Protein Stability and Folding
Supplement 1

Springer-Verlag Berlin Heidelberg GmbH

Wolfgang Pfeil

# Protein Stability and Folding

## Supplement 1

### A Collection of Thermodynamic Data

 Springer

Professor Dr. Wolfgang Pfeil

University of Potsdam
Institute for Biochemistry and Biology/Physical Biochemistry
Karl-Liebknecht-Straße 24–25
14476 Golm
Germany
wpfeil@rz.uni-potsdam.de

Cataloging-in-Publication Data applied for
Die Deutsche Bibliothek - CIP-Einheitsaufnahme

Pfeil, Wolfgang:
Protein stability and folding : a collection of thermodynamic data /
Wolfgang Pfeil. - Berlin ; Heidelberg ; New York ; Barcelona ; Hong Kong ;
London ; Milan ; Paris ; Singapore ; Tokyo : Springer

Suppl. 1 . - (2001)

ISBN 978-3-540-42168-9     ISBN 978-3-642-56462-8 (eBook)
DOI 10.1007/978-3-642-56462-8

http://www.springer.de

© Springer-Verlag Berlin Heidelberg 2001
Originally published by Springer-Verlag Berlin Heidelberg New York in 2001

Cover design: Struve & Partner, Heidelberg
Typesetting: Data conversion by Marianne Schillinger-Dietrich, Berlin

SPIN: 10796750        2/3020 mh – 5 4 3 2 1 0

# Contents

# Preface

In 1998, we published the data compilation PROTEIN STABILITY AND FOLDING which covered the data from the early beginnings of thermodynamic studies of protein folding until 1996. Since then, the amount of available thermodynamic data has increased nearly twice. The data constitute very important additions to the information on the protein folding problem, the construction of mutant protein, and the practical application of proteins in various fields.

The Supplement covers the period 1997–1999 and is designed to make the vast amount of present data accessible to multidisciplinary research where chemistry, physics, biology, and medicine are involved and also biotechnology, pharmaceutical and food research. At the same time the data could be helpful to identify problems unsolved so far, and to avoid unnecessary duplication of scientific work.

The structure of the Supplement is the same as in the previous data compilation. However, some additional data characterizing protein-denaturant interaction and protein unfolding by trifluoroethanol have been added. In that context, some previous data have been reconsidered.

The author wishes to thank everyone who provided data, ideas, or even unpublished results. Furthermore, support by the Deutsche Forschungsgemeinschaft (INK 16 B1-1) is gratefully acknowledged. Finally, I would like to thank the staff of Springer Verlag for their efforts and for excellent assistance during the production of the data collections.

Berlin, 2001                                                                          W. Pfeil

# Abbreviations

| | |
|---|---|
| Abu | α-amino-n-butyric-acid |
| ACES | N-(2-acetamido)-2-aminoethanesulfonic acid |
| ADA | N-(2-acetamido)-iminodiacetic acid |
| ADP | adenosine diphosphate |
| AMP | adenosine monophosphate |
| AMPPNP | 5'-adenylylimidodiphosphate |
| ANS | 1-anilinonaphthalene-8-sulfonic acid |
| ATP | adenosine triphosphate |
| BICINE | N,N-bis(2-hydroxyethyl)glycine |
| bis-ANS | 4,4'-dianilino-1,1'-binaphthyl-5,5'disulfonic acid |
| cal | (superscript) calorimetrically determined value |
| calc | (subscript) calculated value |
| CD | Circular Dichroism |
| conc. | concentration |
| dansyl | 2-(dimethylamino)naphthalene-5-sulfonyl chloride |
| DSC | Differential Scanning Calorimetry |
| DTE | dithioerythritol |
| DTT | dithiothreitol |
| EACA | ε-amino caproic acid |
| EDTA | ethylenediaminetetraacetic acid |
| EGTA | ethyleneglycol-bis(β-aminoethylether)N,N,N',N'-tetraacetic acid |
| EPR | Electron Paramagnetic Resonance |
| EtOH | ethanol |
| FMN | flavin mononucleotide |
| FTIR | Fourier Transform InfraRed spectroscopy |
| GuHCl | guanidine hydrochloride |
| GuHSCN | guanidine isothiocyanate |
| HEPES | N-2(hydroxyethyl)piperazine-N'-2-ethanesulfonic acid |
| HEPPS | N-2(hydroxyethyl)piperazine-N'-(2-hydroxypropanesulfonic acid) |
| HEW | Hen Egg White (Lysozyme) |
| HFIP | 1,1,1,3,3,3-hexafluoro-2-propanol |
| HPLC | High Performance Liquid Chromatography |
| HX | hydrogen exchange |
| IAEDANS | N-(iodoacetyl)-N'-(5-sulfo-1-naphthyl)-ethylenediamine |
| IR | InfraRed Spectroscopy |
| ITC | Isothermal Titration Calorimetry |
| LEM | linear extrapolation method |
| LEM-SB | linear extrapolation to zero denaturant concentration by a method that includes the pre- and postdenaturational baselines for a nonlinear regression of the data according to Refs. 88S4, 88B2, and 92S2: Santoro, M.M., Bolen, D.W.: Biochemistry 27 (1988) 8063-8068, Bolen, D.W., Santoro, M.M.: Biochemistry 27 (1988) 8069-8074, Santoro, M.M., Bolen, D.W.: Biochemistry 31 (1992) 4901-4907. |
| kDa | kilo Dalton |

| | |
|---|---|
| MES | 2-(N-morpholino)ethanesulfonic acid |
| MetOH | methanol |
| MOPS | 3-(N-morpholino)propanesulfonic acid |
| mPa | mega Pascal |
| MW | Molecular Weight |
| n.d. | not determined |
| NMR | Nuclear Magnetic Resonance |
| ORD | Optical Rotatory Dispersion |
| ox. | oxidized |
| PALA | N-(phosphoacetyl)-L-aspartat |
| PBS | phosphate balanced saline |
| PIPES | piperazine-N,N'-bis(2-ethanesulfonic acid) |
| PMSF | phenylmethylsulfonyl fluoride |
| 1-PrOH | 1-propanol, n-propanol |
| 2-PrOH | 2-propanol, isopropanol |
| red. | reduced |
| red., alkyl. | reduced and alkylated |
| SAXS | Small-Angle X-ray Scattering |
| SEC | Size Exclusion Chromatography |
| TCA | trichloroacetate |
| TES | 2-([2-hydroxy-1,1-bis(hydroxymethyl)ethyl]amino)ethane-sulfonic acid |
| TFA | trifluoroacetate |
| TFE | 2,2,2-trifluoroethanol |
| TMAO | trimethylamine-N-oxide |
| Tris | tris(hydroxymethyl)aminomethane |
| TWEEN | polyoxyethylenesorbitan |
| v.H. | (superscript) equilibrium treatment by means of the van't Hoff equation |
| w.t. | wild type |
| w.t.* | pseudo-wild type |

# Symbols

| | |
|---|---|
| A | acid form (intermediate) |
| a | activity |
| $a_{\pm}$ | mean ion activity |
| $c_{1/2}$ | denaturant concentration at which the transition midpoint occurs, i.e., transition midpoint in the linear extrapolation method (LEM) in $Mol_{denaturant}/l$ |
| Cp | heat capacity at constant pressure |
| $\Delta_{unf}Cp$ | heat capacity change at protein unfolding |
| $\Delta_{unf}c_p$ | specific value of heat capacity change at protein unfolding |
| D | denatured state of protein |
| $\Delta G$ | abbreviation for $\Delta_{unf}G$ if not otherwise indicated |
| $\Delta G_{ex}$ | apparent free energy of hydrogen exchange |
| $\Delta G_{op}$ | free energy of opening of protein structure in hydrogen exchange experiments |
| $\Delta_{unf}G$ | Gibbs energy change at protein unfolding |
| $\Delta_{unf}G(T)$ | temperature function of Gibbs energy change at protein unfolding |
| $\Delta_{unf}G_{res}$ | Gibbs energy change at protein unfolding per amino acid residue |
| $\Delta_{unf}G°$ | Gibbs energy change at protein unfolding at standard conditions and in the absence of denaturant |
| $\Delta_{trs}G$ | Gibbs energy change at a structural transition which is different from protein unfolding |
| $\Delta(\Delta G)$ | abbreviation for $\Delta(\Delta_{unf}G)$ |
| $\Delta(\Delta_{unf}G)$ | Gibbs energy change at protein unfolding, difference value, refers to the reference (e.g., wild-type) protein if not otherwise indicated |
| $\Delta g$ | abbreviation for $\Delta_{unf}g$ |
| $\Delta_{unf}g$ | specific value of Gibbs energy change at protein unfolding |
| $\Delta H$ | abbreviation for $\Delta_{unf}H$ if not otherwise indicated |
| $\Delta_{unf}H$ | enthalpy change at protein unfolding |
| $\Delta_{unf}H(T)$ | temperature function of enthalpy change at protein unfolding |
| $\Delta_{unf}H^{cal}$ | enthalpy change at protein unfolding determined by calorimetry |
| $\Delta_{unf}H^{v.H.}$ | enthalpy change at protein unfolding determined by van't Hoff treatment |
| $\Delta_{unf}H°$ | enthalpy change at protein unfolding at standard conditions |
| $\Delta_{unf}H_{res}$ | enthalpy change at protein unfolding per amino acid residue |
| $\Delta h$ | abbreviation for $\Delta_{unf}h$ |
| $\Delta Cp$ | abbreviation for $\Delta_{unf}Cp$ |
| $\Delta c_p$ | abbreviation for $\Delta_{unf}c_p$ |
| $\Delta_{unf}h$ | specific value of enthalpy change at protein unfolding |
| I | ionic strength |
| I or X | intermediate states at protein unfolding |
| I° | standard value of ionic strength, $I = 0.1$ |
| K | abbreviation for equilibrium constant |
| $K_b$ | binding constant |

| | |
|---|---|
| $k_i$ | rate constant |
| $K_{unf}$ | equilibrium constant for protein unfolding |
| $K°_{unf}$ | equilibrium constant for protein unfolding at zero denaturant concentration |
| m | dependence of $\Delta G$ on denaturant concentration, i.e., slope in the linear extrapolation method (LEM) in kJ/mol$_{protein}$/Mol$_{denaturant}$, abbreviation kJ/mol/M |
| N | native state of protein |
| n | number of moles |
| $\Delta n$ | preferential denaturant binding parameter |
| $n_i$ | number of groups of the i-th type |
| pH | pH value |
| pH° | standard pH value, pH = 7.0 |
| R | gas constant, R = 8.3143 J/K/mol |
| $\Delta S$ | abbreviation for $\Delta_{unf}S$ |
| $\Delta_{unf}S$ | entropy change at protein unfolding |
| $\Delta_{unf}S(T)$ | temperature function of entropy change at protein unfolding |
| $\Delta_{unf}S°$ | entropy change at protein unfolding at standard temperature |
| $\Delta_{unf}S$ | specific value of entropy change at protein unfolding |
| T | temperature |
| T° | standard temperature, T = 298.16 K |
| $T_{trs}$ | transition temperature |
| U | unfolded state of protein |
| $\alpha$ | degree of conversion in equilibrium treatment |
| $\varepsilon$ | average degree of exposure of amino acid residues in native protein |
| $\delta g_{tr,i}$ | Gibbs energy change at transfer of the i-th side chain from water to denaturant |
| $\Delta_{unf}\nu$ | difference number of protons between unfolded and native protein states |
| * | asterisk characterizing data which were calculated on the basis of other thermodynamic quantities in the original paper |

# Dimensions

| | |
|---|---|
| $c_{1/2}$ | $Mol_{denaturant}/l$ |
| $C_p$ | $J/g/K$ |
| $\Delta_{unf}c_p$ | $J/g/K$ |
| $Cp$ | $kJ/mol/K$ |
| $\Delta_{unf}Cp$ | $kJ/mol/K$ |
| $\Delta_{unf}G$ | $kJ/mol$ |
| $\Delta_{unf}g$ | $J/g$ |
| $\Delta_{unf}H$ | $kJ/mol$ |
| $\Delta_{unf}h$ | $J/g$ |
| $m$ | $kJ/mol_{protein}/Mol_{denaturant}$, abbreviation $kJ/mol/M$ |
| $\Delta_{unf}S$ | $kJ/mol/K$ |
| $\Delta_{unf}s$ | $J/g/K$ |
| $T$ | $K$ |
| $T(°C)$ | degree centigrade |

# Introduction

# Introduction

There has been a remarkable increase in the number of thermodynamic studies of proteins devoted to the problems of folding and stability. Figure 1 shows the annual number of publications reporting experimentally determined thermodynamic quantities. This has made it necessary to update the compilation of thermodynamic data published "Protein Folding and Stability" (I-98P). This book covers data published from the early beginnings in the 1960s until 1996. The present Supplement covers the years 1997–1999.

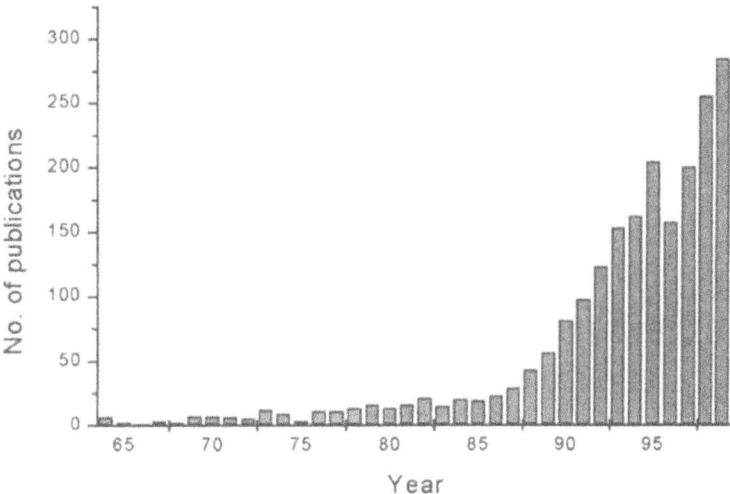

**Fig. 1.** The number of manuscripts published each year which report thermodynamic quantities on protein folding and stability

The tabulated data in the Supplement is organized basically in the same manner as the previously published book "Protein Stability and Folding". However, the data base has been extended in two points. First, parameters of denaturant-induced protein unfolding (m and $c_{1/2}$ values) have been added in Table 1. In that context, previous data on denaturant-induced protein unfolding have been reconsidered. Furthermore, data on TFE-induced unfolding have been arranged in a new Table (see Table 4). Finally, some data have been added which were overlooked during the preparation of the previous data collection (I-98P1).

**Comments on the tabulated data:**

**Table 1** contains Gibbs energy change $\Delta_{unf}G$ and difference values $\Delta(\Delta_{unf}G)$ obtained from protein unfolding experiments using various approaches. $\Delta_{unf}G$ is given as a positive value for reasons of convenience to represent a measure of conformational stability. Similarly, positive $\Delta(\Delta_{unf}G)$ values designate a mutant protein of greater stability than the reference protein, usually the wild-type protein. $\Delta_{unf}G$ and $\Delta(\Delta_{unf}G)$ are given in kJ/mol.

With respect to the widely used linear extrapolation method in denaturant-induced unfolding, the parameters m, representing the slope, and $c_{1/2}$, representing the half transition denaturant concentration, are now included.

**Table 2** contains thermodynamic key values, such as the molar enthalpy change $\Delta_{unf}H$, the molar heat capacity change $\Delta_{unf}Cp$, and transition temperature $T_{trs}$ (or reference temperature $T_{ref}$). $\Delta_{unf}H$ is given in kJ/mol, $\Delta_{unf}Cp$ in kJ/mol/K, and $T_{trs}$ in degrees centigrade if not otherwise indicated.

**Table 3** contains thermodynamic key values, such as the specific enthalpy change $\Delta_{unf}h$, the specific heat capacity change $\Delta_{unf}cp$, and transition temperature $T_{trs}$ (or reference temperature $T_{ref}$). $\Delta_{unf}h$ is given in J/g, $\Delta_{unf}Cp$ in J/g/K, and $T_{trs}$ in degrees centigrade if not otherwise indicated.

**Table 4** has been newly included to represent data concerning protein denaturation by trifluoroethanol (TFE) and other alcohol-based cosolvents. Values of $\Delta G$ and $\Delta H$ have been reported in recent studies of proteins with TFE. The results correspond closely to those of guanidinium hydrochloride (GuHCl) denaturation studies. The similarity is unexpected since the TFE denatured state has a high content of secondary structure in contrast to the GuHCl denatured state. For a recent review refer to I-98B. Basic considerations on the mode of TFE-protein interaction and corresponding equations can be found in Ref. I-94J. For comparison some studies are included in which alcohol-based cosolvents have been used; meaning here only simple alcohols. For the studies in which polyols have been used, also see Table 1.

**Comments on the approaches for the determination of thermodynamic quantitites:**

The introduction to Ref. I-98P1 contained some material necessary for a critical evaluation of the tabulated thermodynamic quantities. In the following section some more recent papers have been added.

a) Calorimetry:

For an overview on calorimetric methods, data treatment, and a wide variety of biological applications the reader is referred to the papers collected in the book "Biocalorimetry" (I-98L). For a recent review on thermodynamic analysis of biomolecular processes, see Ref. I-99C. The calorimetric criterion for a two-state process was newly analyzed in Ref. I-99Z, with implications for the baseline treatment in calorimetry.

Heat capacity changes ($\Delta Cp$ values) have been considered in detail in Ref. I-98M. For an error analysis in case of extremely wide transition regions, see I-99T. The structure-based calculation of thermodynamic quantities including $\Delta Cp$ has achieved importance (see I-95G, I-95L, I-95M, I-97H).

Residual structure in DSC studies was considered in Ref. I-99G1. Proteins may unfold irreversibly thus rendering the determination of thermodynamic quantitites difficult if not impossible. Conditions for folding under inequilibrium conditions were analyzed in Ref. I-98P2.

b) Treatment of solvent-denaturation data:

The enthalpy of transfer of unfolded proteins into solutions of urea and GuHCl has been analyzed in terms of the solvent-exchange model (I-96S1). This model is consistent with the linear extrapolation method (I-98P1, I-00P) which is now widely used. Nevertheless, arguments can be found in favour of both, the linear extrapolation procedure (I-99G3) and the denaturant binding model (I-99W). Proteins with a transition midpoint in the 0–1.5 M GuHCl range need not necessarily give denaturant-concentration independent $\Delta_{unf}G$ values (I-00Y).

The effects of the m-value in solvent denaturation have been discussed in detail by the example of staphylococcal nuclease (I-95S, I-96S2, see also I-00P). Possible deviations from the two-state assumption in staphylococcal-nuclease unfolding have been considered in I-96C. Potential impact of undetected intermediates on m-values has been considered in Ref. I-98S2. The influence of protein concentration of oligomeric proteins on urea and GuHCl unfolding curves has been analyzed in Ref. I-00R.

The development of an automated method for the determination of protein stability has been reported in Ref. I-98S1. The application of capillary zone electrophoresis for monitoring equilibria and kinetics of protein folding/unfolding has been proposed in Ref. I-00V.

c) Hydrogen exchange:
Hydrogen exchange techniques have gained importance in protein folding studies. For a recent review, see Ref. I-99L. For the determination of conformational stability from hydrogen exchange rates, see Ref. I-99H. Specific aspects of native state hydrogen exchange are still disputed (see I-97C, I-98C, I-98E, I-98W, I-00W).

**Other data collections:**

Recently, an electronically available thermodynamic database for proteins and mutants, ProTherm, has been made available (I-99G2, I-00G). ProTherm 2.0 contains currently more than 5500 data entries and has links with other structural and literature databases (I-00G). ProTherm 2.0 is accessible through WWW http://www.rtc.riken.go.jp/protherm.html. A database on experimentally measured heat capacities of amino acids, peptides, and proteins has been made available in Ref. I-98M.

A literature reference list concerning isothermal titration calorimetry and differential scanning calorimetry has been provided by MicroCal Inc. (I-00M).

**Concluding remarks:**

The data contained in Tables 1-4 were adopted from the literature without changes except for conversion into Joules. The trivial names of proteins are cited as suggested by the authors.

It is impossible to prepare the data collection completely free of errors and without overlooking data. I would like to apologize for this in advance. I will gratefully acknowledge any critical remarks and help in updating the data collection.

**References**

(References have been labelled with an I- to distinguish them from the references of the Tables)

I-94J     Jasanoff, A., Fersht, A.R.: Biochemistry 33 (1994) 2129–2135.
I-95G     Gómez, J., Hilser, V.J., Xie, D., Freire, E.: Proteins: Structure, Function, and Genetics 22 (1995) 404–412.
I-95L     Lazaridis, T., Archontis, G., Karplus, M.: Adv. Protein Chem. 47 (1995) 231–305.
I-95M     Makhatadze, G.I., Privalov, P.L.: Adv. Protein Chem. 47 (1995) 307–425.
I-95S     Shortle, D.: Adv. Protein Chem. 46 (1995) 217–248.
I-96C     Carra, J.H., Privalov, P.L.: FASEB J. 10 (1996) 67–74.
I-96S1    Schellman, J.A., Gassner, N.C.: Biophys. Chem. 59 (1996) 259–275.
I-96S2    Shortle, D.: FASEB J. 10 (1996) 27–34.
I-97C     Clarke, J., Itzhaki, L.S., Fersht, A.R.: Trends Biochem. Sci. 22 (1997) 284–287.
I-97H     Hilser, V.J., Townsend, B.D., Freire, E.: Biophys. Chem. 64 (1997) 69–79.
I-98B     Buck, M.: Quart. Rev. Biophys. 31 (1998) 297–355.
I-98C     Clarke, J., Itzhaki, L.S., Fersht, A.R.: Trends Biochem. Sci. 23 (1998) 379–381.
I-98E     Englander, S.W.: Trends Biochem. Sci. 23 (1998) 378.

| | |
|---|---|
| I-98L | Ladbury, J.E., Chowdhry, B.Z. (Eds): in: Biocalorimetry: applications of calorimetry in the biological sciences, John Wiley & Sons, Chichester, New York, Weinheim, Brisbane, Singapore, Toronto, 1998. |
| I-98M | Makhatadze, G.I.: Biophys. Chem. 71 (1998) 133–156. |
| I-98P1 | W. Pfeil: Protein Stability and Folding: a collection of thermodynamic data, Springer-Verlag, Berlin, Heidelberg, New York, 1998. |
| I-98P2 | Potekhin, S.A., Kovrigin, E.L.: Biophys. Chem. 73 (1998) 241–248. |
| I-98S1 | Schwehm, J.M., Stites, W.E.: Methods Enzymol. 295 (1998) 150–170. |
| I-98S2 | Soulages, J.L.: Biophys. J. 75 (1998) 484–492. |
| I-98W | Woodward, C., Li, R.: Trends Biochem. Sci. 23 (1998) 379. |
| I-99C | Cooper, A.: Current Opinion Chem. Biol. 3 (1999) 557–563. |
| I-99G1 | Griko, Y.V.: J. Protein Chem. 18 (1999) 361–369. |
| I-99G2 | Gromiha, M.M., An, J., Kono, H., Oobatake, M., Uedaira, H., Sarai, A.: Nucleic Acids Res. 27 (1999) 286–288. |
| I-99G3 | Gupta, R., Ahmad, F.: Biochemistry 38 (1999) 2471–2479. |
| I-99H | Huyghues-Despointes, B.M.P., Scholtz, J.M., Pace, C.N.: Nature Struct. Biol. 6 (1999) 910–912. |
| I-99L | Li, R., Woodward, C.: Protein Sci. 8 (1999) 1571–1591. |
| I-99T | Taylor, J.W., Greenfield, N.J., Wu, B., Privalov, P.L.: J. Mol. Biol. 291 (1999) 965–976. |
| I-99W | Wu, J.-W., Wang, Z.-X.: Protein Sci. 8 (1999) 2090–2097. |
| I-99Z | Zhou, Y., Hall, C.K., Karplus, M.: Protein Sci. 8 (1999) 1064–1074. |
| I-00G | Gromiha, M.M., An, J., Kono, H., Oobatake, M., Uedaira, H., Prabakaran, P., Sarai, A.: Nucleic Acids Res. 28 (2000) 283–285. |
| I-00M | MicroCal: Ultrasensitive isothermal titration and differential scanning calorimetry, Reference Lists CD-ROM (2000). |
| I-00P | Pace, C.N., Shaw, K.L.: Proteins: Structure, Function, and Genetics Suppl. 4 (2000) 1–7. |
| I-00R | Ragone, R.: Biopolymers 53 (2000) 221–225. |
| I-00V | Verzola, B., Chiti, F., Manao, G., Righetti, P.G.: Analyt. Biochem. 282 (2000) 239–244. |
| I-00W | Wool, J.O., Wrabl, J.O., Hilser, V.J.: J. Mol. Biol. 301 (2000) 247–256. |
| I-00Y | Yang, M., Ferreon, A.C.M., Bolen, D.W.: Proteins: Structure, Function, and Genetics Suppl. 4 (2000) 44–49. |

**Table 1.**
**Gibbs Energy Change – Molar Values**

## Acyl Coenzyme A Binding Protein

Recombinant bovine acyl-coenzyme A binding protein, wild type and mutants

| Mutant | pH | T | ΔG | m | Appr./Rem. | Ref |
|---|---|---|---|---|---|---|
| wild type | 5.3 | 25 | 14.77±0.13 | 33.81±0.33 | GuHCl (1–3) | 99K14 |
| Phe5→Ala | 5.3 | 25 | 14.69±0.04 | 22.80±0.04 | GuHCl (1–3) | 99K14 |
| Ala9→Gly | 5.3 | 25 | 14.23±0.17 | 25.44±0.33 | GuHCl (1–3) | 99K14 |
| Val12→Ala | 5.3 | 25 | 14.52±0.17 | 27.53±0.33 | GuHCl (1–3) | 99K14 |
| Leu15→Ala | 5.3 | 25 | 13.35±0.33 | 18.87±0.46 | GuHCl (1–3) | 99K14 |
| Pro19→Ala | 5.3 | 25 | 13.81±0.21 | 30.00±0.54 | GuHCl (1–3) | 99K14 |
| Asp21→Ala | 5.3 | 25 | 13.89±0.21 | 32.13±0.54 | GuHCl (1–3) | 99K14 |
| Leu25→Ala | 5.3 | 25 | 14.06±0.08 | 25.35±0.17 | GuHCl (1–3) | 99K14 |
| Phe26→Ala | 5.3 | 25 | 15.90±0.13 | 27.41±0.25 | GuHCl (1–3) | 99K14 |
| Tyr28→Ala | 5.3 | 25 | 16.61±0.67 | 23.26±0.84 | GuHCl (1–3) | 99K14 |
| Tyr28→Asn | 5.3 | 25 | 15.27±0.21 | 23.47±0.29 | GuHCl (1–3) | 99K14 |
| Tyr28→Phe | 5.3 | 25 | 14.90±0.08 | 28.53±0.13 | GuHCl (1–3) | 99K14 |
| Tyr31→Asn | 5.3 | 25 | 20.71±0.17 | 29.75±0.21 | GuHCl (1–3) | 99K14 |
| Lys32→Ala | 5.3 | 25 | 14.18±0.08 | 27.07±0.13 | GuHCl (1–3) | 99K14 |
| Lys32→Glu | 5.3 | 25 | 14.31±0.08 | 27.66±0.17 | GuHCl (1–3) | 99K14 |
| Lys32→Arg | 5.3 | 25 | 12.97±0.42 | 24.52±0.67 | GuHCl (1–3) | 99K14 |
| Gln33→Ala | 5.3 | 25 | 18.79±0.63 | 21.13±0.71 | GuHCl (1–3) | 99K14 |
| Ala34→Gly | 5.3 | 25 | 15.27±0.08 | 29.29±0.21 | GuHCl (1–3) | 99K14 |
| Thr35→Ala | 5.3 | 25 | 16.44±0.08 | 28.45±0.75 | GuHCl (1–3) | 99K14 |
| Ile39→Ala | 5.3 | 25 | 14.69±0.13 | 29.37±0.25 | GuHCl (1–3) | 99K14 |
| Pro44→Ala | 5.3 | 25 | 14.64±0.08 | 28.66±0.17 | GuHCl (1–3) | 99K14 |
| Lys52→Met | 5.3 | 25 | 15.52±0.08 | 35.82±0.17 | GuHCl (1–3) | 99K14 |
| Lys54→Ala | 5.3 | 25 | 13.18±0.50 | 27.91±1.09 | GuHCl (1–3) | 99K14 |
| Lys54→Met | 5.3 | 25 | 13.93±0.13 | 32.89±0.33 | GuHCl (1–3) | 99K14 |
| Glu67→Ala | 5.3 | 25 | 14.43±0.04 | 30.46±0.13 | GuHCl (1–3) | 99K14 |
| Ala69→Gly | 5.3 | 25 | 18.70±0.25 | 27.32±0.38 | GuHCl (1–3) | 99K14 |
| Tyr73→Ala | 5.3 | 25 | 16.23±0.33 | 14.94±1.30 | GuHCl (1–3) | 99K14 |
| Tyr73→Phe | 5.3 | 25 | 13.85±0.08 | 33.93±0.17 | GuHCl (1–3) | 99K14 |
| Ile74→Ala | 5.3 | 25 | 16.36±0.08 | 28.20±0.13 | GuHCl (1–3) | 99K14 |
| Val77→Ala | 5.3 | 25 | 15.82±0.08 | 27.36±0.17 | GuHCl (1–3) | 99K14 |
| Leu80→Ala | 5.3 | 25 | 12.89±0.42 | 14.69±0.63 | GuHCl (1–3) | 99K14 |

Remarks:

(1) linear extrapolation, for the procedure see also Ref. 96K7

(2) transition monitored by fluorescence emission at 356 nm

(3) Ref. 99K14 contains data for equilibrium unfolding, kinetics of unfolding and refolding, and Φ-value analysis

Recombinant bovine acyl-coenzyme A binding protein, wild type and mutants

| Mutant | pH | T | ΔG | Approach/Remarks | Ref |
|---|---|---|---|---|---|
| wild type | 5.3 | 5 | 33.0±0.6 | GuHCl (1–3) | 99K15 |
| Phe5→Ala | 5.3 | 5 | 22.5±1.4 | GuHCl (1–3) | 99K15 |
| Ala9→Gly | 5.3 | 5 | 25.4±0.3 | GuHCl (1–3) | 99K15 |
| Val12→Ala | 5.3 | 5 | 25.9±1.7 | GuHCl (1–3) | 99K15 |
| Leu15→Ala | 5.3 | 5 | 20.0±2.1 | GuHCl (1–3) | 99K15 |
| Pro19→Ala | 5.3 | 5 | 28.5±1.0 | GuHCl (1–3) | 99K15 |
| Asp21→Ala | 5.3 | 5 | 34.8±2.0 | GuHCl (1–3) | 99K15 |
| Asp21→His | 5.3 | 5 | 30.7±0.5 | GuHCl (1–3) | 99K15 |
| Leu25→Ala | 5.3 | 5 | 28.7±1.3 | GuHCl (1–3) | 99K15 |
| Tyr28→Ala | 5.3 | 5 | 22.7±1.5 | GuHCl (1–3) | 99K15 |
| Tyr28→Phe | 5.3 | 5 | 28.6±2.3 | GuHCl (1–3) | 99K15 |

Recombinant bovine acyl-coenzyme A binding protein, wild type and mutants (continued)

| Mutant | pH | T | ΔG | Approach/Remarks | | Ref |
|---|---|---|---|---|---|---|
| Tyr28→Asn | 5.3 | 5 | 23.7±2.2 | GuHCl | (1–3) | 99K15 |
| Tyr31→Asn | 5.3 | 5 | 26.7±2.5 | GuHCl | (1–3) | 99K15 |
| Lys32→Ala | 5.3 | 5 | 28.7±3.1 | GuHCl | (1–3) | 99K15 |
| Lys32→Arg | 5.3 | 5 | 25.4±3.3 | GuHCl | (1–3) | 99K15 |
| Lys32→Glu | 5.3 | 5 | 26.0±2.0 | GuHCl | (1–3) | 99K15 |
| Gln33→Ala | 5.3 | 5 | 17.7±1.4 | GuHCl | (1–3) | 99K15 |
| Ala34→Gly | 5.3 | 5 | 26.4±1.0 | GuHCl | (1–3) | 99K15 |
| Thr35→Ala | 5.3 | 5 | 28.5±1.5 | GuHCl | (1–3) | 99K15 |
| Pro44→Ala | 5.3 | 5 | 27.2±1.3 | GuHCl | (1–3) | 99K15 |
| Lys52→Met | 5.3 | 5 | 33.8±1.9 | GuHCl | (1–3) | 99K15 |
| Lys54→Ala | 5.3 | 5 | 29.4±2.3 | GuHCl | (1–3) | 99K15 |
| Lys54→Met | 5.3 | 5 | 34.1±2.2 | GuHCl | (1–3) | 99K15 |
| Glu67→Ala | 5.3 | 5 | 31.5±3.8 | GuHCl | (1–3) | 99K15 |
| Tyr73→Ala | 5.3 | 5 | 12.8±1.1 | GuHCl | (1–3) | 99K15 |
| Tyr73→Phe | 5.3 | 5 | 34.1±2.2 | GuHCl | (1–3) | 99K15 |
| Val77→Ala | 5.3 | 5 | 28.2±1.0 | GuHCl | (1–3) | 99K15 |
| Leu80→Ala | 5.3 | 5 | 17.5±0.8 | GuHCl | (1–3) | 99K15 |
| double mutant (Lys52→Met and Lys54→Met) | | | | | | |
| | 5.3 | 5 | 37.2±3.8 | GuHCl | (1–3) | 99K15 |

Remarks:
(1) linear extrapolation, LEM-SB
(2) transition monitored by fluorescence intensity at 356 nm
(3) buffer: 0.02 M sodium acetate, pH 5.3

Recombinant ($^{15}$N-labeled) bovine acyl-coenzyme A binding protein (ACBP), local $\Delta G_{op}$ from hydrogen exchange

| Residue | pH* | T | $\Delta G_{op}$ | Approach/Remarks | | Ref |
|---|---|---|---|---|---|---|
| Ala8 | 5.2–7.4 | 25 | 20.25±1.38 | HX | (1–5) | 98K14 |
| Ala9 | 5.2–7.4 | 25 | 20.58±0.13 | HX | (1–5) | 98K14 |
| Glu10 | 5.2–7.4 | 25 | 18.16±0.13 | HX | (1–5) | 98K14 |
| Val12 | 5.2–7.4 | 25 | 16.44±1.42 | HX | (1–5) | 98K14 |
| Ser29 | 5.2–7.4 | 25 | 28.03±1.30 | HX | (1–5) | 98K14 |
| His30 | 5.2–7.4 | 25 | 26.28±0.92 | HX | (1–5) | 98K14 |
| Tyr31 | 5.2–7.4 | 25 | 29.79±1.00 | HX | (1–5) | 98K14 |
| Lys32 | 5.2–7.4 | 25 | 27.70±0.71 | HX | (1–5) | 98K14 |
| Ala34 | 5.2–7.4 | 25 | 30.00±0.71 | HX | (1–5) | 98K14 |
| Thr35 | 5.2–7.4 | 25 | 32.30±0.08 | HX | (1–5) | 98K14 |
| Gly37 | 5.2–7.4 | 25 | 26.82±1.21 | HX | (1–5) | 98K14 |
| Trp55 | 5.2–7.4 | 25 | 17.53±1.51 | HX | (1–5) | 98K14 |
| Asp56 | 5.2–7.4 | 25 | 22.51±1.34 | HX | (1–5) | 98K14 |
| Ala57 | 5.2–7.4 | 25 | 25.36±1.05 | HX | (1–5) | 98K14 |
| Trp58 | 5.2–7.4 | 25 | 27.07±0.79 | HX | (1–5) | 98K14 |
| Asn59 | 5.2–7.4 | 25 | 29.66±1.21 | HX | (1–5) | 98K14 |
| Glu60 | 5.2–7.4 | 25 | 28.45±1.09 | HX | (1–5) | 98K14 |
| Lys62 | 5.2–7.4 | 25 | 24.31±0.54 | HX | (1–5) | 98K14 |
| Gly63 | 5.2–7.4 | 25 | 18.58±0.63 | HX | (1–5) | 98K14 |
| Thr64 | 5.2–7.4 | 25 | 27.41±1.05 | HX | (1–5) | 98K14 |
| Met70 | 5.2–7.4 | 25 | 31.59±0.84 | HX | (1–5) | 98K14 |
| Ala72 | 5.2–7.4 | 25 | 26.44±0.13 | HX | (1–5) | 98K14 |
| Ile74 | 5.2–7.4 | 25 | 27.45±1.21 | HX | (1–5) | 98K14 |
| Glu78 | 5.2–7.4 | 25 | 24.35±1.00 | HX | (1–5) | 98K14 |
| Glu79 | 5.2–7.4 | 25 | 29.54±0.92 | HX | (1–5) | 98K14 |

Recombinant ($^{15}$N-labeled) bovine acyl-coenzyme A binding protein (ACBP), local $\Delta G_{op}$ from hydrogen exchange (continued)

| Residue | pH* | T | $\Delta G_{op}$ | Approach/Remarks | | Ref |
|---|---|---|---|---|---|---|
| Lys81 | 5.2–7.4 | 25 | 31.09±1.46 | HX | (1–5) | 98K14 |
| Lys82 | 5.2–7.4 | 25 | 23.18±0.38 | HX | (1–5) | 98K14 |
| Tyr84 | 5.2–7.4 | 25 | 17.78±1.63 | HX | (1–5) | 98K14 |

Remarks:
(1) ACBP is a 86 residue four-α-helix bundle protein
(2) from two data sets: pH* between 5.2 and 8.1 at 298 K, and a second data set between 280 and 300 K
(3) the present analysis was restricted to the pH range from 5.2 to 7.4
(4) reference value from equilibrium denaturation $\Delta G = 23.4\pm1.7$ kJ/mol (GuHCl at pH 7.4, for pH 5.3 see also Ref. 95K12)
(5) pH dependent stability effects are analyzed in Ref. 98K14 based on the pH dependence of chemical shifts and the pH dependence of hydrogen exchange

Acyl-coenzyme A binding protein (ACBP), stability of the protein and ligand-protein complex determined by equilibrium denaturation

| Protein | pH | T | $\Delta G$ | $c_{1/2}$ | Approach/Remarks | | Ref |
|---|---|---|---|---|---|---|---|
| ACBP | 6.65 | 25 | 22.6±2.1 | 1.7 | GuHCl | (1–4) | 95K17 |
| ligated ACBP | 6.65 | 25 | 23.0±1.7 | 1.85 | GuHCl | (1–4) | 95K17 |

Remarks:
(1) linear extrapolation
(2) measured in 0.04 M potassium phosphate
(3) transition monitored by fluorescence at 335 and 356 nm
(4) $c_{1/2}$ was taken from Fig. 1 in Ref. 95K17

Acyl-coenzyme A binding protein (ACBP), difference in local $\Delta G_{ex}$ between protein and ligand-protein complex determined by deuterium exchange kinetics

| Group | pH | T | $\Delta(\Delta G_{ex})$ | Approach/Remarks | Ref |
|---|---|---|---|---|---|
| Ala8 | 6.65 | 25 | −12.05 | HX, NMR | 95K17 |
| Ala9 | 6.65 | 25 | −13.47 | HX, NMR | 95K17 |
| Glu10 | 6.65 | 25 | −12.38 | HX, NMR | 95K17 |
| Glu11 | 6.65 | 25 | −10.04 | HX, NMR | 95K17 |
| Lys13 | 6.65 | 25 | (−13.43) | HX, NMR (1) | 95K17 |
| Met24 | 6.65 | 25 | −12.68 | HX, NMR | 95K17 |
| Leu25 | 6.65 | 25 | (−11.38) | HX, NMR (1) | 95K17 |
| Ile27 | 6.65 | 25 | (−7.61) | HX, NMR (1) | 95K17 |
| Ser29 | 6.65 | 25 | −9.62 | HX, NMR | 95K17 |
| Tyr31 | 6.65 | 25 | −3.77 | HX, NMR | 95K17 |
| Lys32 | 6.65 | 25 | −2.22 | HX, NMR | 95K17 |
| Ala34 | 6.65 | 25 | −3.60 | HX, NMR | 95K17 |
| Thr35 | 6.65 | 25 | −4.94 | HX, NMR | 95K17 |
| Gly37 | 6.65 | 25 | −5.73 | HX, NMR | 95K17 |
| Asp38 | 6.65 | 25 | 0.75 | HX, NMR | 95K17 |
| Ile39 | 6.65 | 25 | −3.35 | HX, NMR | 95K17 |
| Trp55 | 6.65 | 25 | −9.75 | HX, NMR | 95K17 |
| Asp56 | 6.65 | 25 | −11.09 | HX, NMR | 95K17 |
| Ala57 | 6.65 | 25 | −6.74 | HX, NMR | 95K17 |
| Trp58 | 6.65 | 25 | −3.10 | HX, NMR | 95K17 |
| Asn59 | 6.65 | 25 | −6.36 | HX, NMR | 95K17 |
| Glu60 | 6.65 | 25 | −3.72 | HX, NMR | 95K17 |
| Leu61 | 6.65 | 25 | −5.40 | HX, NMR | 95K17 |

Acyl-coenzyme A binding protein (ACBP), difference in local $\Delta G_{ex}$ between protein and ligand-protein complex determined by deuterium exchange kinetics (continued)

| Group | pH | T | $\Delta(\Delta G_{ex})$ | Approach/Remarks | Ref |
|---|---|---|---|---|---|
| Lys62 | 6.65 | 25 | 1.09 | HX, NMR | 95K17 |
| Gly63 | 6.65 | 25 | −0.75 | HX, NMR | 95K17 |
| Thr64 | 6.65 | 25 | −2.05 | HX, NMR | 95K17 |
| Met70 | 6.65 | 25 | −2.64 | HX, NMR | 95K17 |
| Ala72 | 6.65 | 25 | 0.92 | HX, NMR | 95K17 |
| Tyr73 | 6.65 | 25 | −0.50 | HX, NMR | 95K17 |
| Ile74 | 6.65 | 25 | −5.06 | HX, NMR | 95K17 |
| Lys76 | 6.65 | 25 | −1.63 | HX, NMR | 95K17 |
| Val77 | 6.65 | 25 | −0.21 | HX, NMR | 95K17 |
| Glu78 | 6.65 | 25 | 0.46 | HX, NMR | 95K17 |
| Glu79 | 6.65 | 25 | 1.05 | HX, NMR | 95K17 |
| Leu80 | 6.65 | 25 | −3.85 | HX, NMR | 95K17 |
| Lys81 | 6.65 | 25 | −0.25 | HX, NMR | 95K17 |
| Lys82 | 6.65 | 25 | 0.75 | HX, NMR | 95K17 |
| Lys83 | 6.65 | 25 | 1.72 | HX, NMR | 95K17 |
| Tyr84 | 6.65 | 25 | −2.89 | HX, NMR | 95K17 |
| Gly85 | 6.65 | 25 | (−3.05) | HX, NMR (1) | 95K17 |
| Ile86 | 6.65 | 25 | 0.25 | HX, NMR | 95K17 |

Remark:
(1) values in parentheses are based on maximum observable rate constants

## Acylphosphatase

Recombinant muscle acylphosphatase, urea denaturation at different temperatures

| pH | T | $\Delta G$ | $c_{1/2}$ | m | Appr./Rem. | Ref |
|---|---|---|---|---|---|---|
| 5.50 | 15.5 | 28.0 | 5.36 | 5.22 | urea (1–4) | 98C5 |
| 5.50 | 20.2 | 25.3 | 4.91 | 5.14 | urea (1–4) | 98C5 |
| 5.50 | 24.2 | 24.1 | 4.54 | 5.31 | urea (1–4) | 98C5 |
| 5.50 | 28.0 | 22.5 | 4.32 | 5.22 | urea (1–4) | 98C5 |
| 5.50 | 32.6 | 19.5 | 3.70 | 5.26 | urea (1–4) | 98C5 |
| 5.50 | 37.0 | 18.5 | 3.04 | 6.09 | urea (1–4) | 98C5 |
| 5.50 | 41.3 | 12.8 | 2.72 | 4.71 | urea (1–4) | 98C5 |

Remarks:
(1) linear extrapolation, LEM-SB
(2) transition monitored by CD at 222 nm
(3) measured in 50 mM acetate buffer
(4) experimental errors are 0.1 M for $c_{1/2}$, 0.5 kJ/mol/M for m, and ca. 10% for $\Delta G$

Recombinant muscle acylphosphatase, maximal stability obtained from the analysis of stability curves at different urea concentrations

| Urea Conc. | pH | $T_{max}$ | $\Delta G_{max}$ | Approach/Remarks | | Ref |
|---|---|---|---|---|---|---|
| 0.0 M | 5.50 | 4.1 | 28.7 | heat | (1–3) | 98C5 |
| 0.4 M | 5.50 | 3.8 | 26.7 | heat | (1–3) | 98C5 |
| 0.8 M | 5.50 | 4.9 | 24.3 | heat | (1–3) | 98C5 |
| 1.2 M | 5.50 | 5.0 | 22.0 | heat | (1–3) | 98C5 |
| 1.6 M | 5.50 | 2.9 | 20.6 | heat | (1–3) | 98C5 |
| 2.0 M | 5.50 | 3.8 | 18.4 | heat | (1–3) | 98C5 |
| 2.4 M | 5.50 | 4.4 | 16.0 | heat | (1–3) | 98C5 |
| 2.8 M | 5.50 | 1.3 | 15.0 | heat | (1–3) | 98C5 |

Remarks:
(1) calculated from heat denaturation data, see also Table 2
(2) experimental error in $T_{max}$ 2°C, and 5% in $\Delta G_{max}$

Recombinant human muscle acylphosphatase, urea denaturation in the presence of phosphate and glucose

| Solvent | pH | T | $\Delta G$ | $c_{1/2}$ | m | Appr./Rem. | | Ref |
|---|---|---|---|---|---|---|---|---|
| no additives | 5.5 | 28 | 18.8 | 3.97 | 4.56 | urea | (1–3) | 98C7 |
| no additives | 5.5 | 28 | 19.0 | 3.75 | 5.07 | urea | (3,4) | 98C7 |
| 0.24 M glucose | 5.5 | 28 | 21.3 | 4.49 | 4.98 | urea | (1–3) | 98C7 |
| 0.48 M glucose | 5.5 | 28 | 23.1 | 4.88 | 4.80 | urea | (1–3) | 98C7 |
| 0.48 M glucose | 5.5 | 28 | 23.5 | 4.63 | 5.07 | urea | (3,4) | 98C7 |
| 0.72 M glucose | 5.5 | 28 | 25.4 | 5.36 | 4.62 | urea | (1–3) | 98C7 |
| 2 mM phosphate | 5.5 | 28 | 23.2 | 4.90 | 5.00 | urea | (1–3) | 98C7 |
| 2 mM phosphate | 5.5 | 28 | 23.5 | 4.50 | 5.22 | urea | (3,4) | 98C7 |
| 3.4% (v/v) TFE | 5.5 | 28 | 19.1 | 3.88 | 4.92 | urea | (3–5) | 98C7 |

Remarks:
(1) data from equilibrium unfolding, transition monitored by fluorescence at 335 nm
(2) linear extrapolation, LEM-SB
(3) buffer: 50 mM acetate buffer, pH 5.5
(4) data from kinetics of unfolding, transition monitored by fluorescence at 335 nm
(5) TFE = 2,2,2-trifluoroethanol

Recombinant muscle acylphosphatase, wild type and mutants, equilibrium unfolding

| Mutant | pH | T | $\Delta(\Delta G)$ | $c_{1/2}$ | Approach/Remarks | | Ref |
|---|---|---|---|---|---|---|---|
| wild type | 5.5 | 28 | 0.0 | 4.02±0.20 | urea | (1–4) | 99C8 |
| Tyr11→Ile | 5.5 | 28 | −9.3±1.5 | 2.25±0.20 | urea | (2–4) | 99C8 |
| Tyr11→Phe | 5.5 | 28 | 1.8±2.8 | 5.32±0.20 | urea | (2–4) | 99C8 |
| Val13→Ala | 5.5 | 28 | −11.0±1.4 | 1.95±0.20 | urea | (2–4) | 99C8 |
| Val17→Ala | 5.5 | 28 | −7.7±1.7 | 2.57±0.20 | urea | (2–4) | 99C8 |
| Val20→Ala | 5.5 | 28 | −1.2±2.2 | 3.80±0.20 | urea | (2–4) | 99C8 |
| Phe22→Leu | 5.5 | 28 | −5.4±1.8 | 3.00±0.20 | urea | (2–4) | 99C8 |
| Tyr25→Ala | 5.5 | 28 | −1.0±2.2 | 4.36±0.20 | urea | (2–4) | 99C8 |
| Glu29→Asp | 5.5 | 28 | −14.5±2.2 | 1.29±0.20 | urea | (2–4) | 99C8 |
| Ala30→Gly | 5.5 | 28 | −6.9±1.7 | 2.72±0.20 | urea | (2–4) | 99C8 |
| Val36→Ala | 5.5 | 28 | −10.6±1.5 | 2.02±0.20 | urea | (2–4) | 99C8 |
| Trp38→Phe | 5.5 | 28 | −0.1±2.0 | 4.00±0.20 | urea | (2–4) | 99C8 |
| Val39→Ala | 5.5 | 28 | −6.8±1.7 | 2.73±0.20 | urea | (2–4) | 99C8 |
| Thr42→Ala | 5.5 | 28 | −7.7±1.7 | 2.57±0.20 | urea | (2–4) | 99C8 |
| Gly45→Ala | 5.5 | 28 | −7.2±1.8 | 2.66±0.20 | urea | (2–4) | 99C8 |
| Val47→Ala | 5.5 | 28 | −7.8±1.7 | 2.55±0.20 | urea | (2–4) | 99C8 |

Recombinant muscle acylphosphatase, wild type and mutants, equilibrium unfolding (continued)

| Mutant | pH | T | $\Delta(\Delta G)$ | $c_{1/2}$ | Approach/Remarks | | Ref |
|---|---|---|---|---|---|---|---|
| Val51→Ala | 5.5 | 28 | −7.9±1.6 | 2.54±0.20 | urea | (2–4) | 99C8 |
| Pro54→Ala | 5.5 | 28 | −1.9±2.1 | 3.66±0.20 | urea | (2–4) | 99C8 |
| Met61→Ala | 5.5 | 28 | −16.6±2.2 | 0.88±0.40 | urea | (2–4) | 99C8 |
| Trp64→Ala | 5.5 | 28 | −6.3±1.8 | 2.84±0.20 | urea | (2–4) | 99C8 |
| Leu65→Val | 5.5 | 28 | −22.4±1.0 | 0.21±0.40 | urea | (2–4) | 99C8 |
| Pro71→Ala | 5.5 | 28 | −6.7±1.7 | 2.76±0.20 | urea | (2–4) | 99C8 |
| Ile75→Val | 5.5 | 28 | −5.9±1.8 | 2.91±0.20 | urea | (2–4) | 99C8 |
| Thr78→Ser | 5.5 | 28 | −5.9±1.8 | 2.90±0.20 | urea | (2–4) | 99C8 |
| Glu83→Asp | 5.5 | 28 | −6.3±1.8 | 2.83±0.20 | urea | (2–4) | 99C8 |
| Ile86→Val | 5.5 | 28 | −7.4±1.7 | 2.95±0.20 | urea | (2–4) | 99C8 |
| Leu89→Ala | 5.5 | 28 | −7.0±1.7 | 2.70±0.20 | urea | (2–4) | 99C8 |
| Phe94→Leu | 5.5 | 28 | −18.1±1.1 | 0.61±0.40 | urea | (2–4) | 99C8 |

Remarks:

(1) $\Delta G$ for the wild-type protein amounts to $\Delta G = 21.3\pm2.3$ kJ/mol

(2) linear extrapolation, LEM-SB, using an average value for m of m = 5.3 kJ/mol/M

(3) transition monitored by fluorescence emission at 335 nm

(4) buffer: 50 mM acetate, pH 5.5

Recombinant muscle acylphosphatase, wild type and lysine to glutamine mutants

| Mutant | pH | T | $\Delta G$ | $c_{1/2}$ | m | Approach/Remarks | | Ref |
|---|---|---|---|---|---|---|---|---|
| wild type | 5.5 | 25 | 18.5 | 4.61 | 4.0±0.3 | urea | (1–5) | 98C6 |
| Lys32→Gln | 5.5 | 25 | 16.9 | 4.22 | | urea | (1–5) | 98C6 |
| Lys57→Gln | 5.5 | 25 | 16.0 | 4.00 | | urea | (1–5) | 98C6 |
| Lys67→Gln | 5.5 | 25 | 18.7 | 4.68 | | urea | (1–5) | 98C6 |
| Lys84→Gln | 5.5 | 25 | 15.9 | 3.98 | | urea | (1–5) | 98C6 |
| Lys88→Gln | 5.5 | 25 | 18.0 | 4.50 | | urea | (1–5) | 98C6 |

Remarks:

(1) linear extrapolation, LEM-SB

(2) transition monitored by CD at 222 nm

(3) buffer: 50 mM acetate

(4) m = 4.0±0.3 kJ/mol/M is an averaged value that was used for all mutant proteins

(5) estimated error in $c_{1/2}$ ±0.15 M

Recombinant muscle acylphosphatase, mutant Cys21→Ser

| Mutant | pH | T | $\Delta G$ | $c_{1/2}$ | m | Appr./Rem. | Ref |
|---|---|---|---|---|---|---|---|
| Cys21→Ser | 5.5 | 28 | 18.8±1.0 | 3.97±0.2 | 4.75±0.3 | urea (1–3) | 98V2 |
| Cys21→Ser | 5.5 | 28 | 19.0±1.0 | 3.75±0.3 | 5.07±0.3 | urea (3,4) | 98V2 |

Remarks:

(1) data from equilibrium unfolding, linear extrapolation, for details of the procedure see Ref. 98C5

(2) transition monitored by intrinsic fluorescence

(3) measured in 50 mM acetate buffer, pH 5.5

(4) data from kinetics of folding and unfolding, assuming a two-state model and linear dependence of ln(k) on the denaturant concentration

Recombinant muscle acylphosphatase, wild type and mutants, kinetic analysis of folding and unfolding

| Mutant | pH | T | Δ(ΔG) | $c_{1/2}$ | m | Appr./Rem. | Ref |
|--------|-----|-----|--------|-----------|-----|------------|-----|
| wild type | 5.5 | 28 | 0.00 | 3.77±0.10 | 5.45±0.1 | urea (1–4) | 99C8 |
| Tyr11→Ile | 5.5 | 28 | −8.25±0.40 | 2.22±0.09 | 5.60±0.15 | urea (2–4) | 99C8 |
| Tyr11→Phe | 5.5 | 28 | 2.50±0.40 | 5.27±0.18 | 4.40±0.15 | urea (2–4) | 99C8 |
| Val13→Ala | 5.5 | 28 | −10.85±0.55 | 1.91±0.12 | 5.10±0.15 | urea (2–4) | 99C8 |
| Val17→Ala | 5.5 | 28 | −7.40±0.70 | 2.32±0.17 | 5.65±0.30 | urea (2–4) | 99C8 |
| Val20→Ala | 5.5 | 28 | −1.70±0.85 | 3.43±0.19 | 5.50±0.20 | urea (2–4) | 99C8 |
| Phe22→Leu | 5.5 | 28 | −6.15±0.30 | 2.56±0.11 | 5.65±0.20 | urea (2–4) | 99C8 |
| Tyr25→Ala | 5.5 | 28 | −1.50±0.60 | 4.10±0.27 | 4.65±0.25 | urea (2–4) | 99C8 |
| Glu29→Asp | 5.5 | 28 | −15.45±0.35 | | | urea (2–4) | 99C8 |
| Ala30→Gly | 5.5 | 28 | −5.05±0.40 | 2.74±0.13 | 5.70±0.25 | urea (2–4) | 99C8 |
| Val36→Ala | 5.5 | 28 | −8.75±0.65 | 2.15±0.15 | 5.50±0.25 | urea (2–4) | 99C8 |
| Val39→Ala | 5.5 | 28 | −7.80±0.45 | 2.33±0.18 | 5.50±0.40 | urea (2–4) | 99C8 |
| Thr42→Ala | 5.5 | 28 | −9.15±0.40 | 2.09±0.30 | 5.45±0.15 | urea (2–4) | 99C8 |
| Gly45→Ala | 5.5 | 28 | −9.15±0.30 | 2.24±0.20 | 5.15±0.10 | urea (2–4) | 99C8 |
| Val47→Ala | 5.5 | 28 | −8.25±0.35 | 2.41±0.10 | 5.10±0.15 | urea (2–4) | 99C8 |
| Val51→Ala | 5.5 | 28 | −8.25±0.80 | 2.18±0.24 | 5.65±0.50 | urea (2–4) | 99C8 |
| Pro54→Ala | 5.5 | 28 | −2.75±0.50 | 3.35±0.15 | 5.30±0.20 | urea (2–4) | 99C8 |
| Met61→Ala | 5.5 | 28 | −13.40±0.35 | | | urea (2–4) | 99C8 |
| Trp64→Ala | 5.5 | 28 | −5.85±0.30 | 2.63±0.07 | 5.60±0.15 | urea (2–4) | 99C8 |
| Pro71→Ala | 5.5 | 28 | −5.60±0.35 | 2.69±0.08 | 5.60±0.15 | urea (2–4) | 99C8 |
| Ile75→Val | 5.5 | 28 | −3.20±0.60 | 3.12±0.16 | 5.60±0.20 | urea (2–4) | 99C8 |
| Thr78→Ser | 5.5 | 28 | −5.60±0.70 | 2.74±0.18 | 5.45±0.25 | urea (2–4) | 99C8 |
| Glu83→Asp | 5.5 | 28 | −6.00±0.70 | 2.71±0.16 | 5.35±0.20 | urea (2–4) | 99C8 |
| Ile86→Val | 5.5 | 28 | −8.15±0.30 | 2.75±0.07 | 4.50±0.10 | urea (2–4) | 99C8 |
| Leu89→Ala | 5.5 | 28 | −8.65±0.50 | 2.31±0.13 | 5.20±0.20 | urea (2–4) | 99C8 |

Remarks:
(1) ΔG for the wild-type protein amounts to ΔG = 20.6±0.3 kJ/mol
(2) data from the kinetics of protein folding and refolding in the presence of denaturant
(3) transition measured by stopped-flow fluorescence at 335 nm
(4) buffer: 50 mM acetate, pH 5.5

Common-type acylphosphatase (CT AcP) and muscle acylphosphatase (M AcP)

| Protein | pH | T | $\Delta G$ | $c_{1/2}$ | m | Appr./Rem. | Ref |
|---------|-----|-----|------|------|-----|--------------|------|
| CT AcP | 5.5 | 5 | 20.2 | 3.19 | 7.3 | urea (1–5) | 99T1 |
| CT AcP | 5.5 | 10 | 20.1 | 3.18 | 6.5 | urea (1–5) | 99T1 |
| CT AcP | 5.5 | 15 | 20.0 | 3.16 | 6.2 | urea (1–5) | 99T1 |
| CT AcP | 5.5 | 20 | 19.3 | 3.05 | 5.6 | urea (1–5) | 99T1 |
| CT AcP | 5.5 | 25 | 18.3 | 2.89 | 6.2 | urea (1–5,9) | 99T1 |
| CT AcP | 5.5 | 30 | 15.5 | 2.45 | 5.4 | urea (1–5) | 99T1 |
| CT AcP | 5.5 | 35 | 13.0 | 2.05 | 6.6 | urea (1–5) | 99T1 |
| CT AcP | 5.5 | 40 | 9.6 | 1.52 | 6.7 | urea (1–5) | 99T1 |
| CT AcP | 5.5 | 25 | 20.2 | | | heat (6) | 99T1 |
| M AcP | 5.5 | 25 | 28.7 | | | heat (6,7,10) | 99T1 |
| CT AcP | 5.5 | 28 | 18.3 | 2.6 | 7.1 | kin. (3,11) | 99T1 |
| CT AcP | 9.2 | 28 | 13.7 | 3.0 | 4.6 | kin. (11) | 99T1 |
| M AcP | 5.5 | 28 | 19.0 | 3.8 | 5.1 | kin. (11,12) | 99T1 |

Remarks:

(1)  linear extrapolation, LEM-SB, for details see also Ref. 98C5
(2)  transition monitored by CD at 222 nm
(3)  measured in 50 mM acetate buffer
(4)  experimental errors are 0.10 M for $c_{1/2}$, 0.6 kJ/mol/M for m, and ca. 8% for $\Delta G$
(5)  an average value of m = 6.32±0.49 kJ/mol/M was used to calculate $\Delta G$
(6)  see also Table 2
(7)  data from Ref. 98C5
(8)  for the pH dependence of $\Delta G$, m, and $c_{1/2}$ see Fig. 3 in Ref. 99T1
(9)  the dependence of $\Delta G$ on ionic strength (IS) for CT AcP is given by: $\Delta G = 17.2 + 18.3 \times IS$ with $c_{1/2} = 2.89 + 3.08 \times IS$ and m = 5.95 ± 0.54 kJ/mol/M (independent of salt conc.)
(10) the dependence of $\Delta G$ on ionic strength (IS) for M AcP is given by: $\Delta G = 19.2 + 17.7 \times IS$ with $c_{1/2} = 4.00 + 3.69 \times IS$ and m = 4.80 ± 0.45 kJ/mol/M (independent of salt conc.)
(11) data from kinetics of urea-induced unfolding and refolding, transition monitored by fluorescence
(12) the enzyme is the Cys21→Ser mutant of wild-type M AcP


## Adenylate Kinase

Adenylate kinase from *E. coli* ($AK_e$) and a 4-cysteine variant ($AKC_4$) that binds zinc ions

| Mutant | pH | T | $\Delta G$ | $c_{1/2}$ | m | Appr./Rem. | Ref |
|--------|-----|-----|-----------|-----------|---------|-------------|-------|
| $AK_e$ | 7.2 | 20 | 41.0±3.8 | 3.4±0.2 | 8.8±1.3 | urea (1,2) | 98B12 |
| $AKC_4$ | 7.2 | 20 | 25.9±2.1 | 4.5±0.1 | 5.9±0.4 | urea (1,2) | 98B12 |

Remarks:
(1) linear extrapolation
(2) transition monitored by CD at 222 nm

Adenylate kinase (AK) from the archaeon *Sulfolobus acidocaldarius*

| Protein | pH | T | $\Delta G$ | $c_{1/2}$ | m | Appr./Rem. | Ref |
|---------|-----|-----|-----|------|----------|------------|------|
| AK | 7.0 | 32 | 130 | 2.95 | 23.7±4.8 | GuHCl (1–7) | 98B1 |

Remarks:

(1) linear extrapolation, LEM-SB

(2) measured by CD at 222 nm in 50 mM potassium phosphate buffer

(3) Ref. 98B1 contains temperature dependent data for GuHCl-induced unfolding of adenylate kinase from 5 to 70°C treated by linear extrapolation and denaturant binding model

(4) T refers to the maximal stability of adenylate kinase

(5) $\Delta G$ refers to the stability of the trimer

(6) $c_{1/2}$ was taken from Fig. 4a in Ref. 98B1

(7) m is the average value for the temperature range 5 to 70°C

## Adrenodoxin

Recombinant bovine adrenodoxin, wild type, truncated form 4–108 and mutants at the Pro108-Arg14 hydrogen bond

| Protein | pH | T | $\Delta G$ | Approach/Remarks | | Ref |
|---------|-----|-----|-------|------|------|------|
| pseudo-w.t. | 9.2 | 37 | 15.31 | heat | (1–4) | 98G15 |
| Pro108→Ala | 9.2 | 37 | 9.87 | heat | (1–4) | 98G15 |
| Arg14→Ala | 9.2 | 37 | 9.29 | heat | (1–4) | 98G15 |
| Arg14→Glu | 9.2 | 37 | 2.65 | heat | (1–4) | 98G15 |

Remarks:

(1) pseudo-wild type is the truncated protein Adx(4-108)

(2) measured in protectant buffer: 40 mM glycine, pH 9.2, 10 mM $Na_2S$, 1 mM ascorbate, 10 mM 2-mercaptoethanol

(3) transition monitored by CD at 440 nm

(4) $\Delta G$ was calculated using $\Delta Cp = 7.5\pm0.67$ kJ/mol/K from Ref. 95B9

Recombinant bovine adrenodoxin, wild type, truncated form 4-108 with Pro108, and mutants Pro108→X

| Protein | pH | T | $\Delta G$ | Approach/Remarks | | Ref |
|---------|-----|-----|-------|------|------|------|
| wild type | 8.5 | 37 | 9.97 | heat | (1–3) | 98G16 |
| 4–108Pro | 8.5 | 37 | 11.72 | heat | (1–4) | 98G16 |
| 4–108Ala | 8.5 | 37 | 6.72 | heat | (1–4) | 98G16 |
| 4–108Lys | 8.5 | 37 | 1.06 | heat | (1–4) | 98G16 |
| 4–108Ser | 8.5 | 37 | 3.90 | heat | (1–4) | 98G16 |
| 4–108Trp | 8.5 | 37 | 1.62 | heat | (1–4) | 98G16 |

Remarks:

(1) $\Delta G$ was calculated using $\Delta Cp = 10.35$ kJ/mol/K, see Table 2

(2) transition monitored by CD at 440 nm

(3) measured in protectant buffer: 40 mM glycine, pH 8.5, 10 mM $Na_2S$, 1 mM ascorbate, 10 mM 2-mercaptoethanol

(4) mutant description: truncated form consisting of residues 4–108 with Pro or mutants with other residues in position 108

## Aerolysin

Aerolysin, pore-forming toxin from *Aeromonas hydrophila*

| Protein/Transition | | pH | T | $\Delta G$ | $c_{1/2}$ | m | Appr./Rem. | Ref |
|---|---|---|---|---|---|---|---|---|
| wild-type (w.t.) proaerolysin: | | | | | | | | |
| w.t. | trans. (1) | 7.4 | 25 | 31.4 | 2.9 | 10.89 | urea (1–3) | 99L2 |
| | trans. (2) | 7.4 | 25 | | 7.1 | | urea (1–3) | 99L2 |
| w.t. | trans. (1) | 7.4 | 25 | 41.8 | 0.9 | 49.33 | GuHCl (1–3) | 99L2 |
| | trans. (2) | 7.4 | 25 | 19.2 | 2.9 | 6.64 | GuHCl (1–3) | 99L2 |
| w.t. | trans. (1) | 7.4 | 37 | 8.4 | 1.6 | 5.20 | urea (1–3) | 99L2 |
| w.t. | trans. (1) | 8.4 | 25 | 31.8 | 2.4 | 13.17 | urea (1–3) | 99L2 |
| heptamer | | 7.4 | 25 | 83.7 | 3.8 | 22.21 | GuHCl (1–3) | 99L2 |
| mutant (mut.) Cys159→Ser: | | | | | | | | |
| mut. | trans. (1) | 7.4 | 25 | 21.3 | 2.0 | 21.3 | urea (1–3) | 99L2 |
| | trans. (2) | 7.4 | 25 | | 6.9 | | urea (1–3) | 99L2 |

Remarks:
(1) linear extrapolation assuming a three-state model
(2) transition monitored by fluorescence intensity at different wavelengths
(3) buffer: 20 mM HEPES, 150 mM NaCl

## Alkaline Phosphatase

Human placental alkaline phosphatase

| Transition | pH | T | $\Delta G$ | $c_{1/2}$ | m | Appr./Rem. | Ref |
|---|---|---|---|---|---|---|---|
| at protein conc. 131 μg/ml: | | | | | | | |
| N → I | 7.7 | 30 | 56.6±9.3 | 1.48±0.21 | 38.2±6.2 | GuHCl (1,2,3) | 98H12 |
| I → U | 7.7 | 30 | 26.1±5.1 | 2.53±0.46 | 10.5±1.9 | GuHCl (1,2,3) | 98H12 |
| at protein conc. 13.1 μg/ml: | | | | | | | |
| N → I | 7.7 | 30 | 63.1±6.2 | 1.45±0.14 | 43.4±4.1 | GuHCl (1,2,3) | 98H12 |
| N → I | 7.7 | 30 | 43.9±6.3 | 1.37±0.19 | 32.0±4.5 | GuHCl (2,4,5) | 98H12 |
| I → U | 7.7 | 30 | 49.2±7.1 | 2.50±0.35 | 19.7±2.8 | GuHCl (1,2,3) | 98H12 |
| I → U | 7.7 | 30 | 37.8±2.4 | 2.46±0.89 | 15.4±9.5 | GuHCl (2,4,6) | 98H12 |
| at protein conc. 1.31 μg/ml: | | | | | | | |
| N → I | 7.7 | 30 | 15.6±7.1 | 1.24±0.55 | 12.5±5.3 | GuHCl (1,2,3) | 98H12 |
| I → U | 7.7 | 30 | 39.3±30.8 | 2.41±1.88 | 16.3±12.6 | GuHCl (1,2,3) | 98H12 |

Remarks:
(1) linear extrapolation, LEM-SB, treatment by a three-state model
(2) measured in 30 mM Tris-HCl, pH 7.7
(3) transition monitored by fluorescence wavelength
(4) linear extrapolation, LEM-SB, treatment by a two-state model
(5) transition monitored by fluorescence intensity at 346 nm
(6) transition monitored by residual enzyme activity

## α-Amylase

α-Amylase from various species, normalized stability values (per mol of residue)

| Protein | pH | T | $\Delta G_{res}$ | Approach/Remarks | | Ref |
|---|---|---|---|---|---|---|
| AH-A | 7.2 | 17 | 92 | DSC | (1–3) | 99F3 |
| | 7.2 | 0 | 50 | DSC | (1,3) | 99F3 |
| BA-A | 7.2 | 34 | 293 | DSC | (1–3) | 99F3 |
| AO-A | 7.2 | 12 | 331 | DSC | (1–3) | 99F3 |

Explanations:

AH-A = *Alteromonas haloplanctis* α-amylase
BA-A = *Bacillus amyloliquefaciens* α-amylase
AO-A = *Aspergillus oryzae* (Taka amylase) α-amylase

Remarks:
(1) $\Delta G_{res}$ in J per mol of residue
(2) the protein achieves maximum stability at the given temperature
(3) buffer: 30 mM MOPS, 50 mM NaCl, 1 mM $CaCl_2$, pH 7.2

α-Amylase from antarctic psychrophilic *Alteromonas haloplanctis* (AH-A)

| Protein | pH | T | $\Delta G$ | Approach/Remarks | | Ref |
|---|---|---|---|---|---|---|
| AH-A | 7.2 | 25 | 38.8* | DSC | (1) | 99F3 |
| | 7.2 | 17.8 | 42.8* | DSC | (1,2) | 99F3 |

Remarks:
(1) buffer: 30 mM MOPS, 50 mM NaCl, 1 mM $CaCl_2$, pH 7.2
(2) the protein achieves maximum stability near 17°C

α-Amylase from antarctic psychrophile *Alteromonas haloplanctis* A23 (AH-A), $\Delta(\Delta G)$
relative to the more stable porcine α-amylase

| Protein | pH | T | $\Delta(\Delta G)$ | Approach | Remarks | Ref |
|---|---|---|---|---|---|---|
| | 7.0 | 25 | −10 | GuHCl | (1,2) | 94F4 |
| | 7.0 | 50 | −8 | activity | (1,3) | 94F4 |

Remarks:
(1) linear extrapolation
(2) transition monitored by fluorescence intensity
(3) from inactivation kinetics of porcine α-amylase and the enzyme of *Alteromonas haloplanctis* A23

## Amyloid

Soluble fusion protein between serum human amyloid A and staphylococcal nuclease

| Variant | pH | T | $\Delta(\Delta G)$ | $c_{1/2}$ | Approach/Remarks | | Ref |
|---|---|---|---|---|---|---|---|
| nuclease | 7.0 | 20 | 23.4 | 0.82 | GuHCl | (1–3) | 98M13 |
| fusion protein | 7.0 | 20 | 15.1 | 0.72 | GuHCl | (1–3) | 98M13 |

Remarks:
(1) linear extrapolation
(2) transition monitored by fluorescence emission at 325 nm (excitation 295 nm)
(3) buffer: 25 mM sodium phosphate, 0.1 M NaCl, pH 7.0

## Annexin

Annexin I, recombinant porcine protein, and annexin V, recombinant human protein

| Protein | pH | T | $\Delta G$ | $c_{1/2}$ | m | Appr./Rem. | Ref |
|---|---|---|---|---|---|---|---|
| annexin I | 6.0 | 5 | 36.2±7.3 | 1.59 | 22.7±4.6 | GuHCl (1,2) | 99R18 |
| | 6.0 | 12 | 59.9±9.9 | 1.66 | 36.0±6.0 | GuHCl (1,2) | 99R18 |
| | 6.0 | 20 | 47.7±7.2 | 1.57 | 30.3±4.5 | GuHCl (1,2) | 99R18 |
| | 8.0 | 37 | 39.9 | | | DSC  (4) | 99R18 |
| annexin V | 8.0 | 37 | 31.2 | | | DSC  (4) | 99R18 |

Remarks:
(1) linear extrapolation, LEM-SB
(2) measured in 50 mM MES-NaOH, pH 6
(3) transition monitored by CD at 222 nm
(4) from DSC measurements performed in 50 mM sodium phosphate, see also Table 2

Human recombinant annexin V, wild type and mutant Glu112→Gly

| Mutant | pH | T | $\Delta G$ | Approach/Remarks | | Ref |
|---|---|---|---|---|---|---|
| wild type | 8.0 | 20 | 37.4±1.2 | urea | (1–3) | 96L5 |
| Glu112→Gly | 8.0 | 20 | 28.5±1.8 | urea | (1–3) | 96L5 |

Remarks:
(1) linear extrapolation
(2) transition monitored by fluorescence emission of Trp187 at 360 nm
(3) measured at protein conc. of 3 to 5 μM in 20 mM Tris-HCl, pH 8.0

Human recombinant protein annexin V, mutant Glu17→Gly

| Mutant | pH | T | $\Delta G$ | $c_{1/2}$ | m | Appr./Rem. | Ref |
|---|---|---|---|---|---|---|---|
| Glu17→Gly | 7.0 | 25 | 48.3 | | | DSC (1–3) | 97V5 |
| Glu17→Gly | 7.0 | 20 | 53.4 | | | DSC (1–3) | 97V5 |
| Glu17→Gly | 7.0 | –5 | 65.4 | | | DSC (1–3) | 97V5 |
| Glu17→Gly | 8.0 | 20 | 10.4 | 1.74 | 6.0 | GuHCl(1,4,5) | 97V5 |

Remarks:
(1) Ref. 97V5 is a case study for quasi-equilibrium treatment of thermodynamic quantities derived from irreversible denaturation
(2) data were calculated using the data obtained at scan rate 2.34 K/min, i.e., $T_{trs}$ = 54.7°C, $\Delta H$ = 690 kJ/mol, and $\Delta Cp$ = 10.3 kJ/mol/K
(3) buffer: 20 mM sodium phosphate
(4) buffer: 50 mM Tris, 100 mM KCl, pH 8.0
(5) linear extrapolation, LEM-SB

Human recombinant annexin V, isomorphous replacement of Met with aminohexanoic acid, selenomethionine, and telluromethionine

| Mutant | pH | T | $\Delta G$ | $c_{1/2}$ | m | Appr./Rem. | Ref |
|---|---|---|---|---|---|---|---|
| wild type | 7.5 | 25 | 27.38±1.84 | 1.63±0.013 | 16.75±1.13 | GuHCl (1–3) | 98B11 |
| Sem | 7.5 | 25 | 29.98±2.35 | 1.39±0.011 | 21.58±1.68 | GuHCl (1–3) | 98B11 |
| Ahx | 7.5 | 25 | 28.64±1.72 | 1.41±0.009 | 20.32±1.21 | GuHCl (1–3) | 98B11 |
| Tem | 7.5 | 25 | 24.69±3.82 | 0.98±0.021 | 25.36±3.94 | GuHCl  (1–3) | 98B11 |
| wild type | 7.5 | 25 | 28.01±2.77 | 1.84±0.020 | 15.20±1.51 | GuHCl (1,2,4) | 98B11 |
| Sem | 7.5 | 25 | 33.05±3.69 | 1.58±0.017 | 20.70±2.35 | GuHCl (1,2,4) | 98B11 |
| Ahx | 7.5 | 25 | 33.81±3.06 | 1.62±0.013 | 20.91±1.89 | GuHCl (1,2,4) | 98B11 |
| Tem | 7.5 | 25 | 25.53±3.36 | 1.10±0.019 | 23.26±3.06 | GuHCl (1,2,4) | 98B11 |

Human recombinant annexin V, isomorphous replacement of Met with aminohexanoic acid, selenomethionine, and telluromethionine (continued)

| Mutant | pH | T | $\Delta G$ | $c_{1/2}$ | m | Appr./Rem. | Ref |
|---|---|---|---|---|---|---|---|
| wild type | 7.5 | 25 | 50.86±5.29 | 4.84±0.031 | 10.5 ±1.09 | urea (1–3) | 98B11 |
| Sem | 7.5 | 25 | 40.65±4.03 | 4.59±0.035 | 8.86±0.88 | urea (1–3) | 98B11 |
| Ahx | 7.5 | 25 | 41.70±3.44 | 4.62±0.028 | 9.03±0.75 | urea (1–3) | 98B11 |
| Tem | 7.5 | 25 | 26.88±3.65 | 3.56±0.057 | 7.51±1.05 | urea (1–3) | 98B11 |

Remarks:
(1) abbreviations:

       Sem – selenomethionine

       Ahx – 2-aminohexanoic acid

       Tem – telluromethionine

(2) linear extrapolation
(3) transition monitored by fluorescence at 360 nm
(4) transition monitored by CD at 222 nm

## α₁-Antitrypsin

α₁-Antitrypsin (α₁-AT) with fluorescence labels at various positions
Results of a two-state analysis

| Protein | T | $\Delta G$ | $c_{1/2}$ | m | Appr./Rem. | Ref |
|---|---|---|---|---|---|---|
| fluorescence probe: | | | | | | |
| α₁-AT$_{(Cys-232-IANBD)}$ | 25 | 13.97 | 2.84 | 4.98 | GuHCl (1–3,5) | 99J6 |
| α₁-AT$_{(Cys-313-IANBD)}$ | 25 | 5.31 | 1.72 | 3.18 | GuHCl (1–3,6) | 99J6 |
| tryptophan fluorescence and far-UV CD: | | | | | | |
| α₁-AT$_{(P1 = Arg)}$ | 25 | 9.29 | 2.50 | 3.68 | GuHCl (1–4) | 99J6 |
| α₁-AT$_{(P3'Cys-IANBD)}$ | 25 | 9.50 | 2.49 | 3.81 | GuHCl (1–3,7) | 99J6 |
| α₁-AT$_{(Cys-232-IANBD)}$ | 25 | 9.20 | 2.56 | 3.60 | GuHCl (1–3,5) | 99J6 |
| α₁-AT$_{(Cys-313-IANBD)}$ | 25 | 9.12 | 2.45 | 3.72 | GuHCl (1–3,6) | 99J6 |

Remarks:
(1) linear extrapolation
(2) transitions monitored by fluorescence emission and far-UV CD
(3) IANBD = N,N'-dimethyl-(iodoacetyl)-N'-(7-nitrobenz-2-oxa-1,3-diazol-4-yl) ethylenediamine
(4) α₁-AT$_{(P1 = Arg)}$ = control
(5) α₁-AT$_{(Cys-232-IANBD)}$ = unique cysteine residue of α₁-AT, IANBD-modified
(6) α₁-AT$_{(Cys-313-IANBD)}$ = mutant Ser313→Cys, IANBD-modified
(7) α₁-AT$_{(P3'Cys-IANBD)}$ = mutant Pro361→Cys (P3'), IANBD-modified

α₁-Antitrypsin (α₁-AT) with fluorescence labels at various positions
Results of a three-state analysis

| Protein | Transition | T | $\Delta G$ | $c_{1/2}$ | m | Appr./Rem. | Ref |
|---|---|---|---|---|---|---|---|
| fluorescence probe: | | | | | | | |
| α₁-AT$_{(P3'Cys-IANBD)}$ | N→I | 25 | 5.82 | 0.43 | 14.39 | GuHCl (1–3,7) | 99J6 |
| | I→U | 25 | 13.05 | 2.45 | 5.27 | GuHCl (1–3,7) | 99J6 |
| tryptophan fluorescence and far-UV CD: | | | | | | | |
| α₁-AT$_{(P1 = Arg)}$ | N→I | 25 | 17.24 | 0.74 | 23.18 | GuHCl (1–4) | 99J6 |
| | I→U | 25 | 26.44 | 2.70 | 9.79 | GuHCl (1–4) | 99J6 |
| α₁-AT$_{(P3'Cys-IANBD)}$ | N→I | 25 | 18.45 | 0.76 | 24.10 | GuHCl (1–3,7) | 99J6 |
| | I→U | 25 | 27.61 | 2.50 | 11.05 | GuHCl (1–3,7) | 99J6 |

$\alpha_1$-Antitrypsin ($\alpha_1$-AT) with fluorescence labels at various positions
Results of a three-state analysis (continued)

| Protein | Transition | T | $\Delta G$ | $c_{1/2}$ | m | Appr./Rem. | Ref |
|---|---|---|---|---|---|---|---|
| $\alpha_1$-AT$_{(Cys-232-IANBD)}$ | N→I | 25 | 19.37 | 0.79 | 24.43 | GuHCl (1–3,5) | 99J6 |
| | I→U | 25 | 26.78 | 2.66 | 10.04 | GuHCl (1–3,5) | 99J6 |
| $\alpha_1$-AT$_{(Cys-313-IANBD)}$ | N→I | 25 | 17.32 | 0.80 | 21.59 | GuHCl (1–3,6) | 99J6 |
| | I→U | 25 | 25.94 | 2.48 | 10.46 | GuHCl (1–3,6) | 99J6 |

Remarks:
(1) linear extrapolation
(2) transitions monitored by fluorescence emission and far-UV CD
(3) IANBD = N,N'-dimethyl-(iodoacetyl)-N'-(7-nitrobenz-2-oxa-1,3-diazol-4-yl) ethylenediamine
(4) $\alpha_1$-AT$_{(P1 = Arg)}$ = control
(5) $\alpha_1$-AT$_{(Cys-232-IANBD)}$ = unique cysteine residue of $\alpha_1$-AT, IANBD-modified
(6) $\alpha_1$-AT$_{(Cys-313-IANBD)}$ = mutant Ser313→Cys, IANBD-modified
(7) $\alpha_1$-AT$_{(P3'Cys-IANBD)}$ = mutant Pro361→Cys (P3'), IANBD-modified


**Apoflavodoxin**, see Flavodoxin


## Apolipoprotein

Human apolipoprotein A-I (hA-I) and deletion mutants

| Mutant | pH | T | $\Delta G$ | $c_{1/2}$ | m | Appr./Rem. | Ref |
|---|---|---|---|---|---|---|---|
| apo hA-I | 7.4 | 20 | 39.2±2.3 | 2.6 | 14.9±1.4 | urea (1–3) | 97R6 |
| apo hA-I | 7.4 | 20 | 39.2±2.3 | 2.64 | 14.9±1.4 | urea (1–3) | 98R9 |
| apo Δ(1–43)A-I | | | | | | | |
| | 7.4 | 20 | 18.8±1.0 | 1.9 | 10.1±0.5 | urea (1–3) | 97R6 |
| apo Δ(187–243)A-I | | | | | | | |
| | 7.4 | 20 | 31.0±2.3 | 2.6 | 11.9±0.9 | urea (1–3) | 98R9 |

Remarks:
(1) linear extrapolation, the slope of the pre- and postdenaturational baselines was taken into account
(2) transition monitored by CD, fluorescence, and UV absorbance
(3) measured in PBS buffer, pH 7.4

Human plasma apolipoprotein C-1 (apoC-1)

| Protein | pH | T | $\Delta G$ | $c_{1/2}$ | m | Appr./Rem. | Ref |
|---|---|---|---|---|---|---|---|
| apoC-1 | 7.5 | 0 | 13.0±3.3 | 0.84±0.07 | 15.5±2.9 | GuHCl (1–3) | 98G21 |
| | 7.5 | 0 | 11.7±3.3 | | | GuHCl (1–3) | 98G21 |
| | 7.5 | 25 | 7.5±4.2 | | | GuHCl (1–3) | 98G21 |

Remarks:
(1) linear extrapolation
(2) transition monitored by CD at 222 nm
(3) measured in 5 mM sodium phosphate buffer, pH 7.5


**Arc Repressor**, see Repressor Proteins

## Ascorbate Oxidase

Ascorbate oxidase from green zucchini

| Mutant | pH | T | $\Delta G$ | m | Approach/Remarks | | Ref |
|---|---|---|---|---|---|---|---|
| from fluorescence anisotropy: | | | | | | | |
| trans. (1) | 6.0 | 20 | 15.9±3.3 | 7.5±1.7 | urea | (1,2) | 97M6 |
| trans. (2) | 6.0 | 20 | 54.8±5.0 | 3.8±0.8 | urea | (1,2) | 97M6 |
| trans. (1) | 6.0 | 20 | 12.6±1.3 | 12.1±1.3 | GuHCl | (1,2) | 97M6 |
| trans. (2) | 6.0 | 20 | 53.6±2.9 | 7.5±1.3 | GuHCl | (1,2) | 97M6 |
| from fluorescence spectrum: | | | | | | | |
| trans. (1) | 6.0 | 20 | 13.4±1.7 | 6.3±0.8 | urea | (1,3) | 97M6 |
| trans. (2) | 6.0 | 20 | 58.6±3.8 | 5.9±0.8 | urea | (1,3) | 97M6 |
| trans. (1) | 6.0 | 20 | 15.1±1.7 | 13.4±1.3 | GuHCl | (1,3) | 97M6 |
| trans. (2) | 6.0 | 20 | 49.4±2.5 | 6.3±0.8 | GuHCl | (1,3) | 97M6 |
| from circular dichroism (CD): | | | | | | | |
| trans. (2) | 6.0 | 20 | 56.9±2.1 | 3.8±0.4 | urea | (4) | 97M6 |
| trans. (2) | 6.0 | 20 | 53.1±0.8 | 7.5±0.4 | GuHCl | (4) | 97M6 |

Remarks:
(1) the transition was treated as follows: $N_2 \to I_2 \to 2U$, i.e., a transition with a dimeric intermediate in equilibrium with both, native dimer and unfolded monomers, assuming a linear dependence of $\Delta G_1$ and $\Delta G_2$ on denaturant concentration
(2) transition monitored by fluorescence anisotropy
(3) transition monitored by fluorescence spectrum (peak)
(4) transition monitored by CD at 220 nm, the transition was treated by a two-state model

## Ascorbate Peroxidase

Ascorbate peroxidase, wild type and mutants

| Mutant | pH | $T_{trs}$ | $\Delta T$ | $\Delta(\Delta G)$ | Approach/Remarks | | Ref |
|---|---|---|---|---|---|---|---|
| wild type | 7.0 | 58.3±0.5 | 0.0 | 0.0 | DSC | (1–3) | 98M5 |
| Glu112→Ala | 7.0 | 56.0±0.8 | −2.3 | −5.9 | DSC | (1–3) | 98M5 |
| Glu112→Lys | 7.0 | 53.0±0.9 | −5.3 | −11.7 | DSC | (1–3) | 98M5 |

Remarks:
(1) unfolding of a native dimer into two unfolded monomers ($N_2 \to 2U$), values are calculated per monomer
(2) $\Delta(\Delta G)$ was calculated using the wild-type value for $\Delta Cp$ of 0.7 kJ/mol/K
(3) measured in 50 mM potassium phosphate, 0.15 M NaCl, pH 7.0, at a scan rate of 15 K/h

## Azurin

Azurin (Az), effect of metal ions on the conformational stability

| Mutant | pH | T | $\Delta G$ | $c_{1/2}$ | m | Appr./Rem. | Ref |
|---|---|---|---|---|---|---|---|
| Az(Cu$^{2+}$) | 7.0 | 25 | 52.2±4.7 | 3.94±0.02 | 13.3±1.2 | GuHCl (1) | 97L5 |
| Az(Cu$^{2+}$) | 7.0 | 25 | 52.2±4.7 | 3.9 | | GuHCl (1,2) | 97L6 |
| Az(Cu$^{+}$) | 7.0 | 25 | 40.0±6.9 | 2.5 | | GuHCl (1,2) | 97L6 |
| Az(Zn$^{2+}$) | 7.0 | 25 | 44.3±4.1 | 2.75±0.02 | 16.2±1.5 | GuHCl (1) | 97L5 |
| Az(apo)-1 | 7.0 | 25 | | 1.66±0.03 | 24.0±5.0 | GuHCl (1,3) | 97L5 |
| Az(apo)-2 | 7.0 | 25 | | 3.09±0.07 | 10.9±3.2 | GuHCl (1,3) | 97L5 |

Remarks:
(1) linear extrapolation
(2) for a thermodynamic cycle for the redox and unfolding equilibrium of azurin, see Ref. 97L6
(3) apoazurin displays two transitions at equilibrium unfolding

Recombinant wild-type azurin from *Pseudomonas aeruginosa* and mutants

| Protein | pH | T | ΔG | $c_{1/2}$ | m | Appr./Rem. | Ref |
|---|---|---|---|---|---|---|---|
| holo-proteins: | | | | | | | |
| wild type | 7.2 | 20 | 39.3±1.7 | 2.78 | 13.8±1.3 | GuHCl (1–3) | 99M14 |
| Ile7→Ser | 7.2 | 20 | 24.3±1.3 | 1.61 | 15.1±1.3 | GuHCl (1–3) | 99M14 |
| Phe110→Ser | 7.2 | 20 | 20.1±1.3 | 1.24 | 15.1±1.3 | GuHCl (1–3) | 99M14 |
| apo-proteins: | | | | | | | |
| wild type | 7.2 | 20 | 26.8±1.7 | 1.74 | 15.5±1.3 | GuHCl (1–3) | 99M14 |
| Ile7→Ser | 7.2 | 20 | 14.2±1.7 | 1.02 | 14.2±1.3 | GuHCl (1–3) | 99M14 |
| Phe110→Ser | 7.2 | 20 | 11.3±0.8 | 0.86 | 13.4±0.8 | GuHCl (1–3) | 99M14 |

Remarks:

(1) linear extrapolation

(2) transition monitored by CD at 220 nm and fluorescence intensity

(3) buffer: 50 mM Tris-HCl, pH 7.2

Recombinant wild-type azurin from *Pseudomonas aeruginosa*, wild type and double mutant (Cys3→Ala and Cys26→Ala)

| Protein | pH | T | ΔG | Approach/Remarks | | Ref |
|---|---|---|---|---|---|---|
| wild type | | 20 | 59.7±6.5 | DSC | (1) | 99G23 |
| | 7.03 | 20 | 63 | DSC | (4) | 95L14 |
| double mutant (Cys3→Ala and Cys26→Ala) | | | | | | |
| | 7.03 | 20 | 32.6±2.6 | DSC | (2,3) | 99G23 |

Remarks:

(1) data from the authentic protein, presented in Ref. 96M14, cited in Ref. 99G23

(2) measured in 10 mM phosphate buffer, pH 7.03, with 0.1 M NaCl

(3) based on DSC data at infinite scan rate

(4) based on DSC data, the protein achieves maximal stability at about 20°C, for details see Ref. 95L14

Azurin from *Pseudomonas aeruginosa* and $Cu_A$ domain from *Thermus thermophilus*

| Protein | pH | T | ΔG | $c_{1/2}$ | m | Appr./Rem. | Ref |
|---|---|---|---|---|---|---|---|
| azurin ox. | 7.0 | 25 | | 3.9 | | GuHCl (1) | 98W11 |
| azurin red. | 7.0 | 25 | | 2.5 | | GuHCl (1) | 98W11 |
| $Cu_A$ | 7.0 | 25 | | 3.3 | | GuHCl (2) | 98W11 |
| $Cu_A$ | 7.0 | 25 | | 3.1 | | GuHCl (3) | 98W11 |

Remarks:

(1) for more detailed data on azurin, see also Refs. 97L5 and 97L6

(2) fraction of high potential $Cu_A$ from voltammetry

(3) from optical absorption at 360 nm, data taken from Fig. 5 in Ref. 99W14

## Barnase

The data entries are arranged as follows:

a) wild type and mutants, equilibrium unfolding

b) wild type and mutants in the presence of chaperone

c) wild type and mutants, transitions

d) wild type, fragments, and permutants

e) wild type and precursor proteins

*a) wild type and mutants, equilibrium unfolding*

Barnase, Ribonuclease from *Bacillus amyloliquefaciens*, wild type and Gly mutants

| Mutant | pH | $T_{trs}$ | $\Delta T$ | $\Delta(\Delta G)$ | Approach/Remarks | | Ref |
|---|---|---|---|---|---|---|---|
| wild type | 6.3 | 54.1±0.2 | 0.0 | 0.0 | DSC | (1,2) | 99A4 |
| Gly52→Ala | 6.3 | 40.8±0.3 | −13.3 | −22.2 | DSC | (1,2) | 99A4 |
| Gly52→Val | 6.3 | 33.0±0.3 | −21.1 | −35.1 | DSC | (1,2) | 99A4 |
| Gly53→Ala | 6.3 | 45.5±0.2 | −8.6 | −14.2 | DSC | (1,2) | 99A4 |
| Gly53→Val | 6.3 | 34.5±0.4 | −19.6 | −32.6 | DSC | (1,2) | 99A4 |
| Gly53Δ | 6.3 | 41.0±0.5 | −13.1 | −21.8 | DSC | (1–3) | 99A4 |

Remarks:
(1) measured in 50 mM MES buffer, pH 6.3
(2) $\Delta(\Delta G)$ was calculated using $\Delta(\Delta G) = \Delta T_{mut} \times \Delta S_{w.t.}$ with $\Delta S_{w.t.} = 1.67$ kJ/mol/K from Ref. 94M9
(3) deletion mutant

Barnase, Ribonuclease from *Bacillus amyloliquefaciens*, mutants at position 73

| Mutant | pH | T | $\Delta(\Delta G)$ | Approach/Remarks | | Ref |
|---|---|---|---|---|---|---|
| wild type | 6.3 | 25 | 0.0 | urea | (1,2) | 97S7 |
| Glu73→Ala | 6.3 | 25 | −9.6 | urea | (1,2) | 97S7 |
| Glu73→Gln | 6.3 | 25 | −11.3 | urea | (1,2) | 97S7 |
| Glu73→Phe | 6.3 | 25 | −8.8 | urea | (1,2) | 97S7 |
| Glu73→Trp | 6.3 | 25 | −9.2 | urea | (1,2) | 97S7 |

Remarks:
(1) for the procedure, see Ref. 93C6
(2) $\Delta G$ of the wild type, see Ref. 90S3

Barnase, Ribonuclease from *Bacillus amyloliquefaciens*, wild type and mutants with disulfide crosslinks

| Mutant | pH | T | $\Delta G$ | Approach/Remarks | Ref |
|---|---|---|---|---|---|
| wild type | 5 | 25 | 42.2 | (1–4) | 97J4 |
| (85–102) | 5 | 25 | 58.3 | (1–5) | 97J4 |
| (43–80) | 5 | 25 | 45.9 | (1–5) | 97J4 |
| (70–92) | 5 | 25 | 28.3 | (1–5) | 97J4 |

Remarks:
(1) the data were taken from Fig. 6 in Ref. 97J4
(2) Ref. 97J4 contains an extended function of barnase stability ($\Delta G$) on pH, between pH 1 and 7
(3) the data were calculated using thermodynamic quantities obtained by DSC and melting curves
(4) $\Delta C_p$ was taken as 7.1 kJ/mol/K
(5) barnase mutants with crosslinks between residues 85 and 102, 43 and 80, and 70 and 92, respectively

Barnase, Ribonuclease from *Bacillus amyloliquefaciens*, wild type and mutants with disulfide crosslinks, difference values

| Mutant | pH | T | $\Delta(\Delta G)$ | Approach/Remarks | | Ref |
|---|---|---|---|---|---|---|
| (85–102) | 4.4 | 25 | 13.8±1.3 | DSC | (1–3) | 97J4 |
| (85–102)SH | 4.4 | 25 | −0.4±1.3 | DSC | (1,3,4) | 97J4 |
| Ser85→Cys | 4.4 | 25 | −1.7±1.3 | DSC | (1) | 97J4 |
| His102→Cys | 4.4 | 25 | 0.4±1.3 | DSC | (1) | 97J4 |
| (43–80) | 4.4 | 25 | 3.8±0.8 | DSC | (1–3) | 97J4 |
| (43–80)SH | 4.4 | 25 | −4.6±1.3 | DSC | (1,4) | 97J4 |
| Ala43→Cys | 4.4 | 25 | −3.8±1.3 | DSC | (1,3) | 97J4 |
| Ser80→Cys | 4.4 | 25 | −0.8±1.3 | DSC | (1,3) | 97J4 |

Barnase, Ribonuclease from *Bacillus amyloliquefaciens*, wild type and mutants with disulfide crosslinks, difference values (continued)

| Mutant | pH | T | Δ(ΔG) | Approach/Remarks | | Ref |
|---|---|---|---|---|---|---|
| (70–92) | 4.4 | 25 | −15.9±0.8 | DSC | (1–3) | 97J4 |
| (70–92)SH | 4.4 | 25 | −13.0±1.3 | DSC | (1,4) | 97J4 |
| Thr70→Cys | 4.4 | 25 | −5.0±1.3 | DSC | (1,3) | 97J4 |
| Ser92→Cys | 4.4 | 25 | −9.6±1.3 | DSC | (1,3) | 97J4 |

Remarks:

(1) Δ(ΔG) was calculated relative to wild-type barnase using $\Delta Cp = 7.1$ kJ/mol/K

(2) barnase mutants with crosslinks between residues 85 and 102, 43 and 80, and 70 and 92, respectively

(3) multiple measurements at pH 3.4 and/or 4.4 were averaged

(4) (85–102)SH, (43–80)SH, and (70–92)SH are double cysteine mutants under reducing conditions

*b) wild type and mutants in the presence of chaperone*

Barnase, Ribonuclease from *Bacillus amyloliquefaciens*, in the presence and absence of chaperone [remark (6)]

| Mutant | pH | T | ΔG | Approach/Remarks | | Ref |
|---|---|---|---|---|---|---|
| wild type | 6.3 | 25 | 43.1 | heat | (1–3) | 97P7 |
| probarnase | 6.3 | 25 | 41.4 | heat | (1,2) | 97P7 |
| Asn58→Ala | 6.3 | 25 | 32.6 | heat | (1,2) | 97P7 |
| Arg87→Ala | 6.3 | 25 | 29.3 | heat | (1,2) | 97P7 |
| Ser91→Ala | 6.3 | 25 | 36.0 | heat | (1–3) | 97P7 |
| Trp94→Tyr | 6.3 | 25 | 36.0 | heat | (1,2) | 97P7 |
| double mutant (Ile4→Ala and Ile51→Val) | | | | | | |
| | 6.3 | 25 | 26.8 | heat | (1,2) | 97P7 |
| wild type (low-salt) | | | | | | |
| | 6.3 | 25 | 37.7 | heat | (2,4) | 97P7 |
| wild type (high-salt) | | | | | | |
| | 6.3 | 25 | 45.2 | heat | (2,5) | 97P7 |

Remarks:

(1) buffer: 50 mM MES pH 6.3

(2) ΔG was calculated from $T_{trs}$, ΔH (see Table 2), and $\Delta Cp = 6.69$ kJ/mol/K from Ref. 94M9

(3) data from Ref. 96Z

(4) buffer: 5 mM MES, pH 6.3

(5) buffer: 50 mM MES, pH 6.3, with 200 mM NaCl

(6) the data serve as the basis for the interpretation of melting curves of barnase, measured in the presence of GroEL and SecB, see Table 2

*c) wild type and mutants, transitions*

Barnase, Ribonuclease from *Bacillus amyloliquefaciens*, wild type and mutants, transition N → D

| Protein | pH | T | ΔG(N→D) | ∂(ΔG)/∂T | Appr./Rem. | | Ref |
|---|---|---|---|---|---|---|---|
| wild type | 6.3 | 25 | 43.9 | −0.075 | DSC | (1–3) | 98D2 |
| Ile4→Ala | 6.3 | 25 | 36.0 | −0.033 | DSC | (1–3) | 98D2 |
| Thr6→Gly | 6.3 | 25 | 37.2 | 0.038 | DSC | (1–3) | 98D2 |
| Asp8→Ala | 6.3 | 25 | 41.4 | 0.033 | DSC | (1–3) | 98D2 |
| Asp8→Gly | 6.3 | 25 | 39.7 | −0.004 | DSC | (1–3) | 98D2 |
| Asp12→Gly | 6.3 | 25 | 39.7 | −0.017 | DSC | (1–3) | 98D2 |
| Tyr17→Ala | 6.3 | 25 | 36.0 | 0.033 | DSC | (1–3) | 98D2 |
| His18→Asn | 6.3 | 25 | 36.4 | −0.013 | DSC | (1–3) | 98D2 |
| Val36→Ala | 6.3 | 25 | 37.7 | −0.046 | DSC | (1–3) | 98D2 |

Barnase, Ribonuclease from *Bacillus amyloliquefaciens*, wild type and mutants, transition N → D (continued)

| Protein | pH | T | ΔG(N→D) | ∂(ΔG)/∂T | Appr./Rem. | Ref |
|---------|----|----|---------|----------|------------|-----|
| Asn41→Ala | 6.3 | 25 | 27.2 | –0.130 | DSC (1–3) | 98D2 |
| Asn58→Ala | 6.3 | 25 | 35.1 | –0.046 | DSC (1–3) | 98D2 |
| Ile88→Val | 6.3 | 25 | 38.1 | –0.013 | DSC (1–3) | 98D2 |
| Ser91→Ala | 6.3 | 25 | 33.1 | –0.105 | DSC (1–3) | 98D2 |
| Ser92→Ala | 6.3 | 25 | 31.4 | –0.042 | DSC (1–3) | 98D2 |
| Ile96→Val | 6.3 | 25 | 41.0 | 0.025 | DSC (1–3) | 98D2 |
| Thr105→Val | 6.3 | 25 | 34.3 | –0.038 | DSC (1–3) | 98D2 |

Remarks:

(1) ΔG was calculated using ΔCp = 7.11 kJ/mol/K from Ref. 95J1; ΔCp was assumed to be the same for all mutants

(2) measured in 50 mM MES buffer

(3) ∂(ΔG)/∂T is the temperature dependence of ΔG(N→D) in kJ/mol/K

Barnase, Ribonuclease from *Bacillus amyloliquefaciens*, wild type and mutants, temperature dependence of m(N→D) and the dependence of ΔG(N→D) from the urea conc. [see remark (1)]

| Protein | $m(N→D)^{25°C}$ | $m(N→D)^{30°C}$ | $m(N→D)^{35°C}$ | $m(N→D)^{40°C}$ | $m(N→D)^{45°C}$ | Ref |
|---------|------|------|------|------|------|-----|
| wild type | 10.50 | 10.38 | 9.87 | 9.46 | | 98D2 |
| Asp8→Ala | 12.01 | | 10.63 | 10.00 | 9.41 | 98D2 |
| Asp12→Gly | 11.21 | | 10.21 | 9.80 | 9.41 | 98D2 |
| Tyr17→Ala | 12.64 | | 11.00 | 10.29 | 10.33 | 98D2 |
| Asn41→Ala | 11.76 | 11.17 | 10.67 | | | 98D2 |
| Ile96→Val | 11.67 | | 10.63 | 10.21 | | 98D2 |
| Thr105→Val | 11.34 | | 10.13 | 9.62 | | 98D2 |

Remark:

(1) m(N→D) at any temperature was obtained from the slope of a plot ΔG(N→D) versus urea conc., in the range 0 to 1 to 2.5 M (depending on the protein stability) at that temperature, to ±0.2–0.6 kJ/mol/K

Barnase, Ribonuclease from *Bacillus amyloliquefaciens*, wild type and mutants, transition I → D

| Protein | pH | T | ΔG(I→D) | Approach/Remarks | Ref |
|---------|----|----|---------|------------------|-----|
| wild type | 6.3 | 25 | 11.92 | DSC (1,2) | 98D2 |
| Ile4→Ala | 6.3 | 25 | 12.51 | DSC (1,2) | 98D2 |
| Thr6→Gly | 6.3 | 25 | 11.51 | DSC (1,2) | 98D2 |
| Asp8→Ala | 6.3 | 25 | 10.79 | DSC (1,2) | 98D2 |
| Asp8→Gly | 6.3 | 25 | 8.79 | DSC (1,2) | 98D2 |
| Asp12→Gly | 6.3 | 25 | 10.29 | DSC (1,2) | 98D2 |
| Tyr17→Ala | 6.3 | 25 | 9.04 | DSC (1,2) | 98D2 |
| His18→Asn | 6.3 | 25 | 7.53 | DSC (1,2) | 98D2 |
| Val36→Ala | 6.3 | 25 | 9.46 | DSC (1,2) | 98D2 |
| Asn41→Ala | 6.3 | 25 | 8.62 | DSC (1,2) | 98D2 |
| Asn58→Ala | 6.3 | 25 | 3.14 | DSC (1,2) | 98D2 |
| Ile88→Val | 6.3 | 25 | 9.08 | DSC (1,2) | 98D2 |
| Ser91→Ala | 6.3 | 25 | 3.93 | DSC (1,2) | 98D2 |
| Ser92→Ala | 6.3 | 25 | 4.64 | DSC (1,2) | 98D2 |
| Ile96→Val | 6.3 | 25 | 10.50 | DSC (1,2) | 98D2 |
| Thr105→Val | 6.3 | 25 | 10.21 | DSC (1,2) | 98D2 |

Remarks:

(1) data from kinetics and equilibrium unfolding, for details see Ref. 98D2

(2) measured in 50 mM MES buffer

Barnase, Ribonuclease from *Bacillus amyloliquefaciens*, wild type and mutants, temperature dependence of m(I→D) and the [urea] dependence of ΔG(I→D) [see remark (1)]

| Protein | m(I→D)$^{25°C}$ | m(I→D)$^{30°C}$ | m(I→D)$^{35°C}$ | m(I→D)$^{40°C}$ | m(I→D)$^{45°C}$ | Ref |
|---|---|---|---|---|---|---|
| wild type | 4.85 | 4.90 | 4.69 | 4.98 | | 98D2 |
| Asp8→Ala | 7.28 | | 6.61 | 6.11 | 2.30 | 98D2 |
| Asp12→Gly | 6.07 | | 5.15 | 4.60 | 2.47 | 98D2 |
| Tyr17→Ala | 7.11 | | 4.60 | 3.01 | 0.75 | 98D2 |
| Asn41→Ala | 6.78 | 6.65 | 6.44 | | | 98D2 |
| Ile96→Val | 6.32 | | 7.07 | 9.00 | | 98D2 |
| Thr105→Val | 6.61 | | 3.64 | 2.26 | | 98D2 |

Remark:

(1) m(I→D) at any temperature was obtained from the slope of a plot of ΔG(I→D) versus urea conc., in the range 0 to 1 to 2.5 M (depending on the protein stability) at that temperature, to ±0.2–0.6 kJ/mol/K

Barnase, Ribonuclease from *Bacillus amyloliquefaciens*, wild type and mutants, temperature dependence of ΔG(N→D)

| Protein | pH | T | ΔG(N→D) | Approach/Remarks | | Ref |
|---|---|---|---|---|---|---|
| wild type | 6.3 | 25.0 | 43.9 | DSC | (1,2) | 98D1 |
| | 6.3 | 33.0 | 34.3 | DSC | (1,2) | 98D1 |
| | 6.3 | 37.0 | 28.9 | DSC | (1,2) | 98D1 |
| | 6.3 | 45.0 | 16.7 | DSC | (1,2) | 98D1 |
| Cys43-Cys80 | 6.3 | 25.0 | 48.1 | DSC | (1–3) | 98D1 |
| | 6.3 | 33.0 | 39.3 | DSC | (1–3) | 98D1 |
| | 6.3 | 37.0 | 34.7 | DSC | (1–3) | 98D1 |
| | 6.3 | 45.0 | 23.8 | DSC | (1–3) | 98D1 |
| | 6.3 | 50.0 | 16.3 | DSC | (1–3) | 98D1 |
| Asn58→Ala | 6.3 | 25.0 | 35.1 | DSC | (1,2) | 98D1 |
| | 6.3 | 33.0 | 25.5 | DSC | (1,2) | 98D1 |
| | 6.3 | 37.0 | 20.1 | DSC | (1,2) | 98D1 |
| | 6.3 | 45.0 | 8.8 | DSC | (1,2) | 98D1 |
| Ser91→Ala | 6.3 | 25.0 | 33.1 | DSC | (1,2) | 98D1 |
| | 6.3 | 33.0 | 24.3 | DSC | (1,2) | 98D1 |
| | 6.3 | 37.0 | 19.2 | DSC | (1,2) | 98D1 |
| | 6.3 | 45.0 | 7.9 | DSC | (1,2) | 98D1 |
| wild type in D$_2$O | | | | | | |
| | 7.8 | 25.0 | 42.7 | DSC | (1,4) | 98D1 |
| | 7.8 | 33.0 | 33.4 | DSC | (1,4) | 98D1 |
| | 7.8 | 37.0 | 28.3 | DSC | (1,4) | 98D1 |

Remarks:

(1) ΔG(N→D) from DSC, proline residues in *trans* conformation only, ΔG(N→D) was calculated using ΔCp = 7.11 kJ/mol/K from Ref. 95J1; ΔCp was assumed to be the same for all mutants

(2) measured in 50 mM MES buffer/H$_2$O, pH 6.3

(3) Cys43-Cys80 = stabilized disulfide mutant (Ala43→Cys and Ser80→Cys), see also Ref. 93C7

(4) measured in 50 mM imidazole buffer/D$_2$O, pD 7.8

*d) wild type, fragments, and permutants*

Barnase, Ribonuclease from *Bacillus amyloliquefaciens*, wild type and N-terminal fragments

| Fragment | pH | T | ΔG | $c_{1/2}$ | m | Appr./Rem. | Ref |
|---|---|---|---|---|---|---|---|
| wild type | 6.3 | 25 | 36.9 | 4.57 | 8.08 | urea (1–4) | 99N1 |
| B1-105 | 6.3 | 25 | 8.36±0.84 | 1.07±0.05 | 8.24±0.50 | urea (1,3–6) | 99N1 |
| | 6.3 | 25 | | 0.91±0.50 | 4.64±1.17 | urea (1,5–7) | 99N1 |
| | 6.3 | 25 | | 2.66±0.30 | 6.15±1.55 | urea (1,8,9) | 99N1 |

Remarks:

(1) linear extrapolation

(2) data from Ref. 92S8

(3) measured in 50 mM MES buffer, pH 6.3

(4) transition monitored by fluorescence

(5) the following N-terminal barnase fragments were studied: (1–22), (1–36), (1–56), (1–68), (1–79), (1–95), (1–105 = B1–105), and wild type (residues 1–110)

(6) fragments up to (1–95) appeared to be mainly disordered. Thermal denaturation in 5 mM MES, pH 6.3, monitored by ellipticity at 222, 230, and 280 nm gives $T_{trs}$ = 34.7±0.3°C for the fragment B1-105 and $T_{trs}$ = 54.9±0.2°C for the wild type

(7) transition monitored by far-UV CD

(8) transition monitored by size exclusion chromatography

(9) measured in 50 mM MES in the presence of 0.150 M NaCl

Barnase, Ribonuclease from *Bacillus amyloliquefaciens*, wild type, C-terminal fragments

| Fragment | pH | T | ΔG | $c_{1/2}$ | m | Appr./Rem. | Ref |
|---|---|---|---|---|---|---|---|
| wild type | 6.3 | 25 | 36.9 | 4.57 | 8.08 | urea (1–4) | 99N1 |
| B23–110 | 6.3 | 25 | 24.8* | 3.07±0.06 | 8.08±1.21 | urea (1,3–6) | 99N2 |
| | 6.3 | 25 | | 2.94±0.05 | 6.40±0.79 | urea (1,5–7) | 99N2 |
| | 6.3 | 25 | | 3.70±0.07 | 7.90±1.30 | urea (1,8,9) | 99N2 |
| | 6.3 | 25 | | 3.07±0.06 | | urea (1,4,9) | 99N2 |

Remarks:

(1) linear extrapolation

(2) data from Ref. 92S8

(3) measured in 50 mM MES buffer, pH 6.3

(4) transition monitored by fluorescence

(5) the following C-terminal barnase fragments were studied: (96–110), (80–110), (69–110), (57–110), (37–110), (23–110 = B23–110), and wild type (residues 1–110)

(6) fragments up to (37–110) appeared to be mainly disordered. Thermal denaturation in 5 mM MES, pH 6.3, monitored by ellipticity at 230 nm gives $T_{trs}$ = 46°C for the fragment B23–110, i.e., 9°C less than for the wild type

(7) transition monitored by far-UV CD at 222 and 230 nm

(8) transition monitored by size exclusion chromatography

(9) measured in 50 mM MES in the presence of 0.150 M NaCl

Barnase, Ribonuclease from *Bacillus amyloliquefaciens*, permutations of modules and secondary structure units

| Variant | pH | T | ΔG | $c_{1/2}$ | m | Appr./Rem. | Ref |
|---|---|---|---|---|---|---|---|
| wild type | 6.0 | 5 | 43.1±4.6 | 5.7 | 7.53±0.84 | urea (1–5) | 99T15 |
| M3245 | 6.0 | 5 | 5.0±0.8 | 1.3 | 4.02±0.33 | urea (1–6) | 99T15 |
|  | 6.0 | 5 | 7.9±0.4 | 1.6 | 5.02±0.42 | urea (1–5,7) | 99T15 |
| S2543 | 6.0 | 5 | 10.0±0.4 | 0.96 | 10.46±0.42 | urea (1–6) | 99T15 |
|  | 6.0 | 5 | 9.6±0.4 | 0.96 | 10.04±0.42 | urea (1–5,7) | 99T15 |

Remarks:

(1) 44 mutants were constructed containing permutations of internal modules and secondary structure units. Of the 44 mutants, only two showed foldability

(2) for the construction of the mutants, see Ref. 99T15

(3) data from equilibrium unfolding, nonlinear least-squares fit of the two-state transition curve assuming a linear dependence of ΔG on the denaturant conc.

(5) buffer: 5 mM Tris-HCl, pH 6.0

(6) transition monitored by far-UV CD at 230 nm

(7) transition monitored by internal fluorescence at 340 nm

*e) wild type and precursor proteins*

Barnase, Ribonuclease from *Bacillus amyloliquefaciens*, wild type and precursor proteins

| Precursor | pH | T | ΔG | $c_{1/2}$ | m | Appr./Rem. | Ref |
|---|---|---|---|---|---|---|---|
| no | 6.3 | 25 | 36.9 | 4.57 | 8.08 | urea (1–3,5) | 99H15 |
| 35 aa | 6.3 | 25 | 31.3 | 4.4±0.1 | 7.1±0.8 | urea (1–4) | 99H15 |
| 65 aa | 6.3 | 25 | 31.3 | 4.4±0.1 | 7.1±0.4 | urea (1–4) | 99H15 |
| 95 aa | 6.3 | 25 | 30.6 | 4.3±0.2 | 7.1±2.5 | urea (1–4) | 99H15 |

Remarks:

(1) linear extrapolation, for details see also Refs. 89M4, 90M3, 92S8

(2) measured in 45 mM MES buffer in the presence of DTT

(3) transition monitored by fluorescence emission at 320 nm

(4) barnase was converted into a mitochondrial precursor protein by attaching targeting sequences of the first 35, 65, or 95 amino acids of pre-cytochrome $b_2$

(5) data from Refs. 92S8 and 92S9

## Barstar

The data entries are arranged as follows:

a) wild type and tryptophan mutants
b) wild type, cysteine mutants, and mutants derived from the pseudo-wild type
c) chemically modified protein

*a) wild type and tryptophan mutants*

Barstar, wild type

| | pH | T | ΔG | $c_{1/2}$ | m | Approach/Remarks | | Ref |
|---|---|---|---|---|---|---|---|---|
| | 8 | 25 | 4.85 | 1.9 | 2.05 | GuHCl | (1–3) | 97A1 |
| | 8 | 5 | | 1.48 | | GuHCl | (2–4) | 97A1 |

Remarks:

(1) two-state fit by a nonlinear procedure, see Ref. 95A1

(2) data treatment includes kinetic and equilibrium amplitudes of the unfolding of barstar

(3) buffer: 20 mM sodium phosphate, 50 µM EDTA, 0.1 mM DTT

(4) the kinetic data at 5°C reveal the presence of an intermediate during unfolding, using intrinsic Trp fluorescence as a probe

Barstar, analysis of unfolding in terms of a two-state and three-state model

| Transition | pH | T | ΔG | m | Approach/Remarks | | Ref |
|---|---|---|---|---|---|---|---|
| two-state model: | | | | | | | |
| N → U | 8 | 22 | 21.3 | 10.5 | GuHCl | (1–4) | 99B11 |
| | 8 | 22 | 18.8 | 9.6 | GuHCl | (1–3,5) | 99B11 |
| | 8 | 22 | 20.1 | 10.5 | GuHCl | (1–3,6) | 99B11 |
| three-state model: | | | | | | | |
| U → I | 8 | 22 | 8.8 | 6.1 | GuHCl | (1–4,7) | 99B11 |
| | 8 | 22 | 10.0 | 6.5 | GuHCl | (1–3,5,7) | 99B11 |
| | 8 | 22 | 10.5 | 8.8 | GuHCl | (1–3,6,7) | 99B11 |
| N → U$_F$ | 8 | 22 | 23.4 | 9.2 | GuHCl | (1–4,7,8) | 99B11 |
| | 8 | 22 | 22.6 | 9.2 | GuHCl | (1–3,5,7,8) | 99B11 |
| | 8 | 22 | 20.1 | 8.8 | GuHCl | (1–3,6–8) | 99B11 |

Remarks:
(1) nonlinear least-squares fit of the transition curve assuming a linear dependence of ΔG on the denaturant conc.
(2) GuHCl dependence of the initial and final signal values from kinetic experiments
(3) measured in 20 mM phosphate buffer, pH 8, containing 300 μM EDTA and 250 μM DTT
(4) transition monitored by fluorescence at 320 nm
(5) transition monitored by far-UV CD at 222 nm
(6) transition monitored by near-UV CD at 270 nm
(7) GuHCl dependence of the initial and final signal values from kinetic experiments analyzed in terms of a three-state model
(8) U$_F$ represents 31% of fast folding molecules

Barstar, wild type and tryptophan mutants

| Mutant | pH | T | ΔG | c$_{1/2}$ | m | Approach/Remarks | | Ref |
|---|---|---|---|---|---|---|---|---|
| wild type | 7 | 25 | 20.9 | 1.95±0.05 | 10.9 | GuHCl | (1–3) | 97N4 |
| Trp38→Phe | 7 | 25 | | 1.89±0.05 | | GuHCl | (1,3,4) | 97N4 |
| Trp38→Phe | 7 | 25 | | 1.73±0.05 | | GuHCl | (1,3,5) | 97N4 |
| Trp44→Phe | 7 | 25 | | 1.87±0.05 | | GuHCl | (1,3,4) | 97N4 |
| Trp44→Phe | 7 | 25 | | 1.77±0.05 | | GuHCl | (1,3,5) | 97N4 |
| double mutant (Trp38→Phe and Trp44→Phe) | | | | | | | | |
| | 7 | 25 | 18.4 | 1.75±0.05 | 10.5 | GuHCl | (1–3) | 97N4 |

Remarks:
(1) linear extrapolation
(2) transitions monitored by CD at 220 nm and 275 nm as well as by Trp fluorescence (at 334 nm for wild type and 327 nm for the double mutant); the unfolding curves were found to be coincident
(3) the T$_{trs}$ of the thermal transitions monitored by near- and far-UV CD are coincident. For wild type, Trp38→Phe, Trp44→Phe, and the double mutant T$_{trs}$ amounts to 70±0.5, 69.6±0.5, 70.0±0.5, and 68.0±0.5°C, respectively
(4) transition monitored by fluorescence and far-UV CD
(5) transition monitored by near-UV CD

Barstar, wild type and tryptophan double mutant (Trp38→Phe and Trp44→Phe) with Trp53 retained

| Mutant | pH | T | ΔG | m | Approach | Remarks | Ref |
|---|---|---|---|---|---|---|---|
| measurements at low ionic strength [remark (3)]: | | | | | | | |
| wild type | 7 | 10 | 18.0 | 5.02 | urea, CD | (1–4) | 97Z1 |
| wild type | 7 | 10 | 21.8 | 6.28 | urea, FL | (1–3,5) | 97Z1 |
| double mutant | 7 | 10 | 15.5 | 5.44 | urea, CD | (1–4) | 97Z1 |
| double mutant | 7 | 10 | 13.4 | 5.02 | urea, FL | (1–3,5) | 97Z1 |
| wild type | 7 | 25 | 19.7 | 4.60 | urea, CD | (1–4) | 97Z1 |
| wild type | 7 | 25 | 22.2 | 5.02 | urea, FL | (1–3,5) | 97Z1 |
| double mutant | 7 | 25 | 16.7 | 4.60 | urea, CD | (1–4) | 97Z1 |

Barstar, wild type and tryptophan double mutant (Trp38→Phe and Trp44→Phe) with Trp53 retained (continued)

| Mutant | pH | T | ΔG | m | Approach | Remarks | Ref |
|---|---|---|---|---|---|---|---|
| double mutant | 7 | 25 | 20.9 | 5.44 | urea, FL | (1–3,5) | 97Z1 |
| wild type | 7 | 40 | 18.4 | 5.44 | urea, CD | (1–4) | 97Z1 |
| wild type | 7 | 40 | 18.8 | 5.44 | urea, FL | (1–3,5) | 97Z1 |
| double mutant | 7 | 40 | 21.3 | 6.69 | urea, CD | (1–4) | 97Z1 |
| double mutant | 7 | 40 | 20.1 | 6.28 | urea, FL | (1–3,5) | 97Z1 |
| measurements at high ionic strength [remark (6)]: | | | | | | | |
| wild type | 7 | 25 | 29.3 | 5.44 | urea, CD | (1–4) | 97Z1 |
| wild type | 7 | 25 | 31.4 | 6.28 | urea, FL | (1–3,5) | 97Z1 |
| double mutant | 7 | 25 | 20.1 | 4.18 | urea, CD | (1–4) | 97Z1 |
| double mutant | 7 | 25 | 25.9 | 5.02 | urea, FL | (1–3,5) | 97Z1 |

Remarks:

(1) from equilibrium and kinetic amplitudes of urea-induced unfolding of barstar

(2) the paper demonstrates the presence of at least two unfolding intermediates on two competing unfolding pathways

(3) measured in 50 mM sodium phosphate, 0.25 mM EDTA, 0.25 mM DTT, pH 7

(4) CD = transition monitored by relative ellipticity

(5) FL = transition monitored by relative fluorescence

(6) measured in 50 mM sodium phosphate, 0.8 M KCl, 0.25 mM EDTA, 0.25 mM DTT, pH 7

*b) wild type, cysteine mutants, and mutants derived from the pseudo-wild type*

Barstar, wild type and mutant (Cys82→Ala)

| Mutant | pH | T | ΔG | Approach | Remarks | Ref |
|---|---|---|---|---|---|---|
| wild type | 6.4 | 25 | 27.4±2 | DSC | (1,2) | 97S6 |
| wild type | 7.4 | 25 | 23.5±2 | DSC | (1,2) | 97S6 |
| wild type | 8.0 | 25 | 25.7±2 | DSC | (1,2) | 97S6 |
| wild type | 8.3 | 25 | 25.3±2 | DSC | (1,2) | 97S6 |
| wild type | 7.4 | 25 | 23.5±2 | DSC | (1,3) | 97S6 |
| Cys82→Ala | 7.4 | 25 | 25.5±2 | DSC | (1,3) | 97S6 |

Remarks:

(1) Ref. 97S6 contains a detailed consideration of the error propagation in ΔG due to ΔCp

(2) measured in the presence of DTT, buffer: 50 mM sodium phosphate, 1 mM EDTA, 10 mM DTT

(3) measured in the absence of DTT, buffer: 50 mM sodium phosphate, 1 mM EDTA

Barstar, mutant Cys40→Ala, Cys82→Ala, and Pro27→Ala (b*C40A/C82A/P27A)

| Protein | pH | T | ΔG | Approach/Remarks | Ref |
|---|---|---|---|---|---|
| b*C40A/C82A | 8.0 | 25 | 20.25 | urea, Fl (1–3,6,8) | 99G9 |
| | 8.0 | 15 | 20.29 | urea, CD (1,2,4,8) | 99G9 |
| | 8.0 | 25 | 20.25 | urea, CD (1,2,4,6) | 99G9 |
| b*C40A/C82A/P27A | 8.0 | 25 | 13.93 | urea, Fl (1–3,5,8) | 99G9 |
| | 8.0 | 15 | 12.89 | urea, CD (1,2,4,5) | 99G9 |
| | 8.0 | 25 | 13.93 | urea, CD (1,2,4,5) | 99G9 |
| b*w.t.E76A | 8.0 | 25 | 25.48 | urea, Fl (1–3,7,8) | 99G9 |
| | 8.0 | 15 | 21.92 | urea, CD (1,2,4,5) | 99G9 |

Remarks:

(1) linear extrapolation, LEM-SB

(2) buffer: 50 mM sodium phosphate, pH 8.0

(3) transition monitored by fluorescence at 320 (332) nm

(4) transition monitored by CD at 222 nm

(5) ΔG was calculated using the average value of m = 5.27 kJ/mol/M

(6) see also Ref. 93S5

(7) see also Ref. 94S2

(8) Ref. 99G9 deals with the catalysis of the petidyl-prolyl cis/trans isomerization by human cytosolic cyclophilin

Barstar, mutants derived from the pseudo-wild type (Cys40→Ala, Cys82→Ala, and Pro27→Ala)

| Mutant | pH | Δ(ΔG) | Approach | Remarks | Ref |
|---|---|---|---|---|---|
| pseudo-wild type | 8.0 | 0.0 | urea | (1–3) | 97N11 |
| Ile5→Val | 8.0 | −4.2 | urea | (1–3) | 97N11 |
| Ser14→Ala | 8.0 | −2.1 | urea | (1–3) | 97N11 |
| Gln18→Gly | 8.0 | −5.4 | urea | (1–3) | 97N11 |
| Ala25→Gly | 8.0 | −5.4 | urea | (1–3) | 97N11 |
| Leu34→Val | 8.0 | −4.6 | urea | (1–3) | 97N11 |
| Gln58→Gly | 8.0 | −6.7 | urea | (1–3) | 97N11 |
| Ser59→Ala | 8.0 | −5.4 | urea | (1–3) | 97N11 |
| Gln72→Gly | 8.0 | −5.0 | urea | (1–3) | 97N11 |
| Ala77→Gly | 8.0 | −8.4 | urea | (1–3) | 97N11 |

Remarks:

(1) linear extrapolation, LEM-SB

(2) buffer: 50 mM Tris-HCl with 0.1 M KCl, pH 8.0

(3) for details of the approach see Ref. 95N4

Barstar, pseudo-wild type (Cys40→Ala, Cys82→Ala, and Pro27→Ala), *cis/trans* isomerization of Pro48

| Transition | pH | T | ΔG | Approach | Remarks | Ref |
|---|---|---|---|---|---|---|
| N(*trans*) → I(*trans*) | 8 | 10 | −4.2 | urea, kinetics | (1–3) | 97N10 |
| N(*cis*) → N(*trans*) | 8 | 10 | −5.9 | urea, kinetics | (1–3) | 97N10 |

Remarks:

(1) the major folding pathway of barstar is: D(*trans*) ↔ I(*trans*) ↔ N(*trans*) ↔ N(*cis*)

(2) the folding was monitored by fast kinetics from the combined changes in fluorescence of tryptophans 38, 44, and 53

(3) buffer: 50 mM Tris-HCl, 100 mM KCl, pH 8

*c) chemically modified protein*

Carboxyamidomethylated barstar mutant Val73→Ala (CAM-V73A)

| Mutant | pH | T | $\Delta G$ | $c_{1/2}$ | m | Approach/Remarks | | Ref |
|--------|-----|-----|------|------|------|--------------|------|------|
| CAM-V73A | 7 | 25 | 17.6 | 0.89 | 19.7 | GuHCl | (1,2) | 98P3 |

Remarks:
(1) the transition monitored by fluorescence intensities at 320 nm was analyzed in terms of a two-state transition
(2) linear extrapolation, for details see Ref. 95K7

*Bst*HU, see DNA-Binding Protein

## Calbindin

Recombinant apo bovine calbindin D9k, wild type and mutants

| Mutant | pH | T | $\Delta G$ | $c_{1/2}$ | m | Appr./Rem. | Ref |
|--------|-----|-----|------|------|------|--------|------|
| wild type | 7.0 | 25 | 27.4±1.3 | 5.6 | 4.9±0.2 | urea (1–3) | 98J3 |
| Leu6→Val | 7.0 | 25 | 18.2±1.0 | 4.3 | 4.2±0.2 | urea (1–3) | 98J3 |
| Phe10→Ala | 7.0 | 25 | 7.2±0.5 | 2.0 | 3.7±0.2 | urea (1–3) | 98J3 |
| Leu23→Ala | 7.0 | 25 | 11.8±0.6 | 2.7 | 4.3±0.2 | urea (1–3) | 98J3 |
| Leu23→Gly | 7.0 | 25 | 9.5±1.3 | 1.8 | 5.1±0.5 | urea (1–3) | 98J3 |
| Leu28→Ala | 7.0 | 25 | 16.1±0.7 | 3.8 | 4.3±0.2 | urea (1–3) | 98J3 |
| Pro43→Met | | | | 5.3 | | urea (4) | 93L10 |
| Val61→Ala | 7.0 | 25 | 11.6±0.7 | 3.4 | 3.4±0.2 | urea (1–3) | 98J3 |
| Val61→Gly | 7.0 | 25 | 9.2±0.8 | 2.6 | 3.6±0.2 | urea (1–3) | 98J3 |
| Phe66→Trp | 7.0 | 25 | 17.6±1.4 | 6.1 | 2.9±0.2 | urea (1–3) | 98J3 |
| Phe66→Ala | 7.0 | 25 | 6.6±0.5 | 2.0 | 3.3±0.2 | urea (1–3) | 98J3 |
| Val70→Leu | 7.0 | 25 | 22.4±1.0 | 6.6 | 3.4±0.1 | urea (1–3) | 98J3 |
| Ile73→Val | 7.0 | 25 | 21.0±1.1 | 4.4 | 4.8±0.2 | urea (1–3) | 98J3 |
| triple mutant (Leu39→Cys, Pro43→Met, and Ile73→Cys) | | | | | | | |
| | | | | 8.0 | | urea (5) | 93L10 |

Remarks:
(1) linear extrapolation, LEM-SB
(2) buffer: 10 mM potassium phosphate, 0.5 mM EGTA, pH 7.0
(3) transition monitored by CD at 222 nm
(4) thermal transition at 85°C (DSC), pH 7
(5) thermal transition at ≥95°C, pH 7, $\Delta(\Delta G)$ = 8 kJ/mol relative to Pro43→Met, 6.7 M urea

## Calmodulin

Calmodulin (CAM), loss of conformational stability upon methionine oxidation

| Protein | pH | $\Delta T$ | $\Delta(\Delta G)$ | Approach/Remarks | | Ref |
|---------|-----|------|------|-----------|------|------|
| CAM | 7.5 | 0 | 0.0 | heat | (1–3) | 98G2 |
| CAM-ox. | 7.5 | –10 | –1.3 | heat | (1,3,4) | 98G2 |

Remarks:
(1) $\Delta(\Delta G)$ was estimated assuming a two-state transition
(2) thermal transition at $T_{trs}$ = 97±3°C
(3) buffer: 10 mM Tris-HCl, 0.1 M $KClO_4$, 1 mM $Mg(ClO_4)_2$, and 0.1 mM $Ca(ClO_4)_2$
(4) thermal transition at $T_{trs}$ = 87±3°C

## Carbonic Anhydrase

Bovine muscle carbonic anhydrase, isoenzyme III (BCAIII), transitions

| Protein | | pH | T | $\Delta G$ | $c_{1/2}$ | m | Appr./Rem. | Ref |
|---|---|---|---|---|---|---|---|---|
| BCAIII | $N \rightarrow I$ | 7.5 | 23 | | 1.2±0.2 | | GuHCl (1) | 99B16 |
| | $N \rightarrow I$ | 7.5 | 23 | | 1.0±0.2 | | GuHCl (2,5) | 99B16 |
| | $N \rightarrow I$ | 7.5 | 23 | | 1.1±0.1 | | GuHCl (3,5) | 99B16 |
| | $N \rightarrow I$ | 7.5 | 23 | | 1.2±0.1 | | GuHCl (4,5) | 99B16 |
| BCAIII | $I \rightarrow U$ | 7.5 | 23 | | 2.6±0.2 | | GuHCl (3,5) | 99B16 |
| | $I \rightarrow U$ | 7.5 | 23 | | 2.5±0.4 | | GuHCl (4,5) | 99B16 |
| HCAII | $N \rightarrow I$ | 7.5 | 23 | 31.8 | 0.94 | 33.9 | GuHCl (6) | 93M7 |
| | $I \rightarrow U$ | 7.5 | 23 | 24.3 | 2.4 | 10.5 | GuHCl (6) | 93M7 |

Remarks:

(1) transition monitored by enzyme activity, measured in 0.1 M Tris $H_2SO_4$, pH 7.5

(2) transition monitored by near-UV CD at 270 nm

(3) transition monitored by far-UV CD at 218 nm

(4) transition monitored by optical absorption at 292 nm

(5) buffer: 0.01 M Na-phosphate

(6) reference value for human carbonic anhydrase II from Ref. 93M7, based on the transitions monitored by the ratio in absorption $A_{292}/A_{260}$

**Catabolic Activator Protein**, see also CRP, cAMP Receptor Protein

## Cell Cycle Regulatory Protein

Cell cycle regulatory protein p13$^{suc1}$

| Protein | pH | T | $\Delta G$ | $c_{1/2}$ | m | Approach/Remarks | | Ref |
|---|---|---|---|---|---|---|---|---|
| p13$^{suc1}$ | 7.5 | 25 | 30.1 | 4.4 | 11.46 | urea | (1–3) | 98R12 |
| p13$^{suc1}$ | 7.5 | 25 | 30.1 | 4.97±0.01 | 6.07±0.04 | urea | (1,2,4,5) | 98R12 |

Remarks:

(1) from equilibrium unfolding, linear extrapolation to zero denaturant concentration by a method that includes the pre- and postdenaturational baselines for a nonlinear regression of the data

(2) transition monitored by fluorescence intensity

(3) measured in 50 mM Tris buffer

(4) measured in 50 mM phosphate buffer

(5) thermal transition at $T_{trs}$ = 67°C (DSC, in 20 mM phosphate buffer)

## Cell Surface Receptor Protein CD2

Domain 1 of cell surface receptor protein CD2 from rat (CD2.d1), intermediates

| Transition | pH | T | $\Delta G$ | Approach/Remarks | | Ref |
|---|---|---|---|---|---|---|
| $N \rightarrow U$ | 7.5 | 25 | 36.7 | GuHCl | (1–4) | 98P5 |
| $N \rightarrow I$ | 7.5 | 25 | 27.6 | GuHCl | (1–4) | 98P5 |
| $I \rightarrow U$ | 7.5 | 25 | 9.1 | GuHCl | (1–4) | 98P5 |

Remarks:

(1) from folding and unfolding rates, measured by stopped-flow kinetics

(2) transition monitored by fluorescence intensity

(3) buffer: 50 mM triethanolamine, 2 mM DTT, and GuHCl

(4) Ref. 98P5 contains an extended temperature profile of $\Delta G$

Domain 1 of cell surface receptor protein CD2 from rat (CD2.d1, residues 1–98), isotope effects

| Transition/Reaction | | pH | T | $\Delta G$ | m* | Approach/Remarks | | Ref |
|---|---|---|---|---|---|---|---|---|
| protonated CD2: | | | | | | | | |
| $N \to U$ | [NH]CD2-$H_2O$ | 7.5 | 25 | 36.44±1.00 | 8.44±0.10 | GuHCl | (1–3) | 97P5 |
| $N \to U$ | [NH]CD2-$H_2O$ | 7.5 | 25 | 35.52±0.71 | 8.40±0.17 | GuHCl | (1–3,5) | 97P5 |
| $I \to U$ | [NH]CD2-$H_2O$ | 7.5 | 25 | 7.32±0.04 | 5.08±0.14 | GuHCl | (1–3) | 97P5 |
| protonated CD2: | | | | | | | | |
| $N \to U$ | [NH]CD2-$D_2O$ | 7.5 | 25 | 41.50±1.13 | 8.56±0.09 | GuHCl | (1,3,4) | 97P5 |
| $I \to U$ | [NH]CD2-$D_2O$ | 7.5 | 25 | 9.25±0.46 | 5.03±0.12 | GuHCl | (1,3,4) | 97P5 |
| deuterated CD2 | | | | | | | | |
| $N \to U$ | [ND]CD2-$H_2O$ | 7.5 | 25 | 36.28±0.42 | 8.55±0.08 | GuHCl | (1–3) | 97P5 |
| $I \to U$ | [ND]CD2-$H_2O$ | 7.5 | 25 | 7.07±0.25 | 5.00±0.15 | GuHCl | (1–3) | 97P5 |
| deuterated CD2: | | | | | | | | |
| $N \to U$ | [ND]CD2-$D_2O$ | 7.5 | 25 | 41.80±0.79 | 8.51±0.09 | GuHCl | (1,3,4) | 97P5 |
| $N \to U$ | [ND]CD2-$D_2O$ | 7.5 | 25 | 42.26±0.42 | 8.57±0.12 | GuHCl | (1,3–5) | 97P5 |
| $I \to U$ | [ND]CD2-$D_2O$ | 7.5 | 25 | 9.58±0.38 | 5.02±0.09 | GuHCl | (1,3,4) | 97P5 |

Remarks:

(1) from an integrated analysis of equilibrium and kinetic folding reaction, for details of the approach see Ref. 95P11

(2) a specific $C_{1/2}$ value to calculate the molar denaturant activity was set to 7.5

(3) m* values used here describe changes in both equilibrium and rate constants as function of molar denaturant activity relative to the native state (dimension $M^{-1}$)

(4) a specific $C_{1/2}$ value to calculate the molar denaturant activity was set to 9.1

(5) m* was derived from equilibrium unfolding profiles

Domain 1 of cell surface receptor protein CD2 (CD2.d1), interwinded oligomers, wild type and mutant CD2.D1 Δ46Δ47

| Transition | pH | T | $\Delta G$ | m* | Approach/Remarks | Ref |
|---|---|---|---|---|---|---|
| wild-type monomer: | | | | | | |
| $N \to U$ | 7.0 | 25 | 34.3 | 7.4±0.6 | GuHCl (1–5,8) | 99H4 |
| | 7.0 | 25 | 34.3 | | GuHCl (1,2,4,6,8) | 99H4 |
| $I \to U$ | 7.0 | 25 | 8.4 | | GuHCl (1,2,8) | 99H4 |
| $N \to I$ | 7.0 | 25 | 25.9 | 2.9±0.3 | GuHCl (1–4,8) | 99H4 |
| $t \to I$ | 7.0 | 25 | −52.3 | 2.8±0.1 | GuHCl (1–4,6,7,8) | 99H4 |
| wild-type dimer: | | | | | | |
| $N \to U$ | 7.0 | 25 | 64.4 | | GuHCl (1–3,6) | 99H4 |
| mutant monomer (CD2.D1 Δ46Δ47): | | | | | | |
| $N \to U$ | 7.0 | 25 | 20.9 | 6.4±0.6 | GuHCl (1–3,5,9) | 99H4 |
| | 7.0 | 25 | 20.9 | | GuHCl (1,2,6,9) | 99H4 |
| $I \to U$ | 7.0 | 25 | 4.6 | | GuHCl (1,2,9) | 99H4 |
| $N \to I$ | 7.0 | 25 | 16.3 | 3.1±1.1 | GuHCl (1–3,9) | 99H4 |
| $t \to I$ | 7.0 | 25 | −49.8 | 2.2±0.1 | GuHCl (1–3,6,7,9) | 99H4 |
| mutant dimer (CD2.D1 Δ46Δ47): | | | | | | |
| $N \to U$ | 7.0 | 25 | 46.0 | | GuHCl (1–3,6) | 99H4 |

Remarks:

(1) data from an integrated analysis of equilibrium and kinetic folding reaction

(2) equilibrium and kinetic folding/unfolding experiments were monitored by fluorescence, for details of the approach see also Refs. 95P11 and 97P6

(3) m* values used here describe changes in both equilibrium and rate constants as function of molar denaturant activity relative to the native state (dimension $M^{-1}$)

(4) data from Ref. 97P6

(5) $\Delta G_{eq}$, equilibrium data

(6) $\Delta G_{kin}$, data from kinetics

(7) $t \to I$, transition state to intermediate

(8) buffer: 25 mM sodium borate, 25 mM $NaH_2PO_4$ with added $Na_2SO_4$

(9) buffer: 50 mM triethanolamine hydrochloride, containing 0.2 M $Na_2SO_4$

Domain 1 of cell surface receptor protein CD2 (CD2.d1), mutants, $\Delta(\Delta G)$ for the transitions $N \rightarrow U$, $I \rightarrow U$, and $t \rightarrow U$

| Mutant | pH | T | $\Delta(\Delta G)_{N \rightarrow U}$ | $\Delta(\Delta G)_{I \rightarrow U}$ | $\Delta(\Delta G)_{t \rightarrow U}$ | Appr./Rem. | Ref |
|---|---|---|---|---|---|---|---|
| Leu16→Val | 7.5 | 25 | −9.20±1.34 | 0.75±1.09 | 0.46±1.46 | GuHCl (1–4) | 99L9 |
| Ile18→Val | 7.5 | 25 | −4.90±0.92 | 0.75±0.50 | −0.75±1.13 | GuHCl (1–4) | 99L9 |
| Val30→Ala | 7.5 | 25 | −20.42±0.96 | −2.55±0.71 | −8.16±1.05 | GuHCl (1–4) | 99L9 |
| Ala40→Gly | 7.5 | 25 | −7.15±0.84 | −1.26±0.59 | −0.50±0.96 | GuHCl (1–4) | 99L9 |
| Leu50→Val | 7.5 | 25 | 1.55±0.96 | 0.75±0.46 | 0.84±1.21 | GuHCl (1–4) | 99L9 |
| Ile57→Val | 7.5 | 25 | −1.80±0.92 | 1.42±0.50 | 0.54±1.13 | GuHCl (1–4) | 99L9 |
| Val78→Ala | 7.5 | 25 | −12.59±0.71 | −1.72±0.42 | −6.57±0.84 | GuHCl (1–4) | 99L9 |
| Leu89→Val | 7.5 | 25 | −0.59±0.96 | 0.25±0.46 | −1.30±1.21 | GuHCl (1–4) | 99L9 |
| Leu95→Val | 7.5 | 25 | −3.14±0.84 | 0.88±0.59 | 0.67±0.96 | GuHCl (1–4) | 99L9 |
| double mutant (Leu16→Val and Leu95→Val) | | | | | | | |
| | 7.5 | 25 | −8.41±0.79 | 0.88±0.46 | −1.63±0.96 | GuHCl (1–4) | 99L9 |
| double mutant (Ile18→Val and Val78→Ala) | | | | | | | |
| | 7.5 | 25 | −20.17±0.67 | −5.56±0.59 | −9.54±0.79 | GuHCl (1–4) | 99L9 |
| double mutant (Ala40→Gly and Leu50→Val) | | | | | | | |
| | 7.5 | 25 | −3.01±0.96 | −0.42±0.67 | 0.33±1.13 | GuHCl (1–4) | 99L9 |
| double mutant (Val78→Ala and Leu89→Val) | | | | | | | |
| | 7.5 | 25 | −16.99±1.34 | −3.56±1.09 | −8.08±1.42 | GuHCl (1–4) | 99L9 |

Remarks:

(1) data from an integrated analysis of equilibrium and kinetic folding reaction, equilibrium and kinetic folding/unfolding experiments were monitored by fluorescence, for details of the approach see also Refs. 95P11 and 97P6

(2) transitions monitored by fluorescence spectroscopy

(3) buffer: 50 mM triethanolamine hydrochloride with added $Na_2SO_4$

(4) $t \rightarrow u$, transition state to unfolded state

## Cellulase

Cellulase Cel45 from *Humicola insolens*

| Protein | pH | T | $\Delta G$ | $c_{1/2}$ | m | Appr./Rem. | Ref |
|---|---|---|---|---|---|---|---|
| Cel45 | 7.0 | 25 | 61.5±3.2 | 3.05±0.01 | 14.81±1.05 | GuHCl (1–3) | 99O4 |
| Cel45 | 7.0 | 25 | 56.5±6.2 | 3.02±0.11 | 12.30±0.59 | GuHCl (3,4) | 99O4 |
| Cel45, red. | 7.0 | 25 | 67.4±2.6 | 2.60±0.01 | 19.08±1.13 | GuHCl (1–3,5) | 99O4 |

Remarks:

(1) data from equilibrium unfolding, nonlinear least-squares fit of the transition curve assuming a linear dependence of $\Delta G$ on the denaturant conc.

(2) transition monitored by ellipticity at 220 nm and fluorescence

(3) buffer: 50 mM HEPES, pH 7.0

(4) data from stopped-flow kinetics of folding and unfolding in dependence of GuHCl monitored by fluorescence

(5) in the presence of 10 mM DTT

Cellulase Cel45 from *Humicola insolens*, wild type and mutants

| Protein | pH | T | $\Delta(\Delta G)$ | $c_{1/2}$ | m | Appr./Rem. | Ref |
|---|---|---|---|---|---|---|---|
| wild type | 7.0 | 25 | 0.00 | 3.05±0.01 | 14.81±1.05 | GuHCl (1–3) | 99O4 |
| Trp9→Phe | 7.0 | 25 | −16.57±1.30 | 2.46±0.02 | 24.81±2.51 | GuHCl (1–4) | 99O4 |
| Pro14→Ala | 7.0 | 25 | −11.25±0.88 | 2.65±0.02 | 14.56±1.55 | GuHCl (1–4) | 99O4 |
| Ala33→Pro | 7.0 | 25 | −10.67±0.84 | 2.67±0.02 | 26.32±4.35 | GuHCl (1–4) | 99O4 |
| Ser55→Met | 7.0 | 25 | 9.29±0.71 | 3.38±0.02 | 28.03±3.89 | GuHCl (1–4) | 99O4 |
| Trp62→Glu | 7.0 | 25 | −18.28±1.42 | 2.40±0.01 | 25.94±1.46 | GuHCl (1–4) | 99O4 |

Cellulase Cel45 from *Humicola insolens*, wild type and mutants (continued)

| Protein | pH | T | $\Delta(\Delta G)$ | $c_{1/2}$ | m | Appr./Rem. | Ref |
|---|---|---|---|---|---|---|---|
| Ala63→Arg | 7.0 | 25 | −7.57±0.59 | 2.78±0.02 | 21.13±1.55 | GuHCl (1–4) | 99O4 |
| Asn65→Arg | 7.0 | 25 | −13.22±1.00 | 2.58±0.01 | 25.48±1.30 | GuHCl (1–4) | 99O4 |
| Asp67→Arg | 7.0 | 25 | −1.67±0.13 | 2.99±0.02 | 14.90±1.63 | GuHCl (1–4) | 99O4 |
| His119→Gln | 7.0 | 25 | −1.13±0.08 | 3.01±0.02 | 16.19±1.72 | GuHCl (1–4) | 99O4 |
| Asn123→Ala | 7.0 | 25 | 8.70±0.67 | 3.36±0.01 | 24.43±1.76 | GuHCl (1–4) | 99O4 |
| Asp133→Asn | 7.0 | 25 | −9.83±0.75 | 2.70±0.01 | 16.86±2.51 | GuHCl (1–4) | 99O4 |
| Tyr147→Gly | 7.0 | 25 | −6.49±0.50 | 2.82±0.02 | 17.49±1.30 | GuHCl (1–4) | 99O4 |
| Arg158→Glu | 7.0 | 25 | 2.26±0.17 | 3.13±0.01 | 20.08±1.34 | GuHCl (1–4) | 99O4 |
| Arg196→Glu | 7.0 | 25 | −12.38±0.96 | 2.61±0.01 | 21.55±1.30 | GuHCl (1–4) | 99O4 |
| double mutant (Asn65→Arg and Asp67→Arg) | | | | | | | |
|  | 7.0 | 25 | −19.96±1.55 | 2.34±0.02 | 16.28±1.34 | GuHCl (1–3) | 99O4 |
| triple mutant (Ala63→Arg, Asn65→Arg and Asp67→Arg | | | | | | | |
|  | 7.0 | 25 | −23.89±1.88 | 2.20±0.02 | 22.43±3.64 | GuHCl (1–3) | 99O4 |

Remarks:
(1) data from equilibrium unfolding, nonlinear least-squares fit of the transition curve assuming a linear dependence of $\Delta G$ on the denaturant conc.
(2) transition monitored by ellipticity at 220 nm and fluorescence
(3) buffer: 50 mM HEPES, pH 7.0
(4) $\Delta(\Delta G)$ was calculated by $\Delta(\Delta G) = RT \ln(10) <m> (c_{1/2,mutant} - c_{1/2,wild type})$ using the average m value for wild type and the 16 mutants m = (20.67±1.17) kJ/mol/M

Isolated catalytic domain of cellulase E2 from thermophile *Thermomonospora fusca* ($E2_{cd}$) compared with the analogous domain of the cellulase from the mesophile *Cellulomonas fimi* ($CenA_{P30}$)

| Protein | pH | T | $\Delta G$ | $c_{1/2}$ | m | Appr./Rem. | Ref |
|---|---|---|---|---|---|---|---|
| $CenA_{P30}$+ DTT | 6.8 | 30 | 18.0±2.5 | 2.6±0.1 | 7.1±0.4 | urea (1–4) | 99B5 |
| $CenA_{P30}$ native | 6.8 | 30 | 45.2±12.6 | 4.5±0.3 | 10.0±2.1 | urea (1–3) | 99B5 |
| $E2_{cd}$ + DDT | 6.8 | 30 | 46.9±4.2 | 5.2±0.1 | 9.2±0.4 | urea (1–4) | 99B5 |
| $E2_{cd}$ native | 6.8 | 30 |  | 7.4±0.2 |  | urea (1–3) | 99B5 |

Remarks:
(1) linear extrapolation
(2) transition monitored by far-UV CD at 223 nm
(3) buffer: 50 mM potassium phosphate, 100 mM KCl, pH 6.8
(4) in the presence of 2.5 mM DTT

## Chaperone

GroEL, Minichaperones derived from the apical domain of GroEL, residues 191 to 376 and its C-terminally truncated fragment GroEL(191–345), urea-induced equilibrium unfolding

| Fragment | pH | T | $\Delta G$ | $c_{1/2}$ | m | Appr./Rem. | Ref |
|---|---|---|---|---|---|---|---|
| GroEL(191–345) | 7.0 | 25 | 22.6±2.5 | 2.68±0.06 | 8.4±0.8 | urea (1–3) | 98G11 |
|  | 7.0 | 25 | 23.8±1.3 | 2.91±0.02 | 8.4±0.4 | urea (1,2,4) | 98G11 |
| GroEL(191–376) | 7.0 | 25 | 21.8±3.8 | 2.68±0.06 | 7.9±1.3 | urea (1–3) | 98G11 |
|  | 7.0 | 25 | 23.8±3.8 | 2.90±0.06 | 8.4±1.3 | urea (1,2,4) | 98G11 |

Remarks:
(1) linear extrapolation, the slope of the pre- and postdenaturational baselines was taken into account
(2) buffer: 10 mM sodium phosphate, pH 7.0
(3) transition monitored by fluorescence emission at 300–302 nm
(4) transition monitored by CD at 222 nm

GroEL, Minichaperones derived from the apical domain of GroEL, residues 191 to 376 and its C-terminally truncated fragment GroEL(191–345), thermodynamic data from kinetics of folding and unfolding and thermal transition temperatures

| Fragment | Transition | pH | T | $\Delta G$ | $T_{trs}$ | Remarks | Ref |
|---|---|---|---|---|---|---|---|
| GroEL(191–345) | N → I | 7.0 | 25 | 6.7±2.5 | | (1,2) | 98G11 |
| GroEL(191–345) | I → U | 7.0 | 25 | 15.9±1.3 | | (1,2) | 98G11 |
| | | | | | 70.5±0.7 | (2,3) | 98G11 |
| | | | | | 64.5±0.6 | (2,4) | 98G11 |
| | | | | | 67.0±1.0 | (2,5) | 98G11 |
| GroEL(191–376) | N → I | 7.0 | 25 | 6.7±3.8 | | (1,2) | 98G11 |
| GroEL(191–376) | I → U | 7.0 | 25 | 14.6±0.8 | | (1,2) | 98G11 |
| | | | | | 70.6±1.0 | (2,3) | 98G11 |
| | | | | | 66.5±0.8 | (2,4) | 98G11 |
| | trans. (1) | | | | 34.3±1.0 | (2,5,6) | 98G11 |
| | trans. (2) | | | | 67.2±0.2 | (2,5,6) | 98G11 |

Remarks:

(1) data from urea-induced folding and unfolding kinetics

(2) buffer: 10 mM sodium phosphate, pH 7.0

(3) thermal transition monitored by intrinsic fluorescence

(4) thermal transition monitored by near-UV CD (275 nm)

(5) thermal transition monitored by far-UV CD (222 nm)

(6) two transitions were observed by far-UV CD

GroEL, Minichaperone GroEL(193–335), wild type and mutants

| Mutant | pH | T | $\Delta G$ | $c_{1/2}$ | m | Appr./Rem. | Ref |
|---|---|---|---|---|---|---|---|
| wild type | 8.2 | 25 | 25.5±1.3 | 2.81±0.02 | 9.12±0.42 | urea (1–3) | 99W5 |
| Lys207→Asn | 8.2 | 25 | 28.9±3.3 | 2.93±0.03 | 9.92±1.13 | urea (1–3) | 99W5 |
| Pro208→Ser | 8.2 | 25 | 20.9±2.1 | 2.47±0.05 | 8.62±0.84 | urea (1–3) | 99W5 |
| Thr210→Lys | 8.2 | 25 | 26.4±1.7 | 2.69±0.02 | 9.79±0.59 | urea (1–3) | 99W5 |
| Gly211→Met | 8.2 | 25 | 27.3±1.7 | 2.81±0.02 | 9.87±0.59 | urea (1–3) | 99W5 |
| Gly211→Gln | 8.2 | 25 | 25.1±2.1 | 2.80±0.03 | 9.04±0.75 | urea (1–3) | 99W5 |
| Ala212→Glu | 8.2 | 25 | 31.0±1.7 | 3.43±0.02 | 9.04±0.54 | urea (1–3) | 99W5 |

Remarks:

(1) nonlinear least-squares fit of the transition curve to a two-state model, assuming a linear dependence of $\Delta G$ on the denaturant conc.

(2) transition monitored by ellipticity at 218 nm

(3) buffer: 50 mM Tris-HCl, 150 mM NaCl, pH 8.2

Minichaperone GroEL(193–345), wild type and mutants

| Mutant | pH | T | $\Delta G$ | $c_{1/2}$ | m | Appr./Rem. | Ref |
|---|---|---|---|---|---|---|---|
| wild type | 8.2 | 25 | 27.6±1.3 | 3.12±0.01 | 8.87±0.38 | urea (1–4) | 99W5 |
| Asn206→Thr | 8.2 | 25 | 26.8±0.4 | 2.79±0.01 | 9.54±0.21 | urea (1–3) | 99W5 |
| Val213→Ala | 8.2 | 25 | 22.6±0.4 | 2.40±0.01 | 9.46±0.25 | urea (1–3) | 99W5 |
| Ser217→Asp | 8.2 | 25 | 28.5±0.4 | 3.20±0.01 | 8.95±0.17 | urea (1–3) | 99W5 |
| Phe219→Tyr | 8.2 | 25 | 27.6±0.8 | 3.15±0.01 | 8.79±0.21 | urea (1–3) | 99W5 |
| Ala223→Thr | 8.2 | 25 | 31.4±0.8 | 3.84±0.01 | 8.20±0.21 | urea (1–4) | 99W5 |
| Ala223→Val | 8.2 | 25 | 31.0±1.3 | 3.60±0.01 | 8.58±0.33 | urea (1–4) | 99W5 |
| Arg231→Gln | 8.2 | 25 | 28.0±0.8 | 2.95±0.01 | 9.54±0.33 | urea (1–3) | 99W5 |
| Glu232→Asp | 8.2 | 25 | 25.1±0.4 | 2.79±0.01 | 9.08±0.17 | urea (1–3) | 99W5 |
| Met233→Leu | 8.2 | 25 | 29.7±0.8 | 3.95±0.01 | 7.53±0.17 | urea (1–4) | 99W5 |
| Ala239→Gln | 8.2 | 25 | 30.5±0.8 | 3.00±0.01 | 10.25±0.33 | urea (1–3) | 99W5 |
| Lys242→Gln | 8.2 | 25 | 29.7±0.8 | 2.91±0.01 | 10.17±0.29 | urea (1–3) | 99W5 |

Minichaperone GroEL(193–345), wild type and mutants (continued)

| Mutant | pH | T | ΔG | $c_{1/2}$ | m | Appr./Rem. | Ref |
|---|---|---|---|---|---|---|---|
| Ala243→Ser | 8.2 | 25 | 27.2±0.4 | 3.02±0.01 | 8.95±0.21 | urea (1–3) | 99W5 |
| Thr266→Lys | 8.2 | 25 | 28.5±0.4 | 2.97±0.01 | 9.62±0.21 | urea (1–3) | 99W5 |
| Met267→Leu | 8.2 | 25 | 33.1±0.8 | 3.18±0.01 | 10.42±0.33 | urea (1–3) | 99W5 |
| Ile270→Gly | 8.2 | 25 | 27.6±0.4 | 3.01±0.01 | 9.20±0.17 | urea (1–3) | 99W5 |
| Ile270→Thr | 8.2 | 25 | 25.9±0.4 | 2.90±0.01 | 9.00±0.13 | urea (1–3) | 99W5 |
| Val271→Leu | 8.2 | 25 | 27.6±0.4 | 3.27±0.01 | 8.41±0.17 | urea (1–3) | 99W5 |
| Thr294→Arg | 8.2 | 25 | 33.1±1.3 | 3.29±0.01 | 10.04±0.38 | urea (1–3) | 99W5 |
| Thr294→Ile | 8.2 | 25 | 28.9±0.8 | 3.10±0.01 | 9.37±0.25 | urea (1–3) | 99W5 |
| Thr299→Gln | 8.2 | 25 | 25.1±0.4 | 2.64±0.01 | 9.46±0.17 | urea (1–3) | 99W5 |
| Ile305→Leu | 8.2 | 25 | 28.9±0.8 | 3.44±0.01 | 8.45±0.21 | urea (1–3) | 99W5 |
| Met307→Leu | 8.2 | 25 | 25.9±0.4 | 2.86±0.01 | 9.12±0.17 | urea (1–3) | 99W5 |
| Glu308→Lys | 8.2 | 25 | 28.5±1.3 | 3.28±0.01 | 8.70±0.33 | urea (1–3) | 99W5 |
| Glu308→Ser | 8.2 | 25 | 27.6±0.4 | 3.34±0.01 | 8.33±0.17 | urea (1–3) | 99W5 |
| Gln319→Lys | 8.2 | 25 | 28.0±0.8 | 3.14±0.01 | 9.00±0.29 | urea (1–3) | 99W5 |
| Asn326→Thr | 8.2 | 25 | 33.5±0.8 | 3.75±0.01 | 8.87±0.25 | urea (1–4) | 99W5 |
| Thr329→Asn | 8.2 | 25 | 26.8±0.8 | 2.78±0.01 | 9.58±0.25 | urea (1–3) | 99W5 |
| Ile333→Val | 8.2 | 25 | 24.3±0.8 | 2.68±0.01 | 9.00±0.25 | urea (1–3) | 99W5 |
| multiple variant M1: | | | | | | | |
|  | 8.2 | 25 | 49.0±0.8 | 6.37±0.01 | 7.70±0.17 | urea (1–3,5) | 99W5 |
| multiple variant M2: | | | | | | | |
|  | 8.2 | 25 | 44.4±0.8 | 5.98±0.01 | 7.45±0.17 | urea (1–3,6) | 99W5 |

Remarks:

(1) nonlinear least-squares fit of the transition curve to a two-state model assuming a linear dependence of ΔG on the denaturant conc.

(2) transition monitored by ellipticity at 218 nm

(3) buffer: 50 mM Tris-HCl, 150 mM NaCl, pH 8.2

(4) $T_{trs}$ amounts to 67.1°C for wild type, 72.7°C for Ala212→Glu and Ala223→Thr, 68.6°C for Ala223→Val, 73.0°C for Met233→Leu, and 73.9°C for Asn326→Thr

(5) mutant M1 designed for high thermostability consists of Ala212→Glu, Ala223→Thr, Met233→Leu, Ile305→Leu, Glu308→Lys, and Asn326→Thr, and exhibits $T_{trs}$ = 85.7°C (DSC data)

(6) mutant M2 designed for high thermostability consists of Ala212→Glu, Ala223→Val, Met233→Leu, Ile305→Leu, Glu308→Lys, and Asn326→Thr, and exhibits $T_{trs}$ = 81.3°C (DSC data)

GroES, Co-chaperonin GroES

| | pH | T | ΔG | Approach | Remarks | Ref |
|---|---|---|---|---|---|---|
| | 7.0 | 25 | 37.7–42.7 | DSC/urea | (1–4) | 97B11 |

Remarks:

(1) from DSC measurements in the presence and in the absence of urea

(2) the DSC measurements were performed varying the protein conc., see Table 2

(3) buffer: 20 mM sodium phosphate

(4) GroES melting is a spontaneous reversible process involving a highly cooperative transition between folded heptamers and unfolded monomers

GroES, Chaperonin GroES from *Escherichia coli*, three-state unfolding, wild type and Trp mutant (Ile48→Trp)

| Protein/Transition | pH | T | ΔG | $c_{1/2}$ | Approach/Remarks | | Ref |
|---|---|---|---|---|---|---|---|
| wild type | | | | | | | |
| $N_7$ → U | 7.8 | 25 | 47.3 | | GuHCl | (1–5) | 99H9 |
| $N_7$ → X | 7.8 | 25 | | 0.3 | GuHCl | (1,2,4,5) | 99H9 |
| X → U | 7.8 | 25 | 2.5 | 1.2 | GuHCl | (1,4,5) | 99H9 |
| X → U | 7.8 | 25 | | 1.1 | GuHCl | (1,4,6) | 99H9 |

GroES, Chaperonin GroES from *Escherichia coli*, three-state unfolding, wild type and Trp mutant (Ile48→Trp) (continued)

| Protein/Transition | pH | T | $\Delta G$ | $c_{1/2}$ | Approach/Remarks | | Ref |
|---|---|---|---|---|---|---|---|
| Ile48→Trp | | | | | | | |
| $N_7 \to U$ | 7.8 | 25 | 46.0 | | GuHCl | (1–4,6) | 99H9 |
| $N_7 \to U$ | 7.8 | 25 | 39.7 | | GuHCl | (1–5) | 99H9 |
| $N_7 \to X$ | 7.8 | 25 | | 0.3 | GuHCl | (1,4,5) | 99H9 |
| $N_7 \to X$ | 7.8 | 25 | | 0.5 | GuHCl | (1,4,6) | 99H9 |
| $X \to U$ | 7.8 | 25 | 1.7 | 1.2 | GuHCl | (1,4,5) | 99H9 |
| $X \to U$ | 7.8 | 25 | 1.7 | | GuHCl | (1,4,6) | 99H9 |

Remarks:

(1) the data refer to a three-state model with the native heptameric state ($N_7$), the partially folded monomeric state (X), and the unfolded monomer (U)

(2) $\Delta G$ of dissociation ($\Delta G_{diss}$) varies linearly with denaturant conc.

(3) $\Delta G$ from $\Delta G_{diss} + 7 \times \Delta G_{(X \to U)}$

(4) measured in 5 mM Tris-HCl, pH 7.9

(5) transition monitored by size-exclusion experiments

(6) transition monitored by tryptophan fluorescence

## Che Y

Che Y, chemotactic protein from *E. coli*, mutants with enhanced native α-helix propensities, mutants derived from pseudo-wild type Phe14→Asn

| Mutant | pH | T | $\Delta G$ | $c_{1/2}$ | m | Approach/Remarks | Ref |
|---|---|---|---|---|---|---|---|
| Phe14→Asn | 7.0 | 25 | 35.3±1.3 | 4.68 | 7.36±0.13 | urea, kin. (1–4) | 97L12 |
| Phe14→Asn | 7.0 | 25 | 35.1±0.8 | 4.68 | 7.49±0.21 | urea, eq. (1–3,5) | 97L12 |
| Hel1 | 7.0 | 25 | 32.1±0.8 | 4.92 | 6.40±0.13 | urea, kin. (1–4) | 97L12 |
| Hel1 | 7.0 | 25 | 31.0±0.8 | 4.86 | 6.36±0.21 | urea, eq. (1–3,5) | 97L12 |
| Hel2(contr.) | 7.0 | 25 | 30.2±0.8 | 4.38 | 7.03±0.17 | urea, kin. (1–4) | 97L12 |
| Hel2(contr.) | 7.0 | 25 | 31.4±0.8 | 4.42 | 7.07±0.13 | urea, eq. (1–3,5) | 97L12 |
| Hel2 | 7.0 | 25 | 29.3±0.8 | 4.76 | 6.49±0.17 | urea, kin. (1–4) | 97L12 |
| Hel2 | 7.0 | 25 | 30.1±0.4 | 4.77 | 6.32±0.13 | urea, eq. (1–3,5) | 97L12 |
| Hel3 | 7.0 | 25 | 38.3±1.7 | 5.74 | 6.61±0.29 | urea, kin. (1–4) | 97L12 |
| Hel3 | 7.0 | 25 | 34.7±1.3 | 5.35 | 6.49±0.13 | urea, eq. (1–3,5) | 97L12 |
| Hel4 | 7.0 | 25 | 40.6±0.8 | 5.51 | 7.42±0.21 | urea, kin. (1–4) | 97L12 |
| Hel4 | 7.0 | 25 | 37.7±1.3 | 5.39 | 6.99±0.21 | urea, eq. (1–3,5) | 97L12 |
| Hel5 | 7.0 | 25 | 40.9±1.7 | 5.88 | 6.99±0.33 | urea, kin. (1–4) | 97L12 |
| Hel5 | 7.0 | 25 | 38.5±1.7 | 5.65 | 6.78±0.25 | urea, eq. (1–3,5) | 97L12 |

Remarks:

(1) mutant Phe14→Asn is the reference protein

(2) abbreviations:

       Hel1 = (Thr16→Ala, Arg19→Glu, and Asn23→Arg)

       Hel2 (contr.) = control of helix 2 (Gly39→Ala)

       Hel2 = (Gly39→Ala, Asp41→Glu, and Asn44→Arg)

       Hel3 = (Thr71→Arg)

       Hel4 = (Lys91→Asn, Asn94→Ala, and Ile96→Leu)

       Hel5 = (Thr115→Glu, Glu118→Lys, Asn121→Ala, and Lys122→Glu)

(3) buffer: 50 mM PIPES, pH 7

(4) from kinetic data fitted to a three-state model

(5) from equilibrum denaturation fitted to a two-state model with linear dependence of $\Delta G$ on denaturant concentration

## Chitosanase

Chitosanase from *Streptomyces* sp. N174 and tryptophan to phenylalanine mutants

| Protein/Transition | pH | T | $\Delta T$ | $\Delta(\Delta G)$ | Approach/Remarks | | Ref |
|---|---|---|---|---|---|---|---|
| Data obtained by CD measurements: | | | | | | | |
| wild type | | | | | | | |
|     trans. (1) | 7.0 | 42.7 | 0.0 | 0.0 | heat | (1–4) | 99H11 |
|     trans. (2) | 7.0 | 52.2 | 0.0 | 0.0 | heat | (1–4) | 99H11 |
| Trp28→Phe | | | | | | | |
|     trans. (1) | 7.0 | 42.7 | −7.1 | −14.90 | heat | (1–4) | 99H11 |
|     trans. (2) | 7.0 | 52.2 | −7.0 | −7.36 | heat | (1–4) | 99H11 |
| Trp101→Phe | | | | | | | |
|     trans. (1) | 7.0 | 42.7 | −7.1 | −16.36 | heat | (1–4) | 99H11 |
|     trans. (2) | 7.0 | 52.2 | −7.5 | −7.91 | heat | (1–4) | 99H11 |
| Trp227→Phe | | | | | | | |
|     trans. (1) | 7.0 | 42.7 | −6.8 | −12.43 | heat | (1–4) | 99H11 |
|     trans. (2) | 7.0 | 52.2 | −6.4 | −5.48 | heat | (1–4) | 99H11 |
| double mutant (Trp28→Phe and Trp101→Phe) | | | | | | | |
|     trans. (1) | 7.0 | 42.7 | −11.4 | −24.27 | heat | (1–4) | 99H11 |
|     trans. (2) | 7.0 | 52.2 | −8.0 | −6.32 | heat | (1–4) | 99H11 |
| Data obtained by fluorescence measurements: | | | | | | | |
| wild type | 7.0 | 43.2 | 0.0 | 0.0 | heat | (4–7) | 99H11 |
| Trp28→Phe | 7.0 | 43.2 | −7.0 | −18.53 | heat | (4–7) | 99H11 |
| Trp101→Phe | 7.0 | 43.2 | −6.6 | −17.87 | heat | (4–7) | 99H11 |
| Trp227→Phe | 7.0 | 43.2 | −5.9 | −10.50 | heat | (4–7) | 99H11 |
| double mutant (Trp28→Phe and Trp101→Phe) | | | | | | | |
| | 7.0 | 43.2 | −12.0 | −33.60 | heat | (4–7) | 99H11 |

Remarks:

(1) transition monitored by CD at 222 nm

(2) the transition profile characterstic of a three-state transition was treated by a model derived from $N \rightarrow I \rightarrow U$

(3) $\Delta(\Delta G)$ was calculated at $T_{trs}$ of the wild-type protein using $\Delta(\Delta G) = \Delta T_{mut} \times \Delta S_{w.t.}$ with $\Delta S_{w.t.} = 1.5899$ kJ/mol/K for the first transition and $\Delta S_{w.t.} = 1.6736$ kJ/mol/K for the second transition

(4) measured in 50 mM sodium phosphate buffer, pH 7.0

(5) transition monitored by fluorescence at 330 nm

(6) the transition profile was treated by a two-state transition

(7) $\Delta(\Delta G)$ was calculated at $T_{trs}$ of the wild-type protein using $\Delta(\Delta G) = \Delta T_{mut} \times \Delta S_{w.t.}$ with $\Delta S_{w.t.} = 2.1255$ kJ/mol/K

## Chymotrypsinogen

Recombinant rat wild-type chymotrypsinogen and Δ-chymotrypsinogen

| Protein | pH | T | $\Delta G$ | Approach/Remarks | | Ref |
|---|---|---|---|---|---|---|
| chymotrypsinogen | 8.0 | 20 | 34.0 | GuHCl | (1–3) | 99K2 |
| Δ-chymotrypsinogen | 8.0 | 20 | 10.0 | GuHCl | (1–4) | 99K2 |

Remarks:

(1) linear extrapolation

(2) transition monitored by tryptophan fluorescence

(3) buffer: 10 mM Tris-HCl, 10 mM CaCl$_2$, pH 8.0

(4) Δ-chymotrypsinogen, mutant of rat chymotrypsinogen containing the six-amino acid propeptide of rat trypsinogen instead of the 15-amino acid propeptide of rat chymotrypsinogen and with a Cys122→Ser

α-Chymotrypsinogen from bovine pancreas in aqueous polyol solutions

| Cosolvent | Conc. | pH | $T_{trs}$ | $\Delta T$ | $\Delta(\Delta G)$ | Appr./Rem. | Ref |
|---|---|---|---|---|---|---|---|
| control | | 2.5 | 44.9 | 0.0 | 0.00 | heat (1,2) | 98K3 |
| mannitol | 1.00 M | 2.5 | 50.5 | 5.6 | 6.15 | heat (1,2) | 98K3 |
| inositol | 0.75 M | 2.5 | 52.3 | 7.4 | 8.54 | heat (1,2) | 98K3 |
| xylitol | 2.00 M | 2.5 | 52.6 | 7.7 | 8.79 | heat (1,2) | 98K3 |
| adonitol | 2.00 M | 2.5 | 52.5 | 7.6 | 8.24 | heat (1,2) | 98K3 |
| sorbitol | 2.00 M | 2.5 | 54.7 | 9.8 | 10.25 | heat (1,2) | 98K3 |

Remarks:
(1) transition monitored by optical absorption at 287 nm
(2) buffer: 20 mM glycine, pH 2.5

## Chymotrypsin Inhibitor CI2

The data entries are arranged as follows:

a) wild type and mutants, data from equilibrium unfolding and kinetics
b) wild type and mutants, data from exchange experiments
c) circularly permuted variants and complexes

*a) wild type and mutants, data from equilibrium unfolding and kinetics*

Chymotrypsin inhibitor 2 (CI2), wild type and proline mutants

| Mutant | pH | T | $\Delta(\Delta G)$ | $c_{1/2}$ | m | Appr./Rem. | Ref |
|---|---|---|---|---|---|---|---|
| wild type | 3 | 25 | 0.0 | 4.00±0.01 | 7.95±0.13 | GuHCl (1,2) | 97T3 |
| Pro6→Ala | 3 | 25 | −(7.95±0.25) | 3.02±0.01 | 6.82±0.17 | GuHCl (1,3) | 97T3 |
| Pro25→Ala | 3 | 25 | −(7.36±0.21) | 3.09±0.02 | 8.66±0.42 | GuHCl (1,3) | 97T3 |
| Pro33→Ala | 3 | 25 | −(0.71±0.21) | 3.91±0.02 | 7.49±0.29 | GuHCl (1,3) | 97T3 |
| Pro61→Ala | 3 | 25 | −(13.97±0.38) | 2.28±0.04 | 7.53±0.59 | GuHCl (1,3) | 97T3 |
| double mutant (Pro6→Ala and Pro61→Ala) | | | | | | | |
| | 3 | 25 | −(15.35±0.63) | 2.11±0.10 | 6.90±0.13 | GuHCl (1,3) | 97T3 |

Remarks:
(1) nonlinear fit of changes in fluorescence with denaturant concentration assuming a linear dependence of $\Delta G$ on denaturant concentration
(2) $\Delta G$ of the wild type amounts to 31.8±0.4 kJ/mol, see Ref. 95I1
(3) $\Delta(\Delta G)$ was calculated using the average value of m = 8.12±0.14 kJ/mol/M from 124 mutants, see Ref. 95I1

Chymotrypsin inhibitor 2 (CI2), wild type and mutants, data derived from the GuHCl dependence of the unfolding kinetics

| Mutant | pH | T | $\Delta(\Delta G)$ | Approach/Remarks | Ref |
|---|---|---|---|---|---|
| wild type | 6.3 | 25 | 0.0 | GuHCl (1) | 98O5 |
| Leu8→Ala | 6.3 | 25 | −11.2 | GuHCl (1) | 98O5 |
| Leu49→Ala | 6.3 | 25 | −16.1 | GuHCl (1) | 98O5 |
| double mutant (Ile29→Ala and Ile57→Ala) | | | | | |
| | 6.3 | 25 | −17.1 | GuHCl (1) | 98O5 |
| double mutant (Leu32→Ala and Val38→Ala) | | | | | |
| | 6.3 | 25 | −13.2 | GuHCl (1) | 98O5 |
| double mutant (Leu32→Ala and Phe50→Ala) | | | | | |
| | 6.3 | 25 | −20.0 | GuHCl (1) | 98O5 |
| double mutant (Leu32→Val and Phe50→Ala) | | | | | |
| | 6.3 | 25 | −14.3 | GuHCl (1) | 98O5 |

Remark:
(1) from log $k_{unf}$ versus GuHCl concentration, for details see Ref. 98O5

Barley chymotrypsin inhibitor-2 (CI2), multiple lysine substitutions

| Protein | pH | T | ΔG | $c_{1/2}$ | m | Appr./Rem. | Ref |
|---|---|---|---|---|---|---|---|
| wild type | 7.0 | 25 | 29.46±0.17 | 3.97±0.01 | 7.41±0.08 | GuHCl (1–3) | 99R16 |
| BHL1 | 7.0 | 25 | 18.74±1.42 | 2.36±0.04 | 7.99±0.75 | GuHCl (1–4) | 99R16 |
| BHL2 etc. | 7.0 | 25 | 6.53±0.67 | 0.86±0.02 | 7.61±0.79 | GuHCl (1–4) | 99R16 |
| BHL4 | 7.0 | 25 | 20.63±0.79 | 2.59±0.01 | 7.99±0.25 | GuHCl (1–4) | 99R16 |

Remarks:
(1) linear extrapolation
(2) transition monitored by fluorescence emission at 356 nm
(3) buffer: 10 mM sodium phosphate, pH 7.0
(4) mutant with increased Lys content, for the sequence see Ref. 99R16

Chymotrypsin inhibitor CI2, effect of osmolytes on the stability

| Osmolyte Conc. | pH | T | ΔG | m | Approach/Remarks | | Ref |
|---|---|---|---|---|---|---|---|
| none | 4.2 | 21 | 20.42±1.05 | 1.20±0.14 | GuHCl | (1–3) | 98F2 |
| 2 M glycine | 4.2 | 21 | 28.16±1.80 | 1.20±0.14 | GuHCl | (1–3) | 98F2 |

Remark:
(1) linear extrapolation, LEM-SB
(2) m was held to be the same in the absence and presence of glycine
(3) Ref. 98F2 contains the hydrogen exchange rates for more than 20 amino acid residues of CI2 in the absence and presence of glycine

*b) wild type and mutants, data from exchange experiments*

Chymotrypsin inhibitor 2, wild type, GuHCl denaturation parameters obtained from hydrogen exchange measurements

| Residue | pH | T | $\Delta G°_{ex}$ | m | Approach/Remarks | Ref |
|---|---|---|---|---|---|---|
| Glu4 | 5.3 | 33 | 10.5±1.3 | 1.3±0.4 | GuHCl + HX (1–3) | 97I4 |
| Trp5 | 5.3 | 33 | 19.2±0.8 | 9.6±2.1 | GuHCl + HX (1–3) | 97I4 |
| Leu8 | 5.3 | 33 | 20.5±0.8 | 11.7±3.3 | GuHCl + HX (1–3) | 97I4 |
| Val9 | 5.3 | 33 | 25.1±0.4 | 13.0±2.1 | GuHCl + HX (1–3) | 97I4 |
| Gly10 | 5.3 | 33 | 18.8±1.3 | 2.1±0.4 | GuHCl + HX (1–3) | 97I4 |
| Lys11 | 5.3 | 33 | 38.1±1.7 | 11.3±1.3 | GuHCl + HX (1–3) | 97I4 |
| Val13 | 5.3 | 33 | 18.4±0.8 | 2.5±0.4 | GuHCl + HX (1–3) | 97I4 |
| Ala16 | 5.3 | 33 | 18.0±0.8 | 5.4±1.7 | GuHCl + HX (1–3) | 97I4 |
| Lys17 | 5.3 | 33 | 16.7±1.7 | 0.8±0.13 | GuHCl + HX (1–3) | 97I4 |
| Lys18 | 5.3 | 33 | 20.9±0.4 | 10.0±1.7 | GuHCl + HX (1–3) | 97I4 |
| Val19 | 5.3 | 33 | 39.3±3.3 | 11.7±1.3 | GuHCl + HX (1–3) | 97I4 |
| Ile20 | 5.3 | 33 | 37.2±2.5 | 11.7±1.3 | GuHCl + HX (1–3) | 97I4 |
| Leu21 | 5.3 | 33 | 39.3±1.7 | 13.8±0.8 | GuHCl + HX (1–3) | 97I4 |
| Gln22 | 5.3 | 33 | 24.3±0.4 | 13.4±1.3 | GuHCl + HX (1–3) | 97I4 |
| Ala27 | 5.3 | 33 | 17.2±1.3 | 2.1±0.25 | GuHCl + HX (1–3) | 97I4 |
| Gln28 | 5.3 | 33 | 20.9±0.4 | 6.7±2.5 | GuHCl + HX (1–3) | 97I4 |
| Ile30 | 5.3 | 33 | 43.1±1.7 | 15.1±0.08 | GuHCl + HX (1–3) | 97I4 |
| Leu32 | 5.3 | 33 | 41.4±1.7 | 13.4±0.8 | GuHCl + HX (1–3) | 97I4 |
| Val34 | 5.3 | 33 | 13.0±0.4 | 1.7±0.2 | GuHCl + HX (1–3) | 97I4 |
| Arg46 | 5.3 | 33 | 23.0±0.8 | 9.6±0.4 | GuHCl + HX (1–3) | 97I4 |
| Val47 | 5.3 | 33 | 38.9±0.2 | 11.7±0.13 | GuHCl + HX (1–3) | 97I4 |
| Arg48 | 5.3 | 33 | 36.8±2.5 | 10.5±1.3 | GuHCl + HX (1–3) | 97I4 |
| Leu49 | 5.3 | 33 | 36.8±2.5 | 11.3±2.1 | GuHCl + HX (1–3) | 97I4 |
| Phe50 | 5.3 | 33 | 39.7±2.5 | 13.0±1.7 | GuHCl + HX (1–3) | 97I4 |
| Val51 | 5.3 | 33 | 36.4±1.7 | 11.7±0.8 | GuHCl + HX (1–3) | 97I4 |

Chymotrypsin inhibitor 2, wild type, GuHCl denaturation parameters obtained from hydrogen exchange measurements (continued)

| Residue | pH | T | $\Delta G^\circ_{ex}$ | m | Approach/Remarks | Ref |
|---|---|---|---|---|---|---|
| Asp52 | 5.3 | 33 | 21.8±1.3 | 12.1±2.1 | GuHCl + HX (1–3) | 97I4 |
| Leu54 | 5.3 | 33 | 11.3±8.4 | 2.5±1.3 | GuHCl + HX (1–3) | 97I4 |
| Asn56 | 5.3 | 33 | 21.3±0.8 | 6.3±0.8 | GuHCl + HX (1–3) | 97I4 |
| Ile57 | 5.3 | 33 | 25.1±1.7 | 11.7±1.7 | GuHCl + HX (1–3) | 97I4 |
| Ala58 | 5.3 | 33 | 23.0±0.4 | 8.4±2.1 | GuHCl + HX (1–3) | 97I4 |
| Gln59 | 5.3 | 33 | 26.8±0.4 | 10.0±0.4 | GuHCl + HX (1–3) | 97I4 |
| Arg62 | 5.3 | 33 | 25.5±1.3 | 11.7±2.1 | GuHCl + HX (1–3) | 97I4 |
| Val63 | 5.3 | 33 | 18.8±0.4 | 7.5±0.8 | GuHCl + HX (1–3) | 97I4 |
| Gly64 | 5.3 | 33 | 13.0±0.8 | 2.1±0.25 | GuHCl + HX (1–3) | 97I4 |

Remarks:

(1) data from exchange rate at different GuHCl conc., exchange rates were fitted assuming a linear dependence of apparent $\Delta G_{ex}$ on GuHCl concentration

(2) buffer: 50 mM sodium acetate, pH 5.3, the salt concentration was maintained at 2.5 by adding NaCl throughout the range of GuHCl concentration

(3) Ref. 97I4 contains additionally activation energies derived from the temperature dependence of HX

Chymotrypsin inhibitor 2, wild type, apparent free energies of hydrogen-deuterium exchange of amide protons monitored by NMR at various apparent values of pH

| Residue | T | $\Delta G^\circ_{ex}$(pH5.3) | $\Delta G^\circ_{ex}$(pH5.7) | $\Delta G^\circ_{ex}$(pH6.8) | Approach | Ref |
|---|---|---|---|---|---|---|
| Glu4 | 33 | 10.0 | 9.6 | | HX | 97N5 |
| Trp5 | 33 | 21.8 | 24.7 | 24.3 | HX | 97N5 |
| Leu8 | 33 | 20.9 | 22.6 | 23.0 | HX | 97N5 |
| Val9 | 33 | 18.0 | 18.8 | 20.1 | HX | 97N5 |
| Gly10 | 33 | 18.0 | 18.8 | | HX | 97N5 |
| Lys11 | 33 | | 32.6 | 32.6 | HX | 97N5 |
| Val13 | 33 | 17.6 | 19.2 | 15.9 | HX | 97N5 |
| Glu15 | 33 | 7.1 | 6.3 | | HX | 97N5 |
| Ala16 | 33 | 24.3 | 26.8 | 26.8 | HX | 97N5 |
| Lys17 | 33 | 15.5 | 15.9 | | HX | 97N5 |
| Lys18 | 33 | 19.7 | 20.5 | 19.2 | HX | 97N5 |
| Val19 | 33 | 23.4 | 24.3 | 25.1 | HX | 97N5 |
| Leu21 | 33 | | | 29.7 | HX | 97N5 |
| Gln22 | 33 | 25.1 | 26.8 | 26.8 | HX | 97N5 |
| Asp23 | 33 | 13.0 | 14.2 | | HX | 97N5 |
| Lys24 | 33 | 24.7 | 25.9 | 27.6 | HX | 97N5 |
| Ala27 | 33 | 18.0 | 19.2 | 20.5 | HX | 97N5 |
| Gln28 | 33 | 21.8 | 21.8 | 22.6 | HX | 97N5 |
| Ile30 | 33 | | 28.0 | 31.0 | HX | 97N5 |
| Leu32 | 33 | | 26.4 | 28.5 | HX | 97N5 |
| Val34 | 33 | 11.7 | 11.7 | | HX | 97N5 |
| Arg46 | 33 | 25.1 | 26.8 | 27.6 | HX | 97N5 |
| Arg48 | 33 | 28.5 | 26.8 | 28.5 | HX | 97N5 |
| Leu49 | 33 | | | 33.9 | HX | 97N5 |
| Phe50 | 33 | | | 31.0 | HX | 97N5 |
| Asp52 | 33 | 20.9 | 21.8 | 23.4 | HX | 97N5 |
| Leu54 | 33 | 9.6 | | | HX | 97N5 |
| Asp55 | 33 | 10.5 | 9.2 | | HX | 97N5 |
| Asn56 | 33 | | 24.3 | 29.3 | HX | 97N5 |
| Ile57 | 33 | 25.5 | 28.0 | 26.8 | HX | 97N5 |
| Ala58 | 33 | 20.9 | 20.1 | 21.8 | HX | 97N5 |

Chymotrypsin inhibitor 2, wild type, apparent free energies of hydrogen-deuterium exchange of amide protons monitored by NMR at various pD (continued)

| Residue | T | $\Delta G°_{ex}$(pH5.3) | $\Delta G°_{ex}$(pH5.7) | $\Delta G°_{ex}$(pH6.8) | Approach | Ref |
|---------|---|------------|------------|------------|----------|-----|
| Gln59 | 33 | 24.7 | 24.3 | 26.8 | HX | 97N5 |
| Arg62 | 33 | 26.8 | 27.2 | 28.0 | HX | 97N5 |
| Val63 | 33 | 20.5 | 21.8 | 23.0 | HX | 97N5 |
| Gly64 | 33 | 15.1 | 17.6 | 18.8 | HX | 97N5 |

reference value from equilibrium denaturation:

| Protein | pH | T | $\Delta G$ | $c_{1/2}$ | m | Appr./Rem. | Ref |
|---------|----|----|-----------|-----------|---|-----------|-----|
| wild type | 5.3 | 33 | 29.3±0.8 | 3.7±0.01 | 7.95±0.13 | (1,2) | 97N5 |
| wild type | 37 | 32.6±1.7 | | | | HX (3) | 97N5 |

Remarks:

(1) GuHCl-induced unfolding in $D_2O$ monitored by fluorescence at 356 nm

(2) for the procedure see Refs. 93J1, 95I1

(3) average value of slowest exchange rates

Chymotrypsin inhibitor 2, mutants, difference in apparent free energies of hydrogen-deuterium exchange of amide protons relative to the wild type monitored by NMR

| Residue | T | $\Delta(\Delta G°_{ex})$(1) | $\Delta(\Delta G°_{ex})$(2) | $\Delta(\Delta G°_{ex})$(3) | Approach/Remarks | | Ref |
|---------|---|------------|------------|------------|----------|------|-----|
| Trp5 | 37 | 1.7 | 2.1 | 2.9 | HX | (1–3) | 97N5 |
| Leu8 | 37 | 0.0 | −0.4 | 1.7 | HX | (1–3) | 97N5 |
| Val9 | 37 | 1.7 | 1.3 | 2.9 | HX | (1–3) | 97N5 |
| Lys11 | 37 | 3.3 | 5.4 | 5.9 | HX | (1–3) | 97N5 |
| Val13 | 37 | −2.1 | 1.3 | 1.7 | HX | (1–3) | 97N5 |
| Ala16 | 37 | 0.4 | 0.4 | 2.1 | HX | (1–3) | 97N5 |
| Lys17 | 37 | | −0.4 | 1.3 | HX | (1–3) | 97N5 |
| Lys18 | 37 | | | 1.7 | HX | (1–3) | 97N5 |
| Val19 | 37 | 2.5 | 2.5 | 3.8 | HX | (1–3) | 97N5 |
| Ile20 | 37 | 7.1 | 9.2 | 7.5 | HX | (1–3) | 97N5 |
| Leu21 | 37 | 5.0 | 6.7 | 7.1 | HX | (1–3) | 97N5 |
| Gln22 | 37 | 0.0 | | 1.7 | HX | (1–3) | 97N5 |
| Lys24 | 37 | 0.8 | 1.7 | 2.5 | HX | (1–3) | 97N5 |
| Ala27 | 37 | 0.04 | 0.4 | 2.5 | HX | (1–3) | 97N5 |
| Gln28 | 37 | 0.4 | 1.7 | 3.3 | HX | (1–3) | 97N5 |
| Ile30 | 37 | 5.0 | 5.9 | 6.7 | HX | (1–3) | 97N5 |
| Leu32 | 37 | 1.3 | 0.8 | 4.2 | HX | (1–3) | 97N5 |
| Val34 | 37 | 2.5 | | 4.2 | HX | (1–3) | 97N5 |
| Arg46 | 37 | 1.7 | 43.1 | | HX | (1–3) | 97N5 |
| Val47 | 37 | 5.4 | 6.3 | 7.5 | HX | (1–3) | 97N5 |
| Arg48 | 37 | 1.7 | 1.7 | 2.9 | HX | (1–3) | 97N5 |
| Leu49 | 37 | | 1.7 | | HX | (1–3) | 97N5 |
| Phe50 | 37 | 4.6 | 4.6 | | HX | (1–3) | 97N5 |
| Val51 | 37 | 5.4 | 7.5 | 8.4 | HX | (1–3) | 97N5 |
| Asp52 | 37 | −0.8 | −0.4 | | HX | (1–3) | 97N5 |
| Asp56 | 37 | 0.4 | 0.4 | 4.2 | HX | (1–3) | 97N5 |
| Ile57 | 37 | 1.3 | 4.2 | 4.6 | HX | (1–3) | 97N5 |
| Ala58 | 37 | −0.4 | −1.7 | | HX | (1–3) | 97N5 |
| Gln59 | 37 | 0.4 | 0.8 | | HX | (1–3) | 97N5 |
| Arg62 | 37 | 2.1 | 1.7 | 2.9 | HX | (1–3) | 97N5 |
| Val63 | 37 | 0.4 | 0.4 | 2.1 | HX | (1–3) | 97N5 |
| Gly64 | 37 | 2.9 | 0.8 | 5.4 | HX | (1–3) | 97N5 |

Remarks:

(1) mutant Ile20→Val, measured at pH 6.3

(2) triple mutant (Ser12→Gly, Glu14→Ala, and Glu15→Ala), measured at pH 6.5

(3) mutant Ala58→Gly, measured at pH 6.0

*c) circularly permuted variants and complexes*

Chymotrypsin inhibitor 2 (CI2), circular and permuted mutants

| Mutant | pH | T | $\Delta G$ | $c_{1/2}$ | m | Appr./ Rem. | Ref |
|---|---|---|---|---|---|---|---|
| wild type | 6.25 | 25 | 30.21±1.00 | 2.10±0.01 | 14.43±0.46 | GuHSCN(1–3) | 98O7 |
| double mutant (Thr3→Cys and Val63→Cys) reduced | | | | | | | |
| | 6.25 | 25 | 28.74±1.21 | 1.80±0.01 | 15.94±0.67 | GuHSCN(1–4) | 98O7 |
| double mutant (Thr3→Cys and Val63→Cys) oxidized | | | | | | | |
| | 6.25 | 25 | 33.47±1.30 | 3.32±0.01 | 10.08±0.38 | GuHSCN(1–3,5) | 98O7 |
| wild type | 6.25 | 25 | 31.80±0.50 | 4.00±0.01 | 7.95±0.13 | GuHCl    (1–3) | 98O7 |
| double mutant (Thr3→Cys and Val63→Cys) | | | | | | | |
| | 6.25 | 25 | 28.70±0.92 | 3.62±0.02 | 7.95±0.25 | GuHCl    (1–3) | 98O7 |
| permuted (Thr3→Cys and Val63→Cys) | | | | | | | |
| | 6.25 | 25 | 28.91±1.00 | 4.57±0.02 | 6.32±0.21 | GuHCl    (1–3,6) | 98O7 |

Remarks:
(1) linear extrapolation
(2) transition monitored by fluorescence, for the procedure see Ref. 93J1
(3) buffer: 50 mM MES, pH 6.25
(4) in the presence of 20 mM DTT
(5) in the presence of 1 mM oxidized DTT
(6) permuted (Thr3→Cys and Val63→Cys) with intact disulfide bond, but cleaved at Met40

Chymotrypsin inhibitor 2 (CI2), circularized, permuted, and unpermuted mutants, molar denaturant concentration required for half conversion

| Mutant | pH | T | circularized $c_{1/2}$ GuHSCN | permuted $c_{1/2}$ GuHCl | unpermuted $c_{1/2}$ GuHCl | Rem. | Ref |
|---|---|---|---|---|---|---|---|
| reference | 6.25 | 25 | 3.32±0.01 | 4.57±0.02 | 4.00±0.01 | (1–4) | 98O7 |
| Lys2→Ala | 6.25 | 25 | 3.19±0.01 | 4.12±0.02 | 3.72±0.01 | (1–4) | 98O7 |
| Pro6→Ala | 6.25 | 25 | 2.61±0.01 | 3.51±0.02 | 3.19±0.05 | (1–4) | 98O7 |
| Ala16→Gly | 6.25 | 25 | 2.87±0.01 | 3.78±0.01 | 3.44±0.02 | (1–4) | 98O7 |
| Leu32→Ala | 6.25 | 25 | 2.53±0.01 | 3.61±0.01 | 2.78±0.02 | (1–4) | 98O7 |
| Val34→Gly | 6.25 | 25 | 2.53±0.02 | 3.83±0.01 | 2.75±0.01 | (1–4) | 98O7 |
| Val38→Ala | 6.25 | 25 | 2.83±0.01 | 4.48±0.02 | 3.24±0.02 | (1–4) | 98O7 |
| Tyr42→Ala | 6.25 | 25 | 2.79±0.01 | 4.08±0.03 | 3.01±0.01 | (1–4) | 98O7 |
| Arg46→Ala | 6.25 | 25 | 2.34±0.01 | 3.20±0.06 | 2.49±0.02 | (1–4) | 98O7 |
| Phe50→Ala | 6.25 | 25 | 1.99±0.02 | 2.99±0.02 | 2.02±0.03 | (1–4) | 98O7 |
| Ala58→Gly | 6.25 | 25 | 2.63±0.01 | 3.42±0.02 | 3.03±0.04 | (1–4) | 98O7 |
| Val60→Ala | 6.25 | 25 | 2.79±0.01 | 3.78±0.03 | 3.22±0.03 | (1–4) | 98O7 |

Remarks:
(1) reference = double mutant (Thr3→Cys and Val63→Cys) oxidized
(2) unperturbed mutants, data from Ref. 95I1
(3) transition monitored by fluorescence, for the procedure see Ref. 93J1
(4) buffer: 50 mM MES, pH 6.25

Chymotrypsin inbitor-2 (CI2), complex composed of residues (1–40) and (41–64)

| Protein | pH | T | ΔG | Approach/Remarks | | Ref |
|---|---|---|---|---|---|---|
| CI2 complex | 6.0 | 25 | 5.69±0.13 | pressure | (1,2) | 99M19 |
| | 6.0 | 25 | 35.6 | GuHCl | (1,3) | 99M19 |
| | 6.0 | 25 | 41.8 | eq. | (1,4,5) | 99M19 |
| | 6.0 | 25 | 39.7 | kin. | (1,4,6) | 99M19 |
| CI2 native | 6.0 | 25 | 10.5 | pressure | (1,7) | 99M19 |

Remarks:
(1) buffer: 50 mM Bis-Tris, pH 6.0
(2) pressure denaturation, refers to the formation of a high pressure state, ΔV = 35±0.7 ml/mol
(3) transition monitored by fluorescence, the transition refers to dissociation of the complex and unfolding of the constituents
(4) the transition refers to fragment dissociation
(5) equilibrium value
(6) from association kinetics
(7) reference value for native CI2 obtained in the presence of 2.25 M GuHCl, transition monitored by fluorescence


**c-Myb**, see Protooncogene Product


# Coiled-Coil

Coiled-coil, designed leucine zipper to show stabilization by phosphorylation, effect of arginine on change in stability upon phosphorylation

| Coiled-coil | PHOS. | $T_{trs}$ | $\Delta H(T_{trs})$ | ΔG(37°C) | Appr./Rem. | Ref |
|---|---|---|---|---|---|---|
| (g abcdefg ab) | (1) | | | | | |
| (R ARRGTAR VR) | (–) | 33.0±0.11 | 301 | 28.5 | heat (1–4) | 97S17 |
| (R ARRGTAR VR) | (+) | 22.3±0.04 | 176 | 23.4 | heat (1–4) | 97S17 |
| (R ARRGSAR VR) | (–) | 33.1±0.06 | 301 | 28.0 | heat (1–4) | 97S17 |
| (R ARRGSAR VR) | (+) | 38.8±0.06 | 314 | 33.9 | heat (1–4) | 97S17 |
| (A – – – – – –R –R) | (–) | 35.2±0.04 | 310 | 30.5 | heat (1–4) | 97S17 |
| (A – – – – – –R –R) | (+) | 36.4±0.06 | 322 | 31.8 | heat (1–4) | 97S17 |
| (R – – – – – –A –R) | (–) | 23.1±0.04 | 205 | 22.6 | heat (1–4) | 97S17 |
| (R – – – – – –A –R) | (+) | 22.1±0.04 | 184 | 22.6 | heat (1–4) | 97S17 |
| (A – – – – – –A –R) | (–) | 22.2±0.07 | 188 | 23.0 | heat (1–4) | 97S17 |
| (A – – – – – –A –R) | (+) | 22.8±0.07 | 180 | 23.0 | heat (1–4) | 97S17 |
| (R – – – – – –R –A) | (–) | 32.6±0.08 | 297 | 28.0 | heat (1–4) | 97S17 |
| (R – – – – – –R –A) | (+) | 31.9±0.70 | 339 | 26.8 | heat (1–4) | 97S17 |
| (K –KK – – –K –K) | (–) | 24.2±0.09 | 247 | 21.8 | heat (1–4) | 97S17 |
| (K –KK – – –K –K) | (+) | 27.3±0.10 | 255 | 23.8 | heat (1–4) | 97S17 |

Remarks:
(1) column PHOS.:
    (–) unphosphorylated
    (+) phosphorylated
(2) buffer: 10 mM MOPS, 150 mM KCl, 1 mM EDTA, pH 7.4
(3) $T_{trs}$ and $\Delta H(T_{trs})$ from melting curves monitored by CD at 222 nm
(4) ΔG(37°C) was calculated using ΔCp = 5.02 kJ/mol/K from ΔH versus $T_{trs}$ from all the proteins in Ref. 97S17

Coiled-coil, designed leucine zipper to show stabilization by phosphorylation, effect of amino acids in the $a$ and $d$ positions on change in stability upon phosphorylation

| Coiled-coil | PHOS. | $T_{trs}$ | $\Delta H(T_{trs})$ | $\Delta G(37°C)$ | Appr./Rem. | Ref |
|---|---|---|---|---|---|---|
| (g abcdefg ab) | (1) | | | | | |
| (R ARRGSAR VR) | (−) | 36.0±0.06 | 314 | 31.4 | heat (1–4) | 97S17 |
| (R ARRGSAR VR) | (+) | 41.0±0.05 | 297 | 36.0 | heat (1–4) | 97S17 |
| (– G– –G– – – – –) | (−) | 29.6±0.07 | 263 | 25.9 | heat (1–4) | 97S17 |
| (– G– –G– – – – –) | (+) | 35.2±0.25 | 188 | 31.0 | heat (1–4) | 97S17 |
| (– V– –G– – – – –) | (−) | 46.1±0.06 | 285 | 40.2 | heat (1–4) | 97S17 |
| (– V– –G– – – – –) | (+) | 45.7±0.06 | 360 | 42.3 | heat (1–4) | 97S17 |
| (– A– –A– – –  – –) | (−) | 47.4±0.05 | 234 | 39.7 | heat (1–4) | 97S17 |
| (– A– –A– – – – –) | (+) | 47.9±0.02 | 226 | 39.7 | heat (1–4) | 97S17 |
| (– G– –A– – – – –) | (−) | 39.5±0.05 | 326 | 34.7 | heat (1–4) | 97S17 |
| (– G– –A– – – – –) | (+) | 40.0±0.03 | 259 | 34.7 | heat (1–4) | 97S17 |
| (– G– –L– – – – –) | (−) | 56.3±0.02 | 448 | 58.6 | heat (1–4) | 97S17 |
| (– G– –L– – – – –) | (+) | 56.9±0.04 | 439 | 58.6 | heat (1–4) | 97S17 |
| (– G– –L– – – – –) | (−) | 77.2±0.02 | 460 | 84.9 | heat (1–4) | 97S17 |
| (– G– –L– – – – –) | (+) | 77.2±0.04 | 456 | 84.9 | heat (1–4) | 97S17 |

Remarks:
(1) column PHOS.:    (−) unphosphorylated
          (+) phosphorylated
(2) buffer: 10 mM phosphate, 150 mM KCl, 1 mM EDTA, pH 7.4
(3) $T_{trs}$ and $\Delta H(T_{trs})$ from melting curves monitored by CD at 222 nm
(4) $\Delta G(37°C)$ was calculated using $\Delta Cp$ = 5.02 kJ/mol/K from $\Delta H$ versus $T_{trs}$ from all the proteins in Ref. 97S17

Coiled-coil, designed leucine zipper to show stabilization by phosphorylation, effect of buffer on change in stability upon phosphorylation

| Coiled-coil | PHOS. | $T_{trs}$ | $\Delta H(T_{trs})$ | $\Delta G(37°C)$ | Appr./Rem. | | Ref |
|---|---|---|---|---|---|---|---|
| (g abcdefg ab) | (1) | | | | | | |
| (R ARRGSAR VR) | (−) | 36.0±0.06 | 314 | 31.4 | KP | (1–4) | 97S17 |
| (R ARRGSAR VR) | (+) | 41.0±0.05 | 297 | 36.0 | KP | (1–4) | 97S17 |
| (R ARRGSAR VR) | (−) | 33.1±0.06 | 301 | 28.0 | MOPS | (1–4) | 97S17 |
| (R ARRGSAR VR) | (+) | 38.8±0.06 | 314 | 33.9 | MOPS | (1–4) | 97S17 |
| (– – – –A– – –  – –) | (−) | 47.4±0.05 | 234 | 39.7 | KP | (1–4) | 97S17 |
| (– – – –A– – –  – –) | (+) | 47.9±0.02 | 226 | 39.7 | KP | (1–4) | 97S17 |
| (– – – –A– – –  – –) | (−) | 45.6±0.07 | 314 | 40.6 | MOPS | (1–4) | 97S17 |
| (– – – –A– – –  – –) | (+) | 45.2±0.07 | 293 | 39.7 | MOPS | (1–4) | 97S17 |
| (A – – – – – –A – –) | (−) | 23.5±0.59 | 180 | 26.4 | KP | (1–4) | 97S17 |
| (A – – – – – –A – –) | (+) | 20.5±0.60 | 142 | 26.4 | KP | (1–4) | 97S17 |
| (A – – – – – –A – –) | (−) | 20.8±0.63 | 167 | 25.1 | MOPS | (1–4) | 97S17 |
| (A – – – – – –A – –) | (+) | 19.8±0.49 | 155 | 25.1 | MOPS | (1–4) | 97S17 |
| (K –KK– – –K –K) | (−) | 28.0±0.10 | 243 | 25.1 | KP | (1–4) | 97S17 |
| (K –KK– – –K –K) | (+) | 30.3±0.32 | 222 | 27.2 | KP | (1–4) | 97S17 |
| (K –KK– – –K –K) | (−) | 24.2±0.09 | 247 | 23.8 | MOPS | (1–4) | 97S17 |
| (K –KK– – –K –K) | (+) | 27.3±0.10 | 255 | 25.9 | MOPS | (1–4) | 97S17 |

Remarks:
(1) column PHOS.:    (−) unphosphorylated
          (+) phosphorylated
(2) buffer:
          KP  = 10 mM potassium phosphate, 150 mM KCl, 1 mM EDTA, pH 7.4
          MOPS = 10 mM MOPS, 150 mM KCl, 1 mM EDTA, pH 7.4
(3) $T_{trs}$ and $\Delta H(T_{trs})$ from melting curves monitored by CD at 222 nm
(4) $\Delta G(37°C)$ was calculated using $\Delta Cp$ = 5.02 kJ/mol/K from $\Delta H$ versus $T_{trs}$ from all the proteins in Ref. 97S17

Coiled-coil, GCN4 leucine zipper

| Mutant | pH | $T_{trs}$ | $\Delta T$ | $\Delta(\Delta G)$ | Appr./Rem. | Ref |
|---|---|---|---|---|---|---|
| wild type (Ser15, His19, Asn22) | | | | | | |
| | 7.0 | 45 | 0 | 0.0 | heat (1,3,5) | 98S17 |
| Arg,Glu,Arg | 7.0 | 67 | 22 | 7.20 | heat (1–5) | 98S17 |
| Arg,Glu,Ala | 7.0 | 66 | 21 | 6.86 | heat (1–5) | 98S17 |
| Ala,Ala,Arg | 7.0 | 61 | 16 | 5.15 | heat (1–5) | 98S17 |
| Arg,Ala,Arg | 7.0 | 60 | 15 | 4.81 | heat (1–5) | 98S17 |
| Ala,Glu,Arg | 7.0 | 60 | 15 | 4.81 | heat (1–5) | 98S17 |
| Arg,Ala,Ala | 7.0 | 59 | 14 | 4.48 | heat (1–5) | 98S17 |
| Ala,Ala,Ala | 7.0 | 59 | 14 | 4.48 | heat (1–5) | 98S17 |
| Ala,Glu,Ala | 7.0 | 55 | 10 | 3.18 | heat (1–5) | 98S17 |
| Ser,Arg,Glu | 7.0 | 48 | 3 | 0.92 | heat (1–5) | 98S17 |

Remarks:

(1) the GCN4 leucine zipper consists of two 34 amino acid $\alpha$-helices that form a dimer

(2) within the wild-type sequence Val10-Glu-Glu-Leu-Leu-Ser15-Lys-Asn-Tyr-His-Leu20-Glu-Asn-Glu-Val the following residues are replaced: Ser15, His19, and Asn22

(3) mutant description: Arg,Glu,Arg corresponds to the triple mutant (Ser15→Arg, His19→Glu, and Asn22→Arg)

(4) the stability was measured by CD at 222 nm at a fixed monomer conc. of 10 μM peptide in 150 mM NaCl, 50 mM phosphate, pH 7.0

(5) for reference values, $\Delta H$ and $\Delta Cp$, see Ref. 93T4

Coiled-coil, model peptides

| Form | pH | T | $\Delta G$ | Approach/Remarks | | Ref |
|---|---|---|---|---|---|---|
| dKL27ox | 4.7 | 20 | 19.1±0.9 | heat | (1,2,4–6) | 97H7 |
| mKL27ox | 4.7 | 20 | 14.2±0.2 | heat | (1,3–6) | 97H7 |

Remarks:

(1) model peptide: Trp-Cys(SH)-Lys-Ala-Val-Lys-Lys-Leu-Ala-Lys-Ala-Val-Lys-Lys-Leu-Ala-Lys-Ala-Val-Lys-Lys-Leu-Ala-Lys-Ala-Gly-Cys(SH)-NH2

(2) dKL27ox: parallel dimer connected by the Cys residues at the N- and C-termini

(3) mKL27ox: monomeric circular species in which a disulfide bond was formed intramolecularly between the N-terminal and C-terminal Cys residues

(4) buffer: 10 mM sodium acetate, pH 4.7, with 20 mM NaClO4

(5) van't Hoff treatment of a two-state transition $U_2 \rightarrow H_2$ with U = unfolded and H = helical state, for further thermodynamic data see Table 2

(6) the transition was monitored by the ellipticity at 222 nm

Coiled-coil, de novo designed coiled-coils, salt effect on protein stability

| Salt | pH | T | ΔG | $c_{1/2}$ | m | Appr./Rem. | Ref |
|---|---|---|---|---|---|---|---|
| peptide QQx, remark (1): | | | | | | | |
| none | 7 | 20 | 0 | 3.95 | 4.02 | urea (3–5) | 97K8 |
| 1.2 M KCl | 7 | 20 | 8.2 | 5.9 | 4.35 | urea (3–5) | 97K8 |
| 0.4 M MgCl$_2$ | 7 | 20 | 4.8 | 5.1 | 4.35 | urea (3–5) | 97K8 |
| 0.2 M LaCl$_3$ | 7 | 20 | 5.6 | 5.3 | 4.31 | urea (3–5) | 97K8 |
| peptide KEx, remark (2): | | | | | | | |
| none | 7 | 20 | 0 | 4.9 | 4.02 | urea (3–5) | 97K8 |
| 1.2 M KCl | 7 | 20 | 2.2 | 5.5 | 3.89 | urea (3–5) | 97K8 |
| 0.4 M MgCl$_2$ | 7 | 20 | –0.8 | 4.7 | 4.02 | urea (3–5) | 97K8 |
| 0.2 M LaCl$_3$ | 7 | 20 | –12.7 | 1.9 | 4.44 | urea (3–5) | 97K8 |

Remarks:
(1) the designed coiled-coil QQ contains no intra- and interhelical ionic interactions
(2) the designed coiled-coil KE contains Lys in position *e* and Glu in position *g*
(3) linear extrapolation
(4) salt conditions in addition to the standard buffer 50 mM Tris, 100 mM KCl, pH 7
(5) the transition was monitored by CD at 222 nm

Coiled-coil, heterodimeric coiled-coil with altered orientation of helices

| Protein | pH | T | ΔG | Approach/Remarks | | Ref |
|---|---|---|---|---|---|---|
| Acid-a1C-Base-a1N | 7.0 | 25 | 25.5±0.8 | GuHCl | (1–4) | 98O1 |
| Acid-a1C-Base-a1N | 6.0 | 25 | 26.4±2.5 | GuHCl | (1,3–5) | 98O1 |
| Acid-a1N-Base-a1N | 7.0 | 25 | 15.9±2.1 | GuHCl | (1–4) | 98O1 |

Remarks:
(1) linear extrapolation
(2) measured in 10 mM sodium phosphate, 150 mM sodium chloride at 10 μM peptide
(3) transition monitored by CD at 222 nm
(4) for the construction of the peptide, see also Refs. 93O3 and 95L12
(5) measured in 10 mM sodium phosphate, 150 mM sodium chloride at 20 μM peptide in D$_2$O

Coiled-coil, α-helical coiled-coils, effect of interhelical ion pairs

| Peptide | Form | pH | T | Δ(ΔG) | $c_{1/2}$ | m | Approach/Remarks | | Ref |
|---|---|---|---|---|---|---|---|---|---|
| urea denaturation at pH 7, Lys or Glu substitutions: | | | | | | | | | |
| native | ox. | 7.0 | 20 | 0.00 | 6.0 | 3.64 | urea | (1,2,5–7) | 98K11 |
| Glu15 | ox. | 7.0 | 20 | –1.63 | 5.55 | 3.60 | urea | (1–3,5–7) | 98K11 |
| QE(VL) | ox. | 7.0 | 20 | –6.57 | 4.1 | 3.26 | urea | (1,2,4–7) | 98K11 |
| Glu20 | ox. | 7.0 | 20 | –0.79 | 5.8 | 4.18 | urea | (1–3,5–7) | 98K11 |
| EQ(VL) | ox. | 7.0 | 20 | –4.06 | 4.8 | 3.10 | urea | (1,2,4–7) | 98K11 |
| Lys15 | ox. | 7.0 | 20 | –4.06 | 4.95 | 4.10 | urea | (1–3,5–7) | 98K11 |
| QK(VL) | ox. | 7.0 | 20 | –16.82 | 2.1 | 4.98 | urea | (1,2,4–7) | 98K11 |
| Lys20 | ox. | 7.0 | 20 | –1.88 | 5.5 | 3.93 | urea | (1–3,5–7) | 98K11 |
| KQ(VL) | ox. | 7.0 | 20 | –10.21 | 3.6 | 4.85 | urea | (1,2,4–7) | 98K11 |
| urea denaturation at pH 3.2, Lys or Glu substitutions: | | | | | | | | | |
| native | ox. | 3.2 | 20 | 0.00 | 5.7 | 3.60 | urea | (1,2,5,6,8) | 98K11 |
| Glu15 | ox. | 3.2 | 20 | +5.36 | 7.25 | 3.31 | urea | (1–3,5,6,8) | 98K11 |
| QE(VL) | ox. | 3.2 | 20 | | >10 | | urea | (1,2,4,6,8) | 98K11 |
| Glu20 | ox. | 3.2 | 20 | +4.02 | 6.8 | 3.64 | urea | (1–3,5,6,8) | 98K11 |
| EQ(VL) | ox. | 3.2 | 20 | | >10 | | urea | (1,2,4,6,8) | 98K11 |
| Lys15 | ox. | 3.2 | 20 | –3.77 | 4.7 | 3.93 | urea | (1–3,5,6,8) | 98K11 |

Coiled-coil, α-helical coiled-coils, effect of interhelical ion pairs (continued)

| Peptide | Form | pH | T | Δ(ΔG) | $c_{1/2}$ | m | Approach/Remarks | | Ref |
|---------|------|-----|-----|--------|------|------|----------|------------|--------|
| QK(VL) | ox. | 3.2 | 20 | −17.15 | 1.8 | 5.19 | urea | (1,2,4–6,8) | 98K11 |
| Lys20 | ox. | 3.2 | 20 | −2.22 | 5.1 | 3.72 | urea | (1–3,5,6,8) | 98K11 |
| KQ(VL) | ox. | 3.2 | 20 | −10.08 | 3.3 | 4.81 | urea | (1,2,4–6,8) | 98K11 |
| GuHCl denaturation at pH 3.2, Glu substitutions: | | | | | | | | | |
| native | ox. | 3.2 | 20 | 0.00 | 3.1 | 7.53 | GuHCl | (1,2,5,6,8) | 98K11 |
| Glu15 | ox. | 3.2 | 20 | +5.40 | 3.8 | 7.99 | GuHCl | (1–3,5,6,8) | 98K11 |
| QE(VL) | ox. | 3.2 | 20 | +20.21 | 5.6 | 8.28 | GuHCl | (1,2,4–6,8) | 98K11 |
| Glu20 | ox. | 3.2 | 20 | +3.85 | 3.6 | 8.03 | GuHCl | (1–3,5,6,8) | 98K11 |
| EQ(VL) | ox. | 3.2 | 20 | +17.41 | 5.3 | 8.33 | GuHCl | (1,2,4–6,8) | 98K11 |
| urea denaturation at pH 7, ion pair peptides: | | | | | | | | | |
| native | ox. | 7.0 | 20 | 0.00 | 6.0 | 3.64 | urea | (1,2,5–7) | 98K11 |
| Glu15/Lys20 | ox. | 7.0 | 20 | 0.00 | 6.0 | 3.47 | urea | (1–3,5–7) | 98K11 |
| KE(VL) | ox. | 7.0 | 20 | −1.34 | 5.6 | 3.01 | urea | (1,2,4–7) | 98K11 |
| Lys15/Glu20 | ox. | 7.0 | 20 | 0.00 | 6.0 | 3.47 | urea | (1–3,5–7) | 98K11 |
| EK(VL) | ox. | 7.0 | 20 | +0.71 | 6.2 | 3.39 | urea | (1,2,4–7) | 98K11 |
| urea denaturation at pH 3.2, ion pair peptides: | | | | | | | | | |
| native | ox. | 3.2 | 20 | 0.00 | 5.7 | 3.60 | urea | (1,2,5,6,8) | 98K11 |
| Glu15/Lys20 | ox. | 3.2 | 20 | +3.97 | 6.75 | 3.93 | urea | (1–3,5,6,8) | 98K11 |
| KE(VL) | ox. | 3.2 | 20 | +12.80 | 9.1 | 3.89 | urea | (1,2,4–6,8) | 98K11 |
| Lys15/Glu20 | ox. | 3.2 | 20 | +1.13 | 6.0 | 3.85 | urea | (1–3,5,6,8) | 98K11 |
| EK(VL) | ox. | 3.2 | 20 | +3.60 | 6.6 | 4.48 | urea | (1,2,4–6,8) | 98K11 |
| urea denaturation at pH 7, reduced coiled-coils | | | | | | | | | |
| native | red. | 7.0 | 20 | 0.00 | 2.5 | 6.15 | urea | (1,5–7) | 98K11 |
| Glu15 | red. | 7.0 | 20 | −1.21 | 2.3 | 5.77 | urea | (1,3,5–7) | 98K11 |
| Glu20 | red. | 7.0 | 20 | −1.17 | 2.3 | 5.44 | urea | (1,3,5–7) | 98K11 |
| Lys15 | red. | 7.0 | 20 | −5.02 | 1.7 | 6.36 | urea | (1,3,5–7) | 98K11 |
| Lys20 | red. | 7.0 | 20 | −4.98 | 1.7 | 6.28 | urea | (1,3,5–7) | 98K11 |
| Glu15/Lys20 | red. | 7.0 | 20 | 0.00 | 2.5 | 6.07 | urea | (1,3,5–7) | 98K11 |
| Lys15/Glu20 | red. | 7.0 | 20 | 0.00 | 2.5 | 5.56 | urea | (1,3,5–7) | 98K11 |

Remarks:

(1) the sequence of the native peptide: Ac-Gln1-Cys-Gly-Ala-Leu-Gln-Lys-Gln-Val-Gly10-Ala-Leu-Gln-Lys-Gln-Val-Gly-Ala-Leu-Gln20-Lys-Gln-Val-Gly-Ala-Leu-Gln-Lys-Gln-Val30-Gly-Ala-Leu-Gln-Lys-amide

(2) Cys2 (ox. = oxidized) forms two-stranded disulfide-bridged peptides

(3) Glu15, Glu20, Lys15, and Lys20 designate exchanges of Gln in heptad repeat positions *g* and *e* (only the middle heptad), Glu15/Lys20 and Lys15/Glu20 designate the corresponding double mutants

(4) the analogs EQ(VL), QE(VL), KQ(VL), QK(VL), EK(VL), and KE(VL) designate Glu and Lys substitutions in all five heptad repeats; (VL) signifies the presence of Val and Leu at the hydrophobic positions

(5) linear extrapolation

(6) transition monitored by CD at 222 nm

(7) buffer: 50 mM Tris, pH 7.0, 100 mM KCl

(8) buffer: 20 mM glycine, pH 3.2, 100 mM KCl

Coiled-coil, two-stranded α-helical coiled-coils

| Peptide name | pH | T | Δ(ΔG) | $c_{1/2}$ | m | Appr./Rem. | Ref |
|--------------|-----|-----|--------|------|------|------------|--------|
| native | 7.0 | 25 | 0.0 | 3.2 | 7.61 | GuHCl (1–4) | 98K20 |
| control | 7.0 | 25 | 18.83 | 5.6 | 8.08 | GuHCl (1–4) | 98K20 |
| T1 (24) | 7.0 | 25 | −2.26 | 2.9 | 7.03 | GuHCl (1–4) | 98K20 |
| T2a (22,24) | 7.0 | 25 | 2.38 | 3.5 | 7.82 | GuHCl (1–4) | 98K20 |
| T2b (19,24) | 7.0 | 25 | −3.72 | 2.7 | 6.86 | GuHCl (1–4) | 98K20 |
| T3a (19,22,24) | 7.0 | 25 | 0.71 | 3.3 | 6.40 | GuHCl (1–4) | 98K20 |

Coiled-coil, two-stranded α-helical coiled-coils (continued)

| Peptide name | pH | T | Δ(ΔG) | $c_{1/2}$ | m | Appr./Rem. | Ref |
|---|---|---|---|---|---|---|---|
| T3b (18,19,24) | 7.0 | 25 | −6.07 | 2.4 | 7.11 | GuHCl (1–4) | 98K20 |
| T4a (18,19,22,24) | 7.0 | 25 | −1.59 | 3.0 | 7.74 | GuHCl (1–4) | 98K20 |
| T4b (15,18,19,24) | 7.0 | 25 | −7.57 | 2.2 | 7.07 | GuHCl (1–4) | 98K20 |
| T5a (15,18,19,22,24) | 7.0 | 25 | −3.97 | 2.7 | 7.78 | GuHCl (1–4) | 98K20 |
| Ile20→Ala | 7.0 | 25 | −11.97 | 1.6 | 6.86 | GuHCl (1–4) | 98K20 |
| Ile20→Thr | 7.0 | 25 | −13.39 | 1.4 | 7.15 | GuHCl (1–4) | 98K20 |

Remarks:
(1) disulfide-bridged two-stranded homostranded peptides, for the sequences see Ref. 98K20
(2) linear extrapolation
(3) transition monitored by CD at 222 nm
(4) buffer: 50 mM $KH_2PO_4$, 100 mM KCl, pH 7.0

Coiled-coil, de novo designed model peptide, single amino acid substitutions in the hydrophobic core

| Peptide | pH | T | Δ(ΔG) | $c_{1/2}$ | m | Appr./Rem. | Ref |
|---|---|---|---|---|---|---|---|
| Ala19a | 7.0 | 25 | 0.0 | 1.2 | 9.20 | GuHCl (1–5) | 99W1 |
| | | | | 1.0 | | GuHCl (1–4,6) | 99W1 |
| Asn19a | 7.0 | 25 | 3.2 | 1.5 | 7.70 | GuHCl (1–5) | 99W1 |
| | | | | 1.6 | | GuHCl (1–4,6) | 99W1 |
| Glu19a | 7.0 | 25 | −8.8 | 0.52 | 16.32 | GuHCl (1–5) | 99W1 |
| | | | | n.d. | | GuHCl (1–4,6) | 99W1 |
| Ile19a | 7.0 | 25 | 15.5 | 3.3 | 5.48 | GuHCl (1–5) | 99W1 |
| | | | | 3.2 | | GuHCl (1–4,6) | 99W1 |
| Leu19a | 7.0 | 25 | 14.2 | 2.9 | 7.03 | GuHCl (1–5) | 99W1 |
| | | | | 2.7 | | GuHCl (1–4,6) | 99W1 |

Remarks:
(1) model peptide: Ac-Cys-Gly-Gly-Glu-Val-Gly-Ala-Leu-Lys-Ala-Gln-Val-Gly-Ala-Leu-Gln-Ala-Gln-(X19a)-Gly-Ala-Leu-Gln-Lys-Glu-Val-Gly-Ala-Leu-Lys-Lys-Glu-Val-Gly-Ala-Leu-Lys-Lys-amide with Ala, Asn, Glu, Ile, Leu in position (X19a), all peptides in oxidized state
(2) peptide with Ala in position (X19a) is the reference protein
(3) buffer: 50 mM phosphate, 100 mM KCl, pH 7.0
(4) Δ(ΔG) was calculated using $\Delta(\Delta G) = \{(m_{Ala} + m_X)/2\}\times\{(c_{1/2})_X - (c_{1/2})_{Ala}\}$ according to Ref. 91S1
(5) transition monitored by far-UV CD at 222 nm at 50 μM peptide conc.
(6) transition from high performance size-exclusion chromatography (HPSEC) at 2.5 μM peptide conc.

Coiled-coil, model coiled-coil, variation of position *a*

| Amino acid | pH | T | Δ(ΔG) | $c_{1/2}$ | m | Appr./Rem. | Ref |
|---|---|---|---|---|---|---|---|
| Ala | 7.0 | 25 | 0.00 | 2.97 | 3.81 | urea (1–4) | 99W2 |
| | 7.0 | 25 | 0.00 | 1.17 | 10.59 | GuHCl (1–4) | 99W2 |
| Arg | 7.0 | 25 | −5.82 | 1.51 | 4.18 | urea (1–3) | 99W2 |
| | 7.0 | 25 | −3.22 | 0.81 | 7.57 | GuHCl (1–3) | 99W2 |
| Asn | 7.0 | 25 | 2.97 | 3.73 | 4.02 | urea (1–4) | 99W2 |
| | 7.0 | 25 | 3.56 | 1.54 | 8.12 | GuHCl (1–4) | 99W2 |
| Gln | 7.0 | 25 | −0.88 | 2.74 | 3.93 | urea (1–3) | 99W2 |
| | 7.0 | 25 | −0.54 | 1.11 | 9.20 | GuHCl (1–3) | 99W2 |
| Glu | 7.0 | 25 | −15.23 | 0.60 | 9.08 | urea (1–4) | 99W2 |
| | 7.0 | 25 | −8.28 | 0.52 | 15.10 | GuHCl (1–4) | 99W2 |
| Gly | 7.0 | 25 | −7.57 | 1.08 | 4.23 | urea (1–3) | 99W2 |
| | 7.0 | 25 | −10.42 | 0.27 | 12.68 | GuHCl (1–3) | 99W2 |
| His | 7.0 | 25 | −1.38 | 2.64 | 4.64 | urea (1–3) | 99W2 |
| | 7.0 | 25 | −4.81 | 0.77 | 13.56 | GuHCl (1–3) | 99W2 |
| Ile | 7.0 | 25 | 13.77 | 7.40 | 2.43 | urea (1–4) | 99W2 |
| | 7.0 | 25 | 16.23 | 3.26 | 4.94 | GuHCl (1–4) | 99W2 |

Coiled-coil, model coiled-coil, variation of position *a* (continued)

| Amino acid | pH | T | $\Delta(\Delta G)$ | $c_{1/2}$ | m | Appr./Rem. | Ref |
|---|---|---|---|---|---|---|---|
| Leu | 7.0 | 25 | 12.43 | 6.39 | 2.51 | urea (1–4) | 99W2 |
|  | 7.0 | 25 | 14.56 | 2.90 | 6.23 | GuHCl (1–4) | 99W2 |
| Lys | 7.0 | 25 | −3.89 | 2.05 | 4.73 | urea (1–3) | 99W2 |
|  | 7.0 | 25 | −1.67 | 0.98 | 7.70 | GuHCl (1–3) | 99W2 |
| Met | 7.0 | 25 | 13.85 | 6.94 | 3.22 | urea (1–3) | 99W2 |
|  | 7.0 | 25 | 14.35 | 2.85 | 6.40 | GuHCl (1–3) | 99W2 |
| Orn | 7.0 | 25 | −10.59 | 0.68 | 5.48 | urea (1–3) | 99W2 |
|  | 7.0 | 25 | −7.87 | 0.34 | 8.37 | GuHCl (1–3) | 99W2 |
| Phe | 7.0 | 25 | 11.55 | 6.15 | 3.47 | urea (1–3) | 99W2 |
|  | 7.0 | 25 | 12.59 | 2.57 | 7.28 | GuHCl (1–3) | 99W2 |
| Ser | 7.0 | 25 | −4.90 | 1.80 | 4.64 | urea (1–3) | 99W2 |
|  | 7.0 | 25 | −5.36 | 0.67 | 11.13 | GuHCl (1–3) | 99W2 |
| Thr | 7.0 | 25 | 0.75 | 3.18 | 3.60 | urea (1–3) | 99W2 |
|  | 7.0 | 25 | 0.84 | 1.25 | 8.91 | GuHCl (1–3) | 99W2 |
| Trp | 7.0 | 25 | 3.77 | 3.91 | 4.23 | urea (1–3) | 99W2 |
|  | 7.0 | 25 | 3.26 | 1.54 | 6.90 | GuHCl (1–3) | 99W2 |
| Tyr | 7.0 | 25 | 7.91 | 5.02 | 3.89 | urea (1–3) | 99W2 |
|  | 7.0 | 25 | 8.70 | 2.07 | 8.74 | GuHCl (1–3) | 99W2 |
| Val | 7.0 | 25 | 14.69 | 7.56 | 2.59 | urea (1–3) | 99W2 |
|  | 7.0 | 25 | 17.15 | 3.17 | 6.61 | GuHCl (1–3) | 99W2 |

Remarks:
(1) for the sequence (position 19*a* modified) and peptide synthesis, see Ref. 99W2
(2) linear extrapolation, for further details see Ref. 91S1
(3) $\Delta(\Delta G)$ refers to the Ala-substituted analog
(4) data from Ref. 99W1

**Coiled-coil,** see also De Novo Designed Proteins/Peptides

**Coiled-Coil,** see also Leucine Zipper

## Cold Shock Protein

Cold shock protein CspA from *E. coli*, aromatic cluster mutants

| Mutant | pH | T | $\Delta G$ | $c_{1/2}$ | m | Appr./Rem. | Ref |
|---|---|---|---|---|---|---|---|
| Phe,Phe,Phe | 7.5 | 25 | 12.55±0.63 | 4.20±0.10 | 2.97±0.17 | urea (1–3) | 98H8 |
| Phe,Phe,Phe | 7.5 | 10 | 12.22±0.46 | 4.95±0.10 | 2.47±0.21 | urea (1–3) | 98H8 |
| Phe,Phe,Leu | 7.5 | 25 | 9.79±0.42 | 3.27±0.07 | 3.01±0.17 | urea (1–3) | 98H8 |
| Phe,Leu,Phe | 7.5 | 25 | 11.25±0.67 | 3.77±0.03 | 2.97±0.21 | urea (1–3) | 98H8 |
| Leu,Phe,Phe | 7.5 | 25 | 8.37±0.29 | 2.49±0.05 | 3.35±0.17 | urea (1–3) | 98H8 |
| Phe,Leu,Leu | 7.5 | 25 | 10.54±0.21 | 3.62±0.06 | 2.93±<.04 | urea (1–3) | 98H8 |
| Leu,Leu,Phe | 7.5 | 25 | 8.87±0.13 | 2.93±0.02 | 3.01±0.04 | urea (1–3) | 98H8 |
| Leu,Phe,Leu | 7.5 | 25 | 6.02±0.25 | 2.00±0.06 | 3.01±0.04 | urea (1–3) | 98H8 |
| Leu,Leu,Leu | 7.5 | 25 | 8.87±0.25 | 3.22±0.05 | 2.76±0.04 | urea (1–3) | 98H8 |
| Phe,Phe,Ser | 7.5 | 25 | 8.24±0.13 | 2.47±0.02 | 3.35±0.04 | urea (1–3) | 98H8 |
| Phe,Ser,Phe | 7.5 | 25 | 7.70±0.08 | 2.36±0.01 | 3.26±0.04 | urea (1–3) | 98H8 |
| Ser,Phe,Phe | 7.5 | 25 | 6.11±0.08 | 1.91±0.02 | 3.18±0.04 | urea (1–3) | 98H8 |
| Phe,Ser,Ser | 7.5 | 10 | 8.20±0.46 | 2.33±0.01 | 3.51±0.21 | urea (1–3) | 98H8 |
| Ser,Ser,Phe | 7.5 | 10 | 3.97±0.42 | 1.30±0.11 | 3.10±0.04 | urea (1–3) | 98H8 |
| Ser,Phe,Ser | 7.5 | 10 | 6.40±0.38 | 1.84±0.01 | 3.47±0.21 | urea (1–3) | 98H8 |
| Ser,Ser,Ser | 7.5 | 10 | 4.81±0.38 | 1.31±0.13 | 3.68±0.04 | urea (1–3) | 98H8 |

Remarks:
(1) mutants are identified by three sequential three-letter codes designating the identity of the residues at positions 18, 20, and 31. For example, Phe,Phe,Phe denotes Phe18, Phe20, Phe31 of CspA, and Phe,Leu,Leu denotes Phe18, Phe20→Leu, and Phe31→Leu
(2) linear extrapolation, for the procedure see Ref. 98R2
(3) buffer: 50 mM potassium phosphate, 100 mM KCl, pH 7

Cold shock protein CspA from *E. coli*, role of His tags

| Mutant | pH | T | ΔG | $c_{1/2}$ | m | Appr./Rem. | | Ref |
|--------|-----|-----|----------|----------|------------|---------|-------|------|
| wild type | 7.0 | 10 | 12.1±0.4 | 4.9±0.1 | 2.46±0.21 | urea | (1–3) | 98R2 |
| wild type | 7.0 | 25 | 12.6±0.4 | 4.2±0.1 | 2.97±0.17 | urea | (1–3) | 98R2 |
| wild type | 7.0 | 25 | 10.9±2.1 | 4.0±1.1 | 2.51±0.42 | urea | (1,2,4) | 98R2 |
| H6-Xa-CspA | 7.0 | 10 | 13.8±0.8 | 5.0±0.2 | 2.76±0.17 | urea | (1–3) | 98R2 |
| H6-Xa-CspA | 7.0 | 25 | 12.1±0.4 | 4.2±0.1 | 2.89±0.21 | urea | (1–3) | 98R2 |
| H6-Xa-CspA | 7.0 | 25 | 10.9±2.1 | 4.5±1.1 | 2.51±0.42 | urea | (1,2,4) | 98R2 |
| CspA-H6 | 7.0 | 10 | 11.7±0.8 | 4.5±0.2 | 2.68±0.21 | urea | (1–3) | 98R2 |
| CspA-H6 | 7.0 | 25 | 13.4±0.8 | 4.2±0.1 | 3.26±0.25 | urea | (1–3) | 98R2 |
| CspA-H6 | 7.0 | 25 | 13.0±1.7 | 4.1±0.9 | 2.93±0.42 | urea | (1,2,4) | 98R2 |
| wild type | 7.0 | 10 | 13.0±0.8 | 1.8±0.2 | 7.11±0.84 | GuHCl | (1–3) | 98R2 |
| wild type | 7.0 | 25 | 13.0±0.4 | 1.4±0.1 | 9.20±0.42 | GuHCl | (1–3) | 98R2 |
| wild type | 7.0 | 25 | 10.0±2.1 | 1.5±0.4 | 7.11±1.26 | GuHCl | (1,2,4) | 98R2 |
| H6-Xa-CspA | 7.0 | 10 | 15.9±0.8 | 1.6±0.1 | 10.04±0.84 | GuHCl | (1–3) | 98R2 |
| H6-Xa-CspA | 7.0 | 25 | 13.4±0.8 | 1.6±0.1 | 8.37±0.42 | GuHCl | (1–3) | 98R2 |
| H6-Xa-CspA | 7.0 | 25 | 15.5±2.1 | 1.3±0.2 | 12.13±1.67 | GuHCl | (1,2,4) | 98R2 |
| CspA-H6 | 7.0 | 10 | 13.8±0.4 | 1.6±0.1 | 8.79±0.42 | GuHCl | (1–3) | 98R2 |
| CspA-H6 | 7.0 | 25 | 12.6±0.4 | 1.4±0.1 | 9.20±0.42 | GuHCl | (1–3) | 98R2 |
| CspA-H6 | 7.0 | 25 | 13.8±1.7 | 1.6±0.3 | 8.79±1.26 | GuHCl | (1,2,4) | 98R2 |

Explanation:

> H6-Xa-CspA = CspA with six histidine residues at the N-terminus followed by a factor Xa protease cleavage site
> CspA-H6   = CspA with six histidine residues at the C-terminus

Remarks:

(1) linear extrapolation, LEM-SB

(2) buffer: 50 mM potassium phosphate, 100 mM KCl, pH 7.0

(3) transition monitored by fluorescence, excision of the single Trp11 at 280 nm, emission at 349 nm

(4) denaturant-induced unfolding transition monitored by CD at 222 nm, thermal transition monitored by CD at 222 nm yields equivalent $T_{trs}$ = 58.6±1.0°C for all three variants

Cold shock protein CspB from *Bacillus subtilis*, wild type and Phe → Ala mutants

| Mutant | pH | $T_{ref}$ | ΔG | Approach/Remarks | | Ref |
|--------|-----|-----|------------|------|-------|------|
| wild type | 7.0 | 45 | 4.89±0.04 | heat | (1,2) | 99J3 |
| Phe15→Ala | 7.0 | 45 | −3.67±0.04 | heat | (1,2) | 99J3 |
| Phe17→Ala | 7.0 | 45 | −1.46±0.04 | heat | (1,2) | 99J3 |
| Phe27→Ala | 7.0 | 45 | 1.16±0.04 | heat | (1,2) | 99J3 |

Remarks:

(1) transition monitored by CD at 223 nm

(2) buffer: 0.1 M sodium cacodylate/HCl, pH 7.0

Cold shock protein CspB from *Bacillus subtilis*, surface-exposed phenylalanines

| Mutant | pH | T | ΔG | $c_{1/2}$ | m | Appr./Rem. | | Ref |
|--------|-----|-----|----------|------|------------|------|-------|------|
| wild type | 7.0 | 25 | 14.6±0.6 | 4.45 | 3.28±0.13 | urea | (1–3) | 98S2 |
| Phe15→Ala | 7.0 | 25 | 5.1±1.5 | 1.55 | 3.28±0.30 | urea | (1–3) | 98S2 |
| Phe17→Ala | 7.0 | 25 | 8.2±1.0 | 2.57 | 3.19±0.24 | urea | (1–3) | 98S2 |
| Phe27→Ala | 7.0 | 25 | 11.4±0.8 | 3.17 | 3.60±0.19 | urea | (1–3) | 98S2 |
| Phe38→Ala | 7.0 | 25 | 15.2±1.7 | 4.71 | 3.24±0.35 | urea | (1–3) | 98S2 |

Remarks:

(1) linear extrapolation, LEM-SB

(2) transition monitored by fluorescence at 342 nm

(3) buffer: 0.35 M potassium phosphate, pH 7.0

Cold shock protein CspB from *Bacillus subtilis*, surface-exposed phenylalanines

| Mutant | pH | $T_{trs}$ | $\Delta T$ | $\Delta(\Delta G)$ | Approach/Remarks | Ref |
|---|---|---|---|---|---|---|
| wild type | 7.0 | 59.9 | 0.0 | 0.0 | heat, v.H. (1–3) | 98S2 |
| Phe15→Ala | 7.0 | 44.1 | −15.8 | −7.0 | heat, v.H. (1–3) | 98S2 |
| Phe17→Ala | 7.0 | 48.6 | −11.3 | −5.6 | heat, v.H. (1–3) | 98S2 |
| Phe27→Ala | 7.0 | 53.9 | −6.0 | −2.9 | heat, v.H. (1–3) | 98S2 |
| Phe38→Ala | 7.0 | 62.1 | 2.2 | 1.3 | heat, v.H. (1–3) | 98S2 |

Remarks:
(1) $\Delta(\Delta G)$ at $T_{trs}$ of the wild-type protein
(2) transition monitored by absorbance change at 292 nm
(3) buffer: 0.35 M potassium phosphate, pH 7.0

Cold shock protein CspB from mesophile *Bacillus subtilis* (*Bs*-CspB), thermophile *Bacillus caldolyticus* (*Bc*-CspB), and hyperthermophile *Thermotoga maritima* (*Tm*-CspB)

| Mutant | pH | T | $\Delta G$ | $c_{1/2}$ | m | Appr./Rem. | Ref |
|---|---|---|---|---|---|---|---|
| *Bs*-CspB | 7.0 | 25 | 11.3±0.6 | 1.52±0.1 | 7.4±0.3 | GuHCl (1–3) | 98P7 |
| *Bs*-CspB | 7.0 | 25 | 10.5±0.2 | 1.37±0.04 | 7.67±0.20 | GuHCl (1,2,4) | 98P7 |
| *Bs*-CspB | 7.0 | 25 | 11.4±0.6 | 1.37±0.10 | 8.3±0.4 | GuHCl (1,2,5) | 98P7 |
| *Bc*-CspB | 7.0 | 25 | 20.1±1.1 | 2.68±0.2 | 7.5±0.3 | GuHCl (1–3) | 98P7 |
| *Bc*-CspB | 7.0 | 25 | 19.0±0.4 | 2.57±0.09 | 7.39±0.20 | GuHCl (1,2,4) | 98P7 |
| *Bc*-CspB | 7.0 | 25 | 22.2±1.3 | 2.74±0.20 | 8.1±0.5 | GuHCl (1,2,5) | 98P7 |
| *Tm*-CspB | 7.0 | 25 | 26.2±1.2 | 3.32±0.2 | 7.9±0.3 | GuHCl (1–3) | 98P7 |
| *Tm*-CspB | 7.0 | 25 | 25.6±0.5 | 3.36±0.08 | 7.63±0.10 | GuHCl (1,2,4) | 98P7 |
| *Tm*-CspB | 7.0 | 25 | 29.0±1.9 | 3.29±0.31 | 8.8±0.6 | GuHCl (1,2,5) | 98P7 |

Remarks:
(1) linear extrapolation, LEM-SB
(2) buffer: 0.1 M cacodylate/HCl
(3) transition monitored by fluorescence at 343 nm (*Bs*-CspB and *Bc*-CspB) or at 337 nm (*Tm*-CspB)
(4) from kinetics, analysis of λ values
(5) from kinetics, analysis of the endpoints

Cold shock protein *Tm*Csp from *Thermotoga maritima*

| Protein | pH | T | $\Delta G$ | $c_{1/2}$ | m | Appr./Rem. | Ref |
|---|---|---|---|---|---|---|---|
| *Tm*Csp | 7.0 | 12 | 25.2 | 3.52 | 7.16 | GuHCl (1–3) | 99W8 |
|  | 7.0 | 26 | 24.3 | 3.31 | 7.35 | GuHCl (1–3) | 99W8 |
|  | 7.0 | 35 | 21.4 | 2.82 | 7.53 | GuHCl (1–3) | 99W8 |
|  | 7.0 | 45 | 16.8 | 2.29 | 7.35 | GuHCl (1–3) | 99W8 |
|  | 7.0 | 30 | 20.1±2.1 |  |  | DSC (4,5) | 99W6 |
|  | 5.7 | 30 | 24.7±2.5 |  |  | DSC (5) | 99W6 |

Remarks:
(1) linear extrapolation, LEM-SB
(2) transition monitored by fluorescence
(3) buffer: 0.1 M sodium cacodylate, pH 7.0
(4) maximal stability at pH 7 is achieved at 30°C
(5) for buffer conditions and further details, see Table 2

Cold shock protein *Tm*Csp from *Thermotoga maritima*, and mutant proteins

| Mutant | pH | T | ΔG | $c_{1/2}$ | m | Appr./Rem. | Ref |
|---|---|---|---|---|---|---|---|
| *Tm*Csp | 7.0 | 25 | 21.7±1.1 | | 6.7±0.2 | GuHCl (1–3) | 99F8 |
| | 7.0 | 25 | 23.5 | | 7.1 | kin. (2–4) | 99F8 |
| Asp9→Asn | 7.0 | 25 | 21.2±1.8 | | 6.8±0.6 | GuHCl (1–3) | 99F8 |
| | 7.0 | 25 | 23.4 | | 7.2 | kin. (2–4) | 99F8 |
| | 7.0 | 25 | 25.0 | | 7.2±0.9 | kin. (2–5) | 99F8 |

Remarks:
(1) equilibrium unfolding, linear extrapolation, LEM-SB
(2) buffer: 100 mM Na-cacodylate, pH 7.0
(3) transition monitored by fluorescence at 337 nm
(4) data from kinetics of unfolding and folding
(5) from initial and final values of folding and unfolding kinetics varying GuHCl conc.

## Colicin

Soluble colicin E1 channel peptide, wild type and single tryptophan mutants

| Mutant | pH | T | ΔG | $c_{1/2}$ | Approach/Remarks | | Ref |
|---|---|---|---|---|---|---|---|
| wild type | 6.0 | | | 6.7±0.1 | urea | (1,4) | 99S16 |
| wild type | 6.0 | | | 6.4±0.1 | urea | (2,4) | 99S16 |
| wild type | 6.0 | | | 6.2±0.1 | urea | (3,4) | 99S16 |
| Trp355 | 6.0 | | | 4.9±0.1 | urea | (1,4) | 99S16 |
| Trp367 | 6.0 | | | 4.9±0.1 | urea | (1,4) | 99S16 |
| Trp404 | 6.0 | | | 4.8±0.1 | urea | (1,4) | 99S16 |
| Trp413 | 6.0 | | | 5.7±0.1 | urea | (1,4) | 99S16 |
| Trp424 | 6.0 | | | 6.1±0.1 | urea | (1,4) | 99S16 |
| Trp431 | 6.0 | | | 5.8±0.1 | urea | (1,4) | 99S16 |
| Trp443 | 6.0 | | | 5.5±0.2 | urea | (1,4) | 99S16 |
| Trp460 | 6.0 | | | 6.1±0.1 | urea | (1,4) | 99S16 |
| Trp484 | 6.0 | | | 4.6±0.1 | urea | (1,4) | 99S16 |
| Trp495 | 6.0 | | | 5.3±0.1 | urea | (1,4) | 99S16 |
| Trp507 | 6.0 | | | 5.7±0.1 | urea | (1,4) | 99S16 |

Remarks:
(1) transition monitored by fluorescence
(2) transition monitored by far-UV CD at 222 nm
(3) transition monitored by binding of TNS, 2-(p-toluidinyl)naphthalene 6-sulphonic acid
(4) buffer: 100 mM NaCl, 10 mM dimethylglutaric acid, pH 6.0

**Colicin Binding Immunity Protein**, see Immunity Protein

## Complement Receptor Type 1

Complement receptor type 1 (CR1), segment corresponding to the modules 15–17, and fragments

| Transition | pH | T | $\Delta G$ | $c_{1/2}$ | m | Appr./Rem. | Ref |
|---|---|---|---|---|---|---|---|
| segment CR1~15–17 | | | | | | | |
| trans. (1) | 6.4 | 20 | 17.9±1.8 | 1.5 | 11.7±1.1 | GuHCl (1–4) | 99K6 |
| trans. (2) | 6.4 | 20 | 113.0±10 | 3.3 | 34.1±2.9 | GuHCl (1–4) | 99K6 |
| trans. (3) | 6.4 | 20 | 64.4±3.0 | 5.4 | 11.9±0.6 | GuHCl (1–4) | 99K6 |
| fragment CR1~15–16 | | | | | | | |
| trans. (2) | 6.4 | 20 | 45.5±3.3 | 3.8 | 12.7±0.9 | GuHCl (1–4) | 99K6 |
| trans. (3) | 6.4 | 20 | 70.4±6.0 | 5.5 | 12.8±1.1 | GuHCl (1–4) | 99K6 |
| fragment CR1~16 | | | | | | | |
| trans. (2) | 6.4 | 20 | 14.6±1.0 | 3.6 | 3.8±0.4 | GuHCl (1–4) | 99K6 |
| trans. (3) | 6.4 | 20 | 51.2±5.5 | 5.3 | 9.7±1.0 | GuHCl (1–4) | 99K6 |

Remarks:

(1) linear extrapolation assuming independent two-state transitions

(2) transition monitored by fluorescence intensity

(3) measured in 25 mM phosphate buffer, pH 6.4

(4) see also Table 2 for DSC data

## Creatine Kinase

Cytoplasmic creatine kinase

| Transition | pH | T | $\Delta G$ | m | Approach/Remarks | Ref |
|---|---|---|---|---|---|---|
| $N \rightarrow I$ | 8.3 | 25 | 16.4 | 41.5 | GuHCl, FL (1,2) | 98F1 |
| $N \rightarrow I$ | 8.3 | 25 | 16.2 | 41.4 | GuHCl, CD (1,3) | 98F1 |
| $I \rightarrow U$ | 8.3 | 25 | 10.9 | 6.8 | GuHCl, FL (1,2) | 98F1 |
| $I \rightarrow U$ | 8.3 | 25 | 8.7 | 5.7 | GuHCl, CD (1,3) | 98F1 |

Remarks:

(1) data treatment by a three-state model assuming a linear dependence of $\Delta G$ on the denaturant concentration

(2) transition monitored by fluorescence intensity at 335 nm

(3) transition monitored by CD at 222 nm

## CRO Protein

Phage 434 CRO protein, equilibrium unfolding

| | pH | T | $\Delta G$ | $c_{1/2}$ | m | Appr./Rem. | Ref |
|---|---|---|---|---|---|---|---|
| | 6.0 | 20 | 15.1 | 3.50 | 4.27 | urea (1–3) | 97P3 |
| | 6.0 | 20 | 16.3 | 3.48 | 4.77 | urea (1,2,4) | 97P3 |
| | 6.0 | 20 | 14.6 | 3.35 | 4.39 | urea (1,2,5) | 97P3 |
| | 6.0 | 20 | 15.4 | 3.44 | 4.48 | urea (1,2,6) | 97P3 |

Remarks:

(1) linear extrapolation

(2) buffer: 25 mM potassium phosphate, 100 mM KCl, 1 mM DTT, 0.01% sodium azide, pH 6.0

(3) transition monitored by far-UV CD

(4) transition monitored by near-UV CD

(5) transition monitored by fluorescence

(6) mean value

Phage 434 CRO protein, slowest exchanging amide protons

| Residue | pH | T | $\Delta G_{op}$ | Approach | Remarks | Ref |
|---------|-----|-----|------|----------|---------|------|
| Ile13 | 6.0 | 20 | 15.9 | HX | (1) | 97P3 |
| Ile36 | 6.0 | 20 | 13.8 | HX | (1) | 97P3 |
| Ile50 | 6.0 | 20 | 13.8 | HX | (1) | 97P3 |
| Ala53 | 6.0 | 20 | 18.8 | HX | (1) | 97P3 |
| Leu54 | 6.0 | 20 | 15.9 | HX | (1) | 97P3 |
| mean | 6.0 | 20 | 15.5 | HX | (1) | 97P3 |

Remark:

(1) buffer: 25 mM potassium phosphate, 100 mM KCl, 1 mM DTT, 0.01% sodium azide, pH 6.0

Phage 434 Cro protein, comparison of linear extrapolation and denaturant binding model using the same data set
Results of the linear extrapolation method

| Protein | pH | T | $\Delta G$ | $c_{1/2}$ | m | Appr./Rem. | Ref |
|---------|-----|-----|------|------|------|------------|------|
| 434 Cro | 6.0 | 5 | 13.7±1.6 | 3.32±0.08 | 4.14±0.42 | urea (1–3) | 99P2 |
|  | 6.0 | 10 | 14.2±1.0 | 3.31±0.06 | 4.31±0.25 | urea (1–3) | 99P2 |
|  | 6.0 | 15 | 13.5±1.3 | 3.56±0.05 | 3.81±0.21 | urea (1–3) | 99P2 |
|  | 6.0 | 20 | 14.4±2.0 | 3.42±0.09 | 4.23±0.50 | urea (1–3) | 99P2 |
|  | 6.0 | 25 | 12.8±2.1 | 3.07±0.15 | 4.18±0.59 | urea (1–3) | 99P2 |

Results of the denaturant binding model

| Protein | pH | T | $\Delta G$ | k | $\Delta n$ | Appr./Rem. | Ref |
|---------|-----|-----|------|------|------|------------|------|
| 434 Cro | 6.0 | 5 | 17.2±1.8 | 0.086 | 33±2 | urea (1–3) | 99P2 |
|  | 6.0 | 10 | 17.3±1.2 | 0.078 | 36±2 | urea (1–3) | 99P2 |
|  | 6.0 | 15 | 16.2±1.0 | 0.071 | 33±2 | urea (1–3) | 99P2 |
|  | 6.0 | 20 | 16.6±2.1 | 0.065 | 37±4 | urea (1–3) | 99P2 |
|  | 6.0 | 25 | 14.7±2.4 | 0.061 | 37±5 | urea (1–3) | 99P2 |

Remarks:

(1) two state model, for details of the data treatment see Ref. 99P2
(2) transition monitored by far-UV-CD at 222 nm
(3) buffer: 25 mM potassium phosphate, 100 mM KCl, 1 mM DTT

Phage 434 Cro protein, DSC data at various pH values

| Protein | pH | T | $\Delta G$ | Approach/Remarks | | Ref |
|---------|-----|-----|------|----------|---------|------|
| 434 Cro | 2.65 | 25 | −5.69±0.84 | DSC | (1,2) | 99P2 |
|  | 3.00 | 25 | 2.38±0.84 | DSC | (1,2) | 99P2 |
|  | 3.50 | 25 | 2.30±0.42 | DSC | (1,2) | 99P2 |
|  | 3.75 | 25 | 5.40±0.42 | DSC | (1,2) | 99P2 |
|  | 4.00 | 25 | 6.49±0.42 | DSC | (1,2) | 99P2 |
|  | 4.50 | 25 | 8.49±0.42 | DSC | (1,2) | 99P2 |
|  | 5.00 | 25 | 9.20±0.42 | DSC | (1,2) | 99P2 |
|  | 6.00 | 25 | 13.51±0.84 | DSC | (1,2) | 99P2 |

Remarks:

(1) data from individual curve fittings based on linear $\Delta H^{cal}/T_{trs}$ equations except for pH < 3.0 (further expressions are considered in Ref. 99P2)
(2) buffers: 20 or 25 mM sodium glycine, acetate or phosphate, 100 mM KCl

**λ Cro Repressor,** see Repressor Proteins

## CRP, cAMP Receptor Protein

CRP, cAMP receptor protein from *E. coli*

| Process | pH | T | $\Delta G$ | Approach | Remarks | Ref |
|---|---|---|---|---|---|---|
| dissociation | 7.9 | 20 | 46.9±2.5 | kinetics | (1,2) | 97M2 |
| unfolding | 7.9 | 20 | 30.9±1.3 | kinetics | (1,2) | 97M2 |
| unfolding | 7.9 | 20 | 28.0±0.7 | GuHCl | (1,3) | 97M2 |
| unfolding | 7.9 | 20 | 30.5±0.4 | heat/GuHCl | (1,4) | 97M2 |

Remarks:

(1) CRP is a homodimer and each monomer is folded into two structural domains

(2) from folding and unfolding kinetics in the presence of GuHCl, monitored by stopped-flow fluorescence

(3) from equilibrium unfolding, linear extrapolation

(4) from temperature stability curve in the absence of denaturant

## Crystallin

Recombinant αA- and αB-crystallins

| Protein/Transition | pH | T | $\Delta G$ | m | Appr./Rem. | Ref |
|---|---|---|---|---|---|---|
| αA-crystallin | | | | | | |
| $N \rightarrow I$ | | | 12.13±2.09 | 12.09±0.75 | GuHCl (1–3) | 99S19 |
| $I \rightarrow U$ | | | 14.56±3.47 | 16.36±1.88 | GuHCl (1–3) | 99S19 |
| $N \rightarrow U$ | | | 26.69±2.80 | 41.51±1.34 | GuHCl (1–3) | 99S19 |
| αB-crystallin | | | | | | |
| $N \rightarrow I$ | | | 7.07±0.46 | 19.92±1.63 | GuHCl (1–3) | 99S19 |
| $I \rightarrow U$ | | | 14.02±1.38 | 10.59±0.84 | GuHCl (1–3) | 99S19 |
| $N \rightarrow U$ | | | 21.09±0.92 | 30.50±1.26 | GuHCl (1–3) | 99S19 |

Remarks:

(1) three-state fit based on global analysis of data obtained by absorbance at 235 nm, Trp fluorescence at 320 nm, and far-UV CD at 223 nm

(2) the three-state fit is regarded as not perfect in Ref. 99S19

(3) measured at room temperature

γB-Crystallin, stability of the C-terminal domain

| Form | pH | T | $\Delta G$ | $c_{1/2}$ | m | Appr./Rem. | | Ref |
|---|---|---|---|---|---|---|---|---|
| C-domain in γB | 2.0 | 20 | 20.7±3.7 | 2.10 | 9.6±1.8 | urea | (1–3) | 97M5 |
| γBC | 2.0 | 20 | 4.7±0.2 | 0.70 | 7.0±0.3 | urea | (3,4) | 97M5 |
| γBC | 2.5 | 20 | 7.3±0.5 | 0.95 | 8.3±0.6 | urea | (3,4) | 97M5 |
| γBC | 3.0 | 20 | 25.5±1.6 | 1.95 | 13.6±0.8 | urea | (3,4) | 97M5 |
| γBC | 3.5 | 20 | 34.8±2.6 | 2.75 | 13.0±1.0 | urea | (3,4) | 97M5 |
| γBC | 4.0 | 20 | 36.0±1.9 | 4.50 | 8.2±0.4 | urea | (3,4) | 97M5 |

Remarks:

(1) stability of the C-terminal domain as a constituent of the complete γB-crystallin

(2) at neutral pH, the stabilities of the N- and C-terminal domains are closely similar

(3) linear extrapolation, LEM-SB

(4) stability of the isolated (recombinant) C-terminal domain of γB-crystallin ranging from Thr87 to Tyr174

Calf lens γF-crystallin

| Protein | pH | T | ΔG | Approach/Remarks | | Ref |
|---|---|---|---|---|---|---|
| γF-cryst. | 7.4 | 25 | 35.1±0.8 | urea | (1–3) | 98D5 |
| | 7.4 | 25 | 37.2±1.3 | urea | (1,2,4) | 98D5 |

Remarks:
(1) linear extrapolation
(2) measured in 0.1 M sodium phosphate in the presence of 1.5 M GuHCl
(3) transition monitored by CD at 218 nm
(4) transition monitored by Trp fluorescence intensity

## Cytochrome $b_s$

The data entries are arranged as follows:

a) cytochrome $b_s$ from bovine liver
b) cytochrome $b_s$ from rat

*a) cytochrome $b_s$ from bovine liver*

Bovine cytochrome $b_s$, tryptic fragment (Ala3-Lys86), wild type and mutants

| Mutant | pH | T | Δ(ΔG) | $c_{1/2}$ | m | Approach/Remarks | | Ref |
|---|---|---|---|---|---|---|---|---|
| wild type | 7.0 | 30 | 0.0 | 7.7 | 2.3 | urea | (1–3) | 99X1 |
| Val61→Glu | 7.0 | 30 | −4.4 | 6.1 | 3.2 | urea | (1–3) | 99X1 |
| Val61→His | 7.0 | 30 | −6.7 | 5.3 | 3.3 | urea | (1–3) | 99X1 |
| Val61→Lys | 7.0 | 30 | −10.0 | 5.2 | 4.0 | urea | (1–3) | 99X1 |
| Val61→Tyr | 7.0 | 30 | −5.6 | 6.0 | 3.1 | urea | (1–3) | 99X1 |

Remarks:
(1) linear extrapolation
(2) Δ(ΔG) was determined at the transition midpoint
(3) the transition was monitored by absorption at 412 nm
(4) buffer: 100 mM phosphate buffer, pH 7.0

Cytochrome $b_s$, recombinant bovine liver trypsin-solubilized (82 residues in length), wild type and mutants (oxidized)

| Mutant | pH | T | Δ(ΔG) | $c_{1/2}$ | m | Appr./Rem. | Ref |
|---|---|---|---|---|---|---|---|
| wild type | 7.0 | 30 | 0.0 | 7.1 | 3.7 | urea (1–3) | 97Y2 |
| Phe35→Leu | 7.0 | 30 | 4.3 | 8.3 | 3.5 | urea (1–3) | 97Y2 |
| Phe35→Tyr | 7.0 | 30 | −7.8 | 4.9 | 3.4 | urea (1–3) | 97Y2 |

Remarks:
(1) recombinant trypsin-solubilized bovine microsomal cytochrome $b_s$, 82 residues in length
(2) buffer: 0.1 M sodium phosphate, pH 7.0
(3) linear extrapolation

Cytochrome $b_5$, recombinant bovine liver trypsin-solubilized (82 residues in length), wild type and mutants, reduced and oxidized forms

| Mutant | Form | pH | $T_{trs}$ | $\Delta T$ | $\Delta(\Delta G)$ | Approach/Remarks | Ref |
|--------|------|-----|------|------|------|------------------|-----|
| wild type | ox. | 7.0 | 66.2 | 0.0 | 0.0 | heat, v.H. (1–4) | 97Y2 |
| Phe35→His | ox. | 7.0 | 54.9 | −11.3 | −11.8 | heat, v.H. (1–4) | 97Y2 |
| Phe35→Leu | ox. | 7.0 | 59.1 | −7.1 | −7.8 | heat, v.H. (1–4) | 97Y2 |
| Phe35→Tyr | ox. | 7.0 | 69.6 | 3.4 | 3.3 | heat, v.H. (1–4) | 97Y2 |
| wild type | red. | 7.0 | 76.7 | 0.0 | 0.0 | heat, v.H. (1–3,5) | 97Y2 |
| Phe35→His | red. | 7.0 | 63.1 | −13.6 | −16.7 | heat, v.H. (1–3,5) | 97Y2 |
| Phe35→Leu | red. | 7.0 | 68.8 | −7.9 | −8.3 | heat, v.H. (1–3,5) | 97Y2 |
| Phe35→Tyr | red. | 7.0 | 74.4 | −2.3 | −2.3 | heat, v.H. (1–3,5) | 97Y2 |

Remarks:

(1) recombinant trypsin-solubilized bovine microsomal cytochrome $b_5$, 82 residues in length

(2) buffer: 0.1 M sodium phosphate, pH 7.0

(3) $\Delta(\Delta G)$ was calculated from $\Delta(\Delta G) = (\Delta H/T_{trs})*[T_{trs(mutant)}-T_{trs(wild\ type)}] = \Delta(\Delta G) = (\Delta S)*[T_{trs(mutant)}-T_{trs(wild\ type)}]$, for $\Delta H$ see Table 2

(4) $\Delta(\Delta G)$ refers to the oxidized form of the wild type

(5) $\Delta(\Delta G)$ refers to the reduced form of the wild type

Recombinant trypsin-solubilized bovine liver microsomal cytochrome $b_5$ and mutants

| Protein | pH | T | $\Delta(\Delta G)$ | m | Appr./Rem. | Ref |
|---------|-----|-----|------|------|------------|-----|
| wild type | 7.0 | 26 | 0.0 | 7.77±0.01 | urea (1–3) | 98Q1 |
| Glu44→Ala | 7.0 | 26 | 0.21 | 7.83±0.02 | urea (1–3) | 98Q1 |
| Glu56→Ala | 7.0 | 26 | 0.75 | 8.00±0.02 | urea (1–3) | 98Q1 |
| double mutant (Glu44→Ala and Glu56→Ala) | | | | | | |
| | 7.0 | 26 | 0.96 | 8.06±0.03 | urea (1–3) | 98Q1 |

Remarks:

(1) linear extrapolation

(2) transition monitored by optical absorption at 412 nm

(3) buffer: 100 mM phosphate

Recombinant bovine cytochrome $b_5$, Ala1-Ser104 variant

| pH | T | $\Delta G$ | $c_{1/2}$ | m | Approach/Remarks | Ref |
|-----|-----|------|------|------|------------------|-----|
| Results of equilibrium unfolding: | | | | | | |
| 7.0 | 25 | 27.5±3.3 | 3.05±0.1 | 8.95±1.10 | GuHCl, OA (1–3) | 99M6 |
| 7.0 | 25 | 25.5±3.5 | 2.99±0.1 | 8.56±1.20 | GuHCl, FL (1,2,4) | 99M6 |
| Results of stopped flow experiments: | | | | | | |
| 7.0 | 25 | 25.9±2.1 | | 5.38±0.57 | GuHCl, OA fast (1–3) | 99M6 |
| 7.0 | 25 | 31.5±1.0 | | 5.40±0.57 | GuHCl, OA slow (1–3) | 99M6 |
| 7.0 | 25 | 24.0±0.8 | 2.90±0.1 | 5.16±0.12 | GuHCl, FL fast (1,2,4) | 99M6 |
| 7.0 | 25 | 29.3±1.4 | 3.20-3.5 | 5.40±0.22 | GuHCl, FL slow (1,2,4) | 99M6 |

Remarks:

(1) linear extrapolation to zero denaturant concentration by a method that includes the pre- and postdenaturational baselines for a nonlinear regression of the data

(2) measured in 30 mM MOPS, pH 7.0

(3) transition monitored by optical absorbance (OA) at 412 nm

(4) transition monitored by fluorescence emission (FL) at 350 nm

Recombinant bovine cytochrome $b_s$, Ala1-Ser104 variant, kinetic data

| Transition | pH | T | $\Delta G$ | m | Appr./Rem. | Ref |
|---|---|---|---|---|---|---|
| unfolding fast | 7.0 | 25 | 24.0±0.8 | 5.156±0.124 | GuHCl, FL (1,2) | 99M6 |
| unfolding slow | 7.0 | 25 | 29.3±1.4 | 5.404±0.224 | GuHCl, FL (1,2) | 99M6 |
| refolding 1 | 7.0 | 25 | −17.6±0.6 | 8.517±0.518 | GuHCl, FL (1,2) | 99M6 |
| refolding 2 | 7.0 | 25 | −10.1±0.2 | 3.978±0.329 | GuHCl, FL (1,2) | 99M6 |
| equilibrium unf. | 7.0 | 25 | 25.5±3.5 | 8.560±1.200 | GuHCl, FL (2,3) | 99M6 |

Remarks:
(1) data from kinetic rate constants versus GuHCl conc.
(2) transition monitored by fluorescence
(3) reference value

Bovine apocytochrome $b_s$, Ala1-Ser104 variant, data from equilibrium unfolding and stopped flow kinetics

| Method | pH | T | $\Delta G$ | $c_{1/2}$ | m | Appr./Rem. | Ref |
|---|---|---|---|---|---|---|---|
| equilibrium | 7.0 | 10 | 11.6±1.5 | 1.6 | 7.874±0.181 | GuHCl (1–3,5) | 99M7 |
| stopped flow | | | | | | | |
| folding | 7.0 | 10 | −7.0±0.1 | | 2.420±0.420 | GuHCl (2–4) | 99M7 |
| unfolding | 7.0 | 10 | 3.4±0.2 | | 5.589±0.552 | GuHCl (2–4) | 99M7 |
| summation | 7.0 | 10 | 10.4±0.2 | | 8.009±0.486 | GuHCl (2–4) | 99M7 |

Remarks:
(1) linear extrapolation
(2) transition monitored by fluorescence at 349 nm
(3) buffer: 30 mM MOPS, pH 7.0
(4) data from GuHCl dependence of kinetics of folding and unfolding
(5) for the holoprotein $\Delta G$ amounts to 23.5 kJ/mol and $c_{1/2}$ to ~3.0 according to unpublished data of the authors mentioned in Ref. 99M7

*b) cytochrome $b_s$ from rat*

Recombinant rat microsomal cytochrome $b_s$

| Protein | pD | T | $\Delta G$ | $c_{1/2}$ | m | Appr./Rem. | Ref |
|---|---|---|---|---|---|---|---|
| cyt. $b_s$ | 7.0 | 25 | 29±3 | 2.6±0.3 | 11±1 | GuHCl (1–3) | 98A7 |

Remarks:
(1) linear extrapolation, LEM-SB
(2) transition monitored by NMR
(3) measured in 100 mM phosphate buffer in $D_2O$, pD 7.0

Recombinant rat cytochrome $b_5$, wild type and mutants

| Mutant | pH | T | ΔG | $c_{1/2}$ | m | Appr./Rem. | Ref |
|---|---|---|---|---|---|---|---|
| Results of absorption spectroscopy: | | | | | | | |
| wild type | 7 | 25 | 36.4 | 6.9 | 5.27 | urea (1–3) | 99S17 |
| Ser18→Asp | 7 | 25 | 28.9 | 6.5 | 4.06 | urea (1–3) | 99S17 |
| double mutant (Ser18→Cys and Arg47→Cys), oxidized | | | | | | | |
| | 7 | 25 | 21.8 | 6.4 | 3.39 | urea (1–3) | 99S17 |
| Results of fluorescence spectroscopy: | | | | | | | |
| wild type | 7 | 25 | 7.1 | 2.3 | 3.05 | urea (1,4,5) | 99S17 |
| Ser18→Asp | 7 | 25 | 5.9 | 2.5 | 2.34 | urea (1,4,5) | 99S17 |
| double mutant (Ser18→Cys and Arg47→Cys) oxidized | | | | | | | |
| | 7 | 25 | 14.2 | 3.7 | 3.89 | urea (1,4,5) | 99S17 |

Remarks:

(1) linear extrapolation

(2) transition monitored by optical absorption spectroscopy at 412 nm

(3) the thermal transition occurs at 75.2°C, 75.5°C, and 76.0°C for wild type, Ser18→Asp, and double mutant (ox.), respectively

(4) transition monitored by fluorescence spectroscopy

(5) the thermal transition occurs at 50.1°C, 52.7°C, 44.8°C, and 60.8°C for wild type, Ser18→Asp, reduced double mutant, and oxidized double mutant, respectively

Recombinant rat cytochrome $b_5$, wild type and Tyr74→Lys variant, data from equilibrium dissociation of heme

| Mutant | pH | T | ΔG | $c_{1/2}$ | Approach/Remarks | | Ref |
|---|---|---|---|---|---|---|---|
| wild type | 7.25 | 25 | 14.2 | 5.1 | urea | (1–4) | 93V4 |
| Tyr74→Lys | 7.25 | 25 | 8.4 | 3.0 | urea | (1–3,5) | 93V4 |

Remarks:

(1) linear extrapolation

(2) transition monitored by absorbance at 413 nm

(3) buffer: 100 mM sodium phosphate, pH 7.25

(4) thermal transition at $T_{trs}$ = 58.1°C

(5) thermal transition at $T_{trs}$ = 47.1°C

Apocytochrome $b_5$ from rat liver (cyt. $b_5$)

| Protein | pH | T | ΔG | Approach/Remarks | | Ref |
|---|---|---|---|---|---|---|
| cyt. $b_5$ | 6.2 | 25 | 7±1 | heat | (1–3) | 99B9 |
| | 8.0 | 25 | 5±1 | heat | (1–3) | 99B9 |

Remarks:

(1) transition monitored by optical absorption at 285 nm

(2) measured in 20 mM phosphate buffer

(3) see also the data in Ref. 98C10 (Table 2)

## Cytochrome $b_{562}$

Cytochrome $b_{562}$, reduced and oxidized

| Protein | pH | T | ΔG | $c_{1/2}$ | | Approach/Remarks | | Ref |
|---|---|---|---|---|---|---|---|---|
| reduced | 7 | 20 | 43±10 | 6.0 | | GuHCl | (1–3) | 97W6 |
| reduced | 7 | | | 5.7 | | GuHCl | (1) | 99W15 |
| oxidized | 7 | 20 | 18±2 | 1.5 | ˋ | GuHCl | (1–3) | 97W6 |
| oxidized | 7 | | | 1.8 | | GuHCl | (1) | 99W15 |
| oxidized | | 20 | 16±7 | | | GuHCl/heat | (4,5) | 99W13 |

Remarks:

(1) linear extrapolation

(2) transition monitored by far-UV CD and optical absorption in the Soret band

(3) $c_{1/2}$ was taken from the Fig. 1 in Ref. 97W6

(4) transition monitored by far-UV CD at 220 nm

(5) from heat denaturation in the presence of varying GuHCl conc. with extrapolation to zero denaturant conc.

Cytochrome $b_{562}$, stability change upon reduction

| Protein | pH | $T_{trs}$ | ΔT | Δ(ΔG) | Appr./Rem. | | Ref |
|---|---|---|---|---|---|---|---|
| oxidized | 7.0 | 66.7±0.5 | 0.0 | 0.0 | heat | (1) | 91F3 |
| reduced | 7.0 | 81.0±0.5 | 14.3 | 19.2 | heat | (2,3) | 91F3 |

Remarks:

(1) thermal transition monitored by optical absorption at 418 nm

(2) thermal transition monitored by optical absorption at 426 nm

(3) Δ(ΔG) was obtained using Δ(ΔG) = ΔT × ΔS

Cytochrome $b_{562}$ from the periplasm of E. coli, holo- and apoprotein

| Protein | pH | T | ΔG | Approach/Remarks | | Ref |
|---|---|---|---|---|---|---|
| holo | 7.4 | 25 | 43.4* | DSC | (1) | 98R8 |
| apo | 7.4 | 25 | 14.5* | DSC | (1) | 98R8 |

Remark:

(1) buffer: 50 mM sodium phosphate, 100 mM sodium chloride, pH 7.4

Apocytochrome $b_{562}$

| | pH | T | ΔG | $c_{1/2}$ | m | Approach/Remarks | | Ref |
|---|---|---|---|---|---|---|---|---|
| | 4.5 | 25 | 13.8 | 1.1 | 24.3 | GuHCl | (1,2) | 98F6 |
| | 4.5 | 25 | 21 | | | GuHCl | (2,3) | 98F6 |
| | 4.5 | 25 | 24.6±0.5 | | 22.6±1.1 | HX | (2,4) | 98F6 |
| | 4.5 | 25 | 21 | | | DSC | (2,5) | 98F6 |

Remarks:

(1) linear extrapolation

(2) measured in 10 mM acetate buffer, 90% $D_2O$, 0.65 M NaCl, pH* 4.5

(3) denaturant binding model with $K_b$ = 0.69 and Δn = 15.8

(4) hydrogen exchange monitored by NMR, ΔG amounts to 22.2 kJ/mol with correction for proline isomerization

(5) using $\Delta Cp$ = 2.1 kJ/mol

Cytochrome $b_{562}$, heme binding and denaturation of the holo- and apo-protein

| Transition | pH | T | ΔG | Approach | Remarks | Ref |
|---|---|---|---|---|---|---|
| DA → DH | 8.5 | 25 | 31.8±2.5 | calculated | (1) | 97R4 |
| DA → NA | 8.5 | 25 | 13.4±0.8 | experimental | (2) | 97R4 |
| NA → NH | 8.5 | 25 | 46.0±0.8 | ITC | (3) | 97R4 |
| DH → NH | 8.5 | 25 | 27.6±0.8 | experimental | (2) | 97R4 |

Explanations:

DA – denatured apoprotein
NA – native apoprotein
DH – denatured holoprotein
NH – native holoprotein

Remarks:

(1) calculated from a thermodynamic cycle, see Ref. 97R4
(2) from denaturation of native holo- and apoprotein, see Refs. 91F1, 91F3, 93R1
(3) from heme binding to native apoprotein measured by ITC

Apocytochrome $b_{562}$, subglobal and local hydrogen exchange parameters

| Residue | subglobal parameters $(\Delta G)_{sg}$ | $(m)_{sg}$ | local parameters $(\Delta G)_{loc}$ | $(m)_{loc}$ | Appr./Rem. | | Ref |
|---|---|---|---|---|---|---|---|
| Glu8 | | | 11.9±0.8 | 0.0±0.0 | HX | (1,2) | 98F6 |
| Leu10 | | | 9.4±0.7 | 4.6±2.0 | HX | (1,2) | 98F6 |
| Asn11 | | | 14.3±0.6 | 0.0±0.0 | HX | (1,2) | 98F6 |
| Leu14 | 15.1±1.7 | 11.3±3.5 | 14.9±0.6 | 0.0±0.1 | HX | (1,2) | 98F6 |
| Lys15 | 15.1±1.7 | 11.3±3.5 | 12.1±0.5 | 0.0±0.0 | HX | (1,2) | 98F6 |
| Val16 | 15.1±1.7 | 11.3±3.5 | 14.6±0.7 | 0.2±0.7 | HX | (1,2) | 98F6 |
| Ile17 | 15.1±1.7 | 11.3±3.5 | 12.9±0.5 | 2.5±2.2 | HX | (1,2) | 98F6 |
| Glu18 | | | 11.7±0.5 | 1.0±1.9 | HX | (1,2) | 98F6 |
| Lys19 | | | 14.2±0.5 | 0.0±0.1 | HX | (1,2) | 98F6 |
| Ala20 | | | 15.0±0.5 | 0.0±0.0 | HX | (1,2) | 98F6 |
| Gln25 | | | 11.7±0.3 | 1.7±1.2 | HX | (1,2) | 98F6 |
| Val26 | | | 16.7±0.5 | 10.3±1.9 | HX | (1,2) | 98F6 |
| Lys27 | | | 21.8±0.7 | 17.2±3.0 | HX | (1,2) | 98F6 |
| Asp28 | | | 15.5±0.5 | 24.4±2.0 | HX | (1,2) | 98F6 |
| Ala29 | | | 23.8±1.1 | 13.5±6.7 | HX | (1,2) | 98F6 |
| Leu30 | | | 21.8±1.2 | 20.0±3.1 | HX | (1,2) | 98F6 |
| Thr31 | | | 21.7±0.9 | 19.4±2.8 | HX | (1,2) | 98F6 |
| Lys32 | 24.6±0.5 | 22.6±1.1 | 21.1±0.6 | 8.6±4.4 | HX | (1,2) | 98F6 |
| Met33 | 24.6±0.5 | 22.6±1.1 | 26.9±2.3 | 13.1±11.3 | HX | (1,2) | 98F6 |
| Arg34 | 24.6±0.5 | 22.6±1.1 | 25.2±0.9 | 2.3±6.8 | HX | (1,2) | 98F6 |
| Ala35 | 24.6±0.5 | 22.6±1.1 | 26.6±0.7 | 1.7±6.2 | HX | (1,2) | 98F6 |
| Ala36 | 24.6±0.5 | 22.6±1.1 | 28.2±1.8 | 10.4±11.5 | HX | (1,2) | 98F6 |
| Ala37 | 24.6±0.5 | 22.6±1.1 | 26.4±1.1 | 7.5±10.0 | HX | (1,2) | 98F6 |
| Leu38 | | | 23.0±0.8 | 21.0±2.8 | HX | (1,2) | 98F6 |
| Asp39 | | | 20.6±0.6 | 13.9±2.8 | HX | (1,2) | 98F6 |
| Ala40 | | | 20.7±0.3 | 1.9±3.3 | HX | (1,2) | 98F6 |
| Gln41 | | | 19.0±0.4 | 6.5±3.1 | HX | (1,2) | 98F6 |
| Lys42 | | | 15.6±0.4 | 4.3±2.0 | HX | (1,2) | 98F6 |
| Ala43 | | | 12.1±0.3 | 2.8±1.0 | HX | (1,2) | 98F6 |
| Asp66 | | | 11.8±0.5 | 0.0±0.3 | HX | (1,2) | 98F6 |
| Ile67 | | | 7.7±0.4 | 3.4±1.0 | HX | (1,2) | 98F6 |
| Leu68 | | | 17.2±0.5 | 11.5±1.9 | HX | (1,2) | 98F6 |
| Val69 | | | 19.3±1.6 | 14.4±3.6 | HX | (1,2) | 98F6 |
| Gly70 | 24.6±0.5 | 22.6±1.1 | 19.7±0.6 | 4.6±3.4 | HX | (1,2) | 98F6 |

Apocytochrome $b_{562}$, subglobal and local hydrogen exchange parameters (continued)

| Residue | subglobal parameters | | local parameters | | Appr./Rem. | | Ref |
|---|---|---|---|---|---|---|---|
| | $(\Delta G)_{sg}$ | $(m)_{sg}$ | $(\Delta G)_{loc}$ | $(m)_{loc}$ | | | |
| Gln71 | 24.6±0.5 | 22.6±1.1 | 21.8±1.4 | 1.8±3.7 | HX | (1,2) | 98F6 |
| Ile72 | | | 21.5±1.6 | 18.3±3.7 | HX | (1,2) | 98F6 |
| Asp73 | | | 19.5±0.9 | 14.5±2.7 | HX | (1,2) | 98F6 |
| Ala75 | 24.6±0.5 | 22.6±1.1 | 22.3±1.1 | 0.0±0.0 | HX | (1,2) | 98F6 |
| Leu76 | 24.6±0.5 | 22.6±1.1 | 22.3±2.1 | 8.7±8.8 | HX | (1,2) | 98F6 |
| Lys77 | 24.6±0.5 | 22.6±1.1 | 19.0±0.5 | 0.6±1.6 | HX | (1,2) | 98F6 |
| Leu78 | | | 20.4±1.3 | 12.0±4.1 | HX | (1,2) | 98F6 |
| Ala79 | | | 19.7±0.8 | 12.4±2.9 | HX | (1,2) | 98F6 |
| Asn80 | | | 19.9±0.8 | 3.4±3.0 | HX | (1,2) | 98F6 |
| Ala87 | 19.9±1.2 | 11.0±3.4 | 18.7±1.7 | 5.2±8.8 | HX | (1,2) | 98F6 |
| Gln88 | 19.9±1.2 | 11.0±3.4 | 18.1±1.8 | 7.1±6.5 | HX | (1,2) | 98F6 |
| Ala89 | 19.9±1.2 | 11.0±3.4 | 24.2±4.3 | 6.6±14.6 | HX | (1,2) | 98F6 |
| Ala90 | 19.9±1.2 | 11.0±3.4 | 19.0±2.1 | 5.7±9.4 | HX | (1,2) | 98F6 |
| Ala91 | 19.9±1.2 | 11.0±3.4 | 20.1±2.4 | 4.2±10.0 | HX | (1,2) | 98F6 |
| Glu92 | | | 16.3±0.6 | 1.9±2.2 | HX | (1,2) | 98F6 |
| Gln93 | | | 12.0±0.3 | 0.1±0.4 | HX | (1,2) | 98F6 |

Remarks:

(1) measured in 10 mM sodium $d3$-acetate and 0.65 M KCl and/or GuHCl

(2) segregation of individual amino acids into cooperative units on the basis of the dependence of $\Delta G_{(HX)}^{eff.}$ on GuHCl concentration

Apocytochrome $b_{562}$, identification of cooperative units by pressure-dependence of hydrogen exchange

| Group | pH | T | $\Delta G$ | Approach/Remarks | | Ref |
|---|---|---|---|---|---|---|
| global | 4.5 | 25 | 21.63±0.21 | HX | (1,2) | 98F7 |
| subglobal | 4.5 | 25 | 19.08±1.09 | HX | (1,3) | 98F7 |
| subglobal | 4.5 | 25 | 14.98±1.00 to 8.41±0.92 | HX | (1,4) | 98F7 |

Remarks:

(1) measured in 10 mM acetate buffer, 90% $D_2O$, 0.65 M NaCl, pH* 4.5

(2) dominated by global unfolding; the group shows the largest negative $\Delta V°$ of –(102±3) ml/mol concerning residues 34–37 and 75–77

(3) subglobal parameters; the group shows $\Delta V°$ of –(65±5) ml/mol concerning residues 87–91

(4) subglobal parameters; the group shows $\Delta V°$ ranging from –(41±14) ml/mol for Asn13 to –(7±11) ml/mol for Lys15

## Cytochrome $c$

The data entries are arranged as follows:

a) data for various species, mutants, and redox states

b) transitions and intermediates

c) data from exchange techniques

d) data obtained in the presence of cosolutes

*a) data for various species and mutants*

Cytochrome *c* from horse heart, wild type and mutants

| Mutant | pH | T | $\Delta G$ | $c_{1/2}$ | m | Appr./Rem. | Ref |
|---|---|---|---|---|---|---|---|
| wild type | 5.0 | 10 | 33.9±1.3 | 2.61±0.01 | 13.0±0.4 | GuHCl (1–3) | 97C10 |
| His33→Asn | 5.0 | 10 | 30.5±0.8 | 2.44±0.01 | 12.6±0.4 | GuHCl (1–3) | 97C10 |
| His26→Gln | 5.0 | 10 | 33.1±1.3 | 2.54±0.01 | 13.0±0.4 | GuHCl (1–3) | 97C10 |

Remarks:
(1) linear extrapolation, LEM-SB
(2) transition monitored by fluorescence at 350 nm
(3) buffer: 100 mM sodium acetate

Cytochrome *c* from *Thermus thermophilus*, reduced and oxidized forms

| Variant/Form | pH | T | $\Delta G$ | $c_{1/2}$ | m | Appr./Rem. | Ref |
|---|---|---|---|---|---|---|---|
| protein extracted from *Thermus thermophilus*: | | | | | | | |
| oxidized | 7.0 | 20 | 75±5 | 5.8±0.2 | 12.5±0.3 | GuHCl (1,2) | 98W10 |
| reduced | 7.0 | 20 | 105±5 | 6.5±0.2 | 16.0±0.3 | GuHCl (1,2) | 98W10 |
| protein expressed in *E. coli*: | | | | | | | |
| oxidized | 7.0 | 20 | 25±3 | 4.2±0.2 | 7.1±0.3 | GuHCl (1–3) | 98W10 |
| reduced | 7.0 | 20 | 35±3 | 5.0±0.2 | 6.2±0.3 | GuHCl (1–3) | 98W10 |

Remarks:
(1) linear extrapolation
(2) transition monitored by CD at 222 nm
(3) expression of the protein in *E. coli* leads to a molten globule

Cytochrome *c*, reduced form

| | pH | T | $\Delta G$ | $c_{1/2}$ | m | Approach/Remarks | | Ref |
|---|---|---|---|---|---|---|---|---|
| | 6 | 24±1 | 75 | 5.15 | 14.7 | GuHCl | (1–3) | 98B7 |

Remarks:
(1) linear extrapolation
(2) measured in 100 mM phosphate buffer, pH 6, ~500 μM dithionite
(3) transition monitored by CD at 222 nm

Cytochrome *c*, reduced and oxidized

| Protein | pH | T | $\Delta G$ | $c_{1/2}$ | Approach/Remarks | | Ref |
|---|---|---|---|---|---|---|---|
| reduced | 7 | 40 | 61±10 | 4.8 | GuHCl | (1,2) | 96P4 |
| oxidized | 7 | 40 | 30±1 | 2.4 | GuHCl | (1,2) | 96P4 |

Remarks:
(1) linear extrapolation
(2) $c_{1/2}$ was taken from the Fig. 2 in Ref. 96P4
(3) see also Ref. 98T5

Horse and yeast cytochrome $c$ (mutant Cys102→Ser), redox states

| Protein | pH | T | $\Delta G$ | $c_{1/2}$ | m | Appr./Rem. | Ref |
|---|---|---|---|---|---|---|---|
| horse, ox. | 7 | 22.5 | 40±1 | 2.8±0.1 | 14.3±0.4 | GuHCl (1–4) | 96M15 |
| horse, red. | 7 | 22.5 | 74±3 | 5.3±0.1 | 13.8±0.4 | GuHCl (1–4) | 96M15 |
| horse, ox. | 7 | 40.0 | 30±1 | 2.4±0.1 | 12.2±0.4 | GuHCl (1–4) | 96M15 |
| horse, red. | 7 | 40.0 | 61±10 | 4.7±0.1 | 13.1±2.0 | GuHCl (1–4) | 96M15 |
| yeast, ox. | 7 | 22.5 | 24±1 | 1.3±0.1 | 18.9±0.5 | GuHCl (1–5) | 96M15 |
| yeast, red. | 7 | 22.5 | 63±4 | 3.8±0.1 | 16.6±1.0 | GuHCl (1–5) | 96M15 |
| yeast, ox. | 7 | 40.0 | 14±1 | 0.8±0.1 | 17.1±0.1 | GuHCl (1–5) | 96M15 |
| yeast, red. | 7 | 40.0 | 45±3 | 3.3±0.1 | 13.7±0.8 | GuHCl (1–5) | 96M15 |

Remarks:

(1) linear extrapolation

(2) transition monitored by fluorescence

(3) buffer: 100 mM sodium phosphate

(4) Ref. 96M15 contains additional data on the folding kinetics

(5) mutant Cys102→Ser

Cytochrome $c$ from horse heart, combined urea and GuHCl denaturation

| Conditions | pH | T | $\Delta G$ | $c_{1/2}$ | m | Appr./Rem. | Ref |
|---|---|---|---|---|---|---|---|
| 0.1–1.0 M KCl | 2.0 | 25 | 34.23±1.76 | 2.47 | 13.85±0.71 | GuHCl (1–4) | 99G21 |
| 0.1–1.0 M KCl | 2.0 | 25 | 33.56±1.80 | 6.26 | 5.36±0.29 | urea (1–4) | 99G21 |
| 0.60 M GuHCl | 2.0 | 25 | 25.48±1.21 | 4.61 | 5.52±0.25 | urea (1–4) | 99G21 |
| 1.21 M GuHCl | 2.0 | 25 | 16.19±0.54 | 3.10 | 5.23±0.17 | urea (1–4) | 99G21 |
| 2.20 M GuHCl | 2.0 | 25 | 3.31±0.26 | 0.58 | 5.73±0.29 | urea (1–4) | 99G21 |
| 2.35 M GuHCl | 2.0 | 25 | 1.34±0.17 | 0.23 | 5.73±0.21 | urea (1–4) | 99G21 |

Remarks:

(1) linear extrapolation, for details of see Ref. 99G21

(2) transition monitored by optical absorption at 405 nm

(3) measured in 30 mM cacodylic acid, pH 6.0

(4) for comparable results see also lysozyme (HEW) and ribonuclease A (Ref. 99G21)

Cytochrome $c$ from horse heart, stability determined by limited proteolysis

| Form | pH | T | $\Delta(\Delta G)$ | Approach/Remarks | Ref |
|---|---|---|---|---|---|
| oxidized form | 7.5 | 25 | 0.0 | proteolysis (1,2) | 98W5 |
| reduced form | 7.5 | 25 | 8.4 | proteolysis (1,2) | 98W5 |
| azide complex | 7.5 | 25 | −0.3 | proteolysis (1,2) | 98W5 |

Remarks:

(1) $\Delta(\Delta G)$ was determined from the kinetics of proteolysis measured by CD at 222 nm

(2) digestion was performed in 20 mM Tris-HCl, pH 7.5, containing 20 mM NaCl and 1.5 M GuHCl

*b) transitions and intermediates*

Cytochrome $c$ from horse heart, oxidized

| Transition | pH | T | $\Delta G$ | $c_{1/2}$ | m | Approach/Remarks | Ref |
|---|---|---|---|---|---|---|---|
| N → U | 7.0 | 10 | 42.3 | 3.3 | 15.1 | kinetics (1,4) | 98B6 |
| N → U | 7.0 | 10 | 49.0 | | 17.6 | kinetics (2,4) | 98B6 |
| N → I | 7.0 | 10 | 20.5 | | 6.3 | kinetics (1,3–6) | 98B6 |
| N → U | 7.0 | 22 | 32.2 | 3.1 | 11.7 | kinetics (1,4) | 98B6 |
| N → U | 7.0 | 22 | 35.1 | | 12.6 | kinetics (2,4) | 98B6 |

Cytochrome *c* from horse heart, oxidized (continued)

| Transition | pH | T | $\Delta G$ | $c_{1/2}$ | m | Approach/Remarks | Ref |
|---|---|---|---|---|---|---|---|
| N → I | 7.0 | 22 | 42.3 | | 13.8 | kinetics (1,3–5) | 98B6 |
| N → I | 7.0 | 22 | 20.9 | | 5.0 | kinetics (2–5) | 98B6 |
| N → U | 7.0 | 34 | 21.8 | 2.9 | 8.8 | kinetics (1,4) | 98B6 |
| N → U | 7.0 | 34 | 26.8 | | 10.0 | kinetics (2,4) | 98B6 |
| N → I | 7.0 | 34 | 33.1 | | 11.3 | kinetics (1,3–5) | 98B6 |
| N → I | 7.0 | 34 | 18.4 | | 5.4 | kinetics (2–5) | 98B6 |

Remarks:
(1) from GuHCl-induced unfolding monitored by CD at 222 nm
(2) from GuHCl-induced unfolding monitored by fluorescence
(3) from the rate constant of the single observable phase
(4) measured in 0.1 M sodium phosphate buffer, pH 7.0
(5) the results suggest that at least two kinetic unfolding intermediates accumulate during unfolding
(6) no burst phase change in fluorescence occurs at 10°C

Cytochrome *c* from horse heart, stability of the native protein and the early (burst phase) folding intermediate (I)

| Mutant | pH | T | $\Delta G$ | $c_{1/2}$ | m | Appr./Rem. | Ref |
|---|---|---|---|---|---|---|---|
| native | 7 | 10 | 49.79±3.01 | 3.42±0.02 | 14.56±0.88 | GuHCl (1,2) | 98S1 |
| native | 9 | 10 | 36.65±3.97 | 3.62±0.04 | 10.13±1.09 | GuHCl (1,2) | 98S1 |
| native | 11 | 10 | 20.75±0.92 | 3.18±0.02 | 6.53±0.29 | GuHCl (1,2) | 98S1 |
| burst I | 7 | 10 | 4.18±0.92 | 1.26±0.18 | 3.31±0.54 | GuHCl (1,2) | 98S1 |
| burst I | 9 | 10 | 3.64±1.63 | 1.40±0.44 | 2.59±0.84 | GuHCl (1,2) | 98S1 |
| burst I | 11 | 10 | 3.60±1.00 | 1.43±0.26 | 2.51±0.54 | GuHCl (1,2) | 98S1 |

Remarks:
(1) measured by fluorescence-detected stopped-flow technique, for further details, see Ref. 98S1
(2) measured in the presence of 0.4 M $Na_2SO_4$

Cytochrome *c* from horse heart, different unfolded states (U1 and U2) derived from X-ray scattering

| Transition | pH | T | $\Delta G$ | $c_{1/2}$ | m | Appr./Rem. | Ref |
|---|---|---|---|---|---|---|---|
| N → U1 | 7 | | 45.2 | 2.6±0.1 | 17.6 | GuHCl (1–4) | 98S10 |
| N → U1 | 7 | | 39.3 | | 15.1 | GuHCl (1,2,4,5) | 98S10 |
| N → U2 | 7 | | 53.1 | | 19.3 | GuHCl (1,2,4,5) | 98S10 |
| N → U | 7.0 | 20 | | 2.86 | 11.9 | GuHCl (3,6,7) | 97C7 |
| N → U | 7.0 | 25 | | 2.8 | | GuHCl (3,6,8) | 76T2 |
| N → U | 7.0 | 20 | 34.2 | 2.7 | 12.6 | GuHCl (3,6) | 96H1 |
| N → U | 7.0 | 10 | 40.6±4.6 | 2.77±0.02 | 14.6±1.7 | GuHCl (3,6) | 96C14 |

Remarks:
(1) transition monitored by X-ray scattering (SAXS)
(2) result of singular value decomposition (SVD) of the SAXS results
(3) data treatment by a two-state model, linear extrapolation
(4) buffer: 100 mM phosphate buffer
(5) data treatment by a three-state model, linear extrapolation
(6) reference value in Ref. 98S10 obtained from equilibrium unfolding monitored by tryptophan fluorescence
(7) buffer: 0.1 M potassium phosphate, 0.2 M imidazole, pH 7.0
(8) buffer: 0.1 M Tris, pH 7.0

Horse heart cytochrome $c$, on-pathway folding intermediate

| Transition | GuHCl | pH | T | $\Delta G$ | Appr./Rem. | Ref |
|---|---|---|---|---|---|---|
| N → U | 0.0 M | 6.2 | 10 | 44.8±1.3 | GuHCl (1) | 99B1 |
| N → U | 1.0 M | 6.2 | 10 | 28.0±0.8 | GuHCl (1) | 99B1 |
| N → I | 1.0 M | 6.2 | 10 | 17.6±0.4 | GuHCl (2) | 99B1 |
| I → U | 1.0 M | 6.2 | 10 | 10.5 | GuHCl (1,2) | 99B1 |
| I → U | 0.7 M | 6.2 | 10 | 12.6 | GuHCl (3) | 99B1 |
| I → U | 0.7 M | 7.0 | 30 | 9.6 | GuHCl (4) | 99B1 |

Remarks:

(1) data from equilibrium unfolding, linear extrapolation

(2) data from kinetics of unfolding and refolding

(3) data from H/D pulse labeling, cited in Ref. 99B1 (from Elöve, G.A., Roder, H.: ACS Symp. Ser. 470 (1991) 50–63)

(4) data from N hydrogen-exchange, cited in Ref. 99B1 (from Bai, Y., et al.: Science 269 (1995) 192–197)

Cytochrome $c$, see also section d) for sugar-induced molten-globule formation

*c) data from exchange techniques*

Equine cytochrome $c$, oxidized form, hydrogen exchange

| Group | pD | T | $\Delta G$ | Approach/Remarks | Ref |
|---|---|---|---|---|---|
| Lys7 | 7 | 20 | 27.2 | HX, NMR | 98M20 |
| Lys8 | 7 | 20 | 24.3 | HX, NMR | 98M20 |
| Ile9 | 7 | 20 | 28.5 | HX, NMR | 98M20 |
| Phe10 | 7 | 20 | 44.4 | HX, NMR | 98M20 |
| Val11 | 7 | 20 | 29.7 | HX, NMR | 98M20 |
| Gln12 | 7 | 20 | 26.8 | HX, NMR | 98M20 |
| Lys13 | 7 | 20 | 31.4 | HX, NMR | 98M20 |
| Cys14 | 7 | 20 | 33.9 | HX, NMR | 98M20 |
| Ala15 | 7 | 20 | 28.0 | HX, NMR | 98M20 |
| His18 | 7 | 20 | 31.0 | HX, NMR | 98M20 |
| His18 N$\pi$H | 7 | 20 | 30.1 | HX, NMR | 98M20 |
| Thr19 | 7 | 20 | 28.5 | HX, NMR | 98M20 |
| Gly29 | 7 | 20 | 31.0 | HX, NMR | 98M20 |
| Asn31 | 7 | 20 | ~17.2 | HX, NMR | 98M20 |
| Leu32 | 7 | 20 | 35.6 | HX, NMR | 98M20 |
| His33 | 7 | 20 | 34.3 | HX, NMR | 98M20 |
| Leu35 | 7 | 20 | 23.4 | HX, NMR | 98M20 |
| Phe36 | 7 | 20 | 26.8 | HX, NMR | 98M20 |
| Gly37 | 7 | 20 | 25.1 | HX, NMR | 98M20 |
| Arg38 | 7 | 20 | 23.8 | HX, NMR | 98M20 |
| Thr40 | 7 | 20 | 20.5 | HX, NMR | 98M20 |
| Gln42 | 7 | 20 | 20.5 | HX, NMR | 98M20 |
| Ala43 | 7 | 20 | 18.0 | HX, NMR | 98M20 |
| Thr49 | 7 | 20 | ~15.5 | HX, NMR | 98M20 |
| Asn52 | 7 | 20 | 20.5 | HX, NMR | 98M20 |
| Lys53 | 7 | 20 | ~18.8 | HX, NMR | 98M20 |
| Asn54 | 7 | 20 | ~19.2 | HX, NMR | 98M20 |
| Ile57 | 7 | 20 | 17.2 | HX, NMR | 98M20 |
| Trp59 | 7 | 20 | 26.8 | HX, NMR | 98M20 |
| Trp59 N$_i$H | 7 | 20 | 31.4 | HX, NMR | 98M20 |
| Lys60 | 7 | 20 | 32.6 | HX, NMR | 98M20 |
| Leu64 | 7 | 20 | 31.4 | HX, NMR | 98M20 |
| Met65 | 7 | 20 | 36.4 | HX, NMR | 98M20 |

Equine cytochrome *c*, oxidized form, hydrogen exchange (continued)

| Group | pD | T | ΔG | Approach/Remarks | Ref |
|---|---|---|---|---|---|
| Glu66 | 7 | 20 | 22.6 | HX, NMR | 98M20 |
| Tyr67 | 7 | 20 | 29.3 | HX, NMR | 98M20 |
| Leu68 | 7 | 20 | 48.1 | HX, NMR | 98M20 |
| Glu69 | 7 | 20 | 30.5 | HX, NMR | 98M20 |
| Asn70 | 7 | 20 | 25.9 | HX, NMR | 98M20 |
| Lys73 | 7 | 20 | 18.8 | HX, NMR | 98M20 |
| Tyr74 | 7 | 20 | 26.4 | HX, NMR | 98M20 |
| Ile75 | 7 | 20 | 25.1 | HX, NMR | 98M20 |
| Lys79 | 7 | 20 | ~16.7 | HX, NMR | 98M20 |
| Met80 | 7 | 20 | ~15.5 | HX, NMR | 98M20 |
| Ile85 | 7 | 20 | 19.7 | HX, NMR | 98M20 |
| Arg91 | 7 | 20 | 33.5 | HX, NMR | 98M20 |
| Glu92 | 7 | 20 | 33.9 | HX, NMR | 98M20 |
| Asp93 | 7 | 20 | 31.4 | HX, NMR | 98M20 |
| Leu94 | 7 | 20 | ~43.9 | HX, NMR | 98M20 |
| Ile95 | 7 | 20 | >48.5 | HX, NMR | 98M20 |
| Ala96 | 7 | 20 | >46.9 | HX, NMR | 98M20 |
| Tyr97 | 7 | 20 | >47.7 | HX, NMR | 98M20 |
| Leu98 | 7 | 20 | ~56.5 | HX, NMR | 98M20 |
| Lys99 | 7 | 20 | >43.9 | HX, NMR | 98M20 |
| Lys100 | 7 | 20 | 27.2 | HX, NMR | 98M20 |
| Ala101 | 7 | 20 | 26.4 | HX, NMR | 98M20 |
| Thr102 | 7 | 20 | 19.2 | HX, NMR | 98M20 |

Equine cytochrome *c*, characterization of folding intermediates by unfolding free energies of specific marker amino acids

| Residue/Form | | pD | T | ΔG | Approach/Remarks | | Ref |
|---|---|---|---|---|---|---|---|
| Leu98 | reduced | 7–9 | 30 | 74.9 | HX/GuHSCN | (1,2) | 98X2 |
| | oxidized | 7–9 | 30 | 53.6 | HX/GuHCl | (1,2) | 98X2 |
| Leu68 | reduced | 7–9 | 30 | 54.8 | HX/GuHSCN | (1) | 98X2 |
| | oxidized | 7–9 | 30 | 41.0 | HX/GuHCl | (1) | 98X2 |
| Leu64 | reduced | 7–9 | 30 | 46.0 | HX/GuHSCN | (1) | 98X2 |
| | oxidized | 7–9 | 30 | 31.0 | HX/GuHCl | (1) | 98X2 |
| Lys60 | reduced | 7–9 | 30 | 42.3 | HX/GuHSCN | (1,3) | 98X2 |
| Ile75 | reduced | 7–9 | 30 | 38.5 | HX/GuHSCN | (1) | 98X2 |
| | oxidized | 7–9 | 30 | 25.1 | HX/GuHCl | (1) | 98X2 |

Remarks:
(1) from hydrogen exchange rates measured in the presence of denaturant, ΔG extrapolated to zero denaturant concentration
(2) the Met80 to heme-iron binding affinity parallels the Δ(ΔG) values between the oxidized and reduced forms, for details see Ref. 98X2
(3) Lys60 is a marker for the same goups as Leu64

Equine cytochrome *c*, oxidized (ox.) and reduced (red.), ΔG for individual amino acid residues

| Position | pD | T | ΔG$_{ox.}$ | ΔG$_{red.}$ | Appr./Rem. | | Ref |
|---|---|---|---|---|---|---|---|
| a) N/C helix, global unfolding: | | | | | | | |
| Leu98 | 7.0 | 20 | 59 | 79 | H/D | (1–3) | 99M16 |
| Ile9 | 7.0 | 20 | 29 | 29 | H/D | (1,2) | 99M16 |
| Phe10 | 7.0 | 20 | 46 | 46 | H/D | (1,2) | 99M16 |
| Val11 | 7.0 | 20 | 29 | 29 | H/D | (1,2) | 99M16 |
| Glu92 | 7.0 | 20 | 33 | 33 | H/D | (1,2) | 99M16 |
| Asp93 | 7.0 | 20 | 29 | 33 | H/D | (1,2) | 99M16 |

Equine cytochrome $c$, oxidized (ox.) and reduced (red.), $\Delta G$ for individual amino acid residues (continued)

| Position | pD | T | $\Delta G_{ox.}$ | $\Delta G_{red.}$ | Appr./Rem. | Ref |
|---|---|---|---|---|---|---|
| Leu94 | 7.0 | 20 | 46 | 38 | H/D  (1,2) | 99M16 |
| Ala96 | 7.0 | 20 | 50 | 46 | H/D  (1,2) | 99M16 |
| Tyr97 | 7.0 | 20 |  | 54 | H/D  (1,2) | 99M16 |
| Lys99 | 7.0 | 20 | 42 |  | H/D  (1,2) | 99M16 |
| b) 60s helix, subglobal unfolding: | | | | | | |
| Leu68 | 7.0 | 20 | 46 |  | H/D  (1–3) | 99M16 |
| Leu32 | 7.0 | 20 | 33 | 42 | H/D  (1,2) | 99M16 |
| Met65 | 7.0 | 20 | 38 | 42 | H/D  (1,2) | 99M16 |
| Tyr67 | 7.0 | 20 | 29 | 38 | H/D  (1,2) | 99M16 |
| c) omega loop, subglobal unfolding: | | | | | | |
| Leu64 | 7.0 | 20 | 33 | 50 | H/D  (1–3) | 99M16 |
| Lys60 | 7.0 | 20 |  | 46 | H/D  (1–3) | 99M16 |
| Trp59 $N_l H$ | 7.0 | 20 | 33 | 46 | H/D  (1–3) | 99M16 |
| Phe36 | 7.0 | 20 | 25 | 29 | H/D  (1,2) | 99M16 |
| Gly37 | 7.0 | 20 |  | 29 | H/D  (1,2) | 99M16 |
| d) region connecting 60s helix and C helix, subglobal unfolding: | | | | | | |
| Ile75 | 7.0 | 20 | 25 | 38 | H/D  (1–3) | 99M16 |
| Tyr74 | 7.0 | 20 |  | 33 | H/D  (1,2) | 99M16 |

Remarks:

(1) Ref. 99M16 contains data from H/D exchange at varying denaturant conc. and temperature

(2) using GuHCl for studies on oxidized cytochrome $c$, and GuHSCN for the reduced protein

(3) marker proton

*d) data obtained in the presence of cosolutes*

Cytochrome $c$ in aqueous polyol solution

| Cosolvent | Conc. | pH | $T_{trs}$ | $\Delta T$ | $\Delta(\Delta G)$ | Appr./Rem. | Ref |
|---|---|---|---|---|---|---|---|
| control |  | 4.0 | 64.5 | 0.0 | 0.00 | heat (1,2) | 98K3 |
| mannitol | 1.00 M | 4.0 | 69.8 | 5.4 | 3.93 | heat (1,2) | 98K3 |
| inositol | 0.75 M | 4.0 | 70.9 | 6.5 | 3.77 | heat (1,2) | 98K3 |
| xylitol | 2.00 M | 4.0 | 72.5 | 8.1 | 5.44 | heat (1,2) | 98K3 |
| adonitol | 2.00 M | 4.0 | 72.6 | 8.2 | 5.94 | heat (1,2) | 98K3 |
| sorbitol | 2.00 M | 4.0 | 73.8 | 9.3 | 6.36 | heat (1,2) | 98K3 |
| control |  | 7.0 | 48.0 | 0.0 | 0.00 | heat (1,3) | 98K3 |
| mannitol | 1.00 M | 7.0 | 52.7 | 4.7 | 2.09 | heat (1,3) | 98K3 |
| inositol | 0.75 M | 7.0 | 54.0 | 6.0 | 2.72 | heat (1,3) | 98K3 |
| xylitol | 2.00 M | 7.0 | 56.8 | 8.8 | 3.68 | heat (1,3) | 98K3 |
| adonitol | 2.00 M | 7.0 | 56.5 | 8.5 | 3.14 | heat (1,3) | 98K3 |
| sorbitol | 2.00 M | 7.0 | 61.1 | 13.1 | 5.73 | heat (1,3) | 98K3 |

Remarks:

(1) transition monitored by optical absorption at 287 nm

(2) buffer: 40 mM acetate, pH 4.0

(3) buffer: 20 mM phosphate or MOPS, pH 7.0, 1.5 M GuHCl

Equine cytochrome $c$, stability parameters in the presence of cosolvents at relative viscosity of $\eta_{rel} = 2$

| Solvent | pH | T | $\Delta G$ | m | Approach/Remarks | | Ref |
|---|---|---|---|---|---|---|---|
| buffer | 7.0 | 10 | 33.9±1.3 | 11.3±0.4 | GuHCl | (1–4) | 99B10 |
| 27% (w/w) glycol | 7.0 | 10 | 32.6±0.8 | 11.3±0.4 | GuHCl | (1–5) | 99B10 |
| 21% (w/w) glucose | 7.0 | 10 | 43.1±4.6 | 11.3±1.3 | GuHCl | (1–4) | 99B10 |
| 25% (w/w) glycerol | 7.0 | 10 | 39.3±1.3 | 12.1±0.4 | GuHCl | (1–4) | 99B10 |

Remarks:

(1) data from equilibrium measurements, linear extrapolation

(2) transition monitored by fluorescence

(3) buffer: 50 mM imidazole, pH 7.0

(4) Ref. 99B10 contains a deconvolution of viscosity and stability

(5) 27% (w/w) ethylene glycol

Cytochrome $c$ from horse heart (cyt $c$), in the presence and absence of imidazole

| Protein | pH | T | $\Delta G$ | m | Appr./Rem. | Ref |
|---|---|---|---|---|---|---|
| cyt $c$, no imidazole | 7 | 23 | 44.8±3.8 | 17.2±1.3 | GuHCl (1,2) | 99S8 |
| cyt $c$, 200 mM imidazole | 7 | 23 | 30.1±2.1 | 10.5±0.8 | GuHCl (1,2) | 99S8 |

Remarks:

(1) linear extrapolation

(2) transition monitored by small-angle X-ray scattering (SAXS)

Cytochrome $c$, in the presence and absence of glycine

| Osmolyte | Conc. | pH | T | $\Delta G$ | m | Approach/Remarks | | Ref |
|---|---|---|---|---|---|---|---|---|
| none | | 4.2 | 21 | 26.74±1.16 | 2.58±0.26 | GuHCl | (1–3) | 98F2 |
| glycine | 2 M | 4.2 | 21 | 44.02±1.17 | 2.58±0.26 | GuHCl | (1–3) | 98F2 |

Remarks:

(1) linear extrapolation, LEM-SB

(2) m was held to be the same in the absence and presence of glycine

(3) Ref. 98F2 contains the hydrogen exchange rates for about 30 amino acid residues of cytochrome $c$ in the absence and presence of glycine

Cytochrome $c$, in the presence of polyglutamate

| Electrolyte | Conc. | pH | T | $\Delta G$ | Approach/Remarks | | Ref |
|---|---|---|---|---|---|---|---|
| polyglutamate | 0.0 mg/ml | 7.0 | 25 | 22.7±2.7 | DSC | (1) | 97B1 |
| | 1.5 | 7.0 | 25 | 32.0±2.2 | DSC | (1) | 97B1 |

Remark:

(1) measured in 2 mM HEPES, pH 7.0, at 110 μM cytochrome $c$ and 1.5 mg/ml polyglutamate, see also Table 2

Cytochrome $c$ from horse heart, partially unfolded by sodium dodecyl sulphate (SDS)

| pH | T | $\Delta G$ | Approach/Remarks | | Ref |
|---|---|---|---|---|---|
| 7.4 | 25 | 16.8 | SDS | (1,2) | 98D6 |

Remarks:

(1) data treatment by a two-state transition assuming a linear dependence of $\Delta G$ on the denaturant concentration

(2) transition monitored by optical absorption, CD at 220 nm, and Trp fluorescence at 334 nm yields $\Delta G$ from 15 to 19.3 kJ/mol

Cytochrome $c$ from horse heart, sorbitol-induced molten globule formation

| NaCl | pH | T | $\Delta G$ | $c_{1/2}$ | m | Approach/Remarks | Ref |
|------|----|----|------------|-----------|------|------------------|-----|
| 0 mM | 2 | 15 | 7.73 | 1.2 | 6.69 | sorbitol (1–3) | 99K1 |
| | 2 | 20 | 9.20 | 1.4 | 6.56 | sorbitol (1–3) | 99K1 |
| | 2 | 25 | 10.7 | 1.7 | 6.14 | sorbitol (1–3) | 99K1 |
| | 2 | 30 | 12.2 | 2.0 | 6.02 | sorbitol (1–3) | 99K1 |
| | 2 | 35 | 13.3 | 2.4 | 5.60 | sorbitol (1–3) | 99K1 |
| 5 mM | 2 | 25 | 8.70 | 1.4 | 6.08 | sorbitol (1–3) | 99K1 |
| 10 mM | 2 | 25 | 7.36 | 1.2 | 5.96 | sorbitol (1–3) | 99K1 |

Remarks:

(1) linear extrapolation applied to the sorbitol-induced two-state transition AU → MG (acid unfolded → molten globule)

(2) transition monitored by CD at 222 nm

(3) from the temperature dependence of $\Delta G$ it follows $\Delta H = -74.4$ kJ/mol and $\Delta Cp \sim 0$ kJ/mol/K

Equine ferricytochrome $c$, sugar-induced molten globule of the A-state

| Osmolyte | pH | T | $\partial\Delta(\Delta G)/\partial c$ | Approach/Remarks | Ref |
|----------|----|----|----------------------------------------|------------------|-----|
| glycerol | 2 | 1 | 2.9±0.4 | (1,2) | 98D8 |
| glucose | 2 | 1 | 7.5±1.3 | (1,2) | 98D8 |
| galactose | 2 | 1 | 6.3±0.8 | (1,2) | 98D8 |
| sucrose | 2 | 1 | 9.2±0.4 | (1,2) | 98D8 |
| trehalose | 2 | 1 | 10.0±0.4 | (1,2) | 98D8 |
| melizitose | 2 | 1 | 17.2±1.3 | (1–3) | 98D8 |
| stachyose | 2 | 1 | 21.3±0.8 | (1,2) | 98D8 |

Remarks:

(1) transition measured in dependence of osmolyte concentration, assuming a linear dependence of ln K on the osmolyte concentration

(2) osmolyte-induced transition monitored by CD at 222 nm

(3) melizitose-induced transition monitored by NMR in $D_2O$ using fully exchanged samples at 20°C gives
$\Delta G = 9.6\pm1.3$ kJ/mol

# Cytochrome $c_2$

Ferrocytochrome $c_2$ from *Rhodobacter capsulatus*, mutants (Pro35→Ala) and (Tyr75→Phe) relative to the wild-type protein, $\Delta G_{op}$ determined by hydrogen-deuterium exchange

| Group | pH | T | $\Delta(\Delta G_{op})$ (Tyr75→Phe) | $\Delta(\Delta G_{op})$ (Pro35→Ala) | Appr./Rem. | Ref |
|-------|----|----|--------------------------------------|--------------------------------------|------------|-----|
| Glu7 | 6 | 30 | −0.4 | 0.4 | HX, NMR (1) | 92G7 |
| Phe10 | 6 | 30 | 1.3 | ≈−1.3 | HX, NMR (1) | 92G7 |
| Cys13 | 6 | 30 | 0.4 | 0.4 | HX, NMR (1) | 92G7 |
| Lys14 | 6 | 30 | 2.1 | 0.4 | HX, NMR (1) | 92G7 |
| Cys16 | 6 | 30 | −1.7 | <−2.5 | HX, NMR (1) | 92G7 |
| His17 | 6 | 30 | | <−2.5 | HX, NMR (1) | 92G7 |
| His17N$\pi$H | 6 | 30 | 1.3 | −8.8 | HX, NMR (1) | 92G7 |
| Ser18 | 6 | 30 | −0.4 | −2.5 | HX, NMR (1) | 92G7 |
| Ile19 | 6 | 30 | 3.3 | 0.0 | HX, NMR (1) | 92G7 |
| Ile20 | 6 | 30 | −0.4 | −4.2 | HX, NMR (1) | 92G7 |
| Ala21 | 6 | 30 | −0.1 | −1.3 | HX, NMR (1) | 92G7 |
| Ile27 | 6 | 30 | 0.4 | −0.4 | HX, NMR (1) | 92G7 |
| Val28 | 6 | 30 | −0.8 | −0.8 | HX, NMR (1) | 92G7 |
| Gly34 | 6 | 30 | 0.4 | −10.9 | HX, NMR (1) | 92G7 |

Ferrocytochrome $c_2$ from *Rhodobacter capsulatus*, mutants (Pro35→Ala) and (Tyr75→Phe) relative to the wild-type protein, $\Delta G_{op}$ determined by hydrogen-deuterium exchange (continued)

| Group | pH | T | $\Delta(\Delta G_{op})$ (Tyr75→Phe) | $\Delta(\Delta G_{op})$ (Pro35→Ala) | Appr./Rem. | Ref |
|---|---|---|---|---|---|---|
| Leu37 | 6 | 30 | | <−5.4 | HX, NMR (1) | 92G7 |
| Tyr38 | 6 | 30 | −0.8 | −2.1 | HX, NMR (1) | 92G7 |
| Val40 | 6 | 30 | 0.3 | 0.4 | HX, NMR (1) | 92G7 |
| Val41 | 6 | 30 | 0.4 | 0.4 | HX, NMR (1) | 92G7 |
| Gly42 | 6 | 30 | 0.4 | 0.8 | HX, NMR (1) | 92G7 |
| Arg43 | 6 | 30 | −1.7 | −3.8 | HX, NMR (1) | 92G7 |
| Ala45 | 6 | 30 | −3.8 | −6.3 | HX, NMR (1) | 92G7 |
| Gly46 | 6 | 30 | −3.8 | −6.3 | HX, NMR (1) | 92G7 |
| Thr47 | 6 | 30 | −0.8 | −0.4 | HX, NMR (1) | 92G7 |
| Lys54 | 6 | 30 | <−2.5 | 2.1 | HX, NMR (1) | 92G7 |
| Ile57 | 6 | 30 | −5.0 | 0.4 | HX, NMR (1) | 92G7 |
| Val58 | 6 | 30 | −4.2 | −3.3 | HX, NMR (1) | 92G7 |
| Ala59 | 6 | 30 | −0.8 | 1.7 | HX, NMR (1) | 92G7 |
| Leu60 | 6 | 30 | −2.9 | | HX, NMR (1) | 92G7 |
| Gly61 | 6 | 30 | −2.5 | −1.3 | HX, NMR (1) | 92G7 |
| Ala62 | 6 | 30 | −0.8 | 0.1 | HX, NMR (1) | 92G7 |
| Thr68 | 6 | 30 | −0.4 | | HX, NMR (1) | 92G7 |
| Ala73 | 6 | 30 | 1.7 | 0.2 | HX, NMR (1) | 92G7 |
| Thr74 | 6 | 30 | 0.4 | −0.4 | HX, NMR (1) | 92G7 |
| Tyr75 | 6 | 30 | 0.8 | 1.3 | HX, NMR (1) | 92G7 |
| Asp78 | 6 | 30 | −1.7 | −0.4 | HX, NMR (1) | 92G7 |
| Gly80 | 6 | 30 | <−1.3 | −0.4 | HX, NMR (1) | 92G7 |
| Ala81 | 6 | 30 | −0.8 | −0.0 | HX, NMR (1) | 92G7 |
| Phe82 | 6 | 30 | −4.6 | −0.4 | HX, NMR (1) | 92G7 |
| Leu83 | 6 | 30 | −6.3 | −0.4 | HX, NMR (1) | 92G7 |
| Lys84 | 6 | 30 | −6.7 | −0.04 | HX, NMR (1) | 92G7 |
| Glu85 | 6 | 30 | −5.0 | −0.4 | HX, NMR (1) | 92G7 |
| Lys86 | 6 | 30 | −4.6 | −0.08 | HX, NMR (1) | 92G7 |
| Leu87 | 6 | 30 | −6.7 | −0.8 | HX, NMR (1) | 92G7 |
| Phe98 | 6 | 30 | <−3.8 | −0.4 | HX, NMR (1) | 92G7 |
| Leu100 | 6 | 30 | −2.9 | −0.4 | HX, NMR (1) | 92G7 |
| Ala108 | 6 | 30 | −0.4 | −0.8 | HX, NMR (1) | 92G7 |
| Ala109 | 6 | 30 | 0.4 | −0.17 | HX, NMR (1) | 92G7 |
| Tyr110 | 6 | 30 | −1.3 | 0.4 | HX, NMR (1) | 92G7 |
| Ala112 | 6 | 30 | 0.8 | 0.4 | HX, NMR (1) | 92G7 |
| Ser113 | 6 | 30 | 0.08 | −0.4 | HX, NMR (1) | 92G7 |
| Val114 | 6 | 30 | 0.13 | −0.4 | HX, NMR (1) | 92G7 |
| Val115 | 6 | 30 | <−0.4 | <−0.4 | HX, NMR (1) | 92G7 |

Remark:
(1) differences in $\Delta(\Delta G_{op})$ of more than 2 kJ/mol are considered significant in Ref. 92G7

## Cytochrome $c_{551}$

Cytochrome $c_{551}$ from *Pseudomonas aeruginosa* (cyt $c_{551}$)

| Protein | pH | T | $\Delta G$ | $c_{1/2}$ | m | Appr./Rem. | Ref |
|---|---|---|---|---|---|---|---|
| cyt $c_{551}$ | 3.0 | 10 | 13.8 | 1.0 | 13.8 | GuHCl (1–3) | 99T13 |
| | 3.0 | 10 | 7.3 | 0.9 | | kin. (2,4) | 99T13 |
| | 4.0 | 10 | 24.7 | 2.1 | 11.7 | GuHCl (1–3) | 99T13 |
| | 4.0 | 10 | 18.0 | 2.0 | | kin. (2,4) | 99T13 |

Cytochrome $c_{551}$ from *Pseudomonas aeruginosa* (cyt $c_{551}$) (continued)

| Protein | pH | T | $\Delta G$ | $c_{1/2}$ | m | Appr./Rem. | Ref |
|---|---|---|---|---|---|---|---|
| | 4.7 | 10 | 27.6 | 2.1 | 13.0 | GuHCl (1–3) | 99T13 |
| | 4.7 | 10 | 21.3 | 2.3 | | kin. (2,4) | 99T13 |
| | 7.0 | 10 | 25.1 | 1.8 | 13.8 | GuHCl (1–3) | 99T13 |
| | 7.0 | 10 | 19.2 | 1.8 | | kin. (2,4) | 99T13 |

Remarks:

(1) linear extrapolation

(2) buffers: 50 mM sodium phosphate, pH 7.0; 25 mM phosphate and 25 mM acetate, pH 4.0 and 4.7; 15 mM citrate, pH 3.0

(3) transition monitored by fluorescence

(4) kinetic data from stopped-flow fluorescence

Cytochrome $c_{551}$ from *Pseudomonas aeruginosa*, wild type and mutants, GuHCl and acid denaturation

| Protein | pH | T | $\Delta G$ | $c_{1/2}$ | Approach/Remarks | Ref |
|---|---|---|---|---|---|---|
| wild type | 7.0 | 10 | 34.3±0.8 | 1.8 | GuHCl (1) | 98B8 |
| wild type | 7.0 | 10 | 33.5±0.4 | 6.2 | urea (1) | 98B8 |
| Trp56→Phe | 7.0 | 10 | 31.4±0.8 | 1.5 | GuHCl (1) | 98B8 |
| Ile59→Glu | 7.0 | 10 | 31.8±0.4 | 1.5 | GuHCl (1) | 98B8 |
| Pro58→Ala | 7.0 | 10 | 26.4±2.9 | 1.5 | GuHCl (1) | 98B8 |
| Val23→Asp | 7.0 | 10 | 28.5±0.4 | 1.5 | GuHCl (1) | 98B8 |
| wild type | 7.0 | 10 | 26.8±0.8 | | acid (2) | 98B8 |
| Trp56→Phe | 7.0 | 10 | 26.8±0.8 | | acid (2) | 98B8 |
| Ile59→Glu | 7.0 | 10 | 30.1±0.8 | | acid (2) | 98B8 |
| Pro58→Ala | 7.0 | 10 | 32.6±0.8 | | acid (2) | 98B8 |
| Val23→Asp | 7.0 | 10 | 28.5±0.8 | | acid (2) | 98B8 |

Remarks:

(1) data treatment according to Ref. 96C14

(2) data treatment according to Ref. I-70T

Cytochrome $c_{551}$ from *Pseudomonas aeruginosa*, wild type and mutants based on cytochrome $c_{552}$ from thermophilic *Hydrogenobacter thermophilus*

| Mutant | pH | $T_{trs}$ | $\Delta T$ | $\Delta(\Delta G)$ | Appr./Rem. | Ref |
|---|---|---|---|---|---|---|
| wild type | 5.0 | 50.4 | 0.0 | 0.0 | heat (1,2) | 99H3 |
| Phe7→Ala | 5.0 | 59.9 | 9.5 | 4.6 | heat (1,2) | 99H3 |
| Val13→Met | 5.0 | 53.6 | 3.2 | 1.7 | heat (1,2) | 99H3 |
| Phe34→Tyr | 5.0 | 66.4 | 16.0 | 7.9 | heat (1,2) | 99H3 |
| Gln37→Arg | 5.0 | 54.7 | 4.3 | 2.1 | heat (1,2) | 99H3 |
| Glu43→Tyr | 5.0 | 55.5 | 5.1 | 2.5 | heat (1,2) | 99H3 |
| Val78→Ile | 5.0 | 58.8 | 8.4 | 4.2 | heat (1,2) | 99H3 |
| double mutant (Phe7→Ala and Val13→Met) | | | | | | |
| | 5.0 | 62.4 | 12.0 | 5.9 | heat (1,2) | 99H3 |
| double mutant (Phe34→Tyr and Gln37→Arg) | | | | | | |
| | 5.0 | 62.9 | 12.5 | 6.3 | heat (1,2) | 99H3 |
| double mutant (Phe34→Tyr and Glu43→Tyr) | | | | | | |
| | 5.0 | 70.7 | 20.3 | 10.0 | heat (1,2) | 99H3 |
| double mutant (Gln37→Arg and Glu43→Tyr) | | | | | | |
| | 5.0 | 54.7 | 4.3 | 2.1 | heat (1,2) | 99H3 |
| multiple mutant (Phe34→Tyr, Gln37→Arg and Glu43→Tyr) | | | | | | |
| | 5.0 | 57.9 | 17.5 | 8.4 | heat (1,2) | 99H3 |

Remarks:

(1) transition monitored by far-UV CD at 222 nm

(2) measured in water, pH 5.0, adjusted with HCl

(3) $\Delta(\Delta G)$ was calculated using $\Delta(\Delta G) = \Delta T_{mut} \times \Delta H_{w.t.}/T_{trs,w.t.}$, see also Table 2

Cytochrome $c_{551}$ from *Pseudomonas aeruginosa*, wild type and mutants based on cytochrome $c_{552}$ from thermophilic *Hydrogenobacter thermophilus*

| Mutant | pH | T | ΔG | $c_{1/2}$ | m | Appr./Rem. | Ref |
|---|---|---|---|---|---|---|---|
| wild type | 5.0 | 25 | 20.42 | 2.25 | 9.08 | GuHCl (1–3) | 99H3 |
| Val78→Ile | 5.0 | 25 | 20.67 | 2.44 | 8.45 | GuHCl (1–3) | 99H3 |
| double mutant (Phe7→Ala and Val13→Met) | | | | | | | |
| | 5.0 | 25 | 30.71 | 2.87 | 10.71 | GuHCl (1–3) | 99H3 |
| double mutant (Phe34→Tyr and Glu43→Tyr) | | | | | | | |
| | 5.0 | 25 | 33.85 | 2.98 | 11.34 | GuHCl (1–3) | 99H3 |

Remarks:
(1) linear extrapolation
(2) transition monitored by far-UV CD at 222 nm
(3) measured in water with GuHCl, pH 5.0, adjusted with HCl

## Cytochrome $c_{553}$

Cytochrome $c_{553}$ from *Desulfovibrio vulgaris* (Hildenborough), redox states

| Protein | pH | T | ΔG | $c_{1/2}$ | m | Appr./Rem. | Ref |
|---|---|---|---|---|---|---|---|
| $c_{553}$, ox. | 7.0 | 20 | 38±2 | 3.65±0.20 | 10.4±0.4 | GuHCl (1–3) | 99W14 |
| $c_{553}$, red. | 7.0 | 20 | 45±3 | 3.91±0.20 | 11.5±0.7 | GuHCl (1–3) | 99W14 |
| $c_{551}$, ox. | | | 34 | 1.8 | 18.9 | (4) | 99W14 |

Remarks:
(1) linear extrapolation
(2) transition monitored by far-UV CD and Soret absorption
(3) buffer: 5 mM phosphate
(4) calculated value for cytochrome $c_{551}$ (oxidized), reported in Ref. 99W14

## Iso-1-cytochrome $c$

The data entries are arranged as follows:

a) wild type or pseudo-wild type and mutants
b) transitions of iso-1-cytochrome $c$

*a) wild type or pseudo-wild type and mutants*

Iso-1-cytochrome $c$ from yeast, wild type and mutants position 73

| Mutant | pH | T | ΔG | m | Approach/Remarks | | Ref |
|---|---|---|---|---|---|---|---|
| wild type | 4.5 | 25 | 15.5±0.8 | | DSC | (1) | 97H4 |
| wild type | 7.5 | 25 | 23.7±1.7 | 20.38±1.34 | GuHCl | (2–4) | 97H4 |
| Lys73→Ile | 4.5 | 25 | 13.8±0.4 | | DSC | (1) | 97H4 |
| Lys73→Ile | 7.5 | 25 | 18.7±0.5 | 17.95±0.50 | GuHCl | (2,3,5) | 97H4 |
| Lys73→Trp | 4.5 | 25 | 8.8±1.7 | | DSC | (1) | 97H4 |
| Lys73→Trp | 7.5 | 25 | 17.7±1.5 | 17.41±1.42 | GuHCl | (2–4) | 97H4 |
| Lys73→Val | 4.5 | 25 | 15.9±0.4 | | DSC | (1) | 97H4 |
| Lys73→Val | 7.5 | 25 | 17.9±1.3 | 16.78±1.00 | GuHCl | (2,3,5) | 97H4 |

Remarks:
(1) for the experimental data see Table 2, ΔCp from thermal unfolding and m from GuHCl denaturation do not correlate well, as can be concluded from the results
(2) linear extrapolation
(3) measured in 20 mM Tris-HCl, 40 mM NaCl, pH 7.5
(4) data from Ref. 93B6
(5) data from Ref. 95H6

Iso-1-cytochrome *c* from yeast, wild type and mutant (Lys73→His)

| Mutant | pH | T | ΔG | m | Appr./Rem. | Ref |
|---|---|---|---|---|---|---|
| wild type | 7.5 | 25.0±0.1 | 23.7±1.8 | 20.38±1.34 | GuHCl (1–3) | 97G9 |
| wild type | 7.5 | 25.0±0.1 | 22.7±1.3 | 17.07±1.67 | GuHCl (1,2,4) | 97G9 |
| wild type | 7.5 | 25.0±0.1 | 17.8±0.8 | 14.14±0.50 | GuHCl (1,4,5) | 97G9 |
| wild type | 7.5 | 25.0±0.1 | 20.8±1.7 | 6.90±0.08 | urea (1,4,5) | 97G9 |
| Lys73→His | 7.5 | 25.0±0.1 | 18.1±0.5 | 15.02±0.04 | GuHCl (1–3) | 97G9 |
| Lys73→His | 7.5 | 25.0±0.1 | 18.6±1.3 | 14.14±1.17 | GuHCl (1,2,4) | 97G9 |
| Lys73→His | 7.5 | 25.0±0.1 | 18.0±1.0 | 14.23±0.54 | GuHCl (1,4,5) | 97G9 |
| Lys73→His | 7.5 | 25.0±0.1 | 15.2±0.6 | 5.15±0.59 | urea (1,4,5) | 97G9 |

Remarks:

(1) linear extrapolation, LEM–SB

(2) buffer: 20 mM Tris, 40 mM NaCl, pH 7.5

(3) transition monitored by CD

(4) transition monitored by Trp fluorescence

(5) measured in the above buffer (2) in the presence of 200 mM imidazole

Yeast iso-1-cytochrome *c*, wild type and Lys73→His mutant

| Mutant | pH | T | ΔG | $c_{1/2}$ | m | Appr./Rem. | Ref |
|---|---|---|---|---|---|---|---|
| wild type | 7.5 | 25 | 23.68±2.01 | 1.15±0.01 | 20.58±1.72 | GuHCl (1–3) | 99G8 |
| | 7.2 | 25 | 19.58±1.46 | 1.09±0.04 | 18.03±0.84 | GuHCl (1,2) | 99G8 |
| | 6.5 | 25 | 22.76±0.59 | 1.23±0.07 | 18.62±1.09 | GuHCl (1,2) | 99G8 |
| | 6.12 | 25 | 23.68±0.79 | 1.16±0.02 | 20.33±1.00 | GuHCl (1,2) | 99G8 |
| | 5.8 | 25 | 22.09±1.97 | 1.09±0.04 | 20.25±1.55 | GuHCl (1,2) | 99G8 |
| | 5.0 | 25 | 19.96±1.05 | 1.08±0.01 | 18.49±0.88 | GuHCl (1,2) | 99G8 |
| | 4.5 | 25 | 17.20±0.67 | 0.96±0.03 | 17.91±1.05 | GuHCl (1,2) | 99G8 |
| | 4.0 | 25 | 11.38±1.55 | 0.66±0.04 | 17.15±1.55 | GuHCl (1,2) | 99G8 |
| Lys73→His | 7.5 | 25 | 15.40±0.79 | 1.15±0.01 | 13.39±0.75 | GuHCl (1–3) | 99G8 |
| | 7.2 | 25 | 14.27±2.18 | 1.15±0.05 | 12.43±1.55 | GuHCl (1,2) | 99G8 |
| | 6.5 | 25 | 14.73±1.34 | 1.12±0.03 | 13.14±0.79 | GuHCl (1,2) | 99G8 |
| | 6.12 | 25 | 15.06±0.84 | 1.09±0.01 | 13.85±0.88 | GuHCl (1,2) | 99G8 |
| | 5.8 | 25 | 15.48±0.84 | 1.05±0.02 | 14.81±0.84 | GuHCl (1,2) | 99G8 |
| | 5.0 | 25 | 15.06±0.04 | 0.98±0.02 | 15.36±0.33 | GuHCl (1,2) | 99G8 |
| | 4.5 | 25 | 12.97±1.30 | 0.83±0.05 | 15.69±1.21 | GuHCl (1,2) | 99G8 |
| | 4.0 | 25 | 7.15±0.42 | 0.51±0.01 | 13.93±0.71 | GuHCl (1,2) | 99G8 |
| the native-like folding intermediate: | | | | | | | |
| wild type | 4.5 | 25 | | 0.68 | 16.3 | GuHCl (1,4) | 99G8 |
| Lys73→His | 4.5 | 25 | | 0.48 | 16.3 | GuHCl (1,4) | 99G8 |

Remarks:

(1) nonlinear least-squares fit of the transition curve assuming a linear dependence of ΔG on the denaturant conc.

(2) transition monitored by ellipticity at 220 nm

(3) data from Ref. 97G9, refit according to remark (1)

(4) the native-like intermediate due to displacement of Met80 was detected monitoring the unfolding transition by optical absorption at 695 nm

Iso-1-cytochrome $c$ from yeast, wild type and multiple histidine mutants

| Mutant | pH | T | $\Delta G$ | $c_{1/2}$ | m | Approach/Remarks | Ref |
|---|---|---|---|---|---|---|---|
| wild type | 7.5 | 20 | 23.7±1.7 | 1.16 | 20.4±1.3 | GuHCl (1–3) | 97G8 |
| TM (4) | 7.5 | 20 | 16.8±0.7 | 1.008 | 16.7±0.7 | GuHCl (1,2,4) | 97G8 |
| TM + His26 | 7.5 | 20 | 20.9±1.0 | 1.15 | 18.2±0.8 | GuHCl (1,2,4) | 97G8 |
| TM + His54 | 7.5 | 20 | 10.2±0.3 | 0.63 | 16.3±0.5 | GuHCl (1,2,4,5) | 97G8 |

Remarks:
(1) linear extrapolation, weighted least-squares fit
(2) buffer: 20 mM Tris, 40 mM NaCl, pH 7.5
(3) data from Ref. 93B6
(4) TM = triple mutant with all three His residues replaced (His26→Asn, His33→Asn, and His39→Glu)
(5) TM + His54 has residual structure stabilizing its GuHCl-denatured state by 10 kJ/mol relative to TM + His26

Yeast iso-1-cytochrome $c$, pseudo-wild type (Cys102→Thr, w.t.*), and mutants derived from w.t.*, data at reference temperature

| Protein/Form | | pH | $T_{ref}$ | $\Delta T$ | $\Delta(\Delta G)$ | Approach/Remarks | | Ref |
|---|---|---|---|---|---|---|---|---|
| Cys102→Thr | ox. | 6.0 | 57.1 | 3.1 | 3.3 | DSC | (1,2) | 99L3 |
| | red. | 6.0 | 57.1 | 27.6 | 23.8 | DSC | (1,2) | 99L3 |
| | ox. | 4.7 | 51.4 | 2.5 | 2.1 | DSC | (1,2) | 99L3 |
| | red. | 4.7 | 51.4 | 28.5 | 22.2 | DSC | (1,2) | 99L3 |
| Thr69→Glu | ox. | 6.0 | 57.1 | 5.5 | 6.3 | DSC | (1,2) | 99L3 |
| | red. | 6.0 | 57.1 | 29.2 | 30.5 | DSC | (1,2) | 99L3 |
| | ox. | 4.7 | 51.4 | 5.7 | 5.9 | DSC | (1,2) | 99L3 |
| | red. | 4.7 | 51.4 | 30.2 | 28.0 | DSC | (1,2) | 99L3 |
| Thr96→Ala | ox. | 6.0 | 57.1 | 4.0 | 4.2 | DSC | (1,2) | 99L3 |
| | red. | 6.0 | 57.1 | 28.8 | 30.5 | DSC | (1,2) | 99L3 |
| | ox. | 4.7 | 51.4 | 3.4 | 3.3 | DSC | (1,2) | 99L3 |
| | red. | 4.7 | 51.4 | 29.3 | 22.6 | DSC | (1,2) | 99L3 |
| double mutant (Thr69→Glu and Thr96→Ala) | | | | | | | | |
| | ox. | 6.0 | 57.1 | 5.5 | 5.9 | DSC | (1,2) | 99L3 |
| | red. | 6.0 | 57.1 | 29.9 | 31.4 | DSC | (1,2) | 99L3 |
| | ox. | 4.7 | 51.4 | 5.8 | 5.4 | DSC | (1,2) | 99L3 |
| | red. | 4.7 | 51.4 | 31.1 | 27.6 | DSC | (1,2) | 99L3 |
| deletion mutant $\Delta(-5/-1)$ | | | | | | | | |
| | ox. | 6.0 | 57.1 | 0.0 | 0.0 | DSC | (1–3) | 99L3 |
| | red. | 6.0 | 57.1 | 25.9 | 18.4 | DSC | (1–3) | 99L3 |
| | ox. | 4.7 | 51.4 | −1.1 | −0.4 | DSC | (1–3) | 99L3 |
| | red. | 4.7 | 51.4 | 25.9 | 14.6 | DSC | (1–3) | 99L3 |

Remarks:
(1) the reference temperature is defined the by wild-type protein (ox.), i.e., $T_{ref}$ = 57.1°C at pH 6.0 and $T_{ref}$ = 51.4°C at pH 4.7
(2) all yeast iso-1-cytochromes have the Cys102→Thr modification
(3) amino acids are referred to as residues −5 to −1, when using the eukaryotic numbering system for cytochromes $c$

Yeast iso-1-cytochrome $c$, Cys102→Thr mutant

| Protein | pH | T | $\Delta G$ | Approach/Remarks | | Ref |
|---|---|---|---|---|---|---|
| Cys102→Thr | 4.6 | 27 | 20.5±1.3 | heat | (1–4) | 98A3 |

Remarks:
(1) calculated from heat denaturation data (see Table 2), using $\Delta Cp$ = 5.73±0.25 kJ/mol/K from Ref. 94C7
(2) Ref. 98A3 contains an analysis of the error propagation in $\Delta G$ and $\Delta H$ due to removed data points in the pre-denaturational baseline
(3) measured in 50 mM sodium acetate, pH 4.6, transition monitored by CD at 222 nm
(4) for a comparison of $\Delta G$ from equilibrium unfolding with $\Delta G_{op}$ from amide proton exchange, see Ref. 93M19

Iso-1-cytochrome *c* from *Saccharomyces cerevisiae*, Cys102→Thr mutant, stabilization upon reduction

| Mutant/Form | pH | T | ΔG | Approach/Remarks | Ref |
|---|---|---|---|---|---|
| Cys102→Thr, red. | 4.6 | 27 | 49.0±2.9 | DSC | 95C7 |
| Cys102→Thr, ox. | 4.6 | 27 | 21.3±1.7 | DSC (1) | 95C7 |

Remarks:

(1) measured in 100 mM sodium acetate, for further details see Ref. 95C7

(2) data from Ref. 94C7

(3) for differences in ΔG between the oxidized and reduced protein based on hydrogen exchange, see Ref. 99B4

Iso-1-cytochrome *c* from *Saccharomyces cerevisiae*, variants derived from Cys102→Ser, methionine mutations

| Mutant | pH | T | ΔG | m | Approach/Remarks | Ref |
|---|---|---|---|---|---|---|
| wild type* | 7.5 | 25 | 22.46±1.84 | 20.08±1.38 | GuHCl (1–4) | 98H4 |
| Lys4→Met | 7.5 | 25 | 19.54±0.46 | 17.32±0.29 | GuHCl (1–3) | 98H4 |
| Lys11→Met | 7.5 | 25 | 18.12±0.84 | 17.24±0.96 | GuHCl (1–3) | 98H4 |
| Lys22→Met | 7.5 | 25 | 17.95±1.13 | 19.83±1.46 | GuHCl (1–3) | 98H4 |
| Glu44→Met | 7.5 | 25 | 18.03±1.00 | 17.07±0.92 | GuHCl (1–3) | 98H4 |
| Ser47→Met | 7.5 | 25 | 18.70±1.92 | 16.02±1.55 | GuHCl (1–3) | 98H4 |
| Asp50→Met | 7.5 | 25 | 19.00±1.42 | 16.23±1.17 | GuHCl (1–3) | 98H4 |
| Lys54→Met | 7.5 | 25 | 16.74±0.63 | 14.06±0.75 | GuHCl (1–3) | 98H4 |
| TML72→Met | 7.5 | 25 | 18.45±0.88 | 15.77±0.79 | GuHCl (1–3,5) | 98H4 |
| Lys73→Met | 7.5 | 25 | 19.96±1.21 | 19.25±1.30 | GuHCl (1–3) | 98H4 |
| Lys89→Met | 7.5 | 25 | 20.04±1.09 | 17.87±0.88 | GuHCl (1–3) | 98H4 |
| Lys100→Met | 7.5 | 25 | 21.05±1.17 | 16.74±0.96 | GuHCl (1–3) | 98H4 |

Remarks:

(1) linear extrapolation, the slope of the pre- and postdenaturational baselines was taken into account, see also Ref. 993B6

(2) measured in 20 mM Tris, 40 mM NaCl

(3) transition monitored by CD

(4) wild type* refers to the Cys102→Ser mutant

(5) trimethyllysine in the wild-type protein

Iso-1-cytochromne *c* from yeast, wild type (Cys102) and spin-labeled protein

| Form | pH | T | ΔG | $c_{1/2}$ | m | Appr./Rem. | Ref |
|---|---|---|---|---|---|---|---|
| wild type | 5 | 5 | 32.6±2.1 | 1.40±0.13 | 22.6±1.3 | GuHCl (1–3) | 97Q1 |
| wild type | 5 | 5 | 41.8 | | | heat (4) | 97Q1 |
| Cys102-SL | 5 | 5 | 13.0±1.3 | 0.81±0.12 | 15.9±1.7 | GuHCl (1–3,5) | 97Q1 |
| Cys102-SL | 5 | 5 | 12.1±0.4 | 1.00±0.05 | 12.6±0.4 | GuHCl (5,6) | 97Q1 |
| Cys102-SL | 5 | 5 | 11.7 | | | heat (4,5) | 97Q1 |

Remarks:

(1) linear extrapolation, LEM-SB

(2) buffer: 0.1 M acetate, pH 5

(3) transition monitored by CD

(4) from heat denaturation using ΔCp = 5.73 kJ/mol/K from Ref. 94C6 and thermodynamic data listed in Table 2

(5) Cys102-SL = at Cys102 spin-labeled iso-1-cytochrome *c*

(6) transition monitored by EPR

Yeast iso-1-cytochrome $c$, pseudo-wild type (w.t.*) and mutants in the hydrophobic heme pocket region

| Protein | pH | T | $\Delta G$ | Approach/Remarks | | Ref |
|---|---|---|---|---|---|---|
| w.t.* | 6.0 | 25.0 | 23.0±1.7 | DSC | (1,2) | 99L6 |
| Phe82→Ala | 6.0 | 25.0 | 11.3±0.8 | DSC | (2,3) | 99L6 |
| Phe82→Tyr | 6.0 | 25.0 | 20.1±1.7 | DSC | (2,3) | 99L6 |
| double mutant (Phe82→Tyr and Leu85→Ala) | | | | | | |
| | 6.0 | 25.0 | 10.9±0.8 | DSC | (2,3) | 99L6 |

Remarks:

(1) pseudo-wild type iso-1-cytochrome $c$ (Cys102→Thr)

(2) measured in 0.1 M sodium phosphate, pH 6.0

(3) mutants derived from the pseudo-wild type protein (w.t.*)

Iso-1-cytochrome $c$ from yeast, omega loop A replacements, wild type* (Cys102→Thr) and mutants

| Mutant | pH | $T_{trs}$ | $\Delta G(25°C)$ | Approach | Remarks | Ref |
|---|---|---|---|---|---|---|
| wild type* | 4.6 | 51.8±1.0 | 22.2±2.5 | heat | (1,2) | 97F2 |
| RepA2 | 4.6 | 39.6±0.4 | 14.2 | heat | (1–3) | 97F2 |
| RepA2val | 4.6 | 46.3±0.9 | 18.8 | heat | (1,2,4) | 97F2 |
| RepA3 | 4.6 | 38.5 | 13.4 | heat | (1,5,6) | 97F2 |
| RepA5 | 4.6 | 33.1 | 8.8 | heat | (1,6,7) | 97F2 |

Remarks:

(1) $\Delta G(25°C)$ was calculated using $\Delta H$ and $\Delta Cp$ for wild type* (Cys102→Thr) of 345±16 kJ/mol and 5.73±0.25 kJ/mol/K, respectively, from Ref. 94C6

(2) $T_{trs}$ is the average of melting temperatures determined by CD at 222 nm, and UV at 280, 287, and 399 nm

(3) RepA2 = omega loop A replaced by $R.\ rubrum$ cyt $c_2$ residues 18–32 (position 20 Phe)

(4) RepA2val = as RepA2 with Val in position 20

(5) RepA3 = omega loop A replaced by $P.\ denitrificans$ cyt $c_{550}$ residues 19–38

(6) $T_{trs}$ from CD at 222 nm

(7) RepA5 = omega loop A replaced by esterase residues 216–224

*b) transitions of iso-1-cytochrome c*

Iso-1-cytochrome $c$

Yeast iso-1-cytochrome $c$, N → D transition of wild type and mutants of Ala7

| Mutant | pH | $T_{trs}$ | $\Delta T$ | $\Delta G(54.7°C)$ | Approach/Remarks | Ref |
|---|---|---|---|---|---|---|
| wild type | 4.6 | 54.7 | 0.0 | 0.0 | heat, v.H. (1–4) | 98M23 |
| Ala7→Gly | 4.6 | 50.7 | –4.0 | –4.44±1.30 | heat, v.H. (1–4) | 98M23 |
| Ala7→Leu | 4.6 | 53.5 | –1.2 | –1.30±1.21 | heat, v.H. (1–4) | 98M23 |
| Ala7→Phe | 4.6 | 54.4 | –0.3 | –0.33±1.26 | heat, v.H. (1–4) | 98M23 |
| Ala7→Tyr | 4.6 | 55.9 | 1.2 | 1.34±1.21 | heat, v.H. (1–4) | 98M23 |

Remarks:

(1) the Cys102→Thr variant is referred to as the wild-type protein

(2) measured in 50 mM sodium acetate, pH 4.6

(3) transition monitored by CD at 222 nm

(4) $\Delta G$ refers to $T_{trs}$ of the wild-type protein (54.7°C)

Yeast iso-1-cytochrome *c*, A → D transition of the interface variants in 0.33 M Na$_2$SO$_4$/H$_2$SO$_4$, pH 1.8–2.0

| Mutant | pH | $T_{trs}$ | $\Delta G_{ex}$ | $\Delta G_{BS}$ | Appr./Rem. | Ref |
|---|---|---|---|---|---|---|
| Phe10→Trp | 2.0 | 22.7 | −3.68 | −5.19±0.79 | heat (1–5) | 98M23 |
| | 1.9 | 22.3 | −3.01 | −3.56±0.79 | heat (1–5) | 98M23 |
| | 1.8 | 20.4 | −3.22 | −5.06±1.09 | heat (1–5) | 98M23 |
| Phe10→Tyr | 2.0 | 25.5 | −3.51 | −3.72±0.79 | heat (1–5) | 98M23 |
| | 1.9 | 24.8 | −3.05 | −3.93±0.92 | heat (1–5) | 98M23 |
| | 1.8 | 23.2 | −3.85 | −3.72±1.05 | heat (1–5) | 98M23 |
| Leu94→Ala | 2.0 | <−1 | | <−18.0 | heat (1–5) | 98M23 |
| Leu94→Thr | 2.0 | <−10 | | <−12.5 | heat (1–5) | 98M23 |
| | 1.8 | <−1 | | <−15.5 | heat (1–5) | 98M23 |
| Leu94→Val | 2.0 | 30.9 | −2.05 | −1.97±0.79 | heat (1–5) | 98M23 |
| | 1.9 | 27.2 | −1.92 | −2.05±0.84 | heat (1–5) | 98M23 |
| | 1.8 | 25.8 | −2.26 | −2.59±1.05 | heat (1–5) | 98M23 |
| Tyr97→Phe | 2.0 | 30.0 | −1.13 | −1.55±0.84 | heat (1–5) | 98M23 |
| | 1.9 | 31.3 | −0.84 | −1.59±0.88 | heat (1–5) | 98M23 |
| | 1.8 | 27.4 | −1.00 | −1.46±1.05 | heat (1–5) | 98M23 |

Remarks:

(1) mutants were derived from the Cys102→Thr variant which is referred to as the wild-type protein

(2) the A form measured in 0.33 M Na$_2$SO$_4$/H$_2$SO$_4$, pH 1.8–2.0

(3) transition monitored by CD at 222 nm

(4) $\Delta G_{ex}$ was calculated examining −RT ln K$_D$ versus T at T$_{trs}$ of the wild-type protein under the same conditions

(5) $\Delta G_{BS}$ was calculated using $\Delta G_{BS} = (\Delta H_{w.t.}/T_{trs,w.t.})/(T_{trs,mut}-T_{trs,w.t.})$

Yeast iso-1-cytochrome *c*, A → D transition of the non-interface variants in 0.33 M Na$_2$SO$_4$/H$_2$SO$_4$, pH 1.8–2.0

| Mutant | pH | $T_{trs}$ | $\Delta G_{ex}$ | $\Delta G_{BS}$ | Appr./Rem. | Ref |
|---|---|---|---|---|---|---|
| Ala7→Gly | 2.0 | 22.6 | −3.47 | −4.18±0.75 | heat (1–5) | 98M23 |
| | 1.9 | 23.7 | −3.56 | −4.02±0.92 | heat (1–5) | 98M23 |
| | 1.8 | 18.2 | −3.47 | −4.10±0.96 | heat (1–5) | 98M23 |
| Ala7→Leu | 2.0 | 32.9 | −1.13 | −0.96±0.79 | heat (1–5) | 98M23 |
| | 1.9 | 28.6 | −1.34 | −0.96±0.75 | heat (1–5) | 98M23 |
| | 1.8 | 29.6 | −1.09 | −1.09±0.84 | heat (1–5) | 98M23 |
| Ala7→Phe | 2.0 | 29.9 | −1.63 | −0.84±0.75 | heat (1–5) | 98M23 |
| | 1.9 | 30.8 | −0.50 | −0.50±0.84 | heat (1–5) | 98M23 |
| | 1.8 | 27.7 | −0.25 | −0.33±1.17 | heat (1–5) | 98M23 |
| Ala7→Tyr | 2.0 | 33.5 | 0.25 | 0.42±0.79 | heat (1–5) | 98M23 |
| | 1.9 | 32.0 | 0.25 | 0.08±0.84 | heat (1–5) | 98M23 |
| | 1.8 | 31.5 | 0.08 | 0.42±1.00 | heat (1–5) | 98M23 |

Remarks:

(1) mutants were derived from the Cys102→Thr variant which is referred to as the wild-type protein

(2) the A form measured in 0.33 M Na$_2$SO$_4$/H$_2$SO$_4$, pH 1.8–2.0

(3) transition monitored by CD at 222 nm

(4) $\Delta G_{ex}$ was calculated examining −RT ln K$_D$ versus T at T$_{trs}$ of the wild-type protein under the same conditions

(5) $\Delta G_{BS}$ was calculated using $\Delta G_{BS} = (\Delta H_{w.t.}/T_{trs,w.t.})/(T_{trs,mut}-T_{trs,w.t.})$

Yeast iso-1-cytochrome $c$, A $\rightarrow$ D transition of the interface variants at enhanced sodium sulfate concentration (>0.33 M)

| Mutant | $[SO_4]^{2-}$ | pH | $T_{trs}$ | $\Delta G_{ex}$ | $\Delta G_{BS}$ | Appr./Rem. | Ref |
|---|---|---|---|---|---|---|---|
| Phe10$\rightarrow$Trp | 1.00 M | 1.9 | 44.9 | −5.15 | −5.02 | heat (1–5) | 98M23 |
| | 0.68 M | 1.9 | 38.7 | −4.23 | −4.18 | heat (1–5) | 98M23 |
| Leu94$\rightarrow$Ala | 1.00 M | 1.9 | 25.4 | | −19.7±4.2 | heat (1–5) | 98M23 |
| Leu94$\rightarrow$Thr | 1.00 M | 2.0 | 26.2 | | −13.8 | heat (1–5) | 98M23 |
| | 1.00 M | 1.8 | 21.9 | | −16.7 | heat (1–5) | 98M23 |
| | 0.67 M | 2.0 | 14.5 | | −16.3 | heat (1–5) | 98M23 |
| Tyr97$\rightarrow$Ala | 1.00 M | 2.0 | ≤25 | | <−18.4 | heat (1–5) | 98M23 |
| | 1.00 M | 1.9 | ≤25 | | <−19.2 | heat (1–5) | 98M23 |

Remarks:
(1) mutants derived from the Cys102$\rightarrow$Thr variant which is referred to as the wild-type protein
(2) the A form measured in $Na_2SO_4/H_2SO_4$, pH 1.8-2.0, at the indicated sulfate conc.
(3) transition monitored by CD at 222 nm
(4) $\Delta G_{ex}$ was calculated examining $-RT \ln K_D$ versus T at $T_{trs}$ of the wild-type protein under the same conditions
(5) $\Delta G_{BS}$ was calculated using $\Delta G_{BS} = (\Delta H_{w.t.}/T_{trs,w.t.})/(T_{trs,mut}-T_{trs,w.t.})$

Yeast iso-1-cytochrome $c$, mutants in position 68 derived from the pseudo-wild type (Cys102$\rightarrow$Thr), N $\rightarrow$ D and N $\rightarrow$ A transitions

| Protein | pH | T | $\Delta(\Delta G)$ | Approach/Remarks | | Ref |
|---|---|---|---|---|---|---|
| N $\rightarrow$ D transition: | | | | | | |
| wild type | 4.6 | 54 | 0.00±0.21 | heat | (1–3) | 99H14 |
| Leu68$\rightarrow$Ile | 4.6 | 54 | −6.23±0.29 | heat | (1–3) | 99H14 |
| Leu68$\rightarrow$Val | 4.6 | 54 | −10.13±0.33 | heat | (1–3) | 99H14 |
| Leu68$\rightarrow$Met | 4.6 | 54 | −11.21±0.25 | heat | (1–3) | 99H14 |
| N $\rightarrow$ A transition: | | | | | | |
| wild type | 4.6 | 39 | 0.0±0.4 | heat | (2,4,5) | 99H14 |
| Leu68$\rightarrow$Ile | 4.6 | 39 | −4.2±1.3 | heat | (1,4,5) | 99H14 |
| Leu68$\rightarrow$Val | 4.6 | 39 | −7.5±1.7 | heat | (1,4,5) | 99H14 |
| Leu68$\rightarrow$Met | 4.6 | 39 | −5.9±1.3 | heat | (1,4,5) | 99H14 |

Remarks:
(1) measured in 0.05 M sodium acetate
(2) transition monitored by ellipticity at 222 nm
(3) for the procedure see also Table 2 and Ref. 95P3
(4) measured in 0.5 M $Na_2/H_2SO_4$, pH 2.1
(5) $\Delta(\Delta G)$ was calculated using $\Delta(\Delta G) = \Delta T_{mut} \times \Delta H_{w.t.}/T_{trs.w.t.}$

## Iso-2-cytochrome $c$

Yeast iso-2-cytochrome $c$, recombinant, wild type and Asn52 mutant

| Mutant | pH | T | $\Delta G$ | $c_{1/2}$ | m | Appr./Rem. | | Ref |
|---|---|---|---|---|---|---|---|---|
| wild type | 6.0 | 20 | 19.2±1.3 | 1.5±0.2 | 12.6±0.8 | GuHCl | (1–3) | 98M11 |
| Asn52$\rightarrow$Ile | 6.0 | 20 | 35.6±2.1 | 2.2±0.2 | 16.3±0.8 | GuHCl | (1–3) | 98M11 |

Remarks:
(1) linear extrapolation, LEM-SB
(2) transition monitored by fluorescence
(3) buffer: 0.1 M sodium phosphate, pH 6.0

Yeast iso-2-cytochrome *c*, wild type and His double mutant

| Mutant | pH | T | $\Delta G$ | $c_{1/2}$ | m | Appr./Rem. | Ref |
|---|---|---|---|---|---|---|---|
| wild type | 6.0 | 20 | 18.8±1.7 | 1.5±0.2 | 12.6±0.8 | GuHCl (1–3) | 97P11 |
| double mutant (His33→Asn and His39→Lys) | | | | | | | |
| | 6.0 | 20 | 26.4±1.7 | 1.5±0.2 | 18.0±0.8 | GuHCl (1–3) | 97P11 |

Remarks:

(1) linear extrapolation, LEM-SB

(2) buffer: 0.1 M sodium phosphate, pH 6.0

(3) transition monitored by fluorescence emission at 350 nm following excitation at 285 nm

## Cytochrome Oxidase

$Cu_A$ domain of cytochrome oxidase from *Thermus thermophilus*

| Protein | pH | T | $\Delta G$ | $c_{1/2}$ | Approach/Remarks | | Ref |
|---|---|---|---|---|---|---|---|
| reduced | 7.0 | 20 | 65±10 | 6.5 | GuHCl | (1–3) | 98W12 |
| reduced | 7.0 | 20 | | 6.7 | GuHCl | (3,4) | 98W12 |
| reduced | 7.0 | 75 | 18±3 | 3.0 | GuHCl | (1–3) | 98W12 |
| oxidized | 7.0 | 20 | 85±10 | | GuHCl | (5) | 98W12 |
| oxidized | 7.0 | 75 | 45±6 | 6.5 | GuHCl | (1–3) | 98W12 |
| oxidized | 7.0 | 75 | | 6.5 | GuHCl | (6) | 98W12 |

Remarks:

(1) linear extrapolation

(2) transition monitored by CD at 218 nm

(3) $c_{1/2}$ was taken from the Figures in Ref. 98W12

(4) transition monitored by fluorescence

(5) $\Delta G$ for the oxidized form at 20°C was calculated using a thermodynamic cycle based on $\Delta G$ for the reduced protein and the redox potential (see also Ref. I–81P)

(6) transition monitored by optical absorption at 530 nm

## Cytochrome *c* Peroxidase

Cytochrome *c* peroxidase from *E. coli*, recombinant

| | pH | T | $\Delta G$ | $c_{1/2}$ | m | Approach/Remarks | | Ref |
|---|---|---|---|---|---|---|---|---|
| | 7.0 | 25 | 25.9±1.3 | 4.2 | 6.3 | urea | (1,2) | 98T11 |
| | 7.0 | 25 | 26.8±2.1 | 1.9 | 14.2 | GuHCl | (1,2) | 98T11 |

Remarks:

(1) linear extrapolation

(2) transition monitored by CD at 222 nm

## Cytokine

Recombinant human megakaryocyte growth and development factor (rHuMGDF)

| Protein | pH | T | $\Delta G$ | $c_{1/2}$ | Approach/Remarks | Ref |
|---|---|---|---|---|---|---|
| rHuMGDF | 6 | 23 | 39.7±2.1 | 4.9±0.1 | urea, FL  (1) | 98H2 |
|  | 6 | 23 | 21.8±2.1 | 6.9±0.1 | urea, CD (2,3) | 98H2 |

Remarks:

(1) transition monitored by fluorescence wavelength maximum

(2) transition monitored by CD at 224 nm

(3) urea-induced unfolding at pH ~5.5 results in a multiphase transition as detected by CD

## De Novo Designed Proteins/Peptides

Design of a native-like three-helix bundle

| Protein | pH | T | $\Delta G$ | $c_{1/2}$ | m | Appr./Rem. | Ref |
|---|---|---|---|---|---|---|---|
| α3B | 7.3 | 25 | 30.1 | 4.2 | 7.1 | GuHCl  (1–4) | 98B9 |
| α3C | 7.3 | 25 | 23.0 | 2.6 | 8.8 | GuHCl  (1–4) | 98B9 |

Remarks:

(1) for details of the design see Ref. 98B9

(2) linear extrapolation, LEM-SB

(3) transition monitored by CD at 222 nm

(4) buffer: 25 mM MOPS, 100 mM NaCl, pH 7.3

De novo designed ROP-like four-helix bundles

| Form | pH | T | $\Delta G$ | Approach | Remarks | Ref |
|---|---|---|---|---|---|---|
| RLP-1 | 6.97 | 27 | 36.4 | heat, v.H. | (1,2) | 97B8 |
| RLP-1 | 6.97 | 27 | 36.4 | heat, v.H. | (1,3) | 97B8 |
| RLP-2 | 6.97 | 27 | 35.1 | heat, v.H. | (1,2) | 97B8 |
| RLP-2 | 6.97 | 27 | 35.6 | heat, v.H. | (1,3) | 97B8 |
| RLP-3 | 6.97 | 27 | 40.6 | heat, v.H. | (1,2) | 97B8 |
| RLP-3 | 6.97 | 27 | 41.4 | heat, v.H. | (1,3) | 97B8 |

Remarks:

(1) 51-residue helix-turn-helix peptides that form dimers in solution. RLP-1, –2, and –3 designate the number of Val residues in the second helix

(2) van't Hoff treatment, $\Delta G$ was calculated using $\Delta C_p$ from $\Delta H$ versus $T_{trs}$, see Table 2

(3) fit of entire melting curves by the Gibbs-Helmholtz equation, see Table 2

DHP1, 108 amino acid designed four-helix bundle protein, constructed from a reduced alphabet of seven amino acids

| | pH | T | $\Delta G$ | $c_{1/2}$ | m | Appr./Rem. | Ref |
|---|---|---|---|---|---|---|---|
| | 7 | 25 | 38.9±1.7 | 4.63 | 9.16±0.33 | GuHCl  (1,2) | 97S4 |
| | 7 | 7 | 34–42 | | | HX  (3) | 97S4 |

Remarks:

(1) linear extrapolation

(2) transition monitored by CD at 222 nm

(3) from hydrogen/deuterium exchange derived protection factors

MB-1, engineered four-α-helical bundle ('milk-bundle' protein), introduction of potential helix-capping residues

| Protein | pH | T | $\Delta G$ | m | Approach/Remarks | | Ref |
|---|---|---|---|---|---|---|---|
| MB-1 | 7.2 | 4 | 29.2±1.3 | 4.6±0.8 | urea | (1–5) | 98P4 |
| MB-3 | 7.2 | 4 | 36.4±1.7 | 7.1±1.3 | urea | (1–5) | 98P4 |
| MB-5 | 7.2 | 4 | 34.7±2.1 | 6.3±1.7 | urea | (1–5) | 98P4 |
| MB-11 | 7.2 | 4 | 39.3±5.8 | 8.4±4.6 | urea | (1–5) | 98P4 |
| MB-13 | 7.2 | 4 | 40.6±2.5 | 8.8±1.7 | urea | (1–5) | 98P4 |

Remarks:

(1) linear extrapolation, assuming an equilibrium between folded dimer and unfolded monomer

(2) buffer: 100 mM potassium phosphate, pH 7.2

(3) transition monitored by CD at 222 nm

(4) for the mutagenesis procedure and mutant description see Ref. 98P4

(5) the same relative stabilities of the variants were found when a folded trimer to unfolded monomer model was used to calculate $\Delta G$

Synthetic tetrahelix bundle heme protein maquettes

| Protein | pH | T | $\Delta G$ | Approach/Remarks | | Ref |
|---|---|---|---|---|---|---|
| heme-$(\alpha ss\alpha)_2$ | 4.0 | 25 | 81.2 | GuHCl | (1–3) | 98S12 |
| | 8.5 | 25 | 60.7 | GuHCl | (1,2,4) | 98S12 |
| heme-$(\alpha ss\alpha)_2$, mutant (Glu11→Gln) | | | | | | |
| | 4.0 | 25 | 77.8 | GuHCl | (1–3) | 98S12 |
| | 8.5 | 25 | 61.9 | GuHCl | (1,2,4) | 98S12 |
| heme-$(\alpha ss\alpha)_2$, multiple mutant (Glu11→Gln, Glu18→Gln, and Glu25→Gln) | | | | | | |
| | 4.0 | 25 | 62.3 | GuHCl | (1–3) | 98S12 |
| | 8.5 | 25 | 59.8 | GuHCl | (1,2,4) | 98S12 |
| heme-$(\alpha ss\alpha)_2$, mutant (Glu[a11]→Gln) | | | | | | |
| | 4.0 | 25 | 60.2 | GuHCl | (1–3) | 98S12 |
| | 8.5 | 25 | 61.9 | GuHCl | (1,2,4) | 98S12 |

Remarks:

(1) linear extrapolation, dimer-folded to monomer-unfolded equilibrium model

(2) transition monitored by CD at 222 nm

(3) buffer: 50 mM citric acid, 100 mM NaCl

(4) buffer: 50 mM Tris-HCl, 100 mM NaCl

De novo designed globin 1 (DG1), compared with horse-heart myoglobin (Mb)

| Protein | pH | T | $\Delta G$ | $c_{1/2}$ | m | Appr./Rem. | Ref |
|---|---|---|---|---|---|---|---|
| DG1, apo | 8.0 | 25 | 11.7±0.4 | 4.9 | 2.38±0.04 | GuHCl (1–3) | 99I4 |
| DG1, holo | 8.0 | 25 | 13.0±0.4 | | 2.64±0.08 | GuHCl (1–4) | 99I4 |
| Mb, apo | 8.0 | 25 | 12.6±1.3 | 1.0 | 12.55±1.26 | GuHCl (1–3) | 99I4 |
| Mb, holo | 8.0 | 25 | 29.7±1.3 | | 19.20±0.80 | GuHCl (1–3) | 99I4 |
| | 8.0 | 25 | 28.9±2.1 | | 18.80±1.30 | GuHCl (1,3,5) | 99I4 |

Remarks:

(1) linear extrapolation

(2) transition monitored by CD at 222 nm

(3) buffer: 10 mM Tris-HCl, 200 mM NaCl, pH 8.0

(4) the transition monitored by Soret absorption at 412 nm reveals two transitions with $\Delta G_1$ = 3.8±0.4 kJ/mol and $m_1$ = 5.9±0.4 kJ/mol/M, and $\Delta G_2$ = 11.3±0.8 kJ/mol and $m_2$ = 3.2±0.3 kJ/mol/M

(5) transition monitored by Soret absorption at 412 nm

De novo designed four-α-helix bundle, histidine placements at coiled coil heptad *a* positions

| Peptide | pH | T | $\Delta G$ | $c_{1/2}$ | m | Appr./Rem. | Ref |
|---|---|---|---|---|---|---|---|
| [His10,His24]$_2$ | 8.0 | 25 | 72.30 | 5.04 | 8.16 | GuHCl (1–3) | 99G6 |
| [His10,Ala24]$_2$ | 8.0 | 25 | 104.10 | 5.59 | 13.05 | GuHCl (1–3) | 99G6 |
| [Ala10,His24]$_2$ | 8.0 | 25 | 83.89 | 6.36 | 8.28 | GuHCl (1–3) | 99G6 |
| [Ala10,Ala24]$_2$ | 8.0 | 25 | 97.86 | 6.26 | 10.67 | GuHCl (1–3) | 99G6 |
| [His9,Ala24]$_2$ | 8.0 | 25 | >125 | 6.68 | 15.23 | GuHCl (1–3) | 99G6 |
| [His10,Ala24]$_2$ | 8.0 | 25 | 104.10 | 5.59 | 13.05 | GuHCl (1–3) | 99G6 |
| [His11,Ala24]$_2$ | 8.0 | 25 | | 7.53 | | GuHCl (1–3) | 99G6 |
| [His12,Ala24]$_2$ | 8.0 | 25 | >105 | 7.09 | | GuHCl (1–3) | 99G6 |
| [His13,Ala24]$_2$ | 8.0 | 25 | 77.57 | 4.65 | 10.04 | GuHCl (1–3) | 99G6 |
| [His14,Ala24]$_2$ | 8.0 | 25 | 81.96 | 4.99 | 10.21 | GuHCl (1–3) | 99G6 |
| [His15,Ala24]$_2$ | 8.0 | 25 | | >8 | | GuHCl (1–3) | 99G6 |

Remarks:

(1) for the sequence of the heme protein maquette, see Ref. 99G6

(2) linear extrapolation, LEM-SB, applied to a dimer folded to two monomer-unfolded transition

(3) transition monitored by far-UV CD at 222 nm

(4) buffer: 10 mM potassium phosphate, 100 mM KCl, pH 8.0

Four-α-helix bundle protein, with affinity for binding volatile anesthetic

| Protein | pH | T | $\Delta G$ | m | Appr./Rem. | Ref |
|---|---|---|---|---|---|---|
| (Aα2)$_2$ | 7.0 | 25.0±0.1 | 59.8±3.3 | 8.4±0.8 | GuHCl (1–3) | 98J1 |
| (Aα2-MTSSL)$_2$ | 7.0 | 25.0±0.1 | 64.0±2.1 | 7.9±0.1 | GuHCl (1–3) | 98J1 |
| (Aα2-MTSSL)$_2$ | 7.0 | 25.0±0.1 | 66.5±3.3 | 7.0±0.8 | GuHCl (1,2,4) | 98J1 |
| (Lα2)$_2$ | 7.0 | 25.0±0.1 | >125 | | GuHCl (1–3) | 98J1 |

Explanation:

(Aα2)$_2$ is built of two 62-residue peptides, each composed of two 27-residue α-helical segments connected by an eight-residue flexible glycine linker

(Aα2-MTSSL)$_2$ is spin-labeled by (1-oxyl-2,2,5,5-tetramethylpyrroline-3-methyl)methanethiosulfonate

(Lα2)$_2$ is a peptide analogous to (Aα2)$_2$, but contains a core made up predominantly of leucine residues

Remarks:

(1) linear extrapolation based on a model assuming that the native dimer (four-α-helix bundle) unfolds into two denatured monomers

(2) measured in 10 mM potassium phosphate

(3) transition monitored by CD at 222 nm

(4) transition monitored by EPR

De novo designed proteins as models of radical enzymes

| Protein | pH | T | $\Delta G$ | $c_{1/2}$ | m | Appr./Rem. | Ref |
|---|---|---|---|---|---|---|---|
| $\alpha_3$W | 7.0 | 25 | 25.9 | 2.2 | 11.7 | GuHCl (1–3) | 99T11 |
| | 6.85 | 25 | 21.3–26.4 | | | H/D (1,5,6) | 99T11 |
| $\alpha_3$Y | 7.0 | 25 | 22.6 | 1.9 | 11.7 | GuHCl (1–3) | 99T11 |
| | 6.85 | 25 | 22.6–27.2 | | | H/D (1,5,6) | 99T11 |
| $\alpha_3$–1 | 7.0 | 25 | 19.2±1.3 | 2.1 | 9.2±0.8 | GuHCl (1,3,7) | 98J2 |

Remarks:

(1) $\alpha_3$W and $\alpha_3$Y contain 65 residues and fold into thermodynamically stable α-helical conformations. $\alpha_3$W and $\alpha_3$Y differ in position 32 ($\alpha_3$W contains Trp32, and $\alpha_3$Y Tyr32)

(2) linear extrapolation

(3) transition monitored by far-UV CD at 222 nm

(4) buffer: 10 mM potassium phosphate, 10 mM KCl, pH 7.0

(5) transition monitored by NMR, slowest exchanging amide protons in H/D exchange experiments

(6) uncorrected pH-meter reading

(7) $\alpha_3$-1 is the original molecule

Betanova, designed 20 amino acid, three-stranded β-sheet protein

| pH | T | ΔG | m | Approach/Remarks | | Ref |
|---|---|---|---|---|---|---|
| 5.0 | 5 | 2.9 | 1.7 | urea | (1,2) | 98K12 |
| 5.0 | 5 | 2.5 | | heat | (3–5) | 98K12 |

Remarks:

(1) linear extrapolation

(2) transition monitored by fluorescence

(3) thermal transition monitored by CD at 217 nm

(4) buffer: 5 mM sodium acetate with 10% glycerol

(5) ΔG was calculated using $\Delta H^{v.H.} = 13.8$ kJ/mol at 5°C and $\Delta C_p = 0.59$ kJ/mol/K (estimated value)

De novo synthesized proteins albebetin and albeferon

| Protein | pH | T | ΔG | $c_{1/2}$ | m | Approach/Remarks | | Ref |
|---|---|---|---|---|---|---|---|---|
| albebetin | 7.6 | 23 | 12.1±1.3 | 4.6 | 2.34±0.29 | urea | (1–3) | 98A5 |
| albeferon | 7.6 | 23 | 21.8±1.3 | 5.6 | 4.14±0.29 | urea | (1–4) | 98A5 |

Remarks:

(1) linear extrapolation

(2) transition monitored by size exclusion chromatography, see Ref. 96D4

(3) albebetin consists of a four-stranded antiparallel β-sheet with one site screened by two α-helices, for the sequence see Ref. 96D4

(4) albeferon consists of albebetin with an eight amino acid fragment (residues 131–138) of human interferon α2 attached to the N-terminus of albebetin, see Ref. 96D4

Minimal peptides (Min-21, Min-23) that fold like a cysteine-stabilized β-sheet motif

| Protein | pH | ΔG | Approach | Rem. | Ref |
|---|---|---|---|---|---|
| Min-23 | 2.7 | 8.8 | heat, NMR, Cys[21]Hβ | (1,2) | 99H5 |
| | 2.7 | 10.5 | heat, NMR, Leu[6]Hδ | (1,2) | 99H5 |
| | 2.7 | 10.5 | heat, NMR, Gly[25]Hα2 | (1,2) | 99H5 |
| | 2.7 | 7.9 | heat, NMR, Phe[26]H2,H6 | (1,2) | 99H5 |
| | 2.7 | 14.2 | heat, NMR, Phe[26]Hα | (1,2) | 99H5 |
| Min-21 | 2.7 | 6.3 | heat, NMR, Leu[6]Hδ | (1,2) | 99H5 |
| | 2.7 | 5.9 | heat, NMR, Gly[25]Hα2 | (1,2) | 99H5 |
| | 2.7 | 5.0 | heat, NMR, Phe[26]H2,H6 | (1,2) | 99H5 |
| | 2.7 | 5.0 | heat, NMR, Phe[26]Hα | (1,2) | 99H5 |
| EETI II | 2.7 | 11.3 | heat, NMR, Cys[21]Hβ | (1,3) | 99H5 |

Remarks:

(1) data from thermal denaturation followed by 1D NMR

(2) Min-23 and Min-21 = 23 and 21 residue peptide, respectively that corresponds to the cysteine-stabilized β-sheet motif

(3) EETI II = *Ecballium elaterium* trypsin inhibitor II

## Dihydrofolate Reductase

*E. coli* (*Ec*DHFR) and murine (*Mu*DHFR) dihydrofolate reductase

| Protein | pH | T | $\Delta G$ | $c_{1/2}$ | m | Appr./Rem. | Ref |
|---------|-----|----|------|------|------|------------|--------|
| *Mu*DHFR | 7.2 | 8 | 34.7 | 2.5 | 14.0 | urea (1–6) | 99C11 |
| | 7.2 | 12 | 33.7 | | 13.3 | urea (1–6) | 99C11 |
| | 7.2 | 17 | 31.0 | | 11.5 | urea (1–6) | 99C11 |
| | 7.2 | 22 | 27.2 | | 10.2 | urea (1–6) | 99C11 |
| | 7.2 | 26 | 26.4 | | 10.2 | urea (1–6) | 99C11 |
| | 7.2 | 30 | 22.2 | | 8.7 | urea (1–6) | 99C11 |
| | 7.2 | 37 | 19.9 | | 7.9 | urea (1–6) | 99C11 |
| | 7.2 | 42 | 18.2 | 2.5 | 7.0 | urea (1–6) | 99C11 |
| *Ec*DHFR | 7.2 | 8 | 33.5 | 4.0 | 8.7 | urea (1–6) | 99C11 |
| | 7.2 | 12 | 33.5 | | 8.3 | urea (1–6) | 99C11 |
| | 7.2 | 17 | 35.6 | | 9.0 | urea (1–6) | 99C11 |
| | 7.2 | 22 | 35.6 | | 9.0 | urea (1–6) | 99C11 |
| | 7.2 | 26 | 32.4 | | 8.8 | urea (1–6) | 99C11 |
| | 7.2 | 30 | 31.4 | | 9.0 | urea (1–6) | 99C11 |
| | 7.2 | 37 | 23.0 | | 6.9 | urea (1–6) | 99C11 |
| | 7.2 | 42 | 17.8 | 3.9 | 6.4 | urea (1–6) | 99C11 |

Remarks:

(1) linear extrapolation, LEM-SB

(2) transitions monitored by fluorescence at 320 nm (*Mu*DHFR) and 380 nm (*Ec*DHFR)

(3) buffer: 50 mM potassium phosphate, 100 mM KCl, 1 mM DTT

(4) the present data were taken from Figs. 1 and 5 in Ref. 99C11

(5) Ref. 99C11 contains additional kinetic data which characterize folding intermediates

(6) Ref. 99C12 describes the interaction of *Mu*DHFR and *Ec*DHFR folding intermediates with GroEL

Dihydrofolate reductase from the hyperthermophilic bacterium *Thermotoga maritima* (*Tm*DHFR)

| Protein | pH | T | $\Delta G$ | $c_{1/2}$ | m | Appr./Rem. | Ref |
|---------|-----|----|-------|------|------|-------------|------|
| *Tm*DHFR | 7.8 | 5 | 155.5 | 2.23 | 55.4 | GuHCl (1–3) | 99D1 |
| | 7.8 | 15 | 141.0 | 2.75 | 39.3 | GuHCl (1–3) | 99D1 |
| | 7.8 | 15 | 141.0 | 2.85 | 39.3 | GuHCl (1–4) | 99D1 |
| | 7.8 | 15 | 144.4 | 5.45 | 19.7 | urea (1–3) | 99D1 |
| | 7.8 | 25 | 125.9 | 3.07 | 29.9 | GuHCl (1–3) | 99D1 |
| | 7.8 | 37 | 118.5 | 3.36 | 25.0 | GuHCl (1–3) | 99D1 |
| | 7.8 | 55 | 116.2 | 3.01 | 26.1 | GuHCl (1–3) | 99D1 |
| | 7.8 | 55 | | 3.21 | | GuHCl (1–4) | 99D1 |
| | 7.8 | 70 | 114.3 | 2.38 | 31.6 | GuHCl (1–3) | 99D1 |

Remarks:

(1) linear extrapolation for a dimer to monomer transition $N_2 \rightarrow 2U$

(2) transition monitored by far-UV CD at 222 nm, fluorescence emission at 330 nm, and optical absorption at 280 nm

(3) measured at 0.52 μM *Tm*DHFR conc. in 10 mM potassium phosphate buffer, 0.2 mM EDTA, pH 7.8

(4) measured at 5.2 μM *Tm*DHFR conc.

Dihydrofolate reductase from *E. coli*, mutants at Gly67 and Gly121

| Mutant/Form | pH | T | ΔG | m | $c_{1/2}$ | Appr./Rem. | Ref |
|---|---|---|---|---|---|---|---|
| wild type | 7.0 | 15 | 25.44±0.75 | 8.20±0.25 | 3.11 | urea (1–4) | 94G8 |
| Gly67→Ala | 7.0 | 15 | 27.61±0.33 | 9.79±0.13 | 2.86 | urea (1–4) | 96O5 |
| Gly67→Asp | 7.0 | 15 | 24.85±0.50 | 9.50±0.17 | 2.70 | urea (1–4) | 96O5 |
| Gly67→Cys | 7.0 | 15 | 23.93±0.50 | 8.58±0.17 | 2.86 | urea (1–4) | 96O5 |
| Gly67→Leu | 7.0 | 15 | 23.60±0.71 | 8.33±0.25 | 2.78 | urea (1–4) | 96O5 |
| Gly67→Ser | 7.0 | 15 | 25.73±0.75 | 9.12±0.25 | 3.03 | urea (1–4) | 96O5 |
| Gly67→Tyr | 7.0 | 15 | 24.56±0.21 | 9.37±0.21 | 2.72 | urea (1–4) | 96O5 |
| Gly67→Val | 7.0 | 15 | 21.51±0.71 | 7.45±0.21 | 2.89 | urea (1–4) | 96O5 |
| Gly121→Ala | 7.0 | 15 | 27.53±0.71 | 9.62±0.25 | 2.86 | urea (1–4) | 94G8 |
| Gly121→Asp | 7.0 | 15 | 26.23±0.63 | 9.71±0.13 | 2.70 | urea (1–4) | 94G8 |
| Gly121→Cys | 7.0 | 15 | 25.82±0.92 | 9.04±0.29 | 2.86 | urea (1–4) | 94G8 |
| Gly121→His | 7.0 | 15 | 24.98±0.96 | 9.46±0.38 | 2.64 | urea (1–4) | 94G8 |
| Gly121→Leu | 7.0 | 15 | 22.80±0.71 | 8.20±0.25 | 2.78 | urea (1–4) | 94G8 |
| Gly121→Ser | 7.0 | 15 | 25.48±0.67 | 8.41±0.21 | 3.03 | urea (1–4) | 94G8 |
| Gly121→Tyr | 7.0 | 15 | 24.60±0.75 | 9.04±0.25 | 2.72 | urea (1–4) | 94G8 |
| Gly121→Val | 7.0 | 15 | 21.30±0.84 | 8.03±0.33 | 2.65 | urea (1–4) | 94G8 |
| double mutants Gly67→Val (G67V) and Gly121→X: | | | | | | | |
| G67V + Gly121→Ala | 7.0 | 15 | 28.16±0.59 | 8.37±0.17 | 3.36 | urea (1–3) | 98O4 |
| G67V + Gly121→Asp | 7.0 | 15 | 19.96±0.54 | 7.99±0.21 | 2.50 | urea (1–3) | 98O4 |
| G67V + Gly121→Cys | 7.0 | 15 | 20.00±0.54 | 7.41±0.17 | 2.69 | urea (1–3) | 98O4 |
| G67V + Gly121→His | 7.0 | 15 | 21.13±0.54 | 8.41±0.21 | 2.51 | urea (1–3) | 98O4 |
| G67V + Gly121→Leu | 7.0 | 15 | 21.21±0.59 | 7.74±0.21 | 2.73 | urea (1–3) | 98O4 |
| G67V + Gly121→Ser | 7.0 | 15 | 24.52±0.59 | 9.00±0.21 | 2.72 | urea (1–3) | 98O4 |
| G67V + Gly121→Tyr | 7.0 | 15 | 29.62±0.71 | 9.16±0.25 | 3.24 | urea (1–3) | 98O4 |
| G67V + Gly121→Val | 7.0 | 15 | 21.55±0.42 | 8.58±0.17 | 2.51 | urea (1–3) | 98O4 |

Remarks:
(1) linear extrapolation
(2) transition monitored by CD at 222 nm
(3) buffer: 10 mM potassium phosphate, 0.1 mM EDTA, 1.4 mM 2-mercaptoethanol
(4) see also Ref. 98O4

Dihydrofolate reductase from *E. coli*, cysteine-free double mutant (AS DHFR) (Cys85→Ala and Cys152→Ser), and fragment 37–159

| Mutant | pH | T | ΔG | $c_{1/2}$ | m | Appr./Rem. | Ref |
|---|---|---|---|---|---|---|---|
| measurements in the absence of $(NH_4)_2SO_4$ | | | | | | | |
| AS DHFR | 7.8 | 15 | 23.1±2.1 | 2.87±0.35 | 8.03±0.71 | urea (1–3) | 97G4 |
| AS DHFR | 7.8 | 15 | 26.7±0.4 | 2.96±0.06 | 9.04±0.13 | urea (1,2,4) | 97G4 |
| AS DHFR | 7.8 | 15 | 26.3±0.7 | 2.95±0.12 | 8.91±0.25 | urea (1,2,5) | 97G4 |
| measurements in the presence of 0.5 M $(NH_4)_2SO_4$ | | | | | | | |
| AS DHFR | 7.8 | 15 | 44.7±4.9 | 5.54±0.87 | 8.08±0.92 | urea (1–3) | 97G4 |
| AS DHFR | 7.8 | 15 | 48.4±1.1 | 5.64±0.19 | 8.58±0.21 | urea (1,2,4) | 97G4 |
| AS DHFR | 7.8 | 15 | 46.0±2.7 | 5.58±0.48 | 8.24±0.50 | urea (12,5) | 97G4 |
| measurements in the presence of 0.5 M $(NH_4)_2SO_4$ | | | | | | | |
| 37–159 DHFR | 7.8 | 15 | 16.1±2.1 | 2.24±0.40 | 7.20±0.92 | urea (1–3) | 97G4 |
| 37–159 DHFR | 7.8 | 15 | 13.9±1.0 | 2.61±0.26 | 5.35±0.33 | urea (1,2,4) | 97G4 |
| 37–159 DHFR | 7.8 | 15 | 13.8±2.5 | 2.29±0.54 | 6.02±0.92 | urea (1,2,6) | 97G4 |
| 37–159 DHFR | 7.8 | 15 | 13.1±1.0 | 2.31±0.24 | 5.65±0.38 | urea (1,2,5) | 97G4 |

Remarks:
(1) linear extrapolation
(2) buffer: 10 mM potassium phosphate, 0.2 mM $K_2$EDTA, 1 mM 2-mercaptoethanol
(3) transition monitored by far-UV CD (222 nm)
(4) transition monitored by fluorescence spectrum
(5) results of a global fit
(6) transition monitored by near-UV CD (292 nm)

Dihydrofolate reductase (DHFR) from *E. coli*, effect of the length of a glycine linker connecting the N- and C-termini of a circularly permuted DHFR

| Protein | pH | T | $\Delta G$ | $c_{1/2}$ | m | Appr./Rem. | Ref |
|---|---|---|---|---|---|---|---|
| wild type | 7.8 | 15 | 25.5±1.3 | 3.0±0.1 | 8.4±0.4 | urea (1–3) | 98I4 |
| cpM16G3 | 7.8 | 15 | 13.8±1.7 | 2.3±0.1 | 5.9±1.3 | urea (1–5) | 98I4 |
| cpM16G4 | 7.8 | 15 | 17.2±1.3 | 2.4±0.1 | 7.1±0.8 | urea (1–5) | 98I4 |
| cpM16G5 | 7.8 | 15 | 21.8±1.3 | 3.1±0.1 | 7.1±0.8 | urea (1–5) | 98I4 |
| cpM16G6 | 7.8 | 15 | 20.5±1.3 | 2.9±0.1 | 7.1±0.8 | urea (1–5) | 98I4 |
| cpL24G5 | 7.8 | 15 | 20.1±1.3 | 2.5±0.1 | 7.9±0.8 | urea (1–3,6) | 98I4 |
| cpM1G5 | 7.8 | 15 | 25.9±1.3 | 3.1±0.1 | 8.4±0.4 | urea (1–3,6) | 98I4 |

Remarks:

(1) linear extrapolation, LEM-SB

(2) transition monitored by difference UV spectroscopy at 292 nm

(3) measured in 20 mM potassium phosphate, pH 7.8, 0.2 mM EDTA, 1 mM 2-mercaptoethanol

(4) the circularly permuted mutant of DHFR was constructed with Met16 as the new N-terminus

(5) G3, G4, G5, and G6 refer to glycine linkers three to six residues in length

(6) the circularly permuted mutants of DHFR (with G5 linkers) were constructed with Leu24 and Met1 as the new N-termini, respectively

Dihydrofolate reductase from *E. coli*, topological mutations

| Mutant | pH | T | $\Delta G$ | $c_{1/2}$ | m | Approach/Remarks | | Ref |
|---|---|---|---|---|---|---|---|---|
| wild type | 7.8 | 15 | 25.5 | 3.0 | 8.4 | urea | (1–5) | 98I5 |
| GHF33 | 7.8 | 15 | 16.7 | 2.4 | 6.7 | urea | (1–4,6) | 98I5 |
| GHF34 | 7.8 | 15 | 20.1 | 2.4 | 8.4 | urea | (1–4,7) | 98I5 |

Remarks:

(1) linear extrapolation, LEM-SB

(2) transition monitored by difference UV spectroscopy at 292 nm

(3) buffer: 10 mM potassium phosphate, pH 7.8, 0.2 mM EDTA, 1 mM 2-mercaptoethanol

(4) $c_{1/2}$ was taken from Fig. 10 in Ref. 98I5

(5) sequence (1–104)-Pro105–(106–129)–Pro130–(131–159)

(6) mutant GHF33 = sequence (1–104)-(Gly)$_3$-(131–159)-(Gly)$_3$-(106–129) with proline residues in the wild-type sequence (Pro105 and Pro130) replaced by glycine oligomers

(7) mutant GHF34 = sequence (1–104)-(Gly)$_3$-(131–159)-(Gly)$_4$-(106–129) with proline residues in the wild-type sequence (Pro105 and Pro130) replaced by glycine oligomers

## DNA-Binding Protein

The data entries are arranged as follows:

a) HU from *B. stearothermophilus* and *B. subtilis,* wild type and mutants

b) DNA-binding proteins: E2 protein, NikR, Sac7d, Sso7d

*a) BstHU from B. stearothermophilus and B. subtilis, wild type and mutants*

DNA-binding protein HU from *Bacillus stearothermophilus* (*Bst*HU), *Bacillus subtilis* (*Bsu*HU), and mutant proteins

| Protein | pH | $T_{trs}$ | $\Delta T$ | $\Delta(\Delta G)$ | Appr./Rem. | Ref |
|---|---|---|---|---|---|---|
| *Bst*Hu wild type | 7.0 | 63.9 | 0.0 | 0.0 | heat (1) | 96K12 |
| *Bst*Hu-Thr13→Ala | 7.0 | 67.0 | 3.1 | 2.80 | heat (1,2) | 96K12 |
| *Bst*Hu-Gly15→Glu | 7.0 | 54.0 | −9.9 | −8.94 | heat (1,2) | 96K12 |
| *Bst*Hu-Ala27→Ser | 7.0 | 58.4 | −5.5 | −4.97 | heat (1,2) | 98K4 |

DNA-binding protein HU from *Bacillus stearothermophilus* (*Bst*HU), *Bacillus subtilis* (*Bsu*HU), and mutant proteins (continued)

| Protein | pH | $T_{trs}$ | $\Delta T$ | $\Delta(\Delta G)$ | Appr./Rem. | Ref |
|---|---|---|---|---|---|---|
| *Bst*Hu-Ser31→Thr | 7.0 | 65.8 | 1.9 | 1.72 | heat (1,2) | 98K4 |
| *Bst*Hu-Thr33→Leu | 7.0 | 65.6 | 1.7 | 1.54 | heat (1,2) | 96K12 |
| *Bst*Hu-Val42→Ile | 7.0 | 60.1 | −3.8 | −3.43 | heat (1,2) | 98K4 |
| *Bst*Hu-Ala56→Ser | 7.0 | 63.3 | −0.6 | −0.54 | heat (1,2) | 98K4 |
| *Bst*Hu-Met69→Ile | 7.0 | 63.9 | 0.0 | 0.0 | heat (1,2) | 98K4 |
| *Bst*Hu-Lys90→ext | 7.0 | 65.7 | 1.8 | 1.63 | heat (1–3) | 98K4 |
| *Bst*Hu double mutant (Thr13→Ala and Gly15→Glu) | | | | | | |
| | 7.0 | 57.2 | −7.2 | −6.50 | heat (1,2) | 96K12 |
| *Bsu*Hu wild type | 7.0 | 48.6 | 0.0 | 0.0 | heat (1) | 96K12 |
| *Bsu*Hu-Glu15→Gly | 7.0 | 60.4 | 11.8 | 6.47 | heat (1,4) | 96K12 |
| *Bsu*Hu-Ser27→Ala | 7.0 | 54.2 | 5.6 | 3.07 | heat (1,4) | 98K4 |
| *Bsu*Hu-Ile42→Val | 7.0 | 52.6 | 4.0 | 2.19 | heat (1,4) | 98K4 |
| *Bsu*Hu double mutant (Asp34→Glu and Asn38→Lys) | | | | | | |
| | 7.0 | 52.1 | 3.5 | 1.92 | heat (1,4) | 98K4 |
| *Bsu*Hu multiple mutant (Glu15→Gly, Ser27→Ala, Asp34→Glu, Asn38→Lys, Ile42→Val) | | | | | | |
| | 7.0 | 71.3 | 22.7 | 12.44 | heat (1,3,5) | 98K4 |

Remarks:
(1) measured in 0.05 M phosphate buffer, pH 7.0, transition monitored by CD at 222 nm
(2) reference protein is *Bst*Hu ($\Delta T$ and $\Delta(\Delta G)$)
(3) *Bst*Hu-Lys90→ext, protein extended by −Ala90-Gly91-Lys92
(4) reference protein is *Bsu*HU ($\Delta T$ and $\Delta(\Delta G)$)
(5) the five thermostabilizing mutations were simultaneously introduced into the mesophilic protein *Bsu*Hu

*Bst*HU, DNA-binding protein from *Bacillus stearothermophilus*, wild type and mutants

| Mutant | pH | $T_{trs}$ | $\Delta T$ | $\Delta(\Delta G)$ | Appr./Rem. | Ref |
|---|---|---|---|---|---|---|
| wild type | 7.0 | 63.9 | 0.0 | 0.0 | heat (1,2) | 97K3 |
| Glu34→Asp | 7.0 | 61.6 | −2.3 | −2.08 | heat (1,2) | 97K3 |
| Glu34→Gln | 7.0 | 60.9 | −3.0 | −2.71 | heat (1,2) | 97K3 |
| Arg37→Lys | 7.0 | 63.9 | 0.0 | 0.0 | heat (1,2) | 97K3 |
| Lys38→Asn | 7.0 | 61.5 | −2.4 | −2.17 | heat (1,2) | 97K3 |
| double mutant (Glu34→Asp and Lys38→Asn) | | | | | | |
| | 7.0 | 60.5 | −3.4 | −3.07 | heat (1,2) | 97K3 |

Remarks:
(1) buffer: 5 mM sodium phosphate, 0.2 M NaCl
(2) $\Delta(\Delta G)$ was calculated from $\Delta(\Delta G) = \Delta T * \Delta S_{w.t.}$

*Bst*HU, DNA-binding protein from *Bacillus stearothermophilus*, wild type and mutants, salt dependence

| Mutant | Salt | pH | $T_{trs}$ | $\Delta T$ | $\Delta(\Delta G)$ | Appr./Rem. | Ref |
|---|---|---|---|---|---|---|---|
| wild type | 0.0 M | 7.0 | 52.9 | 0.0 | 0.0 | heat (1–3) | 97K3 |
| | 0.1 M | 7.0 | 57.8 | 0.0 | 0.0 | heat (1–3) | 97K3 |
| | 0.2 M | 7.0 | 63.9 | 0.0 | 0.0 | heat (1–3) | 97K3 |
| | 0.5 M | 7.0 | 65.4 | 0.0 | 0.0 | heat (1–3) | 97K3 |
| Glu34→Asp | 0.0 M | 7.0 | 50.7 | −2.2 | −1.46 | heat (1–3) | 97K3 |
| | 0.1 M | 7.0 | 56.3 | −1.5 | −1.22 | heat (1–3) | 97K3 |
| | 0.2 M | 7.0 | 61.6 | −2.3 | −2.08 | heat (1–3) | 97K3 |
| | 0.5 M | 7.0 | 65.0 | −0.4 | −0.31 | heat (1–3) | 97K3 |
| Lys38→Asn | 0.0 M | 7.0 | 49.6 | −3.3 | −2.19 | heat (1–3) | 97K3 |
| | 0.1 M | 7.0 | 54.9 | −2.9 | −2.35 | heat (1–3) | 97K3 |
| | 0.2 M | 7.0 | 61.5 | −2.4 | −2.17 | heat (1–3) | 97K3 |
| | 0.5 M | 7.0 | 65.0 | −0.4 | −0.31 | heat (1–3) | 97K3 |

Remarks:
(1) buffer: 5 mM sodium phosphate with the indicated NaCl concentration
(2) $\Delta(\Delta G)$ was calculated from $\Delta(\Delta G) = \Delta T * \Delta S_{w.t.}$
(3) $\Delta(\Delta G)$ refers to the wild-type protein at the identical NaCl concentration

*Bst*HU, DNA-binding protein from *Bacillus stearothermophilus*, wild type and mutants, pH dependence

| Mutant | pH | $T_{trs}$ | $\Delta T$ | $\Delta(\Delta G)$ | Appr./Rem. | Ref |
|---|---|---|---|---|---|---|
| wild type | 2.0 | 48.4 | 0.0 | 0.0 | heat (1–3) | 97K3 |
| | 3.0 | 61.3 | 0.0 | 0.0 | heat (1–3) | 97K3 |
| | 4.0 | 64.1 | 0.0 | 0.0 | heat (1–3) | 97K3 |
| | 5.0 | 64.3 | 0.0 | 0.0 | heat (1–3) | 97K3 |
| | 7.0 | 63.9 | 0.0 | 0.0 | heat (1–3) | 97K3 |
| Glu34→Asp | 2.0 | 46.7 | −1.7 | −1.21 | heat (1–3) | 97K3 |
| | 3.0 | 59.0 | −2.3 | −2.07 | heat (1–3) | 97K3 |
| | 4.0 | 62.2 | −1.9 | −1.50 | heat (1–3) | 97K3 |
| | 5.0 | 62.3 | −2.0 | −3.00 | heat (1–3) | 97K3 |
| | 7.0 | 61.6 | −2.3 | −2.08 | heat (1–3) | 97K3 |
| Glu34→Gln | 2.0 | 47.1 | −1.3 | −0.93 | heat (1–3) | 97K3 |
| | 7.0 | 60.9 | −3.0 | −2.71 | heat (1–3) | 97K3 |
| Lys38→Asn | 2.0 | 48.1 | −0.3 | −0.21 | heat (1–3) | 97K3 |
| | 3.0 | 58.6 | −2.7 | −2.43 | heat (1–3) | 97K3 |
| | 4.0 | 61.9 | −2.2 | −1.73 | heat (1–3) | 97K3 |
| | 5.0 | 61.7 | −2.6 | −2.73 | heat (1–3) | 97K3 |
| | 7.0 | 61.5 | −2.4 | −2.17 | heat (1–3) | 97K3 |
| double mutant (Glu34→Asp and Lys38→Asn) | | | | | | |
| | 2.0 | 46.5 | −1.9 | −1.36 | heat (1–3) | 97K3 |
| | 7.0 | 60.5 | −3.4 | −3.07 | heat (1–3) | 97K3 |

Remarks:
(1) buffer: 5 mM phosphate and 0.2 M NaCl below pH 4, and 5 mM sodium phosphate and 0.2 M NaCl above pH 5
(2) $\Delta(\Delta G)$ was calculated from $\Delta(\Delta G) = \Delta T * \Delta S_{w.t.}$
(3) $\Delta(\Delta G)$ refers to the wild-type protein at the identical pH

*Bsu*HU, DNA-binding protein from *Bacillus subtilis*, mesophilic protein, wild type and mutants

| Mutant | pH | $T_{trs}$ | $\Delta T$ | $\Delta(\Delta G)$ | Appr./Rem. | Ref |
|---|---|---|---|---|---|---|
| wild type | 7.0 | 48.6 | 0.0 | 0.0 | heat (1,2) | 97K3 |
| Glu15→Gly | 7.0 | 60.4 | 11.8 | 6.47 | heat (1,2) | 97K3 |
| double mutant (Asp34→Glu and Asn38→Lys) | | | | | | |
| | 7.0 | 52.1 | 3.5 | 1.92 | heat (1,2) | 97K3 |
| triple mutant (Glu15→Gly, Asp34→Glu, and Asn38→Lys) | | | | | | |
| | 7.0 | 64.5 | 15.9 | 8.71 | heat (1,2) | 97K3 |

Remarks:
(1) buffer: 5 mM sodium phosphate, 0.2 M NaCl
(2) $\Delta(\Delta G)$ was calculated from $\Delta(\Delta G) = \Delta T * \Delta S_{w.t.}$

*b) DNA-binding proteins: E2 protein, NikR, Sac7d, Sso7d*

81-residue DNA-binding domain of the E2 protein from human papillomavirus, stability under different conditions

| Component | pH | T | $\Delta G$ | $c_{1/2}$ | m | Appr./Rem. | Ref |
|---|---|---|---|---|---|---|---|
| buffer | 7.0 | 25 | 52.3±2.9 | 2.2 | 8.4±1.3 | urea (1–3) | 97L10 |
| NaCl | 7.0 | 25 | 73.2±3.3 | 4.9 | 7.9±0.8 | urea (1,2,4) | 97L10 |
| phosphate | 7.0 | 25 | 70.7±2.5 | 4.3 | 8.4±0.4 | urea (1,2,5) | 97L10 |
| heparin | 7.0 | 25 | | 7.2 | | urea (1,2,6) | 97L10 |
| DNA oligo | 7.0 | 25 | | 7.0 | | urea (1,2,7) | 97L10 |

Remarks:
(1) treatment by a two-state equilibrium with folded dimers and unfolded monomers
(2) transition monitored by fluorescence
(3) buffer: 50 mM Bis-Tris, 1 mM DTT, pH 7
(4) buffer (3) with 0.75 M NaCl,
(5) buffer (3) with 100 mM sodium phosphate
(6) buffer (3) with 2 μg/ml heparin
(7) buffer (3) with 0.25 μM DNA oligonucleotide (36-mer)

NikR, DNA-binding protein from *Escherichia coli*, N-terminal fragment (1–48), wild type and mutant

| Protein | pH | T | $\Delta G$ | $c_{1/2}$ | m | Appr./Rem. | Ref |
|---|---|---|---|---|---|---|---|
| wild type | 7.6 | | 56.9 | 2.97 | 9.20 | GuHCl (1–4) | 99C9 |
| Arg3→Ala | 7.6 | | 55.2 | 2.90 | 8.83 | GuHCl (1–4) | 99C9 |

Remarks:
(1) linear extrapolation
(2) transition monitored by far-UV CD
(3) measured in 50 mM sodium phosphate buffer, pH 7.6
(4) thermal transition at $T_{trs} = 76.9°C$ for the wild type protein, and $T_{trs} = 76.9°C$ for the Arg3→Ala mutant

Sac7d, 7 kDa DNA-binding protein from *Sulfolobus acidocaldarius*, linkage of protonation and anion binding

| KCl Conc. | pH | T | $\Delta G$ | Approach/Remarks | | Ref |
|---|---|---|---|---|---|---|
| 0.3 M | 7 | 20 | 26.4 | global fit (1) | | 98M10 |
| 0.3 M | 7 | 20 | 25.5 | GuHCl | (2) | 98M10 |
| 0.3 M | 4 | 20 | 21.3 | global fit (1) | | 98M10 |
| 0.3 M | 4 | 20 | 20.1 | urea | (2) | 98M10 |

Remarks:
(1) from global fit of complex data to a linkage model, see also Table 2
(2) reference data from Ref. 96M2

Sso7d, 7 kDa DNA-binding protein from thermoacidophilic *Sulfolobus solfataricus*, wild type and mutants, $\Delta G$ at condition of maximal thermostability

| Protein | pH | Tmax | $\Delta Gmax$ | Approach/Remarks | | Ref |
|---|---|---|---|---|---|---|
| Sso7d | 4.5–7.0 | 98.2±0.4 | 32.5* | DSC | (1–3) | 98C2 |
| Trp23→Ala | 4.9–7.2 | 92.2±0.4 | 27.6* | DSC | (1,2) | 98C2 |
| Phe31→Ala | 4.9–7.2 | 74.4±0.5 | 17.2* | DSC | (1,2) | 98C2 |
| Phe31→Tyr | 4.9–7.2 | 88.0±0.3 | 23.7* | DSC | (1,2) | 98C2 |

Remarks:
(1) for reference data see Table 2
(2) Ref. 98C2 contains a temperature profile of $\Delta G$ from 250-380 K
(3) based on data from Ref. 96K5

## DsbA, Disulfide-Bond Forming Enzyme

DsbA, disulfide-bond forming enzyme from *E. coli*, variants

| Mutant/Form | | pH | T | $\Delta G$ | m | Approach/Remarks | | Ref |
|---|---|---|---|---|---|---|---|---|
| DsbA | oxidized | 7.0 | 30 | 48.7 | | DSC | (1,2) | 99M22 |
| | reduced | 7.0 | 30 | 65.93 | | DSC | (1,2) | 99M22 |
| DsbA Pro151→Ala | | | | | | | | |
| | oxidized | 7.0 | 30 | 24.96 | | DSC | (1,2) | 99M22 |
| | reduced | 7.0 | 30 | 37.6 | | DSC | (1,2) | 99M22 |
| DsbA | oxidized | 7.0 | 30 | 34.7 | 20.3±1.1 | GuHCl | (3-6) | 99M22 |
| | reduced | 7.0 | 30 | 52.8 | 24.1±0.9 | GuHCl | (3-6) | 99M22 |
| DsbA Pro151→Ala | | | | | | | | |
| | oxidized | 7.0 | 30 | 11.9 | 8.6±0.4 | GuHCl | (3-6) | 99M22 |
| | reduced | 7.0 | 30 | 14.8 | 10.3±0.4 | GuHCl | (3-6) | 99M22 |

Remarks:
(1) measured in 100 mM phosphate, 1 mM EDTA, pH 7.0
(2) calculated using $\Delta Cp = 9.28±2.09$ kJ/mol/K
(3) linear extrapolation
(4) transition monitored by fluorescence
(5) measured in 100 mM phosphate, 1 mM EDTA, pH 7.0, in the presence of 1 mM DTT for the reduced proteins
(6) the fluorimetrically determined $\Delta G$ value is underestimated, see Ref. 99M22

DsbA, wild type and mutants concerning residues within the active site (Cys30-Pro31-His32-Cys33)

| Mutant | Form | pH | T | $\Delta G$ | $c_{1/2}$ | m | Appr./Rem. | Ref |
|---|---|---|---|---|---|---|---|---|
| wild type | ox. | 7.0 | 30 | 29.3±2.9 | 1.71 | 17.2±1.7 | GuHCl (1) | 95G14 |
| wild type | red. | 7.0 | 30 | 46.4±1.3 | 2.29 | 20.5±0.8 | GuHCl (1) | 95G14 |
| His32→Leu | ox. | 7.0 | 30 | 51.5±4.6 | 2.09 | 22.2±1.7 | GuHCl (1) | 95G14 |
| His32→Leu | red. | 7.0 | 30 | 65.7±7.1 | 2.46 | 29.3±2.1 | GuHCl (1) | 95G14 |
| His32→Ser | ox. | 7.0 | 30 | 51.5±4.6 | | | GuHCl (1) | 97G14 |
| His32→Ser | red. | 7.0 | 30 | 65.7±7.1 | | | GuHCl (1) | 97G14 |
| His32→Tyr | ox. | 7.0 | 30 | 57.7±4.4 | | | GuHCl (1) | 97G14 |
| His32→Tyr | red. | 7.0 | 30 | 48.5±2.0 | | | GuHCl (1) | 97G14 |
| double mutant (Pro31→Leu and His32→Thr) | | | | | | | | |
| | ox. | 7.0 | 30 | 50.2±4.6 | 2.04 | 27.6±2.5 | GuHCl (1) | 95G14 |
| | red. | 7.0 | 30 | 62.8±6.7 | 2.36 | 28.9±2.5 | GuHCl (1) | 95G14 |
| double mutant (Pro31→Ser and His32→Phe) | | | | | | | | |
| | ox. | 7.0 | 30 | 48.1±4.2 | 1.97 | 25.1±0.8 | GuHCl (1) | 95G14 |
| | red. | 7.0 | 30 | 59.8±6.3 | 2.24 | 23.4±0.8 | GuHCl (1) | 95G14 |

DsbA, wild type and mutants concerning residues within the active site (Cys30-Pro31-His32-Cys33) (continued)

| Mutant | Form | pH | T | ΔG | $c_{1/2}$ | m | Appr./Rem. | Ref |
|---|---|---|---|---|---|---|---|---|
| double mutant (Pro31→Ser and His32→Thr) | | | | | | | | |
| | ox. | 7.0 | 30 | 49.0±4.2 | 1.99 | 23.4±0.8 | GuHCl (1) | 95G14 |
| | red. | 7.0 | 30 | 59.8±6.3 | 2.24 | 24.7±1.7 | GuHCl (1) | 95G14 |

Remark:
(1) the folding transitions and the curve fitting were performed as described in Ref. 93W6

DsbA, wild type and mutants concerning charged residues in the vicinity of the acitve-site helix

| Mutant | Form | pH | T | ΔG | $c_{1/2}$ | m | Appr./Rem. | Ref |
|---|---|---|---|---|---|---|---|---|
| wild type | ox. | 7.0 | 25 | 46.0±2.9 | 1.80 | 25.5±1.6 | GuHCl(1–3) | 97H2 |
| wild type | red. | 7.0 | 25 | 58.4±2.6 | 2.37 | 24.6±1.1 | GuHCl(1–3) | 97H2 |
| wild type | ox. | 7.0 | 25 | 47.9±2.7 | 1.82 | 26.3±1.8 | GuHCl(1–3) | 97J1 |
| wild type | red. | 7.0 | 25 | 58.1±2.3 | 2.39 | 24.3±1.0 | GuHCl(1–3) | 97J1 |
| Glu24→Gln | ox. | 7.0 | 25 | 50.2±1.7 | 1.88 | 26.7±0.9 | GuHCl(1–3) | 97J1 |
| Glu24→Gln | red. | 7.0 | 25 | 60.3±2.0 | 2.43 | 24.8±0.8 | GuHCl(1–3) | 97J1 |
| Glu37→Gln | ox. | 7.0 | 25 | 39.1±1.2 | 1.79 | 21.8±0.6 | GuHCl(1–3) | 97H2 |
| Glu37→Gln | red. | 7.0 | 25 | 59.7±1.7 | 2.25 | 26.6±0.7 | GuHCl(1–3) | 97H2 |
| Glu38→Gln | ox. | 7.0 | 25 | 44.2±1.2 | 1.87 | 23.6±0.6 | GuHCl(1–3) | 97H2 |
| Glu38→Gln | red. | 7.0 | 25 | 64.0±2.0 | 2.44 | 26.2±0.8 | GuHCl(1–3) | 97H2 |
| Lys58→Met | ox. | 7.0 | 25 | 39.6±1.5 | 1.83 | 21.6±0.8 | GuHCl(1–3) | 97J1 |
| Lys58→Met | red. | 7.0 | 25 | 61.1±3.4 | 2.38 | 25.7±1.4 | GuHCl(1–3) | 97J1 |
| double mutant (Glu24→Gln and Glu37→Gln) | | | | | | | | |
| | ox. | 7.0 | 25 | 42.5±2.5 | 1.82 | 23.3±1.4 | GuHCl(1–3) | 97J1 |
| | red. | 7.0 | 25 | 50.1±2.8 | 2.26 | 22.2±1.2 | GuHCl(1–3) | 97J1 |
| double mutant (Glu24→Gln and Lys58→Met) | | | | | | | | |
| | ox. | 7.0 | 25 | 51.8±4.3 | 1.90 | 27.2±2.3 | GuHCl(1–3) | 97J1 |
| | red. | 7.0 | 25 | 60.1±2.4 | 2.38 | 25.2±1.0 | GuHCl(1–3) | 97J1 |
| double mutant (Glu37→Gln and Glu38→Gln) | | | | | | | | |
| | ox. | 7.0 | 25 | 38.5±1.0 | 1.79 | 21.5±0.5 | GuHCl(1–3) | 97H2 |
| | red. | 7.0 | 25 | 61.0±1.8 | 2.30 | 26.6±0.8 | GuHCl(1–3) | 97H2 |
| double mutant (Glu37→Gln and Lys58→Met) | | | | | | | | |
| | ox. | 7.0 | 25 | 34.6±1.3 | 1.76 | 19.7±0.7 | GuHCl(1–3) | 97J1 |
| | red. | 7.0 | 25 | 59.5±2.0 | 2.23 | 26.7±0.9 | GuHCl(1–3) | 97J1 |
| triple mutant (Glu24→Gln, Glu37→Gln, and Lys58→Met) | | | | | | | | |
| | ox. | 7.0 | 25 | 41.7±1.8 | 1.85 | 22.5±1.0 | GuHCl(1–3) | 97J1 |
| | red. | 7.0 | 25 | 56.9±2.8 | 2.36 | 24.1±1.2 | GuHCl(1–3) | 97J1 |
| multiple mutant (Glu24→Gln, Glu37→Gln, Glu38→Gln, and Lys58→Met) | | | | | | | | |
| | ox. | 7.0 | 25 | 42.0±2.4 | 1.88 | 22.3±1.3 | GuHCl(1–3) | 97J1 |
| | red. | 7.0 | 25 | 34.8±2.0 | 2.29 | 15.2±0.9 | GuHCl(1–4) | 97J1 |
| deletion mutant (Glu38, Val39, and Leu40 deleted) | | | | | | | | |
| | ox. | 7.0 | 25 | 43.0±2.6 | 1.82 | 23.6±1.4 | GuHCl(1–3) | 97H2 |
| | red. | 7.0 | 25 | 56.3±2.0 | 2.24 | 25.2±0.9 | GuHCl(1–3) | 97H2 |
| deletion mutant (Glu38, Val39, and Leu40 deleted) and mutant His41→Pro | | | | | | | | |
| | ox. | 7.0 | 25 | 34.3±1.6 | 1.72 | 19.9±0.9 | GuHCl(1–3) | 97H2 |
| | red. | 7.0 | 25 | 51.9±3.0 | 2.19 | 23.7±1.3 | GuHCl(1–3) | 97H2 |

Remarks:
(1) linear extrapolation, LEM-SB, for the procedure see also Ref. 93W6
(2) transition monitored by fluorescence at 365 nm and at 327 nm (excitation 280 nm)
(3) buffer: 100 mM sodium phosphate, 0.1 mM EDTA (for reduced DsbA with 1 mM DTT), and for refolding experiments 4 M GuHCl, 100 mM sodium phosphate, 1 mM EDTA
(4) two-state model may not be valid

DsbA, circular permutants

| Mutant | Form | pH | T | ΔG | m | Approach/Remarks | | Ref |
|---|---|---|---|---|---|---|---|---|
| wild type | ox. | 7.0 | 25 | 48.7±1.4 | 10.6±0.3 | urea | (1,2) | 99H7 |
| | red. | 7.0 | 25 | 65.6±2.2 | 11.0±0.4 | urea | (1,2) | 99H7 |
| Leu12-Tyr9 | ox. | 7.0 | 25 | | | urea | (1–3) | 99H7 |
| | red. | 7.0 | 25 | 41.9±1.9 | 8.2±0.4 | urea | (1,2) | 99H7 |
| Gln21-Pro20 | ox. | 7.0 | 25 | 19.0±2.4 | 4.6±0.6 | urea | (1,2,4) | 99H7 |
| | red. | 7.0 | 25 | 33.7±1.1 | 6.7±0.2 | urea | (1,2) | 99H7 |
| His32-Phe29 | red. | 7.0 | 25 | 15.5±0.5 | 5.4±0.1 | urea | (1,2) | 99H7 |
| Tyr34-His32 | red. | 7.0 | 25 | 18.1±1.0 | 5.6±0.3 | urea | (1,2) | 99H7 |
| Glu37-Phe36 | ox. | 7.0 | 25 | 23.8±0.9 | 6.5±0.2 | urea | (1,2) | 99H7 |
| | red. | 7.0 | 25 | 27.3±0.9 | 6.6±0.2 | urea | (1,2) | 99H7 |
| His41-Leu40 | ox. | 7.0 | 25 | 21.3±1.1 | 6.1±0.3 | urea | (1,2) | 99H7 |
| | red. | 7.0 | 25 | 29.9±1.0 | 7.1±0.2 | urea | (1,2) | 99H7 |
| Gly66-Gly65 | ox. | 7.0 | 25 | 33.9±1.7 | 10.9±0.6 | urea | (1,2) | 99H7 |
| | red. | 7.0 | 25 | 50.5±2.1 | 10.9±0.5 | urea | (1,2) | 99H7 |
| Thr89-Val88 | red. | 7.0 | 25 | 14.1±1.3 | 6.2±0.4 | urea | (1,2) | 99H7 |
| Gln100-Thr99 | ox. | 7.0 | 25 | 44.1±2.1 | 13.7±0.7 | urea | (1,2) | 99H7 |
| | red. | 7.0 | 25 | 55.6±3.0 | 12.1±0.6 | urea | (1,2) | 99H7 |
| Ala136-Val135 | ox. | 7.0 | 25 | 20.2±3.6 | 6.5±1.1 | urea | (1,2) | 99H7 |
| | red. | 7.0 | 25 | 5.5±0.9 | 3.9±0.2 | urea | (1,2,4) | 99H7 |
| Ala152-Pro151 | ox. | 7.0 | 25 | 16.0±1.6 | 4.5±0.4 | urea | (1,2,4) | 99H7 |
| | red. | 7.0 | 25 | 18.7±0.9 | 5.1±0.2 | urea | (1,2) | 99H7 |
| Asp172-Met171 | ox. | 7.0 | 25 | | | urea | (1–3) | 99H7 |
| | red. | 7.0 | 25 | 44.4±2.1 | 8.8±0.4 | urea | (1,2) | 99H7 |

Mutant description:
- the first-last natural residue is indicated
- wild type corresponds to Ala1-Lys189

Remarks:
(1) linear extrapolation, LEM-SB
(2) transition monitored fluorimetrically
(3) more than two states for the oxidized form
(4) the two-state model is probably not valid

# Fas

Fas (apo-1/CD95) membrane protein that mediates apoptosis

| Mutant | pH | T | Δ(ΔG) | $c_{1/2}$ | Approach | Remarks | Ref |
|---|---|---|---|---|---|---|---|
| wild type | 7.0 | | 0.0 | 2.4 | GuHCl | (1–3) | 97E1 |
| Val238→Asn | 7.0 | | −5.9 | 2.0 | GuHCl | (1–4) | 97E1 |

Remarks:
(1) linear extrapolation
(2) $c_{1/2}$ was taken from Fig. 4b in Ref. 97E1
(3) the transition was monitored by CD at 222 nm
(4) helix 3 of Fas is unfolded in the mutant protein

## Fatty Acid-Binding Protein

Recombinant intestinal fatty acid-binding protein (IFABP)

| Protein | pH | T | ΔG | $c_{1/2}$ | m | Appr./Rem. | Ref |
|---------|-----|----|-------------|-----------|-----------|-------------|------|
| IFABP | 5.0 | 25 | 19.00±0.96 | 4.05±0.04 | 4.69±0.21 | urea (1–3) | 98D3 |
| | 5.0 | 25 | 25.61±1.76 | 4.21±0.04 | 6.11±0.42 | urea (1,2,4) | 98D3 |
| | 6.0 | 25 | 18.24±0.92 | 4.09±0.03 | 4.44±0.21 | urea (1–3) | 98D3 |
| | 6.0 | 25 | 23.47±1.67 | 4.20±0.04 | 5.56±0.38 | urea (1,2,4) | 98D3 |
| | 7.0 | 25 | 17.91±1.13 | 4.10±0.04 | 4.35±0.25 | urea (1–3) | 98D3 |
| | 7.0 | 25 | 20.04±0.92 | 4.24±0.03 | 4.73±0.29 | urea (1,2,4) | 98D3 |
| | 8.0 | 25 | 20.04±1.26 | 4.12±0.03 | 4.85±0.29 | urea (1–3) | 98D3 |
| | 8.0 | 25 | 21.21±1.46 | 4.20±0.04 | 5.06±0.33 | urea (1,2,4) | 98D3 |
| | 9.0 | 25 | 20.42±0.96 | 4.36±0.03 | 4.69±0.21 | urea (1–3) | 98D3 |
| | 9.0 | 25 | 19.20±2.09 | 4.55±0.07 | 4.23±0.46 | urea (1,2,4) | 98D3 |
| | 10.0 | 25 | 19.20±1.92 | 4.12±0.08 | 4.69±0.50 | urea (1–3) | 98D3 |
| | 10.0 | 25 | 18.66±1.05 | 4.18±0.04 | 4.48±0.25 | urea (1,2,4) | 98D3 |

Remarks:

(1) linear extrapolation, LEM-SB

(2) measured in 25 mM buffer (acetate at pH 5, sodium phosphate at pH 6–8, glycine pH at 9–10), 75 mM NaCl, 0.1 mM EDTA

(3) transition monitored by CD at 218 nm

(4) transition monitored by fluorescence at 327 nm

Intestinal fatty acid-binding protein, wild type and turn mutants

| Mutant | pH | T | ΔG | $c_{1/2}$ | m | Approach/Remarks | Ref |
|--------|-----|----|------|-----------|------|------------------|------|
| wild type | 7.3 | 20 | 21.8 | 1.36 | 16.0 | GuHCl (1–3) | 98K8 |
| Gly44→Val | 7.3 | 20 | 21.3 | 1.0 | 21.3 | GuHCl (1–3) | 98K8 |
| Asn54→Val | 7.3 | 20 | 31.4 | 1.61 | 19.2 | GuHCl (1–3) | 98K8 |
| Gly65→Val | 7.3 | 20 | 3.1 | 0.48 | 6.3 | GuHCl (1–3) | 98K8 |
| Gly75→Val | 7.3 | 20 | 20.5 | 0.91 | 22.6 | GuHCl (1–3) | 98K8 |
| Gly80→Val | 7.3 | 20 | 2.3 | 0.48 | 5.0 | GuHCl (1–3) | 98K8 |
| Gly86→Val | 7.3 | 20 | 22.2 | 1.29 | 17.2 | GuHCl (1–3) | 98K8 |
| Asp97→Val | 7.3 | 20 | 14.2 | 0.89 | 15.9 | GuHCl (1–3) | 98K8 |
| Gly99→Val | 7.3 | 20 | 11.3 | 0.85 | 13.4 | GuHCl (1–3) | 98K8 |
| Gly110→Val | 7.3 | 20 | 18.4 | 1.15 | 15.9 | GuHCl (1–3) | 98K8 |
| Gly121→Val | 7.3 | 20 | 8.8 | 0.89 | 10.0 | GuHCl (1–3) | 98K8 |

Remarks:

(1) linear extrapolation

(2) measured in 20 mM potassium phosphate, 0.25 mM EDTA, pH 7.3

(3) transition monitored by fluorescence at 328 nm

Intestinal fatty acid-binding protein, wild type and mutants pos. 64, 65, 66

| Mutant | pH | T | ΔG | $c_{1/2}$ | m | Appr./Rem. | Ref |
|--------|-----|----|------|-----------|------|-------------|------|
| wild type | 7.4 | 20 | 29.5 | 1.32 | 22.3 | GuHCl (1–3) | 97K5 |
| Leu64→Ala | 7.4 | 20 | 20.9 | 1.01 | 20.8 | GuHCl (1–3) | 97K5 |
| Leu64→Gly | 7.4 | 20 | 18.5 | 0.87 | 21.4 | GuHCl (1–3) | 97K5 |
| Gly65→Ala | 7.4 | 20 | 24.0 | 1.19 | 20.1 | GuHCl (1–3) | 97K5 |
| double mutant (Gly65→Ala and Val66→Arg) | | | | | | | |
| | 7.4 | 20 | 20.5 | 1.12 | 18.3 | GuHCl (1–3) | 97K5 |
| double mutant (Gly65→Ala and Val66→Asn) | | | | | | | |
| | 7.4 | 20 | 12.1 | 0.94 | 12.8 | GuHCl (1–3) | 97K5 |

Intestinal fatty acid-binding protein, wild type and mutants pos. 64, 65, 66 (continued)

| Mutant | pH | T | $\Delta G$ | $c_{1/2}$ | m | Appr./Rem. | Ref |
|---|---|---|---|---|---|---|---|
| double mutant (Gly65→Asp and Val66→His) | | | | | | | |
| | 7.4 | 20 | 26.1 | 1.23 | 21.2 | GuHCl (1–3) | 97K5 |
| double mutant (Gly65→Ala and Val66→Phe) | | | | | | | |
| | 7.4 | 20 | 27.2 | 1.52 | 17.8 | GuHCl (1–3) | 97K5 |
| double mutant (Gly65→Asp and Val66→Phe) | | | | | | | |
| | 7.4 | 20 | 19.2 | 1.03 | 18.7 | GuHCl (1–3) | 97K5 |
| triple mutant (Leu64→Arg, Gly65→Ala, and Val66→Cys) | | | | | | | |
| | 7.4 | 20 | 6.4 | 0.48 | 13.2 | GuHCl (1–3) | 97K5 |
| triple mutant (Leu64→Gly, Gly65→Ala, and Val66→Phe) | | | | | | | |
| | 7.4 | 20 | 5.9 | 0.57 | 10.3 | GuHCl (1–3) | 97K5 |
| triple mutant (Leu64→Lys, Gly65→Val, and Val66→Asp) | | | | | | | |
| | 7.4 | 20 | 8.3 | 0.46 | 17.9 | GuHCl (1–3) | 97K5 |
| triple mutant (Leu64→Ser, Gly65→Ala, and Val66→Asn) | | | | | | | |
| | 7.4 | 20 | 10.0 | 0.60 | 16.7 | GuHCl (1–3) | 97K5 |
| triple mutant (Leu64→Thr, Gly65→Ala, and Val66→Gly) | | | | | | | |
| | 7.4 | 20 | 9.7 | 0.58 | 16.7 | GuHCl (1–3) | 97K5 |
| triple mutant (Leu64→Thr, Gly65→Val, and Val66→Pro) | | | | | | | |
| | 7.4 | 20 | 7.4 | 0.50 | 14.8 | GuHCl (1–3) | 97K5 |

Remarks:

(1) linear extrapolation

(2) transition monitored by fluorescence

(3) buffer: 20 mM potassium phosphate, pH 7.4, 0.25 mM EDTA, and 2 mM DTT for the Cys-containing mutants

Human heart fatty acid-binding protein (H-FABP), wild type and mutants

| Protein | pH | T | $\Delta G$ | $c_{1/2}$ | Approach/Remarks | | Ref |
|---|---|---|---|---|---|---|---|
| wild type | 8.0 | 37 | | 5.9±0.2 | urea | (1,2) | 99Z5 |
| Phe4→Glu | 8.0 | 37 | | 2.4±0.3 | urea | (1,2) | 99Z5 |
| Phe4→Ser | 8.0 | 37 | | 3.5±0.2 | urea | (1,2) | 99Z5 |
| Trp8→Glu | 8.0 | 37 | | 3.4±0.6 | urea | (1,2) | 99Z5 |
| Phe16→Glu | 8.0 | 37 | | 2.6±0.4 | urea | (1,2) | 99Z5 |
| Phe16→Ser | 8.0 | 37 | | 3.3±0.1 | urea | (1,2) | 99Z5 |
| Lys21→Ile | 8.0 | 37 | | 5.9±0.2 | urea | (1,2) | 99Z5 |
| Phe64→Ser | 8.0 | 37 | | 3.4±0.1 | urea | (1,2) | 99Z5 |
| Leu66→Gly | 8.0 | 37 | | 3.5±0.1 | urea | (1,2) | 99Z5 |
| Gly67→Ser | 8.0 | 37 | | 5.1±0.4 | urea | (1,2) | 99Z5 |
| Glu72→Ser | 8.0 | 37 | | 3.5±0.2 | urea | (1,2) | 99Z5 |
| double mutant (Phe4→Ser and Trp8→Glu) | | | | | | | |
| | 8.0 | 37 | | 3.5±0.1 | urea | (1,2) | 99Z5 |

Remarks:

(1) transition monitored by Trp fluorescence

(2) buffer: 10 mM Tris-HCl, pH 8.0

## Ferredoxin

[4Fe-4S] ferredoxin from hyperthermophile *Thermotoga maritima*

| pH | T | $\Delta G$ | Approach | Remarks | Ref |
|----|---|-----------|----------|---------|-----|
| 8.5 | 25 | 36 | DSC and heat | (1,2) | 97P9 |
| 8.5 | 45 | 39 | DSC and heat | (1–3) | 97P9 |

Remarks:

(1) from combined DSC and heat denaturation in the presence of denaturant (see Table 2), data extrapolated to zero denaturant concentration

(2) $\Delta G$ was calculated using nonlinear dependence of $\Delta Cp$ on temperature

(3) temperature at which the protein achieves maximal stability

## Ferritin

Recombinant L subunit of human ferritin

| pH | T | $\Delta G$ | Approach | Ref |
|----|---|-----------|----------|-----|
| 2.0 | 25 | 10.5 | DSC | 98M8 |
| 2.2 | 25 | 25.5 | DSC | 98M8 |
| 2.4 | 25 | 40.2 | DSC | 98M8 |
| 2.8 | 25 | 48.4 | DSC | 98M8 |

## Fetuin

Bovine serum fetuin (BSF)

| Transition | pH | T | $\Delta G$ | $c_{1/2}$ | m | Approach/Remarks | | Ref |
|-----------|----|----|-----------|-----------|---|------------------|---|-----|
| $N \rightarrow I$ | 6.0 | 25 | 17.0±1.0 | | | DSC | (1) | 98W4 |
| $N \rightarrow I$ | 6.0 | 25 | 16.1±1.3 | | | heat | (1,4) | 98W4 |
| $N \rightarrow I$ | 6.0 | 25 | 15.5±1.3 | 1.5±0.2 | 10.0 | GuHCl | (1–3) | 98W4 |
| $N \rightarrow I$ | 6.0 | 25 | | ≤1.5 | | GuHCl | (1,4) | 98W4 |
| $I \rightarrow U$ | 6.0 | 25 | 13.8±1.3 | | | DSC | (1) | 98W4 |
| $I \rightarrow U$ | 6.0 | 25 | 12.6±1.3 | 2.5±0.2 | 5.0 | GuHCl | (1,2,5) | 98W4 |
| $I \rightarrow U$ | 6.0 | 25 | 12.6±0.8 | 2.5±0.1 | 5.0 | GuHCl | (1,2,6) | 98W4 |
| $I \rightarrow U$ | 6.0 | 25 | | ≥2.1 | | GuHCl | (1,7) | 98W4 |

Remarks:

(1) buffer: 50 mM sodium phosphate, pH 6.0

(2) linear extrapolation, LEM-SB

(3) transition monitored by intrinsic tryptophan fluorescence

(4) transition monitored by near-UV CD

(5) transition monitored by fluorescence emission wavelength

(6) transition monitored by far-UV CD

(7) transition monitored by size exclusion chromatography at 25°C

**Fibronectin,** see also Tenascin

**Fibronectin**

Fibronectin type III module, $\Delta G$ from equilibrium unfolding and kinetics

| Fn3 Species | pH | T | $\Delta G$ | $c_{1/2}$ | m | Appr./Rem. | Ref |
|---|---|---|---|---|---|---|---|
| FnFn3(9) | 4.8 | 25 | ≈2.7 | 0.37 | 20.6 | GuHCl (1–3) | 97S14 |
| FnFn3(9) | 4.8 | 25 | 5.9 | | | GuHCl (1,4) | 97S14 |
| FnFn3(9) | 4.8 | 25 | 5.13 | | | GuHCl (1,5) | 97S14 |
| FnFn3(9PG510) | 4.8 | 25 | 11.0 | 0.76 | 14.6 | GuHCl (1–3) | 97S14 |
| FnFn3(9PG510) | 4.8 | 25 | 12.1 | | | GuHCl (1,4) | 97S14 |
| FnFn3(9PG510) | 4.8 | 25 | 9.92 | | | GuHCl (1,5) | 97S14 |
| FnFn3(9PG10) | 4.8 | 25 | 13.7 | 0.90 | 15.4 | GuHCl (1–3) | 97S14 |
| FnFn3(9PG10) | 4.8 | 25 | 14.3 | | | GuHCl (1,4) | 97S14 |
| FnFn3(9PG10) | 4.8 | 25 | 12.47 | | | GuHCl (1,5) | 97S14 |
| FnFn3(9,10) | 4.8 | 25 | 14.3 | 1.11 | 12.9 | GuHCl (1–3) | 97S14 |
| FnFn3(9,10) | 4.8 | 25 | 17.7 | | | GuHCl (1,4) | 97S14 |
| FnFn3(9,10) | 4.8 | 25 | 15.56 | | | GuHCl (1,5) | 97S14 |

Remarks:
(1) abbreviations:

Fn3 = fibronectin type III

FnFn3(9) or (10) = the ninth or tenth fibronectin type III module from
human fibronectin

FnFn3(9,10), FnFn3(9PG10), FnFn3(9PG510) = fragment of human fibronectin
containing FnFn3(9) and FnFn3(10)

(2) equilibrium value obtained by linear extrapolation, entries for double modules refer to the first transition = FnFn3(9) moiety

(3) average error in $\Delta G$ ±10%, except FnFn3(9)

(4) $\Delta G$ from m * $c_{1/2}$

(5) data from kinetics

Fibronectin type III domain of human tenascin

| | pH | T | $\Delta G$ | $c_{1/2}$ | m | Appr./Rem. | Ref |
|---|---|---|---|---|---|---|---|
| | 5.0 | 20 | 22.2±1.7 | 3.8±0.1 | 5.86±0.84 | urea (1,2) | 97C9 |
| | 5.0 | 20 | 24.7±3.3 | 3.7±0.1 | 6.69±1.67 | urea (1,2) | 97C9 |
| | 5.0 | 20 | 22.6±0.8 | 3.8±0.2 | 5.86±0.42 | urea (1,3) | 97C9 |
| | 5.0 | 20 | 18.0±0.8 | 3.9±0.1 | 4.60±0.42 | urea (1,3) | 97C9 |
| | 5.0 | 20 | 21.8±0.8 | 3.7±0.1 | 5.86±0.42 | urea (1,2) | 97C9 |
| | 5.0 | 20 | 22.2±2.1 | | | urea (1,4) | 97C9 |
| | 5.0 | 20 | 21.3 | | 5.86 | urea (1,5) | 97C9 |
| | 5.0 | 20 | 24.3±1.3 | | | DSC (1,6) | 97C9 |
| | 5.0 | 20 | 22.2±1.3 | | | DSC (1,7) | 97C9 |
| | 7.0 | 20 | 11.7±2.1 | 2.5±0.1 | 4.60±1.26 | urea (1,2) | 97C9 |
| | 7.0 | 20 | 12.1±1.3 | 2.4±0.1 | 5.02±0.84 | urea (1,3) | 97C9 |
| | 7.0 | 20 | 12.1±2.1 | 2.4±0.2 | 5.02±1.26 | urea (1,8) | 97C9 |
| | 7.0 | 20 | 13.4±0.8 | 2.3±0.1 | 5.86±0.42 | urea (1,9) | 97C9 |

Remarks:
(1) linear extrapolation, LEM-SB
(2) transition monitored by fluorescence at 320 nm
(3) transition monitored by fluorescence at 350 nm
(4) urea denaturation, mean value
(5) from kinetics
(6) from DSC data (see Table 2) using $\Delta Cp$ = 5.0 kJ/mol/K
(7) from DSC data (see Table 2) using $\Delta Cp$ = 6.3 kJ/mol/K
(8) transition monitored by CD at 222 nm
(9) transition monitored by CD at 230 nm

Third fibronectin type III domain (TNfn3) from human tenascin

| Protein | pH | T | $\Delta G$ | $c_{1/2}$ | Approach/Remarks | | Ref |
|---------|----|----|-----------|-----------|------------------|--|-----|
| TNfn3(1–90) | 5.0 | 20 | 19.92±0.46 | 3.78±0.04 | urea | (2,3) | 98H3 |
| TNfn3(1–92) | 5.0 | 20 | 31.13±0.59 | 5.91±0.04 | urea | (1,2) | 98H3 |
| TNfn3(1–90) | 7.0 | 25 | 13.89±0.71 | | DSC | (4) | 98H3 |
| TNfn3(1–92) | 7.0 | 25 | 23.81±0.54 | 4.52±0.05 | urea | (1,2) | 98H3 |
| TNfn3(1–92) | 5.0 | 25 | 27.95±0.75 | 5.31±0.12 | urea | (1,2) | 98H3 |
| TNfn3(1–92)-(Leu92→Ala) | | | | | | | |
| | 5.0 | 25 | 27.82±0.54 | 5.28±0.03 | urea | (1,2) | 98H3 |

Remarks:

(1) linear extrapolation, using an averaged m-value of 5.27 kJ/mol/M from 14 different experiments

(2) buffer: 50 mM acetate, pH 5.0

(3) data recalculated from Ref. 97C9, using m = 5.27 kJ/mol/M

(4) $\Delta G$ derived from extrapolation of DSC data from Ref. 97C9, $\Delta H^{v.H.}$ = 265 kJ/mol and $\Delta Cp$ = 4.98 kJ/mol/K

(5) Ref. 98H3 contains additional $\Delta(\Delta G)$ values derived from hydrogen exchange of amide protons for single amino acid residues in TNfn3(1–90) and TNfn3(1–92)

(6) thermal stability in the absence of urea as monitored by CD at 230 nm: TNfn3(1–90) = 57.1±0.1°C, TNfn3(1–92) = 64.4 ±0.1°C, and TNfn3(1–92)-(Leu92→Ala) = 64.7±0.1°C

Ninth ($^9$FNIII) and tenth ($^{10}$FNIII) type III modules of human fibronectin

| Domain | pH | T | $\Delta G$ | m | Approach/Remarks | | Ref |
|--------|----|----|-----------|---|------------------|--|-----|
| $^9$FNIII | 5.2 | 25±1 | 5.0±2.1 | 5.94±0.08 | GuHCl | (1–3) | 97P12 |
| $^{10}$FNIII | 5.2 | 25±1 | 25.5±0.4 | 12.6±4.2 | GuHCl | (1–3) | 97P12 |

Remarks:

(1) linear extrapolation

(2) transition monitored by fluorescence at 350 nm

(3) buffer: 20 mM sodium acetate, pH 5.2

## FKBP12, FK506-Binding Protein (Immunophilin)

FKBP12, FK506-binding protein (immunophilin), wild-type FKBP12 and mutants

| Protein | pH | T | $\Delta(\Delta G)$ | Approach/Remarks | | Ref |
|---------|----|----|-------------------|------------------|--|-----|
| wild type | 7.5 | 25 | 0.00 | urea | (1–3) | 99F12 |
| Val2→Ala | 7.5 | 25 | −10.17±0.50 | urea | (1–3) | 99F12 |
| Val4→Ala | 7.5 | 25 | −11.63±0.42 | urea | (1–3) | 99F12 |
| Ile7→Val | 7.5 | 25 | −3.85±0.46 | urea | (1–3) | 99F12 |
| Thr21→Ala | 7.5 | 25 | −6.69±0.46 | urea | (1–3) | 99F12 |
| Thr21→Ser | 7.5 | 25 | −6.02±0.46 | urea | (1–3) | 99F12 |
| Thr21→Val | 7.5 | 25 | 3.60±0.54 | urea | (1–3) | 99F12 |
| Val23→Ala | 7.5 | 25 | −12.43±0.42 | urea | (1–3) | 99F12 |
| Val24→Ala | 7.5 | 25 | −13.35±0.46 | urea | (1–3) | 99F12 |
| Thr27→Ala | 7.5 | 25 | −8.24±0.71 | urea | (1–3) | 99F12 |
| Thr27→Ser | 7.5 | 25 | −6.23±0.46 | urea | (1–3) | 99F12 |
| Thr27→Val | 7.5 | 25 | 0.96±0.50 | urea | (1–3) | 99F12 |
| Phe36→Ala | 7.5 | 25 | −14.81±0.63 | urea | (1–3) | 99F12 |
| Leu50→Ala | 7.5 | 25 | −10.75±0.46 | urea | (1–3) | 99F12 |
| Val55→Ala | 7.5 | 25 | −8.91±0.54 | urea | (1–3) | 99F12 |
| Ile56→Ala | 7.5 | 25 | −10.38±0.50 | urea | (1–3) | 99F12 |
| Ile56→Asp | 7.5 | 25 | −13.22±0.84 | urea | (1–3) | 99F12 |
| Ile56→Thr | 7.5 | 25 | −7.57±0.54 | urea | (1–3) | 99F12 |

FKBP12, FK506-binding protein (immunophilin), wild-type FKBP12 and mutants (continued)

| Protein | pH | T | Δ(ΔG) | Approach/Remarks | | Ref |
|---|---|---|---|---|---|---|
| Arg57→Ala | 7.5 | 25 | −3.39±0.46 | urea | (1–3) | 99F12 |
| Arg57→Gly | 7.5 | 25 | −9.58±0.42 | urea | (1–3) | 99F12 |
| Glu60→Ala | 7.5 | 25 | −8.91±0.46 | urea | (1–3) | 99F12 |
| Glu60→Gly | 7.5 | 25 | −11.88±0.50 | urea | (1–3) | 99F12 |
| Glu61→Ala | 7.5 | 25 | −3.51±0.46 | urea | (1–3) | 99F12 |
| Glu61→Gly | 7.5 | 25 | −10.42±0.42 | urea | (1–3) | 99F12 |
| Val63→Ala | 7.5 | 25 | −12.43±0.42 | urea | (1–3) | 99F12 |
| Thr75→Ala | 7.5 | 25 | −10.88±0.46 | urea | (1–3) | 99F12 |
| Thr75→Val | 7.5 | 25 | −3.39±0.54 | urea | (1–3) | 99F12 |
| Ile76→Ala | 7.5 | 25 | −15.94±0.75 | urea | (1–3) | 99F12 |
| Ile76→Val | 7.5 | 25 | −3.18±0.50 | urea | (1–3) | 99F12 |
| Ile91→Ala | 7.5 | 25 | −6.44±0.42 | urea | (1–3) | 99F12 |
| Ile91→Val | 7.5 | 25 | −1.59±0.46 | urea | (1–3) | 99F12 |
| Leu97→Ala | 7.5 | 25 | −14.90±0.46 | urea | (1–3) | 99F12 |
| Val98→Ala | 7.5 | 25 | −9.04±0.54 | urea | (1–3) | 99F12 |
| Val101→Ala | 7.5 | 25 | −11.51±0.42 | urea | (1–3) | 99F12 |
| Leu106→Ala | 7.5 | 25 | −9.71±0.46 | urea | (1–3) | 99F12 |

Remarks:
(1) nonlinear least-squares fit of the transition curve assuming a linear dependence of ΔG on the denaturant conc., for details see also Ref. 98M1
(2) transition monitored by fluorescence spectroscopy
(3) buffer: 50 mM Tris-HCl, 1 mM DTT, pH 7.5

FKBP12, FK506-binding protein, wild type and mutants

| Mutant | pH | T | Δ(ΔG)$_{50\%}$ | $c_{1/2}$ | m | Approach/Remarks | | Ref |
|---|---|---|---|---|---|---|---|---|
| wild type | 7.5 | 25 | 0.0 | 3.87±0.02 | 5.98±0.21 | urea | (1–4) | 98M1 |
| mutations concerning buried hydrophobic residues: | | | | | | | | |
| Val2→Ala | 7.5 | 25 | −10.17±0.50 | 2.34±0.03 | 6.44±0.38 | urea | (1–4) | 98M1 |
| Val4→Ala | 7.5 | 25 | −11.63±0.42 | 2.12±0.03 | 7.49±0.50 | urea | (1–4) | 98M1 |
| Val23→Ala | 7.5 | 25 | −12.43±0.42 | 2.00±0.03 | 6.82±0.38 | urea | (1–4) | 98M1 |
| Val24→Ala | 7.5 | 25 | −13.35±0.46 | 1.86±0.04 | 5.94±0.33 | urea | (1–4) | 98M1 |
| Phe36→Ala | 7.5 | 25 | −14.81±0.63 | 1.64±0.08 | 7.45±0.75 | urea | (1–4) | 98M1 |
| Leu50→Ala | 7.5 | 25 | −10.75±0.46 | 2.25±0.04 | 6.78±0.38 | urea | (1–4) | 98M1 |
| Val55→Ala | 7.5 | 25 | −8.91±0.54 | 2.53±0.05 | 6.78±0.54 | urea | (1–4) | 98M1 |
| Ile56→Ala | 7.5 | 25 | −10.38±0.50 | 2.31±0.05 | 7.11±0.79 | urea | (1–4) | 98M1 |
| Ile56→Asp | 7.5 | 25 | −13.22±0.84 | 1.88±0.11 | 6.53±0.92 | urea | (1–4) | 98M1 |
| Ile56→Thr | 7.5 | 25 | −7.57±0.54 | 2.55±0.06 | 5.65±0.50 | urea | (1–4) | 98M1 |
| Val63→Ala | 7.5 | 25 | −12.43±0.42 | 2.00±0.03 | 7.07±0.42 | urea | (1–4) | 98M1 |
| Ile76→Val | 7.5 | 25 | −3.18±0.50 | 3.39±0.03 | 6.95±0.42 | urea | (1–4) | 98M1 |
| Ile76→Ala | 7.5 | 25 | −15.94±0.75 | 1.47±0.10 | 7.49±1.05 | urea | (1–4) | 98M1 |
| Ile91→Ala | 7.5 | 25 | −6.44±0.42 | 2.90±0.02 | 6.36±0.17 | urea | (1–4) | 98M1 |
| Ile91→Val | 7.5 | 25 | −1.59±0.46 | 3.63±0.02 | 6.57±0.25 | urea | (1–4) | 98M1 |
| Leu97→Ala | 7.5 | 25 | −14.90±0.46 | 1.63±0.04 | 7.91±0.54 | urea | (1–4) | 98M1 |
| Val98→Ala | 7.5 | 25 | −9.04±0.54 | 2.51±0.06 | 6.19±0.71 | urea | (1–4) | 98M1 |
| Val101→Ala | 7.5 | 25 | −11.51±0.42 | 2.14±0.02 | 6.82±0.17 | urea | (1–4) | 98M1 |
| Leu106→Ala | 7.5 | 25 | −13.89±0.46 | 2.41±0.03 | 6.40±0.38 | urea | (1–4) | 98M1 |
| mutations concerning the α-helix: | | | | | | | | |
| Arg57→Ala | 7.5 | 25 | −3.39±0.46 | 3.36±0.02 | 6.53±0.29 | urea | (1–4) | 98M1 |
| Arg57→Gly | 7.5 | 25 | −9.58±0.42 | 2.43±0.02 | 6.95±0.33 | urea | (1–4) | 98M1 |
| Glu60→Ala | 7.5 | 25 | −8.91±0.46 | 2.53±0.03 | 7.03±0.42 | urea | (1–4) | 98M1 |
| Glu60→Gly | 7.5 | 25 | −11.88±0.50 | 2.08±0.05 | 6.28±0.50 | urea | (1–4) | 98M1 |

FKBP12, FK506-binding protein, wild type and mutants

| Mutant | pH | T | $\Delta(\Delta G)_{50\%}$ | $c_{1/2}$ | m | Approach/Remarks | | Ref |
|---|---|---|---|---|---|---|---|---|
| Glu61→Ala | 7.5 | 25 | $-3.51\pm0.46$ | $3.34\pm0.02$ | $6.15\pm0.21$ | urea | (1–4) | 98M1 |
| Glu61→Gly | 7.5 | 25 | $-10.42\pm0.42$ | $2.30\pm0.02$ | $6.57\pm0.25$ | urea | (1–4) | 98M1 |
| mutations concerning β-sheets: | | | | | | | | |
| Ile7→Val | 7.5 | 25 | $-3.85\pm0.46$ | $3.29\pm0.02$ | $5.90\pm0.29$ | urea | (1–4) | 98M1 |
| Thr21→Ala | 7.5 | 25 | $-6.69\pm0.46$ | $2.86\pm0.03$ | $6.82\pm0.42$ | urea | (1–4) | 98M1 |
| Thr21→Ser | 7.5 | 25 | $-6.02\pm0.46$ | $2.96\pm0.02$ | $6.86\pm0.29$ | urea | (1–4) | 98M1 |
| Thr21→Val | 7.5 | 25 | $3.60\pm0.54$ | $4.03\pm0.04$ | $6.02\pm0.29$ | urea | (1–4) | 98M1 |
| Thr27→Ala | 7.5 | 25 | $-8.24\pm0.71$ | $2.63\pm0.09$ | $6.49\pm0.96$ | urea | (1–4) | 98M1 |
| Thr27→Ser | 7.5 | 25 | $-6.23\pm0.46$ | $2.93\pm0.03$ | $6.99\pm0.46$ | urea | (1–4) | 98M1 |
| Thr27→Val | 7.5 | 25 | $0.96\pm0.50$ | $4.01\pm0.02$ | $6.74\pm0.25$ | urea | (1–4) | 98M1 |
| Thr75→Ala | 7.5 | 25 | $-11.09\pm0.46$ | $2.20\pm0.04$ | $6.95\pm0.50$ | urea | (1–4) | 98M1 |
| Thr75→Val | 7.5 | 25 | $-3.39\pm0.54$ | $3.36\pm0.05$ | $6.15\pm0.63$ | urea | (1–4) | 98M1 |

Remarks:

(1) $c_{1/2}$ and m were obtained by linear extrapolation, for a detailed description of the data analysis see Ref. 93J1

(2) $\Delta(\Delta G)_{50\%}$ was obtained from average m value of 33 mutant proteins, m = 6.65±0.08 kJ/mol/M, and the difference in $c_{1/2}$ between mutant and wild type

(3) transition monitored by fluorescence

(4) buffer: 50 mM Tris-HCl, 1 mM DTT, pH 7.5

FKBP12, FK506-binding protein (immunophilin), unfolding of wild-type FKBP12 by different denaturants

| Protein | pH | T | $\Delta G$ | $c_{1/2}$ | m | Appr./Rem. | Ref |
|---|---|---|---|---|---|---|---|
| data from equilibrium unfolding: | | | | | | | |
| FKBP12 | 7.5 | 25 | $23.14\pm0.50$ | $3.87\pm0.01$ | $5.98\pm0.13$ | urea (1–3) | 99M2 |
| + 3.6% TFE | 7.5 | 25 | $24.69\pm1.30$ | $4.40\pm0.03$ | $5.61\pm0.29$ | urea (1–3) | 99M2 |
| + 9.6% TFE | 7.5 | 25 | $32.59\pm1.76$ | $4.58\pm0.03$ | $7.11\pm0.38$ | urea (1–3) | 99M2 |
| + 17 % TFE | 7.5 | 25 | $25.65\pm0.63$ | $3.65\pm0.01$ | $7.03\pm0.17$ | urea (1–3) | 99M2 |
| FKBP12 | 7.5 | 25 | $21.46\pm0.88$ | $0.78\pm0.01$ | $21.61\pm1.13$ | GuHCl (1–3) | 99M2 |
| FKBP12 + rapamycin | | | | | | | |
|  | 7.5 | 25 | $34.35\pm2.38$ | $2.10\pm0.01$ | $16.36\pm1.13$ | GuHCl (1–3) | 99M2 |
| FKBP12 | 7.5 | 25 | $34.10\pm3.39$ | $26.80\pm0.20$ | $1.26\pm0.13$ | TFE (1–4) | 99M2 |
| data from kinetics of unfolding and refolding: | | | | | | | |
| FKBP12 | 7.5 | 25 | $24.3\pm1.3$ | | $7.1\pm2.1$ | urea (3,5) | 99M2 |
| + 0.0% TFE | 7.5 | 25 | $24.7\pm0.8$ | | $7.1\pm2.5$ | urea (3,5) | 99M2 |
| + 3.6% TFE | 7.5 | 25 | $28.9\pm0.8$ | | $6.3\pm2.5$ | urea (3,5) | 99M2 |
| + 9.6% TFE | 7.5 | 25 | $32.2\pm0.8$ | | $6.3\pm2.9$ | urea (3,5) | 99M2 |
| + 17 % TFE | 7.5 | 25 | $26.8\pm1.3$ | | $6.7\pm2.9$ | urea (3,5) | 99M2 |
| FKBP12 | 7.5 | 15 | $24.3\pm0.3$ | | $7.5\pm2.5$ | urea (3,5) | 99M2 |
|  | 7.5 | 20 | $23.4\pm0.8$ | | $7.1\pm2.5$ | urea (3,5) | 99M2 |
|  | 7.5 | 25 | $23.8\pm0.8$ | | $7.1\pm2.5$ | urea (3,5) | 99M2 |
|  | 7.5 | 30 | $22.2\pm0.8$ | | $7.1\pm2.1$ | urea (3,5) | 99M2 |
|  | 7.5 | 35 | $21.3\pm1.3$ | | $6.7\pm2.1$ | urea (3,5) | 99M2 |
| FKBP12 | 7.5 | 25 | $20.9\pm0.4$ | | $37.2\pm0.8$ | GuHCl (3,5) | 99M2 |
| FKBP12 + rapamycin | | | | | | | |
|  | 7.5 | 25 | $29.3\pm1.3$ | | $13.4\pm0.4$ | GuHCl (3,5) | 99M2 |

Remarks:

(1) nonlinear least-squares fit of the transition curve assuming a linear dependence of $\Delta G$ on the denaturant conc.

(2) transition monitored by fluorescence spectroscopy

(3) buffer: 50 mM Tris-HCl, 1 mM DTT, pH 7.5

(4) TFE denaturation, units for m are here kJ/mol/%

(5) from denaturant dependence of kinetics of unfolding and refolding

FKBP12, FK506-binding protein (immunophilin), unfolding in the presence of trifluoroethanol (TFE)

| Mutant | pH | T | $\Delta G$ | $c_{1/2}$ | m | Appr./Rem. | Ref |
|---|---|---|---|---|---|---|---|
| data at 0% TFE: | | | | | | | |
| wild type | 7.5 | 25 | 25.56±0.33 | 3.87±0.02 | 5.98±0.21 | urea (1–4) | 99M3 |
| Arg57→Ala | 7.5 | 25 | 22.22±0.29 | 3.36±0.02 | 6.53±0.29 | urea (1–4) | 99M3 |
| Arg57→Gly | 7.5 | 25 | 16.07±0.25 | 2.43±0.02 | 6.95±0.33 | urea (1–4) | 99M3 |
| Glu60→Ala | 7.5 | 25 | 16.74±0.29 | 2.53±0.03 | 7.03±0.42 | urea (1–4) | 99M3 |
| Glu60→Gly | 7.5 | 25 | 13.77±0.38 | 2.08±0.05 | 6.28±0.50 | urea (1–4) | 99M3 |
| Glu61→Ala | 7.5 | 25 | 22.09±0.29 | 3.34±0.02 | 6.15±0.21 | urea (1–4) | 99M3 |
| Glu61→Gly | 7.5 | 25 | 15.19±0.25 | 2.30±0.02 | 6.57±0.25 | urea (1–4) | 99M3 |
| Val63→Ala | 7.5 | 25 | 13.22±0.25 | 2.00±0.03 | 7.07±0.42 | urea (1–4) | 99M3 |
| Ile7→Val | 7.5 | 25 | 21.76±0.29 | 3.29±0.02 | 5.90±0.29 | urea (1–4) | 99M3 |
| Ile76→Val | 7.5 | 25 | 22.43±0.33 | 3.39±0.03 | 6.95±0.42 | urea (1–4) | 99M3 |
| Leu97→Ala | 7.5 | 25 | 10.79±0.29 | 1.63±0.04 | 7.91±0.54 | urea (1–4) | 99M3 |
| Val98→Ala | 7.5 | 25 | 16.61±0.46 | 2.51±0.06 | 6.19±0.71 | urea (1–4) | 99M3 |
| Val101→Ala | 7.5 | 25 | 14.14±0.21 | 2.14±0.02 | 6.82±0.17 | urea (1–4) | 99M3 |
| data at 9.6% TFE: | | | | | | | |
| wild type | 7.5 | 25 | 31.80±0.21 | 4.58±0.03 | 7.11±0.38 | urea (1–4) | 99M3 |
| Arg57→Ala | 7.5 | 25 | 29.71±0.17 | 4.28±0.02 | 6.57±0.25 | urea (1–4) | 99M3 |
| Arg57→Gly | 7.5 | 25 | 23.81±0.21 | 3.43±0.03 | 6.49±0.42 | urea (1–4) | 99M3 |
| Glu60→Ala | 7.5 | 25 | 25.15±0.21 | 3.62±0.03 | 6.07±0.38 | urea (1–4) | 99M3 |
| Glu60→Gly | 7.5 | 25 | 21.25±0.29 | 3.06±0.04 | 6.90±0.54 | urea (1–4) | 99M3 |
| Glu61→Ala | 7.5 | 25 | 28.95±0.21 | 4.17±0.03 | 6.44±0.42 | urea (1–4) | 99M3 |
| Glu61→Gly | 7.5 | 25 | 21.59±0.29 | 3.11±0.04 | 6.69±0.54 | urea (1–4) | 99M3 |
| Val63→Ala | 7.5 | 25 | 20.96±0.17 | 3.02±0.02 | 7.61±0.38 | urea (1–4) | 99M3 |
| Ile7→Val | 7.5 | 25 | 27.03±0.21 | 3.89±0.03 | 7.53±0.54 | urea (1–4) | 99M3 |
| Ile76→Val | 7.5 | 25 | 29.37±0.17 | 4.23±0.02 | 7.07±0.29 | urea (1–4) | 99M3 |
| Leu97→Ala | 7.5 | 25 | 16.19±0.59 | 2.33±0.08 | 7.03±0.88 | urea (1–4) | 99M3 |
| Val98→Ala | 7.5 | 25 | 21.97±0.08 | 3.16±0.01 | 7.15±0.25 | urea (1–4) | 99M3 |
| Val101→Ala | 7.5 | 25 | 20.50±0.17 | 2.95±0.02 | 8.08±0.42 | urea (1–4) | 99M3 |
| data at 3.6% TFE: | | | | | | | |
| wild type | 7.5 | 25 | 31.55±0.04 | 4.55±0.02 | 7.11±0.38 | urea (1–4) | 99M3 |
| data at 17% TFE: | | | | | | | |
| wild type | 7.5 | 25 | 25.36±0.04 | 3.65±0.02 | 7.03±0.30 | urea (1–4) | 99M3 |

Remarks:

(1) linear extrapolation, for further details see Ref. 98M1

(2) transition monitored by fluorescence

(3) buffer: 50 mM Tris-HCl, pH 7.5, 1 mM DTT

(4) $\Delta G$ was calculated in 0% TFE, using an average m value of 6.65±0.08, and in 9.6% TFE with an average m value of 6.82±0.17 kJ/mol/M

## Flagellin

Flagellin from *Salmonella typhimurium* and its proteolytic fragments

| Protein/Transition | | pH | T | ΔG | Approach/Remarks | | Ref |
|---|---|---|---|---|---|---|---|
| Flagellin | trans. (1) | 7.0 | 25 | 17 | DSC | (1,2) | 99H13 |
| | trans. (2) | 7.0 | 25 | 47 | DSC | (1,2) | 99H13 |
| | trans. (3) | 7.0 | 25 | 65 | DSC | (1,2) | 99H13 |
| F40$^T$ | trans. (1) | 7.0 | 25 | 15 | DSC | (1,2) | 99H13 |
| | trans. (2) | 7.0 | 25 | 47 | DSC | (1,2) | 99H13 |
| | trans. (3) | 7.0 | 25 | 61 | DSC | (1,2) | 99H13 |
| F27$^S$ | trans. (1) | 7.0 | 25 | 28 | DSC | (1,2) | 99H13 |
| | trans. (2) | 7.0 | 25 | 51 | DSC | (1,2) | 99H13 |

Explanations:

| | |
|---|---|
| flagellin | residues 1–494 |
| F40$^T$ | residues 66–450 |
| F27$^S$ | residues 173–422 |

Remarks:

(1) Ref. 99H13 contains an approach which enables better correlation of the results of deconvolution with molecular masses and morphological domains

(2) see also Table 2

## Flavodoxin

Apoflavodoxin from *Anabaena* PCC7119, fragment comprising residues 1–149 (Fld1–149)

| pH | T | Δ(ΔG) | $c_{1/2}$ | m | Approach/Remarks | | Ref |
|---|---|---|---|---|---|---|---|
| 7.0 | 25.0±0.1 | 5.19±1.00 | 1.46±0.12 | 3.56±0.46 | urea | (1–3,6) | 98M4 |
| 7.0 | 25.0±0.1 | 4.81±0.96 | 1.52±0.13 | 3.14±0.38 | urea | (1–3,7) | 98M4 |
| 7.0 | 25.0±0.1 | 4.39±0.50 | 1.55±0.07 | 2.85±0.21 | urea | (1,2,4,6) | 98M4 |
| 7.0 | 25.0±0.1 | 4.73±0.46 | 1.52±0.07 | 3.56±0.46 | urea | (1,2,4,7) | 98M4 |
| 7.0 | 25.0±2.0 | 4.69±0.25 | 1.53±0.06 | 3.05±0.25 | urea | (1,2,5,8) | 98M4 |

Remarks:

(1) linear extrapolation, LEM-SB

(2) measured in 5 mM sodium phosphate, 0.5 M NaCl

(3) transition monitored by fluorescence

(4) transition monitored by CD

(5) transition monitored by size exclusion chromatography

(6) measured at 0.8 μM Fld1–149

(7) measured at 8.0 μM Fld1–149

(8) measured at 1.2 μM Fld1–149

Apoflavodoxin II from *Azotobacter vinelandii*, pseudo-wild type Cys69→Ala, free energies for local opening calculated from amide proton exchange rates

| Residue | pH* | T | ΔG$_{op}$ | Approach/Remarks | | Ref |
|---|---|---|---|---|---|---|
| Leu5 | 6.2 | 25 | 26.8 | HX | (1,2) | 98S20 |
| Phe6 | 6.2 | 25 | 27.2 | HX | (1,2) | 98S20 |
| Gly8 | 6.2 | 25 | 21.3 | HX | (1,2) | 98S20 |
| Val17 | 6.2 | 25 | 18.8 | HX | (1,2) | 98S20 |
| Ala18 | 6.2 | 25 | 23.8 | HX | (1,2) | 98S20 |
| Lys19 | 6.2 | 25 | 24.7 | HX | (1,2) | 98S20 |
| Ile21 | 6.2 | 25 | 23.4 | HX | (1,2) | 98S20 |

**Table 1. Gibbs Energy Change – Molar Values**

Apoflavodoxin II from *Azotobacter vinelandii*, pseudo-wild type Cys69→Ala, free energies for local opening calculated from amide proton exchange rates

| Residue | pH* | T | $\Delta G_{op}$ | Approach/Remarks | | Ref |
|---|---|---|---|---|---|---|
| Lys22 | 6.2 | 25 | 24.7 | HX | (1,2) | 98S20 |
| Arg24 | 6.2 | 25 | 18.4 | HX | (1,2) | 98S20 |
| Asp43 | 6.2 | 25 | 16.3 | HX | (1,2) | 98S20 |
| Ala45 | 6.2 | 25 | 24.7 | HX | (1,2) | 98S20 |
| Gln46 | 6.2 | 25 | 18.0 | HX | (1,2) | 98S20 |
| Tyr47 | 6.2 | 25 | 22.6 | HX | (1,2) | 98S20 |
| Phe49 | 6.2 | 25 | 28.0 | HX | (1,2) | 98S20 |
| Leu50 | 6.2 | 25 | 31.4 | HX | (1,2) | 98S20 |
| Ile51 | 6.2 | 25 | 29.7 | HX | (1,2) | 98S20 |
| Leu52 | 6.2 | 25 | 29.3 | HX | (1,2) | 98S20 |
| Gly53 | 6.2 | 25 | 28.5 | HX | (1,2) | 98S20 |
| Lys80 | 6.2 | 25 | 22.2 | HX | (1,2) | 98S20 |
| Ile81 | 6.2 | 25 | 27.6 | HX | (1,2) | 98S20 |
| Glu82 | 6.2 | 25 | 20.9 | HX | (1,2) | 98S20 |
| Leu84 | 6.2 | 25 | 22.2 | HX | (1,2) | 98S20 |
| Thr90 | 6.2 | 25 | 28.9 | HX | (1,2) | 98S20 |
| Val91 | 6.2 | 25 | 30.1 | HX | (1,2) | 98S20 |
| Ala92 | 6.2 | 25 | 31.8 | HX | (1,2) | 98S20 |
| Leu93 | 6.2 | 25 | 30.5 | HX | (1,2) | 98S20 |
| Phe94 | 6.2 | 25 | 31.4 | HX | (1,2) | 98S20 |
| Gly95 | 6.2 | 25 | 25.9 | HX | (1,2) | 98S20 |
| Gly97 | 6.2 | 25 | 20.5 | HX | (1,2) | 98S20 |
| Val100 | 6.2 | 25 | 13.8 | HX | (1,2) | 98S20 |
| Leu110 | 6.2 | 25 | 22.6 | HX | (1,2) | 98S20 |
| Gly111 | 6.2 | 25 | 27.6 | HX | (1,2) | 98S20 |
| Leu113 | 6.2 | 25 | 21.8 | HX | (1,2) | 98S20 |
| Tyr114 | 6.2 | 25 | >37.7 | HX | (1,2) | 98S20 |
| Phe117 | 6.2 | 25 | >37.7 | HX | (1,2) | 98S20 |
| Lys118 | 6.2 | 25 | 34.3 | HX | (1,2) | 98S20 |
| Asp119 | 6.2 | 25 | 17.6 | HX | (1,2) | 98S20 |
| Arg120 | 6.2 | 25 | 28.9 | HX | (1,2) | 98S20 |
| Lys123 | 6.2 | 25 | 28.9 | HX | (1,2) | 98S20 |
| Val125 | 6.2 | 25 | 27.6 | HX | (1,2) | 98S20 |
| Trp128 | 6.2 | 25 | 19.2 | HX | (1,2) | 98S20 |
| Ala140 | 6.2 | 25 | 17.6 | HX | (1,2) | 98S20 |
| Val141 | 6.2 | 25 | 16.7 | HX | (1,2) | 98S20 |
| Val142 | 6.2 | 25 | 15.5 | HX | (1,2) | 98S20 |
| Lys145 | 6.2 | 25 | 18.4 | HX | (1,2) | 98S20 |
| Phe146 | 6.2 | 25 | 19.2 | HX | (1,2) | 98S20 |
| Val147 | 6.2 | 25 | 18.4 | HX | (1,2) | 98S20 |
| Gly148 | 6.2 | 25 | 20.5 | HX | (1,2) | 98S20 |
| Leu149 | 6.2 | 25 | 19.2 | HX | (1,2) | 98S20 |
| Ala150 | 6.2 | 25 | 18.8 | HX | (1,2) | 98S20 |
| Leu151 | 6.2 | 25 | 19.2 | HX | (1,2) | 98S20 |
| Ala165 | 6.2 | 25 | 27.2 | HX | (1,2) | 98S20 |
| Trp167 | 6.2 | 25 | 29.7 | HX | (1,2) | 98S20 |
| Leu168 | 6.2 | 25 | 31.4 | HX | (1,2) | 98S20 |
| Ala169 | 6.2 | 25 | 32.2 | HX | (1,2) | 98S20 |
| Gln170 | 6.2 | 25 | 31.4 | HX | (1,2) | 98S20 |
| Ile171 | 6.2 | 25 | 30.1 | HX | (1,2) | 98S20 |
| Ala172 | 6.2 | 25 | 30.5 | HX | (1,2) | 98S20 |
| Leu177 | 6.2 | 25 | 16.7 | HX | (1,2) | 98S20 |

Remarks:
(1) measured in 150 mM pyrophosphate, pH* 6.2, 303 K
(2) residues with $\Delta G_{op} \geq 25.9$ are thought to form the structured part of an apoflavodoxin folding intermediate

Apoflavodoxin II from *Azotobacter vinelandii*, wild type and mutants

| Protein/Transition | pH | T | ΔG | m | Approach/Remarks | | Ref |
|---|---|---|---|---|---|---|---|
| wild type | | | | | | | |
| $\quad$ N → U | 6.0 | 25 | 30.8±1.1 | 17.9±0.5 | GuHCl | (1–4) | 98V1 |
| $\quad$ N → I | 6.0 | 25 | 17.2±1.3 | 10.2±0.9 | GuHCl | (1–4) | 98V1 |
| Cys69→Ala | | | | | | | |
| $\quad$ N → U | 6.0 | 25 | 25.9±0.8 | 15.9±0.4 | GuHCl | (1–5) | 98S20 |
| $\quad$ N → I | 6.0 | 25 | 18.4±1.7 | 10.5±1.3 | GuHCl | (1–5) | 98S20 |
| Cys69→Ala | | | | | | | |
| $\quad$ N → U | 6.0 | 25 | 26.1±0.8 | 16.1±0.5 | GuHCl | (1–4,6) | 98V1 |
| $\quad$ N → I | 6.0 | 25 | 18.5±1.6 | 10.5±1.0 | GuHCl | (1–4,6) | 98V1 |
| Cys69→Ser | | | | | | | |
| $\quad$ N → U | 6.0 | 25 | 26.9±1.5 | 16.6±0.9 | GuHCl | (1–4) | 98V1 |
| $\quad$ N → I | 6.0 | 25 | 18.2±2.2 | 10.5±1.5 | GuHCl | (1–4) | 98V1 |

Remarks:

(1) linear extrapolation, LEM-SB

(2) buffer: 100 mM potassium pyrophosphate, pH 6.0

(3) the transitions monitored by CD at 222 nm and fluorescence at 333 nm do not coincide with GuHCl and thermal denaturation, respectively

(4) ΔG from a three-state model

(5) mutant Cys69→Ala is termed pseudo-wild type in Ref. 98S20

(6) the average $\Delta G_{op}$ of the stable nucleus was calculated from hydrogen exchange to be 29.3 kJ/mol (Ref. 98S20), which is significantly higher than ΔG of the N → U transition of 26.1 kJ/mol from equilibrium denaturation, see Ref. 98V1

# Gelsolin

Gelsolin domain 2, wild type and mutants

| Protein | pH | T | ΔG | $c_{1/2}$ | m | Appr./Rem. | Ref |
|---|---|---|---|---|---|---|---|
| wild type | 7.2 | 25 | 16.48±0.79 | 2.28±0.04 | 7.20±0.50 | urea (1–4) | 99I3 |
| | 7.2 | 15 | 19.71±0.25 | 2.82±0.02 | 6.99±0.13 | urea (1–4) | 99I3 |
| | 7.2 | 25 | 15.36±0.67 | 2.65 | 7.45±0.42 | urea (4,6,7) | 99I3 |
| Asp187→Asn | 7.2 | 25 | 12.80±1.80 | 1.59±0.11 | 8.03±1.34 | urea (1–5) | 99I3 |
| | 7.2 | 25 | 10.17±0.79 | 1.85 | 8.45±0.38 | urea (4–7) | 99I3 |
| Asp187→Tyr | 7.2 | 15 | 11.76±2.01 | 1.60±0.14 | 7.32±1.46 | urea (1–5) | 99I3 |
| | 7.2 | 15 | 10.96±0.79 | 1.82 | 7.53±0.04 | urea (4–7) | 99I3 |

Remarks:

(1) equilibrium data

(2) nonlinear least-squares fit of the transition curve assuming a linear dependence of ΔG on the denaturant conc.

(3) transition monitored by fluorescence emission at 356 nm

(4) buffer: 50 mM Tris-HCl, 150 mM NaCl, pH 7.2

(5) the mutations in domain 2 of the actin-regulating protein gelsolin cause familial amyloidosis Finnish type

(6) data from the kinetics of gelsolin folding and refolding in the presence of urea

(7) transition measured by stopped-flow fluorescence

## Glucanase

Endoglucanase CenC, *Cellulomonas fimi*, β-1,4-glucanase CenC, N1 cellulose-binding domain

| pH | T | ΔG | Approach/Remarks | | Ref |
|---|---|---|---|---|---|
| 5.14 | 25 | 21.1±1.7 | DSC | (1,2) | 98C12 |
| 5.50 | 25 | 21.4 | DSC | (1) | 98C12 |
| 6.10 | 25 | 24.9 | DSC | (1) | 98C12 |
| 7.09 | 25 | 22.5 | DSC | (1) | 98C12 |
| 7.36 | 25 | 20.5 | DSC | (1) | 98C12 |
| 9.06 | 25 | 17.9 | DSC | (1) | 98C12 |
| 10.58 | 25 | 14.7 | DSC | (1,3) | 98C12 |
| 10.86 | 25 | 14.4 | DSC | (1,3) | 98C12 |
| 11.06 | 25 | 9.2 | DSC | (1,3) | 98C12 |
| 6.1 | 1 | 33 | DSC | (1,4) | 98C12 |

Remarks:
(1) calculated using $\Delta Cp$ = 7.5 kJ/mol/K, see also Table 2
(2) thermal unfolding was irreversible at pH 5.14 and below
(3) thermal unfolding was partly reversible at pH 10.58 and above
(4) conditions at which the protein achieves maximal stability

## Glucoamylase

Glucoamylase from *Aspergillus awamori*, proline mutants

| Mutant | pH | T | ΔG | $c_{1/2}$ | m | Approach/Remarks | | Ref |
|---|---|---|---|---|---|---|---|---|
| wild type | 4.5 | 25 | 28.32 | 3.58 | 7.90 | GuHCl | (1) | 97L9 |
| Ala27→Pro | 4.5 | 25 | 17.63 | 3.07 | 5.75 | GuHCl | (1) | 97L9 |
| Ala393→Pro | 4.5 | 25 | 21.97 | 3.00 | 7.32 | GuHCl | (1) | 97L9 |
| Ala435→Pro | 4.5 | 25 | 14.40 | 4.05 | 3.55 | GuHCl | (1) | 97L9 |
| Ser436→Pro | 4.5 | 25 | 14.33 | 4.49 | 3.19 | GuHCl | (1) | 97L9 |
| Ser460→Pro | 4.5 | 25 | 16.50 | 2.83 | 5.84 | GuHCl | (1) | 97L9 |

Remarks:
(1) linear extrapolation
(2) transition monitored by CD at 220 nm
(3) buffer: 50 mM sodium acetate, pH 4.5

## Glutathione Transferase

Human class α glutathione transferase A1-1 (hGST A1-1), dimer of two type-1 subunits

| Protein Conc. | pH | T | ΔG | Appr./Rem. | | Ref |
|---|---|---|---|---|---|---|
| 0.1 μM hGST | 6.5 | 25 | 108.7±13.5 | urea | (1–3) | 98W3 |
| 1.0 μM hGST | 6.5 | 25 | 115.4±15.9 | urea | (1–3) | 98W3 |
| 5.0 μM hGST | 6.5 | 25 | 111.6±17.1 | urea | (1–3) | 98W3 |

Remarks:
(1) linear extrapolation, data normalized to 1 M protein
(2) the coincidence of the results obtained while varying the protein conc. validates the two-state model
(3) buffer: 20 mM sodium phosphate, 1 mM EDTA, 100 mM NaCl, pH 6.5

Human glutathione S-transferase (hGST A1-1), wild type and mutant

| Mutant | pH | T | ΔG | $c_{1/2}$ | m | Appr./Rem. | Ref |
|---|---|---|---|---|---|---|---|
| wild type | 6.5 | | 115.0±15.9 | 4.5 | 17.6±2.9 | urea (1–4) | 98W2 |
| Leu164→Ala | 6.5 | | 70.3±8.4 | 3.6 | 10.9±1.3 | urea (1–4) | 98W2 |

Remarks:

(1) linear extrapolation

(2) transition monitored at room temperature by tryptophan fluorescence at 325 nm for folded, and at 335 nm for unfolded protein

(3) buffer: 20 mM sodium phosphate, 1 mM EDTA, 100 mM NaCl, pH 6.5

(4) data obtained at 1 μM enzyme concentration

Human glutathione transferase A1-1 (hGST A1-1) and deletion of C-terminal helix 9 (a9del hGST A1-1)

| Protein | pH | T | ΔG | $c_{1/2}$ | m | Appr./Rem. | Ref |
|---|---|---|---|---|---|---|---|
| hGST A1-1 | 6.5 | 25 | 115 | 4.6 | 17.6 | urea (1–3) | 99D5 |
| a9del hGST A1-1 | 6.5 | 25 | 105 | 4.6 | 15.9 | urea (1–3) | 99D5 |

Remarks:

(1) linear extrapolation, see also Ref. 98W3

(2) transition monitored by Trp20 fluorescence and ellipticity at 222 nm

(3) buffer: 20 mM sodium phosphate, 0.1 M NaCl, 1 mM EDTA, pH 6.5

26 kDa glutathione S-transferase from *Schistosoma japonicum*

| pH | T | ΔG | $c_{1/2}$ | m | Appr./Rem. | Ref |
|---|---|---|---|---|---|---|
| 6.5 | 23 | 109±7 | 4.5 | 18.8 | urea (1–3) | 97K1 |

Remarks:

(1) treatment according to the linear extrapolation method for an equilibrium between folded dimer and unfolded monomers, LEM-SB

(2) buffer: 20 mM sodium phosphate, 0.1 M NaCl, 1 mM EDTA, 2 mM DTT, 0.02% $NaN_3$, pH 6.5

(3) transition monitored by Trp fluorescence and enzyme activity

Homodimeric squid class sigma glutathione S-transferase (GST S1-1) in the presence of NaCl

| NaCl (M) | pH | T | ΔG | $c_{1/2}$ | m | Appr./Rem. | Ref |
|---|---|---|---|---|---|---|---|
| 0.4 M | 7.0 | 20 | 52.3 | 2.7 | 9.2 | urea (1–5) | 98S21 |
| 1.0 M | 7.0 | 20 | 64.9 | 3.5 | 9.2 | urea (1–5) | 98S21 |

Remarks:

(1) linear extrapolation

(2) GST S1-1 unfolding follows a four-state pathway: $N_2 \to I_2 \to 2M \to 2U$, the present data corresponds to the third unfolding transition

(3) GST S1-1 unfolding was studied by tryptophan fluorescence, anisotropy, enzyme activity, ANS binding, and CD

(4) transition monitored by fluorescence

(5) buffer: 20 mM sodium phosphate, pH 7, 1 mM EDTA, 5 mM DTT, and NaCl as indicated

## Glyceraldehyde-3-Phosphate Dehydrogenase

Glyceraldehyde-3-phosphate dehydrogenase from *Bacillus stearothermophilus*, circular permutation within the coenzyme-binding domain

| Mutant | pH | T | $\Delta(\Delta G)$ | $c_{1/2}$ | m | Appr./Rem. | Ref |
|---|---|---|---|---|---|---|---|
| wild type | 8.2 | 25 | 0.0 | 1.21±0.05 | 10.58±0.46 | GuHCl (1–3) | 95V10 |
| circularized | 8.2 | 25 | −3.10±0.84 | 0.97±0.05 | 9.96±0.54 | GuHCl (1–3) | 95V10 |

Remarks:
(1) linear extrapolation
(2) transition monitored by CD at 220 nm
(3) buffer: 50 mM Tris, 2 mM EDTA, 1 mM DTT, pH 8.2

## Glycodelin A

Human glycodelin A (GdA) and β-lactoglobulin A (β-LgA) in water-alcohol mixtures

| Protein | pH | T | $\Delta G$ | $c_{1/2}$ | m | Appr./Rem. | Ref |
|---|---|---|---|---|---|---|---|
| GdA | | 25 | 22.6±1.3 | 3.5±0.1 | 6.3±0.4 | (1,2) | 99G4 |
| β-LgA | | 25 | 15.9±2.1 | 4.0±0.2 | 3.8±0.8 | (1,2) | 99G4 |

Remarks:
(1) 2-propanol-induced transition monitored by CD at 208 nm
(2) linear extrapolation

## β-Glycosidase

Recombinant β-glycosidase from *Sulfolobus solfataricus* expressed in *E. coli*

| Protein Conc. | pH | T | $\Delta G_1$ | $m_1$ | $\Delta G_2$ | $m_2$ | Appr./Rem. | Ref |
|---|---|---|---|---|---|---|---|---|
| 0.010 mg/ml | 5.0 | 25 | 60.1 | 9.5 | 116.0 | 12.5 | GuHCl (1,2) | 98C3 |
| 0.001 mg/ml | 6.5 | 25 | 67.2 | 9.7 | 125.1 | 13.0 | GuHCl (1,2) | 98C3 |
| 0.005 mg/ml | 6.5 | 25 | 67.7 | 10.0 | 125.9 | 12.9 | GuHCl (1,2) | 98C3 |
| 0.010 mg/ml | 6.5 | 25 | 68.4 | 10.1 | 127.8 | 13.2 | GuHCl (1,2) | 98C3 |
| 0.050 mg/ml | 6.5 | 25 | 68.9 | 10.3 | 128.5 | 13.4 | GuHCl (1,2) | 98C3 |
| 0.100 mg/ml | 6.5 | 25 | 68.2 | 10.5 | 128.3 | 13.0 | GuHCl (1,2) | 98C3 |
| 0.010 mg/ml | 8.0 | 25 | 61.0 | 9.8 | 117.2 | 12.6 | GuHCl (1,2) | 98C3 |

Remarks:
(1) thermodynamic parameters obtained from the nonlinear regression of fluorescence intensity measurements
(2) the transition was resolved into two subtransitions, the mechanism involves a dimeric intermediate
(3) the transition curves obtained by fluorescence did not coincide with those determined by CD
(4) buffer: 50 mM acetate, pH 5.0, 50 mM phosphate, pH 6.5, and 50 mM 2-(N-morpholino)ethanesulfonic acid, pH 8.5

## Granulocyte Colony-Stimulating Factor

Granulocyte colony-stimulating factor (G-CSF) and circularly permuted mutants

| Mutant | pH | T | $\Delta G$ | $c_{1/2}$ | Appr./Rem. | Ref |
|---|---|---|---|---|---|---|
| G-CSF(S17) | 4.0 | 20 | 40.2±1.3 | 5.53±0.36 | urea (1–3) | 99F4 |
| G-CSF[39/38] | 4.0 | 20 | 35.1±2.9 | 4.93±0.02 | urea (1–3) | 99F4 |
| G-CSF(L1)[39/38] | 4.0 | 20 | 29.3±2.5 | 4.45±0.04 | urea (1–3) | 99F4 |
| G-CSF[126/125] | 4.0 | 20 | 33.9±0.8 | 4.81±0.31 | urea (1–3) | 99F4 |
| G-CSF(L1)[126/125] | 4.0 | 20 | 30.1±3.8 | 4.43±0.10 | urea (1–3) | 99F4 |
| G-CSF[133/132] | 4.0 | 20 | 33.1±2.9 | 4.70±0.34 | urea (1–3) | 99F4 |
| G-CSF(L1)[133/132] | 4.0 | 20 | 30.5±1.7 | 4.62±0.29 | urea (1–3) | 99F4 |
| G-CSF[142/141] | 4.0 | 20 | 38.1±4.6 | 4.85±0.27 | urea (1–3) | 99F4 |
| G-CSF(L1)[142/141] | 4.0 | 20 | 24.3±1.3 | 4.43±0.14 | urea (1–3) | 99F4 |
| G-CSF[49/48] | 4.0 | 20 | 30.5±1.3 | 4.46±0.11 | urea (1–3) | 99F4 |

Explanations:

cpG-CSF – circularly permuted G-CSF(S17)

L1 – linker L1, substitution of the N-terminal, 10-residue segment by a 7-residue linker, composed of Gly and Ser

Remarks:

(1) linear extrapolation

(2) transition monitored by far–UV CD at 222 nm

(3) buffer: 35 mM sodium acetate, pH 4.0

**GroEL,** see Chaperone

**GroES,** see Chaperone

## Growth Factor

Human acidic fibroblast growth factor (FGF-1), in the presence of GuHCl

| Protein | pH | T | $\Delta G$ | $c_{1/2}$ | m | Appr./Rem. | Ref |
|---|---|---|---|---|---|---|---|
| FGF-1 | 6.60 | 25 | 21.92±0.26 | 1.13±0.01 | 19.44±0.22 | GuHCl (1–3) | 99B15 |
| | 6.60 | 25 | 20.76±0.39 | 1.11±0.02 | 18.67±0.35 | GuHCl (1,2,4) | 99B15 |
| | 6.60 | 25 | 21.22±1.00 | 1.16±0.05 | 18.26±1.04 | GuHCl (1,2,5) | 99B15 |
| | 6.60 | 25 | 21.30±0.58 | 1.13±0.03 | 18.79±0.60 | GuHCl (1,2,6) | 99B15 |

Remarks:

(1) nonlinear least-squares fit of the transition curve, assuming a linear dependence of $\Delta G$ on the denaturant conc.

(2) measured in 20 mM ADA buffer, 0.1 M NaCl, pH 6.60, in the presence of GuHCl

(3) transition monitored by fluorescence

(4) transition monitored by far-UV CD at 227 nm

(5) from DSC data, measured in the presence of GuHCl, see also Table 2

(6) average value

Human acidic fibroblast growth factor (haFGF), calorimetric results for alanine and serine mutants at positions 16, 83, and 117 of haFGF

| Mutant | pH | $T_{tm}$ | $\Delta T$ | $\Delta(\Delta G)$ | Approach/Remarks | | Ref |
|---|---|---|---|---|---|---|---|
| wild type | 7.3 | 46.9 | 0.0 | 0.0 | DSC | (1) | 98P12 |
| Cys16→Ser | 7.3 | 41.2 | −5.7 | −5.14 | DSC | (1,2) | 98P12 |
| Cys117→Ala | 7.3 | 48.6 | 1.7 | 1.54 | DSC | (1,2) | 98P12 |
| Cys117→Ser | 7.3 | 46.6 | −0.3 | −0.27 | DSC | (1,2) | 98P12 |
| double mutant (Cys16→Ala and Cys117→Ala) | | | | | | | |
| | 7.3 | 42.0 | −4.9 | −4.42 | DSC | (1,2) | 98P12 |
| double mutant (Cys83→Ala and Cys117→Ala) | | | | | | | |
| | 7.3 | 46.1 | −0.8 | −0.72 | DSC | (1,2) | 98P12 |
| double mutant (Cys83→Ser and Cys117→Ser) | | | | | | | |
| | 7.3 | 41.8 | −5.1 | −4.60 | DSC | (1,2) | 98P12 |
| triple mutant (Cys16→Ala, Cys83→Ala and Cys117→Ala) | | | | | | | |
| | 7.3 | 39.1 | −7.8 | −7.04 | DSC | (1,2) | 98P12 |

Remarks:

(1) buffer: 50 mM HEPES, 0.5 mM EDTA, 2.0 mM DTT, 0.15 M NaCl, 10 mM $(NH_4)_2SO_4$

(2) $\Delta(\Delta G)$ was calculated using $\Delta(\Delta G) = \Delta T_{mut} \times \Delta S_{w.t.}$, with $\Delta S_{w.t.} = 0.92$ kJ/mol/K, for haFGF

## Growth Hormone

Human growth hormone

| pH | T | $\Delta G$ | Approach/Remarks | | Ref |
|---|---|---|---|---|---|
| 2.0 | 25 | 12.6 | DSC | (1) | 98G12 |
| 3.01 | 25 | 27.7 | DSC | (1) | 98G12 |

Remark:

(1) thermal two-state transition into an intermediate state

Human growth hormone, recombinant (rhGH), spin-labeled rhGH, surfactant-stabilized molten globule intermediate

| Protein/Excipient | | pH | T | $\Delta G$ | $c_{1/2}$ | m | Approach/Remarks | | Ref |
|---|---|---|---|---|---|---|---|---|---|
| rhGH | none | 6.0 | 25 | 62.3 | | | GuHCl | (1–3) | 96B12 |
| rhGH-L | none | 6.0 | 25 | 61.5 | | | GuHCl | (1–3) | 96B12 |
| rhGH-L | none | 6.0 | 25 | 14.6 | 4.6 | 3.1 | GuHCl | (1,2,4,5) | 96B12 |
| rhGH-L | Tween 20 | 6.0 | 25 | 20.1 | 5.5 | 3.8 | GuHCl | (1,2,4,6) | 96B12 |
| rhGH-L | Tween 40 | 6.0 | 25 | 31.8 | 5.9 | 5.4 | GuHCl | (1,2,4,6) | 96B12 |
| rhGH-L | Tween 80 | 6.0 | 25 | 22.2 | 5.3 | 4.1 | GuHCl | (1,2,4,6) | 96B12 |

Explanations:

      rhGH - reference protein

      rhGH-L rhGH spin labeled using the spin label 4-maleimido-TEMPO

Remarks:

(1) linear extrapolation

(2) measured in 10 mM sodium citrate buffer

(3) unfolding transition monitored by CD at 222 nm

(4) transition monitored by EPR (probably native to molten globule transition)

(5) thermal transition monitored by EPR at 45°C, by CD (222 nm) at 88.8°C

(6) stabilizing effect of Tween at a molar ratio Tween to rhGH 4:1

**GTPase Superfamily**, see also p21[H-ras] Protein


## α-Helix Propensity

α-helix propensity derived from folding studies on model peptides

| Amino acid | pH | T | $(\Delta G_m)$ | Approach/Remarks | | Ref |
|---|---|---|---|---|---|---|
| Ala | 7 | 4 | 1.51 | heat | (1–4) | 97Y1 |
| Arg | 7 | 4 | −0.21 | heat | (1–4) | 97Y1 |
| Asn | 7 | 4 | −2.05 | heat | (1–4) | 97Y1 |
| Asp | 7 | 4 | −1.76 | heat | (1–4) | 97Y1 |
| Cys | 7 | 4 | −3.39 | heat | (1–4) | 97Y1 |
| Gln | 7 | 4 | −1.84 | heat | (1–4) | 97Y1 |
| Glu | 7 | 4 | 0.63 | heat | (1–4) | 97Y1 |
| Gly | 7 | 4 | −2.89 | heat | (1–4) | 97Y1 |
| His | 7 | 4 | −0.73 | heat | (1–4) | 97Y1 |
| Ile | 7 | 4 | −0.08 | heat | (1–4) | 97Y1 |
| Leu | 7 | 4 | 0.00 | heat | (1–4) | 97Y1 |
| Lys | 7 | 4 | −0.94 | heat | (1–4) | 97Y1 |
| Met | 7 | 4 | −0.69 | heat | (1–4) | 97Y1 |
| Phe | 7 | 4 | −3.14 | heat | (1–4) | 97Y1 |
| Pro | 7 | 4 | −9.20 | heat | (1–4) | 97Y1 |
| Ser | 7 | 4 | −1.67 | heat | (1–4) | 97Y1 |
| Thr | 7 | 4 | −2.20 | heat | (1–4) | 97Y1 |
| Trp | 7 | 4 | −3.14 | heat | (1–4) | 97Y1 |
| Tyr | 7 | 4 | −6.63 | heat | (1–4) | 97Y1 |
| Val | 7 | 4 | −1.55 | heat | (1–4) | 97Y1 |

Remarks:
(1) host peptide: succinyl-Tyr-Ser-Glu-Glu-Glu-Glu-Lys-Ala-Lys-Lys-Ala-X-Ala-Glu-Glu-Ala-Glu-Lys-Lys-Lys-Lys-NH$_2$
(2) buffer: 1 mM phosphate, 10 mM KF, pH 7
(3) peptide conc. 37 μM
(4) $\Delta G_m$ is the free energy of α-helix formation, for the approach see 90L4

Ribonuclease T1, Ala21 to X mutations

| Mutant | pH | T | $\Delta(\Delta G)$ | $c_{1/2}$ | m | Appr./Rem. | Ref |
|---|---|---|---|---|---|---|---|
| wild type | 2.5 | 25.0 | 0.0 | 2.78 | 6.61 | urea (1-5) | 97M13 |
| Ala21→Arg | 2.5 | 25.0 | −1.72 | 2.53 | 6.90 | urea (1–4) | 97M13 |
| Ala21→Asn | 2.5 | 25.0 | +1.26 | 2.99 | 7.20 | urea (1–4) | 97M13 |
| Ala21→Asp | 2.5 | 25.0 | +1.38 | 2.98 | 6.65 | urea (1–4) | 97M13 |
| Ala21→Cys | 2.5 | 25.0 | −3.10 | 2.33 | 6.53 | urea (1–4) | 97M13 |
| Ala21→Gln | 2.5 | 25.0 | −1.09 | 2.62 | 6.69 | urea (1–4) | 97M13 |
| Ala21→Glu | 2.5 | 25.0 | +0.21 | 2.81 | 6.36 | urea (1–4) | 97M13 |
| Ala21→Gly | 2.5 | 25.0 | −3.77 | 2.23 | 7.15 | urea (1–4) | 97M13 |
| Ala21→His | 2.5 | 25.0 | −2.34 | 2.44 | 7.49 | urea (1–4) | 97M13 |
| Ala21→Ile | 2.5 | 25.0 | −1.84 | 2.51 | 6.57 | urea (1–4) | 97M13 |
| Ala21→Leu | 2.5 | 25.0 | −0.54 | 2.70 | 6.86 | urea (1–4) | 97M13 |
| Ala21→Lys | 2.5 | 25.0 | −2.13 | 2.47 | 7.11 | urea (1–4) | 97M13 |
| Ala21→Met | 2.5 | 25.0 | −0.63 | 2.69 | 6.69 | urea (1–4) | 97M13 |
| Ala21→Phe | 2.5 | 25.0 | −2.38 | 2.43 | 7.07 | urea (1–4) | 97M13 |
| Ala21→Ser | 2.5 | 25.0 | −2.05 | 2.48 | 6.99 | urea (1–4) | 97M13 |
| Ala21→Thr | 2.5 | 25.0 | −2.38 | 2.43 | 6.69 | urea (1–4) | 97M13 |
| Ala21→Trp | 2.5 | 25.0 | −1.26 | 2.60 | 6.53 | urea (1–4) | 97M13 |
| Ala21→Tyr | 2.5 | 25.0 | −1.63 | 2.54 | 7.07 | urea (1–4) | 97M13 |

Ribonuclease T1, Ala21 to X mutations (continued)

| Mutant | pH | T | $\Delta(\Delta G)$ | $c_{1/2}$ | m | Appr./Rem. | Ref |
|---|---|---|---|---|---|---|---|
| Ala21→Val | 2.5 | 25.0 | −2.76 | 2.38 | 7.41 | urea (1–4) | 97M13 |
| wild type | 7.0 | 25.0 | 0.0 | 4.32 | 5.36 | urea (1–5) | 97M13 |
| Ala21→Asp | 7.0 | 25.0 | −2.97 | 3.73 | 5.19 | urea (1–4) | 97M13 |
| Ala21→Gln | 7.0 | 25.0 | −1.67 | 3.99 | 4.69 | urea (1–4) | 97M13 |
| Ala21→Glu | 7.0 | 25.0 | −2.89 | 3.75 | 4.77 | urea (1–4) | 97M13 |
| Ala21→His | 7.0 | 25.0 | −0.71 | 4.18 | 5.31 | urea (1–4) | 97M13 |
| Ala21→Ser | 7.0 | 25.0 | −1.67 | 3.99 | 5.10 | urea (1–4) | 97M13 |

Remarks:

(1) the approach is based on a comparison of the $\alpha$-helix in RNase T1 (residues 13–29) with a peptide that corresponds to the same residues (see the following table)

(2) transition monitored by fluorescence, treatment of the transition by a two-state model with linear dependence of $\Delta G$ on denaturant conc. and consideration of the pre- and postdenaturational slope

(3) temperature was kept at 25.0±0.1°C

(4) error in $c_{1/2}$ ±0.04, error in m ±0.25 kJ/mol/M, $\Delta G$ was calculated using the average m value for all mutants, m = 6.86± 0.33 kJ/mol/M, pH 2.5 and m = 5.06±0.29 kJ/mol/M, pH 7.0, estimated error in $\Delta G$ ±0.42 kJ/mol

(5) wild type = Ala21

Stability of a peptide, derived from ribonuclease T1 residues 13–29

| Mutant | T | $\Delta(\Delta G)$pH2.5 | $\Delta(\Delta G)$pH7.0 | Approach | Remarks | Ref |
|---|---|---|---|---|---|---|
| wild type | 0 | 0.0 | 0.0 | calc. | (1–3) | 97M13 |
| Ala21→Arg | 0 | −2.34 | −1.59 | calc. | (1–3) | 97M13 |
| Ala21→Asn | 0 | −2.76 | −2.43 | calc. | (1–3) | 97M13 |
| Ala21→Asp | 0 | −2.76 | −2.85 | calc. | (1–3) | 97M13 |
| Ala21→Cys | 0 | −2.22 | −2.13 | calc. | (1–3) | 97M13 |
| Ala21→Gln | 0 | −1.30 | −1.21 | calc. | (1–3) | 97M13 |
| Ala21→Glu | 0 | −0.71 | −1.30 | calc. | (1–3) | 97M13 |
| Ala21→Gly | 0 | −4.10 | −3.97 | calc. | (1–3) | 97M13 |
| Ala21→His | 0 | −5.02 | −2.80 | calc. | (1–3) | 97M13 |
| Ala21→Ile | 0 | −1.59 | −1.21 | calc. | (1–3) | 97M13 |
| Ala21→Leu | 0 | −1.05 | −0.79 | calc. | (1–3) | 97M13 |
| Ala21→Lys | 0 | −1.88 | −1.26 | calc. | (1–3) | 97M13 |
| Ala21→Met | 0 | −0.75 | −0.50 | calc. | (1–3) | 97M13 |
| Ala21→Phe | 0 | −2.55 | −2.47 | calc. | (1–3) | 97M13 |
| Ala21→Pro | 0 | −4.60 | −4.60 | calc. | (1–3) | 97M13 |
| Ala21→Ser | 0 | −2.13 | −1.76 | calc. | (1–3) | 97M13 |
| Ala21→Thr | 0 | −2.97 | −2.47 | calc. | (1–3) | 97M13 |
| Ala21→Trp | 0 | −0.59 | −0.08 | calc. | (1–3) | 97M13 |
| Ala21→Tyr | 0 | −1.88 | −1.30 | calc. | (1–3) | 97M13 |
| Ala21→Val | 0 | −2.76 | −2.55 | calc. | (1–3) | 97M13 |

Remarks:

(1) peptide corresponding to the sequence of the helix in RNase T1 (residues 13–29, Gly instead of Ala23):
    Ser-Ser-Asp-Val-Ser-Thr-Ala-Gln-Ala-Ala-Gly-Tyr-Lys-His-Glu-Asp

(2) $\Delta(\Delta G)$ was calculated using Lifson-Roig helix/coil theory from the fractional helicity

(3) fractional helicity was calculated from the mean residue ellipticity at 222 nm, for details see Ref. 97M13

Helix propensity scale based on experimental studies of proteins and peptides

| Amino acid | Helix propensity (kJ/mol) | Remarks | Ref |
|---|---|---|---|
| Ala | 0.00 | (1,2) | 98P2 |
| Arg+ | 0.88 | (1,2) | 98P2 |
| Asn | 2.72 | (1,2) | 98P2 |
| Aspo | 1.80 | (1,2) | 98P2 |
| Asp- | 2.89 | (1,2) | 98P2 |
| Cys | 2.85 | (1,2) | 98P2 |
| Gln | 1.63 | (1,2) | 98P2 |
| Gluo | 0.67 | (1,2) | 98P2 |
| Glu- | 1.67 | (1,2) | 98P2 |
| Gly | 4.18 | (1,2) | 98P2 |
| Hiso | 2.34 | (1,2) | 98P2 |
| His+ | 2.76 | (1,2) | 98P2 |
| Ile | 1.72 | (1,2) | 98P2 |
| Leu | 0.88 | (1,2) | 98P2 |
| Lys+ | 1.09 | (1,2) | 98P2 |
| Met | 1.00 | (1,2) | 98P2 |
| Phe | 2.26 | (1,2) | 98P2 |
| Pro | 13.2 | (1,2) | 98P2 |
| Ser | 2.09 | (1,2) | 98P2 |
| Thr | 2.76 | (1,2) | 98P2 |
| Trp | 2.05 | (1,2) | 98P2 |
| Tyr | 2.22 | (1,2) | 98P2 |
| Val | 2.55 | (1,2) | 98P2 |

Remarks:

(1) the present scale is based on 11 sets of experimental data reviewed in Ref. 98P2

(2) the scale gives Ala the highest helix propensity, and, excluding Pro, glycine the lowest

$\alpha$-Helix Propensity, for further data see also Che Y and Ribonuclease T1

## Hemolysin

Alpha-hemolysin (alpha-toxin) from *Staphylococcus aureus* ($\alpha$HL)

| Protein | pH | T | $\Delta G$ | $c_{1/2}$ | m | Appr./Rem. | Ref |
|---|---|---|---|---|---|---|---|
| $\alpha$HL | 5.0 | 25 | 17.6 | 1.7 | 10.0 | urea (1–3) | 99B17 |
| | 5.0 | 25 | 17.2 | 1.6 | 10.5 | urea (1,2,4) | 99B17 |
| | 5.0 | 25 | | 1.7 | | urea (2,5) | 99B17 |
| | 5.5 | 25 | 19.7 | 1.7 | 11.3 | urea (1–3) | 99B17 |
| | 6.0 | 25 | 24.7 | 1.9 | 13.0 | urea (1–3) | 99B17 |
| | 6.5 | 25 | 33.5 | 2.3 | 14.2 | urea (1–3) | 99B17 |
| | 7.0 | 25 | 37.2 | 2.4 | 15.5 | urea (1–3) | 99B17 |
| | 7.0 | 25 | 37.2 | 2.4 | 15.5 | urea (1,2,4) | 99B17 |
| | 7.0 | 25 | | 2.6 | | urea (2,5) | 99B17 |
| | 7.5 | 25 | 40.6 | 2.8 | 14.2 | urea (1–3) | 99B17 |

Remarks:

(1) linear extrapolation

(2) buffer: 20 mM acetate (pH 5.0 and 5.5) and 20 mM sodium phosphate (pH 6.0–7.5), in the presence of 150 mM NaCl

(3) transition monitored by intrinsic tryptophan fluorescence emission (ITFE)

(4) transition monitored by CD at 218 nm

(5) transition monitored by size exclusion chromatography (SEC)

**His-Tagged Proteins**, see Arc Repressor, Cold Shock Protein CspA, Protein L, and Trigger Factor

## Histone

Recombinant archaeal histones from the mesophile *Methanobacterium formicicum* (rHFoB), hyperthermophile *Methanothermus fervidus* (rHMfA, rHMfB), and hyperthermophile *Pyrococcus* strain GB-3a (rHPyA1)

| Protein | pH | T | $\Delta G$ | Approach/Remarks | Ref |
|---|---|---|---|---|---|
| rHFoB |  | 32 | 30.1 | DSC/heat (1,2) | 98L4 |
| rHFoB |  | 43 | 28.5 | DSC/heat (1,3) | 98L4 |
| rHMfA |  | 35 | 64.9 | DSC/heat (1,2) | 98L4 |
| rHMfA |  | 83 | 32.6 | DSC/heat (1,3) | 98L4 |
| rHMfB |  | 40 | 61.1 | DSC/heat (1,2) | 98L4 |
| rHMfB |  | 83 | 39.3 | DSC/heat (1,3) | 98L4 |
| rHPyA1 |  | 44 | 72.0 | DSC/heat (1,2) | 98L4 |
| rHPyA1 |  | 95 | 33.5 | DSC/heat (1,3) | 98L4 |

Remarks:
(1) from both DSC and heat denaturation experiments, see also Table 2
(2) temperature of maximal stability ($T_{max}$ and $\Delta G_{max}$)
(3) $\Delta G$ at the optimum growth temperature of the archaeon from which the histone originates

## HIV-1 Protease

Plasmid-encoded mutant HIV-1 protease (Gln7→Lys, Leu33→Ile, Leu63→Ile)

| Protein | pH | T | $\Delta G$ | m | Appr./Rem. | Ref |
|---|---|---|---|---|---|---|
|  | 3.4 | 25 | 44.4±2.9 |  | DSC (1) | 98T9 |
|  | 3.4 | 25 | 38.5 |  | diss. (1,2) | 98T9 |
|  | 3.4 | 25 | 37.2 |  | diss. (1,3) | 98T9 |
|  | 3.5 | 25 | 44.4 | 8.8±0.4 | urea (4,5) | 98T9 |
|  | 5.0 | 25 | 62.3 | 8.8±0.4 | urea (4,5) | 98T9 |
|  | 5.0 | 37 | 49.8 |  | diss. (6,7) | 91Z4 |
|  | 6.0 | 25 | 59.4±5.9 |  | urea (4,6,8) | 92G3 |

Remarks:
(1) measured in 10 mM sodium formate (pH 3.4) and 10 mM sodium acetate (pH 4–5.5)
(2) from concentration dependence of protease stability measured by tryptophan fluorescence
(3) from concentration dependence of protease stability measured by decrease in specific activity
(4) linear extrapolation, $\Delta G$ refers to the unfolding of the dimer into two monomers, i.e. $N_2 \rightarrow 2U$
(5) a general expression describing $\Delta G(T,pH,urea)$ in kcal/mol is:

$$\Delta G(T,pH,urea) = -9000 + 3200 \times (T-298.15) - T \times [-79.15 + 3200 \times \ln(T/298.15)] - R \times T \times 4 \times \ln[(1+10^{4.3-pH})/(1+10^{2.9-pH})] - 2100 \times [urea]$$

with the native state at pH 5.5 and 25°C as the reference state
(6) reference value mentioned in Ref. 98T9
(7) measured in 0.1 M sodium acetate, 1 M NaCl, 1 mM EDTA, pH 5.0
(8) measured in 50 mM MES, 0.2 M NaCl, 1 mM EDTA, 1 mM DTT, and 2 % glycerol

**HPr**, see Phosphocarrier Protein

**HU (*Bsu*HU)**, see DNA-Binding Protein

## Human Serum Albumin (HSA)

Human serum albumin (HSA), fatty acid free

| Protein | pH | T | $\Delta G$ | $c_{1/2}$ | m | Appr./Rem. | Ref |
|---------|-----|----|-----------|-----------|-----------|-----------|------|
| HSA | 9.9 | 20 | 4.2±0.4 | 2.00±0.2 | 2.93±0.38 | GuHCl (1) | 99F2 |
| | 7.4 | 20 | 14.2±1.3 | 2.70±0.1 | 5.27±0.29 | GuHCl (1) | 99F2 |
| | 5.3 | 20 | 15.1±1.7 | 2.80±0.2 | 5.86±0.33 | GuHCl (1) | 99F2 |
| HSA | 9.9 | 20 | 4.6±0.8 | 3.58±0.2 | 2.30±0.13 | urea (1) | 99F2 |
| | 7.4 | 20 | 17.2±4.2 | 4.46±0.1 | 4.11±0.07 | urea (1) | 99F2 |
| | 5.3 | 20 | 14.6±1.3 | 1.10±0.1 | 2.51±0.21 | urea (1) | 99F2 |

Remark:
(1) linear extrapolation

Human serum albumin, fatty acid free, in the presence of polyethylene glycols (PEG) of varying molecular weight (MW)

| PEG | (MW) | pH | T | $\Delta G$ | $c_{1/2}$ | m | Appr./Rem. | Ref |
|------|-------|-----|----|-----------|-----------|-----------|------------|------|
| none | | 7.4 | 20 | 23.0±1.3 | 2.75±0.1 | 8.23±0.42 | GuHCl (1–4) | 97F1 |
| | 1000 | 7.4 | 20 | 4.7±0.4 | 2.07±0.1 | 2.52±0.13 | GuHCl (1–4) | 97F1 |
| | 6000 | 7.4 | 20 | 21.3±0.8 | 3.13±0.05 | 6.75±0.18 | GuHCl (1–4) | 97F1 |
| | 10000 | 7.4 | 20 | 18.8±1.3 | 2.88±0.1 | 6.23±0.37 | GuHCl (1–4) | 97F1 |

Remarks:
(1) linear extrapolation
(2) PEG 12 % (w/w), buffer: 100 mM sodium phosphate
(3) transition monitored by fluorescence, excitation 280 nm, emission 345 nm
(4) thermal transition temperature at $T_{trs}$ = 62.2°C in the absence of PEG, $T_{trs}$ = 53.3, 56.3, 62.4, and 63.5°C for PEG 1000, 6000, 8000, and 10,000, respectively

## Immunity Protein

Immunity protein Im9

| Protein | pH | T | $\Delta G$ | m | Appr./Rem. | Ref |
|---------|-----|----|------------|------------|-------------|------|
| Im9 | 7.0 | 25 | 23.22±0.29 | 10.88±0.13 | GuHCl (1–3) | 99K7 |
| | 7.0 | 25 | 24.22±0.42 | 11.05±0.21 | GuHCl (3,4) | 99K7 |

Remarks:
(1) linear extrapolation, LEM-SB
(2) data from equilibrium unfolding, transition monitored by CD at 225 nm
(3) buffer: 50 mM MOPS, 200 mM NaCl, 2 mM DTT
(4) data from kinetics (chevron plot) measured fluorimetrically at $\lambda > 320$ nm

E colicin binding immunity protein Im9

| Protein | Na$_2$SO$_4$ | pH | T | ΔG | m | Appr./Rem. | Ref |
|---------|--------------|-----|-----|---------|-----------|------------|------|
| Im9 | 0.0 M | 7.0 | 10 | 26.1±0.5 | 4.62±0.09 | urea (1–3) | 99F5 |
| | 0.0 M | 7.0 | 10 | 27.5±0.4 | 4.98±0.06 | urea (2–4) | 99F5 |
| | 0.0 M | 7.0 | 25 | | 4.61±0.11 | urea (1,5)99F5 | |

Remarks:
(1) equilibrium unfolding, nonlinear least-squares fit of the two-state transition curve assuming a linear dependence of ΔG on the denaturant conc.
(2) transition monitored by far-UV CD at 225 nm
(3) measured in 50 mM sodium phosphate, 2 mM DTT, pH 7.0
(4) from urea dependence of kinetics of folding and unfolding
(5) transition monitored by fluorescence

E colicin binding immunity protein Im7

| Protein | Na$_2$SO$_4$ | pH | T | ΔG | m | Appr./Rem. | Ref |
|---------|--------------|-----|-----|---------|-----------|------------|------|
| Im7 | 0.0 M | 7.0 | 10 | 16.8±0.3 | 5.21±0.09 | urea (1–3) | 99F5 |
| | 0.4 M | 7.0 | 10 | 26.5±0.5 | 4.80±0.10 | urea (1–3) | 99F5 |
| | 0.0 M | 7.0 | 10 | 19.7±1.3 | 6.31±0.33 | urea (2–5) | 99F5 |
| | 0.4 M | 7.0 | 10 | 27.1±1.6 | 4.84±0.29 | urea (1–4,6) | 99F5 |

Remarks:
(1) equilibrium unfolding, nonlinear least-squares fit of the two-state transition curve assuming a linear dependence of ΔG on the denaturant conc.
(2) transition monitored by far-UV CD at 225 nm
(3) measured in 50 mM sodium phosphate, 2 mM DTT, pH 7.0
(4) from urea dependence of kinetics of folding and unfolding
(5) Im7 folds according to a three-state mechanism with $m_{IU}$ = 3.45±0.14, $m_{TS-I}$ = 2.42±0.17, and $m_{N-TS}$ = 0.43±0.02 kJ/mol/M in 0.0 M Na$_2$SO$_4$
(6) Im7 folds according to a three-state mechanism with $m_{IU}$ = 3.55±0.09, $m_{TS-I}$ = 0.71±0.09, and $m_{N-TS}$ = 0.58±0.09 kJ/mol/M in 0.4 M Na$_2$SO$_4$

## Immunoglobulin

The data entries are arranged as follows:

a) antibodies
b) domains and single chain antibodies
c) mutants with improved stability

*a) antibodies*

Monoclonal anti-ferritin antibodies

| Protein/Subclass | pH | T | ΔG | Approach/Remarks | | Ref |
|------------------|-----|-----|-----|------------------|-------|------|
| G10, IgG2a | 7.0 | 25 | 433 | DSC | (1,2) | 98C9 |
| | 7.0 | 25 | 430 | DSC | (1,2) | 98K15 |
| F11, IgG2a | 7.0 | 25 | 337 | DSC | (1–3) | 98C9 |
| | 7.0 | 25 | 350 | DSC | (1–3) | 98K15 |
| C5, IgG1 | 7.0 | 25 | 488 | DSC | (1,2) | 98C9 |
| | 7.0 | 25 | 490 | DSC | (1,2) | 98K15 |

Remarks:
(1) measured in 50 mM sodium phosphate, pH 7.0
(2) the transition profile resolved by deconvolution shows individual transitions from 66 to 80°C
(3) antibody F11 lacks the individual subtransition at 74°C

Rabbit immunoglobulin (IgG), conformers

| Transition | pH | T | $\Delta G$ | Approach/Remarks | | Ref |
|---|---|---|---|---|---|---|
| N → D | 7 | 25 | 300±30 | DSC | (1) | 96V6 |
| $N_I$ → D | 7 | 25 | 245±25 | DSC | (1) | 96V6 |
| I → D | 2 | 25 | 82± 8 | DSC | (1) | 96V6 |

Remarks:

(1) conformers of rabbit immunoglobulin (IgG) studied by DSC changing pH from 7.0 to 2.0 and from 2.0 to 7.0, results of deconvolution

(2) for the related thermodynamic quantities see Table 2

*b) domains and single chain antibodies*

Antibody $C_H3$ from human $IgG_1$, single chain (sc) and homodimer (hd)

| Mutant | Form | pH | T | $\Delta G$ | $c_{1/2}$ | m | Appr./Rem. | Ref |
|---|---|---|---|---|---|---|---|---|
| wild type | hd | 7.4 | 20±0.1 | 38.1±1.7 | 3.19±0.14 | 12.22±1.00 | GuHCl (1–3) | 98D4 |
| wild type | sc | 7.4 | 20±0.1 | 38.1±1.3 | 3.25±0.08 | 11.72±0.22 | GuHCl (1–3) | 98D4 |
| double mutant (Phe405→Ala and Phe405'→Ala) | | | | | | | | |
| | hd | 7.4 | 20±0.1 | 4.6±0.8 | 1.20±0.02 | 3.85±0.92 | GuHCl (1–4) | 98D4 |
| | sc | 7.4 | 20±0.1 | 5.0±1.3 | 1.05±0.17 | 4.81±0.29 | GuHCl (1–4) | 98D4 |
| double mutant (Tyr407→Ala and Tyr407'→Ala) | | | | | | | | |
| | hd | 7.4 | 20±0.1 | 20.5±1.7 | 1.66±0.08 | 11.76±0.25 | GuHCl (1–3,5) | 98D4 |
| | sc | 7.4 | 20±0.1 | 18.0±1.7 | 1.74±0.15 | 10.54±0.46 | GuHCl (1–3,5) | 98D4 |

Remarks:

(1) linear extrapolation, LEM-SB

(2) transition monitored by CD at 225 nm

(3) measured in 50 mM sodium phosphate, pH 7.4

(4) (Phe405→Ala and Phe405'→Ala) represents replacements of Phe405 in both domains

(5) (Tyr407→Ala and Tyr407'→Ala) represents replacements of Tyr407 in both domains

Antibody $C_H3$ from human $IgG_1$, single chain $scC_H3$ alanine variants

| Mutant | pH | T | $\Delta G$ | $c_{1/2}$ | m | Appr./Rem. | Ref |
|---|---|---|---|---|---|---|---|
| Gln347→Ala | 7.4 | 20±0.1 | 35.1±1.7 | 2.86±0.06 | 12.26±0.84 | GuHCl (1–3) | 98D4 |
| Tyr349→Ala | 7.4 | 20±0.1 | 35.6±0.8 | 3.00±0.01 | 11.84±0.25 | GuHCl (1–3) | 98D4 |
| Thr350→Ala | 7.4 | 20±0.1 | 36.8±0.4 | 3.21±0.01 | 11.55±0.25 | GuHCl (1–3) | 98D4 |
| Leu351→Ala | 7.4 | 20±0.1 | 29.3±1.7 | 2.76±0.02 | 10.54±0.75 | GuHCl (1–3) | 98D4 |
| Thr366→Ala | 7.4 | 20±0.1 | 26.4±0.8 | 2.38±0.02 | 11.09±0.46 | GuHCl (1–3) | 98D4 |
| Leu368→Ala | 7.4 | 20±0.1 | 25.5±1.3 | 2.40±0.04 | 10.54±0.84 | GuHCl (1–3) | 98D4 |
| Lys370→Ala | 7.4 | 20±0.1 | 28.9±1.3 | 2.83±0.08 | 10.25±0.50 | GuHCl (1–3) | 98D4 |
| Lys392→Ala | 7.4 | 20±0.1 | 35.6±1.3 | 3.11±0.19 | 11.51±0.17 | GuHCl (1–3) | 98D4 |
| Thr394→Ala | 7.4 | 20±0.1 | 33.9±0.4 | 3.02±0.01 | 11.17±0.17 | GuHCl (1–3) | 98D4 |
| Pro395→Ala | 7.4 | 20±0.1 | 22.2±1.7 | 2.08±0.03 | 10.96±0.46 | GuHCl (1–3) | 98D4 |
| Val397→Ala | 7.4 | 20±0.1 | 32.2±1.3 | 3.02±0.01 | 10.75±0.38 | GuHCl (1–3) | 98D4 |
| Leu398→Ala | 7.4 | 20±0.1 | 34.3±1.3 | 3.23±0.04 | 10.67±0.33 | GuHCl (1–3) | 98D4 |
| Asp399→Ala | 7.4 | 20±0.1 | 34.3±1.3 | 2.99±0.02 | 11.38±0.25 | GuHCl (1–3) | 98D4 |
| Phe405→Ala | 7.4 | 20±0.1 | 24.3±0.4 | 2.32±0.03 | 10.46±0.29 | GuHCl (1–3) | 98D4 |
| Tyr407→Ala | 7.4 | 20±0.1 | 27.6±1.7 | 2.49±0.02 | 11.09±0.67 | GuHCl (1–3) | 98D4 |
| Lys409→Ala | 7.4 | 20±0.1 | 21.8±0.4 | 2.37±0.11 | 9.25±0.17 | GuHCl (1–3) | 98D4 |

Remarks:

(1) linear extrapolation, LEM-SB

(2) transition monitored by CD at 225 nm

(3) measured in 50 mM sodium phosphate, pH 7.4

Domain stability of $C_H2$ and $C_H3$ domains from different myeloma IgGs

| Protein | pH | T | $\Delta G$ | Approach/Remarks | | Ref |
|---|---|---|---|---|---|---|
| $C_H2$ (donor 1) | 3.8 | 25 | 25.9 | DSC | (1,2) | 98T7 |
| $C_H2$ (donor 1) | 4.2 | 25 | 45.3 | DSC | (1,2) | 98T7 |
| $C_H2$ (donor 2) | 8.0 | 25 | 66.1 | DSC | (1,2) | 98T7 |
| $C_H3$ (donor 1) | 3.8 | 25 | 75.8 | DSC | (1,2) | 98T7 |
| $C_H3$ (donor 1) | 4.2 | 25 | 48.7 | DSC | (1,2) | 98T7 |
| $C_H3$ (donor 2) | 8.0 | 25 | 52.4 | DSC | (1,2) | 98T7 |

Remarks:
(1) $\Delta G$ refers to a pair of domains
(2) for related thermodynamic quantities see Table 2

Antibody domain $C_H3$

| Protein | pH | T | $\Delta G$ | $c_{1/2}$ | Approach/Remarks | | Ref |
|---|---|---|---|---|---|---|---|
| $C_H3$ | 8.0 | 20 | 66.5±1.5 | 1.0 | GuHCl | (1–4) | 99T10 |

Remarks:
(1) linear extrapolation assuming a two-state transition of folded dimer into unfolded monomers ($N_2 \rightarrow 2U$)
(2) $\Delta G$ from unfolding curves at dimer conc. 0.2 μM, 1.6 μM, and 4.1 μM with $c_{1/2}$ of 0.75 M, 0.9 M, and 1.0 M GuHCl, respectively
(3) transition monitored by fluorescence spectra
(4) buffer: 0.1 M Tris-HCl, pH 8.0

$C_H2$ domains in glycosylated and aglycosylated Fc fragments from mouse IgG2b

| Domain | pH | T | $\Delta G$ | Approach/Remarks | | Ref |
|---|---|---|---|---|---|---|
| $C_H2$ glycosylated | 4.1 | 25 | 19.5 | DSC | (1,2) | 98T8 |
| $C_H2$ aglycosylated | 4.1 | 25 | 14 | DSC | (1,2) | 98T8 |

Remarks:
(1) the $\Delta G$ values were taken from Fig. 19.4 in Ref. 98T8
(2) see also Table 2 for related thermodynamic quantities

Single-chain Fv (scFv) antibody fragments, contributions of highly conserved $v_H/v_L$ hydrogen bonding interactions

| Protein | pH | T | $\Delta G$ | m | Approach/Remarks | | Ref |
|---|---|---|---|---|---|---|---|
| alanine and methionine mutants: | | | | | | | |
| wild type | 7.5 | 22 | 13.39±0.50 | 6.57±0.17 | GuHCl | (1–3) | 98T3 |
| Gln44→Met | 7.5 | 22 | 16.44±0.33 | 7.95±0.08 | GuHCl | (1–3) | 98T3 |
| Gln168→Met | 7.5 | 22 | 16.90±0.96 | 8.24±0.59 | GuHCl | (1–3) | 98T3 |
| double mutant (Gln44→Met and Gln168→Met) | | | | | | | |
| | 7.5 | 22 | 15.48±0.17 | 6.86±0.17 | GuHCl | (1–3) | 98T3 |
| Gln44→Ala | 7.5 | 22 | 15.73±0.54 | 7.95±0.33 | GuHCl | (1–3) | 98T3 |
| Gln168→Ala | 7.5 | 22 | 11.97±0.38 | 6.19±0.42 | GuHCl | (1–3) | 98T3 |
| double mutant (Gln44→Ala and Gln168→Ala) | | | | | | | |
| | 7.5 | 22 | 13.10±0.38 | 6.61±0.25 | GuHCl | (1–3) | 98T3 |
| double mutant (Gln44→Ala and Gln168→Met) | | | | | | | |
| | 7.5 | 22 | 19.54±0.46 | 9.29±0.46 | GuHCl | (1–3) | 98T3 |
| double mutant (Gln44→Met and Gln168→Ala) | | | | | | | |
| | 7.5 | 22 | 18.83±1.09 | 9.20±0.54 | GuHCl | (1–3) | 98T3 |

Single-chain Fv (scFv) antibody fragments, contributions of highly conserved $v_H/v_L$ hydrogen bonding interactions (continued)

| Protein | pH | T | ΔG | m | Approach/Remarks | | Ref |
|---|---|---|---|---|---|---|---|
| salt bridge mutants: | | | | | | | |
| wild type | 7.5 | 22 | 13.39±0.50 | 6.57±0.17 | GuHCl | (1–3) | 98T3 |
| Gln44→Glu | 7.5 | 22 | 13.93±0.84 | 6.53±0.25 | GuHCl | (1–3) | 98T3 |
| Gln168→Lys | 7.5 | 22 | 13.81±0.29 | 6.49±0.17 | GuHCl | (1–3) | 98T3 |
| double mutant (Gln44→Glu and Gln168→Lys) | | | | | | | |
| | 7.5 | 22 | 15.61±0.46 | 7.82±0.21 | GuHCl | (1–3) | 98T3 |
| Gln44→Lys | 7.5 | 22 | 17.36±0.88 | 8.79±0.59 | GuHCl | (1–3) | 98T3 |
| Gln168→Glu | 7.5 | 22 | 14.77±1.34 | 7.45±0.88 | GuHCl | (1–3) | 98T3 |
| double mutant (Gln44→Lys and Gln168→Glu) | | | | | | | |
| | 7.5 | 22 | 22.18±0.84 | 11.09±0.42 | GuHCl | (1–3) | 98T3 |

Remarks:
(1) linear extrapolation, LEM-SB
(2) transition monitored by fluorescence intensity
(3) buffer: 50 mM Tris, 150 mM NaCl, pH 7.5

Single-chain antibodies

| Protein | pH | T | ΔG | $c_{1/2}$ | m | Appr./Rem. | Ref |
|---|---|---|---|---|---|---|---|
| SCA | 8.0 | 25 | 11.9±2.1 | 0.98 | 12.5±1.5 | GuHCl (1–3) | 90B11 |
| ScF11 | 7.0 | 25 | | 2.5 | | GuHCl (4,5) | 98M7 |
| ScF11 | 7.0 | 25 | | 4.5 | | urea (4,5) | 98M7 |

Remarks:
(1) anti-fluorescein single-chain antibody
(2) linear extrapolation
(3) measured in 50 mM Tris/HCl, 150 mM NaCl, pH 8.0, transition monitored by fluorescence
(4) anti-ferritin single-chain antibody
(5) measured in 50 sodium phosphate, pH 7.0, transition monitored by fluorescence

Fv fragment of the phosphorylcholine binding antibody McPC603, a noncovalent heterodimer of the variable domains $V_H$ and $V_L$

| Domain | pH | T | ΔG | $c_{1/2}$ | m | Approach/Remarks | | Ref |
|---|---|---|---|---|---|---|---|---|
| $V_L$ domain | 5.7 | 10 | 27.9 | 1.7 | 15.4 | GuHCl | (1–3) | 99J4 |
| $V_H$ domain | 5.7 | 10 | 27.1 | 1.4 | 18.7 | GuHCl | (1–3) | 99J4 |

Remarks:
(1) linear extrapolation
(2) transition monitored by fluorescence emission intensity at 350 nm ($V_L$) or 370 nm ($V_H$)
(3) buffer: 20 mM MES, pH 5.7

Fv and scFv fragments of an engineered variant of the phosphorylcholine binding antibody McPC603

| Domain | pH | T | ΔG | m | Approach/Remarks | | Ref |
|---|---|---|---|---|---|---|---|
| $V_L$ domain | 6.0 | 20 | 23.5±0.2 | 13.9±0.2 | GuHCl | (1–3) | 99J5 |
| $V_H$ domain | 6.0 | 20 | 18.0±1.1 | 13.6±0.8 | GuHCl | (1–3) | 99J5 |
| Fv (0.1 μM + 50 mM PC) | | | | | | | |
| | 6.0 | 20 | 36.6±1.0 | 22.6±0.6 | GuHCl | (1,3,4,6) | 99J5 |
| | 6.0 | 20 | 30.9±1.1 | 18.0±1.0 | GuHCl | (1,3,5,6) | 99J5 |
| Fv (3.8 μM + 50 mM PC) | | | | | | | |
| | 6.0 | 20 | 43.6±1.0 | 23.9±0.5 | GuHCl | (1,3,4,6) | 99J5 |
| | 6.0 | 20 | 40.8±1.0 | 21.8±0.5 | GuHCl | (1,3,5,6) | 99J5 |

Fv and scFv fragments of an engineered variant of the phosphorylcholine binding antibody McPC603

| Domain | pH | T | $\Delta G$ | m | Approach/Remarks | | Ref |
|---|---|---|---|---|---|---|---|
| scFv (– PC) | | | | | | | |
| | 6.0 | 20 | 43.5±1.0 | 27.1±0.5 | GuHCl | (1,3,4,6) | 99J5 |
| | 6.0 | 20 | 39.1±0.9 | 24.4±0.5 | GuHCl | (1,3,5,6) | 99J5 |
| scFv (+ 50 mM PC) | | | | | | | |
| | 6.0 | 20 | 77.2±1.3 | 37.0±0.6 | GuHCl | (1,3,4,6) | 99J5 |
| | 6.0 | 20 | 71.8±1.7 | 34.1±0.9 | GuHCl | (1,3,5,6) | 99J5 |

Remarks:
(1) linear extrapolation, LEM-SB
(2) transition monitored by fluorescence spectra
(3) buffer: 10 mM MES, pH 6.0
(4) two-state fit of the fluorescence emission maximum data
(5) two-state fit of the fluorescence intensity data at 370 nm
(6) PC = phosphorylcholine

*E. coli* expressed Fv fragments

| Protein | pH | T | $\Delta(\Delta G)$ | $c_{1/2}$ | m | Appr./Rem. | Ref |
|---|---|---|---|---|---|---|---|
| McPC and mutants (engineered turns): | | | | | | | |
| McPC603 | 8.0 | 10 | 0.0 | 2.05 | 5.60±0.33 | urea (1–4) | 95K16 |
| Pro40→Ala | 8.0 | 10 | 0.96±0.8 | 2.05 | 5.23±0.29 | urea (1–4) | 95K16 |
| double mutant (Ser63→Ala and Ala64→Asp) | | | | | | | |
| | 8.0 | 10 | –0.59±0.8 | 1.9 | 6.36±0.42 | urea (1–4) | 95K16 |
| triple mutant (Pro40→Ala, Ser63→Ala and Ala64→Asp) | | | | | | | |
| | 8.0 | 10 | 0.54±0.8 | 1.7 | 6.53±0.42 | urea (1–4) | 95K16 |
| single-chain Fv (scFv) fragment and mutant: | | | | | | | |
| scFv | 7.5 | 20 | | 4.1 | | urea (2,5) | 97N7 |
| Flu4 | 7.5 | 20 | | 4.1 | | urea (2,5,6) | 97N7 |

Remarks:
(1) linear extrapolation
(2) transition monitored by fluorescence
(3) buffer: 200 mM sodium tetraborate, 160 mM NaCl, 5 mM phosphorylcholine, pH 8.0
(4) $c_{1/2}$ was taken from Fig. 4
(5) buffer: 20 mM HEPES, 150 mM NaCl, 1 mM EDTA, pH 7.5
(6) mutant Val84→Asp ($V_H$)

Recombinant human light chain variable domains ($rV_\lambda 6$)

| Protein | pH | T | $\Delta G$ | $c_{1/2}$ | m | Appr./Rem. | Ref |
|---|---|---|---|---|---|---|---|
| $rV_\lambda 6$will | 7.5 | 25 | 9.2±3.8 | 0.83±0.05 | 9.6±3.8 | GuHCl (1–4) | 99W4 |
| | 7.5 | 25 | 12.6 | | | heat (1,4–6) | 99W4 |
| | 7.5 | 25 | 12.6 | | | heat (1,4–6) | 99W4 |
| $rV_\lambda 6$Jto | 7.5 | 25 | 18.0±4.2 | 1.22±0.02 | 14.6±6.3 | GuHCl (1–4) | 99W4 |
| | 7.5 | 25 | 19.2 | | | heat (1,4–6) | 99W4 |
| | 7.5 | 25 | 17.2 | | | heat (1,4–6) | 99W4 |

Remarks:
(1) Ref. 99W4 compares proteins differing in thermodynamic properties and fibrillogenic potential, $rV_\lambda 6$ of two patients
(2) linear extrapolation, LEM-SB
(3) buffer: 10 mM sodium phosphate, pH 7.5
(4) transition monitored by intrinsic tryptophan fluorescence
(5) buffer: 10 mM $Na_2HPO_4$, 10 mM $NaH_2PO_4$, 150 mM NaCl, pH 7.5
(6) data from heat denaturation, see also Table 2

Immunoglobulin V$_\kappa$ domain of the murine antibody McPC603, testing β-turn propensity

| Mutant | pH | T | ΔG | m | Approach/Remarks | | Ref |
|---|---|---|---|---|---|---|---|
| V$_\kappa$ domain (reference): | | | | | | | |
| | 7.4 | 20 | 27.2±0.1 | 13.9±0.1 | GuHCl | (1–3) | 97O3 |
| Leu15→Ala | 7.4 | 20 | 20.2±1.7 | 14.5±1.6 | GuHCl | (2,3) | 97O3 |
| Leu15→Pro | 7.4 | 20 | 24.7±1.8 | 16.9±1.2 | GuHCl | (2,3) | 97O3 |
| Ala15→Pro | 7.4 | 20 | 24.7±1.8 | 16.9±1.2 | GuHCl | (2,3) | 97O3 |
| Pro40→Ser | 7.4 | 20 | 24.8±0.1 | 14.1±0.1 | GuHCl | (2,3) | 97O3 |
| Ser56→Pro | 7.4 | 20 | 30.2±0.5 | 14.8±0.2 | GuHCl | (2,3) | 97O3 |
| Asp60→Ser | 7.4 | 20 | 28.4±0.6 | 14.5±0.3 | GuHCl | (2,3) | 97O3 |
| Asp60→Pro | 7.4 | 20 | 27.7±0.6 | 13.9±0.3 | GuHCl | (2,3) | 97O3 |
| Ser60→Pro | 7.4 | 20 | 27.7±0.6 | 13.9±0.3 | GuHCl | (2,3) | 97O3 |
| Gly68→Ala | 7.4 | 20 | 24.4±1.0 | 13.9±0.5 | GuHCl | (2,3) | 97O3 |
| Gly68→Pro | 7.4 | 20 | 21.2±0.5 | 11.2±0.3 | GuHCl | (2,3) | 97O3 |
| Thr69→Ser | 7.4 | 20 | 28.2±1.0 | 15.3±0.3 | GuHCl | (2,3) | 97O3 |

Remarks:

(1) the V$_\kappa$ domain used as reference contains five stabilizing mutations Ala15→Leu, Met21→Ile, Phe32→Tyr, Thr63→Ser, and Asn90→Gln, see Ref. 94S8

(2) linear extrapolation, LEM-SB

(3) transition monitored by fluorescence at 350±5 nm

*c) mutants with improved stability*

Hyperstable immunoglobulin V$_L$ domains derived from the immunoglobulin variable domain (V$_\kappa$) of the light chain of the murine antibody McPC603, oxidized and reduced proteins

| Mutant/Form | pH | T | ΔG | m | Appr./Rem. | Ref |
|---|---|---|---|---|---|---|
| wild type ox. | 7.4 | 20 | 13.5 | | GuHCl (1–4) | 99O2 |
| V$_L$-500 ox. | 7.4 | 20 | 27.2±0.1 | 13.9±0.1 | GuHCl (1–5) | 99O2 |
| V$_L$-601 ox. | 7.4 | 20 | 34.3±0.6 | 13.6±0.3 | GuHCl (1–5) | 99O2 |
| V$_L$-703 ox. | 7.4 | 20 | 37.9±1.1 | 14.6±0.4 | GuHCl (1–5) | 99O2 |
| V$_L$-705 ox. | 7.4 | 20 | 43.6±1.1 | 14.8±0.4 | GuHCl (1–5) | 99O2 |
| IcaL-14 ox. | 7.4 | 20 | 78.1±1.2 | 20.5±0.5 | GuHCl (1–3,6) | 99O3 |
| V$_L$-500 red. | 7.4 | 20 | 13.3±0.3 | 17.8 | GuHCl (1–5) | 99O2 |
| V$_L$-601 red. | 7.4 | 20 | 18.8±1.1 | 17.7±0.9 | GuHCl (1–5) | 99O2 |
| V$_L$-703 red. | 7.4 | 20 | 21.6±1.1 | 18.9±0.9 | GuHCl (1–5) | 99O2 |
| V$_L$-705 red. | 7.4 | 20 | 24.4±0.7 | 18.0±0.5 | GuHCl (1–5) | 99O2 |
| IcaL-14 red. | 7.4 | 20 | 59.1±2.0 | 20.0±1.2 | GuHCl (1–3,6) | 99O3 |

Remarks:

(1) linear extrapolation, LEB-SB

(2) transition monitored by fluorescence, excitation at λ = 280±1.25 nm, emission at λ = 350±2.5 for the oxidized protein and λ =325±2.5 nm for the reduced protein

(3) for the procedure see also Ref. 97O3

(4) sequences:

    wild type = McPC603 V$_L$

    V$_L$-500   = wild type with the five stabilizing mutations Ala15→Leu,
           Met21→Ile, Phe32→Tyr, Thr63→Ser, and Asn90→Gln

    V$_L$-601   = V$_L$-500 with Ala15→Leu and loop length of CDR1 shortened from 12 to 6 residues

    V$_L$-703   = V$_L$-601 with Tyr32→His and Gly50→Glu

    V$_L$-705   = V$_L$-601 with Tyr32→His and His92→Gln

(5) data from Refs. 94S8 and 97O3

(6) IcaL-14 is a construct of the catalytic intrabody Ica-Fv with the relevant CDR sequences of the light chain of 17E8 placed into the framework of V$_L$-601, the data refer to a N$_2$ → 2U transition

Hyperstable immunoglobulin $V_L$ domains with N-terminal extensions, proteins derived from the immunoglobulin variable domain ($V_\kappa$) of the light chain of the murine antibody McPC603, oxidized and reduced proteins

| Mutant/Ext./Form | | pH | T | ΔG | m | Appr./Rem. | Ref |
|---|---|---|---|---|---|---|---|
| IcaL-μ14 (a) | ox. | 7.4 | 20 | 69.8±1.4 | 17.6±0.6 | GuHCl (1–4) | 99O2 |
| IcaL-α14 (b) | ox. | 7.4 | 20 | 70.7±0.9 | 18.3±0.4 | GuHCl (1–4) | 99O2 |
| IcaL-η14 (c) | ox. | 7.4 | 20 | 74.2±1.1 | 19.1±0.5 | GuHCl (1–4) | 99O2 |
| $V_L$-μ500 (d) | ox. | 7.4 | 20 | 25.2±0.7 | 13.3±0.4 | GuHCl (1–3,5) | 99O2 |
| IcaL-μ14 (a) | red. | 7.4 | 20 | 49.0±2.7 | 13.2±0.6 | GuHCl (1–4) | 99O2 |
| IcaL-α14 (b) | red. | 7.4 | 20 | 62.0±2.8 | 22.1±0.4 | GuHCl (1–4) | 99O2 |
| IcaL-η14 (c) | red. | 7.4 | 20 | 63.3±3.7 | 23.0±0.5 | GuHCl (1–4) | 99O2 |
| $V_L$-μ500 (d) | red. | 7.4 | 20 | 12.4±0.3 | 15.7 | GuHCl (1–3,5) | 99O2 |
| $V_L$-μ601 (d) | red. | 7.4 | 20 | 16.4±1.0 | 14.9±0.4 | GuHCl (1–3,5) | 99O2 |
| $V_L$-μ703 (d) | red. | 7.4 | 20 | 20.3±0.4 | 16.4±0.3 | GuHCl (1–3,5) | 99O2 |
| $V_L$-η703 (e) | red. | 7.4 | 20 | 22.6±0.5 | 16.9±0.4 | GuHCl (1–3,5) | 99O2 |
| $V_L$-μ705 (d) | red. | 7.4 | 20 | 18.5±0.9 | 17.0±0.7 | GuHCl (1–3,5) | 99O2 |

N-terminal extensions:

(a) Met-Asp-Ile-Glu-Leu

(b) Ala-Asp-Ile-Glu-Leu

(c) Ala-Ile-Glu-Leu

(d) Met-Asp-Ile-Val-Met

(e) Ala-Ile-Val-Met

Remarks:

(1) linear extrapolation, LEB-SB

(2) transition monitored by fluorescence, excitation at $\lambda = 280\pm1.25$ nm, emission at $\lambda = 350\pm2.5$ for the oxidized protein and $\lambda = 325\pm2.5$ nm for the reduced protein

(3) for the procedure see also Ref. 97O3

(4) the data refer to a $N_2 \rightarrow 2U$ transition, see also Ref. 99O3

(5) the data refer to a $N \rightarrow U$ transition

Engineered hyperstable $V_H$ domains

| Protein | pH | T | ΔG | $c_{1/2}$ | m | Appr./Rem. | Ref |
|---|---|---|---|---|---|---|---|
| IcaH-01 | 7.4 | 20 | ~22 | 3.63 | | urea (1–3) | 99W12 |
| IcaH-101 | 7.4 | 20 | | 3.85 | | urea (1,2) | 99W12 |
| IcaH-201 | 7.4 | 20 | | 3.88 | | urea (1,2) | 99W12 |
| IcaH-301 | 7.4 | 20 | | 4.11 | | urea (1,2) | 99W12 |
| IcaH-401 | 7.4 | 20 | | 4.34 | | urea (1,2) | 99W12 |
| IcaH-501 | 7.4 | 20 | ~34 | 4.62 | | urea (1–4) | 99W12 |

Explanations:

    IcaH-101 – mutant Ala16→Gly

    IcaH-201 – double mutant Ala16→Gly and Ile58→Thr

    IcaH-301 – multiple mutant Ala16→Gly, Ile58→Thr, and Arg43→Gln

    IcaH-401 – multiple mutant Ala16→Gly, Ile58→Thr, Arg43→Gln, and Pro75→Ser

    IcaH-401 – multiple mutant Pro07→Ser, Ala16→Gly, Ile58→Thr, Arg43→Gln, and Pro75→Ser

Remarks:

(1) $c_{1/2}$ from urea-induced transitions monitored by fluorescence

(2) buffer: PBS, 4 mM $K_2HPO_4$, 16 mM $NaH_2PO_4$, 115 mM NaCl, pH 7.4

(3) nonlinear least-squares fit of the transition curve assuming a linear dependence of ΔG on the denaturant conc., and assuming a two-state transition of a monomeric domain

(4) the contribution of the disulfide bridge amounts to ~15 kJ/mol (from GuHCl-induced unfolding of reduced and oxidized IcaH-501)

Single-chain Fv antibody fragment 4D5Flu, mutants with improved stability selected by phage display

| Mutant | pH | T | $\Delta G$ | $c_{1/2}$ | Approach | Remarks | Ref |
|---|---|---|---|---|---|---|---|
| wild type | 7.4 | 20 | 58.6 | 2.4 | GuHCl, FL, $\lambda_{max}$ | (1–4) | 99J11 |
|  | 7.4 | 20 | 48.5 | 2.3 | GuHCl, FL, I | (1–4) | 99J11 |
| 50 | 7.4 | 20 | 74.1 | 2.8 | GuHCl, FL, $\lambda_{max}$ | (1–6) | 99J11 |
|  | 7.4 | 20 | 60.2 | 2.7 | GuHCl, FL, I | (1–6) | 99J11 |
| 3M | 7.4 | 20 | 63.2 | 2.6 | GuHCl, FL, $\lambda_{max}$ | (1–5) | 99J11 |
|  | 7.4 | 20 | 51.9 | 2.5 | GuHCl, FL, I | (1–5) | 99J11 |
| G1 | 7.4 | 20 | 43.9 | 1.9 | GuHCl, FL, $\lambda_{max}$ | (1–5) | 99J11 |
|  | 7.4 | 20 | 15.9 | 1.7 | GuHCl, FL, I | (1–5) | 99J11 |
| G2 | 7.4 | 20 | 50.2 | 2.2 | GuHCl, FL, $\lambda_{max}$ | (1–5) | 99J11 |
|  | 7.4 | 20 | 60.2 | 2.1 | GuHCl, FL, I | (1–5) | 99J11 |
| T0 | 7.4 | 20 | 54.4 | 2.3 | GuHCl, FL, $\lambda_{max}$ | (1–5) | 99J11 |
|  | 7.4 | 20 | 30.5 | 2.1 | GuHCl, FL, I | (1–5) | 99J11 |

Remarks:
(1) 4D5Flu consists of the 4D5 framework with grafted CDRs from the fluorescein-binding antibody 4-4-20
(2) linear extrapolation
(3) transition monitored by the shift of the wavelength of the fluorescence maximum (FL, $\lambda_{max}$) or the change in fluorescence intensity at 335 nm (FL, I)
(4) buffer: 20 mM HEPES, 150 mM NaCl, pH 7.4
(5) for the detailed mutations see Ref. 99J11
(6) $T_{trs}$ according to DSC is 66.2°C for mutant 50 and 62.3°C for the wild-type protein

## Indole-3-Glycerol Phosphate Synthase

Indole-3-glycerol phosphate synthase from *Sulfolobus solfataricus*

| Transition | pH | T | $\Delta G$ | m | Approach/Remarks | | Ref |
|---|---|---|---|---|---|---|---|
| two-state model: |  |  |  |  |  |  |  |
| $N \rightarrow U$ | 7.0 | 20 | 15.2±1.1 | 6.2±0.6 | GuHCl | (1,2,4) | 97A4 |
| $N \rightarrow U$ | 7.0 | 20 | 14.1±1.3 | 6.6±0.7 | GuHCl | (1,3,4) | 97A4 |
| $N \rightarrow U$ | 9.0 | 20 | 23.3±1.5 | 9.2±0.6 | GuHCl | (1,2,4) | 97A4 |
| $N \rightarrow U$ | 9.0 | 20 | 10.1±1.1 | 5.2±0.6 | GuHCl | (1,3,4) | 97A4 |
| three-state model: |  |  |  |  |  |  |  |
| $N \rightarrow I$ | 7.0 | 20 | 9.9±1.2 | 0.2±0.1 | GuHCl | (2,4,5) | 97A4 |
| $I \rightarrow U$ | 7.0 | 20 | 4.7±1.2 | 5.8±0.7 | GuHCl | (2,4,5) | 97A4 |
| $N \rightarrow I$ | 7.0 | 20 | 12.1±1.2 | 1.0±0.4 | GuHCl | (3–5) | 97A4 |
| $I \rightarrow U$ | 7.0 | 20 | 1.6±1.2 | 5.2±4.1 | GuHCl | (3–5) | 97A4 |
| $N \rightarrow I$ | 9.0 | 20 | 18.7±1.3 | 6.6±1.7 | GuHCl | (2,4,5) | 97A4 |
| $I \rightarrow U$ | 9.0 | 20 | 7.3±1.3 | 4.8±1.4 | GuHCl | (2,4,5) | 97A4 |
| $N \rightarrow I$ | 7.0 | 20 | 18.4±1.2 | 16.5±2.8 | GuHCl | (3–5) | 97A4 |
| $I \rightarrow U$ | 7.0 | 20 | 9.3±1.2 | 4.5±0.4 | GuHCl | (3–5) | 97A4 |

Remarks:
(1) treatment by a two-state transition, LEM-SB
(2) transition monitored by CD at 220 nm
(3) transition monitored by fluorescence
(4) the unfolding mechanism closely approaches a two-state model at pH 7, whereas the unfolding transition obtained by fluorescence at pH 9 required a three-state model
(5) treatment by a three-state model which assumes linear dependence of $\Delta G$ on denaturant conc.

## Insulin Receptor Kinase

Insulin receptor kinase domain (Arg953–Ser1355), wild type and mutant

| Protein/Trans. | pH | T | $\Delta G$ | $c_{1/2}$ | m | Appr./Rem. | Ref |
|---|---|---|---|---|---|---|---|
| apo-wild type | | | | | | | |
|   trans. (1) | 7.0 | 24 | 21.8±7.9 | 0.65±0.25 | 32.6±6.7 | GuHCl (1–3) | 99B14 |
|   trans. (3) | 7.0 | 24 | 37.2±7.1 | 2.70±0.10 | 13.8±6.7 | GuHCl (1–3) | 99B14 |
| phospho-wild type | | | | | | | |
|   trans. (1) | 7.0 | 24 | 11.7±5.4 | 0.40±0.06 | 29.3±5.0 | GuHCl (1–3) | 99B14 |
|   trans. (3) | 7.0 | 24 | 31.0±8.4 | 2.40±0.10 | 13.0±7.9 | GuHCl (1–3) | 99B14 |
| Trp1175→Phe | | | | | | | |
|   trans. (3) | 7.0 | 24 | 39.7±5.0 | 2.80±0.10 | 14.2±4.6 | GuHCl (1–3) | 99B14 |
| truncation mutants | | | | | | | |
|   trans. (1) | 7.0 | 24 | 16.7±13.4 | | | GuHCl (1–4) | 99F7 |
|   trans. (3) | 7.0 | 24 | 41.4±1.7 | | | GuHCl (1–4) | 99F7 |

Remarks:
(1) multistate transition analyzed by linear extrapolation, LEM-SB
(2) transitions monitored by fluorescence and CD
(3) buffer: 50 mM Tris-acetate, 1 mM DTT, pH 7.0
(4) from global fit of apo-wild type, N-terminal mutant lacking the first 25 residues, C-terminal mutant lacking the last 72 residues, and double mutant lacking the first 25 and last 72 residues

## Interferon

Recombinant human γ-interferon

| Buffer Conc. | pH | T | $\Delta G$ | Approach/Remarks | | Ref |
|---|---|---|---|---|---|---|
| 5 mM | 4.2 | 25 | 24.5 | DSC | (1,2) | 99B7 |
| | 4.4 | 25 | 25.2 | DSC | (1,2) | 99B7 |
| | 5.0 | 25 | 31.8 | DSC | (1,2) | 99B7 |
| | 6.8 | 25 | 36.9 | DSC | (1,2) | 99B7 |
| 10 mM | 3.5 | 25 | 9.3 | DSC | (1,2) | 99B7 |
| | 3.8 | 25 | 13.6 | DSC | (1,2) | 99B7 |
| | 4.0 | 25 | 15.5 | DSC | (1,2) | 99B7 |
| | 4.2 | 25 | 20.4 | DSC | (1,2) | 99B7 |
| | 4.4 | 25 | 22.6 | DSC | (1,2) | 99B7 |
| | 4.6 | 25 | 24.2 | DSC | (1,2) | 99B7 |
| | 4.8 | 25 | 26.2 | DSC | (1,2) | 99B7 |
| | 5.0 | 25 | 28.3 | DSC | (1,2) | 99B7 |
| | 5.2 | 25 | 29.6 | DSC | (1,2) | 99B7 |
| | 5.4 | 25 | 31.5 | DSC | (1,2) | 99B7 |
| | 6.0 | 25 | 33.5 | DSC | (1,2) | 99B7 |
| | 6.8 | 25 | 34.4 | DSC | (1,2) | 99B7 |
| | 8.0 | 25 | 33.8 | DSC | (1,2) | 99B7 |
| 20 mM | 3.8 | 25 | 11.3 | DSC | (1,2) | 99B7 |
| | 4.0 | 25 | 16.1 | DSC | (1,2) | 99B7 |
| | 4.2 | 25 | 20.0 | DSC | (1,2) | 99B7 |
| | 4.4 | 25 | 22.0 | DSC | (1,2) | 99B7 |
| | 4.8 | 25 | 25.7 | DSC | (1,2) | 99B7 |
| | 5.0 | 25 | 27.7 | DSC | (1,2) | 99B7 |
| | 6.0 | 25 | 30.8 | DSC | (1,2) | 99B7 |
| | 6.8 | 25 | 32.1 | DSC | (1,2) | 99B7 |
| | 8.0 | 25 | 32.3 | DSC | (1,2) | 99B7 |
| | 9.0 | 25 | 31.4 | DSC | (1,2) | 99B7 |

Recombinant human γ-interferon (continued)

| Buffer Conc. | pH | T | ΔG | Approach/Remarks | | Ref |
|---|---|---|---|---|---|---|
| 50 mM | 4.0 | 25 | 10.2 | DSC | (1,2) | 99B7 |
| | 4.2 | 25 | 12.3 | DSC | (1,2) | 99B7 |
| | 4.4 | 25 | 13.5 | DSC | (1,2) | 99B7 |
| | 5.0 | 25 | 19.3 | DSC | (1,2) | 99B7 |
| | 6.8 | 25 | 27.7 | DSC | (1,2) | 99B7 |
| 100 mM | 4.0 | 25 | 8.1 | DSC | (1,2) | 99B7 |
| | 4.4 | 25 | 10.9 | DSC | (1,2) | 99B7 |
| | 5.8 | 25 | 19.4 | DSC | (1,2) | 99B7 |
| | 6.8 | 25 | 23.1 | DSC | (1,2) | 99B7 |
| | 7.8 | 25 | 24.1 | DSC | (1,2) | 99B7 |

Remarks:

(1) ΔG in kJ/mol of dimer

(2) DSC was carried out in acetic acid/sodium acetate below pH 6.0 and sodium phosphate at and above pH 6.0 (measured at 25°C)

## Interleukin

Murine interleukin-6, recombinant (mIL-6), disulfide bonds cleaved

| Mutant | pH | T | ΔG | $c_{1/2}$ | m | Appr./Rem. | Ref |
|---|---|---|---|---|---|---|---|
| mIL-6 | 4.0 | 25 | 30.1±3.3 | 4.05±0.07 | 7.53±0.84 | urea (1–3) | 97Z5 |
| IAA-IL-6 | 4.0 | 25 | 28.0±5.9 | 3.70±0.13 | 7.53±1.67 | urea (1–4) | 97Z5 |
| IAM-IL-6 | 4.0 | 25 | 20.1±3.8 | 2.81±0.08 | 7.11±1.26 | urea (1–3,5) | 97Z5 |

Remarks:

(1) linear extrapolation, LEM-SB

(2) buffer: 10 mM sodium acetate, pH 4.0

(3) transition monitored by CD at 222 nm

(4) IAA-IL-6 = mIL-6 fully reduced and S-carboxymethylated by iodoacetic acid

(5) IAM-IL-6 = mIL-6 fully reduced and S-carbamidomethylated by iodoacetamide

Human interleukin 10 (huIL-10), homodimer

| Protein | pH | T | ΔG | m | Approach/Remarks | | Ref |
|---|---|---|---|---|---|---|---|
| huIL-10 | 8.5 | 22 | 52.3 | 23.8 | GuHCl | (1–3) | 98S23 |

Remarks:

(1) linear extrapolation assuming a two-state unfolding mechanism with folded dimer and unfolded monomers ($N_2 \rightarrow 2U$), see also Ref. 94N4

(2) transition monitored by ellipticity at 222 nm

(3) buffer: 50 mM Tris, pH 8.5

**Iron-Sulfur Protein,** see also Adrenodoxin, Ferredoxin

## Iron-Sulfur Protein

$Fe_4S_4$-containing high potential iron-sulfur protein from *Chromatium vinosum*

| Transition | pH | T | ΔG | m | Approach/Remarks | | Ref |
|---|---|---|---|---|---|---|---|
| N → I | 6.8 | 25 | 22±3 | 6±1 | GuHCl | (1–4) | 97B7 |

Remarks:

(1) linear extrapolation

(2) the transition was monitored by NMR on the reduced protein

(3) reversible equilibrium between reduced native state and reduced intermediate; the cluster remains associated with the polypeptide chain

(4) no intermediate can be detected in the oxidized form; unfolding of the oxidized protein is irreversible and accompanied by loss of the iron-sulfur cluster

## ISL-1 Protein

Homeodomain of rat ISL-1 protein in the presence and absence of DNA

| Ligand | pH | T | ΔG | m | Approach/Remarks | | Ref |
|---|---|---|---|---|---|---|---|
| none | 6.0 | 23 | 7 | 4.9 | urea | (1,2) | 97B6 |
| DNA | 6.0 | 23 | 29 | 5.2 | urea | (1–4) | 97B6 |

Remarks:

(1) linear extrapolation

(2) measured in 20 mM sodium phosphate, 50 mM NaCl, pH 6.0

(3) measured in the presence of specific 12mer oligonucleotide at a 1:1 ratio

(4) $T_{trs}$ = 31°C in the absence of DNA, increased to $T_{trs}$ = 53°C in the presence of DNA

**Iso-1-Cytochrome** *c*, see Cytochrome

**Iso-2-Cytochrome** *c*, see Cytochrome

## 3-Isopropylmalate Dehydrogenase

3-isopropylmalate dehydrogenase from *Thermus thermophilus* (*Tt*-IPMDH) and *E. coli* (*Ec*-IPMDH)

| Protein Conc. | pH | T | ΔG | m | Approach/Remarks | | Ref |
|---|---|---|---|---|---|---|---|
| *Tt*-IPMDH 0.33 µM | | | | | | | |
|    trans. (1) | 7.0 | 27 | 18.87 | 5.02 | urea | (1–3) | 99M21 |
|    trans. (2) | 7.0 | 27 | 68.2 | 5.90 | urea | (1–3) | 99M21 |
| *Tt*-IPMDH 1.6 µM | | | | | | | |
|    trans. (1) | 7.0 | 27 | 18.91 | 5.65 | urea | (1–3) | 99M21 |
|    trans. (2) | 7.0 | 27 | 77.0 | 6.69 | urea | (1–3) | 99M21 |
| *Ec*-IPMDH 0.74 µM | | | | | | | |
|    trans. (1) | 7.0 | 27 | 61.5 | 23.26 | urea | (1–3) | 99M21 |
|    trans. (2) | 7.0 | 27 | 74.1 | 9.29 | urea | (1–3) | 99M21 |
| *Ec*-IPMDH 3.0 µM | | | | | | | |
|    trans. (1) | 7.0 | 27 | 66.5 | 25.06 | urea | (1–3) | 99M21 |
|    trans. (2) | 7.0 | 27 | 79.5 | 11.46 | urea | (1–3) | 99M21 |

Remarks:

(1) linear extrapolation based on a three-state model that involves a native dimeric state and a dimeric intermediate state and unfolded subunits ($N_2 \rightarrow I_2 \rightarrow D + D$)

(2) transition monitored by fluorescence intensity at 340 and 332 nm

(3) buffer: 50 mM potassium phosphate, 1 mM DTT, 0.01% Tween 20, pH 7.0

## Kinetoplastid Membrane Protein

Kinetoplastid membrane protein-11 from *Leishmania infantum* (KMP-11)

| Protein | pH | T | ΔG | Approach/Remarks | | Ref |
|---------|-----|-----|-------|-------------|------|-------|
| KMP-11 | 7.5 | 20 | 14.63 | urea | (1–3) | 99F10 |

Remarks:
(1) linear extrapolation, LEM-SB
(2) transition monitored by far-UV CD at 222 nm
(3) buffer: 50 mM Tris-HCl, pH 7.5

**Lac repressor**, see Repressor Proteins

## α-Lactalbumin

The data entries are arranged as follows:

a) wild type and mutants
b) chimeric α-lactalbumin
c) transitions and intermediates
d) disulfide variants
e) α-lactalbumin in the presence of osmolytes

*a) wild type and mutants*

Recombinant bovine α-lactalbumin (BLA), modified by an extraneous N-terminal Met residue (Met-BLA)

| Mutant | Form | pH | T | ΔG | Approach/Remarks | Ref |
|--------|------|-----|-----|------|------------------|------|
| BLA | apo | 8.0 | 34 | 0.0 | heat, v.H. (1–3) | 98I3 |
| Met-BLA | apo | 8.0 | 34 | −7.9 | heat, v.H. (1–3) | 98I3 |
| BLA | Ca$^{2+}$ | 8.0 | 64 | 0.0 | heat, v.H. (1,2,4) | 98I3 |
| Met-BLA | Ca$^{2+}$ | 8.0 | 64 | −7.2 | heat, v.H. (1,2,4) | 98I3 |

Remarks:
(1) ΔG was calculated using ΔCp = 4.4 kJ/mol/K from Ref. 94G3
(2) transition monitored by CD at 270 nm
(3) buffer: 10 mM borate, 50 mM NaCl, 1 mM CaCl$_2$
(4) buffer: 10 mM borate, 50 mM NaCl, 1 mM EDTA

Recombinant goat α-lactalbumin, effect of the extra N-terminal Met of α-lactalbumin expressed in *E. coli*

| Mutant | pH | T | ΔG | $c_{1/2}$ | m | Appr./Rem. | Ref |
|--------|-----|-----|----------|-----------|----------|-------------|------|
| authentic | 7.0 | 25 | 57.7±2.9 | 3.15±0.01 | 18.4±0.8 | GuHCl (1–4) | 99C7 |
| reco. Met | 7.0 | 25 | 43.5±2.1 | 2.67±0.01 | 16.3±0.8 | GuHCl (1–6) | 99C7 |

Remarks:
(1) linear extrapolation
(2) transition monitored by CD at 222 and 270 nm
(3) buffer: 50 mM cacodylate, 50 mM NaCl, 1 mM CaCl$_2$, pH 7.0
(4) authentic goat α-lactalbumin prepared from fresh goat milk
(5) *E. coli* expressed recombinant α-lactalbumin with the extra N-terminal methionine residue
(6) the transition curves measured by CD at 222 and 270 nm are coincident with each other in authentic and methionine-free recombinant α-lactalbumin

Recombinant bovine α-lactalbumin (mLA), wild type and mutants

| Mutant | pH | $T_{trs}$ | $\Delta T$ | $\Delta(\Delta G)$ | Appr./Rem. | Ref |
|---|---|---|---|---|---|---|
| mLA | 7.4 | 56.2±0.05 | 0.0 | 0.0 | heat (1–3) | 99G13 |
| functional site: | | | | | | |
| His32→Ala | 7.4 | 44.6±0.07 | −11.6±0.1 | −8.9 | heat (1–3) | 99G13 |
| His32→Tyr | 7.4 | 56.5±0.04 | 0.3±0.1 | 0.3 | heat (1–3) | 99G13 |
| Leu110→Arg | 7.4 | 58.1±0.03 | 1.9±0.1 | 1.8 | heat (1–3) | 99G13 |
| Leu110→Glu | 7.4 | 55.1±0.10 | −1.1±0.1 | −0.8 | heat (1–3) | 99G13 |
| Leu110→His | 7.4 | 62.5±0.04 | 6.3±0.1 | 5.8 | heat (1–3) | 99G13 |
| Gln117→Ala | 7.4 | 52.1±0.04 | −4.1±0.1 | −4.0 | heat (1–3) | 99G13 |
| Trp118→His | 7.4 | 52.9±0.07 | −3.3±0.1 | −2.5 | heat (1–3) | 99G13 |
| Trp118→Tyr | 7.4 | 50.8±0.04 | −5.4±0.1 | −4.9 | heat (1–3) | 99G13 |
| adjacent to site: | | | | | | |
| Ala106→Ser | 7.4 | 50.1±0.04 | −6.1±0.1 | −4.4 | heat (1–3) | 99G13 |
| His107→Ala | 7.4 | 52.1±0.05 | −4.1±0.1 | −3.3 | heat (1–3) | 99G13 |
| His107→Trp | 7.4 | 47.5±0.04 | −8.7±0.1 | −7.2 | heat (1–3) | 99G13 |
| His107→Tyr | 7.4 | 55.2±0.03 | −1.0±0.1 | −0.8 | heat (1–3) | 99G13 |
| Lys114→Asn | 7.4 | 66.9±0.05 | 10.7±0.1 | 11.1 | heat (1–3) | 99G13 |
| Lys114→Gln | 7.4 | 53.5±0.07 | −2.5±0.1 | −2.5 | heat (1–3) | 99G13 |
| Lys114→Glu | 7.4 | 53.0±0.05 | −3.2±0.1 | −2.7 | heat (1–3) | 99G13 |
| other residues: | | | | | | |
| Val42→Ala | 7.4 | 51.7±0.04 | −4.5±0.1 | −3.9 | heat (1–3) | 99G13 |
| Val42→Asn | 7.4 | 55.1±0.03 | −1.1±0.1 | −1.0 | heat (1–3) | 99G13 |
| Val42→Gly | 7.4 | 51.0±0.07 | −5.2±0.1 | −4.8 | heat (1–3) | 99G13 |
| Gln54→Ala | 7.4 | 54.3±0.08 | −1.9±0.1 | −1.7 | heat (1–3) | 99G13 |
| Ile59→Trp | 7.4 | 51.7±0.05 | −4.5±0.1 | −3.9 | heat (1–3) | 99G13 |
| Tyr103→Ala | 7.4 | 45.5±0.05 | −10.7±0.1 | −10.0 | heat (1–3) | 99G13 |
| Tyr103→Pro | 7.4 | 55.1±0.02 | −1.1±0.1 | −0.9 | heat (1–3) | 99G13 |
| Tyr103→Ala | 7.4 | 45.5±0.05 | −10.7±0.1 | −10.0 | heat (1–3) | 99G13 |
| Trp104→Tyr | 7.4 | 43.7±0.03 | −12.5±0.1 | −10.2 | heat (1–3) | 99G13 |
| Reference proteins: | | | | | | |
| bovine LA | 7.4 | 62.7±0.09 | 6.5±0.1 | 5.7 | heat (1–3) | 99G13 |
| goat LA | 7.4 | 62.8±0.04 | 6.6±0.1 | 5.7 | heat (1–3) | 99G13 |

Remarks:
(1) mLA differs from bovine LA in having Val substituted for Met90 and the N-terminal Met extension
(2) buffer: 20 mM Tris-HCl, 1 mM CaCl$_2$, pH 7.4
(3) $\Delta(\Delta G)$ was calculated at 56.2°C ($T_{trs}$ of mLA) using $\Delta Cp$ = 5.6 kJ/mol/K from $\Delta H^{cal}$ versus $T_{trs}$

*b) chimeric α-lactalbumin*

Chimeric protein (HLAEZ), recombinant human α-lactalbumin (α-LA) and equine lysozyme (ELZ)

| Protein/Transition | | pH | T | $\Delta G$ | m | Appr./Rem. | Ref |
|---|---|---|---|---|---|---|---|
| HLAEZ | N → I | 7.0 | 25 | 12.72±1.51 | 8.91±1.17 | GuHCl (1–3) | 99M18 |
| | I → U | 7.0 | 25 | 18.62±0.79 | 4.81±0.21 | GuHCl (1–3) | 99M18 |
| | N → U | 7.0 | 25 | 31.38±2.30 | 13.72±1.80 | GuHCl (1–3) | 99M18 |
| HLAEZ | N → I | 7.0 | 25 | 12.30±0.71 | 10.21±0.59 | GuHCl (1,3,4) | 99M18 |
| | I → U | 7.0 | 25 | 18.87±1.55 | 4.35±0.33 | GuHCl (1,3,4) | 99M18 |
| | N → U | 7.0 | 25 | 31.21±2.26 | 14.60±0.92 | GuHCl (1,3,4) | 99M18 |
| α-LA | N → I | 7.0 | 25 | 12.13±1.21 | 9.00±0.84 | GuHCl (1–4) | 99M18 |
| | I → U | 7.0 | 25 | 14.81±0.46 | 4.10±0.13 | GuHCl (1–4) | 99M18 |
| | N → U | 7.0 | 25 | 26.94±1.67 | 13.10±0.96 | GuHCl (1–4) | 99M18 |

Chimeric protein (HLAEZ), recombinant human α-lactalbumin (α-LA) and equine lysozyme (ELZ) (continued)

| Protein/Transition | | pH | T | ΔG | m | Appr./Rem. | Ref |
|---|---|---|---|---|---|---|---|
| ELZ | N → I | 7.0 | 25 | 23.60±1.34 | 8.37±0.46 | GuHCl (1,3,5) | 99M18 |
| | I → U | 7.0 | 25 | 18.12±2.47 | 4.73±0.79 | GuHCl (1,3,5) | 99M18 |
| | N → U | 7.0 | 25 | 41.71±2.05 | 13.10±0.63 | GuHCl (1,3,5) | 99M18 |

Explanation:

HLAEZ = The chimeric protein HLAEZ is produced by replacing the flexible loop (residues 105–110) in human α-LA with the helix D (residues 109–114) in ELZ

Remarks:

(1) treatment by a three-state model assuming a linear dependence of ΔG on the denaturant concentration, nonlinear fit

(2) transitions monitored by CD at 215 nm

(3) buffer: 50 mM sodium cacodylate (pH 7.0), 50 mM NaCl, 10 mM $CaCl_2$

(4) transition monitored by fluorescence

(5) data from Ref. 98M21

Chimeric protein (LYLA1), kinetic and equilibrium intermediate states

| Protein/Transition | | pH | T | ΔG | $c_{1/2}$ | m | Appr./Rem. | Ref |
|---|---|---|---|---|---|---|---|---|
| LYLA1 | $I_E$ → U | 7.5 | 25 | 33.25 | 2.4 | 11.12 | GuHCl (1–3) | 99H1 |
| | $I_B$ → U | 7.0 | 25 | 8.5 | 2.7 | | GuHCl (1,3,4) | 99H1 |

Explanation:

LYLA1 = The chimeric protein LYLA1 is produced by placing the $Ca^{2+}$-binding loop and the adjacent helix C of bovine α-lactalbumin into the homologous position (residues 76–102) of human lysoyzme

Remarks:

(1) linear extrapolation, LEM-SB

(2) data concerning the equilibrium intermediate (IE) from CD at 270 and 222 nm

(3) measured in 10 mM Tris, 80 mM NaCl, 10 mM EDTA, pH 7.5

(4) data concerning the burst phase intermediate from kinetic measurements followed by CD and fluorescence

*c) transitions and intermediates*

Peptides from α-lactalbumin (α-LA) which define the molten globule

| Protein | pH | T | $c_{1/2}$ | m | Appr./Rem. | Ref |
|---|---|---|---|---|---|---|
| native α-LA + $Ca^{2+}$ | 7.0 | 20 | 7.0±0.5 | 5.27±0.29 | urea (1,2) | 99D3 |
| MG α-LA + EDTA | 2.8 | 20 | 5.8±0.6 | 2.47±0.17 | urea (1,3) | 99D3 |
| AB-D$_{95-120}$ | 2.8 | 20 | 3.5±0.6 | 2.38±0.25 | urea (1,4) | 99D3 |
| AB-D$_{101-120}$ | 2.8 | 20 | 2.7±0.6 | 2.38±0.25 | urea (1,5) | 99D3 |

Remarks:

(1) urea denaturation monitored by far-UV CD at 220 nm

(2) native α-LA at pH 7.0 in the presence of 10 mM $Ca^{2+}$

(3) molten globule (MG) form of α-LA at pH 2.8 in the presence of 1 mM EDTA

(4) AB-D$_{95-120}$ – peptide consisting of residues 1–38 crosslinked via the native disulfide bond to a peptide corresponding to residues 95–120

(5) AB-D$_{101-120}$ – truncated form of AB-D$_{95-120}$

Bovine (Bα-LA) and human (Hα-LA) α-lactalbumin, transitions and pressure denaturation

| Protein/Transition | | pH | T | $\Delta G$ | $c_{1/2}$ | m | Appr./Rem. | Ref |
|---|---|---|---|---|---|---|---|---|
| Bα-LA | N → A | 7.0 | 25 | 16.5 | 2.10 | 7.8 | GuHCl (1–3) | 99K10 |
| | A → U | 7.0 | 25 | 5.8 | 2.72 | 2.1 | GuHCl (1–3) | 99K10 |
| Hα-LA | N → A | 7.0 | 25 | 7.52 | 0.44 | 17.0 | GuHCl (1–3) | 99K10 |
| | A → U | 7.0 | 25 | 14.3 | 3.51 | 4.08 | GuHCl (1–3) | 99K10 |

Remarks:

(1) linear extrapolation

(2) transitions monitored by ellipticity at 270 nm and 290 nm, and first derivative spectrum at 290 nm

(3) buffer: 50 mM cacodylate, 50 mM NaCl

(4) the volume change for the N → A transition from pressure denaturation amounts to $\Delta V = -66\pm6$ cm³/mol for holo-Bα-LA, and $\Delta V = -37$ cm³/mol for apo Bα-LA

(5) the volume change for the N → A transition from pressure denaturation amounts to $\Delta V = -81\pm4$ cm³/mol for holo-Hα-LA

**α-lactalbumin transition,** see also Lysozyme, various Calcium-Binding Lysozymes

*d) disulfide variants*

Human α-lactalbumin, disulfide variants

| Protein | Salt | pH | T | $c_{1/2}$ | m | Appr./Rem. | Ref |
|---|---|---|---|---|---|---|---|
| [all-Ala] | 30 mM NaCl | 4.2 | 20 | 3.42 | 3.55 | urea (1–3) | 99L11 |
| | 20 mM Na$_2$SO$_4$ | 4.2 | 20 | 4.34 | 3.68 | urea (1,2) | 99L11 |
| | 50 mM NaClO$_4$ | 4.2 | 20 | 4.62 | 3.90 | urea (1,2) | 99L11 |
| [28–111] | 30 mM NaCl | 4.2 | 20 | 5.43 | 2.41 | urea (1,2) | 99L11 |
| | 20 mM Na$_2$SO$_4$ | 4.2 | 20 | 5.85 | 2.80 | urea (1,2) | 99L11 |
| | 50 mM NaClO$_4$ | 4.2 | 20 | 6.06 | 1.13 | urea (1,2) | 99L11 |

Explanations:

[all-Ala] – all eight Cys residues of human α-lactalbumin replaced by Ala

[28–111] – human α-lactalbumin with a single disulfide bond 28–111

Remarks:

(1) transition monitored by far-UV CD at 222 nm

(2) measured in 2 mM sodium acetate, pH 4.2 with the salt added

(3) for further information on the all-Ala variant see also Ref. 99R7

Bovine α-lactalbumin (BLA) and a derivative 2CM-3SS-BLA in which the 6–120 disulfide bond is selectively reduced and carboxymethylated

| Variant/Form | pH | T | $\Delta G$ | m | Approach/Remarks | | Ref |
|---|---|---|---|---|---|---|---|
| Apo-BLA | 7.0 | 4.5 | 13.8±0.8 | 9.3±0.5 | GuHCl | (1) | 98I2 |
| Apo-2CM-3SS-BLA | 7.0 | 4.5 | 4.2±0.8 | 9.3±1.7 | GuHCl | (1) | 98I2 |
| Holo-BLA | 7.0 | 4.5 | 31.4±0.9 | 11.5±0.3 | GuHCl | (1,2) | 98I2 |
| Holo-2CM-3SS-BLA | 7.0 | 4.5 | 18.0±3.4 | 11.1±2.0 | GuHCl | (1,2) | 98I2 |

Remarks:

(1) from stopped-flow CD experiments, see also Ref. 92I1

(2) measured in the presence of 1 mM CaCl$_2$

*e) α-lactalbumin in the presence of osmolytes*

Bovine α-lactalbumin (holo-protein) in the presence of noncompatible osmolytes

| Osmolyte | Conc. | pH | ΔT | ΔH | Δ(ΔG)(25°) | Appr./Rem. | Ref |
|---|---|---|---|---|---|---|---|
| arginine | 0.0 M | 7.0 | 0.0 | 0 | 0.0 | heat (1–4) | 98R7 |
| | 0.3 M | 7.0 | −4.3 | −50 | −5.4 | heat (1–4) | 98R7 |
| | 0.5 M | 7.0 | −5.4 | −64 | −6.9 | heat (1–4) | 98R7 |
| | 0.8 M | 7.0 | −6.8 | −68 | −7.2 | heat (1–4) | 98R7 |
| | 1.0 M | 7.0 | −7.9 | −69 | −7.2 | heat (1–4) | 98R7 |
| histidine | 0.00 M | 5.0 | 0.0 | 0 | 0.0 | heat (1–4) | 98R7 |
| | 0.13 M | 5.0 | −1.3 | −30 | −3.3 | heat (1–4) | 98R7 |
| | 0.25 M | 5.0 | −2.3 | −27 | −3.1 | heat (1–4) | 98R7 |
| lysine | 0.0 M | 5.0 | 0.0 | 0 | 0.0 | heat (1–4) | 98R7 |
| | 0.2 M | 5.0 | 0.4 | −7 | 0.9 | heat (1–4) | 98R7 |
| | 0.5 M | 5.0 | 0.9 | 3 | 0.0 | heat (1–4) | 98R7 |
| | 0.7 M | 5.0 | 1.4 | −11 | 1.7 | heat (1–4) | 98R7 |
| | 1.0 M | 5.0 | 3.5 | −4 | 0.8 | heat (1–4) | 98R7 |
| | 2.0 M | 5.0 | 3.9 | −2 | 1.7 | heat (1–4) | 98R7 |

Remarks:
(1) van't Hoff treatment, transition monitored by changes in optical absorption at 295 nm
(2) measured in the presence of 0.1 M KCl
(3) in the absence of osmolytes, the values $T_{trs}$ = 65.3°C and ΔH = 309 kJ/mol were obtained
(4) in the presence of osmolyte ΔG(25°) was calculated using ΔCp = 7.1 kJ/mol/K from Ref. 94G3

Bovine α-lactalbumin (apo-protein) in the presence of noncompatible osmolytes

| Osmolyte | Conc. | pH | ΔT | ΔH | Δ(ΔG)(25°) | Appr./Rem. | Ref |
|---|---|---|---|---|---|---|---|
| arginine | 0.0 M | 7.0 | 0.0 | 0 | 0.0 | heat (1–4) | 98R7 |
| | 0.3 M | 7.0 | −0.3 | −5 | −0.4 | heat (1–4) | 98R7 |
| | 0.5 M | 7.0 | −0.6 | −18 | −1.2 | heat (1–4) | 98R7 |
| | 1.0 M | 7.0 | −2.1 | −23 | −1.8 | heat (1–4) | 98R7 |
| histidine | 0.00 M | 7.0 | 0.0 | 0 | 0.0 | heat (1–4) | 98R7 |
| | 0.13 M | 7.0 | −0.7 | −13 | −0.9 | heat (1–4) | 98R7 |
| | 0.25 M | 7.0 | −1.1 | −26 | −1.7 | heat (1–4) | 98R7 |
| lysine | 0.0 M | 7.0 | 0.0 | 0 | 0.0 | heat (1–4) | 98R7 |
| | 0.1 M | 7.0 | 2.7 | −1 | 0.7 | heat (1–4) | 98R7 |
| | 0.2 M | 7.0 | 5.6 | −4 | 1.1 | heat (1–4) | 98R7 |
| | 0.5 M | 7.0 | 7.2 | −1 | 1.7 | heat (1–4) | 98R7 |
| | 0.7 M | 7.0 | 8.0 | −3 | 1.5 | heat (1–4) | 98R7 |
| | 1.0 M | 7.0 | 9.7 | −5 | 1.6 | heat (1–4) | 98R7 |
| | 2.0 M | 7.0 | 12.1 | −8 | 1.5 | heat (1–4) | 98R7 |

Remarks:
(1) van't Hoff treatment, transition monitored by changes in optical absorption at 295 nm
(2) measured in the presence of 0.1 M KCl, 4 mM EGTA
(3) in the absence of osmolytes, the values $T_{trs}$ = 42.2°C and ΔH = 217 kJ/mol were obtained
(4) in the presence of osmolyte ΔG(25°) was calculated using ΔCp = 6.18 kJ/mol/K from Ref. 94G3

Bovine holo-α-lactalbumin, ethanol-induced transitions

| pH | T | ΔG | Approach/Remarks | | Ref |
|---|---|---|---|---|---|
| 8.0 | 20 | 19.25±0.25 | DSC | (1) | 98G17 |

Remark:
(1) from DSC measurements conducted in the presence of ethanol, ΔG is an extrapolated value for α-lactalbumin unfolding in water

## β-Lactamase

Recombinant β-lactamase AmpC from *E. coli* and mutant Tyr150→Phe in the presence of inhibitors

| Inhibitor | pH | $T_{trs}$ | $\Delta T$ | $\Delta(\Delta G)$ | Appr./Rem. | Ref |
|---|---|---|---|---|---|---|
| wild-type AmpC + inhibitor: | | | | | | |
| none | 6.8 | 54.6 | 0.0 | 0.0 | heat (1–3) | 99B6 |
| Cloxacillin | 6.8 | 60.4±0.1 | 5.8±0.1 | 13.4±0.8 | heat (2–4) | 99B6 |
| Azteonam | 6.8 | 59.5±0.2 | 4.9±0.2 | 11.3±0.8 | heat (2–4) | 99B6 |
| Moxalactam | 6.8 | 51.4±0.1 | –3.2±0.1 | –7.5±0.4 | heat (2–4) | 99B6 |
| Imipenem | 6.8 | 53.4±0.1 | –1.2±0.1 | –2.9±0.3 | heat (2–4) | 99B6 |
| PNPP | 6.8 | 57.1±0.1 | 2.5±0.1 | 5.9±0.4 | heat (2–5) | 99B6 |
| BZBTH2B | 6.8 | 57.6±0.2 | 3.0±0.2 | 7.1±0.4 | heat (2–5) | 99B6 |
| mutant Tyr150→Phe + inhibitor: | | | | | | |
| none | 6.8 | 53.3 | 0.0 | 0.0 | heat (1–3,6) | 99B6 |
| Cloxacillin | 6.8 | 51.5±0.1 | –1.8±0.1 | –5.0±0.4 | heat (2–4) | 99B6 |
| Moxalactam | 6.8 | 49.3±0.1 | –4.0±0.1 | –10.9±0.8 | heat (2–4) | 99B6 |
| PNPP | 6.8 | 53.2±0.2 | –0.1±0.2 | –0.4±0.8 | heat (2–5) | 99B6 |
| BZBTH2B | 6.8 | 53.2±0.3 | –0.1±0.3 | –0.4±1.3 | heat (2–5) | 99B6 |

Remarks:
(1) reference value
(2) buffer: 50 mM potassium phosphate, 200 mM KCl, 38% ethylene glycol, pH 6.8
(3) transition monitored by far-UV CD at 223 nm
(4) $\Delta(\Delta G)$ was calculated using $\Delta(\Delta G) = \Delta T_{mut} \times \Delta S_{apo-enzyme}$
(5) PNPP and BZBTH2B are transition state analogs, see Ref. 99B6
(6) mutant Tyr150→Phe is by $\Delta(\Delta G) = -2.9\pm0.4$ kJ/mol less stable than the wild type, calculated using
$\Delta(\Delta G) = \Delta T_{mut} \times \Delta S_{w.t.}$

Recombinant β-lactamase AmpC from *E. coli* and mutant Tyr150→Phe

| Protein | pH | T | $\Delta G$ | $c_{1/2}$ | m | Appr./Rem. | Ref |
|---|---|---|---|---|---|---|---|
| AmpC w.t. | 6.8 | 25 | 58.6±4.6 | 1.43±0.03 | 41.0±3.3 | GuHCl (1–3) | 99B6 |
| Tyr150→Phe | 6.8 | 25 | 40.6±2.9 | 1.32±0.03 | 31.0±2.1 | GuHCl (1–3) | 99B6 |
| AmpC w.t. + PNPP | | | | | | | |
| | 6.8 | 25 | 56.1±4.2 | 1.56±0.04 | 36.0±2.5 | GuHCl (1–3) | 99B6 |

Remarks:
(1) data from equilibrium unfolding, nonlinear least-squares fit of the two-state transition curve assuming a linear
    dependence of $\Delta G$ on the denaturant conc.
(2) transition monitored by far-UV CD at 223 nm
(3) buffer: 50 mM potassium phosphate, 100 mM KCl, pH 6.8

β-lactamase from *Staphylococcus aureus* PC1, nonproline *cis* peptide bond

| Mutant | pH | T | $\Delta G$ | Approach | Remarks | Ref |
|---|---|---|---|---|---|---|
| wild type | 6.8 | 25 | 16.7 | urea | (1–4) | 97B3 |
| Asn136→Ala | 6.8 | 25 | 8.8 | urea | (2–5) | 97B3 |

Remarks:
(1) the nonproline *cis* peptide bond is present between Glu166 and Ile167
(2) linear extrapolation, LEM-SB
(3) transition monitored by CD at 222 nm
(4) buffer: 0.1 M potassium phosphate, 80 mM ammonium sulfate, pH 6.8
(5) mutant Asn136→Ala to examine the interaction between side chain of Asn136 and main chain of Glu166

Recombinant β-lactamase from *Staphylococcus aureus* PC (rPC1) and Ile167→Pro mutant

| Transition | pH | T | ΔG | $c_{1/2}$ | m | Appr./Rem. | Ref |
|---|---|---|---|---|---|---|---|
| protein rPC1 | | | | | | | |
| N → H | 6.8 | 20 | 14.4±2.2 | 0.64±0.02 | 22.4±3.0 | GuHCl (1–4) | 98W8 |
| H → U | 6.8 | 20 | 17.6±2.0 | 1.31±0.02 | 13.5±1.4 | GuHCl (1–4) | 98W8 |
| mutant (Ile167→Pro) | | | | | | | |
| N → H | 6.8 | 20 | 14.2±1.8 | 0.83±0.02 | 17.2±2.5 | GuHCl (1–4) | 98W8 |
| H → U | 6.8 | 20 | 13.8±2.91 | 1.14±0.07 | 12.15±2.0 | GuHCl (1–4) | 98W8 |

Remarks:
(1) linear extrapolation assuming a three-state model N → H → U
(2) transition monitored by CD
(3) measured in 50 mM sodium phosphate buffer, pH 6.8
(4) the recombinant protein was expressed in *E. coli*

TEM-1 β-lactamase, disulfide bond Cys77-Cys123 removed, transitions

| Mutant | pH | T | ΔG | $c_{1/2}$ | m | Appr./Rem. | Ref |
|---|---|---|---|---|---|---|---|
| wild type | | | | | | | |
| N → I | 7 | 25 | 21.7±1.7 | 0.90±0.02 | 24.2±1.7 | GuHCl (1–3) | 97V2 |
| I → U | 7 | 25 | 23.8±0.8 | 2.24±0.02 | 10.9±0.4 | GuHCl (1–3) | 97V2 |
| Cys77→Ser | | | | | | | |
| N → I | 7 | 25 | 18.0±1.7 | 0.71±0.02 | 25.1±1.7 | GuHCl (1–3) | 97V2 |
| I → U | 7 | 25 | 13.4±0.4 | 1.41±0.06 | 9.6±0.4 | GuHCl (1–3) | 97V2 |

Remarks:
(1) three-state unfolding model assuming a linear dependence of ΔG on denaturant concentration
(2) transition monitored by CD at 220 nm and fluorescence
(3) buffer: 50 mM sodium phosphate, 50 mM NaCl, 10 mM DTT, pH 7

## β-Lactoglobulin

β-lactoglobulin A stability determined by fluorescence and SH-reactivity

| pH | T | ΔG | $c_{1/2}$ | m | Approach/Remarks | Ref |
|---|---|---|---|---|---|---|
| 3 | 25 | 20±0.3 | 4.0 | 5.0±0.1 | urea (1,2) | 98A4 |
| 3 | 25 | 43±6 | 5.0 | 8.5±1.2 | urea (1,3) | 98A4 |
| 3 | 25 | 46.3 | 5.0 | 9.3 | urea (1,4) | 98A4 |

Remarks:
(1) linear extrapolation
(2) transition monitored by ANS fluorescence intensity
(3) transition monitored by SH-reactivity
(4) transition monitored by ORD, data from Ref. 83C

β-lactoglobulin A from bovine milk, urea-induced unfolding transition monitored by various methods

| pH | T | $\Delta G$ | $c_{1/2}$ | m | Appr./Rem. | Ref |
|---|---|---|---|---|---|---|
| 3.2 | 25 | 33±13 | 4.8±0.3 | 6.7±2.1 | urea (1,2) | 98A6 |
| 3.2 | 25 | 38±13 | 4.8±0.1 | 7.5±2.1 | urea (1,3) | 98A6 |
| 3.2 | 25 | 34.7±0.8 | 4.79±0.01 | 7.24±0.17 | urea (1,4) | 98A6 |
| 3.2 | 25 | 43.5±1.7 | 4.70±0.01 | 9.29±0.33 | urea (1,4) | 98A6 |
| 3.2 | 25 | 41.4±0.8 | 4.64±0.01 | 8.91±0.21 | urea (1,4) | 98A6 |

Remarks:

(1) linear extrapolation

(2) transition monitored by X-ray scattering, integral scattering intensity of a limited region

(3) transition monitored by X-ray scattering, integral scattering intensity of a Kratky plot in a limited region

(4) transition monitored by CD at 219 nm

(5) transition monitored by CD at 293 nm

(6) transition monitored by optical absorption at 293 nm

β-lactoglobulin, GuHCl-induced transitions that lead to an intermediate with nonnative α-helical structure

| TFE Conc. | Transition | pH | T | $\Delta G_i$ | m | Appr./Rem. | Ref |
|---|---|---|---|---|---|---|---|
| transition monitored by far-UV CD: | | | | | | | |
| 0% | N → U | 2 | 4 | 41.8±1.2 | 17.1±0.5 | GuHCl (1,2) | 97H1 |
| 0% | I → U | 2 | 4 | 8.0±0.7 | 3.4±0.1 | GuHCl (1,2) | 97H1 |
| 0% | N → I | 2 | 4 | 33.8±1.8 | 13.7±0.6 | GuHCl (1,2) | 97H1 |
| 9.8% | N → U | 2 | 4 | 30.6±0.4 | 12.4±0.5 | GuHCl (1–3) | 97H1 |
| 9.8% | I → U | 2 | 4 | 18.3±0.4 | 6.7±0.1 | GuHCl (1–3) | 97H1 |
| 9.8% | N → I | 2 | 4 | 12.3±0.8 | 5.7±0.3 | GuHCl (1–3) | 97H1 |
| transition monitored by near-UV CD: | | | | | | | |
| 0% | N → I | 2 | 4 | 34.9±2.2 | 14.5±0.9 | GuHCl (1,2) | 97H1 |
| 9.8% | N → I | 2 | 4 | 10.7±1.6 | 5.8±0.8 | GuHCl (1–3) | 97H1 |

Remarks:

(1) the approach assumes a linear combination of the contributions of the three states to the observed ellipticity

(2) the appraoch is based on a linear dependence of $\Delta G_i$ on denaturant concentration

(3) measured in the presence of 2,2,2-trifluoroethanol (TFE)

β-lactoglobulin B (β-LG)

| Protein | pH | T | $\Delta G$ | $c_{1/2}$ | m | Appr./Rem. | Ref |
|---|---|---|---|---|---|---|---|
| β-LG | 2.1 | 37 | | 5.90±0.03 | 6.90±0.57 | urea (1–3) | 99R1 |
| β-LG | 2.1 | 37 | | 6.05±0.05 | 1.86±0.14 | urea (1–3) | 99R1 |

Remarks:

(1) nonlinear least-squares fit of the transition curve assuming a linear dependence of $\Delta G$ on the denaturant conc.

(2) transition monitored by CD at 293 nm

(3) transition monitored by CD at 216 nm

(4) measured in 12 mM phosphate buffer, pH 2.1

Bovine β-lactoglobulin, local stability

| Residue | pH | T | $\Delta G$ | $c_{1/2}$ | m | Approach/Remarks | Ref |
|---|---|---|---|---|---|---|---|
| Met24 | 2.1 | 37 | 19.66 | 4.11 | 4.77 | urea, NMR (1–3) | 99R2 |
| Ala25 | 2.1 | 37 | 28.37 | 4.55 | 6.23 | urea, NMR (1–3) | 99R2 |
| Ala26 | 2.1 | 37 | 31.09 | 4.27 | 7.28 | urea, NMR (1–3) | 99R2 |
| Tyr42 | 2.1 | 37 | 13.39 | 3.49 | 3.85 | urea, NMR (1–3) | 99R2 |
| Val43 | 2.1 | 37 | 27.91 | 5.01 | 5.56 | urea, NMR (1–3) | 99R2 |

Bovine β-lactoglobulin, local stability (continued)

| Residue | pH | T | ΔG | $c_{1/2}$ | m | Approach/Remarks | Ref |
|---------|-----|-----|-------|------|------|------------------|-------|
| Glu44   | 2.1 | 37 | 33.93 | 4.69 | 7.24 | urea, NMR (1–3) | 99R2 |
| Leu46   | 2.1 | 37 | 29.87 | 4.86 | 6.15 | urea, NMR (1–3) | 99R2 |
| Glu55   | 2.1 | 37 | 37.91 | 4.55 | 8.33 | urea, NMR (1–3) | 99R2 |
| Leu54   | 2.1 | 37 | 31.55 | 4.91 | 6.44 | urea, NMR (1–3) | 99R2 |
| Ile56   | 2.1 | 37 | 29.92 | 4.96 | 6.02 | urea, NMR (1–3) | 99R2 |
| Leu57   | 2.1 | 37 | 26.53 | 4.51 | 5.90 | urea, NMR (1–3) | 99R2 |
| Gln59   | 2.1 | 37 | 30.46 | 4.85 | 6.28 | urea, NMR (1–3) | 99R2 |
| Leu58   | 2.1 | 37 | 39.50 | 5.21 | 7.57 | urea, NMR (1–3) | 99R2 |
| Lys60   | 2.1 | 37 | 30.75 | 4.93 | 6.23 | urea, NMR (1–3) | 99R2 |
| Glu62   | 2.1 | 37 | 32.93 | 4.90 | 6.74 | urea, NMR (1–3) | 99R2 |
| Ala67   | 2.1 | 37 | 22.43 | 4.82 | 4.64 | urea, NMR (1–3) | 99R2 |
| Ile71   | 2.1 | 37 | 31.92 | 5.12 | 6.23 | urea, NMR (1–3) | 99R2 |
| Ala73   | 2.1 | 37 | 29.25 | 4.56 | 6.40 | urea, NMR (1–3) | 99R2 |
| Glu74   | 2.1 | 37 | 33.47 | 4.46 | 7.49 | urea, NMR (1–3) | 99R2 |
| Val81   | 2.1 | 37 | 36.61 | 4.72 | 7.74 | urea, NMR (1–3) | 99R2 |
| Phe82   | 2.1 | 37 | 34.43 | 4.75 | 7.24 | urea, NMR (1–3) | 99R2 |
| Lys83   | 2.1 | 37 | 25.02 | 4.64 | 5.40 | urea, NMR (1–3) | 99R2 |
| Ile84   | 2.1 | 37 | 35.52 | 4.96 | 7.15 | urea, NMR (1–3) | 99R2 |
| Asn90   | 2.1 | 37 | 32.17 | 4.81 | 6.69 | urea, NMR (1–3) | 99R2 |
| Lys91   | 2.1 | 37 | 28.58 | 4.92 | 5.82 | urea, NMR (1–3) | 99R2 |
| Val92   | 2.1 | 37 | 22.30 | 4.78 | 4.69 | urea, NMR (1–3) | 99R2 |
| Leu93   | 2.1 | 37 | 36.48 | 5.14 | 7.11 | urea, NMR (1–3) | 99R2 |
| Val94   | 2.1 | 37 | 24.48 | 4.60 | 5.31 | urea, NMR (1–3) | 99R2 |
| Tyr102  | 2.1 | 37 | 40.42 | 5.11 | 7.91 | urea, NMR (1–3) | 99R2 |
| Leu103  | 2.1 | 37 | 43.81 | 4.96 | 8.83 | urea, NMR (1–3) | 99R2 |
| Leu104  | 2.1 | 37 | 31.21 | 5.05 | 6.19 | urea, NMR (1–3) | 99R2 |
| Phe105  | 2.1 | 37 | 32.05 | 5.05 | 6.36 | urea, NMR (1–3) | 99R2 |
| Cys106  | 2.1 | 37 | 29.00 | 4.85 | 5.98 | urea, NMR (1–3) | 99R2 |
| Met107  | 2.1 | 37 | 37.74 | 5.04 | 7.49 | urea, NMR (1–3) | 99R2 |
| Glu108  | 2.1 | 37 | 29.12 | 4.90 | 5.94 | urea, NMR (1–3) | 99R2 |
| Ala118  | 2.1 | 37 | 37.87 | 4.66 | 8.12 | urea, NMR (1–3) | 99R2 |
| Cys119  | 2.1 | 37 | 29.95 | 4.70 | 5.94 | urea, NMR (1–3) | 99R2 |
| Leu122  | 2.1 | 37 | 21.42 | 4.83 | 4.44 | urea, NMR (1–3) | 99R2 |
| Val123  | 2.1 | 37 | 27.41 | 5.02 | 5.44 | urea, NMR (1–3) | 99R2 |
| Glu134  | 2.1 | 37 | 28.24 | 4.32 | 6.53 | urea, NMR (1–3) | 99R2 |
| Ala139  | 2.1 | 37 | 34.14 | 4.32 | 7.91 | urea, NMR (1–3) | 99R2 |
| Leu140  | 2.1 | 37 | 26.82 | 4.42 | 6.07 | urea, NMR (1–3) | 99R2 |

Remarks:

(1) unfolding curves were obtained from pseudo-equilibrium measurements: the $H^{\alpha}$-$H^{N}$ cross-peak volumes in TOCSY spectra were recorded versus the urea conc.

(2) data treatment by linear extrapolation to zero denaturant conc.

(3) measured in 12 mM $H_3PO_4$/NaOH, pH 2.1

Bovine β-lactoglobulin (BLG) and mixed disulfide derivatives

| Protein | pH | T | ΔG | Approach/Remarks | | Ref |
|---|---|---|---|---|---|---|
| BLG | 2.05 | 37 | 20.4±2.6 | DSC | (1,2) | 98B14 |
| BLG-COOH | 2.05 | 37 | 12.2±2.8 | DSC | (1) | 98B14 |
| BLG-OH | 2.05 | 37 | 11.7±1.9 | DSC | (1) | 98B14 |

Explanations:

      BLG-COOH:      mixed disulfide derivative obtained by reaction with mercaptopropionic acid

      BLG-OH:         mixed disulfide derivative obtained by reaction with 2-mercaptoethanol

Remarks:

(1) buffer: 40 mM glycine, pH 2.05

(2) calculated using $\Delta Cp = 9.79 \pm 1.07$ kJ/mol/K, see also Table 2

Equine β-lactoglobulin, transitions

| Transition | pH | T | ΔG | $c_{1/2}$ | m | Appr./Rem. | Ref |
|---|---|---|---|---|---|---|---|
| N → U | 4.0 | 25 | 41.1±5.3 | 3.39±0.57 | 12.1±1.3 | urea (1–3) | 99F11 |
| I → U | 4.0 | 25 | 12.1±4.6 | 3.94±1.84 | 3.1±0.9 | urea (1–3) | 99F11 |
| N → U | 4.0 | 25 | 42.1±3.9 | 2.12±0.28 | 19.8±1.8 | GuHCl (1,4,5) | 99F11 |
| I → U | 4.0 | 25 | 16.4±3.1 | 2.55±0.67 | 6.4±1.2 | GuHCl (1,4,5) | 99F11 |

Remarks:

(1) data were analyzed by a three-state model assuming a linear dependence of ΔG on the denaturant conc.

(2) transition monitored by difference absorption at 287 nm, fluorescence intensity, and CD at 293 and 222 nm

(3) buffer: 50 mM citrate, pH 4.0

(4) transition monitored CD at 293 and 222 nm

(5) buffer: 50 mM formate, pH 4.0

**Lambda Repressor,** see Repressor Proteins

## Lectin

Pea lectin, ΔG from urea denaturation at different temperatures

| Protein | pH | T | ΔG | $c_{1/2}$ | m | Appr./Rem. | Ref |
|---|---|---|---|---|---|---|---|
| pea lectin | 7.2 | 4 | 58.66 | 4.59 | 12.13 | urea (1–4) | 98A2 |
| | 7.2 | 10 | 65.14 | 5.72 | 10.84 | urea (1–4) | 98A2 |
| | 7.2 | 15 | 73.30 | 6.47 | 10.84 | urea (1–4) | 98A2 |
| | 7.2 | 20 | 76.65 | 6.91 | 11.00 | urea (1–4) | 98A2 |
| | 7.2 | 25 | 77.32 | 6.79 | 10.92 | urea (1–5) | 98A2 |
| | 7.2 | 28 | 77.53 | 7.44 | 10.00 | urea (1–4) | 98A2 |
| | 7.2 | 30 | 76.19 | 6.79 | 9.83 | urea (1–4) | 98A2 |
| | 7.2 | 37 | 75.10 | 6.34 | 11.30 | urea (1–4) | 98A2 |
| | 7.2 | 40 | 73.14 | 6.65 | 10.50 | urea (1–4) | 98A2 |
| | 7.2 | 45 | 70.37 | 6.54 | 10.25 | urea (1–4) | 98A2 |
| | 7.2 | 50 | 59.29 | 4.12 | 13.56 | urea (1–4) | 98A2 |
| | 7.2 | 55 | 52.17 | 5.66 | 8.58 | urea (1–4) | 98A2 |

Remarks:

(1) linear extrapolation

(2) treatment by an equilibrium between folded dimer and unfolded monomers ($N_2 \rightarrow 2U$)

(3) transition monitored by fluorescence, and near- and far-UV CD

(4) measured in 50 mM phosphate buffer, pH 7.2

(5) ΔG(25°C) determined at 4 μM, 8 μM, and 20 μM pea-lectin concentration amounts to 76.61, 77.28, and 77.36 kJ/mol, respectively

## Leghemoglobin

Leghemoglobin a from soybean, cyanide complex (Lba.CN)

| Protein | pH | T | $\Delta G$ | Approach | Ref |
|---|---|---|---|---|---|
| Lba.CN | 5.0 | 25 | 31.8 | DSC | 98T12 |
| | 5.9 | 25 | 43.9 | DSC | 98T12 |
| | 7.25 | 25 | 46.9 | DSC | 98T12 |
| | 8.5 | 25 | 44.8 | DSC | 98T12 |
| | 9.2 | 25 | 44.8 | DSC | 98T12 |
| | 9.9 | 25 | 31.8 | DSC | 98T12 |
| | 11.15 | 25 | 29.3 | DSC | 98T12 |

## Leucine Zipper

GCN4-p2', a $\alpha$-helical coiled coil derived from the leucine-zipper region of bZIP transcriptional activator GCN4, stability parameters in the presence of cosolvents

| Solvent/Species | | pH | T | $\Delta G$ | m | Approach/Remarks | Ref |
|---|---|---|---|---|---|---|---|
| buffer | dimeric | 5.5 | 10 | 34.94±0.4 | 7.87±0.17 | GuHCl eq. (1–6) | 99B10 |
| buffer | dimeric | 5.5 | 10 | 35.15±0.54 | 6.28±0.54 | GuHCl kin.(1,5–7) | 99B10 |
| buffer | monomeric | 5.5 | 10 | 27.2 ±0.8 | 6.65±0.21 | GuHCl eq. (1–6) | 99B10 |
| buffer | monomeric | 5.5 | 10 | 25.40±0.54 | 6.57±0.54 | GuHCl kin.(1,5–7) | 99B10 |
| 9% glc | dimeric | 5.5 | 10 | n.d. | n.d. | GuHCl eq. (1–6,8) | 99B10 |
| 9% glc | dimeric | 5.5 | 10 | 47.91±0.92 | 8.41±0.63 | GuHCl kin.(1,5–8) | 99B10 |
| 9% glc | monomeric | 5.5 | 10 | 37.7 ±4.2 | 8.37±0.84 | GuHCl eq. (1–6,8) | 99B10 |
| 9% glc | monomeric | 5.5 | 10 | 36.65±6.3 | 8.08±1.30 | GuHCl kin.(1,5–8) | 99B10 |
| 15.2% | dimeric | 5.5 | 10 | 43.5 ±0.4 | 8.74±0.17 | GuHCl eq. (1–6,9) | 99B10 |
| 15.2% | dimeric | 5.5 | 10 | 43.39±0.92 | 8.62±1.21 | GuHCl kin.(1,5–7,9) | 99B10 |
| 15.2% | monomeric | 5.5 | 10 | 38.5 ±0.4 | 8.8 ±1.7 | GuHCl eq. (1–6,9) | 99B10 |
| 15.2% | monomeric | 5.5 | 10 | 38.58±1.30 | 8.49±0.33 | GuHCl kin.(1,5–7,9) | 99B10 |

Remarks:

(1) Peptide: derived from GCN4-p1' (Ac-Arg-Met-Lys-Gln-Leu-Glu-Asp-Lys-Val-Glu-Glu-Leu-Leu-Ser-Lys-Asn-Trp-His-Leu-Glu-Asn-Glu-Val-Ala-Arg-Leu-Lys-Lys-Leu-Val-Gly-Glu-Arg-NH$_2$) with Tyr17→Trp and an additional tripeptide (Cys-Gly-Gly) at the N-terminus that allows to produce the monomeric form by oxidation and the dimeric form by reduction

(2) equilibrium measurements, linear extrapolation

(3) the data are extrapolated to zero denaturant concentration and, for the dimeric species, 1 M standard peptide conc.

(4) transition monitored by CD at 222 nm

(5) buffer: 20 mM sodium acetate, pH 5.5

(6) Ref. 99B10 contains a deconvolution of viscosity and stability

(7) kinetic data from stopped-flow fluorescence

(8) measured in 9% glycerol, relative viscosity $\eta_{rel}$ = 1.25

(9) measured in 15.2% glycerol, relative viscosity $\eta_{rel}$ = 1.5

Designed leucine zipper by including a structure that cannot form interhelical electrostatic bonds

| pH | T | ΔG | Approach/Remarks | | Ref |
|---|---|---|---|---|---|
| 3.5 | 25 | 52 | Δν | (1) | 99D7 |
| 3.5 | 25 | 55 | DSC | (2) | 99D7 |
| 4.7 | 25 | 41.8 | Δν | (1) | 99D7 |
| 4.7 | 25 | 42.5 | DSC | (2) | 99D7 |

Remarks:
(1) from difference in protonation derived for a two-state transition between folded dimer and unfolded monomers
(2) data from equilibrium unfolding (see also Table 2) using $\Delta C_p = 1.98 \pm 0.60$ kJ/mol/K

## Lipoxygenase

Lipoxygenase-1 (LOX1) from soybean

| Protein/Transition | | pH | T | ΔG | m | Appr./Rem. | Ref |
|---|---|---|---|---|---|---|---|
| LOX1 | N → I | 7.0 | 25 | 59.4±1.2 | 8.28±0.92 | urea (1–3) | 99S18 |
| | I → U | 7.0 | 25 | 49.8±0.5 | 6.69±0.21 | urea (1–3) | 99S18 |

Remarks:
(1) linear extrapolation assuming a three-state model
(2) transition monitored by change in fluorescene emission maximum
(3) buffer: 0.1 M Tris-HCl

## Luciferase

Bacterial luciferase, α-subunit

| Transition | pH | T | ΔG | m | Approach/Remarks | | Ref |
|---|---|---|---|---|---|---|---|
| N → I | 7.0 | 18 | 9.37±1.05 | 10.21±0.92 | urea | (1–5) | 99N4 |
| I → U | 7.0 | 18 | 27.20±3.14 | 9.50±0.67 | urea | (1–5) | 99N4 |

Remarks:
(1) protein from the *Vibrio harveyi luxA* gene expressed in *Escherichia coli*
(2) data analysis by a three-state unfolding model assuming a linear dependence of ΔG on denaturant conc.
(3) the data are mean values from 10 measurements at varying protein conc. from 10 to 30 μg/ml
(4) transition monitored by CD at 222 nm and average fluorescence emission wavelength
(5) buffer: 50 mM phosphate, 1 mM DTT, 0.005% Tween 20, pH 7.0

Firefly *Photinus pyralis* luciferase, transitions

| Transition | pH | T | ΔG | $c_{1/2}$ | m | Appr./Rem. | Ref |
|---|---|---|---|---|---|---|---|
| N → I1 | 7.8 | 10 | 15±3 | 0.3 | 50±6 | GuHCl (1) | 97H3 |
| I1 → I2 | 7.8 | 10 | | 1.7 | | GuHCl (2) | 97H3 |
| I2 → U | 7.8 | 10 | 17±3 | 3.8 | 4.5±1 | GuHCl (2) | 97H3 |
| N → U | 7.8 | 10 | ~50 | | | GuHCl (3) | 97H3 |

Remarks:
(1) transition monitored by luciferase activity
(2) transition monitored by CD
(3) estimated value for the three unfolding steps

## Lysyl-tRNA Synthetase

LysN, the anticodon binding domain of the *E. coli* lysyl-tRNA synthetase, comparison of $\Delta G$ from hydrogen exchange and equilibrium unfolding

| Protein | pH | T | $\Delta G$ | $c_{1/2}$ | m | Appr./Rem. | | Ref |
|---|---|---|---|---|---|---|---|---|
| LysN | 6.0 | 20 | 25.9±2.8 | 1.98±0.02 | 13.0±1.3 | GuHCl (1–3) | | 99A1 |
| | 6.0 | 20 | 26.8±1.0 | 2.03±0.01 | 13.2±0.5 | GuHCl (1,2,4) | | 99A1 |
| $\Delta$A51-LysN | 6.0 | 20 | 12.0±2.3 | 0.87±0.06 | 13.9±1.8 | GuHCl (1–3,5) | | 99A1 |
| Data for various structure elements of LysN (hydrogen exchange/NMR): | | | | | | | | |
| all | 6.0 | 20 | 31.5±6.6 | | 8.8±3.4 | H/D | (6) | 99A1 |
| all $\alpha$ | 6.0 | 20 | 24.0±4.6 | | 6.6±1.4 | H/D | (6–8) | 99A1 |
| all $\beta$ | 6.0 | 20 | 34.3±4.9 | | 9.7±3.3 | H/D | (6–8) | 99A1 |
| all $\beta$-trimmed | 6.0 | 20 | 35.4±4.4 | | 10.3±3.2 | H/D | (6–8) | 99A1 |
| loops ($L_{\beta_3\alpha}$) | 6.0 | 20 | 23.3±4.9 | | 4.3±2.4 | H/D | (6–8) | 99A1 |
| ($\alpha$1,$\alpha$2) | 6.0 | 20 | 21.5±3.3 | | 6.4±1.5 | H/D | (6–8) | 99A1 |
| $\alpha$3 | 6.0 | 20 | 26.6±4.6 | | 6.9±1.5 | H/D | (6–8) | 99A1 |
| ($L_{\beta_3\alpha}$,all $\alpha$) | 6.0 | 20 | 23.9±4.4 | | 6.2±1.8 | H/D | (6–8) | 99A1 |
| ($L_{\beta_3\alpha}$,all $\alpha$,$\beta$4, $\beta$5) | | | | | | | | |
| | 6.0 | 20 | 29.4±6.3 | | 7.9±2.9 | H/D | (6–8) | 99A1 |
| OB-fold: (all $\beta$+$\alpha$3) | | | | | | | | |
| | 6.0 | 20 | 34.1±5.2 | | 9.8±3.2 | H/D | (6–8) | 99A1 |
| $\beta$1-$\beta$3 | 6.0 | 20 | 36.6±5.1 | | 11.7±2.9 | H/D | (6–8) | 99A1 |
| $\beta$4-$\beta$5 | 6.0 | 20 | 34.4±3.5 | | 9.2±3.0 | H/D | (6–8) | 99A1 |
| ($\beta$1,$\beta$2,$\beta$4) | 6.0 | 20 | 37.6±3.8 | | 11.3±3.4 | H/D | (6–8) | 99A1 |
| ($\beta$3,$\beta$5) | 6.0 | 20 | 32.3±3.1 | | 8.9±2.4 | H/D | (6–8) | 99A1 |

Remarks:

(1) equilibrium unfolding, linear extrapolation, LEM-SB

(2) buffer: 20 mM potassium phosphate, pH 6.6

(3) transition monitored by fluorescence at 350 nm

(4) transition monitored by CD at 222 nm

(5) the deletion fragment retains cooperative structure

(6) data from hydrogen exchange in the presence of GuHCl measured in 99,98% $D_2O$ at pH 5.8 to 6.0

(7) the data from NMR/hydrogen exchange manifest a distribution of $\Delta G$ and indicate subglobal unfolding

(8) the largest $\Delta G$ values from NMR/hydrogen exchange are from $\beta$-sheet and exceed the equilibrium unfolding values by approx. 12 to 17 kJ/mol

## Lysozyme, Equine

Equine lysozyme, holo- and apo-protein, transitions

| Protein/Transition | | pH | T | $\Delta G$ | m | Appr./Rem. | Ref |
|---|---|---|---|---|---|---|---|
| holo | N → I | 7.0 | 25 | 23.60±1.34 | 8.37±0.46 | GuHCl (1–3) | 98M21 |
| | I → U | 7.0 | 25 | 18.12±2.47 | 4.73±0.79 | GuHCl (1–3) | 98M21 |
| | N → U | 7.0 | 25 | 41.71±2.05 | 13.10±0.63 | GuHCl (1–3) | 98M21 |
| apo | N → I | 7.0 | 25 | 1.72±0.29 | 8.24±0.84 | GuHCl (1,2,4) | 98M21 |
| | I → U | 7.0 | 25 | 18.28±1.38 | 4.81±1.21 | GuHCl (1,2,4) | 98M21 |
| | N → U | 7.0 | 25 | 20.00±1.38 | 13.05±0.88 | GuHCl (1,2,4) | 98M21 |

Remarks:

(1) treatment by a three-state model assuming a linear dependence of $\Delta G$ on the denaturant concentration, nonlinear fit

(2) transitions monitored by CD at 230 and 292.5 nm

(3) buffer: 50 mM sodium cacodylate (pH 7.0), 50 mM NaCl, 10 mM $CaCl_2$

(4) buffer: 50 mM sodium cacodylate (pH 7.0), 50 mM NaCl, 1 mM EGTA

**Lysozyme, HEW** (Hen Egg White Lysozyme)

The data entries are arranged as follows:

a) wild type, data obtained by various approaches
b) wild type and mutants
c) wild type and mutants, transitions
d) wild type and mutants in the presence of solutes

*a) wild type, data obtained by various approaches*

Hen egg white lysozyme

| pH | T | $\Delta G$ | m | Approach/Remarks | Ref |
|----|---|------------|---|------------------|-----|
| 5.2 | 20 | 42.9±4.5 | 10.5±1.5 | GuHCl   (1–3) | 97K4 |

Remarks:
(1) linear extrapolation, LEM-SB
(2) buffer: 20 mM sodium acetate, pH 5.2
(3) the paper describes an nucleated state which is separated from the folding intermediate by $\Delta(\Delta G) = 13.7\pm3$ kJ/mol at zero denaturant conc.

Hen egg white lysozyme

| pH | T | $\Delta G$ | $c_{1/2}$ | m | Appr./Rem. | Ref |
|----|---|------------|-----------|---|------------|-----|
| 4.0 | 10 | 34.3 | 4.19 | 8.37 | GuHCl (1–4) | 97L3 |

Remarks:
(1) linear extrapolation, LEM-SB
(3) buffer: 10 mM sodium acetate, 100 mM NaCl, pH* 4.0
(2) measured in $D_2O$
(4) transition monitored by CD at 222 nm

Hen egg white lysozyme, combined urea and GuHCl denaturation

| Conditions | pH | T | $\Delta G$ | $c_{1/2}$ | m | Appr./Rem. | Ref |
|------------|----|---|------------|-----------|---|------------|-----|
| 0.1–1.0 M KCl | 2.0 | 25 | 17.49±0.88 | 2.06 | 8.49±0.42 | GuHCl (1–4) | 99G21 |
| 0.1 M KCl | 2.0 | 25 | 17.11±1.09 | 3.32 | 5.15±0.33 | urea (1–4) | 99G21 |
| 0.30–0.50 M KCl | 2.0 | 25 | 18.32±1.09 | 3.98 | 4.60±0.25 | urea (1–4) | 99G21 |
| 0.75–1.00 M KCl | 2.0 | 25 | 22.13±1.84 | 4.68 | 4.73±0.38 | urea (1–4) | 99G21 |
| 0.90 M GuHCl | 2.0 | 25 | 10.71±0.59 | 2.33 | 4.60±0.25 | urea (1–4) | 99G21 |
| 1.23 M GuHCl | 2.0 | 25 | 7.20±0.25 | 1.72 | 4.18±0.13 | urea (1–4) | 99G21 |
| 1.84 M GuHCl | 2.0 | 25 | 2.93±0.13 | 0.66 | 4.44±0.13 | urea (1–4) | 99G21 |
| 1.98 M GuHCl | 2.0 | 25 | 1.26±0.21 | 0.27 | 4.69±0.42 | urea (1–4) | 99G21 |

Remarks:
(1) linear extrapolation, for details of see Ref. 99G21
(2) transition monitored by optical absorption at 300 nm
(3) measured in KCl-HCl, pH 2.0
(4) for comparable results see also ribonuclease A and cytochrome *c* (Ref. 99G21)

Hen egg white lysozyme, exclusion of equilibrium intermediates in urea-induced unfolding

| Protein Conc. | pH | T | $c_{1/2}$ | m | Approach | Remarks | Ref |
|---|---|---|---|---|---|---|---|
| 7 mg/ml | 2.9 | 20 | 5.58±0.04 | 6.7±0.7 | urea, kinetics | (1,2) | 97I1 |
| 10 mg/ml | 2.9 | 20 | 5.69±0.14 | | urea, kinetics | (1,2) | 97I1 |
| 7 mg/ml | 2.9 | 20 | 5.51±0.07 | 5.3±0.8 | urea, kinetics | (1,3) | 97I1 |
| 10 mg/ml | 2.9 | 20 | 5.53±0.04 | | urea, kinetics | (1,3) | 97I1 |
| 0.046 mg/ml | 2.9 | 20 | 5.75±0.23 | 6.0±0.5 | urea, chevron | (1,4) | 97I1 |
| 0.046 mg/ml | 2.9 | 20 | 5.69±0.06 | 5.6±0.8 | urea, fluores. | (1,5) | 97I1 |
| 0.046 mg/ml | 2.9 | 20 | 5.63±0.06 | 6.3±1.0 | urea, fluores. | (1,6) | 97I1 |
| (see 96C10) | 2.9 | 20 | 5.51±0.06 | 5.3±0.4 | urea, X-ray+CD | (1,7) | 97I1 |

Remarks:

(1) buffer: sodium citrate, 100 mM NaCl, pH 2.9

(2) double-jump unfolding assay

(3) double-jump refolding assay

(4) chevron plot of folding-unfolding rate constant versus GuHCl concentration

(5) fluorescence with linear unfolded baseline

(6) fluorescence with nonlinear unfolded baseline

(7) global analysis of radius of gyration and circular dichroism data from Ref. 96C10

Hen egg white lysozyme, pressure denaturation

see Ref. 99S2: profiles of $\Delta G$ versus pressure at varying temperatures and GuHCl conc. are provided in Ref. 99S2

*b) wild type and mutants*

Hen egg white lysozyme, wild type and mutations of the N-cap residue Gly4

| Mutant | pH | T | $\Delta G$ | $c_{1/2}$ | m | Appr. | Rem. | Ref |
|---|---|---|---|---|---|---|---|---|
| wild type | 5.5 | 35 | 43.1 | 3.62±0.01 | 12.3 | GuHCl | (1–3) | 97M10 |
| Gly4→Ala | 5.5 | 35 | 42.0 | 3.53±0.01 | 11.1 | GuHCl | (1–3) | 97M10 |
| Gly4→Asp | 5.5 | 35 | 43.7 | 3.67±0.02 | 11.9 | GuHCl | (1–3) | 97M10 |
| Gly4→Glu | 5.5 | 35 | 42.0 | 3.53±0.01 | 11.4 | GuHCl | (1–3) | 97M10 |
| Gly4→Pro | 5.5 | 35 | 34.3 | 2.88±0.03 | 13.0 | GuHCl | (1–3) | 97M10 |
| Gly4→Ser | 5.5 | 35 | 43.1 | 3.62±0.01 | 12.3 | GuHCl | (1–3) | 97M10 |

Remarks:

(1) linear extrapolation

(2) the transition was monitored by fluorescence at 360 nm

(3) buffer: 0.1 M sodium acetate, pH 5.5

(4) $\Delta G$ was determined using the mean value of m = 11.9 kJ/mol/M

Hen egg white lysozyme, wild type and mutants position 15

| Mutant | pH | T | $\Delta G$ | $c_{1/2}$ | Approach/Remarks | | Ref |
|---|---|---|---|---|---|---|---|
| wild type | 5.5 | 35 | 42.3±0.5 | 3.62±0.03 | GuHCl | (1–3) | 97O4 |
| His15→Gly | 5.5 | 35 | 38.9±0.5 | 3.34±0.03 | GuHCl | (1–3) | 97O4 |
| His15→Ala | 5.5 | 35 | 41.4±0.5 | 3.56±0.03 | GuHCl | (1–3) | 97O4 |
| His15→Val | 5.5 | 35 | 36.8±0.5 | 3.15±0.03 | GuHCl | (1–3) | 97O4 |
| His15→Phe | 5.5 | 35 | 40.6±0.5 | 3.48±0.03 | GuHCl | (1–3) | 97O4 |

Remarks:

(1) linear extrapolation

(2) transition monitored by fluorescence at 360 nm

(3) buffer: 0.1 M sodium acetate adjusted to pH 5.5 with HCl

# 146     Table 1. Gibbs Energy Change – Molar Values

Hen egg white lysozyme, wild type and mutants that affect Asx in Pos. 18 and 27

| Mutant | pH | T | ΔG | $c_{1/2}$ | m | Appr./Rem. | Ref |
|---|---|---|---|---|---|---|---|
| wild type | 5.5 | 35 | 44.5 | 3.62±0.01 | 12.3 | GuHCl (1–3) | 97M9 |
| Asp18→Asn | 5.5 | 35 | 42.2 | 3.43±0.01 | 12.3 | GuHCl (1–3) | 97M9 |
| Asn27→Asp | 5.5 | 35 | 46.0 | 3.74±0.01 | 12.2 | GuHCl (1–3) | 97M9 |
| double mutant (Asp18→Asn and Asn27→Asp) | | | | | | | |
| | 5.5 | 35 | 43.7 | 3.55±0.01 | 12.5 | GuHCl (1–3) | 97M9 |

Remarks:
(1) linear extrapolation (LEM) using the mean value of $c_{1/2}$ = 12.3
(2) the transition was monitored by fluorescence at 360 nm
(3) wild type: Asp18 and Asn27

Hen egg white lysozyme (LYZ), wild type and 3-SS variants

| Protein | pH | T | $c_{1/2}$ | m | Approach/Remarks | | Ref |
|---|---|---|---|---|---|---|---|
| wild-type LYZ | 3.0 | 4.0 | 3.21 | 15 | GuHCl | (1,2,5) | 99Y2 |
| Met-LYZ | 3.0 | 4.0 | 2.49 | 11 | GuHCl | (1–3,5) | 99Y2 |
| RCM-LYZ | 3.0 | 4.0 | 1.09 | 13 | GuHCl | (1,2,4,5) | 99Y2 |
| $\Delta1_{SA}$ = double mutant (Cys6→Ser and Cys127→Ala) | | | | | | | |
| | 3.0 | 4.0 | 0.43 | 11 | GuHCl | (1,2,5) | 99Y2 |
| $\Delta2_A$ = double mutant (Cys30→Ala and Cys115→Ala) | | | | | | | |
| | 3.0 | 4.0 | 0.82 | 9.0 | GuHCl | (1,2,5) | 99Y2 |
| $\Delta3_A$ = double mutant (Cys64→Ala and Cys80→Ala) | | | | | | | |
| | 3.0 | 4.0 | 0.90 | 9.5 | GuHCl | (1,2,5) | 99Y2 |
| $\Delta4_A$ = double mutant (Cys76→Ala and Cys94→Ala) | | | | | | | |
| | 3.0 | 4.0 | 1.18 | 12 | GuHCl | (1,2,5) | 99Y2 |

Remarks:
(1) linear extrapolation
(2) transition monitored by far-UV CD at 222 nm
(3) Met-LYZ = recombinant lysozyme containing the four authentic disulfide bridges and the extra N-terminal Met
(4) RCM-LYZ = wild-type lysozyme with Cys6 and Cys127 replaced by carboxymethylated Cys residues
(5) the thermal transition temperature at pH 4.0 in the absence of GuHCl amounts to $T_{trs}$ = 75.9°C for wild type, 68.7°C for Met-Lyz, 49.3°C for RCM-Lyz, 38.1°C for $\Delta1_{SA}$, 47.9°C for $\Delta2_A$, 46.3°C for $\Delta3_A$, and 51.3°C for $\Delta4_A$

*c) wild type and mutants, transitions*

Hen egg white lysozyme, transient intermediate

| Transition | pH | T | ΔG | Approach/Remarks | Ref |
|---|---|---|---|---|---|
| I → D | 5.2 | 20 | 4 | see (1,2) | 99S7 |
| N → I | 5.2 | 20 | 36 | see (1,2) | 99S7 |

Remarks:
(1) characterization of transient intermediates in lysozyme folding with time resolved small-angle X-ray scattering, stopped-flow and continous-flow techniques, and time resolved fluorescence spectra
(2) measured in 50 mM acetate, pH 5.2, with 0.6 M GuHCl

Hen egg white lysozyme, transitions

| Transition | pH | T | $\Delta G$ | m* | Approach/Remarks | | Ref |
|---|---|---|---|---|---|---|---|
| data based on GuHCl activity: | | | | | | | |
| N $\rightarrow$ I* | 7.5 | 25 | 34.3 | 2.9 | GuHCl | (1–3) | 95P11 |
| I* $\rightarrow$ I | 7.5 | 25 | 13.8 | 6.2 | GuHCl | (1–3) | 95P11 |
| I $\rightarrow$ U | 7.5 | 25 | 29.3 | 12.9 | GuHCl | (1–3) | 95P11 |
| I $\rightarrow$ U | 7.5 | 25 | | 12.7 | GuHCl | (1,3,4) | 95P11 |
| data based on GuHCl concentration: | | | | | | | |
| N $\rightarrow$ I* | 7.5 | 25 | 27.6 | 1.3 | GuHCl | (1–3) | 95P11 |
| I* $\rightarrow$ I | 7.5 | 25 | 11.3 | 3.4 | GuHCl | (1–3) | 95P11 |
| I $\rightarrow$ U | 7.5 | 25 | 20.5 | 6.1 | GuHCl | (1–3) | 95P11 |
| I $\rightarrow$ U | 7.5 | 25 | | 6.7 | GuHCl | (1,3,4) | 95P11 |

Remarks:

(1a)  from an integrated analysis of equilibrium and kinetic folding reaction, for details of the approach see Ref. 95P11

(1b)  buffer: 50 mM triethanolamine-HCl, 2 mM DTT, pH 7.5, transition monitored by fluorescence

(2)  a specific $C_{1/2}$ value to calculate the molar denaturant activity was set to 7.5

(3)  m* values used here describe changes in both equilibrium and rate constants as function of molar denaturant activity relative to the native state (dimension $M^{-1}$)

(4)  m* refers to kinetics

*d) wild type and mutants in the presence of solutes*

Hen egg white lysozyme in the presence and absence of $Na_2SO_4$

| Salt Conc. | pH | T | $\Delta G$ | m | Approach/Remarks | | Ref |
|---|---|---|---|---|---|---|---|
| 0.0 M | 5.2 | 20 | 46.4 | 11.0 | GuHCl | (1) | 97W5 |
| 0.0 M | 5.2 | 20 | 46.6 | | | (2) | 97W5 |
| 0.5 M | 5.2 | 20 | 84.5 | 17.3 | GuHCl | (1) | 97W5 |

Remarks:

(1) linear extrapolation, LEM-SB

(2) calculated from kinetic data

Hen egg white lysozyme in the presence and absence of NaCl

| NaCl Conc. | pH | T | $\Delta G$ | m | Appr./Rem. | Ref |
|---|---|---|---|---|---|---|
| 0.00 M | 5.2 | 20 | 46.4±1.5 | 11.0±0.6 | GuHCl (1–3) | 99B12 |
| 0.85 M | 5.2 | 20 | 49.5±2.0 | 11.7±0.7 | GuHCl (1–3) | 99B12 |
| 0.85 M | 5.2 | 10 | 51.5±1.9 | 11.6±0.5 | GuHCl (1–3) | 99B12 |

Remarks:

(1) linear extrapolation

(2) transition monitored by CD at 225 nm, CD at 289 nm gave identical results

(3) buffer: 20 mM sodium acetate, pH 5.2

Hen egg white lysozyme in the presence of noncompatible osmolytes

| Osmolyte | Conc. | pH | $\Delta T$ | $\Delta H$ | $\Delta(\Delta G)(25°)$ | Appr./Rem. | Ref |
|---|---|---|---|---|---|---|---|
| arginine | 0.00 M | 2.5 | 0.0 | 0 | 0.0 | heat (1–4) | 98R7 |
| | 0.25 M | 2.5 | −0.9 | −50 | −5.4 | heat (1–4) | 98R7 |
| | 0.75 M | 2.5 | −4.5 | −109 | −12.1 | heat (1–4) | 98R7 |
| | 1.00 M | 2.5 | −6.1 | −130 | −14 | heat (1–4) | 98R7 |
| histidine | 0.00 M | 2.5 | 0.0 | 0 | 0.0 | heat (1–4) | 98R7 |
| | 0.05 M | 2.5 | −0.8 | −13 | −1.7 | heat (1–4) | 98R7 |

Hen egg white lysozyme in the presence of noncompatible osmolytes (continued)

| Osmolyte | Conc. | pH | $\Delta T$ | $\Delta H$ | $\Delta(\Delta G)(25°)$ | Appr./Rem. | Ref |
|---|---|---|---|---|---|---|---|
| lysine | 0.15 M | 2.5 | −3.8 | −84 | −9.6 | heat (1–4) | 98R7 |
| | 0.0 M | 4.74 | 0.0 | 0 | 0.0 | heat (1,5–7) | 98R7 |
| | 0.2 M | 4.74 | 2.0 | 5 | 1.3 | heat (1,5–7) | 98R7 |
| | 0.5 M | 4.74 | 3.7 | 4 | 1.3 | heat (1,5–7) | 98R7 |
| | 0.7 M | 4.74 | 4.3 | 4 | 1.3 | heat (1,5–7) | 98R7 |
| | 1.0 M | 4.74 | 6.7 | 5 | 1.8 | heat (1,5–7) | 98R7 |

Remarks:

(1) van't Hoff treatment, transition monitored by changes in optical absorption at 292 nm

(2) measured in the presence of 0.1 M KCl

(3) in the absence of osmolytes, the values $T_{trs}$ = 62.3°C and $\Delta H$ = 380 kJ/mol were obtained

(4) $\Delta G(25°)$ in the presence of osmolyte was calculated using $\Delta Cp$ = 6.56 kJ/mol/K from Ref. 88P3

(5) heat denaturation of lysozyme in the presence of lysine was found to be irreversible at pH 2.5; heat denaturation was, therefore, conducted at pH 4.74 in the presence of 1.5 M GuHCl

(6) $T_{trs}$ = 63.1°C and $\Delta H$ = 334 kJ/mol values were obtained at pH 4.74 in the presence of 1.5 M GuHCl (in the absence of lysine)

(7) $\Delta G(25°)$ in the presence of lysine and GuHCl was calculated using $\Delta Cp$ = 6.68 kJ/mol/K from Ref. 76P1

Hen egg white lysozyme in aqueous polyol solution

| Cosolvent | Conc. | pH | $T_{trs}$ | $\Delta T$ | $\Delta(\Delta G)$ | Appr./Rem. | Ref |
|---|---|---|---|---|---|---|---|
| control | | 2.5 | 61.1 | 0.0 | 0.00 | heat (1,2) | 98K3 |
| mannitol | 1.00 M | 2.5 | 67.1 | 7.0 | 7.11 | heat (1,2) | 98K3 |
| inositol | 0.75 M | 2.5 | 66.8 | 6.4 | 6.23 | heat (1,2) | 98K3 |
| xylitol | 2.00 M | 2.5 | 68.7 | 8.0 | 9.12 | heat (1,2) | 98K3 |
| adonitol | 2.00 M | 2.5 | 69.1 | 9.1 | 9.25 | heat (1,2) | 98K3 |
| sorbitol | 2.00 M | 2.5 | 71.0 | 9.9 | 11.63 | heat (1,2) | 98K3 |
| control | | 4.0 | 73.1 | 0.0 | 0.00 | heat (1,3) | 98K3 |
| mannitol | 1.00 M | 4.0 | 77.9 | 4.8 | 5.52 | heat (1,3) | 98K3 |
| inositol | 0.75 M | 4.0 | 77.7 | 4.6 | 5.52 | heat (1,3) | 98K3 |
| xylitol | 2.00 M | 4.0 | 80.1 | 7.0 | 7.91 | heat (1,3) | 98K3 |
| adonitol | 2.00 M | 4.0 | 79.6 | 6.5 | 7.45 | heat (1,3) | 98K3 |
| sorbitol | 2.00 M | 4.0 | 82.5 | 9.4 | 10.75 | heat (1,3) | 98K3 |
| control | | 7.0 | 58.5 | 0.0 | 0.00 | heat (1,4) | 98K3 |
| mannitol | 1.00 M | 7.0 | 63.5 | 5.0 | 5.73 | heat (1,4) | 98K3 |
| inositol | 0.75 M | 7.0 | 64.5 | 6.0 | 6.82 | heat (1,4) | 98K3 |
| xylitol | 2.00 M | 7.0 | 66.5 | 8.0 | 9.41 | heat (1,4) | 98K3 |
| adonitol | 2.00 M | 7.0 | 67.0 | 8.5 | 10.00 | heat (1,4) | 98K3 |
| sorbitol | 2.00 M | 7.0 | 70.1 | 11.6 | 13.76 | heat (1,4) | 98K3 |

Remarks:

(1) transition monitored by optical absorption at 287 nm

(2) buffer: 20 mM glycine, pH 2.5

(3) buffer: 40 mM acetate, pH 4.0

(4) buffer: 20 mM phosphate or MOPS, pH 7.0, 1.5 M GuHCl

Hen egg white lysozyme in the presence of polyols

| Comp. | Conc. | pH | T | ΔT | ΔG | Approach/Remarks | | Ref |
|---|---|---|---|---|---|---|---|---|
| none | | 2.50 | 25 | 0.0 | 34.4* | DSC | (1) | 97J2 |
| glycerol | 2.0 | 2.50 | 25 | 2.3 | 38.5* | DSC | (1–3) | 97J2 |
| i-erythritol | 2.0 | 2.50 | 25 | 3.6 | 43.3* | DSC | (1–3) | 97J2 |
| adonitol | 2.0 | 2.50 | 25 | 5.5 | 43.8* | DSC | (1–3) | 97J2 |
| arabitol | 2.0 | 2.50 | 25 | 5.0 | 45.0* | DSC | (1–3) | 97J2 |

Remarks:

(1) ΔG was calculated from thermodynamic data given in Ref. 97J2, see also the temperature profile of ΔG in Fig. 2 of Ref. 97J2 and data in Table 2

(2) ΔG at 25°C, ΔT is the difference in thermal transition temperature

(3) polyol concentration is the molal concentration


## Lysozyme, Human

The data entries are arranged as follows:

a) wild type and mutants
b) mutants that create a calcium binding site

*a) wild type and mutants*

Human lysozyme, Val to Ala mutants, thermodynamic parameters at the denaturation temperature (64.9°C) of the wild-type protein at pH 2.7

| Mutant | pH | $T_{trs}$ | ΔT | Δ(ΔG) | Approach/Remarks | | Ref |
|---|---|---|---|---|---|---|---|
| wild type | 2.7 | 64.9±0.5 | 0.0 | 0.0 | DSC | (1,2) | 97T2 |
| Val2→Ala | 2.7 | 60.3±0.2 | −4.6 | −6.3±0.2 | DSC | (1) | 97T2 |
| Val74→Ala | 2.7 | 63.8±0.1 | −1.1 | −1.5±0.2 | DSC | (1) | 97T2 |
| Val93→Ala | 2.7 | 62.6±0.2 | −2.3 | −3.1±0.3 | DSC | (1) | 97T2 |
| Val99→Ala | 2.7 | 61.9±0.02 | −3.0 | −4.1±0.1 | DSC | (1) | 97T2 |
| Val100→Ala | 2.7 | 64.1±0.2 | −0.8 | −1.1±0.3 | DSC | (1) | 97T2 |
| Val110→Ala | 2.7 | 66.4±0.4 | +1.5 | +2.2±0.6 | DSC | (1) | 97T2 |
| Val121→Ala | 2.7 | 60.4±0.2 | −4.5 | −6.0±0.3 | DSC | (1) | 97T2 |
| Val125→Ala | 2.7 | 60.9±0.1 | −4.0 | −5.5±0.1 | DSC | (1) | 97T2 |
| Val130→Ala | 2.7 | 62.3±0.1 | −2.6 | −3.5±0.1 | DSC | (1) | 97T2 |

Remarks:
(1) for reference data see also Table 2
(2) data from Ref. 95T1

Human lysozyme, Ile to Ala/Gly mutants compared at the denaturation temperature of 64.9°C of the wild-type protein at pH 2.7

| Mutant | pH | $T_{trs}$ | ΔT | Δ(ΔG) | Approach/Remarks | | Ref |
|---|---|---|---|---|---|---|---|
| wild type | 2.7 | 64.9±0.5 | 0.0 | 0.0 | DSC | (1) | 97T1 |
| Ile23→Ala | 2.7 | 56.8 | −8.1 | −10.6 | DSC | | 97T1 |
| Ile56→Ala | 2.7 | 52.4±0.4 | −12.5 | −15.5±0.5 | DSC | | 97T1 |
| Ile59→Ala | 2.7 | 59.7±0.2 | −5.2 | −7.2±0.3 | DSC | | 97T1 |
| Ile89→Ala | 2.7 | 56.2±0.3 | −8.7 | −11.3±0.4 | DSC | | 97T1 |
| Ile106→Ala | 2.7 | 61.9±0.3 | −3.0 | −3.9±0.4 | DSC | | 97T1 |
| Ile59→Gly | 2.7 | 52.2±0.6 | −12.7 | −16.0±0.8 | DSC | | 97T1 |

Remark:
(1) data from Ref. 95T1

Human lysozyme, proline mutants

| Mutant | pH | T | ΔG | $c_{1/2}$ | m | Appr./Rem. | RefF |
|---|---|---|---|---|---|---|---|
| wild type | 3.0 | 25 | 40.6±2.1 | 2.97±0.06 | 23.10±0.46 | GuHCl (1) | 91H5 |
| Ala47→Pro | 3.0 | 25 | 45.2±2.1 | 3.30±0.07 | 23.01±0.42 | GuHCl (1) | 91H5 |
| Pro71→Gly | 3.0 | 25 | 35.6±1.7 | 2.63±0.05 | 22.72±0.46 | GuHCl (1) | 91H5 |
| Asp91→Pro | 3.0 | 25 | 37.2±2.1 | 2.78±0.06 | 22.68±0.46 | GuHCl (1) | 91H5 |
| Pro103→Gly | 3.0 | 25 | 40.6±2.1 | 2.95±0.06 | 23.22±0.46 | GuHCl (1) | 91H5 |
| Val110→Pro | 3.0 | 25 | 50.6±2.5 | 3.05±0.06 | 27.99±0.54 | GuHCl (1) | 91H5 |
| double mutant (Pro71→Gly and Pro103→Gly) | | | | | | | |
|  | 3.0 | 25 | 36.0±1.7 | 2.64±0.05 | 22.93±0.46 | GuHCl (1) | 91H5 |

Remark:
(1) linear extrapolation, for details see Ref. 90M3

Human lysozyme, Tyr to Phe mutants

| Mutant | pH | $T_{ref}$ | ΔT | Δ(ΔG) | Approach/Remarks | | Ref |
|---|---|---|---|---|---|---|---|
| wild type | 2.7 | 64.9 | 0.0 | 0.0 | DSC | (1,2) | 98Y1 |
| Tyr20→Phe | 2.7 | 64.9 | −1.5 | −2.1±0.1 | DSC | (1,2) | 98Y1 |
| Tyr38→Phe | 2.7 | 64.9 | −0.6 | −0.8±0.2 | DSC | (1,2) | 98Y1 |
| Tyr45→Phe | 2.7 | 64.9 | 0.2 | 0.3±0.4 | DSC | (1,2) | 98Y1 |
| Tyr54→Phe | 2.7 | 64.9 | −3.0 | −4.0±0.2 | DSC | (1,2) | 98Y1 |
| Tyr63→Phe | 2.7 | 64.9 | −0.7 | −1.0±0.2 | DSC | (1,2) | 98Y1 |
| Tyr124→Phe | 2.7 | 64.9 | −1.1 | −1.5±0.2 | DSC | (1,2) | 98Y1 |

Remarks:
(1) Δ(ΔG) refers to $T_{trs}$ of the wild-type protein $T_{trs}$ = 64.9±0.5°C
(2) for further thermodynamic quantities see Table 2

Human lysozyme, mutants derived from the wild-type protein 4SS and from the 3SS variant (Cys77→Ala and Cys95→Ala), Ile to Val and Val to Ala mutants, values normalized to $T_{trs}$ (64.9°C and 49.2°C) of the 4SS and 3SS protein, respectively, at pH 2.7

| Protein | ΔT-(4SS) | ΔT-(3SS) | Δ(ΔG)-(4SS) | Δ(ΔG)-(3SS) | Appr./Rem. | | Ref |
|---|---|---|---|---|---|---|---|
| Ile23→Val | −1.1 | −1.6 | −1.5 | −1.7 | DSC | (1) | 98T1 |
| Ile56→Val | −3.6 | −6.3 | −5.0 | −5.6 | DSC | (1) | 98T1 |
| Ile59→Val | −3.4 | −4.4 | −4.6 | −3.9 | DSC | (1) | 98T1 |
| Ile89→Val | −1.4 | −2.7 | −2.0 | −2.7 | DSC | (1) | 98T1 |
| Ile106→Val | −2.2 | −4.7 | −3.0 | −4.1 | DSC | (1) | 98T1 |
| Val2→Ala | −4.6 | −7.9 | −6.3 | −6.3 | DSC | (1) | 98T1 |
| Val74→Ala | −1.1 | −1.9 | −1.5 | −1.8 | DSC | (1) | 98T1 |
| Val93→Ala | −2.3 | −4.4 | −3.1 | −4.2 | DSC | (1) | 98T1 |
| Val99→Ala | −3.0 | −4.4 | −4.1 | −3.7 | DSC | (1) | 98T1 |
| Val100→Ala | −0.8 | −2.6 | −1.1 | −2.4 | DSC | (1) | 98T1 |
| Val110→Ala | +1.5 | +0.3 | +2.2 | +0.3 | DSC | (1) | 98T1 |
| Val121→Ala | −4.5 | −9.3 | −6.0 | −7.3 | DSC | (1) | 98T1 |
| Val125→Ala | −4.0 | −7.0 | −5.5 | −6.7 | DSC | (1) | 98T1 |
| Val130→Ala | −2.6 | −4.8 | −3.5 | −4.7 | DSC | (1) | 98T1 |

Remarks:
(1) primary data for mutants derived from 3SS, see also Table 2
(2) data for mutants derived from 4SS are taken from Refs. 95T1 and 97T2

Human lysozyme, wild type and Thr to Ala and Thr to Val mutants at the denaturation temperature (64.9°C) of the wild-type protein at pH 2.7

| Mutant | pH | $T_{trs}$ | $\Delta T$ | $\Delta(\Delta G)$ | Appr./Rem. | Ref |
|---|---|---|---|---|---|---|
| wild type | 2.7 | 64.9±0.5 | 0.0 | 0.0 | DSC (1,2) | 99T4 |
| Thr11→Ala | 2.7 | 66.1±0.6 | 1.2 | 1.6 | DSC (1) | 99T4 |
| Thr11→Val | 2.7 | 65.9±0.3 | 1.0 | 1.3 | DSC (1) | 99T4 |
| Thr40→Ala | 2.7 | 60.2±0.6 | −4.7 | −6.3 | DSC (1) | 99T4 |
| Thr40→Val | 2.7 | 60.7±0.3 | −4.2 | −5.6 | DSC (1) | 99T4 |
| Thr43→Ala | 2.7 | 63.8±0.2 | −1.1 | −1.5 | DSC (1) | 99T4 |
| Thr43→Val | 2.7 | 68.1±0.3 | 3.2 | 4.0 | DSC (1) | 99T4 |
| Thr52→Ala | 2.7 | 62.0±0.2 | −2.9 | −3.8 | DSC (1) | 99T4 |
| Thr52→Val | 2.7 | 62.1±0.3 | −2.8 | −3.6 | DSC (1) | 99T4 |
| Thr70→Ala | 2.7 | 60.2±0.5 | −4.7 | −6.2 | DSC (1) | 99T4 |
| Thr70→Val | 2.7 | 62.8±0.4 | −2.1 | −2.9 | DSC (1) | 99T4 |

Remarks:
(1) see also Table 2
(2) wild-type data from Ref. 95T1

Human lysozyme, Ser to Ala mutants, thermodynamic parameters at the denaturation temperature (64.9°C) of the wild-type protein at pH 2.7

| Mutant | $T_{trs}$ | $\Delta T_{trs}$ | $\Delta(\Delta G)$ | Approach/Remarks | | Ref |
|---|---|---|---|---|---|---|
| wild type | 64.9±0.5 | 0.0 | 0.0 | DSC | (1,2) | 99T5 |
| Ser24→Ala | 63.3±0.2 | −1.6 | −2.2 | DSC | (1) | 99T5 |
| Ser36→Ala | 61.4±0.1 | −3.5 | −4.7 | DSC | (1) | 99T5 |
| Ser51→Ala | 64.2±0.2 | −0.7 | −1.0 | DSC | (1) | 99T5 |
| Ser61→Ala | 60.6±0.2 | −4.3 | −5.7 | DSC | (1) | 99T5 |
| Ser80→Ala | 66.3±0.1 | 1.4 | 2.0 | DSC | (1) | 99T5 |
| Ser82→Ala | 66.0±0.2 | 1.1 | 1.6 | DSC | (1) | 99T5 |

Remarks:
(1) see also Table 2
(2) data from Ref. 95T1

Human lysozyme mutants, thermodynamic parameters at $T_{trs}$ (64.9°C) of the wild–type protein at pH 2.7

| Mutant | pH | $T_{trs}$ | $\Delta T$ | $\Delta(\Delta G)$ | Appr./Rem. | Ref |
|---|---|---|---|---|---|---|
| wild type | 2.70 | 64.9±0.5 | 0.0 | 0.0 | DSC (1) | 99T2 |
| Asn27→Leu | 2.70 | 65.3 | 0.4 | | DSC | 99T2 |
| Ala32→Leu | 2.70 | 64.6±0.2 | −0.3 | −0.4 | DSC (1) | 99T2 |
| Glu35→Leu | 2.70 | 63.1 | −1.8 | −2.2 | DSC (1) | 99T2 |
| Gly37→Gln | 2.70 | 64.1±0.0 | −0.8 | −1.1 | DSC (1) | 99T2 |
| Arg50→Gly | 2.70 | 65.8±0.1 | 0.9 | 1.1 | DSC (1) | 99T2 |
| Gln58→Gly | 2.70 | 70.6±0.5 | 5.7 | 7.8 | DSC (1) | 99T2 |
| Val74→Ser | 2.70 | 63.7±0.5 | −1.2 | −1.6 | DSC (1) | 99T2 |
| His78→Gly | 2.70 | 64.4±0.0 | −0.5 | −0.5 | DSC (1) | 99T2 |
| Ala96→Met | 2.70 | 65.0±0.4 | 0.1 | 0.1 | DSC (1) | 99T2 |
| Val100→Phe | 2.70 | 58.4±0.5 | −6.5 | −6.9 | DSC (1) | 99T2 |

Remark:
(1) normalized data at $T_{trs}$ of the wild-type protein

**Table 1. Gibbs Energy Change – Molar Values**

Human lysozyme, mutants at positions 56 and 59, parameters at the denaturation temperature (64.9°C) of the wild-type human lysozyme at pH 2.7

| Mutant | pH | $T_{trs}$ | $\Delta T$ | $\Delta(\Delta G)$ | Appr./Rem. | Ref |
|--------|-----|-----------|------------|--------------------|------------|------|
| wild type | 2.7 | 64.9±0.5 | 0.0 | 0.0 | DSC (1,2) | 99F13 |
| Ile56→Ala | | 52.4±0.4 | −12.5 | −15.5 | DSC (1,3) | 99F13 |
| Ile56→Gly | | 62.2 | −2.7 | | DSC (3) | 99F13 |
| Ile56→Leu | | 64.6±0.3 | −0.3 | −0.4±0.4 | DSC (1) | 99F13 |
| Ile56→Met | | 59.1±0.1 | −5.8 | −7.4±0.1 | DSC (1) | 99F13 |
| Ile56→Phe | | 49.9 | −15.0 | −17.1 | DSC (1) | 99F13 |
| Ile56→Thr | | 52.4 | −12.5 | −15.2 | DSC (1,4) | 99F13 |
| Ile56→Val | | 61.3±0.3 | −3.6 | −5.0±0.4 | DSC (1,2) | 99F13 |
| Ile59→Ala | | 59.7±0.2 | −5.2 | −7.2±0.3 | DSC (1,3) | 99F13 |
| Ile59→Gly | | 52.2±0.6 | −12.7 | −16.0±0.8 | DSC (1,3) | 99F13 |
| Ile59→Leu | | 64.9±0.5 | 0.0 | 0.0±0.7 | DSC (1) | 99F13 |
| Ile59→Met | | 60.7±0.4 | −4.2 | −5.4±0.5 | DSC (1) | 99F13 |
| Ile59→Phe | | 62.3±0.1 | −2.6 | −3.4±0.1 | DSC (1) | 99F13 |
| Ile59→Ser | | 53.1±0.1 | −11.8 | −15.0±0.2 | DSC (1) | 99F13 |
| Ile59→Thr | | 58.0±0.3 | −6.9 | −9.3±0.5 | DSC (1) | 99F13 |
| Ile59→Tyr | | 51.1±0.3 | −13.8 | −15.8±0.3 | DSC (1) | 99F13 |
| Ile59→Val | | 61.5±0.4 | −3.4 | −4.6±0.4 | DSC (1,2) | 99F13 |

Remarks:
(1) see also Table 2
(2) data from Ref. 95T1
(3) data from Ref. 97T1
(4) data from Ref. 96F6

Recombinant human lysozyme, modification of the N-terminus

| Mutant | pH | $T_{trs}$ | $\Delta T$ | $\Delta(\Delta G)$ | Appr./Rem. | Ref |
|--------|-----|-----------|------------|--------------------|------------|------|
| wild type | 2.7 | 64.9 | 0.0 | 0.0 | DSC (1) | 99T3 |
| Lys1→Ala | 2.7 | 62.7 | −2.2 | −2.5 | DSC (2) | 99T3 |
| Lys1→Met | 2.7 | 64.4 | −0.5 | −0.5 | DSC (2) | 99T3 |
| Gly(−1) | 2.7 | 55.1 | −9.8 | −12.2 | DSC (2) | 99T3 |
| Met(−1) | 2.7 | 56.9 | −8.0 | −9.1 | DSC (2) | 99T3 |
| Pro(−1) | 2.7 | 55.9 | −9.0 | −10.2 | DSC (2) | 99T3 |

Explanations:
   the N-terminus of authentic human lysozyme is Lys1-Val2-Phe3-
   Xaa(−1) = additional N-terminal residue

Remarks:
(1) data from Ref. 95T1
(2) $\Delta(\Delta G)$ at the denaturation temperature of the wild-type protein at pH 2.7, see also Table 2

*b) mutants that create a calcium binding site*

Human lysozyme, mutants with partially introduced Ca²⁺ binding site

| Mutant | pH | $T_{trs}$ | ΔT | Δ(ΔG) | Approach/Remarks | | Ref |
|---|---|---|---|---|---|---|---|
| wild type | 4.5 | 73.31 | 0.0 | 0.0 | DSC | (1) | 98K13 |
| Gln86→Asp | 4.5 | 72.94 | −0.37 | −0.4 | DSC | (1–3) | 98K13 |
| Ala92→Asp | 4.5 | 70.34 | −2.97 | −4.6 | DSC | (1–3) | 98K13 |
| double mutant (Gln86→Asp and Ala92→Asp) | | | | | | | |
| | 4.5 | 69.51 | −3.80 | −5.9 | DSC | (1–3) | 98K13 |
| triple mutant (Gln86→Asp, Asp91→Gln and Ala92→Asp) | | | | | | | |
| | 4.5 | 69.27 | −4.04 | −5.4 | DSC | (1–3) | 98K13 |

Remarks:
(1) measured in 50 mM sodium acetate buffer
(2) calculated using ΔCp = 7.15 kJ/mol/K of the wild type, see Table 2
(3) Δ(ΔG) refers to the wild-type protein at 73°C

Human lysozyme, mutants that create a calcium binding site

| Protein | Ca²⁺ | pH | $T_{ref}$ | ΔT | Δ(ΔG) | Appr./Rem. | | Ref |
|---|---|---|---|---|---|---|---|---|
| wild type | 0.0 mM | 4.5 | 80 | 0.3 | 0.4 | DSC | (1–3) | 98K19 |
| Gln86→Asp | 0.0 mM | 4.5 | 80 | 0.3 | 0.4 | DSC | (1,3,4) | 98K19 |
| Ala92→Asp | 0.0 mM | 4.5 | 80 | −3.0 | −4.6 | DSC | (1,3,4) | 98K19 |
| double mutant (Gln86→Asp and Ala92→Asp) | | | | | | | | |
| | 0.0 mM | 4.5 | 80 | −3.5 | −5.9 | DSC | (1,3,4) | 98K19 |
| triple mutant (Ala83→Lys, Gln86→Asp and Ala92→Asp) | | | | | | | | |
| | 0.0 mM | 4.5 | 80 | −3.0 | −5.0 | DSC | (1,3,4) | 98K19 |
| wild type 10.0 mM | | 4.5 | 80 | 0.0 | 0.0 | DSC | (1–3) | 98K19 |
| Gln86→Asp | 10.0 mM | 4.5 | 80 | 0.3 | 0.4 | DSC | (1,3) | 98K19 |
| Ala92→Asp | 10.0 mM | 4.5 | 80 | −2.8 | −5.0 | DSC | (1,3) | 98K19 |
| double mutant (Gln86→Asp and Ala92→Asp) | | | | | | | | |
| | 10.0 mM | 4.5 | 80 | 9.2 | 14.2 | DSC | (1,3,5) | 98K19 |
| triple mutant (Ala83→Lys, Gln86→Asp and Ala92→Asp) | | | | | | | | |
| | 10.0 mM | 4.5 | 80 | 9.3 | 14.6 | DSC | (1,3,5) | 98K19 |

Remarks:
(1) measured in 0.05 M sodium acetate buffer, pH 4.5, with and without 10 mM CaCl₂
(2) data from Ref. 92K12
(3) for the corresponding calorimetric data, see Table 2
(4) ΔCp = 4.06 kJ/mol/K of the wild-type protein (see Ref. 92K12) was used for all mutants
(5) ΔCp = 6.44 kJ/mol/K of the double mutant (see Ref. 92K12) was used also for the triple mutant

**Human lysozyme,** see also Lysozyme, various Calcium-Binding Lysozymes

## Lysozyme, Phage Lambda

Lysozyme phage lambda, wild type and histidine mutants

| Protein | pH | T | ΔG | Approach/Remarks | | Ref |
|---|---|---|---|---|---|---|
| wild type | 7.0 | 25 | 29.7 | heat | (1,2) | 99E4 |
| His31→Asp | 7.0 | 25 | 23.0 | heat | (1,2) | 99E4 |
| His48→Asn | 7.0 | 25 | 8.8 | heat | (1,2) | 99E4 |
| double mutant (His31→Asn and His48→Asn) | | | | | | |
| | 7.0 | 25 | 37.2 | heat | (1,2) | 99E4 |

Remarks:
(1) for the procedure see also Ref. 98S15
(2) buffer: 100 mM phosphate, pH 7.0

Lysozyme phage lambda (λL), double mutant (His31→Asn and His137→Asn) with the remaining His48 replaced by 1,2,4-triazole-3-alanine (Taz)

| Mutant | pH | T | Δ(ΔG) | Approach/Remarks | | Ref |
|---|---|---|---|---|---|---|
| His48-λL | 7 | 25 | 0.0 | pH | (1,2) | 98S15 |
| Taz48-λL | 7 | 25 | −14.6 | pH | (1,2) | 98S15 |

Remarks:
(1) from differences in $pK_a$ between His and Taz
(2) for thermal denaturation data see Table 2

## Lysozyme, Phage T4

The data entries are arranged as follows:

a) wild type and mutants
b) data from more complex mutations
c) data from H/D exchange

*a) wild type and mutants*

Lysozyme phage T4, alanine mutants derived from pseudo-wild type w.t.*

| Protein | pH | $T_{trs}$ | Δ(ΔG) | Approach/Remarks | | Ref |
|---|---|---|---|---|---|---|
| Met6→Ala | 5.4 | 60.8 | −6.7 | heat | (1–4) | 99G3 |
| Leu7→Ala | 5.4 | 59.0 | −9.6 | heat | (1–4) | 99G3 |
| Ile17→Ala | 5.4 | 58.9 | −9.6 | heat | (1–4) | 99G3 |
| Leu33→Ala | 5.4 | 56.5 | −12.1 | heat | (1–4) | 99G3 |
| Ile50→Ala | 5.4 | 61.1 | −6.7 | heat | (1–4) | 99G3 |
| Leu66→Ala | 5.4 | 55.2 | −13.8 | heat | (1–4) | 99G3 |
| Ile78→Ala | 5.4 | 61.8 | −5.0 | heat | (1–4) | 99G3 |
| Leu84→Ala | 5.4 | 54.8 | −15.5 | heat | (1–4) | 99G3 |
| Val87→Ala | 5.4 | 61.0 | −6.3 | heat | (1–4) | 99G3 |
| Leu91→Ala | 5.4 | 57.9 | −10.9 | heat | (1–4) | 99G3 |
| Leu99→Ala | 5.4 | 53.6 | −17.2 | heat | (1–4) | 99G3 |
| Met102→Ala | 5.4 | 57.1 | −12.1 | heat | (1–4) | 99G3 |
| Ile100→Ala | 5.4 | 58.2 | −10.5 | heat | (1–4) | 99G3 |
| Met102→Ala/Met106→Ala | | | | | | |
| | 5.4 | 54.5 | −15.5 | heat | (1–4) | 99G3 |
| Val103→Ala | 5.4 | 60.9 | −6.7 | heat | (1–4) | 99G3 |

Lysozyme phage T4, alanine mutants derived from pseudo-wild type w.t.* (continued)

| Protein | pH | $T_{trs}$ | $\Delta(\Delta G)$ | Approach/Remarks | | Ref |
|---|---|---|---|---|---|---|
| Phe104→Ala | 5.4 | 57.8 | −11.3 | heat | (1–4) | 99G3 |
| Met106→Ala | 5.4 | 60.1 | −7.9 | heat | (1–4) | 99G3 |
| Val111→Ala | 5.4 | 62.4 | −4.2 | heat | (1–4) | 99G3 |
| Leu118→Ala | 5.4 | 56.3 | −13.4 | heat | (1–4) | 99G3 |
| Leu121→Ala | 5.4 | 59.1 | −9.2 | heat | (1–4) | 99G3 |
| w.t.* (Ala129) | 5.4 | 65.3 | 0.0 | heat | (1–5) | 99G3 |
| Val149→Ala | 5.4 | 56.3 | −13.4 | heat | (1–4) | 99G3 |
| Phe153→Ala | 5.4 | 55.6 | −14.2 | heat | (1–4) | 99G3 |

Remarks:

(1) w.t.* is the cysteine-free pseudo-wild-type protein

(2) measured in 0.10 M sodium chloride, 1.4 mM acetic acid, 8.6 mM sodium acetate, pH 5.4, for the procedure see Ref. 93E2

(3) uncertainty in $T_{trs}$ ±0.14°C

(4) $\Delta Cp$ = 10.46 kJ/mol/K was used at an isotherm of 59°C to calculate $\Delta G$

(5) data from Ref. 96G2

Lysozyme phage T4, methionine mutants derived from pseudo-wild type w.t.*

| Protein | pH | $T_{trs}$ | $\Delta(\Delta G)$ | Approach/Remarks | | Ref |
|---|---|---|---|---|---|---|
| w.t.*(Met6) | 5.4 | 65.3 | 0.0 | heat | (1–5) | 99G3 |
| Ile17→Met | 5.4 | 59.4 | −9.2 | heat | (1–4) | 99G3 |
| Ile27→Met | 5.4 | 55.2 | −13.0 | heat | (1–4) | 99G3 |
| Ile17→Met/Leu33→Met | | | | | | |
| | 5.4 | 55.0 | −13.0 | heat | (1–4) | 99G3 |
| Leu33→Met | 5.4 | 60.0 | −8.4 | heat | (1–4) | 99G3 |
| Ile50→Met | 5.4 | 64.7 | −1.7 | heat | (1–4) | 99G3 |
| Leu66→Met | 5.4 | 62.6 | −4.2 | heat | (1–4) | 99G3 |
| Ile78→Met | 5.4 | 61.6 | −6.3 | heat | (1–5) | 99G3 |
| Leu84→Met | 5.4 | 60.4 | −7.9 | heat | (1–5) | 99G3 |
| Val87→Met | 5.4 | 59.0 | −9.6 | heat | (1–4) | 99G3 |
| Leu91→Met | 5.4 | 63.3 | −3.3 | heat | (1–5) | 99G3 |
| Leu99→Met | 5.4 | 64.0 | −1.7 | heat | (1–4,6) | 99G3 |
| w.t.*(Met102) | 5.4 | 65.3 | 0.0 | heat | (1–5) | 99G3 |
| Ile100→Met | 5.4 | 60.8 | −6.7 | heat | (1–4) | 99G3 |
| w.t.*(Met102→Met/Met106→Met) | | | | | | |
| | 5.4 | 65.3 | 0.0 | heat | (1–4) | 99G3 |
| Val103→Met | 5.4 | 62.2 | −5.0 | heat | (1–5) | 99G3 |
| Phe104→Met | 5.4 | 64.5 | −1.7 | heat | (1–4) | 99G3 |
| w.t.*(Met106) | 5.4 | 65.3 | 0.0 | heat | (1–4) | 99G3 |
| Val111→Met | 5.4 | 63.3 | −2.9 | heat | (1–4) | 99G3 |
| Leu118→Met | 5.4 | 63.5 | −2.9 | heat | (1–5) | 99G3 |
| Leu121→Met | 5.4 | 63.2 | −3.3 | heat | (1–5) | 99G3 |
| Ala129→Met | 5.4 | 60.1 | −7.9 | heat | (1–4) | 99G3 |
| Val149→Met | 5.4 | 57.7 | −11.7 | heat | (1–4) | 99G3 |
| Phe153→Met | 5.4 | 63.7 | −2.5 | heat | (1–4,6) | 99G3 |

Remarks:

(1) w.t.* is the cysteine-free pseudo-wild type protein

(2) measured in 0.10 M sodium chloride, 1.4 mM acetic acid, 8.6 mM sodium acetate, pH 5.4, for the procedure see Ref. 93E2

(3) uncertainty in $T_{trs}$ ±0.14°C

(4) $\Delta Cp$ = 10.46 kJ/mol/K was used at an isotherm of 59°C was used to calculate $\Delta G$

(5) data from Ref. 96G2

(6) data from Ref. 93E2

Lysozyme phage T4, methionine mutants derived from pseudo-wild type w.t.*

| Mutant | pH | $T_{ref}$ | ΔT | Δ(ΔG) | Approach/Remarks | | Ref |
|---|---|---|---|---|---|---|---|
| w.t.* | 3.0 | 51.7 | 0.0 | 0.0 | heat | (1–3) | 98L5 |
| Met6→Leu | 3.0 | 51.7 | −10.6 | −11.7 | heat | (2–5) | 98L5 |
| Met102→Leu | 3.0 | 51.7 | − 2.4 | − 3.8 | heat | (2–4) | 98L5 |
| Met106→Leu | 3.0 | 51.7 | 1.7 | 2.1 | heat | (2–4) | 98L5 |
| Met120→Leu | 3.0 | 51.7 | 1.7 | 2.1 | heat | (2–4) | 98L5 |
| Met102→Lys | 3.0 | 51.7 | −35 | −37.7 | heat | (2–4,6) | 98L5 |
| Met106→Lys | 3.0 | 51.7 | −10.5 | −14.2 | heat | (2–4) | 98L5 |
| Met120→Lys | 3.0 | 51.7 | − 4.8 | − 6.7 | heat | (2–4) | 98L5 |

Remarks:

(1) w.t.* = cysteine-free pseudo-wild type (Cys54→Thr and Cys97→Ala)

(2) $T_{trs}$ of w.t.* = 51.7°C is the reference temperature for the Met mutants, see also Ref. 95Z1

(3) buffer: 25 mM potassium chloride, 3 mM phosphoric acid, 17 mM mono-potassium phosphate, pH 3.0

(4) Δ(ΔG) was calculated from heat denaturation (see Table 2) using ΔCp = 7.53 kJ/mol/K

(5) data from Ref. 92H10

(6) data from Ref. 91D3

*b) data from more complex mutations*

Lysozyme phage T4, large-to-small amino acid substitutions within the core

| Protein | pH | ΔT | Δ(ΔG)(44°C) | Approach/Remarks | Ref |
|---|---|---|---|---|---|
| wild type* | 3.0 | 0.0 | 0.0 | heat, v.H. (1–6) | 98X1 |
| Ile17→Ala | 3.0 | −8.4 | −11.3 | heat, v.H. (1–6) | 98X1 |
| Ile27→Ala | 3.0 | −10.1 | −13.0 | heat, v.H. (1–6) | 98X1 |
| Ile29→Ala | 3.0 | −8.2 | −10.9 | heat, v.H. (1–6) | 98X1 |
| Ile50→Ala | 3.0 | −5.8 | −8.4 | heat, v.H. (1–6) | 98X1 |
| Ile58→Ala | 3.0 | −10.4 | −13.4 | heat, v.H. (1–6) | 98X1 |
| Ile78→Ala | 3.0 | −4.7 | −6.7 | heat, v.H. (1–6) | 98X1 |
| Ile100→Ala | 3.0 | −10.7 | −14.2 | heat, v.H. (1–6) | 98X1 |
| Val71→Ala | 3.0 | −4.7 | −6.3 | heat, v.H. (1–6) | 98X1 |
| Val87→Ala | 3.0 | −4.9 | −7.1 | heat, v.H. (1–6) | 98X1 |
| Val94→Ala | 3.0 | −5.0 | −7.5 | heat, v.H. (1–6) | 98X1 |
| Val103→Ala | 3.0 | −6.6 | −9.2 | heat, v.H. (1–6) | 98X1 |
| Val111→Ala | 3.0 | −3.7 | −5.4 | heat, v.H. (1–6) | 98X1 |
| Val149→Ala | 3.0 | −11.0 | −13.4 | heat, v.H. (1–6) | 98X1 |
| Met6→Ala | 3.0 | −5.7 | −7.9 | heat, v.H. (1–6) | 98X1 |
| Met102→Ala | 3.0 | | −13.8 | heat, v.H. (1–6) | 98X1 |
| Met106→Ala | 3.0 | −7.1 | −9.6 | heat, v.H. (1–6) | 98X1 |
| Phe67→Ala | 3.0 | −5.7 | −7.9 | heat, v.H. (1–6) | 98X1 |
| Phe104→Ala | 3.0 | −9.7 | −13.0 | heat, v.H. (1–6) | 98X1 |
| Val7→Ala | 3.0 | −8.1 | −10.9 | heat, v.H. (1–6) | 98X1 |
| Val33→Ala | 3.0 | −12.3 | −15.1 | heat, v.H. (1–6) | 98X1 |
| Val66→Ala | 3.0 | −13.4 | −16.3 | heat, v.H. (1–6) | 98X1 |
| Val84→Ala | 3.0 | −13.4 | −16.3 | heat, v.H. (1–6) | 98X1 |
| Val91→Ala | 3.0 | −9.7 | −13.0 | heat, v.H. (1–6) | 98X1 |

Remarks:

(1) all mutants were constructed in the cysteine-free pseudo-wild-type lysozyme, wild type*

(2) measured in 25 mM KCl, 20 mM potassium phosphate, pH 3.0, see also Ref. 92E3

(3) ΔT refers to $T_{trs}$ = 51.65°C of the pseudo-wild type at the above buffer conditions

(4) Δ(ΔG) was estimated at 44°C using a thermodynamic model

(5) thermodynamic calculations in Ref. 98X1 are based on ΔCp = 7.5 kJ/mol/K

(6) the estimated error in ΔT is about 0.2°C and in Δ(ΔG) 0.6 kJ/mol (increasing to 1.3 for the least stable proteins)

(7) unpublished data of E. Baldwin and B.W. Matthews, cited in Ref. 98X1 (Table 1)

Lysozyme phage T4, circularly permuted mutants

| Mutant | pH | $T_{trs}$ | T | $\Delta(\Delta G)$ | $c_{1/2}$ | Approach/Remarks | | Ref |
|---|---|---|---|---|---|---|---|---|
| wild type* | 5.75 | 69.9 | 4 | 59.0±3.3 | 2.6 | GuHCl | (1–3) | 98L6 |
| 13cpT4L* | 5.75 | 64.0 | 4 | 46.4±2.1 | 2.2 | GuHCl | (1–3) | 98L6 |
| H31N 13cpT4L* | 5.75 | 60.2 | 4 | 30.5±1.7 | 2.0 | GuHCl | (1–3) | 98L6 |
| 75cpT4L* | 5.75 | 48.4 | 4 | 19.2±2.5 | 1.2 | GuHCl | (1–3) | 98L6 |
| C-domain | 5.75 | 31.5 | 4 | 8.8±1.3 | 2.1 | GuHCl | (1–3) | 98L6 |

Explanations:

all mutants were constructed in the cysteine-free pseudo-wild-type lysozyme, wild type*

13cpT4L* = residues 13–164 of wild type*, six-amino acid linker-Ser-(Gly)$_4$-Ala-, and residues 1-12
  H31N 13cpT4L* = 13cpT4L* with replacement His31→Asn

75cpT4L* = residues 75–164 of wild type*, six-amino acid linker-Ser-(Gly)$_4$-Ala-, and residues 1–75

C-domain = C-terminal subdomain consisting of an initiator Met followed by Val75-Leu164 from wild type* and
  the -Ser-(Gly)$_4$-Ala- linker ending in Met1-Leu13 from wild type*

Remarks:

(1) $T_{trs}$ refers to thermal denaturation measured in $H_2O$ at pH 4.5

(2) T, $\Delta(\Delta G)$, and $c_{1/2}$ refer to denaturant-induced unfolding measured in 50 mM potassium phosphate, pH 5.75, data
treatment by linear extrapolation, LEM-SB

(3) transitions were monitored by CD at 222 nm

Lysozyme phage T4, incorporation of α-hydroxy acids in α-helix 39–50

| Mutant | pH | $T_{trs}$ | $\Delta T$ | $\Delta(\Delta G)$ | Appr./Rem. | Ref |
|---|---|---|---|---|---|---|
| wild type (Leu39) | 2.5 | 45.19±0.04 | 0.0 | 0.0 | heat (1–3) | 97K7 |
| Leu39→leucic acid | 2.5 | 41.95±0.06 | −3.24±0.10 | −3.72±0.13 | heat (1–3) | 97K7 |
| Ser44→Leu | 2.5 | 47.19±0.19 | | | heat (1–3) | 97K7 |
| Ser44→leucic acid | 2.5 | 40.87±0.10 | −6.31±0.30 | −7.24±0.33 | heat (1–3) | 97K7 |
| wild type (Ile50) | 2.5 | 45.19±0.04 | | | heat (1–3) | 97K7 |
| Ile50→isoleucic acid | | | | | | |
| | 2.5 | 42.55±0.10 | −2.64±0.10 | −3.01±0.13 | heat (1–3) | 97K7 |

Remarks:

(1) $\Delta(\Delta G)$ was obtained by $\Delta(\Delta G) = \Delta T \times \Delta S$ using $\Delta S = 1.146\pm0.050$ kJ/mol/K from Ref. 91Z2

(2) transition monitored by ellipticity at 223 nm

(3) buffer: 20 mM potassium phosphate, 25 mM KCl

*c) data from H/D exchange*

Lysozyme phage T4 w.t.*, $\Delta G$ of unfolding ($\Delta G_{unf}(H_2O)$) and $\Delta G$ of local fluctations ($\Delta G_{fl}$) from exchange experiments

| Residue | pH | T | $\Delta G_{unf}(H_2O)$ | m | $\Delta G_{fl}$ | Appr/Rem. | Ref |
|---|---|---|---|---|---|---|---|
| Phe4 | 6.0 | 25 | 46.4 | 15.1 | 18.8 | H/D (1–5) | 99L7 |
| Glu5 | 6.0 | 25 | 62.8 | 23.0 | 30.5 | H/D (1–5) | 99L7 |
| Met6 | 6.0 | 25 | 61.9 | 21.3 | 32.2 | H/D (1–5) | 99L7 |
| Ile7 | 6.0 | 25 | 64.9 | 23.8 | 33.1 | H/D (1–5) | 99L7 |
| Arg8 | 6.0 | 25 | 58.2 | 19.7 | | H/D (1–5) | 99L7 |
| Ile9 | 6.0 | 25 | 60.7 | 22.2 | | H/D (1–5) | 99L7 |
| Asp10 | 6.0 | 25 | 66.1 | 24.7 | 29.7 | H/D (1–5) | 99L7 |
| Glu11 | 6.0 | 25 | 61.9 | 23.4 | 31.8 | H/D (1–5) | 99L7 |
| Gly12 | 6.0 | 25 | | | 18.0 | H/D (1–5) | 99L7 |
| Arg14 | 6.0 | 25 | 46.4 | 13.8 | 21.3 | H/D (1–5) | 99L7 |
| Lys16 | 6.0 | 25 | 39.3 | 11.3 | 28.9 | H/D (1–5) | 99L7 |
| Ile17 | 6.0 | 25 | 28.5 | 5.9 | 21.3 | H/D (1–5) | 99L7 |
| Tyr18 | 6.0 | 25 | 45.2 | 14.6 | 23.0 | H/D (1–5) | 99L7 |

Lysozyme phage T4 w.t.*, $\Delta G$ of unfolding ($\Delta G_{unf}(H_2O)$) and $\Delta G$ of local fluctuations ($\Delta G_{fl}$) from exchange experiments (continued)

| Residue | pH | T | $\Delta G_{unf}(H_2O)$ | m | $\Delta G_{fl}$ | Appr/Rem. | Ref |
|---------|-----|----|------|------|------|----------|------|
| Asp20 | 6.0 | 25 | | | 14.6 | H/D (1–5) | 99L7 |
| Tyr25 | 6.0 | 25 | 23.4 | 5.9 | 18.4 | H/D (1–5) | 99L7 |
| Thr26 | 6.0 | 25 | 34.7 | 9.2 | | H/D (1–5) | 99L7 |
| Ile27 | 6.0 | 25 | 34.3 | 8.4 | | H/D (1–5) | 99L7 |
| Gly28 | 6.0 | 25 | 34.7 | 8.8 | | H/D (1–5) | 99L7 |
| Ile29 | 6.0 | 25 | 32.2 | 8.8 | 20.1 | H/D (1–5) | 99L7 |
| His31 | 6.0 | 25 | 39.7 | 11.3 | 23.8 | H/D (1–5) | 99L7 |
| Leu33 | 6.0 | 25 | 32.2 | 9.6 | 26.8 | H/D (1–5) | 99L7 |
| Thr34 | 6.0 | 25 | | | 11.7 | H/D (1–5) | 99L7 |
| Ala42 | 6.0 | 25 | 31.8 | 7.5 | 18.0 | H/D (1–5) | 99L7 |
| Lys43 | 6.0 | 25 | 22.6 | 4.2 | | H/D (1–5) | 99L7 |
| Ser44 | 6.0 | 25 | 39.7 | 10.9 | 20.9 | H/D (1–5) | 99L7 |
| Glu45 | 6.0 | 25 | 31.0 | 7.9 | 22.6 | H/D (1–5) | 99L7 |
| Leu46 | 6.0 | 25 | 36.0 | 10.9 | 25.5 | H/D (1–5) | 99L7 |
| Asp47 | 6.0 | 25 | 27.6 | 6.3 | | H/D (1–5) | 99L7 |
| Lys48 | 6.0 | 25 | 41.0 | 10.5 | 28.5 | H/D (1–5) | 99L7 |
| Ala49 | 6.0 | 25 | 32.2 | 6.7 | 26.4 | H/D (1–5) | 99L7 |
| Ile50 | 6.0 | 25 | 35.6 | 9.6 | 28.0 | H/D (1–5) | 99L7 |
| Gly51 | 6.0 | 25 | 17.6 | 2.1 | | H/D (1–5) | 99L7 |
| Val57 | 6.0 | 25 | 39.3 | 11.7 | 18.0 | H/D (1–5) | 99L7 |
| Ile58 | 6.0 | 25 | 35.6 | 9.6 | 29.3 | H/D (1–5) | 99L7 |
| Thr59 | 6.0 | 25 | 36.8 | 10.5 | 32.6 | H/D (1–5) | 99L7 |
| Glu62 | 6.0 | 25 | 38.5 | 11.7 | 18.4 | H/D (1–5) | 99L7 |
| Ala63 | 6.0 | 25 | 43.9 | 12.1 | 34.7 | H/D (1–5) | 99L7 |
| Glu64 | 6.0 | 25 | 41.8 | 12.1 | 23.8 | H/D (1–5) | 99L7 |
| Lys65 | 6.0 | 25 | 40.2 | 10.9 | 19.7 | H/D (1–5) | 99L7 |
| Leu66 | 6.0 | 25 | 66.5 | 24.7 | 31.4 | H/D (1–5) | 99L7 |
| Phe67 | 6.0 | 25 | 54.0 | 18.4 | 32.6 | H/D (1–5) | 99L7 |
| Asn68 | 6.0 | 25 | | | 23.0 | H/D (1–5) | 99L7 |
| Gln69 | 6.0 | 25 | | | 20.1 | H/D (1–5) | 99L7 |
| Asp70 | 6.0 | 25 | 53.1 | 17.2 | 31.8 | H/D (1–5) | 99L7 |
| Val71 | 6.0 | 25 | 63.6 | 23.4 | 29.7 | H/D (1–5) | 99L7 |
| Asp72 | 6.0 | 25 | 66.9 | 25.5 | 25.1 | H/D (1–5) | 99L7 |
| Ala73 | 6.0 | 25 | 72.0 | 26.8 | 26.4 | H/D (1–5) | 99L7 |
| Ala74 | 6.0 | 25 | 68.6 | 24.7 | 35.1 | H/D (1–5) | 99L7 |
| Val75 | 6.0 | 25 | 61.5 | 22.2 | 28.0 | H/D (1–5) | 99L7 |
| Arg76 | 6.0 | 25 | 67.8 | 24.3 | | H/D (1–5) | 99L7 |
| Gly77 | 6.0 | 25 | 66.5 | 22.6 | 31.4 | H/D (1–5) | 99L7 |
| Ile78 | 6.0 | 25 | 70.7 | 26.8 | 28.9 | H/D (1–5) | 99L7 |
| Leu79 | 6.0 | 25 | 68.2 | 26.8 | | H/D (1–5) | 99L7 |
| Arg80 | 6.0 | 25 | 59.4 | 20.5 | 25.9 | H/D (1–5) | 99L7 |
| Asn81 | 6.0 | 25 | 45.6 | 11.3 | 28.0 | H/D (1–5) | 99L7 |
| Leu84 | 6.0 | 25 | 61.1 | 21.8 | 24.3 | H/D (1–5) | 99L7 |
| Lys85 | 6.0 | 25 | 63.2 | 22.2 | 33.5 | H/D (1–5) | 99L7 |
| Val87 | 6.0 | 25 | | | 13.8 | H/D (1–5) | 99L7 |
| Tyr88 | 6.0 | 25 | 59.8 | 20.9 | | H/D (1–5) | 99L7 |
| Asp89 | 6.0 | 25 | | | 28.0 | H/D (1–5) | 99L7 |
| Ser90 | 6.0 | 25 | | | 20.9 | H/D (1–5) | 99L7 |
| Leu91 | 6.0 | 25 | 58.6 | 20.5 | 28.5 | H/D (1–5) | 99L7 |
| Val94 | 6.0 | 25 | | | 10.0 | H/D (1–5) | 99L7 |
| Arg96 | 6.0 | 25 | 70.7 | 25.1 | 38.1 | H/D (1–5) | 99L7 |
| Ala97 | 6.0 | 25 | 69.5 | 24.3 | | H/D (1–5) | 99L7 |

Lysozyme phage T4 w.t.*, ΔG of unfolding (ΔG$_{unf}$(H$_2$O)) and ΔG of local fluctations (ΔG$_n$) from exchange experiments (continued)

| Residue | pH | T | ΔG$_{unf}$(H$_2$O) | m | ΔG$_n$ | Appr/Rem. | Ref |
|---------|-----|----|------|------|------|-----------|------|
| Ala98   | 6.0 | 25 | 74.1 | 27.2 |      | H/D (1–5) | 99L7 |
| Leu99   | 6.0 | 25 | 67.4 | 25.1 |      | H/D (1–5) | 99L7 |
| Ile100  | 6.0 | 25 | 76.6 | 30.5 |      | H/D (1–5) | 99L7 |
| Asn101  | 6.0 | 25 | 68.6 | 24.3 |      | H/D (1–5) | 99L7 |
| Met102  | 6.0 | 25 | 70.3 | 25.1 |      | H/D (1–5) | 99L7 |
| Val103  | 6.0 | 25 | 69.0 | 25.9 |      | H/D (1–5) | 99L7 |
| Phe104  | 6.0 | 25 | 71.5 | 27.2 |      | H/D (1–5) | 99L7 |
| Gln105  | 6.0 | 25 | 63.2 | 21.8 | 26.4 | H/D (1–5) | 99L7 |
| Met106  | 6.0 | 25 | 65.3 | 22.6 | 28.5 | H/D (1–5) | 99L7 |
| Gly107  | 6.0 | 25 |      |      | 20.9 | H/D (1–5) | 99L7 |
| Val111  | 6.0 | 25 |      |      | 16.7 | H/D (1–5) | 99L7 |
| Ala112  | 6.0 | 25 | 58.2 | 20.5 | 21.3 | H/D (1–5) | 99L7 |
| Leu121  | 6.0 | 25 | 64.0 | 23.4 | 29.3 | H/D (1–5) | 99L7 |
| Gln122  | 6.0 | 25 | 56.5 | 19.2 | 25.5 | H/D (1–5) | 99L7 |
| Gln123  | 6.0 | 25 | 64.0 | 21.3 | 25.5 | H/D (1–5) | 99L7 |
| Lys124  | 6.0 | 25 |      |      | 19.2 | H/D (1–5) | 99L7 |
| Arg125  | 6.0 | 25 | 43.5 | 10.9 | 24.7 | H/D (1–5) | 99L7 |
| Trp126  | 6.0 | 25 | 42.3 | 12.1 | 21.8 | H/D (1–5) | 99L7 |
| Ala129  | 6.0 | 25 | 50.6 | 15.5 | 27.6 | H/D (1–5) | 99L7 |
| Ala130  | 6.0 | 25 | 51.9 | 16.7 | 24.3 | H/D (1–5) | 99L7 |
| Val131  | 6.0 | 25 | 59.0 | 20.9 | 23.0 | H/D (1–5) | 99L7 |
| Asn132  | 6.0 | 25 |      |      | 15.9 | H/D (1–5) | 99L7 |
| Leu133  | 6.0 | 25 |      |      | 16.7 | H/D (1–5) | 99L7 |
| Ala134  | 6.0 | 25 |      |      | 19.2 | H/D (1–5) | 99L7 |
| Lys135  | 6.0 | 25 |      |      | 16.7 | H/D (1–5) | 99L7 |
| Trp138  | 6.0 | 25 |      |      | 16.3 | H/D (1–5) | 99L7 |
| Tyr139  | 6.0 | 25 |      |      | 18.0 | H/D (1–5) | 99L7 |
| Asn140  | 6.0 | 25 |      |      | 20.5 | H/D (1–5) | 99L7 |
| Gln141  | 6.0 | 25 |      |      | 16.7 | H/D (1–5) | 99L7 |
| Thr142  | 6.0 | 25 |      |      | 18.4 | H/D (1–5) | 99L7 |
| Ala146  | 6.0 | 25 |      |      | 27.2 | H/D (1–5) | 99L7 |
| Lys147  | 6.0 | 25 | 67.4 | 24.3 | 24.3 | H/D (1–5) | 99L7 |
| Arg148  | 6.0 | 25 |      |      | 22.2 | H/D (1–5) | 99L7 |
| Val149  | 6.0 | 25 | 68.2 | 25.9 | 28.0 | H/D (1–5) | 99L7 |
| Ile150  | 6.0 | 25 | 60.7 | 22.6 |      | H/D (1–5) | 99L7 |
| Thr151  | 6.0 | 25 | 55.6 | 19.2 | 24.7 | H/D (1–5) | 99L7 |
| Thr152  | 6.0 | 25 | 54.4 | 16.7 |      | H/D (1–5) | 99L7 |
| Phe153  | 6.0 | 25 | 62.8 | 22.2 |      | H/D (1–5) | 99L7 |
| Arg154  | 6.0 | 25 | 56.5 | 18.0 |      | H/D (1–5) | 99L7 |
| Thr155  | 6.0 | 25 | 52.3 | 16.3 | 23.8 | H/D (1–5) | 99L7 |
| Gly156  | 6.0 | 25 | 59.0 | 18.8 | 38.1 | H/D (1–5) | 99L7 |
| Thr157  | 6.0 | 25 | 54.4 | 17.6 | 27.2 | H/D (1–5) | 99L7 |
| Ala160  | 6.0 | 25 |      |      | 16.7 | H/D (1–5) | 99L7 |
| Tyr161  | 6.0 | 25 | 60.2 | 20.9 | 28.9 | H/D (1–5) | 99L7 |
| Lys162  | 6.0 | 25 |      |      | 13.4 | H/D (1–5) | 99L7 |

Remarks:

(1) data for w.t.*, the cysteine-free pseudo-wild type (Cys54→Thr and Cys97→Ala)

(2) data from hydrogen-deuterium exchange monitored by NMR

(3) the exchange was measured as a function of denaturant conc. (0 M to 2 M GuHCl)

(4) ΔG$_{unf}$(H$_2$O) is the extrapolated free energy of unfolding in the absence of denaturant, and m is the dependence on denaturant conc.

(5) ΔG$_n$ of local fluctuations under native conditions (0 M GuHCl)

## Lysozyme, Pigeon

Pigeon lysozyme, intermediate states

| Transition | pH | T | ΔG | $c_{1/2}$ | m | Approach/Remarks | | Ref |
|---|---|---|---|---|---|---|---|---|
| N → U | 6 | 25 | 39±1 | 3.2 | 12.1±0.5 | GuHCl | (1–3) | 98H1 |
| N → U | 6 | 25 | 40.5±3 | | | heat | | 98H1 |
| I → U | | 25 | 18.5±7 | | 6.6±0.2 | GuHCl | (4) | 98H1 |
| N → I | 6 | 25 | 21±2 | | 5.5±0.7 | GuHCl | (5) | 98H1 |
| N → I | | 25 | 24±1 | | 4.4±0.7 | GuHCl | (6) | 98H1 |

Remarks:
(1) linear extrapolation
(2) transition monitored by CD at 276 nm
(3) buffer: 20 mM MES, 80 mM NaCl, pH 6
(4) from final values of the fluorescence intensity on the unfolding-refolding kinetics
(5) from equilibrium unfolding
(6) from activation free energy

**Lysozyme, Pigeon,** see also various Calcium-Binding Lysozymes

## Lysozyme, various Calcium-Binding Lysozymes

Unfolding parameters of calcium-binding lysozymes compared with α-lactalbumins

| Protein/Transition | pH | T | ΔG | m | Approach/Remarks | | Ref |
|---|---|---|---|---|---|---|---|
| canine lysozyme, apo-form | | | | | | | |
| N → A | 7.2 | 25 | 7.1±0.5 | 3.94±0.08 | GuHCl | (1–3,5) | 98K7 |
| A → U | 7.2 | 25 | 28.0±4.0 | 5.52±0.16 | GuHCl | (1,2,4,5) | 98K7 |
| pigeon lysozyme, apo-form in 1 mM EDTA | | | | | | | |
| N → U | | | 12.1±0.06 | 6.63±0.05 | GuHCl | (6) | 98K7 |
| pigeon lysozyme, holo-form in 0.18 mM $CaCl_2$ | | | | | | | |
| N → U | | | 30.9±0.7 | 9.73±0.07 | GuHCl | (6) | 98K7 |
| pigeon lysozyme, holo-form in 10 mM $CaCl_2$ | | | | | | | |
| N → U | | | 36.2±0.5 | 9.73±0.04 | GuHCl | (6) | 98K7 |
| equine lysozyme, apo- or holo-form | | | | | | | |
| A → U | | | 14.7±0.3 | 3.55±0.02 | GuHCl | (6) | 98K7 |
| bovine α-lactalbumin | | | | | | | |
| N → A | | | 14.1–15.1 | 4.0–6.1 | GuHCl | (7) | 98K7 |
| A → U | | | 8.6 | 3.7 | GuHCl | (7) | 98K7 |
| human α-lactalbumin | | | | | | | |
| N → A | | | 9.5 | 8.3 | GuHCl | (8) | 98K7 |
| A → U | | | 13.1 | 4.0 | GuHCl | (8) | 98K7 |

Remarks:
(1) linear extrapolation
(2) measured in 0.1 M TES buffer with 1 mM EDTA
(3) transition monitored by CD at 294 nm
(4) transition monitored by CD at 222 nm
(5) Ref. 98K7 contains further transition curves for the holo-form of canine lysozyme, and apo- and hololysozyme from echidna milk lysozyme (*Tachyglossus aculeatus multiaculeatus*)
(6) reference value in Ref. 98K7, taken from Ref. 93N2
(7) reference value in Ref. 98K7, taken from Refs. 76K1 and 76K2
(8) reference value in Ref. 98K7, taken from Ref. 78N2

## Macrophage Migration Inhibitory Factor

Recombinant human macrophage migration inhibitory factor (MIF)

| Protein | pH | T | $\Delta G$ | $c_{1/2}$ | m | Appr./Rem. | Ref |
|---------|----|----|-----------|-----------|-----|------------|-----|
| MIF | 7 | 25 | 24.5±2.5 | 1.20±0.05 | 21.1±2.0 | GuHCl (1,3,5) | 99Z3 |
| | 7 | 25 | 32.1±3 | 1.64±0.05 | 19.5±1.5 | GuHCl (1,4,5) | 99Z3 |
| | 7 | 25 | 27.9±3 | 1.69±0.05 | 16.5±1.5 | GuHCl (2,4,5) | 99Z3 |
| | 7 | 25 | 31.5±4 | 1.50 | 21.0 | GuHCl (6,7) | 99Z3 |
| | 7 | 25 | 31.5±3 | 3.20±0.1 | 10.1±1.0 | urea (1,3,6) | 99Z3 |
| | 7 | 25 | 28.8±3 | 4.15±0.1 | 6.9±1.0 | urea (1,4,6) | 99Z3 |
| | 7 | 25 | 33.9±3 | 4.13±0.1 | 8.2±1.0 | urea (2,4,6) | 99Z3 |

Remarks:

(1) linear extrapolation

(2) nonlinear curve fit, LEM-SB

(3) transition monitored by fluorescence

(4) transition monitored by far-UV CD

(5) buffer: 0.02 M phosphate, 0.15 M NaCl, pH 7

(6) buffer: 0.02 M phosphate, pH 7

(7) measurements based on far-UV CD and fluorescence, linear extrapolation with corrections for populated intermediates

## Malic Enzyme

Malic enzyme, (S)-malate:NADP$^+$ oxidoreductase, EC 1.1.1.40, recombinant pigeon liver enzyme

| Mutant/Trans. | pH | T | $\Delta G$ | $c_{1/2}$ | m | Appr./Rem. | | Ref |
|---------------|----|----|-----------|-----------|-----|------------|-----|-----|
| wild type | | | | | | | | |
| N → I | 7.4 | 30 | 39.3±3.3 | 2.52±0.1 | 15.5±0.8 | urea | (1) | 98H11 |
| I → U | 7.4 | 30 | 32.2±36.0 | 3.22±0.2 | 10.0±9.6 | urea | (1) | 98H11 |
| Arg9→Glu | | | | | | | | |
| N → I | 7.4 | 30 | 25.1±7.9 | 1.41±0.4 | 18.0±5.9 | urea | (1) | 98H11 |
| I → U | 7.4 | 30 | 33.9±2.5 | 2.25±0.1 | 15.1±1.3 | urea | (1) | 98H11 |
| Met17→Lys | | | | | | | | |
| N → I | 7.4 | 30 | 30.5±2.9 | 2.48±0.1 | 12.6±1.3 | urea | (1) | 98H11 |
| I → U | 7.4 | 30 | 20.5±6.3 | 3.27±0.1 | 6.3±1.7 | urea | (1) | 98H11 |
| double mutant (Arg9→Glu and Met17→Lys) | | | | | | | | |
| N → I | 7.4 | 30 | 16.3±6.7 | 2.19±0.1 | 7.5±3.3 | urea | (1) | 98H11 |
| I → U | 7.4 | 30 | 31.4±5.9 | 3.54±0.2 | 8.8±1.7 | urea | (1) | 98H11 |

Remarks:

(1) three-state unfolding model assuming a linear dependence on denaturant concentration, fit of the whole data set

(2) transition monitored by fluorescence

(3) buffer: 30 mM Tris-HCl, pH 7.4

## Maltose-Binding Protein

Cytoplasmatic (cmMBP) and periplasmatic (pmMBP) maltose-binding protein

| Protein | pH | T | $\Delta G$ | $c_{1/2}$ | m | Appr./Rem. | Ref |
|---------|----|----|-----------|-----------|-----|------------|-----|
| pmMBP | 7.3 | 25 | 64.4±5.0 | 3.15 | 20.5±1.7 | urea (1,2) | 99G1 |
| cmMBP | 7.3 | 25 | 49.8±3.8 | 2.96 | 16.7±1.3 | urea (1,2) | 99G1 |

Remarks:

(1) linear extrapolation, LEM-SB

(2) transition monitored by fluorescence

Maltose-binding protein from *E. coli*, recombinant

| pH | T | $\Delta G$ | $c_{1/2}$ | m | Appr./Rem. | | Ref |
|---|---|---|---|---|---|---|---|
| 7.4 | 28 | 38.1±4.2 | 1.0 | 38.5±4.2 | GuHCl | (1,2) | 97G1 |
| 7.4 | 28 | 42.7 | | | DSC | | 97G1 |
| 7.4 | 2 | 27.6±3.3 | 0.75 | 35.6±4.2 | GuHCl | (1,2) | 97G1 |
| 7.4 | 34±5.5 | 44.8±9.2 | | | DSC | (3) | 97G1 |

Remarks:
(1) linear extrapolation
(2) transition monitored by fluorescence
(3) the protein achieves maximal stability at the given temperature

Maltose-binding protein from *E. coli*

| pH | T | $\Delta G$ | Approach | Remarks | Ref |
|---|---|---|---|---|---|
| 3.4 | 25 | 34.5* | DSC | (1,2) | 97N13 |
| 3.0 | 25 | 25.0* | DSC | (1,3) | 97N13 |

Remarks:
(1) for reference data see Table 2
(2) heat denaturation at $T_{trs} = 56.1°C$, cold denaturation at about 0°C
(2) heat denaturation at $T_{trs} = 50.7°C$, cold denaturation at about 4°C

Maltose-binding protein (MalE), misfolding substitutions in loops

| Protein | pH | T | $\Delta G$ | $c_{1/2}$ | m | Appr./Rem. | Ref |
|---|---|---|---|---|---|---|---|
| MalE wild type | 7.5 | 25 | 42±4 | 1.05±0.05 | 38±8 | GuHCl (1–3) | 98R1 |
| MalE31 | 7.5 | 25 | | 0.75±0.05 | | GuHCl (2–5) | 98R1 |
| | 7.5 | 25 | | 0.35±0.05 | | GuHCl (2–4,6) | 98R1 |
| MalE219 | 7.5 | 25 | 14.6±2.1 | 0.35±0.01 | 38±8 | GuHCl (2,3,7) | 98R1 |

Remarks:
(1) linear extrapolation
(2) transition monitored by fluorescence emission at 345 nm
(3) buffer: 50 mM HEPES, pH 7.5
(4) MalE31 = double mutant (Gly32→Asp and Ile33→Pro) in the αI/βB loop of the N-domain which is critical for in vivo folding
(5) $c_{1/2}$ for the unfolding reaction
(6) $c_{1/2}$ for the refolding reaction
(7) MalE219 = double mutant of an equivalent α/β loop in the C-domain

Maltose-binding protein (MBP), native and molten globule state

| Protein | pH | T | $\Delta G$ | $c_{1/2}$ | m | Appr./Rem. | Ref |
|---|---|---|---|---|---|---|---|
| MBP | 7.1 | 4 | 37.7±1.7 | 3.20 | 11.7±0.4 | urea (1–4) | 99S9 |
| | 7.1 | 9 | 41.4±3.3 | 3.09 | 13.4±1.3 | urea (1–4) | 99S9 |
| | 7.1 | 14 | 49.0±5.0 | 3.34 | 14.6±1.7 | urea (1–4) | 99S9 |
| | 7.1 | 20 | 61.1±5.0 | 3.24 | 18.8±1.3 | urea (1–4) | 99S9 |
| | 7.1 | 26 | 61.5±5.0 | 3.27 | 18.8±1.3 | urea (1–4) | 99S9 |
| | 7.1 | 30 | 66.1±3.8 | 3.36 | 19.7±2.1 | urea (1–4) | 99S9 |
| | 7.1 | 33 | 62.3±5.4 | 3.24 | 19.2±1.7 | urea (1–4) | 99S9 |
| | 7.1 | 34 | 61.5±2.9 | 3.13 | 19.7±0.8 | urea (1–4) | 99S9 |
| | 7.1 | 37 | 50.6±5.4 | 2.95 | 17.2±1.7 | urea (1–4) | 99S9 |
| | 7.1 | 39 | 50.6±5.4 | 2.88 | 17.6±2.1 | urea (1–4) | 99S9 |
| | 7.1 | 40 | 42.7±6.3 | 2.91 | 14.6±2.1 | urea (1–4) | 99S9 |
| | 7.1 | 46 | 30.1±2.5 | 2.12 | 14.2±1.3 | urea (1–4) | 99S9 |
| | 7.1 | 55 | 11.7±2.1 | 1.13 | 10.5±1.3 | urea (1–4) | 99S9 |

Maltose-binding protein (MBP), native and molten globule state (continued)

| Protein | pH | T | $\Delta G$ | $c_{1/2}$ | m | Appr./Rem. | Ref |
|---------|-----|----|-----------|-----------|-----------|-------------|------|
|         | 7.1 | 56 | 11.7±1.7  | 1.04      | 11.3±1.3  | urea (1–4)  | 99S9 |
| MBP     | 7.1 | 1  | 24.3±2.5  | 0.73      | 33.5±3.8  | GuHCl (1–3) | 99S9 |
|         | 7.1 | 11 | 43.9±2.5  | 0.97      | 44.8±2.5  | GuHCl (1–3) | 99S9 |
|         | 7.1 | 20 | 50.2±2.5  | 1.06      | 47.3±2.1  | GuHCl (1–3) | 99S9 |
|         | 7.1 | 25 | 52.7±5.9  | 1.03      | 51.0±5.0  | GuHCl (1–3) | 99S9 |
|         | 7.1 | 31 | 50.2±3.3  | 1.03      | 48.5±3.3  | GuHCl (1–3) | 99S9 |
|         | 7.1 | 36 | 38.9±3.3  | 0.87      | 45.2±3.8  | GuHCl (1–3) | 99S9 |
|         | 7.1 | 40 | 35.1±1.7  | 0.84      | 41.8±2.1  | GuHCl (1–3) | 99S9 |
|         | 7.1 | 45 | 20.5±1.7  | 0.69      | 29.7±3.3  | GuHCl (1–3) | 99S9 |
| MBP     | 3.0 | 3  | 6.7±1.3   | 0.80      | 8.4±0.8   | urea (1–3)  | 99S9 |
|         | 3.0 | 9  | 10.0±1.7  | 1.09      | 9.2±1.7   | urea (1–3)  | 99S9 |
|         | 3.0 | 14 | 14.6±1.3  | 1.12      | 13.0±0.8  | urea (1–3)  | 99S9 |
|         | 3.0 | 18 | 16.7±1.7  | 1.08      | 15.5±2.1  | urea (1–3)  | 99S9 |
|         | 3.0 | 23 | 14.6±1.3  | 0.83      | 17.2±1.3  | urea (1–3)  | 99S9 |
|         | 3.0 | 26 | 13.4±2.1  | 0.82      | 16.3±2.5  | urea (1–3)  | 99S9 |
|         | 3.0 | 27 | 11.7±1.7  | 0.76      | 15.5±1.7  | urea (1–3)  | 99S9 |
|         | 3.0 | 30 | 9.2±1.3   | 0.60      | 14.2±1.7  | urea (1–3)  | 99S9 |

Remarks:

(1) linear extrapolation, for details see also Ref. 95A1

(2) transition monitored by far-UV CD at 222 nm

(3) buffer: CGH10, 10 mM each of citrate, glycine, and HEPES

(4) the buffer contained additionally 150 mM KCl


# Mannitol Permease

Mannitol permease of *E. coli*, enzyme II, isolated domains and phosphorylated forms

| Protein | pH | T | $\Delta G$ | $c_{1/2}$ | m | Appr./Rem. | Ref |
|---------|-----|----|-----------|-----------|-----------|-------------|-------|
| IIA$^{mtl}$    | 7.6 | 25 | 27.6±1.6 | 1.21 | 22.8±1.3 | GuHCl (1–3)  | 96M13 |
| P-IIA$^{mtl}$  | 7.6 | 25 | 15.8±1.7 | 0.83 | 19.1±2.0 | GuHCl (1–3)  | 96M13 |
| IIB$^{mtl}$    | 7.6 | 25 | 12.3±0.6 | 0.95 | 13.1±0.6 | GuHCl (1–3)  | 96M13 |
| P-IIB$^{mtl}$  | 7.6 | 25 | 13.9±1.2 | 0.86 | 15.7±1.4 | GuHCl (1,2,5)| 96M13 |
| IIBA$^{mtl}$   | 7.6 | 25 |          | 1.07 |          | GuHCl (4,6)  | 96M13 |
|                | 7.6 | 25 |          | 1.17 |          | GuHCl (5,6)  | 96M13 |
| P-IIBA$^{mtl}$ | 7.6 | 25 |          | 0.89 |          | GuHCl (4,6)  | 96M13 |
|                | 7.6 | 25 |          | 0.92 |          | GuHCl (5,6)  | 96M13 |

Explanation:

| | |
|---|---|
| EII$^{mtl}$    | mannitol-specific enzyme II |
| IIA$^{mtl}$    | A domain of the mannitol-specific enzyme II |
| IIB$^{mtl}$    | B domain of the mannitol-specific enzyme II |
| IIBA$^{mtl}$   | BA domain of the mannitol-specific enzyme II |
| P-IIA$^{mtl}$  | A domain of the mannitol-specific enzyme II, phosphorylated |
| P-IIB$^{mtl}$  | B domain of the mannitol-specific enzyme II, phosphorylated |
| P-IIBA$^{mtl}$ | BA domain of the mannitol-specific enzyme II, phosphorylated |

Remarks:

(1) treatment by a two-state model, linear extrapolation, LEM-SB

(2) measured in 10 mM HEPES, pH 7.6, 2 mM 2-mercaptoethanol

(3) transition monitored by CD at 222 nm and fluorescence at 305 nm

(4) transition monitored by CD at 222 nm

(5) transition monitored by fluorescence at 305 nm

(6) the transition exhibits two stages, see also Table 2

**Mannose Transporter**

Mannose transporter of *Escherichia coli*, IIAB$^{Man}$ subunit, tryptophan to phenylalanine substitutions

| Protein | pH | T | ΔG | $c_{1/2}$ | Approach/Remarks | | Ref |
|---|---|---|---|---|---|---|---|
| IIAB$^{Man}$ wild type | 7.4 | 22 | 44±11 | 2.3 | GuHCl | (1–3) | 99M9 |
| IIAB$^{Man}$ Trp12→Phe | 7.4 | 22 | 59±15 | 2.6 | GuHCl | (1–3) | 99M9 |
| IIAB$^{Man}$ Trp33→Phe | 7.4 | 22 | 35±6 | 2.2 | GuHCl | (1–3) | 99M9 |
| IIAB$^{Man}$ Trp69→Phe | 7.4 | 22 | 26±3 | 2.1 | GuHCl | (1–3) | 99M9 |
| IIAB$^{Man}$ double mutant (Trp12→Phe and Trp33→Phe) | | | | | | | |
| | 7.4 | 22 | 45±7 | 2.2 | GuHCl | (1–3) | 99M9 |
| IIAB$^{Man}$ double mutant (Trp12→Phe and Trp69→Phe) | | | | | | | |
| | 7.4 | 22 | 46±5 | 2.1 | GuHCl | (1–3) | 99M9 |
| IIAB$^{Man}$ double mutant (Trp33→Phe and Trp69→Phe) | | | | | | | |
| | 7.4 | 22 | 22±4 | 1.8 | GuHCl | (1–3) | 99M9 |

Remarks:
(1) linear extrapolation, LEM-SB
(2) buffer: 20 mM sodium phosphate, pH 7.4
(3) transition monitored by fluorescence

**MerP**

MerP, recombinant protein

| pH | T | ΔG | $c_{1/2}$ | m | Appr./Rem. | Ref |
|---|---|---|---|---|---|---|
| 7.5 | 25 | 14.1±0.7 | 8.9±0.4 | 1.57±0.03 | GuHCl (1–4) | 97A6 |
| 7.5 | 25 | 17.4±1.6 | 10.7±0.9 | 1.63±0.03 | GuHCl (1,2,5) | 97A6 |
| 4.9 | 25 | 15.9±1.5 | 10.6±0.9 | 1.50±0.03 | GuHCl (1,2,5) | 97A6 |
| 9.0 | 25 | 10.6±1.8 | 9.6±1.2 | 1.10±0.07 | GuHCl (1,2,5) | 97A6 |

Remarks:
(1) MerP is a water-soluble protein of 72 amino acids, located in the periplasm of Gram-negative bacteria
(2) linear extrapolation
(3) transition monitored by CD at 222 nm
(4) the stability of the protein towards temperature is highest at pH 6.5 with $T_{trs}$ = 69.5±0.5°C
(5) transition monitored by Tyr fluorescence at 304 nm

**MyoD**

Muscle-specific bHLH (basic helix-loop-helix) transcription factor MyoD and its partner E47, MyoD-E47 heterodimer

| Dimer | pH | T | ΔG | $c_{1/2}$ | Approach/Remarks | | Ref |
|---|---|---|---|---|---|---|---|
| E47 | 7 | 25 | 44.0±1.5 | 3.0 | urea | (1–3) | 98W6 |
| MyoD | 7 | 25 | 33.4±3.1 | 0.8 | urea | (1–3) | 98W6 |
| MyoEL | 7 | 25 | 35.2±0.4 | 1.75 | urea | (1–3) | 98W6 |

Remarks:
(1) linear extrapolation
(2) buffer: 20 mM ammonium acetate, 100 mM KCl, 1 mM DTT, 0.1 mM EDTA
(3) transition monitored by far-UV CD
(4) ΔG is the free energy change associated with folding and dimerization of bHLH domains

## Myoglobin

The data entries are arranged as follows:

a) *Aplysia limacina* apomyoglobin
b) sperm whale myoglobin, wild type and mutants
c) apomyoglobin, transitions and intermediate
d) horse heart myoglobin in the presence of osmolytes

*a) Aplysia limacina apomyoglobin*

Apomyoglobin from *Aplysia limacina*, GuHCl-induced unfolding transitions

| Transition | pH | T | $\Delta G$ | $c_{1/2}$ | $n_{as}$ | Approach/Remarks | | Ref |
|---|---|---|---|---|---|---|---|---|
| I → U | 2 | 4 | 17.2 | 1.75 | 21.6 | GuHCl | (1–4) | 98S19 |
| I → U | 3 | 4 | 20.9 | 1.86 | 24.9 | GuHCl | (1–4) | 98S19 |
| I → U | 4 | 4 | 22.4 | 1.77 | 27.9 | GuHCl | (1–4) | 98S19 |
| N → I | 4.5 | 4 | 1.9 | 0.1 | 39.8 | GuHCl | (1–4) | 98S19 |
| I → U | 4.5 | 4 | 23.9 | 1.69 | 28.5 | GuHCl | (1–4) | 98S19 |
| N → I | 5 | 4 | 3.4 | 0.17 | 36.0 | GuHCl | (1–4) | 98S19 |
| I → U | 5 | 4 | 20.3 | 1.57 | 27.8 | GuHCl | (1–4) | 98S19 |
| N → I | 6 | 4 | 7.5 | 0.45 | 33.2 | GuHCl | (1–4) | 98S19 |
| I → U | 6 | 4 | 17.2 | 1.47 | 25.1 | GuHCl | (1–4) | 98S19 |
| N → I | 7 | 4 | 11.7 | 0.62 | 35.1 | GuHCl | (1–4) | 98S19 |
| I → U | 7 | 4 | 12.1 | 1.30 | 19.1 | GuHCl | (1–4) | 98S19 |

Remarks:
(1) the parameter $n_{as}$ indicates the number of amino acid side-chains that become exposed on unfolding, for details of the approach see Ref. 93S9
(2) buffer: 2 mM citrate (pH 1.5 to 7)
(3) transitions monitored by tryptophan fluorescence (340 nm) and CD (222 nm)
(4) estimated error ±1.3 in $\Delta G$ and ±2 in $n_{as}$

Apomyoglobin from *Aplysia limacina*, urea-induced unfolding transitions in the presence and absence of KCl

| Transition | [KCl] | pH | T | $\Delta G$ | $c_{1/2}$ | $n_{as}$ | Appr./Rem. | Ref |
|---|---|---|---|---|---|---|---|---|
| I → U | 0.25 | 2 | 4 | 10.7 | 1.95 | 29.8 | urea (1–4) | 98S19 |
| I → U | 0.5 | 2 | 4 | 15.8 | 3.36 | 26.75 | urea (1–4) | 98S19 |
| I → U | 1.0 | 2 | 4 | 13.4 | 4.23 | 18.6 | urea (1–4) | 98S19 |
| I → U | 0.0 | 3 | 4 | 14.2 | 2.21 | 35.2 | urea (1–4) | 98S19 |
| I → U | 0.5 | 3 | 4 | 12.1 | 3.19 | 21.5 | urea (1–4) | 98S19 |
| I → U | 1.0 | 3 | 4 | 10.7 | 4.06 | 15.5 | urea (1–4) | 98S19 |
| I → U | 1.5 | 3 | 4 | 10.2 | 4.34 | 13.9 | urea (1–4) | 98S19 |
| I → U | 0.0 | 4.5 | 4 | 28.5 | 2.9 | 55 | urea (1–4) | 98S19 |
| I → U | 0.1 | 4.5 | 4 | 26.4 | 2.96 | 50 | urea (1–4) | 98S19 |
| I → U | 0.5 | 4.5 | 4 | 25.9 | 3.35 | 44.4 | urea (1–4) | 98S19 |
| I → U | 1.0 | 4.5 | 4 | 20.3 | 4.27 | 28 | urea (1–4) | 98S19 |
| I → U | 2.0 | 4.5 | 4 | 16.2 | 5.04 | 19.5 | urea (1–4) | 98S19 |
| N → U | 0.0 | 5 | 4 | 13.1 | 2.82 | 28 | urea (1–4) | 98S19 |
| N → U | 0.0 | 6 | 4 | 15.1 | 2.38 | 32 | urea (1–4) | 98S19 |
| N → U | 0.0 | 7 | 4 | 23.1 | 3.03 | 41.75 | urea (1–4) | 98S19 |
| N → I | 0.5 | 5 | 4 | 8.8 | 1.04 | 80 | urea (1–4) | 98S19 |
| I → U | 0.5 | 5 | 4 | 22.5 | 3.35 | 38.25 | urea (1–4) | 98S19 |
| N → I | 1.0 | 5 | 4 | 6.3 | 0.96 | 36 | urea (1–4) | 98S19 |
| I → U | 1.0 | 5 | 4 | 25.5 | 4.1 | 36.5 | urea (1–4) | 98S19 |
| N → I | 1.0 | 6 | 4 | 16.6 | 1.5 | 59 | urea (1–4) | 98S19 |

Apomyoglobin from *Aplysia limacina*, urea-induced unfolding transitions in the presence and absence of KCl (continued)

| Transition | [KCl] | pH | T | $\Delta G$ | $c_{1/2}$ | $n_{as}$ | Appr./Rem. | Ref |
|---|---|---|---|---|---|---|---|---|
| I → U | 1.0 | 6 | 4 | 23.2 | 3.5 | 38 | urea (1–4) | 98S19 |
| N → I | 1.0 | 7 | 4 | 8.8 | 2.25 | 27 | urea (1–4) | 98S19 |
| I → U | 1.0 | 7 | 4 | 23.9 | 3.93 | 35.5 | urea (1–4) | 98S19 |
| N → I | 2.0 | 7 | 4 | 13.6 | 2.4 | 31 | urea (1–4) | 98S19 |
| I → U | 2.0 | 7 | 4 | 26.8 | 5.11 | 31.7 | urea (1–4) | 98S19 |

Remarks:

(1) the parameter $n_{as}$ indicates the number of amino acid side-chains that become exposed on unfolding, for details of the approach see Ref. 93S9

(2) buffer: 2 mM citrate (pH 1.5 to 7)

(3) transitions monitored by tryptophan fluorescence (340 nm) and CD (222 nm)

(4) estimated error ±1.3 in $\Delta G$ and ±2 in $n_{as}$

*b) sperm whale myoglobin, wild type and mutants*

Myoglobin, various species and mutants derived from sperm whale myoglobin, transitions

| Protein | pH | T | $\Delta G(N{\to}I)$ | $\Delta G(I{\to}U)$ | Approach/Remarks | | Ref |
|---|---|---|---|---|---|---|---|
| sperm whale myoglobin | | | | | | | |
| native | 7.0 | 25 | 15.4 | 13.7 | GuHCl | (1–5) | 94H12 |
| wild type (Asn122) | | | | | | | |
| | 7.0 | 25 | 11.2±0.7 | 15.4±0.3 | GuHCl | (1–5) | 94H12 |
| wild type (Asp122) | | | | | | | |
| | 7.0 | 25 | 11.4 | 16.1 | GuHCl | (1–5) | 94H12 |
| Leu29→Asn | 7.0 | 25 | ~–10 | 11.4 | GuHCl | (1–5) | 94H12 |
| Leu29→Phe | 7.0 | 25 | 11.4 | 17.1 | GuHCl | (1–5) | 94H12 |
| Phe43→Ile | 7.0 | 25 | 10.0 | 6.1 | GuHCl | (1–5) | 94H12 |
| Phe43→Val | 7.0 | 25 | 6.1 | 17.1 | GuHCl | (1–5) | 94H12 |
| His64→Ala | 7.0 | 25 | 14.4 | 17.1 | GuHCl | (1–5) | 94H12 |
| His64→Gln | 7.0 | 25 | 11.4 | 6.1 | GuHCl | (1–5) | 94H12 |
| His64→Leu | 7.0 | 25 | 14.7 | 18.8 | GuHCl | (1–5) | 94H12 |
| His64→Phe | 7.0 | 25 | 16.0 | 19.4 | GuHCl | (1–5) | 94H12 |
| Val68→Ala | 7.0 | 25 | 9.8 | 15.4 | GuHCl | (1–5) | 94H12 |
| Val68→Asn | 7.0 | 25 | 0.0 | 13.1 | GuHCl | (1–5) | 94H12 |
| Val68→Gln | 7.0 | 25 | 4.5 | 14.4 | GuHCl | (1–5) | 94H12 |
| Val68→Phe | 7.0 | 25 | 15.4 | 17.1 | GuHCl | (1–5) | 94H12 |
| Val68→Ser | 7.0 | 25 | 1.3 | 15.4 | GuHCl | (1–5) | 94H12 |
| Val68→Thr | 7.0 | 25 | 6.7 | 15.4 | GuHCl | (1–5) | 94H12 |
| Ile110→Phe | 7.0 | 25 | 14.4 | 17.1 | GuHCl | (1–5) | 94H12 |
| Ile110→Thr | 7.0 | 25 | 8.1 | 11.7 | GuHCl | (1–5) | 94H12 |
| Ile110→Val | 7.0 | 25 | 11.4 | 13.1 | GuHCl | (1–5) | 94H12 |
| pig myoglobin | | | | | | | |
| wild type | 7.0 | 25 | 7.1 | 8.8 | GuHCl | (1–5) | 94H12 |
| human myoglobin | | | | | | | |
| wild type | 7.0 | 25 | 9.8 | 9.0 | GuHCl | (1–5) | 94H12 |

Remarks:

(1) three-state fit assuming a linear dependence of $\Delta G$ in denaturant conc.

(2) transition monitored by fluorescence emission spectra

(3) buffer: 0.2 M potassium phosphate, pH 7

(4) the values for $\Delta G(N{\to}I)$ and $\Delta G(I{\to}U)$ were calculated from the corresponding equilibrium constants given in Ref. 94H12

(5) Ref. 94H12 contains additional stability data derived from rates of hemin loss

Sperm whale myoglobin (cyanoMB), wild type and mutants concerning putative interhelix ion pairs (mutated residue in italics)

| Paired residues | pH | T | Δ(ΔG) | $c_{1/2}$ | Approach/Remarks | | Ref |
|---|---|---|---|---|---|---|---|
| wild type | 5.0 | 25 | 0.0 | 4.20±0.01 | urea | (1–5) | 99R3 |
| *Glu04*-Lys79 | 5.0 | 25 | −2.09±0.63 | 3.89±0.08 | urea | (1–3) | 99R3 |
| Glu06-*Lys133* | 5.0 | 25 | 0.10±0.31 | 4.22±0.06 | urea | (1–3) | 99R3 |
| Lys16-*Asp122* | 5.0 | 25 | −0.31±0.10 | 4.17±0.03 | urea | (1–3) | 99R3 |
| *Glu18*-Lys77 | 5.0 | 25 | −4.29±0.31 | 3.56±0.03 | urea | (1–3) | 99R3 |
| Glu18-*Lys77* | 5.0 | 25 | 0.84 | 4.32±0.01 | urea | (1–3) | 99R3 |
| *Asp20*-Arg118 | 5.0 | 25 | −1.88±0.21 | 3.92±0.05 | urea | (1–3) | 99R3 |
| Asp20-*Arg118* | 5.0 | 25 | −2.72 | 3.79±0.01 | urea | (1–3) | 99R3 |
| *Asp44*-Lys47 | 5.0 | 25 | 0.63±0.42 | 4.30±0.04 | urea | (1–3) | 99R3 |
| Arg45-*Asp60* | 5.0 | 25 | −0.84±0.21 | 4.09±0.03 | urea | (1–3) | 99R3 |
| Glu52-*Lys56* | 5.0 | 25 | −1.26±0.20 | 4.01±0.04 | urea | (1–3) | 99R3 |
| Glu105-*Arg139* | 5.0 | 25 | −1.88 | 3.92±0.01 | urea | (1–3) | 99R3 |

Remarks:
(1) linear extrapolation, LEM-SB
(2) transition monitored by CD at 222 nm
(3) buffer: 10 mM Na-acetate, 0.5 mM KCN, pH 5.0
(4) CD at 222 nm, absorbance at 423 nm, and fluorescence at 320 nm gave identical results
(5) ΔG for the wild-type protein is 28.03 kJ/mol by CD and 28.24 kJ/mol by fluorescence

*c) apomyoglobin, transitions and intermediate*

Sperm whale apomyoglobin (apo-Mb), pH 4 intermediate

| Protein | pH | T | ΔG | $c_{1/2}$ | Approach/Remarks | | Ref |
|---|---|---|---|---|---|---|---|
| apo-Mb | 4 | 20 | 11.3 | 2.06 | urea | (1,2) | 99J7 |

Remarks:
(1) data from kinetics of folding and unfolding reaction at 0 M urea
(2) transition monitored by fluorescence

Sperm whale apomyoglobin, recombinant, wild type and mutants, stability expresed by the midpoint concentration at urea-induced unfolding in 30 mM NaCl, 20 mM $Na_2SO_4$, and 50 mM $NaClO_4$

| Mutant | pH | T | $c_{1/2}$ (NaCl) | $c_{1/2}$ ($Na_2SO_4$) | $c_{1/2}$ ($NaClO_4$) | Appr./Rem. | Ref |
|---|---|---|---|---|---|---|---|
| wild type | 4.2 | 4 | 1.42 | 2.08 | 3.14 | urea (1) | 97L14 |
| | 4.2 | 4 | 1.45 | 2.09 | 3.24 | urea (2) | 97L14 |
| Gln8→Gly | 4.2 | 4 | 0.80 | 1.85 | 2.84 | urea (1) | 97L14 |
| | 4.2 | 4 | 0.75 | 1.86 | 2.84 | urea (2) | 97L14 |
| Glu109→Gly | 4.2 | 4 | n.m. | 1.73 | 2.64 | urea (1,3) | 97L14 |
| | 4.2 | 4 | n.m. | 1.01 | 1.94 | urea (2,3) | 97L14 |

Remarks:
(1) measured by tryptophan fluorescence emission at 330 nm (with excitation at 280 nm), buffer: 2 mM sodium acetate
(2) measured by CD at 222 nm
(3) n.m. = not measurable

Sperm whale apomyoglobin, wild type and double mutant (Asn132→Gly and Glu136→Gly)

| Protein | pH | T | $\Delta G$ | m | Approach/Remarks | | Ref |
|---|---|---|---|---|---|---|---|
| Data from equilibrium unfolding: | | | | | | | |
| wild type | 5.74 | 25 | 19.2±1.7 | 6.3±0.4 | urea | (1,2) | 99C3 |
| double mutant (Asn132→Gly and Glu136→Gly) | | | | | | | |
| | 5.74 | 25 | 15.1±1.3 | 5.9±0.4 | urea | (1–3) | 99C3 |
| Data from urea-dependent refolding kinetics: | | | | | | | |
| wild type | 5 | | 8.4±0.8 | 5.19±0.25 | kinetics | (4,5) | 99C3 |
| double mutant (Asn132→Gly and Glu136→Gly) | | | | | | | |
| | 5 | | 7.1±0.8 | 4.60±0.08 | kinetics | (3–5) | 99C3 |

Remarks:

(1) data from equilibrium unfolding, transition monitored by CD at 222 nm

(2) linear extrapolation, LEM-SB

(3) the double mutant was designed to destabilize residual secondary structure in the H helix region

(4) thermodynamic parameters derived from the urea-dependence of apomyoglobin refolding kinetics

(5) the data correspond to the on-pathway intermediate (I → U transition)

Sperm whale myoglobin, intermediate, wild type and mutants concerning putative interhelix ion pairs

| Mutant | pH | T | $\Delta(\Delta G)$ | $c_{1/2}$ | Approach/Remarks | | Ref |
|---|---|---|---|---|---|---|---|
| wild type | 4.2 | 4 | 0.0 | 1.54±0.04 | urea | (1–4) | 99R3 |
| Glu04→Ala | 4.2 | 4 | −2.62±0.31 | 1.20 | urea | (1–3) | 99R3 |
| Glu18→Ala | 4.2 | 4 | −2.51±0.42 | 1.22±0.01 | urea | (1–3) | 99R3 |
| Asp44→Ala | 4.2 | 4 | −0.73±0.10 | 1.45±0.05 | urea | (1–3) | 99R3 |
| Lys56→Ala | 4.2 | 4 | −0.94±0.10 | 1.42±0.02 | urea | (1–3) | 99R3 |
| Asp60→Ala | 4.2 | 4 | −0.52±0.10 | 1.48±0.05 | urea | (1–3) | 99R3 |
| Lys77→Ala | 4.2 | 4 | −0.13±0.05 | 1.53±0.03 | urea | (1–3) | 99R3 |

Remarks:

(1) linear extrapolation, LEM-SB

(2) transition monitored by CD at 222 nm, and fluorescence at 320 nm

(3) buffer: 4 mM citrate, pH 4.2

(4) $\Delta G$ for the wild-type protein is 12.13 kJ/mol by CD and 11.51 kJ/mol by fluorescence

Apomyoglobin, unfolding transitions of the native protein and partially folded intermediates

| Transitions | pH | T | $\Delta G$ | $c_{1/2}$ | m | Approach/Remarks | | Ref |
|---|---|---|---|---|---|---|---|---|
| N → U | 7.0 | 20 | 18.4 | 2.5 | 6.7 | urea | (1,2) | 97F3 |
| $A_3$ → U | 2.0 | 20 | 10.5 | 2.4 | 4.2 | urea | (1–3) | 97F3 |
| $A_2$ → U | 2.0 | 20 | 7.5 | 1.9 | 3.85 | urea | (1–3) | 97F3 |
| $A_1$ → U | 2.0 | 20 | 6.3 | 1.8 | 3.56 | urea | (1–3) | 97F3 |

Remarks:

(1) linear extrapolation

(2) transition monitored by CD at 222nm

(3) $A_3$ was obtained in the presence of 30 mM trichloroacetate at pH 2.0

(4) $A_2$ was obtained in the presence of 100 mM trifluoroacetate at pH 2.0

(5) $A_1$ was obtained in the presence of 500 mM KCl at pH 2.0

Apomyoglobin intermediate (I), wild type and electrostatic mutants

| Mutant | pH | T | Δ(ΔG) | $c_{1/2}$ | m | Approach/Remarks | | Ref |
|---|---|---|---|---|---|---|---|---|
| wild type | 4.2 | 4 | 0.0 | 1.56 | 6.276 | urea | (1–3) | 98K5 |
| Asp20→Ala | 4.2 | 4 | 0.25 | 1.52 | | urea | (2–4) | 98K5 |
| Asp20→Ala | 4.2 | 4 | 0.88 | 1.42 | | urea | (2,4,5) | 98K5 |
| Arg31→Ala | 4.2 | 4 | 1.63 | 1.82 | | urea | (2–4) | 98K5 |
| His36→Gln | 4.2 | 4 | −0.42 | 1.49 | | urea | (2–4) | 98K5 |
| Arg118→Ala | 4.2 | 4 | 0.96 | 1.71 | | urea | (2–4) | 98K5 |
| Asp122→Ala | 4.2 | 4 | 0.0 | 1.56 | | urea | (2–4) | 98K5 |
| Asp122→Ala | 4.2 | 4 | 0.29 | 1.51 | | urea | (2,4,5) | 98K5 |
| Lys133→Ala | 4.2 | 4 | 2.05 | 1.89 | | urea | (2–4) | 98K5 |
| Arg139→Ala | 4.2 | 4 | 3.14 | 2.06 | | urea | (2–4) | 98K5 |
| Lys140→Ala | 4.2 | 4 | 1.26 | 1.76 | | urea | (2–4) | 98K5 |
| Lys147→Ala | 4.2 | 4 | 1.80 | 1.85 | | urea | (2–4) | 98K5 |

Remarks:

(1) linear extrapolation

(2) stability of the intermediate (I) measured in 4 mM citrate, pH 4.2

(3) transition monitored by CD at 4°C at 222 nm

(4) Δ(ΔG) from wild-type value for average m-value of 6.276 kJ/mol/M

(5) transition monitored by fluorescence

Stability of hydrophobic mutants in native (N = cyano-metmyoglobin) and intermediate state (I) of myoglobin

| Mutant | pH | T | $c_{1/2}$ (1) | $c_{1/2}$ (2) | Δ(ΔG) | Appr./Rem. | | Ref |
|---|---|---|---|---|---|---|---|---|
| data for N: | | | | | | | | |
| wild type | 5 | 25 | 4.21 | 4.17 | 0.0 | urea | (1–4) | 99K4 |
| Leu11→Ala | 5 | 25 | 3.93 | 3.68 | −2.5 | urea | (1–4) | 99K4 |
| Val13→Ala | 5 | 25 | 4.15 | 4.22 | 0.0 | urea | (1–4) | 99K4 |
| Ala15→Leu | 5 | 25 | 4.26 | 4.20 | 0.4 | urea | (1–4) | 99K4 |
| Phe106→Ala | 5 | 25 | 3.71 | 3.75 | −2.9 | urea | (1–4) | 99K4 |
| Leu115→Ala | 5 | 25 | 3.33 | 3.26 | −5.9 | urea | (1–4) | 99K4 |
| Phe123→Ala | 5 | 25 | 3.49 | 3.57 | −4.6 | urea | (1–4) | 99K4 |
| Ala125→Leu | 5 | 25 | 4.56 | 4.57 | 2.5 | urea | (1–4) | 99K4 |
| Leu135→Ala | 5 | 25 | 2.96 | 3.32 | −7.1 | urea | (1–4) | 99K4 |
| Leu137→Ala | 5 | 25 | 4.04 | 3.98 | −1.3 | urea | (1–4) | 99K4 |
| Ala144→Leu | 5 | 25 | 4.44 | 4.45 | 1.7 | urea | (1–4) | 99K4 |
| data for I: | | | | | | | | |
| wild type | 4 | 4 | 1.60 | 1.56 | 0.0 | urea | (1–3,5) | 99K4 |
| Leu11→Ala | 4 | 4 | 1.20 | 1.31 | −2.1 | urea | (1–3,5) | 99K4 |
| Val13→Ala | 4 | 4 | 1.58 | 1.49 | −0.4 | urea | (1–3,5) | 99K4 |
| Ala15→Leu | 4 | 4 | 1.72 | 1.66 | 0.8 | urea | (1–3,5) | 99K4 |
| Phe106→Ala | 4 | 4 | 1.35 | 1.51 | −0.8 | urea | (1–3,5) | 99K4 |
| Leu115→Ala | 4 | 4 | 1.06 | 0.74 | −4.2 | urea | (1–3,5) | 99K4 |
| Phe123→Ala | 4 | 4 | 1.46 | 1.27 | −1.3 | urea | (1–3,5) | 99K4 |
| Ala125→Leu | 4 | 4 | 1.81 | 1.84 | 1.7 | urea | (1–3,5) | 99K4 |
| Leu135→Ala | 4 | 4 | 1.28 | 0.81 | −3.3 | urea | (1–3,5) | 99K4 |
| Leu137→Leu | 4 | 4 | 1.55 | 1.48 | −0.4 | urea | (1–3,5) | 99K4 |
| Ala144→Leu | 4 | 4 | 1.74 | 1.60 | 0.4 | urea | (1–3,5) | 99K4 |

Remarks:

(1) $c_{1/2}$ determined by CD at 222 nm

(2) $c_{1/2}$ determined by Soret absorption at 423 nm for N, and by fluorescence for I

(3) linear extrapolation, LEM-SB; Δ(ΔG) is the difference between the average $c_{1/2}$ given by CD and fluorescence and the $c_{1/2}$ of wild type, multiplied by the average m-value for all mutants, 6.276 kJ/mol/M for I and 6.694 kJ/mol/M for N

(4) N = cyano-metmyoglobin, measured in 10 mM acetate buffer, 0.5 mM sodium cyanide, pH 5, at 25°C

(5) I = apomyoglobin, measured in 4 mM citrate buffer, pH 4, at 4°C

*d) horse heart myoglobin in the presence of osmolytes*

Horse heart myoglobin in the presence of noncompatible osmolytes

| Osmolyte | Conc. | pH | $\Delta T$ | $\Delta H$ | $\Delta(\Delta G)(25°)$ | Appr./Rem. | Ref |
|---|---|---|---|---|---|---|---|
| arginine | 0.00 M | 5.95 | 0.0 | 0 | 0.0 | heat (1–4) | 98R7 |
| | 0.25 M | 5.95 | –3.8 | –54 | –4.2 | heat (1–4) | 98R7 |
| | 0.50 M | 5.95 | –6.6 | –75 | –5.0 | heat (1–4) | 98R7 |
| | 0.75 M | 5.95 | –8.4 | –100 | –6.7 | heat (1–4) | 98R7 |
| | 1.00 M | 5.95 | –10.7 | –117 | –7.1 | heat (1–4) | 98R7 |
| histidine | 0.00 M | 5.95 | 0.0 | 0 | 0.0 | heat (1–4) | 98R7 |
| | 0.05 M | 5.95 | –3.7 | –54 | –4.2 | heat (1–4) | 98R7 |
| | 0.10 M | 5.95 | –6.6 | –75 | –5.0 | heat (1–4) | 98R7 |
| | 0.15 M | 5.95 | –7.4 | –88 | –5.9 | heat (1–4) | 98R7 |
| lysine | 0.00 M | 5.95 | 0.0 | 0 | 0.0 | heat (1–4) | 98R7 |
| | 0.05 M | 5.95 | 0.3 | 1 | 0.0 | heat (1–4) | 98R7 |
| | 0.15 M | 5.95 | 0.1 | 1 | 0.1 | heat (1–4) | 98R7 |
| | 0.25 M | 5.95 | 0.1 | 2 | 0.3 | heat (1–4) | 98R7 |
| | 0.50 M | 5.95 | 0.2 | 5 | 0.4 | heat (1–4) | 98R7 |

Remarks:

(1) van't Hoff treatment, transition monitored by changes in optical absorption at 409 nm

(2) measured in 0.05 M citrate, 0.1 M KCl, in the presence of 0.6 M GuHCl

(3) in the absence of osmolytes, $T_{trs}$ = 66.8°C and $\Delta H$ = 339 kJ/mol was obtained

(4) $\Delta G$(25°) in the presence of osmolyte was calculated using $\Delta Cp$ = 11.57 kJ/mol/K from Ref. 88P3

**NikR**, see DNA-Binding Protein

## Nuclease from *Staphylococcus aureus* (Staphylococcal nuclease)

The data entries are arranged as follows:

a) comparison of various approaches
b) pressure denaturation
c) wild type and mutants
d) stabilization by xylose

*a) comparison of various approaches*

Staphylococcal nuclease (SN), wild type, temperature dependence of $\Delta G$

| Protein | pH | T | $\Delta G$ | $c_{1/2}$ | m | Appr./Rem. | Ref |
|---|---|---|---|---|---|---|---|
| SN | 7.0 | 15.2 | 21.34±0.17 | 0.90 | 23.64±0.17 | GuHCl (1–4) | 99Y1 |
| | 7.0 | 20.0 | 20.04±0.08 | 0.84 | 23.93±0.08 | GuHCl (1–4) | 99Y1 |
| | 7.0 | 25.0 | 19.96±0.17 | 0.80±0.01 | 25.10±0.17 | GuHCl (1–4) | 99Y1 |
| | 7.0 | 25.0 | 21.00±1.84 | 0.77±0.09 | 27.20±2.18 | ITC (1,4,5) | 99Y1 |
| | 7.0 | 30.0 | 17.36±0.13 | 0.72 | 24.18±0.17 | GuHCl (1–4) | 99Y1 |
| | 7.0 | 35.0 | 15.10±0.08 | 0.61 | 24.60±0.13 | GuHCl (1–4) | 99Y1 |

Remarks:

(1) linear extrapolation, LEM-SB

(2) Ref. 99Y1 is aimed to determine the ratio $\Delta H^{v.H.}/\Delta H^{cal}$ for SN in GuHCl, see also Table 2

(3) transition monitored by fluorescence (excitation at 295 nm, emission at 335 nm)

(4) buffer: 25 mM phosphate, 0.1 M NaCl, pH 7.0

(5) transition monitored by heat of mixing of SN with GuHCl (ITC)

Staphylococcal nuclease A, wild type and mutants, global fit of data over two perturbation axes

| Mutant | Case | T | ΔG | m | Approach/Remarks | Ref |
|---|---|---|---|---|---|---|
| wild type | 1 | 20 | 26.19±1.02 | 11.13±0.44 | urea/pH (1–4) | 97I3 |
| | 2 | 20 | 27.07±1.53 | 12.51±0.61 | urea/pH (1–3,5) | 97I3 |
| | 3 | 20 | 21.63±0.82 | 16.65±0.38 | urea/pH (1–3,6) | 97I3 |
| | 4 | 20 | 25.77±1.22 | 11.30±0.64 | urea/pH (1–3,7) | 97I3 |
| | 5 | 20 | 27.74±0.75 | 10.67±0.21 | urea/pH (1–3,8) | 97I3 |
| Val66→Ala' | | 20 | 6.65±0.84 | 4.98±0.19 | urea/pH (1–3) | 97I3 |
| Val66→Ala | f | 20 | 8.20±0.75 | 4.94±0.50 | urea/pH (1–3,9) | 97I3 |
| N → I | a | 20 | 11.30±0.85 | 4.77±0.29 | urea/pH (1–3,10) | 97I3 |
| I → U | a | 20 | 6.69 | 4.98 | urea/pH (1–3,10) | 97I3 |
| N → I | b | 20 | 11.63±1.05 | 5.23±0.35 | urea/pH (1–3,11) | 97I3 |
| I → U | a | 20 | 9.25±1.70 | 5.86±0.75 | urea/pH (1–3,11) | 97I3 |

Remarks:

(1)  abbreviations:

wild type = full-length sequence, 149 residues, with single Trp at pos. 140

Val66→Trp = full-length mutant with Trp residues at positions 66 and 140

Val66→Trp' = variant that contains residues 1–136 with single Trp at pos. 66

(2)  global fit over two perturbation axes: acid-induced unfolding as a function of urea conc., and urea-induced unfolding as a function of pH

(3)  transitions monitored by CD at 222, 228, and 235 nm, and fluorescence at 340 nm

(4)  case 1 = global two-state fit, no additional parameters

(5)  case 2 = global two-state fit, concentration dependence of m as an additional parameter, $\partial m/\partial[\text{urea}] = -0.42$ kJ/mol/M

(6)  case 3 = global two-state fit, pH dependence of m as an additional parameter, $\partial m/\partial pH = -1.13$ kJ/mol/M, given is m at pH 0

(7)  case 4 = global two-state fit, additional parameter is the difference $pK_{a,U} - pK_{a,N} = 1.06$

(8)  case 5 = global two-state fit, additional parameter is p $K_{a,U_{His}} = 6.74$, case 5 is the best fit

(9)  case f = formal two-state treatment

(10)  case a = three-state fit, parameters of the transition I → U fixed

(11)  case b = three-state fit, no parameters fixed, most reliable result for mutant Val66→Trp

Staphylococcal nuclease (SN), wild type and mutant, unfolding analyzed by fluorescence and size-exclusion chromatography

| Mutant | pH | T | ΔG | $c_{1/2}$ | m | Appr./Rem. | Ref |
|---|---|---|---|---|---|---|---|
| SN wild type | 7.00 | 23 | 23.39±0.21 | 2.42±0.03 | 9.67±0.08 | urea (1–3) | 98B4 |
| | 7.00 | 23 | 29.37±2.09 | 2.48±0.15 | 11.84±1.09 | urea (1,4,5) | 98B4 |
| SN Ala69→Thr | 7.00 | 23 | 10.75±0.21 | 1.05±0.04 | 10.25±0.17 | urea (1–3) | 98B4 |
| | 7.00 | 23 | 15.82±1.46 | 1.07±0.1 | 14.77±1.05 | urea (1,4,5) | 98B4 |
| RNase A | 3.0 | 25 | 21.09±0.63 | 3.02±0.13 | 6.99±0.21 | urea (1,6–8) | 98B4 |
| | 3.0 | 25 | 20.25±2.64 | 2.81±0.32 | 7.20±1.09 | urea (1,4,6) | 98B4 |

Remarks:

(1) linear extrapolation, LEM-SB

(2) transition monitored by fluorescence intensity at 335 nm at 23.0±0.1°C

(3) buffer: 30 mM MOPS, 0.1 M NaCl, pH 7.00

(4) transition from the analysis of size-exclusion chromatography profiles recorded at 23.0±0.5°C

(5) buffer: 30 mM MOPS, 0.2 M NaCl, pH 7.00

(6) buffer: 0.1 M β-alanine, 0.421 M NaCl, pH 3.0

(7) transition monitored by optical absorption at 278 nm

(8) data from Ref. 95Y5

Staphylococcal nuclease, unfolding monitored by urea gradient electrophoresis

| pH | T | $\Delta G$ | $c_{1/2}$ | m | Appr./Rem. | Ref |
|---|---|---|---|---|---|---|
| 8.0 | 10 | 26.4±1.7 | 2.5 | 10.5 | urea (1) | 94C12 |
| 7.0 | 2 | 25.5 | 2.56 | 10.0 | urea (2) | 86S3 |

Remarks:

(1) urea gradient electrophoresis

(2) linear extrapolation, reference value from Ref. 86S3

Staphylococcal nuclease, wild type and mutants in $H_2O$ and $D_2O$

| Protein | | pH | T | $\Delta G$ | m | Appr./Rem. | Ref |
|---|---|---|---|---|---|---|---|
| wild type | $H_2O$ | 7.0 | 20 | 23.0±0.4 | 28.5±0.1 | GuHCl (1–4) | 99W16 |
| | $H_2O$ | 5.6 | 25 | 15.9±0.4 | 24.7±0.4 | GuHCl (1–3,5) | 99W16 |
| | $D_2O$ | 5.2 | 25 | 16.7±0.4 | 23.0±0.4 | GuHCl (1–3,6) | 99W16 |
| Glu75→Ala | $H_2O$ | 7.0 | 20 | 13.8±0.4 | 22.2±0.1 | GuHCl (1–4) | 99W16 |
| | $H_2O$ | 5.6 | 25 | 8.8±0.4 | 20.9±0.0 | GuHCl (1–3,5) | 99W16 |
| | $D_2O$ | 5.2 | 25 | 8.4±0.4 | 18.0±0.8 | GuHCl (1–3,6) | 99W16 |
| Asp77→Ala | $H_2O$ | 7.0 | 20 | 10.0±0.4 | 21.8±0.1 | GuHCl (1–4) | 99W16 |
| | $H_2O$ | 5.6 | 25 | 8.4±0.0 | 21.3±0.0 | GuHCl (1–3,5) | 99W16 |
| | $D_2O$ | 5.2 | 25 | 9.6±0.4 | 18.8±0.4 | GuHCl (1–3,6) | 99W16 |
| Met26→Gly | $H_2O$ | 7.0 | 20 | 13.8±0.4 | 31.8±0.2 | GuHCl (1–4) | 99W16 |
| | $H_2O$ | 5.6 | 25 | 10.5±0.4 | 33.1±0.8 | GuHCl (1–3,5) | 99W16 |
| | $D_2O$ | 5.2 | 25 | 13.0±0.4 | 31.0±0.4 | GuHCl (1–3,6) | 99W16 |
| Val23→Ala | $H_2O$ | 7.0 | 20 | 10.9±0.4 | 33.9±0.2 | GuHCl (1–4) | 99W16 |
| | $H_2O$ | 5.6 | 25 | 7.1±0.4 | 34.3±2.1 | GuHCl (1–3,5) | 99W16 |
| | $D_2O$ | 5.2 | 25 | 9.2±0.8 | 32.2±1.3 | GuHCl (1–3,6) | 99W16 |

Remarks:

(1) linear extrapolation

(2) transition monitored by intrinsic Trp fluorescence (excitation at 295 nm, emission at 325 nm)

(3) Ref. 99W16 contains additionally H/D exchange measurements

(4) 50 μg/ml protein in 100 mM NaCl, 25 mM sodium phosphate, pH 7.0, 20°C

(5) 50 μg/ml protein in 100 mM NaCl, 50 mM sodium acetate, pH 5.6, 25°C

(6) 50 μg/ml protein in 100 mM NaCl, 50 mM sodium acetate, pH* 5.2 (uncorrected meter reading), 25°C

Staphylococcal nuclease (SNase) and deuterated SNase ($^D$SNase)

| Protein | pH | T | $\Delta G$ | $c_{1/2}$ | m | Appr./Rem. | Ref |
|---|---|---|---|---|---|---|---|
| $^H$SNase/$H_2O$ | 7 | 20 | 25.8±1.4 | 2.65±0.06 | 9.79±0.75 | urea (1–3) | 98F4 |
| $^D$SNase/$D_2O$ | 7 | 20 | 27.0±0.5 | 3.10±0.06 | 8.70±0.04 | urea (1–3) | 98F4 |
| $^D$SNase/$D_2O$ | 7 | 20 | | 2.8 | | urea (4,5a) | 98F4 |
| $^D$SNase/$D_2O$ | 7 | 20 | | 2.95 | | urea (4,5b) | 98F4 |
| $^D$SNase/$D_2O$ | 7 | 20 | | 2.94 | | urea (4,5c) | 98F4 |

Remarks:

(1) linear extrapolation

(2) transition monitored by CD at 220 nm

(3) buffer: 25 mM sodium phosphate, 100 mM sodium chloride

(4) using deuterated $^{13}$C-urea and $^D$SNase in $D_2O$

(5) transition monitored by Fourier Transform Infrared Spectroscopy (FTIR) (a) at 1630 cm$^{-1}$, (b) at 1627 cm$^{-1}$, and (c) at 1642 cm$^{-1}$

Staphylococcal nuclease, mutant Thr62→Pro, trimethylamine N-oxide-induced (TMAO) folding transition

| Mutant | pH | T | ΔG | m | Approach/Remarks | Ref |
|---|---|---|---|---|---|---|
| Thr62→Pro | 8.8 | 25 | −16.82 | −10.67 | activity (1–3) | 98B3 |
| Thr62→Pro | 8.8 | 25 | +12.13 | | TMAO/act.(1–4) | 98B3 |

Remarks:

(1) the staphylococcal nuclease mutant Thr62→Pro is thermodynamically unstable in buffer solution (25 mM Tris-HCl, 0.1 M NaCl, 10 mM CaCl$_2$, pH 8.8)

(2) transition monitored by activity, data normalized to the wild-type protein

(3) data from TMAO-induced folding, data treatment by linear extrapolation, LEM-SB

(4) ΔG refers to 2.7 M TMAO

Staphylococcal nuclease, characterization of multiple equilibrium partially folded intermediates

| State | Conditions | pH | T | ΔG | m | Appr./Rem. | Ref |
|---|---|---|---|---|---|---|---|
| N state | 0.1 M KCl | 7.3 | 23 | 20.5±0.8 | 8.33±0.33 | urea (1) | 98U2 |
| A$_3$-state | 50 mM TCA | 2.5 | 23 | 13.4±0.4 | 35.15±1.67 | urea (1,2) | 98U2 |
| A$_3$-state unfolding intermediate | | | | | | | |
| | 50 mM TCA | 2.5 | 23 | 1.88±0.17 | 3.26±0.08 | urea (1,3) | 98U2 |
| A$_2$-state | 0.27 M TFA | 2.5 | 23 | 8.79±0.84 | 6.28±0.84 | urea (1,4) | 98U2 |
| A$_2$-state unfolding intermediate | | | | | | | |
| | 0.27 M TFA | 2.5 | 23 | 3.35±0.42 | 2.13±0.13 | urea (1,5) | 98U2 |
| A$_1$-state | 0.25 M Na$_2$SO$_4$ | 2.5 | 23 | 4.81±0.33 | 3.26±0.17 | urea (1,4) | 98U2 |
| A$_1$-state | 0.9 M KCl | 2.5 | 23 | 2.76±0.21 | 2.05±0.04 | urea (1,4) | 98U2 |

Explanations:

TCA: trichloroacetate

TFA: trifluoroacetate

Remarks:

(1) linear extrapolation

(2) second transition (0.9–6 M urea)

(3) first transition (0–0.9 M urea)

(4) second transition (1.9–6 M urea)

(5) first transition (0–1.9 M urea)

*b) pressure denaturation*

Staphylococcal nuclease, pressure denaturation, wild type and Val66 mutants

| Protein | pH | T | ΔG | ΔV | Approach/Remarks | Ref |
|---|---|---|---|---|---|---|
| wild type | 5.5 | 21 | 13.4±0.8 | −77.0±8.0 | pressure (1–3) | 98F5 |
| Val66→Ala | 7.0 | 21 | 9.2±0.8 | −105.7±7.6 | pressure (2,3,5) | 98F5 |
| Val66→Ala | 7.0 | 21 | 8.8±0.8 | −103.8±6.1 | pressure (2,4,5) | 98F5 |
| Val66→Gly | 7.0 | 21 | 8.4±0.8 | −112.6±8.2 | pressure (2,3,6) | 98F5 |
| Val66→Gly | 7.0 | 21 | 8.4±0.8 | −111.9±7.7 | pressure (2,4,6) | 98F5 |
| Val66→Leu | 5.5 | 21 | 10.0±1.3 | −95.7±4.1 | pressure (2,3,7) | 98F5 |
| Val66→Leu | 5.5 | 21 | 9.6±1.7 | −94.2±11.7 | pressure (2,4,7) | 98F5 |

Remarks:

(1) reference value in Ref. 98F5 taken from Ref. 95V4

(2) transition monitored by fluorescence

(3) average value

(4) data from global fit

(5) buffer: 10 mM Bis-Tris, pH 7.0

(6) buffer: 10 mM Bis-Tris with 19 mM xylose, pH 7.0

(7) buffer: 10 mM Bis-Tris with 0.5 M GuHCl, pH 5.5

Staphylococcal nuclease, recombinant (SNase), pressure denaturation

| Protein | pH | T | $\Delta G$ | $\Delta V$ | Approach/Remarks | Ref |
|---------|-----|------|-----------|-----------|------------------|------|
| Results from fluorescence-detected high-pressure experiments: | | | | | | |
| SNase | 5.5 | 2 | 10.9±4.2 | 90±15 | pressure (1,2,5) | 99P4 |
| | 5.5 | 10 | 13.4±4.2 | 84±10 | pressure (1,2,5) | 99P4 |
| | 5.5 | 21 | 13.4±2.5 | 70±10 | pressure (1,2,5) | 99P4 |
| | 5.5 | 30 | 8.8±0.8 | 58± 6 | pressure (1,2,5) | 99P4 |
| | 5.5 | 40 | 5.0±0.4 | 52± 4 | pressure (1,2,5) | 99P4 |
| Results from FTIR-detected high-pressure experiments: | | | | | | |
| SNase | 5.5 | −4.5 | 8.4±3.1 | 92±20 | pressure (3–5) | 99P4 |
| | 5.5 | 0.5 | 9.2±3.8 | 90±20 | pressure (3–5) | 99P4 |
| | 5.5 | 5.0 | 12.6±6.3 | 87±20 | pressure (3–5) | 99P4 |
| | 5.5 | 10.0 | 17.4±7.1 | 87±25 | pressure (3–5) | 99P4 |
| | 5.5 | 15.0 | 18.8±7.1 | 84±20 | pressure (3–5) | 99P4 |
| | 5.5 | 20.0 | 19.2±7.1 | 80±20 | pressure (3–5) | 99P4 |
| | 5.5 | 25.0 | 16.7±6.7 | 77±20 | pressure (3–5) | 99P4 |
| | 5.5 | 36.0 | 13.0±4.6 | 67±20 | pressure (3–5) | 99P4 |
| | 5.5 | 40.0 | 8.8±3.3 | 55±25 | pressure (3–5) | 99P4 |
| | 5.5 | 45.0 | 2.6±2.1 | 35±15 | pressure (3–5) | 99P4 |

Remarks:
(1) from high-pressure fluorescence measurements
(2) buffer: 10 mM Bis-Tris at pH 5.5
(3) from high-pressure Fourier Transform Infrared spectroscopy (FTIR) measurements
(4) buffer: 50 mM Bis-Tris at pH 5.5 in $D_2O$
(5) $\Delta V$ in ml/mol

Staphylococcal nuclease, wild type and mutants, pressure denaturation

| Protein | pH | T | $\Delta G$ | $\Delta V$ | Approach/Remarks | Ref |
|---------|-----|-----|-----------|-----------|------------------|------|
| wild type | 5.5 | 21 | 15.7 | −92.2±10 | pressure (1,2) | 99R19 |
| | 4.5 | 21 | 12.8 | | pressure (1,2) | 99R19 |
| | 3.5 | 21 | 5.6 | | pressure (1,2) | 99R19 |
| double mutant (His124→Leu and Gly79→Ser) | | | | | | |
| | 5.5 | 21 | 13.1 | −84.6±7 | pressure (1,2) | 99R19 |
| | 4.5 | 21 | 11.6 | | pressure (1,2) | 99R19 |
| | 3.5 | 21 | 6.2 | | pressure (1,2) | 99R19 |
| double mutant (His124→Leu and Asp77→Ala) | | | | | | |
| | 7.0 | 21 | 9.6 | | pressure (1,2) | 99R19 |
| | 5.5 | 21 | 2.3 | −89.9±8 | pressure (1,2) | 99R19 |
| | 4.5 | 21 | 0.7 | | pressure (1,2) | 99R19 |
| double mutant (His124→Leu and Phe76→Val) | | | | | | |
| | 5.5 | 21 | 4.3 | −56.7±8 | pressure (1,2) | 99R19 |
| | 4.5 | 21 | 2.9 | | pressure (1,2) | 99R19 |

Remarks:
(1) pressure-induced transition monitored by fluorescence
(2) buffer: 10 mM Bis-Tris

Staphylococcal nuclease, pressure denaturation

| pH | T | ΔG | Approach/Remarks | Ref |
|---|---|---|---|---|
| 6.0 | 37.7 | 9.5±0.3 | pressure (1–3) | 98P14 |
| 6.0 | 37.7 | 10.9±1.3 | pressure (1,2,4) | 98P14 |

Remarks:
(1) pressure denaturation monitored by fluorescence
(2) ΔG at 1 bar
(3) curve fit assuming a zero compressibility change, $\Delta V = -(72\pm2)$ ml/mol
(4) curve fit assuming compressibility ($\Delta K$) variable, $\Delta K = 0.02\pm0.02$ ml/mol/bar, and $\Delta V = -(94\pm19)$ ml/mol

Staphylococcal nuclease, wild type and mutants, results of high-pressure denaturation

| Protein | pH* | T | ΔG | ΔV | $p_{1/2}$ | Approach/Rem. | Ref |
|---|---|---|---|---|---|---|---|
| wild type | 5.1 | 37 | 5.0 | −53±11 | 925 | pressure (1–3) | 99R19 |
| His124→Leu | 5.1 | 37 | 20.5 | −65±7 | 3109 | pressure (1–3) | 99R19 |
| double mutant (His124→Leu and Gly79→Ser) | | | | | | | |
| | 5.1 | 37 | 5.9 | −47±3 | 1195 | pressure (1–3) | 99R19 |
| wild type | 5.5 | 37 | 9.8 | −57±2 | 1550 | pressure (2–4) | 99R19 |
| double mutant (His124→Leu and Gly79→Ser) | | | | | | | |
| | 5.5 | 37 | 9.6 | −55±2 | 1550 | pressure (2–4) | 99R19 |

Remarks:
(1) data from pressure-induced unfolding monitored by ¹H-NMR
(2) pressure in bar
(3) buffer: 10 mM Bis-Tris in $D_2O$
(4) data from pressure-induced unfolding monitored by fluorescence

*c) wild type and mutants*

Staphylococcal nuclease, wild type and mutants at position 27

| Mutant | pH | T | ΔG | $c_{1/2}$ | m (1) | Approach/Remarks | | Ref |
|---|---|---|---|---|---|---|---|---|
| wild type | 7.0 | 20.0 | 20.9 | 0.84 | 4.18 | GuHCl | (1–3) | 97B9 |
| wild type | 7.0 | 20.0 | 22.6 | 0.84 | 4.60 | GuHCl | (1–4) | 97B9 |
| Tyr27→Ala | 7.0 | 20.0 | 9.2 | 0.29 | 4.48 | GuHCl | (1–3) | 97B9 |
| Tyr27→Arg | 7.0 | 20.0 | 8.8 | 0.30 | 5.40 | GuHCl | (1–3) | 97B9 |
| Tyr27→Asn | 7.0 | 20.0 | 4.2 | 0.22 | 4.81 | GuHCl | (1–3) | 97B9 |
| Tyr27→Asp | 7.0 | 20.0 | −1.7 | | | renat. | (6) | 97B9 |
| Tyr27→Cys | 7.0 | 20.0 | 10.5 | 0.41 | 4.27 | GuHCl | (1–4) | 97B9 |
| Tyr27→Cys | 7.0 | 20.0 | 8.8 | | | GuHCl | (5) | 97B9 |
| Tyr27→Gln | 7.0 | 20.0 | 7.5 | 0.26 | 5.31 | GuHCl | (1–3) | 97B9 |
| Tyr27→Glu | 7.0 | 20.0 | 0.0 | | | renat. | (6) | 97B9 |
| Tyr27→Gly | 7.0 | 20.0 | −0.4 | | | renat. | (6) | 97B9 |
| Tyr27→His | 7.0 | 20.0 | 14.2 | 0.48 | 4.98 | GuHCl | (1–3) | 97B9 |
| Tyr27→Ile | 7.0 | 20.0 | 10.5 | 0.49 | 3.39 | GuHCl | (1–3) | 97B9 |
| Tyr27→Leu | 7.0 | 20.0 | 14.6 | 0.57 | 4.48 | GuHCl | (1–3) | 97B9 |
| Tyr27→Lys | 7.0 | 20.0 | 4.6 | 0.22 | 4.48 | GuHCl | (1–3) | 97B9 |
| Tyr27→Lys | 7.0 | 20.0 | 5.0 | | | renat. | (6) | 97B9 |
| Tyr27→Met | 7.0 | 20.0 | 12.6 | 0.52 | 4.10 | GuHCl | (1–3) | 97B9 |
| Tyr27→Phe | 7.0 | 20.0 | 20.9 | 0.77 | 4.60 | GuHCl | (1–3) | 97B9 |
| Tyr27→Pro | 7.0 | 20.0 | −4.6 | | | renat. | (6) | 97B9 |

Staphylococcal nuclease, wild type and mutants at position 27 (continued)

| Mutant | pH | T | ΔG | $c_{1/2}$ | m (1) | Approach/Remarks | | Ref |
|---|---|---|---|---|---|---|---|---|
| Tyr27→Ser | 7.0 | 20.0 | 8.4 | 0.26 | 5.52 | GuHCl | (1–3) | 97B9 |
| Tyr27→Thr | 7.0 | 20.0 | 7.1 | 0.32 | 4.48 | GuHCl | (1–3) | 97B9 |
| Tyr27→Trp | 7.0 | 20.0 | 18.4 | 0.70 | 4.52 | GuHCl | (1–3) | 97B9 |
| Tyr27→Val | 7.0 | 20.0 | 8.4 | 0.45 | 3.14 | GuHCl | (1–3) | 97B9 |

Remarks:

(1) linear extrapolation, LEM-SB, all m values are normalized to the wild-type value of 24.7 kJ/mol/M

(2) buffer: 25 mM sodium phosphate, 100 mM NaCl

(3) average estimated error in ΔG ±0.4 kJ/mol

(4) buffer contained 0.1 mM DTT

(5) assumed true value in the absence of DTT as calculated from the wild type with and without DTT

(6) values obtained by sulfate renaturation as described in Ref. 90S4

Staphylococcal nuclease, phenylalanine substitutions mutants

| Mutant | pH | T | ΔG | $c_{1/2}$ | $(m)_{rel}$ | Approach/Remarks | | Ref |
|---|---|---|---|---|---|---|---|---|
| wild type | 7.0 | 20 | 23.0 | 0.83 | 1.00 | GuHCl | (1–3) | 98S8 |
| Lys9→Phe | 7.0 | 20 | 18.0 | 0.70 | 0.93 | GuHCl | (1–3) | 98S8 |
| Pro11→Phe | 7.0 | 20 | 20.5 | 0.79 | 0.94 | GuHCl | (1–3) | 98S8 |
| Lys16→Phe | 7.0 | 20 | 20.9 | 0.79 | 0.96 | GuHCl | (1–3) | 98S8 |
| Asp19→Phe | 7.0 | 20 | 16.7 | 0.67 | 0.90 | GuHCl | (1–3) | 98S8 |
| Lys24→Phe | 7.0 | 20 | 20.5 | 0.80 | 0.93 | GuHCl | (1–3) | 98S8 |
| Lys28→Phe | 7.0 | 20 | 19.2 | 0.71 | 0.99 | GuHCl | (1–3) | 98S8 |
| Gly29→Phe | 7.0 | 20 | 17.2 | 0.57 | 1.09 | GuHCl | (1–3) | 98S8 |
| Pro47→Phe | 7.0 | 20 | 21.8 | 0.83 | 0.95 | GuHCl | (1,4,5) | 98S8 |
| Lys48→Phe | 7.0 | 20 | 20.9 | 0.83 | 0.91 | GuHCl | (1–3) | 98S8 |
| Lys49→Phe | 7.0 | 20 | 23.0 | 0.89 | 0.94 | GuHCl | (1–3) | 98S8 |
| Gly50→Phe | 7.0 | 20 | 19.7 | 0.75 | 0.94 | GuHCl | (1–3) | 98S8 |
| Glu52→Phe | 7.0 | 20 | 20.1 | 0.78 | 0.94 | GuHCl | (1,4,5) | 98S8 |
| Glu57→Phe | 7.0 | 20 | 20.1 | 0.72 | 1.01 | GuHCl | (1,4,5) | 98S8 |
| Ser59→Phe | 7.0 | 20 | 23.0 | 1.00 | 0.81 | GuHCl | (1–3) | 98S8 |
| Ala60→Phe | 7.0 | 20 | 19.7 | 0.71 | 1.00 | GuHCl | (1–3) | 98S8 |
| Lys63→Phe | 7.0 | 20 | 14.6 | 0.58 | 0.93 | GuHCl | (1–3) | 98S8 |
| Lys64→Phe | 7.0 | 20 | 21.8 | 0.85 | 0.92 | GuHCl | (1–3) | 98S8 |
| Met65→Phe | 7.0 | 20 | 16.3 | 0.58 | 1.03 | GuHCl | (1–3) | 98S8 |
| Glu67→Phe | 7.0 | 20 | 17.2 | 0.64 | 0.97 | GuHCl | (1–3) | 98S8 |
| Lys70→Phe | 7.0 | 20 | 22.2 | 0.83 | 0.96 | GuHCl | (1–3) | 98S8 |
| Lys71→Phe | 7.0 | 20 | 23.0 | 0.84 | 1.00 | GuHCl | (1–3) | 98S8 |
| Glu73→Phe | 7.0 | 20 | 18.4 | 0.73 | 0.92 | GuHCl | (1–3) | 98S8 |
| Gln80→Phe | 7.0 | 20 | 20.1 | 0.76 | 0.96 | GuHCl | (1,3,5) | 98S8 |
| Asp83→Phe | 7.0 | 20 | 1.3 | 0.06 | 0.88 | GuHCl | (1–3) | 98S8 |
| Lys84→Phe | 7.0 | 20 | 19.2 | 0.74 | 0.95 | GuHCl | (1–3) | 98S8 |
| Tyr85→Phe | 7.0 | 20 | 22.2 | 0.82 | 0.98 | GuHCl | (1–3) | 98S8 |
| Gly86→Phe | 7.0 | 20 | 14.6 | 0.53 | 1.01 | GuHCl | (1–3) | 98S8 |
| Asp95→Phe | 7.0 | 20 | 0.8 | 0.06 | 0.59 | GuHCl | (1–3) | 98S8 |
| Gly96→Phe | 7.0 | 20 | 10.5 | 0.51 | 0.75 | GuHCl | (1–3) | 98S8 |
| Lys97→Phe | 7.0 | 20 | 20.1 | 0.76 | 0.95 | GuHCl | (1–3) | 98S8 |
| Glu101→Phe | 7.0 | 20 | 10.0 | 0.47 | 0.78 | GuHCl | (1–3) | 98S8 |
| Ala102→Phe | 7.0 | 20 | 15.1 | 0.60 | 0.91 | GuHCl | (1–3) | 98S8 |
| Arg105→Phe | 7.0 | 20 | 10.0 | 0.44 | 0.84 | GuHCl | (1–3) | 98S8 |
| Ala112→Phe | 7.0 | 20 | 16.3 | 0.64 | 0.93 | GuHCl | (1–3) | 98S8 |
| Tyr113→Phe | 7.0 | 20 | 23.0 | 0.84 | 0.98 | GuHCl | (1–3) | 98S8 |
| Tyr115→Phe | 7.0 | 20 | 22.6 | 0.83 | 0.97 | GuHCl | (1–3) | 98S8 |

Staphylococcal nuclease, phenylalanine substitutions mutants (continued)

| Mutant | pH | T | ΔG | $c_{1/2}$ | $(m)_{rel}$ | Approach/Remarks | Ref |
|---|---|---|---|---|---|---|---|
| Lys116→Phe | 7.0 | 20 | 22.6 | 0.91 | 0.91 | GuHCl (1–3) | 98S8 |
| Pro117→Phe | 7.0 | 20 | 24.3 | 0.96 | 0.92 | GuHCl (1–3) | 98S8 |
| Glu122→Phe | 7.0 | 20 | 19.7 | 0.82 | 0.86 | GuHCl (1–3) | 98S8 |
| Gln123→Phe | 7.0 | 20 | 20.1 | 0.78 | 0.94 | GuHCl (1–3) | 98S8 |
| His124→Phe | 7.0 | 20 | 24.3 | 0.94 | 0.93 | GuHCl (1–3) | 98S8 |
| Lys127→Phe | 7.0 | 20 | 21.8 | 0.81 | 0.97 | GuHCl (1–3) | 98S8 |
| Ser128→Phe | 7.0 | 20 | 18.8 | 0.68 | 1.00 | GuHCl (1–3) | 98S8 |
| Glu129→Phe | 7.0 | 20 | 6.3 | 0.26 | 0.89 | GuHCl (1–3) | 98S8 |
| Gln131→Phe | 7.0 | 20 | 20.1 | 0.77 | 0.95 | GuHCl (1–3) | 98S8 |
| Lys133→Phe | 7.0 | 20 | 17.2 | 0.66 | 0.93 | GuHCl (1–3) | 98S8 |
| Lys134→Phe | 7.0 | 20 | 20.9 | 0.82 | 0.93 | GuHCl (1–3) | 98S8 |
| Glu135→Phe | 7.0 | 20 | 18.0 | 0.68 | 0.96 | GuHCl (1–3) | 98S8 |
| Lys136→Phe | 7.0 | 20 | 18.0 | 0.72 | 0.91 | GuHCl (1–3) | 98S8 |
| Asp143→Phe | 7.0 | 20 | 22.6 | 0.83 | 0.98 | GuHCl (1,4,5) | 98S8 |

Remarks:

(1) linear extrapolation LEM-SB, $(m)_{rel}$ is expressed relative to the wild-type value of 27.61 kJ/mol/M, error in m is estimated to be ±0.08 kJ/mol/M, error in $c_{1/2}$ is estimated to be ±0.01 M, error in ΔG is estimated to be ±0.4 kJ/mol

(2) for the procedure and data treatment see Ref. 95S14

(3) buffer: 25 mM sodium phosphate, 100 mM NaCl, see also Ref. 95S14

(4) buffer: 25 mM sodium phosphate, 600 mM NaCl, pH 7.0

(5) modified data treatment, see Ref. 98S8

Staphylococcal nuclease, wild type and proline mutants

| Mutant | pH | T | ΔG | $c_{1/2}$ | m | Appr./Rem. | Ref |
|---|---|---|---|---|---|---|---|
| wild type | 7.0 | 20 | 21.3±0.6 | 2.44±0.01 | 8.74±0.25 | urea (1) | 97I2 |
| Pro47→Ala | 7.0 | 20 | 20.8±0.8 | 2.42±0.01 | 8.58±0.33 | urea (1) | 97I2 |
| Pro47→Thr | 7.0 | 20 | 23.4±0.5 | 2.45±0.01 | 9.54±0.21 | urea (1) | 97I2 |
| Pro117→Gly | 7.0 | 20 | 25.9±1.3 | 3.10±0.02 | 8.37±0.42 | urea (1) | 97I2 |
| double mutant (Pro47→Ala and Pro117→Gly) | | | | | | | |
| | 7.0 | 20 | 26.6±0.7 | 2.99±0.01 | 8.87±0.25 | urea (1) | 97I2 |
| double mutant (Pro47→Thr and Pro117→Gly) | | | | | | | |
| | 7.0 | 20 | 25.3±0.7 | 3.02±0.01 | 8.37±0.25 | urea (1) | 97I2 |
| Pro-free mutant | | | | | | | |
| | 5.3 | 15 | | 0.98 | 25.5 | GuHCl (2,3) | 97W1 |

Remarks:

(1) linear extrapolation, LEM-SB

(2) Pro-free mutant with the following amino acid replacements: Pro11→Ala, Pro31→Ala, Pro42→Ala, Pro47→Gly, Pro56→Ala, and Pro117→Gly

(3) from GuHCl-induced folding and refolding kinetics, for reference data see also Ref. 86S4

Staphylococcal nuclease, wild type and proline mutants

| Mutant | pH | T | ΔG | $c_{1/2}$ | m | Appr./Rem. | Ref |
|---|---|---|---|---|---|---|---|
| wild type | 7.0 | 20 | 22.17±1.42 | 2.44±0.02 | 9.04±0.54 | urea (1–4) | 99M4 |
| Pro11→Ala | 7.0 | 20 | 22.51±0.92 | 2.35±0.01 | 9.58±0.38 | urea (1–4) | 99M4 |
| Pro31→Ala | 7.0 | 20 | 19.87±0.88 | 2.34±0.02 | 8.49±0.04 | urea (1–4) | 99M4 |
| Pro42→Ala | 7.0 | 20 | 23.56±3.14 | 2.41±0.04 | 9.75±1.21 | urea (1–4) | 99M4 |
| Pro56→Ala | 7.0 | 20 | 24.14±1.92 | 2.65±0.02 | 9.08±0.67 | urea (1–4) | 99M4 |
| double mutant (Pro47→Thr and Pro117→Gly) | | | | | | | |
| | 7.0 | 20 | 25.27±0.71 | 3.02±0.01 | 8.37±0.25 | urea (1–5) | 99M4 |

Staphylococcal nuclease, wild type and proline mutants (continued)

| Mutant | pH | T | $\Delta G$ | $c_{1/2}$ | m | Appr./Rem. | Ref |
|---|---|---|---|---|---|---|---|
| multiple mutant (Pro11→Ala, Pro47→Thr and Pro117→Gly) | | | | | | | |
| | 7.0 | 20 | 26.40±0.75 | 2.91±0.01 | 9.08±0.25 | urea (1–4) | 99M4 |
| multiple mutant (Pro11→Ala, Pro31→Ala, Pro47→Thr and Pro117→Gly) | | | | | | | |
| | 7.0 | 20 | 26.57±0.46 | 2.85±0.01 | 9.33±0.17 | urea (1–4) | 99M4 |
| multiple mutant (Pro11→Ala, Pro31→Ala, Pro42→Ala, Pro47→Thr and Pro117→Gly) | | | | | | | |
| | 7.0 | 20 | 24.18±1.09 | 2.86±0.02 | 8.45±0.38 | urea (1–4) | 99M4 |
| multiple mutant (Pro11→Ala, Pro31→Ala, Pro42→Ala, Pro47→Thr, Pro56→Ala and Pro117→Gly) | | | | | | | |
| | 7.0 | 20 | 26.90±1.05 | 3.08±0.01 | 8.74±0.33 | urea (1–4) | 99M4 |

Remarks:

(1) nonlinear least-squares fit of the transition curve assuming a two-state transition and a linear dependence of $\Delta G$ on the denaturant conc.

(2) transition monitored by CD at 225 nm

(3) buffer: 50 mM sodium cacodylate, 50 mM sodium chloride, 1 mM EGTA, pH 7.0

(4) Ref. 99M4 concerns the effects of proline isomerizations on the equilibrium unfolding and refolding kinetics of SNase, studied by far-UV CD and tryptophan fluorescence

(5) data from Ref. 97I2

Staphylococcal nuclease, biosynthetically incorporated Trp analogues

| Trp-analogue | pH | T | $\Delta G$ | m | Approach/Remarks | | Ref |
|---|---|---|---|---|---|---|---|
| Trp140 | 7.3 | 20 | 20.96±1.30 | 24.81±1.42 | GuHCl | (1,2) | 97W7 |
| 5HW-containing | 7.3 | 20 | 21.25±1.42 | 23.77±1.46 | GuHCl | (1–3) | 97W7 |
| 7AW-containing | 7.3 | 20 | 9.12±0.96 | 18.83±1.30 | GuHCl | (1,2,4,5) | 97W7 |

Remarks:

(1) global two-state fit of CD (at 222 and 228 nm) and fluorescence assuming a linear dependence of $\Delta G$ on denaturant concentration

(2) buffer: 0.02 M sodium phosphate, pH 7.3

(3) 5HW = 5-hydroxytryptophan

(4) 7AW = 7-azatryptophan

(5) assuming a mixture of 7AW and Trp containing nuclease, $\Delta G$ = 11.7 kJ/mol and m = 24.48 kJ/mol/M is obtained, for details see Ref. 97W7

Staphylococcal nuclease, biosynthetically incorporated Trp analogues

| Trp-analogue | pH | T | $\Delta G$ | m | $c_{1/2}$ | Approach/Remarks | | Ref |
|---|---|---|---|---|---|---|---|---|
| variants derived from the wild-type protein: | | | | | | | | |
| Trp-140 | 7.3 | 25 | 20.96 | 24.81 | 0.84 | GuHCl | (1,2,4,5) | 98W13 |
| 5HW-containing | 7.3 | 25 | 21.25 | 23.77 | 0.89 | GuHCl | (1,2,4,5) | 98W13 |
| 7AW-containing | 7.3 | 25 | 9.12 | 18.83 | 0.49 | GuHCl | (1–5) | 98W13 |
| 4FW-containing | 7.3 | 25 | 23.14 | 24.43 | 0.95 | GuHCl | (1,2,4) | 98W13 |
| 5FW-containing | 7.3 | 25 | 21.84 | 24.27 | 0.90 | GuHCl | (1,2,4) | 98W13 |
| 6FW-containing | 7.3 | 25 | 23.01 | 25.02 | 0.92 | GuHCl | (1,2,4) | 98W13 |
| variants derived from nuclease V66W´: | | | | | | | | |
| Trp-66 | 7.3 | 25 | 9.92 | 11.55 | 0.86 | GuHCl | (1,2,4) | 98W13 |
| 5HW-containing | 7.3 | 25 | 4.14 | 4.23 | 0.98 | GuHCl | (1,2,4) | 98W13 |
| 7AW-containing | 7.3 | 25 | 5.98 | 11.55 | 0.52 | GuHCl | (1–4) | 98W13 |
| 4FW-containing | 7.3 | 25 | 10.04 | 10.96 | 0.92 | GuHCl | (1,2,4) | 98W13 |
| 5FW-containing | 7.3 | 25 | 7.66 | 8.95 | 0.86 | GuHCl | (1,2,4) | 98W13 |
| 6FW-containing | 7.3 | 25 | 8.54 | 9.58 | 0.89 | GuHCl | (1,2,4) | 98W13 |

Staphylococcal nuclease, biosynthetically incorporated Trp analogues (continued)

| Trp-analogue | pH | T | ΔG | m | $c_{1/2}$ | Approach/Remarks | | Ref |
|---|---|---|---|---|---|---|---|---|
| variants derived from nuclease V66W: | | | | | | | | |
| Trp-66, Trp-140 | 7.3 | 25 | 5.44 | 16.19 | 0.34 | GuHCl | (1–4) | 98W13 |
| 5HW-containing | 7.3 | 25 | 16.78 | 25.77 | 0.65 | GuHCl | (1–4) | 98W13 |
| 7AW-containing | 7.3 | 25 | 9.33 | 12.13 | 0.77 | GuHCl | (1–4) | 98W13 |
| 4FW-containing | 7.3 | 25 | 11.38 | 21.88 | 0.52 | GuHCl | (1–4) | 98W13 |
| 5FW-containing | 7.3 | 25 | 7.78 | 20.79 | 0.37 | GuHCl | (1–4) | 98W13 |
| 6FW-containing | 7.3 | 25 | 8.37 | 19.04 | 0.44 | GuHCl | (1–4) | 98W13 |

Explanations:

    5HW = 5-hydroxytryptophan
    7AW = 7-azatryptophan
    4FW = 4-fluorotryptophan
    5FW = 5-fluorotryptophan
    6FW = 6-fluorotryptophan
    V66W′ = deletion mutant lacking residues 137–149, Val66 replaced by Trp
    V66W = mutant of nuclease with Val66 replaced by Trp

Remarks:

(1) global two-state fit of CD (at 222 and 228 nm) and fluorescence assuming a linear dependence of ΔG on denaturant concentration
(2) buffer: 0.02 M sodium phosphate, pH 7.3
(3) Ref. 98W13 contains additional data for a three-state fit
(4) Ref. 98W13 contains additional data derived from thermal denaturation
(5) see also Ref. 97W7

Staphylococcal nuclease, wild type and unstable mutant, ΔG at the temperature of maximum stability ($T_{max}$)

| Mutant | pH | $T_{max}$ | ΔG | Approach | Remarks | Ref |
|---|---|---|---|---|---|---|
| wild type | 7.0 | 20 | 15.2 | heat | (1,2) | 97E2 |
| NCA | 7.0 | 9 | 6.3 | heat | (1–3) | 97E2 |
| NCA-s. | 7.0 | 6 | 3.9 | heat | (1,2,4) | 97E2 |

Remarks:

(1) thermally induced transitions were simultaneously monitored by fluoresecence and CD at various wavelengths, for the primary data see Table 2
(2) buffer: 0.01 M Tris-HCl, 0.1 M NaCl, pH 7.0
(3) NCA = hybrid of nuclease having the hexapeptide Ser-Ser-Asn-Gly-Ser-Pro at positions 27–31 (a type I β turn)
(4) NCA-s. = hybrid nuclease with additional substitution Ser28→Gly

Staphylococcal nuclease insertion mutants with randomly located internal tandem duplications of random size

| Mutant | pH | T | $\Delta G$ | $c_{1/2}$ | m | Appr./Rem. | Ref |
|---|---|---|---|---|---|---|---|
| wild type | 7 | 20 | 35.61 | 1.41 | 25.27 | GuHCl (1–4) | 98N2 |
| (31–80) | 7 | 20 | 25.94 | 1.05 | 24.60 | GuHCl (1–4) | 98N2 |
| (38–43) | 7 | 20 | 33.64 | 1.31 | 25.61 | GuHCl (1–4) | 98N2 |
| (42–56) | 7 | 20 | 30.42 | 1.22 | 24.94 | GuHCl (1–4) | 98N2 |
| (43–109) | 7 | 20 | 24.52 | 1.08 | 22.72 | GuHCl (1–4) | 98N2 |
| (45–61) | 7 | 20 | 28.16 | 1.23 | 22.89 | GuHCl (1–4) | 98N2 |
| (46–69) | 7 | 20 | 24.73 | 1.11 | 22.30 | GuHCl (1–4) | 98N2 |
| (51–55) | 7 | 20 | 31.84 | 1.36 | 23.43 | GuHCl (1–4) | 98N2 |
| (51–55)' | 7 | 20 | 33.81 | 1.33 | 25.40 | GuHCl (1–4) | 98N2 |
| (55–62) | 7 | 20 | 25.56 | 1.12 | 22.76 | GuHCl (1–4) | 98N2 |
| (55–76) | 7 | 20 | 28.16 | 1.16 | 24.35 | GuHCl (1–4) | 98N2 |
| (62–62) | 7 | 20 | 19.87 | 0.75 | 26.57 | GuHCl (1–4) | 98N2 |

Remarks:
(1) linear extrapolation
(2) the mutants were derived from wild-type staphylococcal nuclease PHS (Pro117→Gly, His124→Leu, and Ser128→Ala)
(3) buffer: 25 mM sodium phosphate, 100 mM NaCl, pH 7
(4) transition monitored by fluorescence emission spectra

*d) stabilization by xylose*

Staphylococal nuclease, stabilization by xylose

| Xylose Conc. | pH | T | $\Delta G$ | Approach | Remarks | Ref |
|---|---|---|---|---|---|---|
| 0.0015 | 4.5 | 21 | 9.2±0.8 | pressure, eq. | (1) | 97F5 |
| 0.0015 | 4.5 | 21 | 10.0±1.7 | pressure, kin. | (2) | 97F5 |
| 0.003 | 4.5 | 21 | 10.5±0.8 | pressure, eq. | (1) | 97F5 |
| 0.003 | 4.5 | 21 | 10.0±2.1 | pressure, kin. | (2) | 97F5 |
| 0.006 | 4.5 | 21 | 13.4±1.3 | pressure, eq. | (1) | 97F5 |
| 0.006 | 4.5 | 21 | 12.6±2.5 | pressure, kin. | (2) | 97F5 |
| 0.009 | 4.5 | 21 | 13.8±1.7 | pressure, eq. | (1) | 97F5 |
| 0.009 | 4.5 | 21 | 15.1 | pressure, kin. | (2,3) | 97F5 |

Remarks:
(1) pressure denaturation, equilibrium unfolding
(2) pressure denaturation, kinetics, pressure jump relaxation
(3) error 15.1+1.8/–0.8

## Oncomodulin

Recombinant rat oncomodulin and mutants, apo- and $Ca^{2+}$-bound proteins

| Protein | pH | T | $\Delta G$ | $c_{1/2}$ | m | Appr./Rem. | Ref |
|---|---|---|---|---|---|---|---|
| $Ca^{2+}$-free form: | | | | | | | |
| Tyr57→Trp | 6.5 | 20 | 16.3±1.3 | 1.2±0.1 | 13.4±0.8 | GuHCl (1–3) | 98Z6 |
| Tyr57→Trp | 6.5 | 20 | 10.0±1.3 | | | heat (2,3,7) | 98Z6 |
| Tyr65→Trp | 6.5 | 20 | 2.5±0.4 | 0.3±0.1 | 7.9±1.7 | GuHCl (1–3) | 98Z6 |
| Tyr65→Trp | 6.5 | 20 | 5.9±0.8 | | | heat (2,3,7) | 98Z6 |
| Phe102→Trp | 6.5 | 20 | 7.9±0.4 | 0.8±0.1 | 9.2±0.8 | GuHCl (1–3) | 98Z6 |
| Phe102→Trp | 6.5 | 20 | 19.2±2.1 | | | heat (2,3,7) | 98Z6 |
| CDOM33 | 6.5 | 20 | 12.6±1.3 | 1.2±0.1 | 11.3±1.3 | GuHCl (1–4) | 98Z6 |
| CDOM33 | 6.5 | 20 | 12.6±1.3 | | | heat (2–4,7) | 98Z6 |

Recombinant rat oncomodulin and mutants, apo- and Ca²⁺-bound proteins (continued)

| Protein | pH | T | ΔG | $c_{1/2}$ | m | Appr./Rem. | Ref |
|---|---|---|---|---|---|---|---|
| Ca²⁺-loaded form: | | | | | | | |
| Tyr57→Trp | 6.5 | 20 | 28.0±2.1 | 2.0±0.1 | 13.8±1.3 | GuHCl (1,2,5) | 98Z6 |
| Tyr57→Trp | 6.5 | 20 | 12.6±1.7 | | | heat  (2,3,8) | 98Z6 |
| Tyr65→Trp | 6.5 | 20 | 27.6±2.1 | 1.9±0.1 | 14.2±1.3 | GuHCl (1–3) | 98Z6 |
| Tyr65→Trp | 6.5 | 20 | 8.4±1.3 | | | heat  (2,3,8) | 98Z6 |
| Phe102→Trp | 6.5 | 20 | 21.8±2.1 | 1.8±0.1 | 12.1±1.3 | GuHCl (1–3) | 98Z6 |
| Phe102→Trp | 6.5 | 20 | 52.3±5.0 | | | heat  (2,3,8) | 98Z6 |
| CDOM33 | 6.5 | 20 | 28.9±1.7 | 2.2±0.1 | 13.0±1.3 | GuHCl (1–4) | 98Z6 |
| CDOM33 | 6.5 | 20 | 46.9±5.0 | | | heat  (2–4,8) | 98Z6 |
| Tb³⁺-loaded form: | | | | | | | |
| Tyr57→Trp | 6.5 | 20 | 32.6±2.1 | 2.8±0.1 | 11.7±0.8 | GuHCl (1,2,6) | 98Z6 |
| Tyr57→Trp | 6.5 | 20 | 41.4±4.2 | | | heat  (2,3,8) | 98Z6 |
| Tyr65→Trp | 6.5 | 20 | 28.0±2.9 | 2.9±0.1 | 9.6±1.3 | GuHCl (1–3) | 98Z6 |
| Tyr57→Trp | 6.5 | 20 | 28.5±2.9 | | | heat  (2,3,8) | 98Z6 |
| Phe102→Trp | 6.5 | 20 | 33.5±3.3 | 2.6±0.2 | 13.0±1.3 | GuHCl (1–3) | 98Z6 |
| Phe102→Trp | 6.5 | 20 | 48.5±4.6 | | | heat  (2,3,8) | 98Z6 |
| Phe102→Trp | 6.5 | 20 | 46.4±6.3 | 2.6±0.3 | 17.6±2.1 | GuHCl (1–3) | 98Z6 |
| CDOM33 | 6.5 | 20 | 35.6±3.8 | 4.0±0.1 | 8.8±0.8 | GuHCl (1–4) | 98Z6 |
| CDOM33 | 6.5 | 20 | 62.3±7.1 | | | heat  (2–4,8) | 98Z6 |

Remarks:

(1) linear extrapolation, LEM-SB

(2) buffer: 10 mM PIPES, 100 mM KCl, pH 6.5, and EGTA for apo- proteins

(3) transition monitored by fluorescence intensity changes

(4) CDOM33 represents a CD loop prepared by insertion of a 12 amino acid sequence which has a significantly higher affinity for Ca²⁺

(5) transition monitored by fluorescence emission maximum

(6) transition monitored by Tb³⁺ luminescence

(7) ΔG was calculated using ΔCp = 5.56 kJ/mol/K from Ref. 78F reported for the apo- form of carp parvalbumin (see also Ref. 96H8)

(8) ΔG was calculated using ΔCp = 4.60 kJ/mol/K from Ref. 78F reported for the Ca²⁺-bound form of carp parvalbumin (see also Ref. 96H8)

Recombinant rat oncomodulin and mutants (rOM), apo- and Ca²⁺-bound proteins

| Protein | pH | T | ΔG | Approach/Remarks | | Ref |
|---|---|---|---|---|---|---|
| Apo- proteins: | | | | | | |
| rOM | 7.4 | 25 | 16.3±1.7 | DSC | (1,2) | 96H8 |
| Ser55→Asp | 7.4 | 25 | 10.9±1.3 | DSC | (1,2) | 96H8 |
| Gly98→Asp | 7.4 | 25 | 13.4±1.3 | DSC | (1,2) | 96H8 |
| double mutant (Ser55→Asp and Gly98→Asp) | | | | | | |
|  | 7.4 | 25 | 10.0±1.3 | DSC | (1,2) | 96H8 |
| Ca²⁺-bound proteins: | | | | | | |
| rOM | 7.4 | 25 | 52.7±4.2 | DSC | (3,4) | 96H8 |
| Ser55→Asp | 7.4 | 25 | 54.8±4.2 | DSC | (3,4) | 96H8 |
| Gly98→Asp | 7.4 | 25 | 62.8±4.2 | DSC | (3,4) | 96H8 |
| double mutant (Ser55→Asp and Gly98→Asp) | | | | | | |
|  | 7.4 | 25 | 66.5±4.2 | DSC | (3,4) | 96H8 |

Remarks:

(1) buffer: 25 mM HEPES, 0.15 M NaCl, 5 mM EDTA

(2) calculated using the data from Table 2, and ΔCp = 5.56 kJ/mol/K from Ref. 78F reported for the apo- form of carp parvalbumin

(3) buffer: 25 mM HEPES, 0.15 M NaCl, 5 mM CaCl₂

(4) calculated using the data from Table 2, and ΔCp = 4.60 kJ/mol/K from Ref. 78F reported for the Ca²⁺-bound form of carp parvalbumin

Recombinant rat oncomodulin and mutants (rOM), $Mg^{2+}$-bound proteins

| Mutant | pH | $T_{trs}$ | $\Delta T$ | $\Delta(\Delta G)$ | Approach/Remarks | | Ref |
|---|---|---|---|---|---|---|---|
| $Mg^{2+}$-bound proteins | | | | | | | |
| rOM | 7.4 | 68.5 | 0.0 | 0.0 | DSC | (1,2) | 96H8 |
| Ser55→Asp | 7.4 | 79.0 | 10.5 | 11.3 | DSC | (1,2) | 96H8 |
| Gly98→Asp | 7.4 | 69.0 | 0.5 | 0.4 | DSC | (1,2) | 96H8 |
| double mutant (Ser55→Asp and Gly98→Asp) | | | | | | | |
| | 7.4 | 79.4 | 10.9 | 11.7 | DSC | (1,2) | 96H8 |

Remarks:
(1) buffer: 25 mM HEPES, 0.15 M NaCl, 20 mM $MgCl_2$, 1.0 mM EGTA
(2) calculated using $\Delta(\Delta G) = \Delta T \times \Delta H_{rOM} / T_{trs,rOM}$

## Oncoprotein

BTB/POZ domain from the PLZF oncoprotein

| Protein | pH | T | $\Delta G$ | Approach/Remarks | | Ref |
|---|---|---|---|---|---|---|
| BTB/POZ domain | 8.5 | 25 | 53.6±1.7 | GuHCl | (1–3) | 97L8 |

Remark:
(1) linear extrapolation
(2) $\Delta G$ is an average from two measurements at 21 μM and 125 μM protein
(3) measured in 20 mM boric acid, pH 8.5, 100 mM NaCl

$p21^{H-ras}$ protein, stability as function of ligand concentration

| [GDP] | [$Mg^{2+}$] | [P] | pH | T | $\Delta G$ | $c_{1/2}$ | m | Appr./Rem. | | Ref |
|---|---|---|---|---|---|---|---|---|---|---|
| ternary complex: | | | | | | | | | | |
| 17.5 | 5.00 | 12.5 | 7.5 | 25 | 57.3±2.9 | 3.0 | 5.69±0.84 | urea | (1–6) | 98Z4 |
| 17.5 | 5.00 | 16.5 | 7.5 | 25 | 58.6±1.3 | 3.0 | 5.61±0.42 | urea | (1–6) | 98Z4 |
| 23.6 | 5.00 | 12.5 | 7.5 | 25 | 56.5±2.1 | 3.2 | 5.44±0.84 | urea | (1–6) | 98Z4 |
| 37.4 | 5.00 | 12.5 | 7.5 | 25 | 58.6±0.8 | 3.5 | 5.77±0.42 | urea | (1–6) | 98Z4 |
| 112.5 | 5.00 | 12.5 | 7.5 | 25 | 58.2±2.1 | 4.0 | 5.73±0.42 | urea | (1–6) | 98Z4 |
| 27.0 | 1.02 | 17.0 | 7.5 | 25 | 58.6±1.7 | 2.8 | 5.61±0.25 | urea | (1–6) | 98Z4 |
| 27.0 | 5.02 | 17.0 | 7.5 | 25 | 58.6±0.8 | 3.4 | 5.82±0.21 | urea | (1–6) | 98Z4 |
| 23.65 | 5.00 | 12.5 | 7.5 | 25 | 59.0±2.1 | 3.2 | 5.56±0.84 | urea | (1–6) | 98Z4 |
| 27.06 | 1.02 | 17.0 | 7.5 | 25 | 61.1±1.7 | 2.7 | 5.90±0.42 | urea | (1–6) | 98Z4 |
| 27.0 | 5.02 | 17.0 | 7.5 | 25 | 60.7±2.1 | 3.3 | 5.82±0.42 | urea | (1–6) | 98Z4 |
| 37.4 | 5.00 | 12.5 | 7.5 | 25 | 59.4±2.1 | 3.4 | 5.56±0.84 | urea | (1–6) | 98Z4 |
| | | | | | 59.0±0.8 | | 5.73±0.21 | urea | (1–5,7) | 98Z4 |
| binary complex: | | | | | | | | | | |
| 22.0 | 0.001 | 13.5 | 7.5 | 25 | 31.8±1.3 | 1.4 | 4.60±0.84 | urea | (1–6) | 98Z4 |
| 26.0 | 0.001 | 13.5 | 7.5 | 25 | 31.8±1.3 | 1.4 | 4.18±0.84 | urea | (1–6) | 98Z4 |
| 100.0 | 0.001 | 13.5 | 7.5 | 25 | 31.8±0.8 | 2.2 | 4.14±1,26 | urea | (1–6) | 98Z4 |
| 27.0 | 0.001 | 17.0 | 7.5 | 25 | 32.2±4.2 | 1.4 | 4.10±0.42 | urea | (1–6) | 98Z4 |
| 26.0 | 0.001 | 13.5 | 7.5 | 25 | 31.4±3.3 | 1.4 | 4.60±0.84 | urea | (1–6) | 98Z4 |
| 27.0 | 0.001 | 17.0 | 7.5 | 25 | 31.0±3.8 | 1.4 | 3.97±1.26 | urea | (1–6) | 98Z4 |
| | | | | | 31.4±1.7 | | 4.18±0.42 | urea | (1–5,8) | 98Z4 |

p21$^{H\text{-ras}}$ protein, stability as function of ligand concentration (continued)

| [GDP] | [Mg$^{2+}$] | [P] | pH | T | $\Delta G$ | $c_{1/2}$ | m | Appr./Rem. | Ref |
|---|---|---|---|---|---|---|---|---|---|
| apo-protein: | | | | | | | | | |
| | | | 7.5 | 25 | 7.5±0.8 | | 3.26±0.08 | urea (2,3,9) | 98Z4 |
| burst-phase intermediate: | | | | | | | | | |
| | | | 7.5 | 25 | 6.7±0.8 | | 3.05±0.08 | urea (2,3,9) | 98Z5 |

Remarks:
(1) data fit by a two-state model: N → U + Mg$^{2+}$ + GDP, assuming a linear dependence of $\Delta G$ on the denaturant concentration (similar to LEM-SB), for the detailed equations see Ref. 98Z4
(2) transition monitored by CD at 222 nm and UV absorbance at 287 nm
(3) buffer: 20 mM Tris-HCl, pH 7.5, 67–100 mM NaCl, 2 mM 2-mercaptoethanol; MgCl$_2$ and GDP concentrations were adjusted as necessary
(4) concentrations: [GDP] in μM, [Mg$^{2+}$] in mM, [P] in μM
(5) error in $c_{1/2}$ ±0.06 M
(6) individual fit of the transition
(7) global fit including 11 data sets
(8) global fit including 6 data sets
(9) data fit by a two-state model (N → U)


## Organophosphorus Hydrolase

Organophosphorus hydrolase (OPH), recombinant, three-state unfolding

| Transition | pH | T | $\Delta G$ | $c_{1/2}$ | m | Appr./Rem. | Ref |
|---|---|---|---|---|---|---|---|
| trans. (1) | 8.3 | 25 | 18.1±1.1 | | 4.10±0.33 | urea (1–4) | 97G12 |
| trans. (1) | 8.3 | 25 | 18.1±0.1 | | 3.97±0.08 | urea (1–3,5) | 97G12 |
| trans. (1) | 8.3 | 25 | 18.0±2.1 | 3.77±.13 | 4.77±0.54 | urea (1–3,6) | 97G12 |
| trans. (2) | 8.3 | 25 | 151±8 | | 18.0±1.5 | urea (1–4) | 97G12 |
| trans. (2) | 8.3 | 25 | 158±9 | | 19.0±0.2 | urea (1–3,5) | 97G12 |

Remarks:
(1) linear extrapolation based on a three-state mechanism N$_2$ → I$_2$ → 2U in which a homodimeric intermediate is populated
(2) buffer: 10 mM Tris, pH 8.3
(3) T$_{trs}$ at thermal denatation of OPH in the absence of urea occurs around 75°C
(4) average of eight measurements monitored by fluorescence at 320 nm and CD at 230 nm, protein conc. from 13 to 125 μg/ml
(5) from global analysis of four CD monitored (230 nm) denaturation curves, protein conc. from 13 to 125 μg/ml
(6) from enzymatic activity versus urea concentration, data treatment by a two-state transition


## Ovomucoid

Ovomucoid third domain (OMTKY3) from turkey egg white, H/D exchange of amide protons

| Residue | pH | T | $\Delta G_{op}$ | Approach/Remarks | | Ref |
|---|---|---|---|---|---|---|
| Leu23 | 6–10 | 30 | 28.0 | HX | (1–3) | 97A7 |
| Cys24 | 6–10 | 30 | 31.4 | HX | (1,2,4) | 97A7 |
| Gly25 | 6–10 | 30 | 32.2 | HX | (1–3) | 97A7 |
| Asn28 | 6–10 | 30 | 29.3 | HX | (1,2,4) | 97A7 |
| Lys29 | 6–10 | 30 | 31.0 | HX | (1–3) | 97A7 |
| Tyr31 | 6–10 | 30 | 30.5 | HX | (1–3) | 97A7 |
| Asn33 | 6–10 | 30 | 28.9 | HX | (1,2,4) | 97A7 |
| Phe37 | 6–10 | 30 | 29.7 | HX | (1–3) | 97A7 |

Ovomucoid third domain (OMTKY3) from turkey egg white, H/D exchange of amide protons (continued)

| Residue | pH | T | $\Delta G_{op}$ | Approach/Remarks | | Ref |
|---|---|---|---|---|---|---|
| Cys38 | 6–10 | 30 | 35.1 | HX | (1–3) | 97A7 |
| Asn39 | 6–10 | 30 | 34.7 | HX | (1–3) | 97A7 |
| Ala40 | 6–10 | 30 | 35.1 | HX | (1–3) | 97A7 |
| Val41 | 6–10 | 30 | 30.5 | HX | (1–3) | 97A7 |
| Ser51 | 6–10 | 30 | 28.0 | HX | (1–3) | 97A7 |

Remarks:

(1) the stability of the ovomucoid third domain shows little change from pH 6–10

(2) the midpoint of the thermal transition in $H_2O$ at pH 9 and 10 is 80°C according to Ref. 97F4

(3) the estimated error for $\Delta G_{op}$ is ±2.1 kJ/mol

(4) modified approach, for details see Ref. 97A7

Ovomucoid third domain (OMTKY3) from turkey egg white, H/D exchange for the
14 slowest-exchanging hydrogen atoms

| Residue | pH | T | $\Delta G_{HX}$ | Approach/Remarks | | Ref |
|---|---|---|---|---|---|---|
| Leu23 | 3.5 | 30 | 27.6 | H/D | (1,2) | 99A3 |
| Cys24 | 3.5 | 30 | 31.0 | H/D | (1,2) | 99A3 |
| Gly25 | 3.5 | 30 | 31.4 | H/D | (1,2) | 99A3 |
| Ser26 | 3.5 | 30 | 29.7 | H/D | (1,2) | 99A3 |
| Asn28 | 3.5 | 30 | 28.5 | H/D | (1,2) | 99A3 |
| Lys29 | 3.5 | 30 | 30.1 | H/D | (1,2) | 99A3 |
| Tyr31 | 3.5 | 30 | 29.7 | H/D | (1,2) | 99A3 |
| Asn33 | 3.5 | 30 | 28.9 | H/D | (1,2) | 99A3 |
| Phe37 | 3.5 | 30 | 29.3 | H/D | (1,2) | 99A3 |
| Cys38 | 3.5 | 30 | 34.3 | H/D | (1,2) | 99A3 |
| Asn39 | 3.5 | 30 | 34.3 | H/D | (1,2) | 99A3 |
| Ala40 | 3.5 | 30 | 33.9 | H/D | (1,2) | 99A3 |
| Val41 | 3.5 | 30 | 29.7 | H/D | (1,2) | 99A3 |
| Ser51 | 3.5 | 30 | 28.0 | H/D | (1,2) | 99A3 |

Remarks:

(1) results from NMR and electrospray ionization mass spectrometry (ESI-MS)

(2) estimated error for $\Delta G_{HX}$ ±2.1 kJ/mol

## Oxytocin Precursor

Bovine oxytocin precursor and bovine oxytocin-associated neurophysin (NP–1)

| Protein | pH | T | $\Delta G$ | m | Appr./Rem. | Ref |
|---|---|---|---|---|---|---|
| precursor | 6.0 | 25 | 28.5±1.7 | 6.7±0.4 | GuHCl (1–3) | 99E3 |
| | 8.0 | 25 | 20.5±0.4 | 6.3±0.0 | GuHCl (1–3) | 99E3 |
| | 8.0 | 25 | 18.4±1.3 | 2.9±0.13 | urea (1–3) | 99E3 |
| | 10.0 | 25 | 11.3±0.4 | 3.8±0.4 | GuHCl (1–3) | 99E3 |
| NP–1 | 6.0 | 25 | 13.0±0.4 | 4.6±0.4 | GuHCl (1–3) | 99E3 |
| | 8.0 | 25 | 12.6±1.5 | 4.6±0.4 | GuHCl (1–3) | 99E3 |
| | 8.0 | 25 | 10.0±0.0 | 1.7±0.04 | urea (1–3) | 99E3 |
| | 10.0 | 25 | 10.9±0.4 | 4.2±0.4 | GuHCl (1–3) | 99E3 |

Remarks:

(1) linear extrapolation

(2) transition monitored by CD at 250 nm

(3) measured in 0.1 M ammonium acetate

## Parvalbumin

Rat α- and β-parvalbumin (α-PV and β-PV), wild type and mutants

| Protein | pH | T | $\Delta G$ | Approach/Remarks | | Ref |
|---|---|---|---|---|---|---|
| α-PV | 7.4 | 25 | 14.6 | DSC | (1,2) | 99H8 |
| β-PV | 7.4 | 25 | 19.7 | DSC | (1,2) | 99H8 |

Remarks:
(1) measured in the presence of 0.20 M NaCl, 10 mM EDTA
(2) $\Delta G$ was calculated using $\Delta Cp = 5.56$ kJ/mol/K from Ref. 78F

## Peripheral Subunit-Binding Protein

Peripheral subunit-binding domain (psbd) from the dihydrolipoamide acetyltransferase, component of the *Bacillus stearothermophilus* pyruvate dehydrogenase multienzyme complex, truncation mutant of a protein minidomain

| Protein | pH | T | $\Delta G$ | $m_{urea}$ | $m_{GuHCl}$ | Appr./Rem. | Ref |
|---|---|---|---|---|---|---|---|
| psbd41 | 8.0 | 25 | 9.16 | 1.42±0.46 | 2.80±0.33 | (1–6) | 99S13 |
| psbd36 | 8.0 | 25 | 7.57 | 1.34±0.29 | 2.68±0.25 | (1–5) | 99S13 |
| psbd33 | 8.0 | 25 | 4.31 | 1.72±0.42 | 2.09±0.67 | (1–5) | 99S13 |

Remarks:
(1) $\Delta G$ was obtained from thermal denaturation monitored by CD at 222 nm and 280 nm using the Gibbs-Helmholtz equation, see also Table 2
(2) the $m_{urea}$-values were obtained from the global analysis of thermal denaturation at various urea concentrations
(3) the $m_{GuHCl}$-values were obtained from isothermal GuHCl denaturations
(4) psbd41 corresponds to residues 3–43, psbd36 to residues 6–41, and psbd33 to residues 7–39
(5) buffer: 2 mM phosphate, 2 mM borate, 2 mM citrate, 50 mM NaCl, pH 8.0
(6) see also Ref. 99S14

## Peroxidase

Horseradish peroxidase isoenzyme C, ferric form

| | pH | T | $\Delta G$ | $c_{1/2}$ | m | Approach/Remarks | | Ref |
|---|---|---|---|---|---|---|---|---|
| | 7.0 | 25 | 16.7±2.5 | 2.0 | 8.4 | GuHCl | (1,2) | 98T11 |

Remarks:
(1) linear extrapolation
(2) transition monitored by CD at 222 nm

**Phage 434 Cro Protein**, see CRO Protein

**434-Phage Repressor**, see Repressor Proteins

## Phenylalanine Hydroxylase

Recombinant human phenylalanine hydroxylase (PAH) in the presence and absence of phenylalanine

| Protein/Transition | pH | T | $\Delta G$ | $c_{1/2}$ | m | Appr./Rem. | Ref |
|---|---|---|---|---|---|---|---|
| PAH, no Phe | | | | | | | |
| trans. (1) | 7.5 | 25 | 21±2 | 1.40±0.04 | 15±1 | urea (1–3) | 99K8 |
| trans. (2) | 7.5 | 25 | 57±5 | 5.70±0.08 | 10±1 | urea (1–3) | 99K8 |
| PAH, 0.5 mM L-Phe | | | | | | | |
| trans. (1) | 7.5 | 25 | 17±1 | 2.70±0.04 | 6.3±0.4 | urea (1–3) | 99K8 |
| trans. (2) | 7.5 | 25 | 57±3 | 5.84±0.08 | 9.7±0.6 | urea (1–3) | 99K8 |

Remarks:
(1) data were analyzed by a three-state model assuming a linear dependence of $\Delta G$ on the denaturant conc.
(2) transition monitored by fluorescence intensity
(3) buffer: 0.1 M Na-HEPES, 0.2 M NaCl, 1.25 mM EDTA, pH 7.5

**Phosphatase**, see Alkaline Phosphatase

## Phosphatidylethanolamine-Binding Protein

Bovine brain phosphatidylethanolamine-binding protein (PEBP)

| Protein | pH | T | $\Delta G$ | $c_{1/2}$ | m | Appr./Rem. | Ref |
|---|---|---|---|---|---|---|---|
| PEBP | 7.4 | 25 | 18.8 | 4.8 | 3.97 | urea (1–3) | 99V1 |
| | 4.0 | 25 | 20.9 | | | DSC (4) | 99V1 |

Remarks:
(1) linear extrapolation
(2) transition monitored by fluorescence at 330 nm
(3) buffer: 10 mM Tris-HCl, 2 mM 2-mercaptoethanol, pH 7.4
(4) buffer: 10 mM sodium acetate, 2 mM 2-mercaptoethanol, pH 4.0

**Phosphatidylinositol 3'-Kinase**, see SH3 Domain

## Phosphocarrier Protein

Histidine-containing phosphocarrier protein HPr from *E. coli*

| | pH | T | $\Delta G$ | $c_{1/2}$ | m | | Appr./Rem. | Ref |
|---|---|---|---|---|---|---|---|---|
| | 7.0 | 20 | 19.1±0.5 | 2.09±0.09 | 9.1±0.3 | | GuHCl (1–3) | 98V3 |
| | 7.0 | 20 | 19.8±1.7 | 2.19±0.27 | 9.0±0.8 | | GuHCl (1,2,4) | 98V3 |
| | 7.0 | 20 | 20.0±0.8 | 2.14±0.13 | 9.3±0.4 | | GuHCl (1,2,5) | 98V3 |
| | 7.0 | 20 | 23.3±0.8 | 2.01±0.10 | 11.6±0.4 | | GuHCl (1,2,6) | 98V3 |
| | 7.0 | 20 | 21.6±0.4 | 2.01±0.06 | 10.8±0.24 | | GuHCl (1,2,7) | 98V3 |

Remarks:

(1) linear extrapolation, LEM-SB

(2) measured in 100 mM sodium phosphate, pH 7.0

(3) transition monitored by CD at 222 nm, and difference in ellipticity at 261 and 258 nm

(4) transition monitored by NMR, mean of 56 native state amide $^1$H-$^{15}$N cross peaks

(5) transition monitored by NMR, mean of 13 denatured state amide $^1$H-$^{15}$N cross peaks

(6) transition monitored by ANS fluorescence at 472 nm, equilibrium data

(7) from kinetics, transition monitored by ANS fluorescence

Histidine-containing phosphocarrier protein HPr, wild type and mutant Lys49→Glu

| Mutant | pH | T | $\Delta(\Delta G)$ | $c_{1/2}$ | m | Appr./Rem. | Ref |
|---|---|---|---|---|---|---|---|
| wild type | 7.0 | 30 | 0.0 | 4.34±0.05 | 4.44±0.21 | urea (1–3) | 99P7 |
| Lys49→Glu | 7.0 | 30 | 8.8±0.8 | 6.24±0.06 | 3.97±0.21 | urea (1–3) | 99P7 |
| | 7.0 | 30 | 8.0±0.8 | | | DSC (2–4) | 99P7 |

Remarks:

(1) linear extrapolation, for the procedure see also Ref. 96N2

(2) transition monitored by far-UV CD at 222 nm

(3) measured in 10 mM potassium phosphate

(4) calculated using $\Delta C_p$ of the wild-type protein (5.94 kJ/mol/K) from Ref. 96N2

## Phosphoglycerate Kinase

Phosphoglycerate kinase from yeast, wild type and deletion mutant

| Mutant | pH | T | $\Delta G$ | $c_{1/2}$ | m | Remarks | Ref |
|---|---|---|---|---|---|---|---|
| wild type | 7.5 | 25 | 45.6±6.3 | 0.78±0.01 | 58.2±7.9 | (1,2,4,5) | 97S9 |
| wild type | 7.5 | 25 | 38.9±3.3 | 0.77±0.01 | 50.6±4.6 | (1,3–5) | 97S9 |
| 198–400 del. | 7.5 | 25 | 13.8±2.1 | 0.78±0.03 | 17.6±2.5 | (1,2,4,6) | 97S9 |
| 198–400 del. | 7.5 | 25 | 13.4±2.9 | 0.80±0.05 | 16.7±2.9 | (1,3,4,6) | 97S9 |

Remarks:

(1) GuHCl denaturation, linear extrapolation, LEM-SB

(2) transition monitored by fluorescence, for details see Ref. 95S4

(3) transition monitored by CD at 220 nm

(4) buffer: 20 mM sodium phosphate, pH 7.5

(5) reported by Mas et al., Ref. 95M7

(6) 198–400 del. = deletion mutant in which the C-terminal domain is deleted

Phosphoglycerate kinase (PGK) from yeast, wild type and mutant (His388→Gln) in context with previous data considering the cooperativity of the transition (m value)

| Mutant | pH | T | $\Delta G$ | $c_{1/2}$ | m | Approach/Remarks | | Ref |
|---|---|---|---|---|---|---|---|---|
| wild type | 6.5 | 25 | 16.0 | 0.66 | 24.4 | GuHCl | (1,2–4) | 91J3 |
| wild type | 6.5 | 25 | 16.0 | 0.59 | 27.3 | GuHCl | (1,2,5) | 91J3 |
| His388→Gln | 6.5 | 25 | 12.5 | 0.49 | 25.4 | GuHCl | (1–4) | 91J3 |
| His388→Gln | 6.5 | 25 | 11.8 | 0.40 | 29.7 | GuHCl | (1,2,5) | 91J3 |
| wild type | 6.5 | 20 | 17.9 | 0.729 | 24.4 | GuHCl | (1,6,7,9) | 85A2 |
| wild type | 6.5 | 20 | 27.5 | 0.865 | 31.6 | GuHCl | (1,6,8,9) | 85A2 |
| wild type | 7.5 | 25 | 15.2±0.8 | 0.62 | 24.5 | GuHCl | (1,10,11,13) | 77N |
| *Thermus th.* | 7.5 | 25 | 26.4±0.7 | 2.34 | 11.3 | GuHCl | (1,10,12,13) | 77N |

Remarks:

(1)  linear extrapolation

(2)  measured in 20 mM sodium phosphate, 0.25 mM EDTA

(3)  transition monitored by CD at 225 nm

(4)  DSC of the wild type shows a single transition at 54°C, and mutant His388→Gl shows two transitions at 42.5°C and 51.5°C (buffer: 50 mM PIPES, 0.1 mM DTT, 7.5 mM sodium azide)

(5)  transition monitored by activity

(6)  measured in 10 mM MES, 100 mM NaCl

(7)  transition monitored by CD at 222 nm

(8)  transition monitored by fluorescence at 345 nm

(9)  Ref. 85A2 contains additional data that refer to the denaturant binding model and to the transfer model

(10) measured in 20 mM Tris/HCl, 10 mM EDTA

(11) transition monitored by CD at 220 nm and fluorescence at 340 nm

(12) transition of PGK from *Thermus thermophilus* monitored by CD at 225 nm and fluorescence at 328 nm

(13) $\Delta G$ = 22.3±0.5 and 49.7±0.9 kJ/mol for yeast and *Thermus th.* PGK, respectively, obtained by the denaturant binding model (k = 1.20) are regarded as more reliable in Ref. 77N

Recombinant phosphoglycerate kinase from the hyperthermophilic bacterium *Thermotoga maritima*, temperature dependence of $\Delta G$

| | pH | T | $\Delta G$ | $c_{1/2}$ | m | Approach/Remarks | | Ref |
|---|---|---|---|---|---|---|---|---|
| | 7.0 | 10 | 100±8 | 2.1±0.05 | 47±4 | GuHCl | (1–5) | 98G13 |
| | 7.0 | 20 | 121±9 | 2.5±0.05 | 48±4 | GuHCl | (1–5) | 98G13 |
| | 7.0 | 40 | 124±9 | 2.7±0.05 | 46±4 | GuHCl | (1–5) | 98G13 |
| | 7.0 | 50 | 94±7 | 2.6±0.05 | 37±3 | GuHCl | (1–5) | 98G13 |
| | 7.0 | 60 | 76±6 | 2.4±0.05 | 32±3 | GuHCl | (1–5) | 98G13 |
| | 7.0 | 70 | 47±4 | 1.9±0.05 | 25±2 | GuHCl | (1–5) | 98G13 |

Remarks:

(1) linear extrapolation, LEM-SB

(2) transition monitored by far-UV CD at 222 nm

(3) buffer: 50 mM sodium phosphate (pH 7.0), 1 mM EDTA

(4) at zero GuHCl concentration and pH 7.0, optimum stability is observed at approximately 30°C

(5) the extrapolated $T_{trs}$ of cold and heat denaturation are about −10° and 85°C

N-terminal domain of phophoglycerate kinase from *Bacillus stearothermophilus*, intermediates

| Transition | pH | T | $\Delta G$ | Approach/Remarks | | Ref |
|---|---|---|---|---|---|---|
| $N \rightarrow U$ | 7.5 | 25 | 36.0 | GuHCl | (1–4) | 98P5 |
| $N \rightarrow I$ | 7.5 | 25 | 16.7 | GuHCl | (1–4) | 98P5 |
| $I \rightarrow U$ | 7.5 | 25 | 19.3 | GuHCl | (1–4) | 98P5 |

Remarks:
(1) from folding and unfolding rates measured by stopped-flow kinetics
(2) transition monitored by fluorescence intensity
(3) buffer: 50 mM triethanolamine, 2 mM DTT, and GuHCl
(4) Ref. 98P5 contains an extended temperature profile of $\Delta G$

N-terminal domain of phosphoglycerate kinase from *Bacillus stearothermophilus*

| Transition | pH | T | $\Delta G$ | m* | Approach/Remarks | | Ref |
|---|---|---|---|---|---|---|---|
| data based on GuHCl activity: | | | | | | | |
| $N \rightarrow I$ | 7.5 | 25 | 15.9 | 4.7 | GuHCl | (1–3) | 95P11 |
| $I \rightarrow U$ | 7.5 | 25 | 21.3 | 16.6 | GuHCl | (1–3) | 95P11 |
| $I \rightarrow U$ | 7.5 | 25 | | 15.8 | GuHCl | (1–4) | 95P11 |
| data based on GuHCl concentration: | | | | | | | |
| $N \rightarrow I$ | 7.5 | 25 | 14.2 | 3.7 | GuHCl | (1,3) | 95P11 |
| $I \rightarrow U$ | 7.5 | 25 | 21.8 | 12.8 | GuHCl | (1,3) | 95P11 |
| $I \rightarrow U$ | 7.5 | 25 | | 13.4 | GuHCl | (1,3,4) | 95P11 |

Remarks:
(1a)  from an integrated analysis of equilibrium and kinetic folding reaction, for details of the approach see Ref. 95P11
(1b)  buffer: 50 mM triethanolamine-HCl, 2 mM DTT, pH 7.5, transition monitored by fluorescence
(2)  a specific $C_{1/2}$ value to calculate the molar denaturant activity was set to 7.5
(3)  m* values used here describe changes in both equilibrium and rate constants as function of molar denaturant activity relative to the native state (dimension $M^{-1}$)
(4)  m* refers to kinetics

## Plant Seed Proteins

11 S globulins from three seeds

| Species | pH | T | $\Delta G$ | Approach | Remarks | Ref |
|---|---|---|---|---|---|---|
| soya bean | 8.0 | 20 | 94±10 | DSC | (1,2) | 89G6 |
| faba bean | 8.0 | 20 | 60±14 | DSC | (1,3) | 89G6 |
| sunflower | 8.0 | 20 | 76±28 | DSC | (1,4) | 89G6 |

Remarks:
(1) measured in at pH 8.0 in the presence of 0.3 M NaCl
(2) soya bean = *Glycine max* L
(3) faba bean = *Vicia faba* L
(4) sunflower = *Helianthus annuus* L

## Plastocyanin

Plastocyanin from spinach chloroplasts

| | pH | T | $\Delta G$ | Approach/Remarks | | Ref |
|---|---|---|---|---|---|---|
| | 7.03 | 25 | 15.8* | DSC | (1,2) | 98M17 |

Remarks:
(1) from scan rate dependent DSC measurements, extrapolated to infinite scan rate, see also Table 2
(2) $\Delta G$ was calculated using $\Delta Cp$ = 5.7 kJ/mol/K (estimated value)


## Prion Protein

Recombinant human prion protein PrP(90–231)

| Transition | pH | $\Delta G$ | $c_{1/2}$ | Approach | Remarks | Ref |
|---|---|---|---|---|---|---|
| two-state | 7.2 | 19 | 2.2 | GuHCl | (1–3) | 97S16 |
| three-state | | | | | | |
| trans. (1) | 3.6 | 13 | 0.5 | GuHCl | (1–3) | 97S16 |
| trans. (2) | 3.6 | 21 | 2.4 | GuHCl | (1–3) | 97S16 |

Remarks:
(1) linear extrapolation
(2) measured by CD at 222 nm at room temperature
(3) buffer: 50 mM sodium phosphate or sodium acetate

Recombinant human prion protein PrP(90–231), wild type and mutants

| Mutant | pH | T | $\Delta G$ | $c_{1/2}$ | m | Appr./Rem. | Ref |
|---|---|---|---|---|---|---|---|
| wild type | 7.2 | 30 | 26.3±1.7 | 2.19±0.02 | 12.0±0.3 | GuHCl (1,2) | 98S22 |
| Pro102→Leu | 7.2 | 30 | 24.9±2.4 | 2.18±0.02 | 11.4±1.0 | GuHCl (1,2) | 98S22 |
| Glu200→Lys | 7.2 | 30 | 22.2±0.9 | 2.04±0.01 | 10.9±0.3 | GuHCl (1,2) | 98S22 |

Remarks:
(1) linear extrapolation, LEM-SB
(2) transition monitored by CD at 222 nm

Recombinant human prion protein PrP (91–231)

| Transition | pH | T | $\Delta G$ | m | Approach/Remarks | | Ref |
|---|---|---|---|---|---|---|---|
| N → U | 8.0 | 25 | 23.6 | 20.0 | GuHCl | (1–4) | 99J1 |
| N → U | 4.0 | 25 | 21.1 | 17.9 | GuHCl | (1–3) | 99J1 |
| N → I | 4.0 | 25 | 11.0 | 34.9 | GuHCl | (1–3) | 99J1 |

Remarks:
(1) linear extrapolation based on a two-state unfolding model for pH 8.0 and a three-state unfolding model for pH 4.0
(2) transition monitored by far-UV CD at 222 nm
(3) buffer: 10 mM sodium acetate + 10 mM Tris at either pH 8.0 or 4.0
(4) at pH 8.0 thermal transition at $T_{trs}$ = 56°C (12.5 µM PrP) and 52°C (25 µM PrP)

Recombinant murine prion proteins $m$PrP(121–231), wild type and variants, and full-length $m$PrP(23–231)

| Mutant | pH | T | $\Delta G$ | $c_{1/2}$ | m | Appr./Rem. | Ref |
|---|---|---|---|---|---|---|---|
| wild-type proteins: | | | | | | | |
| $m$PrP(121–231)w.t. | 7.0 | 25 | 29.7±1.0 | 6.2 | 4.8±0.2 | urea (1–3) | 99L5 |
| $m$PrP(23–231)w.t. | 7.0 | 25 | 25.5±1.0 | 6.3 | 4.1±0.2 | urea (1–3) | 99L5 |
| variants derived from $m$PrP(121–231)w.t.: | | | | | | | |
| Met129→Val | 7.0 | 25 | 28.2±1.0 | 6.3 | 4.5±0.2 | urea (1–3) | 99L5 |
| Asp178→Asn/Met129 | 7.0 | 25 | 22.5±0.7 | 4.8 | 4.7±0.1 | urea (1–3) | 99L5 |
| Asp178→Asn/Val129 | 7.0 | 25 | 21.7±0.9 | 4.9 | 4.5±0.2 | urea (1–3) | 99L5 |
| Val180→Ile | 7.0 | 25 | 27.6±0.8 | 6.2 | 4.5±0.1 | urea (1–3) | 99L5 |
| Thr183→Ala | 7.0 | 25 | 10.4±2.1 | 3.0 | 3.5±0.5 | urea (1–3) | 99L5 |
| Thr190→Val | 7.0 | 25 | 30.3±1.5 | 6.4 | 4.8±0.2 | urea (1–3) | 99L5 |
| Phe198→Ser | 7.0 | 25 | 19.4±0.8 | 4.5 | 4.4±0.2 | urea (1–3) | 99L5 |
| Glu200→Lys | 7.0 | 25 | 29.1±1.5 | 5.9 | 4.9±0.3 | urea (1–3) | 99L5 |
| Arg208→His | 7.0 | 25 | 23.7±1.5 | 5.6 | 4.3±0.3 | urea (1–3) | 99L5 |
| Val210→Ile | 7.0 | 25 | 28.6±1.6 | 6.3 | 4.5±0.3 | urea (1–3) | 99L5 |
| Gln217→Arg | 7.0 | 25 | 20.8±0.8 | 4.7 | 4.5±0.2 | urea (1–3) | 99L5 |

Remarks:

(1) linear extrapolation, LEM-SB

(2) buffer: 50 mM sodium phosphate, pH 7.0

(3) transition monitored by CD at 222 nm

Recombinant murine prion proteins $m$PrP(121–231), the structured 111-residue domain of the murine cellular prion protein

| Mutant | pH | T | $\Delta G$ | $c_{1/2}$ | m | Appr./Rem. | Ref |
|---|---|---|---|---|---|---|---|
| wild type | 7.0 | 22 | 28.95±0.94 | 6.21 | 4.66±0.16 | urea (1,2,5) | 99W11 |
| Phe175→Trp | 7.0 | 22 | 28.74±1.16 | 6.47 | 4.44±0.19 | urea (1,2,5) | 99W11 |
| Phe175→Trp | 7.0 | 22 | 28.42±1.88 | 6.40 | 4.44±0.31 | urea (1,3,5) | 99W11 |
| Phe175→Trp | 7.0 | 4 | 35.14±1.30 | 6.60 | 5.35±0.20 | urea (1,2,5) | 99W11 |
| Phe175→Trp | 7.0 | 4 | 32.90±3.34 | 6.96 | 4.72±0.48 | urea (1,4–6) | 99W11 |

Remarks:

(1) linear extrapolation

(2) equilibrium data, transition monitored by CD at 222 nm

(3) equilibrium data, transition monitored by protein fluorescence at 360 nm

(4) kinetic data from stopped-flow fluorescence

(5) buffer: 50 mM phosphate, pH 7.0

(6) m = $m_{NU}$ – $m_{UN}$, with $m_{UN}$ = –1.41±0.28 kJ/mol/M and $m_{NU}$ = 3.31±0.44 kJ/mol/M

Recombinant Syrian hamster prion protein, $\alpha$-helical state $\alpha$-rPrP

| Transition | pH | T | $\Delta G$ | Approach | Remarks | Ref |
|---|---|---|---|---|---|---|
| trans. (1) | 5 | | 7.9±1.7 | GuHCl | (1,2) | 97Z4 |
| trans. (2) | 5 | | 27.2±5.0 | GuHCl | (1,2) | 97Z4 |

Remarks:

(1) the ellipticity at 222 nm as function of GuHCl conc. was fitted by a three-state model assuming a linear dependence on denaturant concentration

(2) $\beta$-rPrP is thermodynamically more stable than $\alpha$-rPrP according to melting curves, see Ref. 97Z4

Yeast prion protein determinant Ure2

| Protein | pH | T | $\Delta G$ | $c_{1/2}$ | m | Appr./Rem. | Ref |
|---|---|---|---|---|---|---|---|
| Ure2-His (native purification): | | | | | | | |
| 0.3 μM | 8.4 | 25 | 55.2 | 3.01 | 18.4 | GuHCl (1–3) | 99P6 |
| 1.0 μM | 8.4 | 25 | 47.7 | 2.94 | 16.3 | GuHCl (1–3) | 99P6 |
| 3.0 μM | 8.4 | 25 | 52.7 | 2.99 | 17.6 | GuHCl (1–3) | 99P6 |
| Ure2 (GuHCl purification): | | | | | | | |
| 1.0 μM | 8.4 | 25 | 47.3 | 3.15 | 15.5 | GuHCl (1–3) | 99P6 |
| Ure2-His (urea purification): | | | | | | | |
| 1.0 μM | 8.4 | 25 | 51.0 | 3.13 | 16.3 | GuHCl (1–3) | 99P6 |
| 90Ure2-His | | | | | | | |
| 0.2 μM | 8.4 | 25 | 43.9 | 3.00 | 14.6 | GuHCl (1–3) | 99P6 |
| 90Ure2 | | | | | | | |
| 1.0 μM | 8.4 | 25 | 50.2 | 2.86 | 17.6 | GuHCl (1–3) | 99P6 |
| 74Ure2 | | | | | | | |
| 1.0 μM | 8.4 | 25 | 55.6 | 2.90 | 19.2 | GuHCl (1–3) | 99P6 |
| Δ15–42Ure2-His | | | | | | | |
| 0.2 μM | 8.4 | 25 | 68.2 | 3.08 | 22.2 | GuHCl (1–3) | 99P6 |
| 1.0 μM | 8.4 | 25 | 54.8 | 2.90 | 18.8 | GuHCl (1–3) | 99P6 |
| 3.0 μM | 8.4 | 25 | 49.0 | 2.92 | 16.7 | GuHCl (1–3) | 99P6 |
| 10.0 μM | 8.4 | 25 | 46.0 | 2.98 | 15.5 | GuHCl (1–3) | 99P6 |
| 15Ure2-His | | | | | | | |
| 1.0 μM | 8.4 | 25 | 51.5 | 2.93 | 17.6 | GuHCl (1–3) | 99P6 |
| mean values | 8.4 | 25 | 51.9±1.7 | 2.98±0.03 | 17.6±0.8 | GuHCl (1–3) | 99P6 |

Remarks:
(1) nonlinear least-squares fit of the transition curve assuming a linear dependence of $\Delta G$ on the denaturant conc.
(2) transitions monitored by intrinsic fluorescence, far and near-UV CD at 222 nm and 280 nm
(3) measured in 50 mM Tris-HCl, pH 8.4, 0.2 M NaCl
(4) Ure2 comprises residues 1–354, 90Ure2 res. 90–354, 74Ure2 res. 74–354, 15Ure2 res. 15–354, and Δ15–42Ure2 res. 1–14 and 43–354

## Procarboxypeptidase

Recombinant human procarboxypeptidase, wild type and mutants

| Mutant | pH | T | $\Delta G$ | $c_{1/2}$ | m | Appr./Rem. | Ref |
|---|---|---|---|---|---|---|---|
| wild type | 7.0 | 25 | 17.07±0.42 | 4.3 | 3.97±0.42 | urea (1–4) | 98V6 |
| Val12→Ala | 7.0 | 25 | 12.47±0.88 | 2.9 | 4.35±0.29 | urea (1–3) | 98V6 |
| Glu14→Ala | 7.0 | 25 | 15.82±0.46 | 4.1 | 3.89±0.13 | urea (1–3) | 98V6 |
| Ile15→Val | 7.0 | 25 | 15.27±0.46 | 3.9 | 3.89±0.13 | urea (1–3) | 98V6 |
| Glu20→Gly | 7.0 | 25 | 11.13±0.42 | 3.2 | 3.47±0.13 | urea (1–3) | 98V6 |
| Ile23→Val | 7.0 | 25 | 18.83±0.50 | 4.3 | 4.39±0.13 | urea (1–3) | 98V6 |
| Leu26→Val | 7.0 | 25 | 13.01±0.71 | 3.2 | 4.10±0.21 | urea (1–3) | 98V6 |
| Ala31→Gly | 7.0 | 25 | 12.05±0.38 | 3.5 | 3.43±0.08 | urea (1–3) | 98V6 |
| Asp38→Ala | 7.0 | 25 | 18.58±0.84 | 4.5 | 4.10±0.17 | urea (1–3) | 98V6 |
| Phe39→Leu | 7.0 | 25 | 9.20±0.08 | 2.5 | 3.68±0.33 | urea (1–3) | 98V6 |
| Lys41→Ala | 7.0 | 25 | 8.79±0.84 | 2.9 | 3.05±0.42 | urea (1–3) | 98V6 |
| Ala50→Gly | 7.0 | 25 | 11.21±0.13 | 2.8 | 3.97±0.17 | urea (1–3) | 98V6 |
| His51→Ala | 7.0 | 25 | 14.52±0.79 | 3.5 | 4.18±0.21 | urea (1–3) | 98V6 |
| Val52→Ala | 7.0 | 25 | 12.55±1.38 | 3.1 | 4.06±0.42 | urea (1–3) | 98V6 |
| Asn58→Ala | 7.0 | 25 | 16.74±0.84 | 4.1 | 4.10±0.25 | urea (1–3) | 98V6 |
| Gln60→Gly | 7.0 | 25 | 15.02±0.42 | 3.7 | 4.02±0.13 | urea (1–3) | 98V6 |

Recombinant human procarboxypeptidase, wild type and mutants (continued)

| Mutant | pH | T | ΔG | $c_{1/2}$ | m | Appr./Rem. | Ref |
|---|---|---|---|---|---|---|---|
| Val64→Gly | 7.0 | 25 | 12.47±0.42 | 3.5 | 3.56±0.13 | urea (1–3) | 98V6 |
| Phe65→Ala | 7.0 | 25 | 12.97±0.67 | 2.8 | 4.60±0.21 | urea (1–3) | 98V6 |
| Ile71→Val | 7.0 | 25 | 11.59±0.42 | 2.8 | 4.18±0.13 | urea (1–3) | 98V6 |
| Tyr73→Leu | 7.0 | 25 | 5.4 ±0.4 | 1.4 | 3.8 | urea (1–3) | 98V6 |
| Ile75→Ala | 7.0 | 25 | 11.59±0.8 | 2.8 | 4.10±0.21 | urea (1–3) | 98V6 |
| α1 stab. | 7.0 | 25 | 22.2 ±0.8 | 5.9 | 3.8 ±0.1 | urea (1,5) | 98V6 |
| α2 stab. | 7.0 | 25 | 23.8 ±0.4 | 6.3 | 3.8 ±0.1 | urea (1,5) | 98V6 |
| wild type | 7.0 | 25 | 18.4 ±0.4 | | 4.14±0.08 | kinet.(2–4,6) | 98V6 |
| Val12→Ala | 7.0 | 25 | 10.9 ±0.8 | | 4.56±0.29 | kinet.(2–3,6) | 98V6 |
| Glu14→Ala | 7.0 | 25 | 16.7 ±0.4 | | 4.14±0.25 | kinet.(2–3,6) | 98V6 |
| Ile15→Val | 7.0 | 25 | 16.3 ±0.4 | | 4.06±0.13 | kinet.(2–3,6) | 98V6 |
| Glu20→Gly | 7.0 | 25 | 14.2 ±0.4 | | 4.14±0.33 | kinet.(2–3,6) | 98V6 |
| Ile23→Val | 7.0 | 25 | 18.0 ±0.4 | | 4.18±0.13 | kinet.(2–3,6) | 98V6 |
| Leu26→Val | 7.0 | 25 | 13.0 ±0.4 | | 4.56±0.21 | kinet.(2–3,6) | 98V6 |
| Ala31→Gly | 7.0 | 25 | 16.3 ±0.4 | | 4.23±0.25 | kinet.(2–3,6) | 98V6 |
| Asp38→Ala | 7.0 | 25 | 18.4 ±0.8 | | 4.02±0.17 | kinet.(2–3,6) | 98V6 |
| Phe39→Leu | 7.0 | 25 | 9.6 ±0.8 | | 3.68±0.33 | kinet.(2–3,6) | 98V6 |
| Lys41→Ala | 7.0 | 25 | 8.8 ±0.8 | | 3.05±0.42 | kinet.(2–3,6) | 98V6 |
| Ala50→Gly | 7.0 | 25 | 11.3 ±0.4 | | 4.23±0.17 | kinet.(2–3,6) | 98V6 |
| His51→Ala | 7.0 | 25 | 15.1 ±0.4 | | 4.39±0.21 | kinet.(2–3,6) | 98V6 |
| Val52→Ala | 7.0 | 25 | 13.8 ±0.8 | | 4.18±0.42 | kinet.(2–3,6) | 98V6 |
| Asn58→Ala | 7.0 | 25 | 16.7 ±0.8 | | 4.02±0.25 | kinet.(2–3,6) | 98V6 |
| Gln60→Gly | 7.0 | 25 | 16.7 ±0.8 | | 4.39±0.21 | kinet.(2–3,6) | 98V6 |
| Val64→Gly | 7.0 | 25 | 16.3 ±0.4 | | 4.31±0.21 | kinet.(2–3,6) | 98V6 |
| Phe65→Ala | 7.0 | 25 | 11.7 ±0.4 | | 4.81±0.13 | kinet.(2–3,6) | 98V6 |
| Ile71→Val | 7.0 | 25 | 12.1 ±0.4 | | 4.73±0.13 | kinet.(2–3,6) | 98V6 |
| Tyr73→Leu | 7.0 | 25 | 5.4 ±0.8 | | 3.77±0.42 | kinet.(2–3,6) | 98V6 |
| Ile75→Ala | 7.0 | 25 | 10.5 ±0.4 | | 4.06±0.21 | kinet.(2–3,6) | 98V6 |
| α1 stab. | 7.0 | 25 | 24.3 ±0.4 | | 4.02±0.13 | kinet.(5,6) | 98V6 |
| α2 stab. | 7.0 | 25 | 25.1 ±0.4 | | 3.77±0.08 | kinet.(5,6) | 98V6 |

Remarks:

(1) data from equilibrium unfolding using linear extrapolation of ΔG to zero denaturant concentration by a method that includes the pre- and postdenaturational baselines for a nonlinear regression of the data, see also Ref. 95V6

(2) transition monitored by fluorescence

(3) buffer: 50 mM sodium phosphate, pH 7.0

(4) data from Ref. 95V6

(5) stabilized α-helix 1 and α-helix 2, respectively, see Refs. 95V7 and 97V3

(6) data from kinetics of folding and unfolding

Activation domain of procarboxypeptidase A2, role of helix-stabilizing mutants

| Mutant | pH | T | ΔG | m | Approach | Remarks | Ref |
|---|---|---|---|---|---|---|---|
| wild type | 7.0 | 25 | 18.4±0.4 | 4.2±0.13 | urea | (1–4) | 97V3 |
| | 7.0 | 25 | 17.6±0.4 | 9.6±0.33 | urea | (1–3,5) | 97V3 |
| | 7.0 | 25 | 18.8±0.4 | 4.2±0.21 | urea | (1–3,6) | 97V3 |
| | 7.0 | 25 | 15.9 | | urea | (1–3,7) | 97V3 |
| mutant M1 | 7.0 | 25 | 22.2±0.8 | 3.8±0.13 | urea | (1–3,5) | 97V3 |
| | 7.0 | 25 | 22.2±0.4 | 10.0±0.29 | urea | (1–3,5) | 97V3 |
| | 7.0 | 25 | 24.3±1.3 | 4.2±0.21 | urea | (1–3,6) | 97V3 |
| mutant M2 | 7.0 | 25 | 23.8±0.4 | 3.8±0.08 | urea | (1–4) | 97V3 |
| | 7.0 | 25 | 25.1±0.4 | 8.8±0.4 | urea | (1–3,5) | 97V3 |
| | 7.0 | 25 | 25.1±0.8 | 3.8±0.08 | urea | (1–3,6) | 97V3 |

Activation domain of procarboxypeptidase A2, role of helix-stabilizing mutants (continued)

| Mutant | pH | T | ΔG | m | Approach | Remarks | Ref |
|---|---|---|---|---|---|---|---|
| double mutant DM | | | | | | | |
| | 7.0 | 25 | 29.3±0.8 | 3.8 | urea | (1–4) | 97V3 |
| | 7.0 | 25 | 28.9±2.1 | 9.2±0.4 | urea | (1–3,5) | 97V3 |
| | 7.0 | 25 | 31.4±1.3 | 3.8 | urea | (1–3,6) | 97V3 |

Explanations:
For the helix-stabilizing mutants see Ref. 95V7
      M1 with the following replacements in helix 1: Asn25→Lys, Gln32→Lys, and Glu33→Lys
      M2 with the following replacements in helix 2: Gln60→Glu, Val64→Ala,
      Ser68→Ala, and Gln69→His

Remarks:
(1) linear extrapolation, LEM-SB
(2) buffer: 50 mM sodium phosphate
(3) transition monitored by the fluorescence spectrum of Trp40
(4) from equilibrium urea denaturation
(5) from equilibrium GuHCl denaturation
(6) from kinetics of urea denaturation
(7) from thermal denaturation with $T_{trs}$ 77.8±0.2, 85.0±0.5, and 85.3±0.5 for wild type, M1, and M2, respectively. ΔH for the
    wild type amounts to 205±3 kJ/mol from DSC, and 188±8 kJ/mol from van't Hoff treatment, see also Ref. 95V6

## Profilin

Profilin from *Saccharomyces cerevisiae*, *Acanthamoebia* and mutants

| Protein | pH | T | ΔG | $c_{1/2}$ | m | Appr./Rem. | | Ref |
|---|---|---|---|---|---|---|---|---|
| *Sacharomyces* | 8 | 20 | 24.7 | 3.4 | 7.5 | urea | (1–3) | 98E1 |
| *Acanthamoebia* | | | | | | | | |
| PI native | 7.8 | 20 | ~31 | 3.5 | | urea | (4–6) | 96K11 |
| PII native | 7.8 | 20 | ~31 | 3.5 | | urea | (4–6) | 96K11 |
| rPI 122QGF | 7.8 | 20 | ~31 | 3.5 | | urea | (4–6) | 96K11 |
| rPII 122QGF | 7.8 | 20 | 31.8 | 3.5 | | urea | (4–6) | 96K11 |
| rPII 122PLV | 7.8 | 20 | 13.4 | 1.5 | | urea | (4–6) | 96K11 |
| rPII 122PSSLD | 7.8 | 20 | 12.6 | 1.5 | | urea | (4–6) | 96K11 |
| rPII 117PLV | 7.8 | 20 | 7.5 | 0.9 | | urea | (4–6) | 96K11 |
| rPI Ser76→Cys | 7.8 | 20 | ~31 | 3.5 | | urea | (4–6) | 96K11 |
| rPI Ser92→Cys | 7.8 | 20 | ~31 | 3.5 | | urea | (4–6) | 96K11 |

Explanations:
    PI           *Acanthamoebia* profilin I
    PII         *Acanthamoebia* profilin II
    rPI         recombinant *Acanthamoebia* profilin I
    rPII        recombinant *Acanthamoebia* profilin II
    rPI 122QGF    wild-type rPI with C-terminus (Gln123-Gly124-Phe125)
    rPII 122QGF   wild-type rPII with C-terminus (Gln123-Gly124-Phe125)
    rPII 122PLV    rPII triple mutant (Gln123→Pro, Gly124→Leu, Phe125→Val)
    rPII 122PSSLD  rPII with (Pro-Ser-Ser-Leu-Asp) substituted for (Gln123-Gly124-Phe125)
    rPII 117PLV    rPII with (Pro-Leu-Val) substituted for (Asp118-Tyr119-Leu120-Ile121-Gly122-Gln-123-
                   Gly124-Phe125)

Remarks:
(1) linear extrapolation
(2) measured in 10 mM Tris (pH 8), 40 mM KCl, 1 mM DTT
(3) transition monitored by fluorescence intensity at 370 nm
(4) linear extrapolation, LEM-SB
(5) measured in 25 mM Tris-HCl, pH 7.8
(6) transition monitored by fluorescence intensity at 350 nm

## Protease (α-Lytic Protease)

α-Lytic Protease (αLP), recombinant

| Protein | pH | T | ΔG | Approach/Remarks | | Ref |
|---|---|---|---|---|---|---|
| αLP, I → N | 5 | 4 | 16.7±0.4 | GuHCl | (1–3) | 98S14 |
| αLP, I → N | 5 | 25 | 24.7±2.1 | GuHCl | (1–3) | 98S14 |

Remarks:
(1) measured by folding and refolding kinetics
(2) both I and U state are more stable than the folded N state
(3) at 10°C the native conformation is favoured enthalpically by 75±6 kJ/mol over the I state

α-Lytic protease precursor (αLP), recombinant

| Protein | pH | T | ΔG | m | Approach/Remarks | | Ref |
|---|---|---|---|---|---|---|---|
| αLP | 6.4 | 5 | 9.2±1.3 | 8.8±1.3 | urea | (1–3) | 99A2 |
| αLP + glycerol | 6.4 | 5 | 13.8±0.8 | 5.4±0.4 | urea | (1–4) | 99A2 |
| pro region of αLP | 6.4 | 5 | 9.6±1.3 | 8.8±1.3 | urea | (1–3) | 99A2 |

Remarks:
(1) linear extrapolation
(2) measured in 10 mM MES, pH 6.4, 0.1 M NaCl
(3) transition monitored by far-UV CD
(4) in the presence of 15% glycerol

## Protein A

Staphylococcal protein A, B-domain, 58-residue proteine with a three-helix bundle structure

| Protein | pH | T | ΔG | Approach/Remarks | | Ref |
|---|---|---|---|---|---|---|
| protein A | | | 29.3±4.1 | GuHCl | (1) | 97B2 |
| protein A | 7.0 | 20 | 31.0±2.1 | HX | (2,3) | 97B2 |
| Ile16→Trp | 6.0 | 20 | 5.8 | GuHCl | (1) | 97B2 |

Remarks:
(1) linear extrapolation, LEM-SB
(2) buffer: 100 mM potassium phosphate, pH* 7.0
(3) largest values of $\Delta G_{op}$

Staphylococcal protein A, B-domain, designated Z protein, wild type and mutants

| Mutant | pH | T | ΔG | $c_{1/2}$ | m | Appr./Rem. | Ref |
|---|---|---|---|---|---|---|---|
| Z protein | 7.5 | 23 | 27.6 | 3.9 | 6.99 | GuHCl (1–3) | 93C8 |
| Z-Asn28→Ala | 7.5 | 23 | 28.9 | 4.0 | 7.24 | GuHCl (1–3) | 93C8 |
| Z-Phe30→Ala | 7.5 | 23 | 17.6 | 2.1 | 8.16 | GuHCl (1–3) | 93C8 |

Remarks:
(1) linear extrapolation
(2) transition monitored by CD at 222 nm
(3) measured in 20 mM phosphate buffer, pH 7.5

## Protein DmpM

Stimulatory protein DmpM of phenol hydroxylase from *Pseudomonas sp.* Strain CF600, native and recombinant protein

| Protein | pH | T | $\Delta G$ | Approach/Remarks | Ref |
|---|---|---|---|---|---|
| native DmpM | 7.5 | 25 | 15.36±1.88 | urea, CD (1–3) | 99C1 |
| | 7.5 | 25 | 15.65±1.67 | urea, FL (1,2,4) | 99C1 |
| reco. DmpM | 7.5 | 25 | 18.12±2.26 | urea, CD (1–3) | 99C1 |
| | 7.5 | 25 | 18.58±1.55 | urea, FL (1,2,4) | 99C1 |

Remarks:
(1) linear extrapolation, LEM-SB
(2) buffer: 50 mM sodium phosphate, pH 7.5, with 10% glycerol
(3) transition monitored by far-UV CD
(4) transition monitored by fluorescence

## Protein G (Immunoglobulin-Binding Protein)

Streptococcal protein G, IgG-binding domain B1

| Form | $Na_2SO_4$ | pH | T | $\Delta G$ | $c_{1/2}$ | m | Appr./Rem. | Ref |
|---|---|---|---|---|---|---|---|---|
| GB1–57 | 0.0 M | 4.0 | 20 | 19.2±1.7 | 2.7±0.1 | 7.11±0.42 | GuHCl (1–3) | 97P4 |
| GB1–57 | 0.0 M | 4.0 | 20 | 20.5 | 2.6 | 7.95 | GuHCl (1,4–6) | 97P4 |
| GB1–57 | 0.4 M | 4.0 | 20 | 28.5±2.1 | 3.6±0.1 | 7.95±0.42 | GuHCl (1–3) | 97P4 |
| GB1–57 | 0.4 M | 4.0 | 20 | 29.3 | 3.3 | 8.79 | GuHCl (1,4–6) | 97P4 |
| GB1–66 | 0.0 M | 4.0 | 20 | 18.4±1.7 | 2.6±0.1 | 7.11±0.42 | GuHCl (1–3) | 97P4 |
| GB1–66 | 0.0 M | 4.0 | 20 | 19.7 | 2.5 | 7.95 | GuHCl (1,4–6) | 97P4 |
| GB1–66 | 0.4 M | 4.0 | 20 | 25.5±2.1 | 3.4±0.1 | 7.53±0.42 | GuHCl (1–3) | 97P4 |
| GB1–66 | 0.4 M | 4.0 | 20 | 27.2 | 3.2 | 8.37 | GuHCl (1,4–6) | 97P4 |

Remarks:
(1) GB1–57 = 57-residue IgG-binding domain of streptococcal protein G GB1–66 = 66-residue fragment of GB1 with an N-terminal extension containing five apolar side chains
(2) equilibrium unfolding monitored by fluorescence emission at 340 nm
(3) linear extrapolation
(4) from kinetics
(5) treatment by a three-state model
(6) m, $\Delta G$, and $c_{1/2}$ are defined as $m = m_{eq} = m_{UI} - m_{IN}\ddagger + m_{NI}\ddagger$, $\Delta G = \Delta_{GU}^{\circ} = \Delta G_{IU}^{\circ} + \Delta G_{NI}^{\circ}$, $c_{1/2} = \Delta G_{U}^{\circ}/m_{eq}$.

Streptococcal protein G, IgG-binding domain B1, mutants in the antiparallel β-sheet, positions 44 and 53

| Mutant | pH | T | $\Delta G$ | Approach/Remarks | | Ref |
|---|---|---|---|---|---|---|
| Ala44, Ala53 | 5.2 | 25 | 15.56±0.38 | heat | (1–3) | 98M16 |
| Ala44, Gly53 | 5.2 | 25 | 10.5 ±0.4 | heat | (1–3) | 98M16 |
| Phe44, Ala53 | 5.2 | 25 | 15.5 ±0.8 | heat | (1–3) | 98M16 |
| Phe44, Gly53 | 5.2 | 25 | 12.1 ±0.4 | heat | (1–3) | 98M16 |

Remarks:
(1) transition monitored by CD at 222 nm
(2) measured in 50 mM sodium acetate buffer, pH 5.2
(3) $\Delta G$ was calculated using $\Delta Cp$ = 2.6 kJ/mol/K from Refs. 92A2 and 94M15

Immunoglobulin G (IgG) binding protein G (β1), variants with charged residues at the parallel cross-strand site (positions 6 and 53)

| Mutant | pH | $T_{trs}$ | $\Delta(\Delta G)^{25°C}$ | Approach/Remarks | | Ref |
|---|---|---|---|---|---|---|
| Ala6, Ala53 | 7.2 | 38.19±0.09 | 0.0 | heat | (1–3) | 99M15 |
| Ala6, Glu53 | 7.2 | 40.3 ±0.1 | 1.05±0.08 | heat | (1,2) | 99M15 |
| Ala6, Arg53 | 7.2 | 46.62±0.08 | 3.51±0.29 | heat | (1,2) | 99M15 |
| Glu6, Ala53 | 7.2 | 39.55±0.06 | 0.67±0.17 | heat | (1,2) | 99M15 |
| Glu6, Glu53 | 7.2 | 32.55±0.06 | −2.47±0.08 | heat | (1,2) | 99M15 |
| Glu6, Arg53 | 7.2 | 52.25±0.09 | 7.28±0.08 | heat | (1,2) | 99M15 |
| Lys6, Ala53 | 7.2 | 50.9 ±0.2 | 6.7 ±0.4 | heat | (1,2) | 99M15 |
| Lys6, Glu53 | 7.2 | 55.19±0.01 | 9.33±0.29 | heat | (1,2) | 99M15 |
| Lys6, Arg53 | 7.2 | 52.9 ±0.1 | 8.8 ±0.4 | heat | (1,2) | 99M15 |

Remarks:
(1) transition monitored by far-UV CD at 222 nm
(2) buffer: 50 mM phosphate buffer, pH 7.2
(3) $\Delta G$ for Ala6, Ala53 amounts to $\Delta G = 9.54±0.21$ kJ/mol at 25°C

Immunoglobulin G (IgG) binding protein G (β1), variants with polar and noncharged residues at the parallel cross-strand site (positions 6 and 53)

| Mutant | pH | $T_{trs}$ | $\Delta(\Delta G)^{25°C}$ | Approach/Remarks | | Ref |
|---|---|---|---|---|---|---|
| Ile6, Tyr53 | 5.2 | 79.3 ±0.2 | 17.2 ±1.3 | heat | (1–3) | 99M15 |
| Ile6, Phe53 | 5.2 | 77.78±0.03 | 19.29±0.21 | heat | (1–3) | 99M15 |
| Ile6, Ile53 | 5.2 | 77.2 ±0.2 | 13.8 ±0.4 | heat | (1–3) | 99M15 |
| Val6, Phe53 | 5.2 | 75.96±0.03 | 14.6 ±0.4 | heat | (1–3) | 99M15 |
| Ile6, Thr53 | 5.2 | 75.9 ±0.3 | 15.1 ±0.4 | heat | (1–3) | 99M15 |
| Ile6, Val53 | 5.2 | 74.3 ±0.2 | 11.7 ±0.4 | heat | (1–3) | 99M15 |
| Leu6, Ile53 | 5.2 | 73.96±0.06 | 11.7 ±0.8 | heat | (1–3) | 99M15 |
| Val6, Ile53 | 5.2 | 73.8 ±0.3 | 12.18±0.38 | heat | (1–3) | 99M15 |
| Phe6, Tyr53 | 5.2 | 72.2 ±0.2 | 11.7 ±1.3 | heat | (1–3) | 99M15 |
| Val6, Val53 | 5.2 | 71.5 ±0.1 | 10.5 ±0.8 | heat | (1–3) | 99M15 |
| Phe6, Phe53 | 5.2 | 70.7 ±0.3 | 12.43±0.33 | heat | (1–3) | 99M15 |
| Phe6, Ile53 | 5.2 | 69.42±0.06 | 8.8 ±0.8 | heat | (1–3) | 99M15 |
| Phe6, Val53 | 5.2 | 67.1 ±0.1 | 8.8 ±0.4 | heat | (1–3) | 99M15 |
| Thr6, Tyr53 | 5.2 | 62.75±0.05 | 7.1 ±0.8 | heat | (1–3) | 99M15 |
| Asn6, Thr53 | 5.2 | 62.7 ±0.1 | 8.66±0.13 | heat | (1–3) | 99M15 |
| Thr6, Thr53 | 5.2 | 62.3 ±0.1 | 8.4 ±0.8 | heat | (1–3) | 99M15 |
| Thr6, Phe53 | 5.2 | 61.62±0.01 | 6.3 ±0.8 | heat | (1–3) | 99M15 |
| Thr6, Ile53 | 5.2 | 60.9 ±0.1 | 5.36±0.08 | heat | (1–3) | 99M15 |
| Thr6, Val53 | 5.2 | 58.61±0.03 | 4.64±0.29 | heat | (1–3) | 99M15 |

Remarks:
(1) transition monitored by far-UV CD at 222 nm
(2) buffer: 50 mM phosphate buffer, pH 7.2
(3) $\Delta G$ for the reference protein Ala6, Ala53 amounts to $\Delta G = 9.54±0.21$ kJ/mol at 25°C

Immunoglobulin-binding domain GB1 of Streptococcal protein G, replacement of an α-helix 23-36 by a β-hairpin

| Mutant | pH | T | $\Delta G$ | m | Approach/Remarks | | Ref |
|---|---|---|---|---|---|---|---|
| $G_{B1}$ | 7.0 | 25 | 29.7 | | heat | (1–3) | 99C14 |
| $G_{In}$ | 7.0 | 25 | 21.3 | | heat | (1–4) | 99C14 |
| | | 20 | 18.0±0.4 | 3.22±0.13 | urea | (4–6) | 99C14 |
| | 7.0 | 20 | 17.2±0.4 | | urea, kin. | (4,7,8) | 99C14 |
| $G_{Out}$ | 7.0 | 25 | 10.9 | | heat | (1–4) | 99C14 |

Immunoglobulin-binding domain GB1 of Streptococcal protein G, replacement of an α-helix 23-36 by a β-hairpin (continued)

| Mutant | pH | T | ΔG | m | Approach/Remarks | Ref |
|---|---|---|---|---|---|---|
| | | 20 | 10.5±0.4 | 3.31±0.17 | urea (4–6) | 99C14 |
| | 7.0 | 20 | 10.0±0.2 | | urea, kin. (4,7,8) | 99C14 |
| $G_Y$ | 7.0 | 25 | 8.8 | | heat (1–4) | 99C14 |
| | | 20 | 5.9±0.4 | 2.89±0.17 | urea (4–6) | 99C14 |
| | 7.0 | 20 | 6.3±0.2 | | urea, kin. (4,7,8) | 99C14 |
| $G_{Hel}$ | 7.0 | 25 | –0.04 | | heat (1–4) | 99C14 |
| | | 20 | 1.7±1.3 | 2.80±0.25 | urea (4–6) | 99C14 |

Remarks:
(1) from heat denaturation monitored by far-UV CD, see also Table 2
(2) ΔG was calculated using ΔCp from Ref. 92A2
(3) measured in 10 mM sodium phosphate, pH 7
(4) mutant with the α-helix 23-36 replaced by a β-hairpin derived from residues 43–55 (with point mutations described in Ref. 99C14)
(5) linear extrapolation to zero denaturant concentration by a method that includes the pre- and postdenaturational baselines for a nonlinear regression of the data
(6) transition monitored by fluorescence
(7) data from kinetics of folding and unfolding
(8) measured in 50 mM sodium phosphate, pH 7

Streptococcal protein G, B1 domain (B1G) and modified domain (DBF-B1G)

| Protein | pH | T | ΔG | Approach/Remarks | Ref |
|---|---|---|---|---|---|
| B1G | 3.5 | 5 | 13.9±0.5 | heat (1) | 99O1 |
| DBF-B1G | 3.5 | 5 | 13.5±0.6 | heat (1,2) | 99O1 |

Remarks:
(1) data from heat denaturation using ΔCp = 2.6 kJ/mol/K of the recombinant protein
(2) DBF-B1G = the residues 10 and 11 of B1G are replaced by dibenzofuran (DFB)

Streptococcal protein G, β1 domain, design of a hyperthermophilic protein variant

| Protein | pH | T | ΔG | $c_{1/2}$ | m | Approach/Remarks | Ref |
|---|---|---|---|---|---|---|---|
| Gβ1 | 5.5 | 50 | 11.7±0.4 | 1.8 | 6.69±0.21 | GuHCl (1,2) | 98M3 |
| Gβ1-c3 | 5.5 | 50 | 13.8±1.3 | 2.6 | 5.86±0.42 | GuHCl (1,3) | 98M3 |
| Gβ1-c3b1 | 5.5 | 50 | 17.2±0.8 | 3.0 | 5.86±0.25 | GuHCl (1,4) | 98M3 |
| Gβ1-c3b2 | 5.5 | 50 | 25.5±1.3 | 4.4 | 5.86±0.29 | GuHCl (1,5) | 98M3 |
| Gβ1-c3b4 | 5.5 | 5= | 29.7±1.3 | 5.1 | 5.86±0.29 | GuHCl (1,6) | 98M3 |

Remarks:
(1)  linear extrapolation LEM-SB, transition measured at 50°C by CD at 218 nm, buffer: 50 mM sodium phosphate
(2a) thermal transition in the absence of GuHCl at 83°C
(2b) Gβ1 contains Tyr3, Leu7, Thr16, Thr18, Thr25, Val29, Val39, and Trp43
(3a) thermal transition in the absence of GuHCl at 91°C
(3b) Gβ1-c3 contains Phe3, Ile7, Thr16, Thr18, Thr25, Val29, Ile39, and Trp43
(4a) thermal transition in the absence of GuHCl at 93°C
(4b) Gβ1-c3b1 contains Phe3, Ile7, Thr16, Thr18, Glu25, Val29, Ile39, and Trp43
(5a) thermal transition in the absence of GuHCl at >99°C
(5b) Gβ1-c3b2 contains Phe3, Ile7, Ile16, Ile18, Thr25, Val29, Ile39, and Trp43
(6a) thermal transition in the absence of GuHCl at >99°C
(6b) Gβ1-c3b4 corresponds to the computed ground state, Gβ1-c3b4 contains Phe3, Ile7, Ile16, Ile18, Glu25, Ile29, Val39, and Trp43

## Protein L

IgG-binding domain of peptostreptococcal protein L

| Mutant | pH | T | ΔG | m | Approach | Remarks | Ref |
|---|---|---|---|---|---|---|---|
| wild type | 7.0 | 22 | 19.7±0.8 | 7.11±0.17 | GuHCl | (1–3) | 97S3 |
| Tyr43→Trp | 7.0 | 22 | 19.2±0.8 | 7.95±0.29 | GuHCl | (1–3) | 97S3 |
| Tyr43→Trp | 7.0 | 22 | 19.7±0.4 | 8.37±0.08 | kinetics | (1,3,4) | 97S3 |

Remarks:

(1) linear extrapolation

(2) transitions monitored by CD at 220 nm and fluorescence emission at 328 nm

(3) wild type and mutant represent the 62 residue IgG-binding domain

(4) from stopped-flow fluorescence measurements in GuHCl, see also Ref. 97S2

IgG-binding domain of peptostreptococcal protein L, mutant (Tyr45→Trp), comparison of various approaches

| Mutant | pH | T | ΔG | m | Approach/Remarks | | Ref |
|---|---|---|---|---|---|---|---|
| Tyr45→Trp | 7.0 | 22 | 26.8±0.4 | 10.0±0.4 | heat | (1–3) | 97Y3 |
| Tyr45→Trp | 7.0 | 22 | 29.3±0.4 | 14.6±0.4 | H/D | (4,5) | 97Y3 |
| Tyr45→Trp | 7.0 | 22 | 20.5±0.4 | 7.1±0.4 | GuHCl | (6) | 97Y3 |

Remarks:

(1) data from global fit of protein melting carried out at different GuHCl conc.

(2) buffer: 100 mM sodium phosphate, pH 7.0, 0.5 M KCl, varying GuHCl conc. from 0 to 3.0 M

(3) transition monitored by CD at 220 nm

(4) measured in $D_2O$, the protein is slightly more stable in $D_2O$ ($\Delta(\Delta G)$~1.3 kJ/mol)

(5) the rates of the global opening and closing reactions are identical at pH 5.0 and pH 7.0

(6) application of linear extrapolation to the H/D experiments underestimates ΔG due to a nonlinear dependence of the Gibbs energy on the denaturant concentration, see also Ref. 97S3

IgG-binding domain of peptostreptococcal protein L, 62 residue B1 IgG binding domain

| Mutant | pH | T | Δ(ΔG) | m | Approach/Remarks | | Ref |
|---|---|---|---|---|---|---|---|
| wild type | 7.0 | 22 | 0.0 | 7.9 | GuHCl | (1–3) | 98G18 |
| combinatorial mutants: | | | | | | | |
| plt1_a | 7.0 | 22 | −14.6 | 10.0 | GuHCl | (1–4) | 98G18 |
| plt1_b | 7.0 | 22 | −6.7 | 6.3 | GuHCl | (1–4) | 98G18 |
| plt2_a | 7.0 | 22 | −0.8 | 7.1 | GuHCl | (1–4) | 98G18 |
| plt2_b | 7.0 | 22 | −13.8 | 9.2 | GuHCl | (1–4) | 98G18 |
| plt2_c | 7.0 | 22 | −6.7 | 8.8 | GuHCl | (1–4) | 98G18 |
| point mutants: | | | | | | | |
| protein L | 7.0 | 22 | 0.0 | 7.9 | GuHCl | (1–3) | 98G18 |
| Ala13→Pro | 7.0 | 22 | −0.46 | 7.9 | GuHCl | (1–3) | 98G18 |
| Asn14→Ala | 7.0 | 22 | −7.1 | 7.9 | GuHCl | (1–3) | 98G18 |
| Gly15→Ala | 7.0 | 22 | −6.3 | 8.8 | GuHCl | (1–3) | 98G18 |
| Gly55→Ala | 7.0 | 22 | −8.4 | 8.8 | GuHCl | (1–3) | 98G18 |

Remarks:

(1) Δ(ΔG) refers to the change in ΔG between wild type and mutant at 2.0 M GuHCl

(2) ΔG appears to be a nonlinear function of denaturant concentration

(3) for the procedure see Ref. 97S3

(4) for the sequences see Ref. 98G18

IgG-binding domain of peptostreptococcal protein L, 62 residue B1 IgG binding domain

| Mutant | pH | T | $\Delta(\Delta G)$ | m | Approach/Remarks | | Ref |
|---|---|---|---|---|---|---|---|
| wild type | 7.0 | 22 | 0.0 | 7.9 | GuHCl | (1–3) | 98K9 |
| combinatorial mutants: | | | | | | | |
| h-c | 7.0 | 22 | −6.7 | 9.2 | GuHCl | (1–4) | 98K9 |
| h-d | 7.0 | 22 | −18.4 | 13.0 | GuHCl | (1–4) | 98K9 |
| triple mutants: | | | | | | | |
| (Ala29→Leu, Thr30→Tyr, Ala33→Ile) | | | | | | | |
| | 7.0 | 22 | −2.9 | 9.2 | GuHCl | (1–3) | 98K9 |
| (Glu32→Gly, Ala35→Gly, Thr39→Gly) | | | | | | | |
| | 7.0 | 22 | −12.1 | 8.8 | GuHCl | (1–3) | 98K9 |
| helix destabilized mutants: | | | | | | | |
| Lys28→Gly | 7.0 | 22 | +0.4 | 7.1 | GuHCl | (1–3) | 98K9 |
| Glu32→Gly | 7.0 | 22 | −2.9 | 7.9 | GuHCl | (1–3) | 98K9 |
| Glu32→Ile | 7.0 | 22 | −3.8 | 8.4 | GuHCl | (1–3) | 98K9 |
| Ala35→Gly | 7.0 | 22 | −5.4 | 8.4 | GuHCl | (1–3) | 98K9 |
| Thr39→Gly | 7.0 | 22 | −0.8 | 7.9 | GuHCl | (1–3) | 98K9 |

Remarks:
(1) $\Delta(\Delta G)$ refers to the change in $\Delta G$ between wild type and mutant at 2.0 M GuHCl
(2) $\Delta G$ appears to be a nonlinear function of denaturant concentration
(3) for the procedure see Ref. 97S3
(4) for the sequences see Ref. 98G18

IgG-binding domain of peptostreptococcal protein L, 62 residue B1 IgG binding domain, solvent viscosity varied by glucose

| Protein | pH | T | $\Delta G$ | m | Approach/Remarks | | Ref |
|---|---|---|---|---|---|---|---|
| protein L | 7.0 | 22.5±1.0 | 20.8±1.3 | 7.82±0.46 | GuHCl | (1–3,5) | 98P10 |
| | 7.0 | 22.5±1.0 | 18.4±5.0 | 7.50±1.30 | GuHCl | (1,3–5) | 98P10 |

Remarks:
(1) protein L was used without cleaving of the His tag
(2) data from equilibrium unfolding, linear extrapolation
(3) transition monitored by CD at 220 nm in 50 mM sodium phosphate buffer
(4) data from kinetics
(5) general expression $\Delta G = \Delta G° - m_{(GuHCl)} + m_{(glucose)}$, ($\Delta G$ and m are figured out here in positive values) with
$m_{(glucose)} = 4.2±0.4$ kJ/mol/M from equilibrium unfolding, and
$m_{(glucose)} = 5.0±0.8$ kJ/mol/M from kinetics

IgG-binding domain of peptostreptococcal protein L, 62 residue B1 IgG binding domain

| Protein | pH | T | $\Delta G$ | m | Approach/Remarks | | Ref |
|---|---|---|---|---|---|---|---|
| protein L | 7.0 | 5 | 23.0±3.3 | 8.4±1.3 | GuHCl | (1,2) | 99P8 |

Remarks:
(1) linear extrapolation
(2) transition monitored by small angle X-ray scattering (SAXS)
(3) the radius of gyration amounts to $R_g = 16.2±0.2$ Å for the native protein and $R_g = 26.0±0.3$ Å for the fully denatured protein

IgG binding domain of peptostreptococcal protein L, wild type and mutants

| Protein | pH | T | $\Delta G$ | m | Approach/Remarks | | Ref |
|---|---|---|---|---|---|---|---|
| wild type | 7.0 | 23 | 19.2 | 7.9 | GuHCl | (1) | 99G18 |
| 2Phe | 7.0 | 23 | 15.5 | 8.4 | GuHCl | (1,2) | 99G18 |
| 4Phe | 7.0 | 23 | 15.1 | 6.7 | GuHCl | (1,3,4) | 99G18 |
| Lys7→Phe | 7.0 | 23 | 22.2 | 7.9 | GuHCl | (1,3) | 99G18 |
| Ala13→Val | 7.0 | 23 | 18.4 | 9.2 | GuHCl | (1,3) | 99G18 |
| Asn14→Ala | 7.0 | 23 | 12.1 | 7.9 | GuHCl | (1,3,6,7) | 99G18 |
| Gly15→Ala | 7.0 | 23 | 15.1 | 8.8 | GuHCl | (1,3,6,7) | 99G18 |
| Gly15→Val | 7.0 | 23 | 13.0 | 10.9 | GuHCl | (1,3) | 99G18 |
| Glu32→Ile | 7.0 | 23 | 16.3 | 8.4 | GuHCl | (1,3,5) | 99G18 |

Remarks:

(1) for details of the procedure and experimental conditions see Ref. 97S3

(2) mutant 2Phe has two Phe substitutions on the fourth β-strand

(3) mutant 4Phe has paired Phe substitutions on the third and fourth β-strand

(4) the values are calculated from kinetic data

(5) data from Ref. 98K9

(6) data from Ref. 97G13

(7) Ref. 97G13 contains $\Delta(\Delta G)$ values for various mutants at 2 M GuHCl

## Protein S

Recombinant protein S, spore coat protein from *Myxococcus xanthus*, wild type and isolated domains

| Protein | pH | T | $\Delta G$ | $c_{1/2}$ | Approach/Remarks | | Ref |
|---|---|---|---|---|---|---|---|
| protein S | 7.0 | 20 | | 4.8 | urea | (1–4,6) | 98W7 |
| N domain | 7.0 | 20 | 30 | 5.2 | urea | (1–4,6) | 98W7 |
| C domain | 7.0 | 20 | 20 | 2.5 | urea | (1–3,5,6) | 98W7 |

Remarks:

(1) protein S is a two-domain protein consisting of 172 amino acid residues with a 88 residues N-terminal domain and a 91 residues C-terminal domain

(2) data from kinetics of unfolding and folding

(3) measured in 25 mM MOPS, 3 mM $CaCl_2$, pH 7.0

(4) kinetics followed by optical absorbance at 286 nm

(5) kinetics followed by CD at 222 nm

(6) data treatment by a two-state model assuming a linear depedence of ln(k) on the urea conc.

Protein S from *Myxococcus xanthus*, isolated N- and C-terminal domains (NPS and CPS), data from denaturant-induced unfolding

| Protein | pH | T | $\Delta G$ | $c_{1/2}$ | m | Appr./Rem. | | Ref |
|---|---|---|---|---|---|---|---|---|
| NPS ($-Ca^{2+}$) | 7.0 | 20 | 21±2 | 4.4±0.1 | 5.0±0.4 | urea | (1–3) | 99W9 |
| NPS ($+Ca^{2+}$) | 7.0 | 20 | 31±3 | 5.0±0.1 | 6.0±0.5 | urea | (1–3) | 99W9 |
| NPS | 2.0 | 20 | 26±4 | 2.2±0.1 | 12.0±1.0 | urea | (1–3) | 99W9 |
| CPS ($-Ca^{2+}$) | 7.0 | 20 | 15±2 | 1.8±0.1 | 8.0±0.5 | urea | (1–3) | 99W9 |
| CPS ($+Ca^{2+}$) | 7.0 | 20 | 21±2 | 2.6±0.1 | 8.0±1.0 | urea | (1–3) | 99W9 |

Remarks:

(1) linear exrapolation, LEM-SB

(2) transition monitored by optical absorption at 286 nm, far-UV CD at 222 nm, and fluorescence spectroscopy at 305 nm

(3) buffer: 25 mM MOPS-NaOH, 1 mM EDTA ($-Ca^{2+}$), or 3 mM $CaCl_2$ ($+Ca^{2+}$)

Protein S (PS) from *Myxococcus xanthus*, a βγ-crystallin homolog, and the C-terminal (CPS) and N-terminal (NPS) fragments
Data from DSC measurements

| Protein | pH | T | ΔG | Approach/Remarks | | Ref |
|---|---|---|---|---|---|---|
| data in the absence of $Ca^{2+}$: | | | | | | |
| PS trans. (1) | 7.0 | 20 | 29±3 | DSC | (1) | 99W10 |
| trans. (2) | 7.0 | 20 | 16±3 | DSC | (1) | 99W10 |
| NPS | 7.0 | 20 | 18±3 | DSC | (1) | 99W10 |
| | 7.0 | 20 | 21±2 | urea | (3) | 99W10 |
| CPS | 7.0 | 20 | 16±4 | DSC | (1) | 99W10 |
| | 7.0 | 20 | 15±2 | urea | (3) | 99W10 |
| data in the presence of $Ca^{2+}$: | | | | | | |
| PS trans. (1) | 7.0 | 20 | 39±4 | DSC | (2) | 99W10 |
| trans. (2) | 7.0 | 20 | 25±3 | DSC | (2) | 99W10 |
| NPS | 7.0 | 20 | 26±3 | DSC | (2) | 99W10 |
| | 7.0 | 20 | 31±3 | urea | (3) | 99W10 |
| CPS | 7.0 | 20 | 20±4 | DSC | (2) | 99W10 |
| | 7.0 | 20 | 20±2 | urea | (3) | 99W10 |

Remarks:

(1) DSC measurements in the presence of 1 mM EDTA

(2) DSC measurements in the presence of 3 mM $CaCl_2$

(3) data from Ref. 99W9

## Protooncogene Product (c-Myb)

c-myb protooncogene product, R2 subdomain of mouse c-Myb with natural and nonnatural amino acid replacements at position 103

| Protein | pH | T | ΔG | Approach/Remarks | | Ref |
|---|---|---|---|---|---|---|
| Ala-R2 | 7.5 | 25 | –3.9±0.2 | heat | (1–3) | 99M20 |
| Abu-R2 | 7.5 | 25 | 6.0±0.3 | heat | (1–3) | 99M20 |
| Val-R2 | 7.5 | 25 | 5.8±0.2 | heat | (1–4) | 99M20 |
| Cha-R2 | 7.5 | 25 | 3.8±0.3 | heat | (1–3) | 99M20 |
| Ail-R2 | 7.5 | 25 | 8.9±0.3 | heat | (1–3) | 99M20 |
| Ile-R2 | 7.5 | 25 | 8.7±0.5 | heat | (1–3) | 99M20 |
| Nle-R2 | 7.5 | 25 | 10.7±0.5 | heat | (1–3) | 99M20 |
| Chg-R2 | 7.5 | 25 | 10.3±0.5 | heat | (1–3) | 99M20 |
| Nva-R2 | 7.5 | 25 | 13.1±0.7 | heat | (1–3) | 99M20 |
| Leu-R2 | 7.5 | 25 | 16.0±1.0 | heat | (1–3) | 99M20 |

Explanations:

[Xaa103]-Myb(90–141) changing Val to Ala, Leu, Ile, 2-aminobutyric acid (Abu), norvaline (Nva), norleucine (Nle), *allo*-isoleucine (Ail), cyclohexylglycine (Chg), and cyclohexylalanine (Cha)

Remarks:

(1) measured in 50 mM potassium phosphate, 50 mM potassium chloride, 10 mM DTT, pH 7.5

(2) transition monitored by CD at 222 nm

(3) ΔG was calculated using $\Delta C_p$ = 1.35 kJ/mol/K for all mutants, see also Table 2

(4) Val-R2 = wild type

## Pseudoazurin

Apopseudoazurin from *Thiosphaera pantotropha*

| Transition | pH | T | $\Delta G$ | $c_{1/2}$ | m | Appr./Rem. | Ref |
|---|---|---|---|---|---|---|---|
| $N \rightarrow U_{eq}$ | 7.0 | 15 | 28.8±0.7 | | 7.84±0.10 | urea (1–3) | 99C2 |
| $N \rightarrow U_{eq}$ | 7.0 | 15 | 32.0±2.1 | | 8.5 ±0.6 | urea (2–4) | 99C2 |
| $N \rightarrow U_{nat}$ | 7.0 | 15 | 30.0±1.7 | | 7.0 ±0.4 | urea (2,4,5) | 99C2 |

Remarks:

(1) data from equilibrium denaturation monitored by far-UV CD at 220 nm

(2) measured in 0.5 M $Na_2SO_4$

(3) $U_{eq}$ = unfolded state with relaxed equilibrium forms of the eight proline residues

(4) data from kinetic experiments, two-state fit; kinetics shows the rapid formation of an intermediate that is stabilized by 25 kJ/mol

(5) $U_{nat}$ = unfolded state with all native isomeric forms of proline residues

## Pullulanase

Heat-stable pullulanase from *Pyrococcus wosei* in the presence and absence of $Ca^{2+}$

| $Ca^{2+}$ Conc. | pH | T | $\Delta G$ | $c_{1/2}$ | m | Appr./Rem. | Ref |
|---|---|---|---|---|---|---|---|
| 0 µM | 7.4 | 20 | 25.29±2.72 | 4.88±0.53 | 5.180±0.556 | GuHCl (1–3) | 99S5 |
| 0 µM | 7.4 | 20 | 26.40±1.42 | 4.85±0.26 | 5.439±0.293 | GuHCl (1,2,4) | 99S5 |
| 0 µM | 7.4 | 20 | 28.59±1.81 | 4.90±0.31 | 5.828±0.368 | GuHCl (1,2,5) | 99S5 |
| 0 µM | 7.4 | 20 | 26.30±1.68 | 4.86±0.31 | 5.414±0.347 | GuHCl (1,2,6) | 99S5 |
| 0 µM | 7.4 | 20 | 29.50±1.76 | 4.86±0.29 | 6.067±0.360 | GuHCl (1,2,7) | 99S5 |
| 120 µM | 7.4 | 20 | 35.38±4.60 | 5.02±0.65 | 7.046±0.916 | GuHCl (1–3) | 99S5 |
| 120 µM | 7.4 | 20 | 37.45±2.25 | 5.04±0.30 | 7.427±0.448 | GuHCl (1,2,4) | 99S5 |
| 120 µM | 7.4 | 20 | 35.23±2.31 | 5.04±0.33 | 6.987±0.456 | GuHCl (1,2,5) | 99S5 |
| 120 µM | 7.4 | 20 | 35.91±2.31 | 5.06±0.31 | 7.092±0.439 | GuHCl (1,2,6) | 99S5 |
| 120 µM | 7.4 | 20 | 36.35±2.28 | 5.10±0.32 | 7.133±0.448 | GuHCl (1,2,7) | 99S5 |

Remarks:

(1) linear extrapolation, LEM-SB

(2) buffer: 25 mM Tris-HCl, pH 7.4, and 100 µM EGTA, with and without 120 µM $CaCl_2$

(3) transition based on the molar ellipticity at 222 nm

(4) transition based on the singular value decomposition algorithm (SVD) applied to the far-UV CD spectra (210–250 nm)

(5) transition from CD spectra (210–250 nm), SVD and matrix decomposition ($V_1$ matrix)

(6) transition monitored by fluorescence intensity at 340 nm

(7) transition monitored by the intensity-averaged emission wavelength

## Pyrophosphatase

Pyrophosphatases from *Sulfolobus acidocaldarius* (S.ac.), *Escherichia coli* (E.coli), and *Thermus thermophilus* (Th.th.), data from GuHCl-induced unfolding

| Protein | | pH | T | $\Delta G$ | $c_{1/2}$ | m | Approach/Remarks | Ref |
|---|---|---|---|---|---|---|---|---|
| *S.ac.* | $Mg^{2+}$ | 7.0 | 25 | 247±11 | 3.2 | 34±4 | GuHCl, FL (1–3) | 99H2 |
| | $Mg^{2+}$ | 7.0 | 25 | 250±3 | 2.9 | 29±1 | GuHCl, FL (1–3) | 99H2 |
| | EDTA | 7.0 | 25 | 246±6 | 3.1 | 25±2 | GuHCl, FL (1–3) | 99H2 |
| | EDTA | 7.0 | 25 | 240±2 | 3.1 | 23±1 | GuHCl, CD (1–3) | 99H2 |
| | $Mg^{2+}$ | 7.0 | 50 | 243±8 | 2.8 | 29±3 | GuHCl, CD (1–3) | 99H2 |
| | EDTA | 7.0 | 50 | 236±10 | 2.9 | 24±4 | GuHCl, CD (1–3) | 99H2 |

Pyrophosphatases from *Sulfolobus acidocaldarius* (S.ac.), *Escherichia coli* (E.coli), and *Thermus thermophilus* (Th.th.), data from GuHCl-induced unfolding (continued)

| Protein | | pH | T | $\Delta G$ | $c_{1/2}$ | m | Approach/Remarks | Ref |
|---|---|---|---|---|---|---|---|---|
| *E.coli* | $Mg^{2+}$ | 7.0 | 25 | 236±3 | 4.9 | 14±1 | GuHCl, FL (1–3) | 99H2 |
| | $Mg^{2+}$ | 7.0 | 25 | 236±6 | 4.9 | 14±2 | GuHCl, FL (1–3) | 99H2 |
| | EDTA | 7.0 | 25 | 220±7 | 4.5 | 11±1 | GuHCl, FL (1–3) | 99H2 |
| *Th.th.* | $Mg^{2+}$ | 7.0 | 25 | 290±6 | 5.1 | 24±2 | GuHCl, FL (1–3) | 99H2 |
| | EDTA | 7.0 | 25 | 283±6 | 4.7 | 24±2 | GuHCl, FL (1–3) | 99H2 |
| | EDTA | 7.0 | 25 | 281±10 | 4.7 | 24±4 | GuHCl, CD (1–3) | 99H2 |

Remarks:

(1) fit of the transition curve by a two-state process comprising dissociation and denaturation of the folded hexamer into six monomers assuming a linear dependence of $\Delta G$ on the denaturant conc.

(2) transition monitored by far-UV CD at 226 (CD) or fluorescence intensity at 340 nm (FL)

(3) buffer: 50 mM Tris-HCl, pH 7.0

Pyrophosphatases from *Sulfolobus acidocaldarius* (S.ac.), *Escherichia coli* (E.coli), and *Thermus thermophilus* (Th.th.), data from heat denaturation

| Protein | | pH | $T_{trs}$ | $\Delta G$ | Appr./Rem. | Ref |
|---|---|---|---|---|---|---|
| *S.ac.* | + 5 mM EDTA | 7.0 | 89 | 188 | heat (1,2) | 99H2 |
| | + 5 mM $Mg^{2+}$ | 7.0 | 98 | 193 | heat (1,2) | 99H2 |
| | + 5 mM $Mn^{2+}$ | 7.0 | >100 | 195 | heat (1,2) | 99H2 |
| *E.coli* | + 5 mM EDTA | 7.0 | 58 | 172 | heat (1,2) | 99H2 |
| | + 5 mM $Mg^{2+}$ | 7.0 | 84 | 185 | heat (1,2) | 99H2 |
| | + 5 mM $Mn^{2+}$ | 7.0 | 93 | 190 | heat (1,2) | 99H2 |
| *Th.th.* | + 5 mM EDTA | 7.0 | 86 | 186 | heat (1,2) | 99H2 |
| | + 5 mM $Mg^{2+}$ | 7.0 | 99 | 194 | heat (1,2) | 99H2 |
| | + 5 mM $Mn^{2+}$ | 7.0 | 96 | 192 | heat (1,2) | 99H2 |

Remarks:

(1) for the approach see Ref. 99H2

(2) treatment of the transition curve by a two-state process comprising dissociation and denaturation of the folded hexamer into six monomers

## Pyrrolidone Carboxyl Peptidase

Pyrrolidone carboxyl peptidase from hyperthermophilic archaeon *Pyrococcus furiosus* (*Pf*PCP) and from mesophile *Bacillus amyloliquefaciens* (*Ba*PCP)

| Protein | pH | T | $\Delta G$ | $c_{1/2}$ | m | Appr./Rem. | Ref |
|---|---|---|---|---|---|---|---|
| *Pf*PCP | 7 | 60 | 54.4±1.7 | 3.2 | 17.3±0.5 | GuHCl (1–3) | 98O3 |
| *Pf*PCP double mutant (Cys142→Ser and Cys188→Ser) | | | | | | | |
| | 7 | 60 | 56.6±4.1 | 3.4 | 17.4±1.3 | GuHCl (1–3) | 98O3 |
| | 7 | 60 | 57.8 | | | GuHCl (2–4) | 98O3 |
| | 7 | 25 | 78.6 | | | GuHCl (2–4) | 98O3 |
| *Ba*PCP | 7 | 40 | 7.6±0.8 | 0.4 | 19.5±1.8 | GuHCl (1–3) | 98O3 |
| *Ba*PCP | 7 | 25 | 41.8 | | | GuHCl (2–4) | 98O3 |

Remarks:

(1) linear extrapolation

(2) transition monitored by CD at 222 nm

(3) buffer: 20 mM Tris-HCl, 2 mM EDTA, 0.1 mM DTE for *Pf*PCP, and the same buffer without EDTA and DTE for *Ba*PCP

(4) from kinetics of folding and unfolding in the presence of GuHCl

## Pyruvate Dehydrogenase

Minidomain derived from peripheral subunit-binding domain of the dihydrolipoamide acetyltransferase component (E2) of the pyruvate dehygrogenase multienzyme complex from *Bacillus stearothermophilus*, 41 residue peptide (psbd41) and mutant Asp34→Asn (psbd41Asn)

| Mutant | pH | T | $\Delta G$ | $c_{1/2}$ | m | Approach/Remarks | | Ref |
|---|---|---|---|---|---|---|---|---|
| psbd41 | 8.0 | 25 | 13.0±2.1 | 4.1 | 3.18±0.54 | GuHCl | (1–3) | 98S16 |
| psbd41Asn | 8.0 | 25 | 5.9 | 1.9 | 3.10 | GuHCl | (1,2,4) | 98S16 |

Remarks:

(1) linear extrapolation

(2) transition monitored by CD at 222 nm

(3) thermal denaturation monitored by far-UV CD (222 nm), near-UV CD (280 nm), and NMR chemical shift of Tyr10 yielded coincident transitions with $T_{trs} = 54°C$ and $\Delta H$ from 109 to 176 kJ/mol, depending on the assumed $\Delta Cp$ value (2.70 and 1.26 kJ/mol/K)

(4) thermal transition at $T_{trs} = 31°C$


## Repressor Proteins

The data entries are arranged as follows:

a) Arc repressor
b) Lac repressor
c) λ repressor
d) λ Cro repressor
e) 434-phage repressor
f) Tet repressor
g) tryptophan repressor

*a) Arc repressor*

Arc *st11* repressor dimer, wild type and MYL mutant, transitions

| Transition | pH | T | $\Delta G$ | Approach | Remarks | Ref |
|---|---|---|---|---|---|---|
| wild type, two-state: | | | | | | |
| $N_2 \rightarrow 2D$ | 7.5 | 25 | 43.9 | heat | (1–4) | 97R5 |
| wild type, three-state: | | | | | | |
| $N_2 \rightarrow D_2$ | 7.5 | 25 | 20.1 | heat | (1,2) | 97R5 |
| $D_2 \rightarrow 2D$ | 7.5 | 25 | 23.0 | heat | (1,2) | 97R5 |
| MYL mutant, two-state: | | | | | | |
| $N_2 \rightarrow 2D$ | 7.5 | 25 | 62.3 | heat | (1,2) | 97R5 |
| MYL mutant, three-state: | | | | | | |
| $N_2 \rightarrow D_2$ | 7.5 | 25 | 45.6 | heat | (1,2) | 97R5 |
| $D_2 \rightarrow 2D$ | 7.5 | 25 | 16.7 | heat | (1,2) | 97R5 |

Remarks:

(1) from thermal denaturation, varying protein concentrations from 5 to 334 µM

(2) buffer: 50 mM Tris-HCl, 250 mM KCl, 0.1 mM EDTA, pH 7.5

(3) poor fit at protein concentration above 160 µM

(4) formation of a denatured dimer limits the thermal stability of Arc repressor

(5) MYL mutant = triple mutant (Arg31→Met, Glu36→Tyr, and Arg40→Leu) of Arc *st11* which has a C-terminal extension (His)$_6$-Lys-Asn-Gln-His-Glu

Arc repressor, multiple alanine substitutions

| Mutant | pH | $T_{trs}$ | $\Delta T$ | $\Delta G(25°C)$ | Approach/Remarks | Ref |
|---|---|---|---|---|---|---|
| Arc native | 7.5 | 57.9±0.6 | 0.0 | 46.0±2.1 | heat/GuHCl (1–3) | 99B19 |
| 3 Ala | 7.5 | 57.7±0.6 | –0.2 | | heat/GuHCl (1–3) | 99B19 |
| 7 Ala | 7.5 | 59.8±0.6 | 1.9 | | heat/GuHCl (1–3) | 99B19 |
| 11 Ala | 7.5 | 69.2±0.6 | 11.3 | | heat/GuHCl (1–3) | 99B19 |
| 12 Ala | 7.5 | 71.1±0.6 | 13.2 | 66.9±2.1 | heat/GuHCl (1–3) | 99B19 |
| 15 Ala | 7.5 | 72.3±0.6 | 14.4 | 63.6±2.1 | heat/GuHCl (1–3) | 99B19 |

Remarks:
(1) $T_{trs}$ from melting curves, $\Delta G(25°C)$ from GuHCl denaturation
(2) transition monitored by CD at 222 nm
(3) buffer: 50 mM Tris-HCl, 250 mM KCl, 0.2 mM EDTA, pH 7.5

Arc repressor (Arc) in the presence of polyvinylsulfate (PVS)

| Protein | pH | T | $\Delta G$ | m | Appr./Rem. | Ref |
|---|---|---|---|---|---|---|
| Arc | 7.5 | 25 | 9.98 | 1.41 | urea (1–3) | 99R10 |
| Arc + PVS | 7.5 | 25 | 14.03 | 1.43 | urea (1–4) | 99R10 |

Remarks:
(1) linear extrapolation, $\Delta G$ per mol of dimer
(2) transition monitored by far-UV CD at 230 nm
(3) measured in 50 mM Tris, 50 mM KCl, pH 7.5
(4) in the presence of 500 µM PVS

*b) Lac repressor*

Lactose repressor, integrated analysis of unfolding and dissociation

| Transition | pH | T | $\Delta G$ | Approach/Remarks | | Ref |
|---|---|---|---|---|---|---|
| $M_4 \rightarrow 4U$ | 7.4 | | ~253 | urea | (1–3) | 99B2 |
| $M \rightarrow U$ | 7.4 | | 20.1 | urea | (4) | 99B2 |
| $M_2 \rightarrow 2M$ | 7.4 | | 40.6 | urea | (5) | 99B2 |
| $M_2 \rightarrow 2M$ | 7.4 | | 41.8 | urea | (1,2,6) | 99B2 |
| $M_4 \rightarrow 2M_2$ | 7.4 | | ~85 | urea | (1,2,7) | 99B2 |
| $M_4 \rightarrow 4U$ | 7.4 | | ~249 | urea | (1,2,8) | 99B2 |

Remarks:
(1) Ref. 99B2 presents an integrated analysis of lactose repressor unfolding/renaturation and dissociation based on CD and fluorescence spectroscopy, operator/inducer binding, and sedimentation equilibria
(2) measured in 0.01 M Tris-HCl, 0.1 M $K_2SO_4$, pH 7.4
(3) from global analysis varying protein conc., the data support the $M_4 \rightarrow U_4 \rightarrow 4U$ model
(4) data are based on a monomeric lac repressor mutant Tyr28→Asp, see also Ref. 94C6
(5) data are based on a dimeric repressor (deletion mutant), see also Ref. 94C6
(6) present study
(7) present study using the Lys84→Leu mutant
(8) total free energy given by the sum of the following contributions: $\Delta G(M_4 \rightarrow 4U) = \Delta G(M_4 \rightarrow 2M_2) + 2 \times \Delta G(M_2 \rightarrow 2M) + 4 \times \Delta G(M \rightarrow U)$

Lactose repressor, substitutions that alter the monomer-monomer interaction

| Mutant | pH | ΔG | m | Approach/Remarks | | Ref |
|---|---|---|---|---|---|---|
| –11 deletion | 7.5 | 74.9±0.8 | 28.9±0.8 | urea | (1–3) | 97N6 |
| –11 deletion | 7.5 | 80.3±3.3 | | urea | (1–3,4) | 94C6 |
| –11 deletion | 7.5 | 80.8±5.9 | | urea | (1–3,5) | 94C6 |
| double mutant (Lys84→Leu and Tyr282→Asp)/–11 deletion | | | | | | |
| | 7.5 | 65.7±2.9 | 26.8±1.7 | urea | (1–3) | 97N6 |
| double mutant (Lys84→Ala and Tyr282→Asp)/–11 deletion | | | | | | |
| | 7.5 | 58.6±0.4 | 19.2±0.4 | urea | (1–3) | 97N6 |
| R3 | 7.5 | 100.0±4.2 | | urea | (2,3,5,6) | 94C6 |
| Tyr282→Asp | 7.5 | 20.1±1.3 | | urea | (3,5,7) | 94C6 |
| Lys84→Ala/–11 deletion | | | | | | |
| | 7.5 | ~88 | | urea | (1–3) | 97N6 |

Remarks:

(1) –11 deletion = 11 C-terminal residues deleted

(2) ΔG refers to transition of a dimer to unfolded monomers, data treatment assuming a linear dependence of ΔG on denaturant concentration

(3) buffer: 10 mM Tris-HCl, 0.1 M potassium sulfate, pH 7.5

(4) transition monitored by CD at 222 nm

(5) transition monitored by fluorescence

(6) R3 = mutant with the C-terminal Leu heptad repeats replaced by the GCN4 dimerization sequence

(7) Tyr282→Asp = monomeric mutant, equilibrium treatment by a monomer unfolding model

*c) λ repressor*

λ repressor, monomeric version of N-terminal domain of lambda repressor $\lambda_{6-85}$, wild type (w.t.), wild type* (w.t.*, see remark (1)), and variants

| Protein | pH | T | ΔG | $c_{1/2}$ | m | Appr./Remarks | | Ref |
|---|---|---|---|---|---|---|---|---|
| data from CD urea titration in H₂O: | | | | | | | | |
| w.t.* | 8.0 | 25 | 22.43 | 5.20 | 4.31 | urea | (1–5) | 99M23 |
| Asp14→Ala* | 8.0 | 25 | 15.86 | 3.96 | 4.02 | urea | (1–6) | 99M23 |
| Ser77→Ala* | 8.0 | 25 | 17.82 | 4.06 | 4.39 | urea | (1–6) | 99M23 |
| double mutant (Asp14→Ala* and Ser77→Ala*) | | | | | | | | |
| | 8.0 | 25 | 16.48 | 4.34 | 3.81 | urea | (1–6) | 99M23 |
| w.t. | 8.0 | 25 | 15.27 | 3.54 | 4.31 | urea | (1–5) | 99M23 |
| Asp14→Ala | 8.0 | 25 | 8.08 | 2.19 | 3.68 | urea | (1–5) | 99M23 |
| Ser77→Ala | 8.0 | 25 | 10.33 | 2.45 | 4.23 | urea | (1–5) | 99M23 |
| double mutant (Asp14→Ala and Ser77→Ala) | | | | | | | | |
| | 8.0 | 25 | 10.00 | 2.51 | 3.97 | urea | (1–5) | 99M23 |
| data from NMR line shape analysis in D₂O: | | | | | | | | |
| w.t.* | 8.0 | 25 | 26.15 | 5.53 | 4.73 | urea | (1,2,7–9) | 99M23 |
| Asp14→Ala* | 8.0 | 25 | 19.20 | 4.25 | 4.52 | urea | (2,6–9) | 99M23 |
| Ser77→Ala* | 8.0 | 25 | 22.47 | 4.55 | 4.94 | urea | (2,6–9) | 99M23 |
| double mutant (Asp14→Ala* and Ser77→Ala*) | | | | | | | | |
| | 8.0 | 25 | 22.50 | 4.54 | 4.52 | urea | (2,6–9) | 99M23 |
| w.t. | 8.0 | 25 | 17.57 | 3.82 | 4.60 | urea | (2,7–10) | 99M23 |
| Asp14→Ala | 8.0 | 25 | 10.75 | 2.54 | 4.23 | urea | (2,7–10) | 99M23 |
| Ser77→Ala | 8.0 | 25 | 13.39 | 2.67 | 5.02 | urea | (2,7–10) | 99M23 |
| double mutant (Asp14→Ala and Ser77→Ala) | | | | | | | | |
| | 8.0 | 25 | 12.93 | 2.86 | 4.52 | urea | (2,7–10) | 99M23 |

data concerning the transition state:

| Protein | pH | T | $\Delta(\Delta G_{TS-D})$ | $\Delta(\Delta G_{TS-N})$ | Appr./Remarks | Ref |
|---|---|---|---|---|---|---|
| w.t.* | 8.0 | 25 | –1.13±0.84 | –7.28±0.75 | urea (11,12) | 99M23 |
| Asp14→Ala* | 8.0 | 25 | 0.33±0.96 | –4.02±0.79 | urea (11,12) | 99M23 |
| Ser77→Ala* | 8.0 | 25 | –1.42±0.96 | –6.44±0.84 | urea (11,12) | 99M23 |
| double mutant (Asp14→Ala* and Ser77→Ala*) | | | | | | |
| | 8.0 | 25 | 0.63±1.59 | 4.85±1.38 | urea (11,12) | 99M23 |
| w.t. | 8.0 | 25 | –0.50±0.25 | 6.28±0.29 | urea (11,12) | 99M23 |
| Asp14→Ala | 8.0 | 25 | 1.05±0.25 | 4.35±0.21 | urea (11,12) | 99M23 |
| Ser77→Ala | 8.0 | 25 | –1.09±0.25 | 5.40±0.29 | urea (11,12) | 99M23 |
| double mutant (Asp14→Ala and Ser77→Ala) | | | | | | |
| | 8.0 | 25 | 1.63±0.42 | 5.23±0.46 | urea (11,12) | 99M23 |

Remarks:

(1)  w.t.* is a more stable variant of w.t. produced by Gly46→Ala and Gly48→Ala double mutation in helix 3

(2)  linear extrapolation, LEM-SB

(3)  transition monitored by CD

(4)  buffer: 20 mM potassium phosphate, pH 8.0, 100 mM NaCl

(5)  errors are estimated to be ±0.1 M in $c_{1/2}$, ±0.21 in m, and ±0.84 in $\Delta G$

(6)  mutant derived from w.t.*

(7)  transition monitored by NMR line shape analysis

(8)  buffer: 20 mM potassium phoshate, pD 8.0, 100 mM NaCl, 1 mM sodium azide, 1 mM TMSP (3-(trimethylsilyl)-tetradeutero-propiomic acid)

(9)  errors are estimated to be ±1.25 in $\Delta G$

(10) mutants derived from w.t.

(11) transition state relative to the denatured and native state, respectively

(12) calculated from equilibrium stabilities and kinetic parameters from NMR line shape analysis in 3 M urea

*d) λ Cro repressor*

λ Cro repressor, wild type and mutant

| Mutant | pH | T | $\Delta G$ | Approach | Remarks | Ref |
|---|---|---|---|---|---|---|
| wild type | 7 | | 46.8±2.5 | GuHCl | (1–3) | 97J3 |
| Phe58→Trp | 7 | | 56.9±2.1 | GuHCl | (1–3) | 97J3 |
| monomeric reference proteins: | | | | | | |
| natural monomer | | | 8.8 | GuHCl | | 97J3 |
| engineered monomer | | | 12.6 | GuHCl | (4) | 97M7 |

Remarks:

(1) linear extrapolation procedure for a concerted dimer-dissociation and unfolding reaction $F_2 \rightarrow 2U$

(2) transition monitored by CD at 222 nm

(3) buffer: 20 mM potassium phosphate, 200 mM potassium chloride and EDTA

(4) engineered monomer in which the C-terminal extension Lys56-[Asp-Gly-Glu-Val-Lys] forms a β-hairpin, and destabilizes the dimer interface, see Ref. 97J3

*e) 434-phage repressor*

434-phage repressor, N-terminal domain, residues 1–69 (R69)

| Protein | Salt | pH | T | $\Delta G$ | Appr./Rem. | Ref |
|---|---|---|---|---|---|---|
| R69 | L | 7.0 | 25 | 22.4±1.5 | DSC (1) | 99R21 |
| | H | 7.0 | 25 | 22.3±1.5 | DSC (1) | 99R21 |
| | L | 4.5 | 25 | 18.4±1.0 | DSC (1) | 99R21 |
| | H | 4.5 | 25 | 18.1±1.0 | DSC (1) | 99R21 |
| | L | 4.0 | 25 | 15.0±0.8 | DSC (1) | 99R21 |
| | H | 4.0 | 25 | 15.8±0.9 | DSC (1) | 99R21 |
| | L | 3.0 | 25 | 10.8±0.8 | DSC (1) | 99R21 |

434-phage repressor, N-terminal domain, residues 1–69 (R69) (continued)

| Protein | Salt | pH | T | ΔG | Appr./Rem. | | Ref |
|---|---|---|---|---|---|---|---|
| | H | 3.0 | 25 | 9.5±0.8 | DSC | (1) | 99R21 |
| | L | 2.0 | 25 | 0.9±0.5 | DSC | (1) | 99R21 |
| | H | 2.0 | 25 | 4.9±0.6 | DSC | (1) | 99R21 |

Remark:

(1) measured in 20 mM buffer: PIPES (pH 7.0), sodium acetate (pH 4.5 and 4.0), and glycine (pH 3.0 and 2.0), L = low salt, H = high salt, the same buffer with 200 mM NaCl

*f) Tet repressor*

Tet repressor (TetR) dimers, variants

| Mutant | pH | T | ΔG | $c_{1/2}$ | Approach/Remarks | | Ref |
|---|---|---|---|---|---|---|---|
| TetR variant: | | | | | | | |
| (B) | 7.5 | 25 | 74.6±4 | 4.7 | urea | (1–3) | 99S4 |
| (B/D)128 | 7.5 | 25 | 77.2±3 | 4.9 | urea | (1–3) | 99S4 |
| (B/D)179–184 | 7.5 | 25 | 77.9±1 | 4.8 | urea | (1–3) | 99S4 |
| (B/D)128,179–184 | 7.5 | 25 | 86.7±3 | 5.6 | urea | (1–3) | 99S4 |

Remarks:

(1) linear extrapolation

(2) for the sequences of the TetR variants see Ref. 99S4

(3) transition monitored by CD at 220 nm

(4) buffer: 50 mM Tris-HCl, 150 mM NaCl, 5 mM MgCl₂, 1 mM EDTA, 1 mM DTT, pH 7.5

*g) tryptophan repressor*

Tryptophan repressor from *E. coli*

| Mutant | pH | T | ΔG | m | Approach/Remarks | Ref |
|---|---|---|---|---|---|---|
| wild type | 7.6 | 25 | 96±8 | 12.6±1.3 | urea CD  (1–3,6) | 97G7 |
| wild type | 7.6 | 25 | 86±2 | 10.5±0.4 | urea IAEW (1,2,4,6) | 97G7 |
| wild type | 7.6 | 25 | 96±13 | 12.6±2.1 | urea FL  (1,2,5,6) | 97G7 |
| wild type | 7.6 | 25 | 100±8 | 13.0±1.3 | urea glob.(1,2,7) | 97G7 |
| [2–66]₂ (8) | 7.6 | 25 | 52±4 | 7.5±0.8 | urea CD  (1–3,6) | 97G7 |
| [2–66]₂ (8) | 7.6 | 25 | 60±1 | 9.2±1.7 | urea IAEW (1,2,4,6) | 97G7 |
| [2–66]₂ (8) | 7.6 | 25 | 54±8 | 8.8±1.3 | urea FL  (1,2,5,6) | 97G7 |
| [2–66]₂ (8) | 7.6 | 25 | 60±1 | 8.4±0.4 | urea glob.(1,2,7) | 97G7 |
| [NHIS–7–66]₂ (9) | 7.6 | 25 | 59±8 | 8.8±2.1 | urea CD  (1–3,6) | 97G7 |
| [NHIS–7–66]₂ (9) | 7.6 | 25 | 64±3 | 10.0±1.7 | urea IAEW (1,2,4,6) | 97G7 |
| [NHIS–7–66]₂ (9) | 7.6 | 25 | 54±4 | 8.4±2.1 | urea FL  (1,2,5,6) | 97G7 |
| [NHIS–7–66]₂ (9) | 7.6 | 25 | 61±2 | 9.2±0.4 | urea glob.(1,2,7) | 97G7 |

Remarks:

(1) two-state dimer model for native dimer in equilibrium with unfolded monomers, assuming a linear dependence on denaturant concentration

(2) buffer: 10 mM potassium phosphate, 0.1 mM EDTA, pH 7.6

(3) CD = transition monitored by CD at 222 nm

(4) IAEW = transition monitored by fluorescence intensity-averaged emission wavelength

(5) FL = transition monitored by fluorescence intensity at 330 nm

(6) local fit according to model (1)

(7) glob. = global fit, i.e., simultaneous fit of multiple data sets collected by fluorescence and CD at varying concentrations of monomer

(8) [2–66]₂ dimeric Trp repressor fragment containing residues 2–66 of wild type

(9) [NHIS-7–66]₂ dimeric Trp repressor fragment containing residues 7–66 of wild type and six additional His residues at the N-terminus

Tryptophan repressor (TR) from *E. coli*, comparison of wild-type dimeric TR and monomeric Leu39→Glu TR, and the burst-phase intermediate species

| Mutant/Conc. | pH | T | $\Delta G$ | $c_{1/2}$ | m | Appr./Rem. | Ref |
|---|---|---|---|---|---|---|---|
| wild type, dimeric TR | | | | | | | |
| | 7.6 | 25 | 93.7±2.9 | 5.6±0.1 | 11.72±0.42 | urea (1,2,4) | 97S8 |
| mutant (Leu39→Glu) monomeric TR | | | | | | | |
| 4.3 µM | 7.6 | 25 | 10.8±1.6 | 2.69±0.12 | 4.02±0.50 | urea (1,3,4) | 97S8 |
| 4.3 µM | 7.6 | 25 | 10.1±1.5 | 2.69±0.11 | 3.77±0.46 | urea (1,3,5) | 97S8 |
| 19.3 µM | 7.6 | 25 | 10.2±2.3 | 2.75±0.17 | 3.72±0.71 | urea (1,3,4) | 97S8 |
| 19.3 µM | 7.6 | 25 | 9.7±0.9 | 2.81±0.09 | 3.47±0.25 | urea (1,3,5) | 97S8 |
| | 7.6 | 25 | 9.9±0.6 | 2.77±0.05 | 3.60±0.17 | urea (1,3,6) | 97S8 |
| burst-phase species | | | | | | | |
| | 7.6 | 25 | 15.1±1.3 | 3.6 | 4.2 ±0.4 | urea (2,4) | 97S8 |

Remarks:

(1) linear extrapolation, LEM-SB

(2) data from Ref. 93M6

(3) buffer: 10 mM sodium phosphate, 0.1 mM $Na_2$EDTA, pH 7.6

(4) transition monitored by CD at 222 nm

(5) transition monitored by tryptophan fluorescence

(6) global fit of signals monitored by far-UV CD, tryptophan fluorescence, near-UV CD at 300 nm, and absorbance at 292 nm

Monomeric tryptophan repressor, mutants that replace a single Trp residue

| Protein/Conc. | pH | T | $\Delta G$ | m | Approach/Remarks | Ref |
|---|---|---|---|---|---|---|
| double mutant (Trp19→Phe and Leu39→Glu) | | | | | | |
| 2.6 µM | 7.6 | 25 | 7.49±1.42 | 3.39±0.38 | urea, CD (1–3) | 98S11 |
| 2.6 µM | 7.6 | 25 | 6.36±1.13 | 3.10±0.38 | urea, FL (1,2,4) | 98S11 |
| 14.4 µM | 7.6 | 25 | 7.66±0.79 | 3.47±0.21 | urea, CD (1–3) | 98S11 |
| 14.4 µM | 7.6 | 25 | 8.41±0.42 | 3.81±0.13 | urea, FL (1,2,4) | 98S11 |
| | 7.6 | 25 | 8.08±0.33 | 3.77±0.08 | urea, glob.(1,2,5) | 98S11 |
| 2.23 µM | 7.6 | 25 | 7.57±0.13 | 3.60±0.08 | urea, anis.(1,2,6) | 98S11 |
| double mutant (Trp99→Phe and Leu39→Glu) | | | | | | |
| 2.6 µM | 7.6 | 25 | 6.57±1.46 | 3.31±0.33 | urea, CD (1–3) | 98S11 |
| 2.6 µM | 7.6 | 25 | 7.57±0.42 | 3.81±0.13 | urea, FL (1,2,4) | 98S11 |
| 14.4 µM | 7.6 | 25 | 7.87±1.21 | 3.68±0.33 | urea, CD (1–3) | 98S11 |
| 14.4 µM | 7.6 | 25 | 7.49±0.79 | 3.56±0.25 | urea, FL (1,2,4) | 98S11 |
| | 7.6 | 25 | 7.70±0.33 | 3.81±0.08 | urea, glob.(1,2,5) | 98S11 |
| 2.23 µM | 7.6 | 25 | 7.36±0.21 | 3.72±0.08 | urea, anis.(1,2,6) | 98S11 |
| Leu39→Glu | 7.6 | 25 | 9.92±0.63 | 3.60±0.17 | urea (1,7) | 98S11 |

Remarks:

(1) linear extrapolation, for the procedure see also Refs. 98S11 and 97S8

(2) measured in 10 mM sodium phosphate, 0.1 mM $Na_2$EDTA, pH 7.6

(3) transition monitored by CD at 222 nm

(4) transition monitored by fluorescence emission, the fluorescence emission spectra were decomposed into spectral and urea-dependent basis vectors

(5) global fit of data sets obtained at different protein concentration by CD and fluorescence

(6) transition monitored by fluorescence anisotropy

(7) data from Ref. 97S8

Dimeric core domain of *E. coli* Tryptophan repressor ([2–66]$_2$ TR), transitions

| Transition | pH | T | $\Delta G$ | m | Approach/Remarks | | Ref |
|---|---|---|---|---|---|---|---|
| I$_2$ → N$_2$ | 7.6 | 25 | 10.4±2.0 | 3.22±0.29 | kinet. | (1,3) | 98G7 |
| | 7.6 | 25 | 10.6±0.1 | 3.26±0.04 | kinet. | (2,3) | 98G7 |
| 2U → I$_2$ | 7.6 | 25 | 43.1±1.3 | 5.02±0.33 | kinet. | (1,3) | 98G7 |
| 2U → N$_2$ | 7.6 | 25 | 53.6±2.1 | 8.28±0.63 | kinet. | (1,3) | 98G7 |
| | 7.6 | 25 | 55.6±0.8 | 8.37±0.21 | urea | (3,4) | 98G7 |

Remarks:

(1) from kinetic data, local fit

(2) from kinetic data, global fit

(3) measured in 10 mM potassium phosphate, pH 7.6, 1 mM EDTA

(4) reference value in Ref. 98G7 from equilibrium unfolding, taken from Ref. 97G7

(5) $\Delta H$ and $\Delta Cp$ for the I$_2$ → N$_2$ transition amount to –9.6 kJ/mol at 25°C and ~2.9 kJ/mol/K, respectively, see Ref. 98G8

Tryptophan repressor, wild type and temperature-sensitive mutant Leu75→Phe

| Protein | pH | T | $\Delta G$ | m | Approach/Remarks | | Ref |
|---|---|---|---|---|---|---|---|
| wild type | 7.6 | 25 | 84.9±2.1 | 10.0±0.4 | urea | (1–4) | 99J10 |
| Leu75→Phe | 7.6 | 25 | 92.0±2.1 | 10.5±0.4 | urea | (1–4) | 99J10 |

Remarks:

(1) linear extrapolation

(2) data per mol of dimer

(3) transition monitored by CD at 222 nm

(4) buffer: 10 mM sodium phosphate, 0.1 M NaCl, 0.1 mM EDTA, pH 7.6

Tryptophan repressor (TR), pressure denaturation

| TR Dimer Conc. | pH | T | $\Delta G_{linked}$ | $\Delta G_{unlinked}$ | Approach/Remarks | Ref |
|---|---|---|---|---|---|---|
| 5 μM | 7.6 | 21 | 50.2 | 51.0 | pressure (1,2,4,5) | 99D4 |
| 1 μM | 7.6 | 21 | 48.1 | 47.7 | pressure (1,3–5) | 99D4 |

Remarks:

(1) measured in 10 mM Tris, pH 7.6, 200 mM KCl, and 2.5 M GuHCl

(2) half transition at about 1.3 kbar

(3) half transition at about 0.9 kbar

(4) the raw intensity data were fit to a two-state dimer to unfolded monomer model, for details see 99D4

(5) the kinetics of pressure-jump relaxation yields $\Delta G$ = 49.4 kJ/mol and 50.2 kJ/mol at 5 μM and 1 μM, respectively

Tryptophan repressor (TR), GuHCl dependence of high pressure denaturation

| GuHCl Conc. | pH | T | $\Delta G_{linked}$ | $\Delta G_{unlinked}$ | Approach/Remarks | Ref |
|---|---|---|---|---|---|---|
| 2.1 M | 7.6 | 21 | 51.0 | 51.0 | pressure (1) | 99D4 |
| 2.3 M | 7.6 | 21 | 54.0 | 55.2 | pressure (1) | 99D4 |
| 2.5 M | 7.6 | 21 | 50.2 | 51.0 | pressure (1) | 99D4 |
| 2.6 M | 7.6 | 21 | 46.9 | 47.3 | pressure (1) | 99D4 |
| 2.7 M | 7.6 | 21 | 41.0 | 41.0 | pressure (1) | 99D4 |

Remark:

(1) measured in 10 mM Tris, pH 7.6, 200 mM KCl

Tryptophan repressor (TR), temperature dependence of high pressure denaturation

| GuHCl Conc. | pH | T | ΔG | Approach/Remarks | Ref |
|---|---|---|---|---|---|
| 2.5 M | 7.6 | 10 | 40.5±1.3 | pressure (1,2) | 99D4 |
| 2.5 M | 7.6 | 15 | 41.2±1.3 | pressure (1) | 99D4 |
| 2.5 M | 7.6 | 21 | 50.3±2.9 | pressure (1) | 99D4 |
| 2.5 M | 7.6 | 30 | 43.3±1.7 | pressure (1) | 99D4 |
| 2.5 M | 7.6 | 40 | 40.8±0.8 | pressure (1) | 99D4 |

Remarks:
(1) measured in 10 mM Tris, pH 7.6, 200 mM KCl, and 2.5 M GuHCl
(2) the half transition occurs at about 1.3 kbar at 10°C and decreases to about 0.6 kbar at 40°C

## Retinoic Acid-Binding Protein

Cellular retinoic acid-binding protein I (CRABP I) and II (CRABP II) compared with intestinal fatty acid-binding protein (IFABP)

| Protein | pH | T | ΔG | $c_{1/2}$ | m | Appr./Rem. | Ref |
|---|---|---|---|---|---|---|---|
| CRABP I: | | | | | | | |
| | 8.0 | 25 | 26.7±2.5 | 3.78±0.04 | 7.07±0.63 | urea (1–3) | 98B13 |
| | 8.0 | 25 | 21.4±6.2 | 3.78±0.14 | 5.65±1.55 | urea (1,2,4) | 98B13 |
| | 8.0 | 25 | 27.7±1.4 | 3.76±0.02 | 7.36±0.38 | urea (1,2,5) | 98B13 |
| | 8.0 | 25 | 22.1±4.3 | 3.61±0.09 | 6.11±1.17 | urea (1,2,6) | 98B13 |
| CRABP II: | | | | | | | |
| | 8.0 | 25 | 34.3±1.7 | 4.83±0.02 | 7.11±0.38 | urea (1–3) | 98B13 |
| | 8.0 | 25 | 29.5±6.4 | 4.81±0.09 | 6.15±1.30 | urea (1,2,4) | 98B13 |
| | 8.0 | 25 | 35.3±1.2 | 4.78±0.01 | 7.36±0.25 | urea (1,2,5) | 98B13 |
| | 8.0 | 25 | 39.3±2.6 | 4.73±0.03 | 8.33±0.54 | urea (1,2,6) | 98B13 |
| IFABP: | | | | | | | |
| | 8.0 | 25 | 20.0±1.3 | 4.12±0.03 | 4.85±0.29 | urea (1–3) | 98B13 |
| | 8.0 | 25 | 21.4±2.5 | 4.03±0.06 | 5.31±0.63 | urea (1,2,4) | 98B13 |
| | 8.0 | 25 | 21.2±1.5 | 4.20±0.04 | 5.06±0.33 | urea (1,2,5) | 98B13 |
| | 8.0 | 25 | 24.6±1.5 | 4.13±0.03 | 5.98±0.38 | urea (1,2,6) | 98B13 |

Remarks:
(1) linear extrapolation, LEM-SB
(2) measured in 25 mM sodium phosphate, 75 mM NaCl, 0.1 mM EDTA, 0.1 mM DTT
(3) data from ellipticity at 218 nm
(4) data from stopped flow monitored by ellipticity at 218 nm
(5) data from fluorescence intensity at 355 nm for CRABP I, 327 nm for CRABP II, and 340 nm for IFABP, respectively
(6) data from stopped flow monitored by fluorescence

## Rhodanese

Rhodanese expressed in *E. coli*, wild type and sequential N-terminal deletion mutants

| Mutant | pH | T | ΔG | $c_{1/2}$ | m | Appr./Rem. | Ref |
|---|---|---|---|---|---|---|---|
| wild type | 7.6 | 23 | 52.7±9.2 | 3.6 | 14.64±1.46 | urea (1–3) | 98T10 |
| Δ 1–3 | 7.6 | 23 | 48.5±2.9 | 3.6 | 13.47±0.46 | urea (1–3) | 98T10 |
| Δ 1–7 | 7.6 | 23 | 32.2±1.3 | 2.9 | 11.13±0.46 | urea (1–3) | 98T10 |
| Δ 1–9 | 7.6 | 23 | 27.2±1.7 | 2.45 | 11.09±0.67 | urea (1–3) | 98T10 |

Remarks:
(1) linear extrapolation
(2) transition monitored by loss of activity at urea denaturation
(3) measured using 1.5 μM protein in 50 mM $Na_2S_2O_3$, 50 mM Tris-HCl, pH 7.6, 200 mM 2-mercaptoethanol

Bovine rhodanese, N-terminal sequence truncation

| Mutant | pH | T | $\Delta G$ | $c_{1/2}$ | Appr./Rem. | Ref |
|---|---|---|---|---|---|---|
| wild type | 7.8 | 25 | 25.9±0.6 | 3.98±0.02 | urea (1–3) | 99T14 |
| Δ1–3 | 7.8 | 25 | 26.4±0.4 | 4.02±0.03 | urea (1–3) | 99T14 |
| Δ1–7 | 7.8 | 25 | 21.8±1.7 | 3.27±0.02 | urea (1–3) | 99T14 |
| Δ1–9 | 7.8 | 25 | 17.6±0.8 | 2.74±0.03 | urea (1–3) | 99T14 |

Remarks:

(1) linear extrapolation

(2) transition monitored by tryptophan fluorescence and bis-ANS fluorescence

(3) buffer: 50 mM Tris-HCl, 50 mM $Na_2S_2O_3$, 200 mM 2-mercaptoethanol, pH 7.8

## Ribonuclease A

The data entries are arranged as follows:

a) comparison of various approaches
b) data from H/D exchange techniques
c) wild type and mutants
d) data obtained in the presence of cosolutes

*a) comparison of various approaches*

Bovine pancreatic ribonuclease A (RNase A)

| Protein | pH | T | $\Delta G$ | Approach/Remarks | | Ref |
|---|---|---|---|---|---|---|
| RNase A | 2.8 | 17.1 | 23.4 | DSC | (1) | 99P1 |
| | 2.8 | 17.1 | 22.6 | urea | (2) | 99P1 |
| | 2.8 | 21.1 | 20.5 | DSC | (1) | 99P1 |
| | 2.8 | 21.1 | 20.5 | urea | (2) | 99P1 |
| | 2.8 | 24.9 | 18.0 | DSC | (1) | 99P1 |
| | 2.8 | 24.9 | 18.0 | urea | (2) | 99P1 |
| | 2.8 | 27.8 | 15.5 | DSC | (1) | 99P1 |
| | 2.8 | 27.8 | 14.6 | urea | (2) | 99P1 |
| | 2.8 | 25.0 | 18.0 | DSC | (1) | 99P1 |
| | 2.8 | 25.0 | 18.0 | urea | (3) | 99P1 |
| | 3.0 | 25.0 | 21.8 | DSC | (1) | 99P1 |
| | 3.0 | 25.0 | 21.8 | urea | (3) | 99P1 |
| | 3.6 | 25.0 | 28.0 | DSC | (1) | 99P1 |
| | 3.6 | 25.0 | 26.8 | urea | (3) | 99P1 |
| | 4.0 | 25.0 | 30.1 | DSC | (1) | 99P1 |
| | 4.0 | 25.0 | 30.5 | urea | (3) | 99P1 |
| | 5.0 | 25.0 | 33.9 | DSC | (1) | 99P1 |
| | 5.0 | 25.0 | 33.1 | urea | (3) | 99P1 |
| | 6.0 | 25.0 | 35.6 | DSC | (1) | 99P1 |
| | 6.0 | 25.0 | 36.0 | urea | (3) | 99P1 |
| | 7.0 | 25.0 | 37.7 | DSC | (1) | 99P1 |
| | 7.0 | 25.0 | 38.1 | urea | (3) | 99P1 |

Remarks:

(1) data calculated using $\Delta Cp$ = 4.56 kJ/mol/K, see also Table 2

(2) reconsideration of data from Ref. 89P2

(3) reconsideration of data from Ref. 90P1

Bovine pancreatic ribonuclease A (RNase A), comparison of the denaturant binding model and the linear extrapolation method

| pH | T | $\Delta G$ | K | $\Delta n$ | Appr./Rem. | Ref |
|----|---|------------|---|------------|------------|-----|
| data treatment by the denaturant binding model: | | | | | | |
| 5.0 | 25 | 48.765±1.828 | 0.150±0.009 | 31.616±1.170 | urea (1–3) | 99W17 |
| 5.3 | 25 | 50.162±1.577 | 0.143±0.008 | 32.715±1.052 | urea (1–3) | 99W17 |
| 5.8 | 25 | 47.735±0.791 | 0.131±0.007 | 32.598±0.552 | urea (1–3) | 99W17 |
| 6.5 | 25 | 48.170±0.987 | 0.129±0.006 | 31.827±0.667 | urea (1–3) | 99W17 |
| 7.0 | 25 | 44.396±1.527 | 0.121±0.002 | 30.701±1.064 | urea (1–3) | 99W17 |
| 7.5 | 25 | 48.158±1.556 | 0.116±0.001 | 31.806±1.033 | urea (1–3) | 99W17 |
| data treatment by the denaturant binding model with K fixed: | | | | | | |
| 5.0 | 25 | 45.170±1.695 | 0.10 | 40.127±1.494 | urea (2–4) | 99W17 |
| 5.3 | 25 | 45.685±1.502 | 0.10 | 39.367±1.324 | urea (2–4) | 99W17 |
| 5.8 | 25 | 45.468±0.745 | 0.10 | 38.292±0.639 | urea (2–4) | 99W17 |
| 6.5 | 25 | 45.965±0.916 | 0.10 | 36.984±0.756 | urea (2–4) | 99W17 |
| 7.0 | 25 | 46.191±1.477 | 0.10 | 35.470±1.147 | urea (2–4) | 99W17 |
| 7.5 | 25 | 46.806±1.523 | 0.10 | 34.530±1.128 | urea (2–4) | 99W17 |

data treatment by the linear extrapolation model:

| pH | T | $\Delta G$ | m | Appr./Rem. | Ref |
|----|---|------------|---|------------|-----|
| 5.0 | 25 | 38.572±0.230 | 6.150±1.452 | urea (2,3,5) | 99W17 |
| 5.3 | 25 | 39.238±0.205 | 6.038±1.293 | urea (2,3,5) | 99W17 |
| 5.8 | 25 | 39.357±0.096 | 5.866±0.640 | urea (2,3,5) | 99W17 |
| 6.5 | 25 | 39.928±0.117 | 5.674±0.808 | urea (2,3,5) | 99W17 |
| 7.0 | 25 | 40.539±0.180 | 5.460±1.305 | urea (2,3,5) | 99W17 |
| 7.5 | 25 | 41.489±0.172 | 5.343±1.331 | urea (2,3,5) | 99W17 |

Remarks:

(1) denaturant binding model, nonlinear curve fit

(2) transition monitored by fluorescence

(3) buffer: acetate (pH 5.0 and 5.3), MES (pH 5.8 and 6.5), MOPS (pH 7.0 and 7.5), all buffer solutions were of I = 0.20, the inert electrolyte being NaCl

(4) denaturant binding model, nonlinear curve fit, K fixed at 0.10 (see also Ref. 86P1)

(5) linear extrapolation

Bovine pancreatic ribonuclease A, combined urea and GuHCl denaturation

| Conditions | pH | T | $\Delta G$ | $c_{1/2}$ | m | Appr./Rem. | Ref |
|------------|----|---|------------|-----------|---|------------|-----|
| 0.1–1.0 M KCl | 2.0 | 25 | 21.05±0.88 | 1.78 | 11.84±0.50 | GuHCl (1–4) | 99G21 |
| 0.10 M KCl | 2.0 | 25 | 17.03±0.67 | 2.05 | 8.28±0.33 | urea (1–4) | 99G21 |
| 0.50 M KCl | 2.0 | 25 | 17.15±1.05 | 2.40 | 7.15±0.42 | urea (1–4) | 99G21 |
| 1.00 M KCl | 2.0 | 25 | 18.79±0.79 | 2.82 | 6.65±0.29 | urea (1–4) | 99G21 |
| 1.25 M KCl | 2.0 | 25 | 20.88±0.54 | 3.02 | 6.90±0.21 | urea (1–4) | 99G21 |
| 1.50 M KCl | 2.0 | 25 | 22.01±0.88 | 3.41 | 6.44±0.25 | urea (1–4) | 99G21 |
| 0.63 M GuHCl | 2.0 | 25 | 12.59±0.46 | 1.90 | 6.61±0.25 | urea (1–4) | 99G21 |
| 0.90 M GuHCl | 2.0 | 25 | 10.25±0.46 | 1.39 | 7.36±0.33 | urea (1–4) | 99G21 |
| 1.30 M GuHCl | 2.0 | 25 | 5.19±0.21 | 0.79 | 6.57±0.25 | urea (1–4) | 99G21 |
| 1.50 M GuHCl | 2.0 | 25 | 3.43±0.17 | 0.47 | 7.24±0.21 | urea (1–4) | 99G21 |
| 1.70 M GuHCl | 2.0 | 25 | 0.67±0.13 | 0.09 | 7.45±0.29 | urea (1–4) | 99G21 |

Remarks:

(1) linear extrapolation, for details of see Ref. 99G21

(2) transition monitored by optical absorption at 287 nm

(3) measured in 50 mM glycine-HCl, pH 3.0

(4) for comparable results see also lysozyme (HEW) and cytochrome $c$ (Ref. 99G21)

Bovine pancreatic ribonuclease A, pressure denaturation

| His signal | pH* | T | ΔG | Approach/Remarks | Ref |
|---|---|---|---|---|---|
| His12 | 2.0 | 22 | 13.8±1.3 | pressure (1–3) | 98P14 |
| His48 | 2.0 | 22 | 10.5±0.8 | pressure (1–3) | 98P14 |
| His119,105 | 2.0 | 22 | 11.7±0.4 | pressure (1–3) | 98P14 |
| denatured | 2.0 | 22 | 11.6±0.3 | pressure (1–3) | 98P14 |
| average | 2.0 | 22 | 11.7±1.3 | pressure (1–3) | 98P14 |

Remarks:

(1) pressure denaturation monitored by NMR using histidine signals

(2) ΔG at 1 bar

(3) curve fit assuming compressibility (ΔK) variable, average values ΔK = 0.015± 0.002 ml/mol/bar, ΔV = –(21±2) ml/mol, and ΔG = 11.7±1.3 kJ/mol

Bovine pancreatic ribonuclease A, wild type and mutants, pressure denaturation compared with thermal denaturation (see below)

| Mutant | pH | T | ΔG | ΔV | $p_{1/2}$ | Appr./Rem. | Ref |
|---|---|---|---|---|---|---|---|
| wild type | 5.0 | 40 | 23.24±0.44 | –46.5±3.3 | 500 | press. (1–3) | 99T12 |
| rec. w.t. | 5.0 | 40 | 24.61±0.70 | –46.5 | 529 | press. (1–4) | 99T12 |
| Tyr115→Trp | 5.0 | 40 | 18.12±1.39 | –43.2±4.4 | 419 | press. (1–3) | 99T12 |
| Tyr115→Trp | 5.0 | 40 | 19.66±0.44 | –47.6±1.2 | 413 | press. (1,3,5) | 99T12 |
| Ile106→Ala | 5.0 | 40 | 2.50±0.50 | –63.4±3.8 | 39 | press. (1–3) | 99T12 |
| Lle106→Leu | 5.0 | 40 | 12.72±0.58 | –47.7±1.7 | 267 | press. (1–3) | 99T12 |
| Ile106→Val | 5.0 | 40 | 17.42±0.91 | –52.8±2.7 | 330 | press. (1–3) | 99T12 |
| Ile107→Ala | 5.0 | 40 | 9.89±0.35 | –47.4±2.3 | 209 | press. (1–3) | 99T12 |
| Lle107→Leu | 5.0 | 40 | 11.60±0.36 | –47.5±1.0 | 244 | press. (1–3) | 99T12 |
| Ile107→Val | 5.0 | 40 | 23.75±2.30 | –52.2±6.9 | 455 | press. (1–3) | 99T12 |
| Val108→Ala | 5.0 | 40 | 3.13±0.30 | –55.9±2.7 | 56 | press. (1–3) | 99T12 |
| Ala109→Gly | 5.0 | 40 | 23.72±2.3 | –53.2±3.4 | 446 | press. (1–3) | 99T12 |
| Val116→Ala | 5.0 | 40 | 20.04±0.52 | –53.6±1.5 | 374 | press. (1–3) | 99T12 |
| Val116→Gly | 5.0 | 40 | 18.41±1.30 | –57.9±3.8 | 318 | press. (1–3) | 99T12 |
| Val118→Ala | 5.0 | 40 | 16.25±1.31 | –51.2±3.9 | 317 | press. (1–3) | 99T12 |
| Val118→Gly | 5.0 | 40 | 11.90±1.48 | –55.1±4.8 | 216 | press. (1–3) | 99T12 |

Remarks:

(1) data from pressure denaturation, pressure at half transition in mPa

(2) transition monitored by absorbance, 4th derivative spectroscopy

(3) buffer: 50 mM MES, pH 5.0

(4) rec. w.t. = recombinant wild-type protein

(5) transition monitored by fluorescence spectroscopy

Bovine pancreatic ribonuclease A, wild type and mutants, thermal denaturation compared with pressure denaturation (see above)

| Protein | pH | T | ΔG | Approach/Remarks | | Ref |
|---|---|---|---|---|---|---|
| wild type | 5.0 | 40 | 24.76±0.52 | heat | (1–3) | 99T12 |
| rec. w.t. | 5.0 | 40 | 21.16±0.67 | heat | (1–4) | 99T12 |
| Tyr115→Trp | 5.0 | 40 | 16.39±0.82 | heat | (1–3) | 99T12 |
| Tyr115→Trp | 5.0 | 40 | 15.67±0.81 | heat | (1,3,5) | 99T12 |
| Ile106→Ala | 5.0 | 40 | 2.87±0.21 | heat | (1–3) | 99T12 |
| Lle106→Leu | 5.0 | 40 | 13.68±0.54 | heat | (1–3) | 99T12 |
| Ile106→Val | 5.0 | 40 | 17.81±0.94 | heat | (1–3) | 99T12 |
| Ile107→Ala | 5.0 | 40 | 9.26±0.32 | heat | (1–3) | 99T12 |
| Lle107→Leu | 5.0 | 40 | 11.98±0.58 | heat | (1–3) | 99T12 |

Bovine pancreatic ribonuclease A, wild type and mutants, thermal denaturation compared with pressure denaturation (see above) (continued)

| Protein | pH | T | ΔG | Approach/Remarks | | Ref |
|---|---|---|---|---|---|---|
| Ile107→Val | 5.0 | 40 | 20.83±0.96 | heat | (1–3) | 99T12 |
| Val108→Ala | 5.0 | 40 | 3.59±0.14 | heat | (1–3) | 99T12 |
| Val108→Gly | 5.0 | 40 | −9.31±0.37 | heat | (1–3) | 99T12 |
| Ala109→Gly | 5.0 | 40 | 19.36±0.98 | heat | (1–3) | 99T12 |
| Val116→Ala | 5.0 | 40 | 18.38±1.22 | heat | (1–3) | 99T12 |
| Val116→Gly | 5.0 | 40 | 16.21±0.57 | heat | (1–3) | 99T12 |
| Val118→Ala | 5.0 | 40 | 13.12±0.60 | heat | (1–3) | 99T12 |
| Val118→Gly | 5.0 | 40 | 9.56±0.36 | heat | (1–3) | 99T12 |

Remarks:
(1) data from thermal denaturation, ΔG at 40°C, see also Table 2
(2) transition monitored by absorbance, 4th derivative spectroscopy
(3) buffer: 50 mM MES, pH 5.0
(4) rec. w.t. = recombinant wild-type protein
(5) transition monitored by fluorescence spectroscopy

b) data from H/D exchange techniques

Bovine pancreatic ribonuclease A and ribonuclease S in $D_2O$

| Protein | pH | T | ΔG | Approach/Remarks | | Ref |
|---|---|---|---|---|---|---|
| RNase A | 6.0 | 25 | 37 | DSC | (1,2) | 99N3 |
| RNase S | 6.0 | 25 | 20 | DSC | (1–3) | 99N3 |

Remarks:
(1) data from Fig. 3 in Ref. 99N3
(2) Ref. 99N3 contains a profile of ΔG versus T from 15 to 45°C
(3) for the calculation of ΔG see Ref. 99N3

Bovine pancreatic ribonuclease A, $\Delta G_{ex}$ from H/D exchange experiments at pH 6.0

| Residue | $\Delta G_{ex}$(25°C) | $\Delta G_{ex}$(35°C) | $\Delta G_{ex}$(40°C) | $\Delta G_{ex}$(45°C) | Appr./Rem. | | Ref |
|---|---|---|---|---|---|---|---|
| Arg10 | 31.0 | 28.9 | | | H/D | (1) | 99N3 |
| Gln11 | >40.6 | 38.5 | 36.8 | 35.6 | H/D | (1) | 99N3 |
| His12 | 32.6 | 34.7 | 32.6 | 29.3 | H/D | (1) | 99N3 |
| Met13 | >35.6 | 34.7 | 32.6 | 31.4 | H/D | (1) | 99N3 |
| Asp14 | | 20.1 | | | H/D | (1) | 99N3 |
| Met29 | | | | | | | 99N3 |
| Met30 | >33.9 | >37.2 | | 28.5 | H/D | (1) | 99N3 |
| Lys31 | | | | | | | 99N3 |
| Val43 | 26.8 | 25.1 | | | H/D | (1) | 99N3 |
| Phe46 | | | | | | | 99N3 |
| Val47 | | | | | | | 99N3 |
| His48 | | | | | | | 99N3 |
| Glu49 | 25.1 | 24.3 | 25.1 | 25.5 | H/D | (1) | 99N3 |
| Val54 | >36.8 | >40.6 | 38.1 | 36.8 | H/D | (1) | 99N3 |
| Gln55 | >38.9 | >42.3 | | 39.3 | H/D | (1) | 99N3 |
| Ala56 | 30.5 | 31.0 | 32.2 | 30.5 | H/D | (1) | 99N3 |
| Val57 | >36.0 | >39.7 | 38.1 | 39.3 | H/D | (1) | 99N3 |
| Cys58 | >39.7 | >43.1 | 39.7 | 39.7 | H/D | (1) | 99N3 |
| Ser59 | 30.1 | 28.5 | | | H/D | (1) | 99N3 |
| Gln60 | 32.0 | 30.1 | | | H/D | (1) | 99N3 |
| Lys61 | 25.5 | 28.0 | | | H/D | (1) | 99N3 |

Bovine pancreatic ribonuclease A, $\Delta G_{ex}$ from H/D exchange experiments at pH 6.0 (continued)

| Residue | $\Delta G_{ex}(25°C)$ | $\Delta G_{ex}(35°C)$ | $\Delta G_{ex}(40°C)$ | $\Delta G_{ex}(45°C)$ | Appr./Rem. | Ref |
|---|---|---|---|---|---|---|
| Val63 | >37.2 | 41.0 | 38.9 | 38.5 | H/D (1) | 99N3 |
| Cys72 | 31.0 | 33.1 | 33.9 | 31.8 | H/D (1) | 99N3 |
| Tyr73 | >38.9 | 42.7 | 41.0 | 39.7 | H/D (1) | 99N3 |
| Gln74 | >38.9 | 42.7 | 41.0 | 39.7 | H/D (1) | 99N3 |
| Met79 | >40.2 | 43.9 | | 40.6 | H/D (1) | 99N3 |
| Ile81 | | | | | | 99N3 |
| Thr82 | | | | | | 99N3 |
| Cys84 | >41.4 | 39.7 | 39.7 | 38.9 | H/D (1) | 99N3 |
| Arg85 | 32.6 | 34.7 | 35.6 | 34.7 | H/D (1) | 99N3 |
| Tyr97 | | | | | | 99N3 |
| Lys98 | 36.0 | 38.1 | 38.5 | 37.2 | H/D (1) | 99N3 |
| Thr100 | 36.4 | 38.1 | | 37.7 | H/D (1) | 99N3 |
| Ala102 | >41.0 | >44.8 | | 39.7 | H/D (1) | 99N3 |
| Lys104 | >34.7 | >38.5 | 34.7 | 31.4 | H/D (1) | 99N3 |
| Ile106 | >39.7 | >43.5 | 37.2 | 36.4 | H/D (1) | 99N3 |
| Ile107 | >28.0 | >31.4 | 28.5 | | H/D (1) | 99N3 |
| Val108 | >35.6 | >38.9 | 41.4 | 40.6 | H/D (1) | 99N3 |
| Ala109 | >36.4 | >39.7 | 41.0 | 39.7 | H/D (1) | 99N3 |
| Cys110 | >36.4 | >39.7 | 25.9 | | H/D (1) | 99N3 |
| Glu111 | 34.3 | 34.7 | 36.0 | 34.3 | H/D (1) | 99N3 |
| Val116 | >33.5 | >37.2 | 37.7 | 37.2 | H/D (1) | 99N3 |
| Val118 | >34.3 | >37.7 | 38.9 | 40.6 | H/D (1) | 99N3 |
| His119 | >35.1 | >38.5 | 35.1 | 32.6 | H/D (1) | 99N3 |

Remark:

(1) for further details see Ref. 99N3

Bovine pancreatic ribonuclease A, $\Delta G_{ex}$ from H/D exchange experiments at pH 2.5, in the presence of 0.2 M NaCl

| Residue | $\Delta G_{ex}(15°C)$ | $\Delta G_{ex}(25°C)$ | $\Delta G_{ex}(35°C)$ | Approach/Remarks | | Ref |
|---|---|---|---|---|---|---|
| Thr3 | 0.4 | | | H/D | (1) | 99N3 |
| Ala6 | 14.6 | 13.4 | 12.6 | H/D | (1) | 99N3 |
| Lys7 | >18.8 | 13.4 | 10.0 | H/D | (1) | 99N3 |
| Phe8 | | | 11.3 | H/D | (1) | 99N3 |
| Glu9 | >20.1 | 22.6 | 18.8 | H/D | (1) | 99N3 |
| Arg10 | >18.8 | >21.8 | 18.0 | H/D | (1) | 99N3 |
| Gln11 | >19.7 | >22.6 | 18.8 | H/D | (1) | 99N3 |
| His12 | | 14.6 | 11.3 | H/D | (1) | 99N3 |
| Met13 | >16.3 | >19.2 | 12.6 | H/D | (1) | 99N3 |
| Asp14 | >24.3 | 17.6 | 15.1 | H/D | (1) | 99N3 |
| Thr17 | 2.1 | 2.9 | | H/D | (1) | 99N3 |
| Asn24 | | | | | | 99N3 |
| Cys26 | >15.9 | >18.8 | 11.7 | H/D | (1) | 99N3 |
| Gln28 | 15.9 | 15.1 | | H/D | (1) | 99N3 |
| Met29 | >15.9 | >18.8 | >21.8 | H/D | (1) | 99N3 |
| Met30 | >15.9 | >18.8 | >21.8 | H/D | (1) | 99N3 |
| Lys31 | >16.3 | >19.2 | 19.2 | H/D | (1) | 99N3 |
| Ser32 | | | | | | 99N3 |
| Arg33 | | | | | | 99N3 |
| Asn34 | | 15.9 | 14.6 | H/D | (1) | 99N3 |
| Leu35 | 7.9 | 7.9 | 5.4 | H/D | (1) | 99N3 |
| Thr36 | 6.3 | 5.0 | | H/D | (1) | 99N3 |
| Val43 | 10.5 | 10.9 | 12.6 | H/D | (1) | 99N3 |
| Asn44 | | >19.2 | 14.6 | H/D | (1) | 99N3 |

Bovine pancreatic ribonuclease A, $\Delta G_{ex}$ from H/D exchange experiments at pH 2.5, in the presence of 0.2 M NaCl (continued)

| Residue | $\Delta G_{ex}(15°C)$ | $\Delta G_{ex}(25°C)$ | $\Delta G_{ex}(35°C)$ | Approach/Remarks | | Ref |
|---|---|---|---|---|---|---|
| Phe46 | | >20.1 | 18.8 | H/D | (1) | 99N3 |
| Val47 | >15.9 | >18.8 | >21.8 | H/D | (1) | 99N3 |
| His48 | >15.9 | >18.8 | 12.1 | H/D | (1) | 99N3 |
| Glu49 | | 11.3 | 10.9 | H/D | (1) | 99N3 |
| Leu51 | 6.3 | | | H/D | (1) | 99N3 |
| Ala52 | 10.0 | 9.2 | | H/D | (1) | 99N3 |
| Asp53 | 10.5 | 8.8 | 5.9 | H/D | (1) | 99N3 |
| Val54 | >15.1 | >18.0 | 17.2 | H/D | (1) | 99N3 |
| Gln55 | >19.7 | >22.6 | 20.1 | H/D | (1) | 99N3 |
| Ala56 | >18.4 | >21.3 | 17.1 | H/D | (1) | 99N3 |
| Val57 | >14.6 | >17.6 | >20.5 | H/D | (1) | 99N3 |
| Cys58 | >17.6 | >20.5 | 18.8 | H/D | (1) | 99N3 |
| Ser59 | >18.4 | 15.5 | 15.5 | H/D | (1) | 99N3 |
| Gln60 | >18.4 | >21.3 | 20.5 | H/D | (1) | 99N3 |
| Lys61 | >18.4 | 12.1 | 13.0 | H/D | (1) | 99N3 |
| Asn62 | 6.3 | | | H/D | (1) | 99N3 |
| Val63 | >15.5 | >18.4 | 19.2 | H/D | (1) | 99N3 |
| Cys65 | 5.4 | 5.4 | 6.7 | H/D | (1) | 99N3 |
| Gln69 | 9.2 | | | H/D | (1) | 99N3 |
| Thr70 | 4.6 | 4.6 | | H/D | (1) | 99N3 |
| Asn71 | | 5.0 | | H/D | (1) | 99N3 |
| Cys72 | >17.6 | >20.5 | 19.2 | H/D | (1) | 99N3 |
| Tyr73 | >16.7 | >19.7 | 20.1 | H/D | (1) | 99N3 |
| Gln74 | >17.6 | >20.5 | 20.1 | H/D | (1) | 99N3 |
| Ser75 | >17.6 | >20.1 | 4.6 | H/D | (1) | 99N3 |
| Met79 | >18.0 | >20.5 | 19.2 | H/D | (1) | 99N3 |
| Ser80 | | 9.2 | 7.5 | H/D | (1) | 99N3 |
| Ile81 | >16.3 | >19.2 | 20.9 | H/D | (1) | 99N3 |
| Thr82 | >17.2 | >20.1 | >23.0 | H/D | (1) | 99N3 |
| Asp83 | >16.3 | 15.9 | 13.4 | H/D | (1) | 99N3 |
| Cys84 | >18.8 | >21.8 | 19.7 | H/D | (1) | 99N3 |
| Arg85 | >17.6 | >20.5 | 18.8 | H/D | (1) | 99N3 |
| Glu86 | | | | | | 99N3 |
| Thr87 | >13.8 | 10.5 | 7.9 | H/D | (1) | 99N3 |
| Lys91 | 5.0 | | | H/D | (1) | 99N3 |
| Asn94 | | | | | | 99N3 |
| Cys95 | 8.4 | | | H/D | (1) | 99N3 |
| Ala96 | 6.3 | | | H/D | (1) | 99N3 |
| Tyr97 | >16.7 | >19.7 | 16.3 | H/D | (1) | 99N3 |
| Lys98 | >17.6 | >20.9 | 20.5 | H/D | (1) | 99N3 |
| Thr99 | 4.2 | 2.1 | | H/D | (1) | 99N3 |
| Thr100 | >19.7 | >22.6 | 21.8 | H/D | (1) | 99N3 |
| Gln101 | 7.1 | | | H/D | (1) | 99N3 |
| Ala102 | | 19.2 | 18.0 | H/D | (1) | 99N3 |
| Asn103 | 6.3 | 8.4 | | H/D | (1) | 99N3 |
| Lys104 | | 33.5 | 16.7 | H/D | (1) | 99N3 |
| His105 | 17.2 | 7.5 | | H/D | (1) | 99N3 |
| Ile106 | >17.2 | >20.1 | 20.5 | H/D | (1) | 99N3 |
| Ile107 | >12.6 | >15.5 | | H/D | (1) | 99N3 |
| Val108 | >13.8 | >16.7 | >19.7 | H/D | (1) | 99N3 |
| Ala109 | >16.3 | >19.2 | 21.3 | H/D | (1) | 99N3 |
| Cys110 | >17.2 | >20.1 | | H/D | (1) | 99N3 |
| Glu111 | >18.4 | >21.3 | 17.6 | H/D | (1) | 99N3 |

Bovine pancreatic ribonuclease A, $\Delta G_{ex}$ from H/D exchange experiments at pH 2.5, in the presence of 0.2 M NaCl (continued)

| Residue | $\Delta G_{ex}(15°C)$ | $\Delta G_{ex}(25°C)$ | $\Delta G_{ex}(35°C)$ | Approach/Remarks | | Ref |
|---|---|---|---|---|---|---|
| Gly112 | 7.9 | 7.5 | | H/D | (1) | 99N3 |
| Asn113 | 8.8 | | | H/D | (1) | 99N3 |
| Val116 | >14.2 | >17.2 | 17.6 | H/D | (1) | 99N3 |
| Val118 | >15.5 | >18.4 | >21.3 | H/D | (1) | 99N3 |
| His119 | >16.3 | >19.2 | 12.6 | H/D | (1) | 99N3 |
| Phe120 | >16.3 | >18.8 | | H/D | (1) | 99N3 |
| Asp121 | 10.9 | 9.6 | 7.5 | H/D | (1) | 99N3 |
| Ser123 | 6.3 | 7.5 | | H/D | (1) | 99N3 |
| Val124 | 10.9 | 10.9 | | H/D | (1) | 99N3 |

Remark:

(1) for further details see Ref. 99N3

*c) wild type and mutants*

Bovine pancreatic ribonuclease A, Tyr to Phe mutants

| Mutant | pH | T | $\Delta G$ | Approach | Remarks | Ref |
|---|---|---|---|---|---|---|
| wild type | 4.0 | 15 | 32.3±2.4 | heat | (1) | 97J5 |
| Tyr25→Phe | 4.0 | 15 | 25.1±3.8 | heat | (1) | 97J5 |
| Tyr92→Phe | 4.0 | 15 | 24.1±4.0 | heat | (1) | 97J5 |
| Tyr97→Phe | 4.0 | 15 | 21.0±4.7 | heat | (1) | 97J5 |

Remark:

(1) $\Delta G$ was calculated from $\Delta H$ and $T_{trs}$ (see Table 2) with $\Delta Cp$ fixed at its calorimetric value (Ref. 95M3)

Bovine pancreatic ribonuclease A, Tyr to Phe mutants

| Mutant | pH | T | $\Delta G$ | $c_{1/2}$ | m | Appr./Rem. | Ref |
|---|---|---|---|---|---|---|---|
| wild type | 5.0 | 15 | 0.0 | 2.99±0.04 | 16.3±2.9 | GuHCl (1) | 97J5 |
| Tyr25→Phe | 5.0 | 15 | −7.7 | 2.38±0.02 | 15.0±2.0 | GuHCl (1) | 97J5 |
| Tyr92→Phe | 5.0 | 15 | 0.5 | 3.03±0.02 | 12.0±1.2 | GuHCl (1) | 97J5 |
| Tyr97→Phe | 5.0 | 15 | −11.5 | 2.07±0.02 | 14.6±2.2 | GuHCl (1) | 97J5 |

Remark:

(1) linear extrapolation, LEM-SB, with m fixed at 12.55 kJ/mol/Mol (Ref. 89P1)

Ribonuclease A, proline mutants, note that the peptide bonds preceding Pro93 and Pro114 are in *cis* conformation

| Mutant | pH | $T_{trs}$ | $\Delta T$ | $\Delta(\Delta G)$ | $c_{1/2}$ | Appr./Rem. | Ref |
|---|---|---|---|---|---|---|---|
| wild type | 6.0 | 61.6±2 | 0.0 | 0.0 | | heat (1,2) | 98S5 |
| wild type | 6.0 | 61.8±1 | 0.0 | 0.0 | | DSC (3) | 98S5 |
| Pro93→Gly | 6.0 | 55.4±2 | −6.2 | −9.2 | | heat (1,2) | 98S5 |
| Pro93→Gly | 6.0 | 54.5±1 | −7.3 | −10.9 | | DSC (3) | 98S5 |
| Pro114→Gly | 6.0 | 52.1±2 | −9.5 | −13.8 | | heat (1,2) | 98S5 |
| Pro114→Gly | 6.0 | 51.6±1 | −10.2 | −15.1 | | DSC (3) | 98S5 |
| wild type | 4.2 | 54.5 | | | | heat (4,5) | 92S6 |
| pseudo w.t. | 4.2 | 56.0 | 0.0 | 0.0 | | heat (4–6) | 92S6 |
| Pro93→Ala | 4.2 | 46.5 | −9.5 | −11.3 | | heat (4–6) | 92S6 |
| Pro93→Ser | 4.2 | 47.5 | −8.5 | −8.8 | | heat (4–6) | 92S6 |
| Pro114→Ala | 4.2 | 45.5 | −10.5 | −13.4 | | heat (4–6) | 92S6 |
| Pro114→Gly | 4.2 | 45.5 | −10.5 | −11.7 | | heat (4–6) | 92S6 |

Ribonuclease A, proline mutants, note that the peptide bonds preceding Pro93 and Pro114 are in *cis* conformation (continued)

| Mutant | pH | $T_{trs}$ | $\Delta T$ | $\Delta(\Delta G)$ | $c_{1/2}$ | Appr./Rem. | Ref |
|---|---|---|---|---|---|---|---|
| double mutant (Pro93→Ala and Pro114→Gly) | | | | | | | |
| | 4.2 | 36.0 | −20 | −17.6 | | heat (4–6) | 92S6 |
| wild type | 5 | 44.4 | 0.0 | | 3.1 | heat (7–9) | 96D3 |
| Pro42→Ala | 5 | 43.0 | −1.4 | | 3.0 | heat (7–9) | 96D3 |
| Pro93→Ala | 5 | 33.7 | −10.7 | | 2.2 | heat (7–9) | 96D3 |
| Pro114→Ala | 5 | 33.8 | −10.6 | | 2.2 | heat (7–9) | 96D3 |
| Pro117→Ala | 5 | 37.7 | −6.7 | | 2.5 | heat (7–9) | 96D3 |

Remarks:

(1) transition monitored by UV spectroscopy at 287 nm, $\Delta(\Delta G)$ at the reference temperature $T_{trs} = 61.6°C$

(2) buffer: 30 mM sodium acetate, pH 6.0, 0.1 M NaCl

(3) DSC experiments were performed in $H_2O$, $\Delta(\Delta G)$ at the reference temperature $T_{trs} = 61.6°C$

(4) transition monitored by CD spectroscopy

(5) buffer: 0.01 M sodium acetate, pH 4.2, 0.1 M NaCl

(6) pseudo-wild type containing an additional N-terminal Met residue, $\Delta(\Delta G)$ at the reference temperature $T_{trs} = 56.0°C$

(7) transition monitored by UV absorption at 287 nm

(8) thermal transition measured in 50 mM acetate buffer in the presence of 1.3 M GuHCl

(9) $c_{1/2}$ from GuHCl denaturation studies at 15°C in 50 mM acetate buffer

Bovine pancreatic ribonuclease A, wild type and mutants

| Mutant | pH | T | $\Delta G$ | $c_{1/2}$ | m | Approach/Remarks | Ref |
|---|---|---|---|---|---|---|---|
| wild type | 5.0 | 15 | 3.03 | 15.56 | | GuHCl (1–3) | 98J4 |
| Pro42→Ala | 5.0 | 15 | 3.02 | 10.13 | | GuHCl (1–3) | 98J4 |
| Tyr92→Phe | 5.0 | 15 | 3.05 | 12.13 | | GuHCl (1–3) | 98J4 |
| Pro93→Ala | 5.0 | 15 | 2.20 | 11.67 | | GuHCl (1–3) | 98J4 |
| Pro114→Ala | 5.0 | 15 | 2.18 | 18.41 | | GuHCl (1–3) | 98J4 |
| Tyr115→Phe | 5.0 | 15 | 3.22 | 13.39 | | GuHCl (1–3) | 98J4 |
| Pro117→Ala | 5.0 | 15 | 2.52 | 11.13 | | GuHCl (1–3) | 98J4 |
| double mutant (Tyr92→Phe and Pro114→Ala) | | | | | | | |
| | 5.0 | 15 | 2.28 | 17.99 | | GuHCl (1–3) | 98J4 |
| double mutant (Tyr92→Phe and Tyr115→Phe) | | | | | | | |
| | 5.0 | 15 | 3.23 | 11.05 | | GuHCl (1–3) | 98J4 |

Remarks:

(1) linear extrapolation, LEM-SB

(2) transition monitored by absorbance at 287 nm

(3) buffer: 100 mM sodium acetate, pH 5.0

Bovine pancreatic Ribonuclease A, wild type and mutants at positions 106–118

| Mutant | pH | $T_{trs}$ | $\Delta T$ | $\Delta(\Delta G)$ | Appr./Rem. | Ref |
|---|---|---|---|---|---|---|
| wild type | 5.0 | 60.8 | 0.0 | 0.0 | DSC (1,2) | 99C13 |
| Ile106→Ala | 5.0 | 46.6 | −14.2 | −20.50 | DSC (1,2) | 99C13 |
| Ile106→Leu | 5.0 | 54.0 | −6.8 | −9.83 | DSC (1,2) | 99C13 |
| Ile106→Val | 5.0 | 56.5 | −4.3 | −6.19 | DSC (1,2) | 99C13 |
| Ile107→Ala | 5.0 | 50.6 | −10.2 | −14.73 | DSC (1,2) | 99C13 |
| Ile107→Leu | 5.0 | 51.8 | −9.0 | −13.01 | DSC (1,2) | 99C13 |
| Ile107→Val | 5.0 | 60.4 | −0.4 | −0.59 | DSC (1,2) | 99C13 |
| Val108→Ala | 5.0 | 46.8 | −14.0 | −20.21 | DSC (1,2) | 99C13 |
| Val108→Gly | 5.0 | 33.8 | −27.0 | −38.99 | DSC (1,2) | 99C13 |
| Ala109→Gly | 5.0 | 58.1 | −2.7 | −3.89 | DSC (1,2) | 99C13 |
| Val116→Ala | 5.0 | 57.5 | −3.3 | −4.77 | DSC (1,2) | 99C13 |

Bovine pancreatic Ribonuclease A, wild type and mutants at positions 106–118 (continued)

| Mutant | pH | $T_{trs}$ | $\Delta T$ | $\Delta(\Delta G)$ | Appr./Rem. | Ref |
|--------|-----|------|-------|--------|-----------|------|
| Val116→Gly | 5.0 | 55.5 | −5.3 | −7.66 | DSC (1,2) | 99C13 |
| Val118→Ala | 5.0 | 54.3 | −6.5 | −9.37 | DSC (1,2) | 99C13 |
| Val118→Gly | 5.0 | 50.0 | −10.8 | −15.61 | DSC (1,2) | 99C13 |

Remarks:
(1) $\Delta(\Delta G)$ was calculated using $\Delta(\Delta G) = \Delta T_{mut} \times \Delta H_{w.t.}/T_{trs,w.t.}$
(2) measured in 50 mM sodium acetate buffer, pH 5.0

Bovine pancreatic ribonuclease and variants

| Mutant | pH | T | $\Delta G$ | Approach/Remarks | | Ref |
|--------|-----|-----|-----------|---------|---|------|
| wild type | 6 | 25 | 37.7±0.4 | heat | (1) | 98Q2 |
| Asp121→Asn | 6 | 25 | 29.7±0.4 | heat | (1) | 98Q2 |
| Asp121→Ala | 6 | 25 | 28.5±0.4 | heat | (1) | 98Q2 |
| His119→Ala | 6 | 25 | 38.5±0.4 | heat | (1) | 98Q2 |
| wild type | 1.2 | 25 | 4.2±0.4 | heat | (1) | 98Q2 |
| Asp121→Asn | 1.2 | 25 | 1.00±0.04 | heat | (1) | 98Q2 |
| Asp121→Ala | 1.2 | 25 | 1.17±0.04 | heat | (1) | 98Q2 |
| His119→Ala | 1.2 | 25 | 8.8±0.4 | heat | (1) | 98Q2 |

Remark:
(1) see also Table 2

Bovine pancreatic ribonuclease A, wild type and three-disulfide mutants

| Protein | pH | T | $\Delta G$ | $c_{1/2}$ | m | Appr./Rem. | Ref |
|---------|-----|-----|-----------|--------|-------|-----------|------|
| wild type | 8.0 | 25 | 47.3±2.9 | 3.00±0.20 | 15.9±0.8 | GuHCl (1–3) | 99I5 |
| double mutant (Cys65→Ser and Cys72→Ser) | | | | | | | |
| | 8.0 | 25 | 20.9±2.1 | 1.30±0.10 | 15.9±1.7 | GuHCl (1–3) | 99I5 |
| double mutant (Cys40→Ala and Cys95→Ala) | | | | | | | |
| | 8.0 | 25 | 12.6±0.8 | 0.77±0.05 | 16.3±0.8 | GuHCl (1–3) | 99I5 |

Remarks:
(1) linear extrapolation, LEM-SB
(2) transition monitored by optical absorption at 287 nm
(3) measured in 100 mM Tris buffer, pH 8.0

Bovine pancreatic ribonuclease A and three-disulfide mutant (Cys40→Ala and Cys95→Ala)

| Protein | pH | T | $\Delta G$ | $c_{1/2}$ | m | Appr./Rem. | Ref |
|---------|-----|-----|-----------|--------|-------|-----------|------|
| wild type | 4.60 | 20 | 33.18±3.35 | 2.96±0.15 | 11.21±1.21 | GuHCl (1–3,5) | 99L1 |
| | 4.60 | 25 | 30.21±3.39 | 2.82±0.16 | 10.71±1.21 | GuHCl (1–3) | 99L1 |
| | 4.60 | 30 | 27.57±2.09 | 2.56±0.11 | 11.09±0.84 | GuHCl (1–3) | 99L1 |
| | 4.60 | 35 | 23.43±2.55 | 2.25±0.15 | 10.38±1.09 | GuHCl (1–4) | 99L1 |
| double mutant (Cys40→Ala and Cys95→Ala) | | | | | | | |
| | 4.60 | 12 | 20.04±2.26 | 1.00±0.17 | 19.50±2.47 | GuHCl (1–4) | 99L1 |
| | 4.60 | 16 | 17.82±2.09 | 0.95±0.17 | 18.70±2.34 | GuHCl (1–3) | 99L1 |
| | 4.60 | 20 | 14.14±2.30 | 0.83±0.23 | 17.07±2.72 | GuHCl (1–4) | 99L1 |
| | 4.60 | 24 | 10.04±2.09 | 0.63±0.26 | 16.02±2.43 | GuHCl (1–3) | 99L1 |

Remarks:
(1) linear extrapolation, LEM-SB
(2) transition monitored by absorbance at 287 nm
(3) buffer: 100 mM sodium acetate, pH 4.60±0.03
(4) Ref. 99L1 contains additional data for H/D exchange
(5) not including 7.1 kJ/mol correction for proline isomerization

Bovine pancreatc ribonuclease A (RNase A) and mutants containing the minimal structural requirements for dimerization and the N-terminal swapping of bovine seminal ribonuclease (BS-RNase, see also Table 2)

| Protein | pH | T | $\Delta G$ | Approach/Remarks | | Ref |
|---|---|---|---|---|---|---|
| RNase A | 5.0 | 37 | 28 | heat | (1,2) | 98C1 |
| P-RNase A | 5.0 | 37 | 24 | heat | (1,3) | 98C1 |

Explanations:

      RNase A         – bovine pancreatic ribonuclease A

      P-RNase A     – monomeric mutant (Ala19→Pro) of RNase A

Remarks:

(1) buffer: 10 mM acetate buffer, pH 5.0

(2) $\Delta G$ was calculated using $\Delta Cp = 5.5$ kJ/mol for RNase A from Ref. 97C4

(3) $\Delta G$ was calculated using $\Delta Cp = 5.2$ kJ/mol for P-RNase A from Ref. 97C4

*d) data obtained in the presence of cosolutes*

Bovine pancreatic ribonuclease A in aqueous polyol solutions

| Cosolvent | Conc. | pH | $T_{trs}$ | $\Delta T$ | $\Delta(\Delta G)$ | Appr./Rem. | Ref |
|---|---|---|---|---|---|---|---|
| control | | 2.5 | 38.3 | 0.0 | 0.00 | heat (1,2) | 98K3 |
| mannitol | 1.00 M | 2.5 | 46.6 | 8.3 | 8.87 | heat (1,2) | 98K3 |
| inositol | 0.75 M | 2.5 | 47.0 | 8.7 | 9.41 | heat (1,2) | 98K3 |
| xylitol | 2.00 M | 2.5 | 48.5 | 10.2 | 10.88 | heat (1,2) | 98K3 |
| adonitol | 2.00 M | 2.5 | 50.2 | 11.9 | 12.47 | heat (1,2) | 98K3 |
| sorbitol | 2.00 M | 2.5 | 51.5 | 13.2 | 14.14 | heat (1,2) | 98K3 |
| control | | 4.0 | 54.2 | 0.0 | 0.00 | heat (1,3) | 98K3 |
| mannitol | 1.00 M | 4.0 | 59.0 | 4.8 | 5.86 | heat (1,3) | 98K3 |
| inositol | 0.75 M | 4.0 | 59.8 | 5.6 | 6.86 | heat (1,3) | 98K3 |
| xylitol | 2.00 M | 4.0 | 61.4 | 7.2 | 8.83 | heat (1,3) | 98K3 |
| adonitol | 2.00 M | 4.0 | 63.3 | 9.1 | 11.13 | heat (1,3) | 98K3 |
| sorbitol | 2.00 M | 4.0 | 66.0 | 11.8 | 14.64 | heat (1,3) | 98K3 |
| control | | 7.0 | 46.0 | 0.0 | 0.00 | heat (1,4) | 98K3 |
| mannitol | 1.00 M | 7.0 | 50.2 | 3.9 | 5.27 | heat (1,4) | 98K3 |
| inositol | 0.75 M | 7.0 | 50.7 | 4.5 | 6.02 | heat (1,4) | 98K3 |
| xylitol | 2.00 M | 7.0 | 52.9 | 6.5 | 8.95 | heat (1,4) | 98K3 |
| adonitol | 2.00 M | 7.0 | 53.4 | 6.8 | 9.00 | heat (1,4) | 98K3 |
| sorbitol | 2.00 M | 7.0 | 56.1 | 10.3 | 10.46 | heat (1,4) | 98K3 |

Remarks:

(1) transition monitored by optical absorption at 287 nm

(2) buffer: 20 mM glycine, pH 2.5

(3) buffer: 40 mM acetate, pH 4.0

(4) buffer: 20 mM phosphate or MOPS, pH 7.0, 1.5 M GuHCl

Ribonuclease A, preferential interactions in aqueous cosolvent systems, the sorbitol-water system

| Cosolvent | pH | T | $\Delta G$ | Approach/Remarks | Ref |
|---|---|---|---|---|---|
| 0 % sorbitol (w/v) | 2.0 | 20 | 10.9±1.3 | heat, v.H. (1) | 97X2 |
| 10 % sorbitol (w/v) | 2.0 | 20 | 12.1 | heat, v.H. (1) | 97X2 |
| 20 % sorbitol (w/v) | 2.0 | 20 | 15.6 | heat, v.H. (1) | 97X2 |
| 30 % sorbitol (w/v) | 2.0 | 20 | 19.3±1.3 | heat, v.H. (1) | 97X2 |
| 40 % sorbitol (w/v) | 2.0 | 20 | 23.2 | heat, v.H. (1) | 97X2 |
| 0 % sorbitol (w/v) | 5.5 | 20 | 56.9±2.1 | heat, v.H. (1) | 97X2 |
| 10 % sorbitol (w/v) | 5.5 | 20 | 59.0 | heat, v.H. (1) | 97X2 |
| 20 % sorbitol (w/v) | 5.5 | 20 | 64.0 | heat, v.H. (1) | 97X2 |
| 30 % sorbitol (w/v) | 5.5 | 20 | 70.6±1.3 | heat, v.H. (1) | 97X2 |

Ribonuclease A, preferential interactions in aqueous cosolvent systems, the sorbitol-water system (continued)

| Co-solvent | pH | T | ΔG | Approach/Remarks | Ref |
|---|---|---|---|---|---|
| 40 % sorbitol (w/v) | 5.5 | 20 | 76.5 | heat, v.H. (1) | 97X2 |
| 30 % sorbitol (w/v) | 1.5 | 20 | 15.9±0.8 | heat, v.H. (1) | 97X2 |
| 30 % sorbitol (w/v) | 2.0 | 20 | 19.2±1.3 | heat, v.H. (1) | 97X2 |
| 30 % sorbitol (w/v) | 3.0 | 20 | 35.1 | heat, v.H. (1) | 97X2 |
| 30 % sorbitol (w/v) | 5.5 | 20 | 70.7±1.3 | heat, v.H. (1) | 97X2 |

Remark:
(1) data treatment by means of a truncated form of the integrated van't Hoff equation with inherent $\Delta Cp$

Bovine pancreatic ribonuclease A, stabilization by trehalose

| Trehalose | pH | T | ΔG | Approach/Remarks | Ref |
|---|---|---|---|---|---|
| 0.0 M | 2.8 | 20 | 23.05±1.26 | heat, v.H. (1,2) | 97X1 |
| 0.1 M | 2.8 | 20 | 23.93 | heat, v.H. (1,2) | 97X1 |
| 0.2 M | 2.8 | 20 | 24.77 | heat, v.H. (1,2) | 97X1 |
| 0.3 M | 2.8 | 20 | 25.90 | heat, v.H. (1,2) | 97X1 |
| 0.4 M | 2.8 | 20 | 26.90 | heat, v.H. (1,2) | 97X1 |
| 0.5 M | 2.8 | 20 | 27.99 | heat, v.H. (1,2) | 97X1 |
| 0.6 M | 2.8 | 20 | 29.46 | heat, v.H. (1,2) | 97X1 |
| 0.7 M | 2.8 | 20 | 30.88 | heat, v.H. (1,2) | 97X1 |
| 0.8 M | 2.8 | 20 | 32.43 | heat, v.H. (1,2) | 97X1 |
| 0.9 M | 2.8 | 20 | 33.72 | heat, v.H. (1,2) | 97X1 |
| 0.0 M | 5.5 | 20 | 55.02±1.67 | heat, v.H. (1,3) | 97X1 |
| 0.1 M | 5.5 | 20 | 55.23 | heat, v.H. (1,3) | 97X1 |
| 0.2 M | 5.5 | 20 | 56.36 | heat, v.H. (1,3) | 97X1 |
| 0.3 M | 5.5 | 20 | 56.48 | heat, v.H. (1,3) | 97X1 |
| 0.4 M | 5.5 | 20 | 57.36 | heat, v.H. (1,3) | 97X1 |
| 0.5 M | 5.5 | 20 | 57.66 | heat, v.H. (1,3) | 97X1 |
| 0.6 M | 5.5 | 20 | 57.99 | heat, v.H. (1,3) | 97X1 |
| 0.7 M | 5.5 | 20 | 59.16 | heat, v.H. (1,3) | 97X1 |
| 0.8 M | 5.5 | 20 | 59.83 | heat, v.H. (1,3) | 97X1 |
| 0.9 M | 5.5 | 20 | 62.72 | heat, v.H. (1,3) | 97X1 |

Remarks:
(1) transition monitored by optical absorption at 287 nm, see also Table 2
(2) measured in 0.04 M glycine at pH 2.8
(3) measured in 0.04 M sodium acetate at pH 5.5

Bovine pancreatic ribonuclease A in the presence of β-hydroxyectoine and betaine

| Cosolvent | Conc. | pH | T | ΔG | Approach/Remarks | | Ref |
|---|---|---|---|---|---|---|---|
| β-hydroxyectoine: | | | | | | | |
| | 0.0 M | 5.5 | 25 | 29.8 | DSC | (1,2) | 99K9 |
| | 1.5 M | 5.5 | 25 | 36.3 | DSC | (1,2) | 99K9 |
| | 3.0 M | 5.5 | 25 | 39.2 | DSC | (1,2) | 99K9 |
| betaine: | | | | | | | |
| | 0.0 M | 5.5 | 25 | 29.8 | DSC | (1,2) | 99K9 |
| | 2.0 M | 5.5 | 25 | 38.0 | DSC | (1,2) | 99K9 |
| | 3.0 M | 5.5 | 25 | 38.7 | DSC | (1,2) | 99K9 |
| | 4.0 M | 5.5 | 25 | 39.7 | DSC | (1,2) | 99K9 |
| | 5.0 M | 5.5 | 25 | 37.8 | DSC | (1,2) | 99K9 |

Remarks:
(1) measured in 50 mM phosphate buffer, 200 mM sodium chloride, and the indicated osmolyte conc.
(2) Ref. 99K9 contains further data and general expressions for the concentration dependence of the osmolytes at varying pH

Bovine pancreatic ribonuclease A in the presence of noncompatible osmolytes

| Osmolyte | Conc. | pH | $\Delta T$ | $\Delta H$ | $\Delta(\Delta G)(25°)$ | Appr./Rem. | Ref |
|----------|-------|-----|------|------|------|------------|------|
| arginine | 0.0 M | 5.0 | 0.0 | 0 | 0.0 | heat (1–4) | 98R7 |
|          | 0.2 M | 5.0 | –3.6 | –21 | –4.7 | heat (1–4) | 98R7 |
|          | 0.5 M | 5.0 | –4.5 | –59 | –9.3 | heat (1–4) | 98R7 |
|          | 0.7 M | 5.0 | –4.5 | –67 | –10.2 | heat (1–4) | 98R7 |
|          | 1.0 M | 5.0 | –4.5 | –63 | –9.8 | heat (1–4) | 98R7 |
| histidine | 0.0 M | 5.0 | 0.0 | 0 | 0.0 | heat (1–4) | 98R7 |
|          | 0.1 M | 5.0 | –1.5 | –20 | –3.5 | heat (1–4) | 98R7 |
|          | 0.2 M | 5.0 | –4.3 | –23 | –5.5 | heat (1–4) | 98R7 |
| lysine | 0.0 M | 5.0 | 0.0 | 0 | 0.0 | heat (1–4) | 98R7 |
|        | 0.1 M | 5.0 | –1.7 | 0 | –1.1 | heat (1–4) | 98R7 |
|        | 0.2 M | 5.0 | –2.2 | 4 | –1.1 | heat (1–4) | 98R7 |
|        | 0.35 M | 5.0 | –1.3 | –13 | 0.2 | heat (1–4) | 98R7 |
|        | 0.5 M | 5.0 | –1.2 | –8 | –1.7 | heat (1–4) | 98R7 |
|        | 0.7 M | 5.0 | 0.4 | 0 | 0.3 | heat (1–4) | 98R7 |
|        | 1.0 M | 5.0 | 1.6 | –15 | –0.8 | heat (1–4) | 98R7 |
|        | 2.0 M | 5.0 | 4.2 | –13 | 0.4 | heat (1–4) | 98R7 |
| arginine | 0.0 M | 6.0 | 0.0 | 0 | 0.0 | heat (1–4) | 98R7 |
|          | 0.2 M | 6.0 | –1.3 | –25 | –2.9 | heat (1–4) | 98R7 |
|          | 0.5 M | 6.0 | –4.3 | –50 | –8.5 | heat (1–4) | 98R7 |
|          | 0.7 M | 6.0 | –4.3 | –63 | –9.9 | heat (1–4) | 98R7 |
|          | 1.0 M | 6.0 | –4.3 | –63 | –9.9 | heat (1–4) | 98R7 |
| arginine | 0.0 M | 3.0 | 0.0 | 0 | 0.0 | heat (1–4) | 98R7 |
|          | 0.2 M | 3.0 | 0.0 | –8 | –0.6 | heat (1–4) | 98R7 |
|          | 0.5 M | 3.0 | 0.2 | –29 | –1.8 | heat (1–4) | 98R7 |
|          | 0.7 M | 3.0 | 1.2 | –42 | –2.1 | heat (1–4) | 98R7 |
|          | 1.0 M | 3.0 | 1.9 | –50 | –2.2 | heat (1–4) | 98R7 |
| arginine | 0.0 M | 2.0 | 0.0 | 0 | 0.0 | heat (1,3–5) | 98R7 |
|          | 0.2 M | 2.0 | 1.6 | –8 | 1.0 | heat (1,3–5) | 98R7 |
|          | 0.5 M | 2.0 | 3.1 | –25 | 1.5 | heat (1,3–5) | 98R7 |
|          | 0.7 M | 2.0 | 3.5 | –17 | 2.1 | heat (1,3–5) | 98R7 |
|          | 1.0 M | 2.0 | 3.5 | –17 | 1.8 | heat (1,3–5) | 98R7 |
| arginine | 0.0 M | 1.5 | 0.0 | 0 | 0.0 | heat (1,3–5) | 98R7 |
|          | 0.2 M | 1.5 | 5.4 | –4 | 4.6 | heat (1,3–5) | 98R7 |
|          | 0.5 M | 1.5 | 7.2 | –8 | 5.8 | heat (1,3–5) | 98R7 |
|          | 0.7 M | 1.5 | 8.7 | –13 | 6.1 | heat (1,3–5) | 98R7 |
|          | 1.0 M | 1.5 | 8.7 | –2 | 6.1 | heat (1,3–5) | 98R7 |

Remarks:

(1) van't Hoff treatment, transition monitored by changes in optical absorption at 287 nm

(2) measured in 0.05 M citrate, 0.1 M KCl

(3) in the absence of osmolytes, $T_{trs} = 65.0°C$ and $\Delta H = 481$ kJ/mol was obtained, and in the presence of 0.2 M His $T_{trs} = 60.7°C$ and $\Delta H = 458$ kJ/mol

(4) $\Delta G(25°)$ in the presence of osmolyte was calculated using $\Delta Cp = 5.14$ kJ/mol/K from Ref. 88P3; the value agrees with $\Delta Cp$ derived from $\Delta H^{v.H.}$ versus $T_{trs}$ from the above measurements in the presence of arginine (see also Table 2)

(5) measured in 0.05 M citrate, 0.05 M KCl

Bovine pancreatic ribonuclease A, preferential interactions in aqueous cosolvent systems, magnesium salts-containing aqueous solutions

| Salt Conc. | pH | T | ΔG | Approach/Remarks | | Ref |
|---|---|---|---|---|---|---|
| buffer | 1.5 | 20 | 6.7±2.5 | heat | (1,2) | 97X3 |
| | 2.0 | 20 | 10.9±1.3 | heat | (1,2) | 97X3 |
| | 2.8 | 20 | 23.0±1.3 | heat | (1,2) | 97X3 |
| | 3.0 | 20 | 26.4 | heat | (1,2) | 97X3 |
| | 3.2 | 20 | 28.5 | heat | (1,2) | 97X3 |
| | 5.5 | 20 | 56.9±2.1 | heat | (1,2) | 97X3 |
| | 5.8 | 20 | 58.2 | heat | (1,2) | 97X3 |
| 0.6 M MgCl$_2$ | 1.5 | 20 | 24.3±0.4 | heat | (1,2) | 97X3 |
| | 2.0 | 20 | 27.6 | heat | (1,2) | 97X3 |
| | 3.0 | 20 | 37.7 | heat | (1,2) | 97X3 |
| | 5.5 | 20 | 56.1±1.3 | heat | (1,2) | 97X3 |
| 0.6 M MgSO$_4$ | 1.5 | 20 | 33.1±0.8 | heat | (1,2) | 97X3 |
| | 2.0 | 20 | 37.2 | heat | (1,2) | 97X3 |
| | 3.0 | 20 | 51.5 | heat | (1,2) | 97X3 |
| | 5.5 | 20 | 70.7±1.3 | heat | (1,2) | 97X3 |

Remarks:

(1) data treatment by means of a truncated form of the integrated van't Hoff equation with inherent ΔCp

(2) buffer: 0.04 M glycine at pH 1.5 and 2.0, 0.04 M acetate at pH 5.5

## Ribonuclease B

Ribonuclease B, comparison with ribonuclease A:

Ribonuclease B and ribonuclease A possess identical protein structures, but differ by the presence of a carbohydrate chain attached to Asn34

| Protein | pH | T | ΔG | Approach | Remarks | Ref |
|---|---|---|---|---|---|---|
| RNase A | 8.0 | 25 | 42.6 | heat/GuHCl | (1) | 97A5 |
| RNase B | 8.0 | 25 | 44.9 | heat/GuHCl | (1) | 97A5 |

Remark:

(1) Ref. 97A5 contains a ΔG profile from 0 to 70°C obtained by combined thermal and GuHCl denaturation, see also Table 2

## Ribonuclease H

The data entries are arranged as follows:

a) ribonuclease H, wild type and mutants

b) comparison of RNase H, RNase HI, and HIV RNase H domains

*a) ribonuclease H, wild type and mutants*

Ribonuclease H from *E. coli*, stabilized by introduction of an artificial disulfide bond

| Protein | pH | T | $\Delta G$ | $c_{1/2}$ | m | Appr./Rem. | Ref |
|---|---|---|---|---|---|---|---|
| wild type | 5.5 | 25 | 37.4±1.2 | 1.90±0.05 | 19.71±0.25 | GuHCl (1–4) | 91K7 |
| wild type | 5.5 | 25 | 37.8±1.2 | 1.76±0.05 | 21.51±0.25 | GuHCl (1–3,5) | 91K7 |
| Asn44→Cys with disulfide bond Cys13-Cys44, oxidized protein [remark (6)]: | | | | | | | |
| | 5.5 | 25 | 49.1±1.2 | 2.26±0.05 | 21.67±0.25 | GuHCl (1–4) | 91K7 |
| | 5.5 | 25 | 46.7±1.2 | 2.24±0.05 | 20.84±0.25 | GuHCl (1–3,5) | 91K7 |
| Asn44→Cys, reduced protein: | | | | | | | |
| | 5.5 | 25 | 35.6±1.2 | 1.95±0.05 | 18.28±0.25 | GuHCl (1–4) | 91K7 |
| | 5.5 | 25 | 35.4±1.2 | 1.87±0.05 | 18.91±0.25 | GuHCl (1–3,5) | 91K7 |

Remarks:
(1) RNase H has three free Cys residues at positions 13, 63, and 133, and no intramolecular disulfide bond
(2) linear extrapolation
(3) buffer: 10 mM sodium acetate, pH 5.5, 0.1 M NaCl, 1 mM DTT
(4) transition monitored by CD at 220 nm
(5) transition monitored by the intrinsic fluorescence at 324 nm
(6) the artificial disulfide bond Cys13-Cys44 increases the thermal stability of *E. coli* RNase H by 11.8°C

Ribonuclease H, *E. coli* ribonuclease H1 with three free cysteines replaced by alanine, molten globule unfolding monitored by hydrogen exchange at pH 1 in urea

| Residue | $\Delta G_{unf}(H_2O)$ | m | Protection factors at urea conc. | | | | Appr./Rem. | Ref |
|---|---|---|---|---|---|---|---|---|
| | | | 0.0 M | 0.2 M | 1.0 M | 2.0 M | | |
| Ile7 | 2.1 | (3.8) | 3.4 | 3.0 | 1.5 | 0.8 | HX | (1,2) | 98C4 |
| Phe8 | 1.7 | (5.9) | 3.1 | 2.2 | 1.2 | 0.6 | HX | (1,2) | 98C4 |
| Gly20 | 3.8 | (3.3) | 6.2 | 4.3 | 2.1 | 0.8 | HX | (1,2) | 98C4 |
| Gly21 | 6.7 | (3.8) | 18 | 15 | 6.8 | 1.7 | HX | (1,2) | 98C4 |
| Tyr22 | 6.3 | (3.3) | 14 | 11 | 5.3 | 1.6 | HX | (1,2) | 98C4 |
| Ala24 | 7.9 | 2.9 | 34 | 22 | 12 | 3.6 | HX | (1) | 98C4 |
| Ile25 | 3.3 | (2.1) | 5.7 | 5.0 | 2.2 | 1.7 | HX | (1,2) | 98C4 |
| Leu26 | 2.1 | (4.2) | 3.6 | 2.9 | 1.4 | 0.7 | HX | (1,2) | 98C4 |
| Arg27 | 4.6 | (3.3) | 8.4 | 6.0 | 2.7 | 1.4 | HX | (1,2) | 98C4 |
| Phe35 | 2.9 | (3.8) | 4.5 | 3.5 | 1.6 | 0.9 | HX | (1,2) | 98C4 |
| Arg46 | 9.6 | 2.9 | 68 | 44 | 22 | 5.5 | HX | (1) | 98C4 |
| Ala51 | 13.8 | 4.2 | 400 | 340 | 69 | 11 | HX | (1) | 98C4 |
| Ala52 | 13.4 | 4.2 | 300 | 270 | 58 | 9.5 | HX | (1) | 98C4 |
| Val54 | 16.7 | 5.9 | | 620 | 210 | 7.4 | HX | (1) | 98C4 |
| Ala55 | 14.2 | 4.6 | 440 | 370 | 90 | 10 | HX | (1) | 98C4 |
| Leu56 | 14.6 | 4.6 | 480 | 400 | 99 | 11 | HX | (1) | 98C4 |
| Ile66 | 6.3 | (3.3) | 16 | 12 | 4.3 | 1.8 | HX | (1,2) | 98C4 |
| Leu67 | 5.9 | (3.3) | 15 | 11 | 4.6 | 1.8 | HX | (1,2) | 98C4 |
| Ser68 | 7.5 | 2.9 | 25 | 17 | 10 | 2.6 | HX | (1) | 98C4 |
| Trp104 | 8.4 | 2.9 | 46 | 27 | 15 | 4.1 | HX | (1) | 98C4 |
| Gln105 | 6.3 | (3.8) | 16 | 10 | 4.4 | 1.5 | HX | (1,2) | 98C4 |
| Leu107 | 7.5 | 3.3 | 31 | 17 | 7.1 | 2.2 | HX | (1) | 98C4 |
| Ala109 | 9.2 | 2.9 | 61 | 42 | 21 | 6.1 | HX | (1) | 98C4 |
| Ala110 | 7.9 | 2.9 | 38 | 27 | 9.1 | 3.5 | HX | (1) | 98C4 |
| Lys117 | 2.9 | (3.8) | 5.2 | 3.4 | 1.8 | 0.9 | HX | (1,2) | 98C4 |
| Glu119 | 1.3 | (6.3) | 2.7 | 2.1 | 1.1 | 0.4 | HX | (1,2) | 98C4 |
| Leu136 | 4.6 | (2.1) | 8.5 | 6.3 | 3.7 | 2.0 | HX | (1,2) | 98C4 |
| Ala139 | 5.9 | (3.3) | 14 | 9.4 | 4.6 | 1.6 | HX | (1,2) | 98C4 |
| Ala140 | 7.1 | (3.8) | 21 | 15 | 6.7 | 1.8 | HX | (1,2) | 98C4 |

Remarks:
(1) lyophilized, $^{15}$N-labeled ribonuclease H was dissolved in deuterated buffer (10 mM phosphoric acid, 50 mM KCl, pD 0.9–1.0) with 0.0, 0.2, 1.0, or 2.0 M urea and 110, 220, 370, or 700 mM DCl, respectively
(2) m values in parentheses were poorly determined due to low protection factors, P < 2 in either 1.0 or 2.0 M urea

| Mutant/Transition | pH | T | $\Delta G$ | m | Approach/Remarks | Ref |
|---|---|---|---|---|---|---|
| wild type* | | | | | | |
| $N \to U$ | 5.5 | 25 | 40.6±1.7 | 8.79±0.38 | urea, eq. (1–4,6) | 99R5 |
| $I \to U$ | 5.5 | 25 | 14.85 | 5.19 | urea, kin. (2–5) | 99R5 |
| $N \to I \to U$ | 5.5 | 25 | 41.3 | 8.87 | urea, kin. (2–5) | 99R5 |
| Asp10→Ala | | | | | | |
| $N \to U$ | 5.5 | 25 | 54.4±4.2 | 8.79±0.84 | urea, eq. (1–4,6) | 99R5 |
| $I \to U$ | 5.5 | 25 | 18.33 | 6.07 | urea, kin. (2–5) | 99R5 |
| $N \to I \to U$ | 5.5 | 25 | 61.9 | 10.79 | urea, kin. (2–5) | 99R5 |
| Arg27→Ala | | | | | | |
| $N \to U$ | 5.5 | 25 | 31.4±3.3 | 8.37±0.84 | urea, eq. (1–4,6) | 99R5 |
| $I \to U$ | 5.5 | 25 | 16.65 | 5.48 | urea, kin. (2–5) | 99R5 |
| $N \to I \to U$ | 5.5 | 25 | 38.0 | 10.21 | urea, kin. (2–5) | 99R5 |
| Ile53→Ala | | | | | | |
| $N \to U$ | 5.5 | 25 | 31.8±1.7 | 8.79±0.42 | urea, eq. (1–4,6) | 99R5 |
| $I \to U$ | 5.5 | 25 | 6.95 | 5.19 | urea, kin. (2–5) | 99R5 |
| $N \to I \to U$ | 5.5 | 25 | 33.6 | 10.04 | urea, kin. (2–5) | 99R5 |
| Gln105→Gly | | | | | | |
| $N \to U$ | 5.5 | 25 | 31.4±1.3 | 8.79±0.33 | urea, eq. (1–4,6) | 99R5 |
| $I \to U$ | 5.5 | 25 | 7.45 | 3.55 | urea, kin. (2–5) | 99R5 |
| $N \to I \to U$ | 5.5 | 25 | 36.4 | 10.13 | urea, kin. (2–5) | 99R5 |

Remarks:

(1) equilibrium measurements, linear extrapolation

(2) transition monitored by far-UV CD

(3) buffer: 20 mM sodium acetate, 50 mM potassium chloride, pH 5.5

(4) wild type* is the triple cysteine to alanine variant, used as the template

(5) data from kinetics of folding and unfolding in dependence of the denaturant concentration

(6) the thermal transition (in 1 M GuHCl) takes place at $T_{trs}$ = 47.8±0.3°C for wild type*, at 62.6±0.1°C for Asp10→Ala, at 42.2±0.2°C for Arg27→Ala, at 35.7±0.2°C for Ile53→Ala, and at 38.2±0.2°C for Gln105→Gly

Ribonuclease H core fragment (eABCD), residues Arg41-Gln113

| Protein | pH | T | $\Delta G$ | Approach/Remarks | Ref |
|---|---|---|---|---|---|
| eABCD | 5.5 | 25 | 50.2 | urea    (1,2) | 99C5 |

Remarks:

(1) linear extrapolation assuming a $D \to 2U$ model

(2) transition monitored by far-UV CD at 222 nm

*b) comparison of RNase H, RNase HI, and HIV RNase H domains*

Ribonuclease H domain of Moloney murine leukemia virus (MMLV) reverse transcriptase compared with *E. coli* RNase HI, and HIV RNase H domain

| Protein | pH | $T_{trs}$ | $\Delta G$ | $c_{1/2}$ | Approach/Remarks | Ref |
|---|---|---|---|---|---|---|
| MMLV RNase H domain (MRH-175) | 8.0 | 30 | 18.0 | 1.1 | GuHCl    (1–3) | 98G10 |
| HIV RNase H domain | 8.0 | | 16.5±0.2 | 1.18 | GuHCl    (1,4) | 95K15 |
| | 8.0 | 45.6 | | | heat    (6) | 97G10 |
| HIV RNase H chimera | 8.0 | | 21.5±0.2 | 2.17 | GuHCl    (1,4,5) | 95K15 |

Ribonuclease H domain of Moloney murine leukemia virus (MMLV) reverse transcriptase compared with *E. coli* RNase HI, and HIV RNase H domain (continued)

| Protein | pH | $T_{trs}$ | $\Delta G$ | $c_{1/2}$ | Approach/Remarks | | Ref |
|---|---|---|---|---|---|---|---|
| *E. coli* RNase HI | 8.0 | | 30.1±2.9 | 1.60 | GuHCl | (1,8) | 94D1 |
| | 8.0 | | 31.0±4.2 | 3.84 | urea | (1,8) | 94D1 |
| | 8.0 | | 31.8±5.4 | 3.93 | urea | (1,9) | 94D1 |
| | 8.0 | 50.9 | | | heat | (6) | 97G10 |
| RNHΔE | 8.0 | | 12.6 | 1.3 | GuHCl | (7) | 97G10 |
| | 8.0 | 42.5 | | | heat | (6,7) | 97G10 |

Remarks:

(1) linear extrapolation, LEM-SB

(2) $\Delta G$ was determined in 100 mM potassium phosphate and 20 mM potassium chloride at 25°C by CD at 222 nm

(3) $T_{trs}$ was determined in 5 mM HEPES, pH 8.0, in the presence of 0.8 M GuHCl by CD at 222 nm

(4) $\Delta G$ was determined in 50 mM potassium phosphate and 50 mM potassium chloride at 25°C by CD at 222 nm

(5) HIV RNase H chimera is composed of the HIV RNase sequence from Tyr427 to Ile506, *E. coli* RNase HI sequence from Thr79 to Asp102, and HIV RNase H sequence from Leu517 to Leu560

(6) $T_{trs}$ was determined in 100 mM potassium phosphate and 20 mM potassium chloride in the presence of 0.8 M GuHCl by CD at 222 nm

(7) RNHΔE is a deletion mutant which lacks the E-helix and adjacent residues (the C-terminal 33 of 155 residues)

(8) $\Delta G$ was determined in 20 mM potassium phosphate and 50 mM potassium chloride at 25°C by CD at 222 nm

(9) $\Delta G$ was determined in 20 mM potassium phosphate and 50 mM potassium chloride at 25°C by Trp fluorescence

Recombinant ribonuclease H (RNase H) from *Thermus thermophilus* HB8, compared with RNase HI from *E. coli*

| Protein | pH | T | $\Delta G$ | $c_{1/2}$ | m | Appr./Rem. | Ref |
|---|---|---|---|---|---|---|---|
| RNase HI from *E. coli*: | | | | | | | |
| | 5.5 | 25 | 38.1±1.5 | 1.86±0.05 | 20.5±0.8 | GuHCl (1–3) | 92K13 |
| | 5.5 | 50 | 36.1±1.4 | 1.15±0.04 | 31.4±1.3 | GuHCl (1–3) | 92K13 |
| RNase H from *Thermus thermophilus*: | | | | | | | |
| | 5.5 | 25 | 87.4±3.5 | 3.73±0.10 | 23.4±0.8 | GuHCl (1–3) | 92K13 |
| | 5.5 | 50 | 94.9±3.8 | 3.49±0.08 | 27.2±1.3 | GuHCl (1–3) | 92K13 |

Remarks:

(1) linear extrapolation

(2) buffer: 20 mM sodium acetate, pH 5.5, 0.1 M NaCl

(3) transition monitored by CD at 220 nm

Ribonuclease H domain from HIV-1 (HIV RNase H) and fragment HIVΔE

| Transition | pH | T | $\Delta G$ | $c_{1/2}$ | m | Approach/Remarks | Ref |
|---|---|---|---|---|---|---|---|
| HIV RNase H | | | | | | | |
| N → U | 5.5 | 25 | 15.5 | 1.25 | 12.1 | GuHCl, eq. (1,2) | 98K6 |
| N → U | 5.5 | 25 | 14.6 | 1.25 | 12.5 | GuHCl, kin. (2–4) | 98K6 |
| I → U | 5.5 | 25 | 6.3 | 1.05 | 5.9 | GuHCl, kin. (2,3,5) | 98K6 |
| fragment HIVΔE | | | | | | | |
| N → U | 5.5 | 25 | 10.9 | 1.0 | 10.9 | GuHCl, eq. (1–3,6) | 98K6 |

Remarks:

(1) data from equilibrium unfolding, transition monitored by CD at 222 nm and trytophan fluorescence

(2) linear extrapolation, LEM-SB

(3) buffer: 50 mM sodium acetate

(4) data from kinetics, final amplitudes of stopped-flow CD

(5) data from kinetics, burst-phase amplitudes measured by CD

(6) fragment HIVΔE containing a C-terminal deletion of helix E, residues 427–539

Comparison of two ribonucleases H, one from mesophile *Escherichia coli* and one from thermophile *Thermus thermophilus*

| Protein | pH | T | ΔG | Approach/Remarks | | Ref |
|---|---|---|---|---|---|---|
| ribonuclease H from *E. coli*: | | | | | | |
| | 5.5 | 24 | 31.4 | GuHCl | (1–4) | 99H10 |
| | 5.5 | 37 | 28.5 | GuHCl | (1–3,5) | 99H10 |
| ribonuclease H from *T. thermophilus*: | | | | | | |
| | 5.5 | 20 | 53.1 | GuHCl | (1–4) | 99H10 |
| | 5.5 | 68.5 | 23.4 | GuHCl | (1–3,5) | 99H10 |

Remarks:

(1) data from GuHCl denaturation at 5, 10, 15, 20, 25, 30, 40, 50, and 60°C assuming a two-state model and a linear relationship between ΔG and GuHCl conc.

(2) measured in 5 mM sodium acetate, 50 mM KCl

(3) transition monitored by CD at 225 nm

(4) temperature at which maximal stability is observed

(5) optimal growth temperature

## Ribonuclease H*

Ribonuclease H*, a triple mutant derived from ribonuclease HI, alanine has been substituted for all the three native cysteines

| Transition | pH | T | ΔG | $c_{1/2}$ | m | Approach/Remarks | | Ref |
|---|---|---|---|---|---|---|---|---|
| N → D | 5.5 | 25 | 41.34±0.21 | 4.66 | 8.87±0.38 | urea | (1,2) | 97R2 |
| N → I | 5.5 | 25 | 14.85±0.13 | 2.86 | 5.19±0.33 | urea | (2,3) | 97R2 |

Remarks:

(1) equilibrium unfolding, LEM

(2) the transition was monitored by CD at 222 nm

(3) kinetics, from burst-phase amplitude

## Ribonuclease HI

Ribonuclease HI, wild type and mutants, position 52

| Mutant | pH | $T_{trs}$ | ΔT | Δ(ΔG) | Appr./Rem. | | Ref |
|---|---|---|---|---|---|---|---|
| wild type | 3.2 | 53.0 | 0.0 | 0.0 | heat | (1) | 97A2 |
| Ala52→Asn | 3.2 | 47.1 | −5.9 | −7.49 | heat | (1,2) | 97A2 |
| Ala52→Asp | 3.2 | 46.9 | −6.1 | −7.74 | heat | (1,2) | 97A2 |
| Ala52→Cys | 3.2 | 55.5 | 2.5 | 3.18 | heat | (1,2) | 97A2 |
| Ala52→Gln | 3.2 | 49.1 | −3.9 | −4.98 | heat | (1,2) | 97A2 |
| Ala52→Glu | 3.2 | 48.0 | −5.0 | −6.36 | heat | (1,2) | 97A2 |
| Ala52→Gly | 3.2 | 44.1 | −8.9 | −11.34 | heat | (1,2) | 97A2 |
| Ala52→His | 3.2 | 41.2 | −11.8 | −15.02 | heat | (1,2) | 97A2 |
| Ala52→Ile | 3.2 | 59.2 | 6.2 | 7.87 | heat | (1,2) | 97A2 |
| Ala52→Leu | 3.2 | 57.3 | 4.3 | 5.48 | heat | (1,2) | 97A2 |
| Ala52→Lys | 3.2 | 33.5 | −19.5 | −24.81 | heat | (1,2) | 97A2 |
| Ala52→Met | 3.2 | 54.6 | 1.6 | 2.05 | heat | (1,2) | 97A2 |
| Ala52→Phe | 3.2 | 51.5 | −1.5 | −1.92 | heat | (1,2) | 97A2 |
| Ala52→Pro | 3.2 | 47.6 | −5.4 | −6.86 | heat | (1,2) | 97A2 |

Ribonuclease HI, wild type and mutants, position 52 (continued)

| Mutant | pH | $T_{trs}$ | $\Delta T$ | $\Delta(\Delta G)$ | Appr./Rem. | | Ref |
|---|---|---|---|---|---|---|---|
| Ala52→Ser | 3.2 | 47.2 | −5.8 | −7.36 | heat | (1,2) | 97A2 |
| Ala52→Thr | 3.2 | 50.3 | −2.7 | −3.43 | heat | (1,2) | 97A2 |
| Ala52→Tyr | 3.2 | 45.4 | −7.6 | −9.67 | heat | (1,2) | 97A2 |
| Ala52→Val | 3.2 | 58.5 | 5.5 | 6.99 | heat | (1,2) | 97A2 |

Remarks:
(1) heat denaturation, transition monitored by CD at 220 nm
(2) $\Delta(\Delta G)$ was calculated from $\Delta(\Delta G) = \Delta T * \Delta S_{w.t.}$, where $\Delta S_{w.t.} = 1.272$ kJ/mol/K was taken from Ref. 92K8

Ribonuclease HI, wild type and double mutants, positions 52 and 74

| Mutant | pH | $T_{trs}$ | $\Delta T$ | $\Delta(\Delta G)$ | Appr./Rem. | | Ref |
|---|---|---|---|---|---|---|---|
| wild type | 3.2 | 53.0 | 0.0 | 0.0 | heat | (1) | 97A2 |
| Val74→Ala | 3.2 | 45.4 | −7.6 | −9.67 | heat | (1–3) | 97A2 |
| Val74→Leu | 3.2 | 56.7 | 3.7 | 4.69 | heat | (1–3) | 97A2 |
| double mutant (Ala52→Ile and Val74→Ala) | | | | | | | |
| | 3.2 | 56.9 | 3.9 | 4.98 | heat | (1,2) | 97A2 |
| double mutant (Ala52→Leu and Val74→Ala) | | | | | | | |
| | 3.2 | 54.2 | 1.2 | 1.50 | heat | (1,2) | 97A2 |
| double mutant (Ala52→Phe and Val74→Ala) | | | | | | | |
| | 3.2 | 48.4 | −4.6 | −5.86 | heat | (1,2) | 97A2 |
| double mutant (Ala52→Val and Val74→Ala) | | | | | | | |
| | 3.2 | 51.3 | −1.7 | −2.18 | heat | (1,2) | 97A2 |
| double mutant (Ala52→Gly and Val74→Leu) | | | | | | | |
| | 3.2 | 48.4 | −4.6 | −5.86 | heat | (1,2) | 97A2 |
| double mutant (Ala52→Ile and Val74→Leu) | | | | | | | |
| | 3.2 | 56.9 | 3.9 | 4.98 | heat | (1,2) | 97A2 |
| double mutant (Ala52→Leu and Val74→Leu) | | | | | | | |
| | 3.2 | 57.9 | 4.9 | 6.23 | heat | (1,2) | 97A2 |
| double mutant (Ala52→Phe and Val74→Leu) | | | | | | | |
| | 3.2 | 53.1 | 0.1 | 0.13 | heat | (1,2) | 97A2 |
| double mutant (Ala52→Val and Val74→Leu) | | | | | | | |
| | 3.2 | 61.0 | 8.0 | 10.17 | heat | (1,2) | 97A2 |

Remarks:
(1) heat denaturation, transition monitored by CD at 220 nm
(2) $\Delta(\Delta G)$ was calculated from $\Delta(\Delta G) = \Delta T * \Delta S_{w.t.}$, $\Delta S_{w.t.} = 1.272$ kJ/mol/K was taken from Ref. 92K8
(3) reference protein, data from Ref. 93I2

Ribonuclease HI from *Thermus thermophilus* HB8, proteins with C-terminal truncations

| Protein | pH | $T_{trs}$ | $\Delta T$ | $\Delta(\Delta G)$ | Appr./Rem. | Ref |
|---|---|---|---|---|---|---|
| wild type | 5.5 | 83.7 | 0.0 | 0.0 | heat (1–3) | 98H9 |
| TRNH[Δ156–161] | 5.5 | 83.3 | −0.4 | −0.7 | heat (1–3) | 98H9 |
| TRNH[Δ154–161] | 5.5 | 82.0 | −1.7 | −3.1 | heat (1–3) | 98H9 |
| TRNH[Δ152–161] | 5.5 | 80.0 | −3.7 | −6.8 | heat (1–3) | 98H9 |
| TRNH[Δ150–161] | 5.5 | 80.9 | −2.8 | −5.1 | heat (1–3) | 98H9 |
| TRNH[Δ149–161] | 5.5 | 72.0 | −11.7 | −21.4 | heat (1–3) | 98H9 |
| TRNH[Δ148–161] | 5.5 | 68.1 | −15.6 | −28.5 | heat (1–3) | 98H9 |
| TRNH[Δ146–161] | 5.5 | 69.1 | −14.6 | −26.7 | heat (1–3) | 98H9 |
| TRNH[Δ144–161] | 5.5 | 66.7 | −17.0 | −31.1 | heat (1–3) | 98H9 |

Remarks:
(1) measured in 20 mM sodium acetate (pH 5.5) in the presence of 1 M GuHCl
(2) transition monitored by CD at 220 mM, see also Table 2
(3) $\Delta(\Delta G)$ was calculated using $\Delta(\Delta G) = \Delta T_{mutant} \times \Delta S_{w.t.}$, with $\Delta S_{w.t.} = 1.828$ kJ/mol/K

Ribonuclease HI from *Thermus thermophilus* HB8, proteins with C-terminal truncations and cysteine residues replaced

| Protein | pH | $T_{trs}$ | $\Delta T$ | $\Delta(\Delta G)$ | Appr./Rem. | Ref |
|---|---|---|---|---|---|---|
| wild type | 5.5 | 83.7 | 0.0 | 0.0 | heat (1,2) | 98H9 |
| wild type, reduced | 5.5 | 77.4 | −6.3 | −11.5 | heat (1,2) | 98H9 |
| Cys149→Ala RNase HI | 5.5 | 74.7 | −9.0 | −16.4 | heat (1–4) | 98H9 |
| TRNH[Δ150–161] | 5.5 | 80.9 | | | heat (5) | 98H9 |
| Cys13→Ala TRNH[Δ150–161] | 5.5 | 80.1 | −0.8 | −1.5 | heat (1–4) | 98H9 |
| Cys41→Ala TRNH[Δ150–161] | 5.5 | 71.7 | −9.2 | −16.8 | heat (1–4) | 98H9 |
| Cys63→Ala TRNH[Δ150–161] | 5.5 | 78.8 | −2.1 | −3.9 | heat (1–4) | 98H9 |
| Cys149→Ala TRNH[Δ150–161] | 5.5 | 72.5 | −8.4 | −15.4 | heat (1–4) | 98H9 |
| Cys149→Ile TRNH[Δ150–161] | 5.5 | 74.6 | −6.3 | −11.5 | heat (1–4) | 98H9 |
| Cys149→Ser TRNH[Δ150–161] | 5.5 | 73.1 | −7.8 | −14.3 | heat (1–4) | 98H9 |
| Cys149→Thr TRNH[Δ150–161] | 5.5 | 75.0 | −5.9 | −10.8 | heat (1–4) | 98H9 |
| Cys149→Val TRNH[Δ150–161] | 5.5 | 75.8 | −5.1 | −9.3 | heat (1–4) | 98H9 |
| double mutant (Cys13→Ala and Cys63→Ala) TRNH[Δ150–161] | | | | | | |
| | 5.5 | 78.5 | −2.4 | −4.4 | heat (1–4) | 98H9 |
| oxidized form | 5.5 | 78.5 | −2.4 | −4.4 | heat (1–4) | 98H9 |
| reduced form | 5.5 | 71.6 | −9.3 | −17.0 | heat (1–4) | 98H9 |
| multiple mutant (Cys13→Ala, Cys41→Ala, Cys63→Ala, and Cys149→Ala) TRNH[Δ150–161] | | | | | | |
| | 5.5 | 68.9 | −12.0 | −21.9 | heat (1–4) | 98H9 |

Remarks:
(1) measured in 20 mM sodium acetate (pH 5.5) in the presence of 1 M GuHCl
(2) transition monitored by CD at 220 mM, see also Table 2
(3) $\Delta(\Delta G)$ was calculated using $\Delta(\Delta G) = \Delta T_{mutant} \times \Delta S_{w.t.}$ with $\Delta S_{w.t.} = 1.828$ kJ/mol/K
(4) RNase HI from *Thermus thermophilus* HB8 contains four cysteine residues: Cys13, Cys41, Cys63, and Cys149
(5) reference protein for the following data entries

# Ribonuclease P2

Ribonuclease P2 from *Sulfolobus solfataricus*, mutants, pressure denaturation

| Mutant | pH | T | $\Delta G$ | Approach/Remarks | Ref |
|---|---|---|---|---|---|
| Phe31→Ala | 5 | 35 | 10.5±2.2 | pressure (1–3) | 97M8 |
| Phe31→Tyr | 5 | 35 | 4.6±0.3 | pressure (1,2,4) | 97M8 |

Remarks:
(1) buffer: 50 mM MES
(2) the transition was monitored by forth derivative spectroscopy of the Tyr bands
(3) transition midpoint at pm = 256 mPa, $\Delta V = -43\pm10$ ml/mol
(4) transition midpoint at pm = 200 mPa, $\Delta V = -23\pm2$ ml/mol

Ribonuclease P2, see also DNA-binding protein Sso7d from the
hyperthermophile *Sulfolobus solfataricus*

**Ribonuclease S**

Bovine pancreatic ribonuclease S, $\Delta G_{ex}$ from H/D exchange experiments at pH 6.0

| Residue | $\Delta G_{ex}(15°C)$ | $\Delta G_{ex}(25°C)$ | $\Delta G_{ex}(35°C)$ | $\Delta G_{ex}(40°C)$ | Appr./Rem. | Ref |
|---|---|---|---|---|---|---|
| Arg10 | | | | | | 99N3 |
| Gln11 | | | | | | 99N3 |
| His12 | | | | | | 99N3 |
| Met13 | 20.1 | | | | H/D (1) | 99N3 |
| Asp14 | | | | | | 99N3 |
| Met29 | 23.8 | 26.4 | | | H/D (1) | 99N3 |
| Met30 | >30.1 | 25.9 | 29.7 | | H/D (1) | 99N3 |
| Lys31 | 26.4 | | | | H/D (1) | 99N3 |
| Val43 | 23.4 | 23.8 | 25.1 | | H/D (1) | 99N3 |
| Phe46 | >36.0 | >39.4 | 31.0 | | H/D (1) | 99N3 |
| Val47 | >34.3 | 27.2 | | | H/D (1) | 99N3 |
| His48 | >31.0 | 20.1 | | | H/D (1) | 99N3 |
| Glu49 | 24.7 | 23.4 | 20.5 | | H/D (1) | 99N3 |
| Val54 | >33.1 | >36.8 | 31.0 | 27.2 | H/D (1) | 99N3 |
| Gln55 | >35.1 | 38.9 | 32.6 | | H/D (1) | 99N3 |
| Ala56 | >34.7 | 27.6 | | | H/D (1) | 99N3 |
| Val57 | >32.2 | >36.0 | 28.5 | | H/D (1) | 99N3 |
| Cys58 | >35.6 | >39.7 | 30.5 | 29.3 | H/D (1) | 99N3 |
| Ser59 | 25.9 | 26.8 | | | H/D (1) | 99N3 |
| Gln60 | 33.1 | 31.8 | 29.3 | | H/D (1) | 99N3 |
| Lys61 | 27.6 | 26.8 | 27.2 | | H/D (1) | 99N3 |
| Val63 | >33.5 | 34.3 | 31.0 | 26.8 | H/D (1) | 99N3 |
| Cys72 | >36.0 | 33.1 | 29.7 | | H/D (1) | 99N3 |
| Tyr73 | >35.6 | >38.9 | 30.1 | 26.8 | H/D (1) | 99N3 |
| Gln74 | >35.1 | >38.9 | 30.1 | 28.0 | H/D (1) | 99N3 |
| Met79 | >36.4 | 35.6 | | | H/D (1) | 99N3 |
| Ile81 | >35.1 | 35.6 | 29.3 | 25.9 | H/D (1) | 99N3 |
| Thr82 | >35.6 | 30.5 | 28.0 | 26.4 | H/D (1) | 99N3 |
| Cys84 | >37.7 | 37.2 | 31.0 | 29.3 | H/D (1) | 99N3 |
| Arg85 | >36.4 | 34.3 | 30.5 | 28.0 | H/D (1) | 99N3 |
| Tyr97 | >34.3 | >39.3 | 28.9 | | H/D (1) | 99N3 |
| Lys98 | >36.4 | 33.9 | 30.1 | 27.6 | H/D (1) | 99N3 |
| Thr100 | >38.5 | 36.0 | 33.1 | 34.7 | H/D (1) | 99N3 |
| Ala102 | >37.2 | 33.1 | 30.1 | | H/D (1) | 99N3 |
| Lys104 | | 28.5 | 23.8 | | H/D (1) | 99N3 |
| Ile106 | >36.0 | 35.1 | 30.1 | 27.6 | H/D (1) | 99N3 |
| Ile107 | >24.3 | >28.0 | | | H/D (1) | 99N3 |
| Val108 | >31.8 | >35.6 | 28.9 | 25.5 | H/D (1) | 99N3 |
| Ala109 | >32.6 | >36.4 | 29.3 | 25.9 | H/D (1) | 99N3 |
| Cys110 | 25.9 | 25.9 | | | H/D (1) | 99N3 |
| Glu111 | >36.8 | 34.3 | 32.6 | | H/D (1) | 99N3 |
| Val116 | >30.1 | >33.5 | 28.5 | 24.7 | H/D (1) | 99N3 |
| Val118 | >31.0 | >34.3 | 28.5 | 23.8 | H/D (1) | 99N3 |
| His119 | 25.9 | 24.3 | 22.6 | | H/D (1) | 99N3 |

Remark:
(1) for further details see Ref. 99N3

## Ribonucleases Sa, Sa2, and Sa3

Ribonucleases (RNases) Sa, Sa2, and Sa3 from *Streptomyces aureofaciens* compared with selected ribonucleases at pH 7.0, heat denaturation

| Protein | pH | T | $\Delta G$ | Approach/Remarks | | Ref |
|---|---|---|---|---|---|---|
| RNase Sa | 7.0 | 25 | 24.3 | heat | (1) | 98P1 |
| RNase Sa2 | 7.0 | 25 | 12.6 | heat | (1) | 98P1 |
| RNase Sa3 | 7.0 | 25 | 22.2 | heat | (1) | 98P1 |
| RNase T1 | 7.0 | 25 | 28.5 | heat | (2) | 98P1 |
| Barnase | 7.0 | 25 | 36.4 | heat | (3) | 98P1 |
| RNase A | 7.0 | 25 | 38.5 | heat | (4) | 98P1 |
| RNase A | 7.0 | 25 | 43.9 | heat | (5) | 98P1 |

Remarks:
(1) $\Delta G$ was calculated using $T_{trs}$, $\Delta H$, and $\Delta Cp$ from Ref. 98P1, see also Table 2
(2) $\Delta G$ was calculated using $T_{trs}$ and $\Delta H$ averaged from Refs. 89S8, 92S11, and 94Y2, and $\Delta Cp$ from Ref. 89P2, see Table 2
(3) $\Delta G$ was calculated using $T_{trs}$ and $\Delta H$ from Ref. 98P1, and averaged $\Delta Cp$ from Refs. 94O1 and 95J1, see Table 2
(4) $\Delta G$ was calculated using $T_{trs}$ and $\Delta H$ from Ref. 96C6, and averaged $\Delta Cp$ from 13 literature values, see Table 2
(5) $\Delta G$ was calculated using $T_{trs}$, $\Delta H$, and $\Delta Cp$ from Ref. 96C6, see Table 2

Ribonucleases (RNases) Sa, Sa2, and Sa3 from *Streptomyces aureofaciens* compared with selected ribonucleases at pH 7.0, urea denaturation

| Protein | pH | T | $\Delta G$ | $c_{1/2}$ | m | Approach/Remarks | | Ref |
|---|---|---|---|---|---|---|---|---|
| RNase Sa | 7.0 | 25 | 26.8 | 6.44 | 4.14 | urea | (1,2) | 98P1 |
| RNase Sa2 | 7.0 | 25 | 11.7 | 2.10 | 5.48 | urea | (1,2) | 98P1 |
| RNase Sa3 | 7.0 | 25 | 24.7 | 5.63 | 4.39 | urea | (1,2) | 98P1 |
| RNase T1 | 7.0 | 25 | 26.8 | 5.30 | 5.06 | urea | (3) | 98P1 |
| Barnase | 7.0 | 25 | 34.4 | 4.49 | 8.12 | urea | (4) | 98P1 |
| RNase A | 7.0 | 25 | 38.5 | 6.92 | 5.65 | urea | (3) | 98P1 |

Remarks:
(1) linear extrapolation, LEM-SB
(2) transition monitored by CD at 234 nm
(3) data from Ref. 90P4 and Ref. 90P5, see also Refs. 89S8, 89T2, and 90P1
(4) data from Ref. 92P1

Ribonucleases (RNases) Sa, Sa2, and Sa3 from *Streptomyces aureofaciens* compared with selected ribonucleases by thermal denaturation at pH 7.0 in the presence of 0 M and 0.5 M NaCl

| Protein | NaCl | pH | $T_{trs}$ | $\Delta T$ | $\Delta(\Delta G)$ | Appr./Rem. | | Ref |
|---|---|---|---|---|---|---|---|---|
| RNase Sa | 0.0 M | 7.0 | 48.4 | 0.0 | 0.0 | heat | (1–3) | 98P1 |
| | 0.5 M | 7.0 | 52.4 | 4.0 | 5.0 | heat | (1–3) | 98P1 |
| RNase Sa2 | 0.0 M | 7.0 | 41.1 | 0.0 | 0.0 | heat | (1–3) | 98P1 |
| | 0.5 M | 7.0 | 41.2 | 0.1 | 0.0 | heat | (1–3) | 98P1 |
| RNase Sa3 | 0.0 M | 7.0 | 47.2 | 0.0 | 0.0 | heat | (1–3) | 98P1 |
| | 0.5 M | 7.0 | 54.4 | 7.2 | 8.8 | heat | (1–3) | 98P1 |
| RNase T1 | 0.0 M | 7.0 | 51.6 | 0.0 | 0.0 | heat | (1–3) | 98P1 |
| | 0.5 M | 7.0 | 59.4 | 7.8 | 10.5 | heat | (1–3) | 98P1 |
| Barnase | 0.0 M | 7.0 | 53.2 | 0.0 | 0.0 | heat | (1–3) | 98P1 |
| | 0.5 M | 7.0 | 55.8 | 2.6 | 4.2 | heat | (1–3) | 98P1 |

Remarks:
(1) $\Delta(\Delta G)$ refers to $T_{trs}$ at 0 M NaCl, $\Delta(\Delta G)$ was determined by $\Delta(\Delta G) = \Delta T \times \Delta S$, with $\Delta T = T_{trs(0.5\,M\,NaCl)} - T_{trs(0\,M\,NaCl)}$
(2) measured in 30 mM MOPS buffer without and with NaCl as indicated
(3) transition monitored by CD at 234 nm

Ribonucleases (RNases) Sa, Sa2, and Sa3 from *Streptomyces aureofaciens*, contribution of the disulfide bond to the stability of the RNases

| Protein | S-S bond | pH | $T_{trs}$ | $\Delta T$ | $\Delta(\Delta G)$ | Appr./Rem. | Ref |
|---------|----------|----|-----------|------------|--------------------|------------|-----|
| RNase Sa | intact | 7.0 | 48.4 | 0.0 | 0.0 | heat (1–3) | 98P1 |
| | reduced | 7.0 | 28.4 | −20.0 | −19.2 | heat (1–3) | 98P1 |
| RNase Sa2 | intact | 5.0 | 48.9 | 0.0 | 0.0 | heat (1–3) | 98P1 |
| | reduced | 5.0 | 17.4 | −31.5 | −24.7 | heat (1–3) | 98P1 |
| RNase Sa3 | intact | 3.0 | 50.6 | 0.0 | 0.0 | heat (1–3) | 98P1 |
| | reduced | 3.0 | 23.6 | −27.0 | −31.4 | heat (1–3) | 98P1 |

Remarks:

(1) RNase Sa was measured in 30 mM PIPES, pH 7.0

(2) RNase Sa2 was measured in 30 mM acetate, pH 5.0

(3) RNase Sa3 was measured in 30 mM glycylglycine, pH 3.0

(4) $\Delta(\Delta G)$ was calculated at 25°C using $T_{trs}$, $\Delta H$, and $\Delta C_p$ from Ref. 98P1, see Table 2

Ribonuclease Sa from *Streptomyces aureofaciens* (RNase Sa), wild type and mutants

| Mutant | pH | $T_{trs}$ | $\Delta T$ | $\Delta(\Delta G)$ | Appr./Rem. | Ref |
|--------|----|-----------|------------|--------------------|------------|-----|
| wild type | 7.0 | 47.2±0.3 | 0.0 | 0.0 | DSC (1–3) | 99G17 |
| Asp25→Lys | 7.0 | 50.2±0.3 | 3.0 | 3.8 | DSC (1–3) | 99G17 |
| Glu74→Lys | 7.0 | 51.1±0.3 | 3.9 | 4.6 | DSC (1–3) | 99G17 |

Remarks:

(1) data from thermal denaturation

(2) $\Delta(\Delta G)$ was calculated using $\Delta(\Delta G) = \Delta T_{mut} \times \Delta S_{w.t.}$

(3) buffer: 30 mM MOPS, pH 7.0

Ribonuclease Sa from *Streptomyces aureofaciens* (RNase Sa), mutants concerning Asn39

| Mutant | pH | T | $\Delta T$ | $\Delta(\Delta G)$ | Approach/Remarks | | Ref |
|--------|----|---|------------|--------------------|------------------|---|-----|
| wild type | 7.0 | 48.4 | 0.0 | 0.0 | heat | (1–4) | 98H7 |
| Asn39→Ala | 7.0 | 40.8 | −7.6 | −9.2 | heat | (1–4) | 98H7 |
| Asn39→Asp | 7.0 | 43.2 | −5.2 | −6.3 | heat | (1–4) | 98H7 |
| Asn39→Ser | 7.0 | 40.4 | −8.0 | −9.6 | heat | (1–4) | 98H7 |

Remarks:

(1) RNase Sa was expressed in *E. coli*

(2) the conserved Asn39 corresponds to position 44 in RNase T1, and 58 in barnase, see also RNase T1, Ref. 98H7

(3) measured by CD at 234 nm in 30 mM MOPS, pH 7.0

(4) $\Delta(\Delta G)$ from $\Delta(\Delta G) = \Delta T \times \Delta S$, see also Table 2

## Ribonuclease, Bovine Seminal

Ribonuclease, bovine seminal, step-wise mutation of ribonuclease A toward the dimeric bovine seminal ribonuclease

| Form | pH | T | $\Delta G$ | $c_{1/2}$ | m | Appr./Rem. | Ref |
|------|----|---|------------|-----------|---|------------|-----|
| RNase A | 5.0 | 25 | 36.9±2.5 | 5.2 | 7.1±0.3 | urea (1,2) | 97C4 |
| | 5.0 | 25 | 39.2 | | | DSC (2) | 97C4 |
| P-RNase A | 5.0 | 25 | 32.3±2.6 | 4.9 | 6.6±0.4 | urea (1,2) | 97C4 |
| | 5.0 | 25 | 33.5 | | | DSC (2) | 97C4 |
| PL-RNase A | 5.0 | 25 | 31.7±2.8 | 4.8 | 6.5±0.4 | urea (1,2) | 97C4 |
| | 5.0 | 25 | 33.0 | | | DSC (2) | 97C4 |

Ribonuclease, bovine seminal, step-wise mutation of ribonuclease A toward the dimeric bovine seminal ribonuclease (continued)

| Form | pH | T | ΔG | $c_{1/2}$ | m | Appr./Rem. | Ref |
|---|---|---|---|---|---|---|---|
| MCAM-PLCC-RNase A | | | | | | | |
| | 5.0 | 25 | 27.9±3.0 | 4.5 | 6.2±0.5 | urea (1,2) | 97C4 |
| | 5.0 | 25 | 29.7 | | | DSC  (2) | 97C4 |
| MCAM-BS-RNase | | | | | | | |
| | 5.0 | 25 | 26.7±2.7 | 4.3 | 6.2±0.4 | urea (1,2) | 97C4 |
| | 5.0 | 25 | 28.2 | | | DSC  (2) | 97C4 |

Explanations:

    RNase A:        bovine pancreatic ribonuclease
    P-RNase A:      monomeric RNase A, mutant Ala19→Pro
    PL-RNase A:     monomeric RNase A, double mutant (Ala19→Pro and Gln28→Leu)
    PLCC-RNase A: monomeric RNase, multiple mutant (Ala19→Pro, Gln28→Leu, Lys31→Cys, and Ser32→Cys)
    MCAM-PLCC-RNase A: monomeric PLCC-RNase A, carboxyamidomethylated at Cys31 and Cys32
    MCAM-BS-RNase: monomeric bovine seminal ribonuclease, carboxyamidomethylated at Cys31 and Cys32

Remarks:
(1) linear extrapolation
(2) buffer: 100 mM acetate

## Ribonuclease T1

The data entries are arranged as follows:

a) reduced and carboxyamidated ribonuclease T1
b) wild type and mutants
c) data from H/D exchange
d) data obtained in the presence of osmolytes

*a) reduced and carboxyamidated ribonuclease T1*

Ribonuclease T1 (RNase T1) reduced and carboxyamidated (RCAM-T1), trimethylamine N-oxide-induced (TMAO) folding transition

| Mutant | pH | T | ΔG | m | Approach/Remarks | Ref |
|---|---|---|---|---|---|---|
| RCAM-T1 | 7.0 | 25 | −9.33 | −7.41 | TMAO   (1–3) | 98B3 |
| RCAM-T1 | 7.0 | 25 | −10.38 | −7.11 | TMAO   (1,2,4) | 98B3 |

Remarks:
(1) RCAM-T1 is thermodynamically unstable in buffer solution (30 mM MOPS, 0.1 M NaCl, 2 mM EDTA, pH 7.0)
(2) data from TMAO-induced folding, data treatment by linear extrapolation, LEM-SB
(3) transition monitored by fluorescence at 319 nm
(4) transition monitored by activity

Ribonuclease T1 (RNase T1) reduced and carboxyamidated (RCAM-T1), ΔG determined from urea-induced denaturation and osmolyte-induced folding

| Solute | pH | T | ΔG(278 nm) | ΔG(295 nm) | Remarks | Ref |
|---|---|---|---|---|---|---|
| urea | 7.0 | 0 | 5.48±0.33 | 5.52±0.38 | (1–4) | 99B3 |
| urea | 7.0 | 2 | 5.02±0.38 | 6.74±0.33 | (1–4) | 99B3 |
| urea | 7.0 | 5 | 3.56±0.29 | 3.31±0.29 | (1–4) | 99B3 |
| TMAO | 7.0 | 5 | 3.85±0.17 | 3.60±0.13 | (1–5) | 99B3 |
| TMAO | 7.0 | 8 | 2.30±0.13 | 2.26±0.13 | (1–5) | 99B3 |

Ribonuclease T1 (RNase T1) reduced and carboxyamidated (RCAM-T1), ΔG determined from urea-induced denaturation and osmolyte-induced folding (continued)

| Solute | pH | T | ΔG(278 nm) | ΔG(295 nm) | Remarks | Ref |
|---|---|---|---|---|---|---|
| TMAO | 7.0 | 12 | 0.25±0.05 | 0.16±0.04 | (1–5) | 99B3 |
| TMAO | 7.0 | 16 | –2.05±0.08 | 2.13±0.08 | (1–5) | 99B3 |
| TMAO | 7.0 | 20 | –4.94±0.25 | –5.10±0.25 | (1–5) | 99B3 |
| TMAO | 7.0 | 25 | –9.20±0.63 | –9.37±0.54 | (1–5) | 99B3 |
| TMAO | 7.0 | 30 | –13.35±1.34 | –14.14±1.30 | (1–5) | 99B3 |
| TMAO | 7.0 | 35 | –18.07±2.01 | –19.08±1.67 | (1–5) | 99B3 |
| sarcosine | 7.0 | 20 | –5.40±0.17 | –5.15±0.21 | (1–4) | 99B3 |
| sarcosine | 7.0 | 25 | –8.95±0.38 | –8.95±0.42 | (1–4) | 99B3 |
| sucrose | 7.0 | 20 | –5.48±0.21 | –5.36±0.17 | (1–4) | 99B3 |
| sucrose | 7.0 | 25 | –9.46±0.63 | –9.62±0.50 | (1–4) | 99B3 |

Remarks:

(1) linear extrapolation, LEM-SB, see also Ref. 98B3

(2) transition monitored by intrinsic fluorescence at 319 nm emission with excitation at 278 nm and 295 nm

(3) buffer: 30 mM MOPS, pH 7.0, 0.1 M NaCl, 2 mM EDTA

(4) average m value for RCAM-T1 was m = 8.16±0.42 kJ/mol/M, and for RNase T1 m = 5.19±0.21 kJ/mol/M

(5) TMAO – trimethylamine-N-oxide

*b) wild type and mutants*

Ribonuclease T1, wild type and Ala→Gly substitutions, equilibrium unfolding

| Protein | pD | T | ΔG | $c_{1/2}$ | m | Appr./Rem. | Ref |
|---|---|---|---|---|---|---|---|
| wild type | 7.4 | 25 | 34.3 | 6.42 | 5.23 | urea (1–3) | 99H16 |
|  | 7.7 | 25 | 33.1 | 6.15 | 5.15 | urea (1–3) | 99H16 |
| Ala21→Gly | 7.4 | 25 | 29.7 | 5.53 | 5.44 | urea (1–3) | 99H16 |
| Gly23→Ala | 7.4 | 25 | 29.3 | 5.49 | 5.27 | urea (1–3) | 99H16 |
| double mutant (Ala21→Gly and Gly23→Ala) | | | | | | | |
|  | 7.4 | 25 | 25.1 | 4.70 | 5.73 | urea (1–3) | 99H16 |

Remarks:

(1) linear extrapolation, ΔG was calculated using the average m value for all mutants, m = 5.36±0.29 kJ/mol/M

(2) measured in $D_2O$ with 50 mM sodium acetate, pD 7.4 or 7.7

(3) transition monitored by Trp fluorescence intensity at 320 nm

Ribonuclease T1, mutants concerning Asn44

| Mutant | pH | T | ΔG | $c_{1/2}$ | m | Appr./Rem. | Ref |
|---|---|---|---|---|---|---|---|
| wild type | 7.0 | 25 | 25.48±0.96 | 5.20±0.02 | 4.90±0.18 | urea (1–3) | 98H7 |
| Asn44→Ala | 7.0 | 25 | 17.24±0.67 | 3.51±0.04 | 4.91±0.18 | urea (1–3) | 98H7 |
| Asn44→Asp | 7.0 | 25 | 18.53±1.13 | 3.54±0.04 | 5.23±0.25 | urea (1–3) | 98H7 |
| Asn44→Ser | 7.0 | 25 | 18.37±0.79 | 3.60±0.02 | 5.10±0.24 | urea (1–3) | 98H7 |
| wild type | 2.7 | 25 | 25.65 | 3.75 | 6.85 | urea (1,2,4) | 98H7 |
| Asn44→Asp | 2.7 | 25 | 23.89 | 3.45 | 6.93 | urea (1,2,4) | 98H7 |

Remarks:

(1) linear extrapolation to zero denaturant concentration by a method that includes the pre- and postdenaturational baselines for a nonlinear regression of the data, see also Ref. 97P2

(2) the conserved Asn44 corresponds to position 39 in RNase Sa and 58 in barnase, see also Δ(ΔG) values for RNase T1 and RNase Sa, Ref. 98H7

(3) measured by fluorescence intensity at 320 nm in 30 mM MOPS buffer

(4) measured by fluorescence intensity at 320 nm in 30 mM glycine buffer

Ribonuclease T1, mutants concerning Asn44

| Mutant | pH | T | ΔT | Δ(ΔG) | Approach/Remarks | | Ref |
|---|---|---|---|---|---|---|---|
| wild type | 7.0 | 25 | | 0.0 | urea | (1,2,4) | 98H7 |
| Asn44→Ala | 7.0 | 25 | | −8.4±0.8 | urea | (1,2,4) | 98H7 |
| Asn44→Asp | 7.0 | 25 | | −8.4±0.8 | urea | (1,2,4) | 98H7 |
| Asn44→Ser | 7.0 | 25 | | −8.0±0.8 | urea | (1,2,4) | 98H7 |
| wild type | 2.7 | 25 | | 0.0 | urea | (1,3,5) | 98H7 |
| Asn44→Asp | 2.7 | 25 | | −2.1 | urea | (1,3,5) | 98H7 |
| wild type | 7.0 | 50.8 | 0.0 | 0.0 | heat | (1,4,6,7) | 98H7 |
| Asn44→Ala | 7.0 | | −5.2 | −6.7 | heat | (1,4,6) | 98H7 |
| Asn44→Asp | 7.0 | | −5.5 | −6.7 | heat | (1,4,6) | 98H7 |
| Asn44→Ser | 7.0 | | −5.0 | −6.3 | heat | (1,4,6) | 98H7 |
| reference values: | | | | | | | |
| Asn44→Ala | 7.0 | 25 | | −8.49 | urea | (8) | 98H7 |
| Asn44→Ala | 7.0 | 25 | | −8.70 | urea | (8) | 92S11 |
| Asn44→Ala | 7.0 | 25 | | −6.49 | heat | (8) | 98H7 |
| Asn44→Ala | 7.0 | 25 | | −7.78 | heat | (8) | 92S11 |
| Asn44→Ala | 7.0 | 25 | | −6.86 | DSC | (8) | 97P2 |
| Asn44→Ala | 7.0 | 25 | | −7.66±0.96 | | (8,9) | 98G1 |

Remarks:
(1) the conserved Asn44 corresponds to position 39 in RNase Sa and 58 in barnase, see also Δ(ΔG) values for RNase Sa, Ref. 98H7
(2) Δ(ΔG) was calculated using an average m = 5.04±0.16 kJ/mol/M
(3) Δ(ΔG) was calculated using an average m = 6.89±0.06 kJ/mol/M
(4) measured in 30 mM MOPS buffer
(5) measured in 30 mM glycine buffer
(6) Δ(ΔG) from Δ(ΔG) = ΔT × ΔS, see also Table 2
(7) T = 50.8°C is the transition temperature of the wild-type protein
(8) reference values taken from Table 5 in Ref. 98H7
(9) average of the reference values

Ribonuclease T1, wild type and Asp49→Ala mutant (equilibrium denaturation)

| Mutant | pH | T | Δ(ΔG) | $c_{1/2}$ | m | Appr./Rem. | Ref |
|---|---|---|---|---|---|---|---|
| wild type | 7.0 | 25 | 0.0 | 5.23±0.03 | 5.15±0.25 | urea (1–3) | 99G17 |
| Asp49→Ala | 7.0 | 25 | 2.1 | 5.67±0.03 | 5.06±0.25 | urea (1–3) | 99G17 |
| wild type | 2.5 | 25 | 0.0 | 3.66±0.03 | 6.65±0.25 | urea (1–3) | 99G17 |
| Asp49→Ala | 2.5 | 25 | −1.7 | 3.42±0.03 | 7.07±0.25 | urea (1–3) | 99G17 |

Remarks:
(1) linear extrapolation
(2) transition monitored by fluorescence intensity or CD
(3) buffer: 30 mM glycine (pH 2.5), 30 mM MOPS (pH 7.0)

Ribonuclease T1, wild type, Asp49→Ala, and Asp49→His mutant (thermal denaturation)

| Mutant | pH | $T_{trs}$ | ΔT | Δ(ΔG) | Appr./Rem. | Ref |
|---|---|---|---|---|---|---|
| wild type | 7.0 | 52.3±0.3 | 0.0 | 0.0 | DSC (1) | 99G17 |
| Asp49→Ala | 7.0 | 54.0±0.3 | 1.7 | 2.1 | DSC (1,2) | 99G17 |
| wild type | 6.0 | 55.5±0.3 | 0.0 | 0.0 | heat (1) | 99G17 |
| Asp49→His | 6.0 | 58.9±0.3 | 3.4 | 4.6 | heat (1,2) | 99G17 |

Remarks:
(1) buffer: 30 mM MOPS (pH 7.0), 30 mM MES (pH 6.0)
(2) Δ(ΔG) was calculated using Δ(ΔG) = $\Delta T_{mut} \times \Delta S_{w.t.}$

Ribonuclease T1, Asp49 mutants (DSC data)

| Mutant | pH | $T_{trs}$ | $\Delta T$ | $\Delta(\Delta G)$ | Appr./Rem. | Ref |
|---|---|---|---|---|---|---|
| Asp49→Ala | 7.0 | 54.0±0.3 | 0.0 | 0.0 | DSC (1–3) | 99G17 |
| Asp49→Phe | 7.0 | 52.7±0.3 | −1.3 | −1.7 | DSC (1–3) | 99G17 |
| Asp49→Trp | 7.0 | 51.2±0.3 | −1.8 | −3.8 | DSC (1–3) | 99G17 |
| Asp49→Tyr | 7.0 | 52.4±0.3 | −1.6 | −2.1 | DSC (1–3) | 99G17 |

Remarks:

(1) mutant Asp49→Ala is the reference protein

(2) $\Delta(\Delta G)$ was calculated using $\Delta(\Delta G) = \Delta T_{mut} \times \Delta S_{(Asp49 \to Ala)}$

(3) buffer: 30 mM MOPS, pH 7.0

Ribonuclease T1, Asp76 mutants

| Mutant | pH | T | $\Delta(\Delta G)$ | $c_{1/2}$ | m | Approach/Remarks | | Ref |
|---|---|---|---|---|---|---|---|---|
| RNase Sa | 7.0 | 15 | 0.0 | 6.43 | 4.77 | urea | (1–3) | 99G7 |
| Asp76→Ala | 7.0 | 15 | −15.5 | 2.57 | 5.94 | urea | (1–3) | 99G7 |
| Asp76→Asn | 7.0 | 15 | −13.0 | 3.01 | 5.90 | urea | (1–3) | 99G7 |
| Asp76→Ser | 7.0 | 15 | −13.8 | 2.99 | 5.69 | urea | (1–3) | 99G7 |
| RNase Sa | 7.0 | 15 | 0.0 | | | heat | (2) | 99G7 |
| Asp76→Ala | 7.0 | 15 | −15.9 | | | heat | (2) | 99G7 |
| Asp76→Asn | 7.0 | 15 | −13.0 | | | heat | (2) | 99G7 |
| Asp76→Ser | 7.0 | 15 | −13.0 | | | heat | (2) | 99G7 |

Remarks:

(1) linear extrapolation to zero denaturant concentration by a method that includes the pre- and postdenaturational baselines for a nonlinear regression of the data

(2) measured by fluorescence intensity at 320 nm in 30 mM MOPS buffer, for the procedure see also Refs. 98H7 and 97P2

(3) average errors are ±0.17 for m, and ±0.03 for $c_{1/2}$

Ribonuclease T1 and mutant Asp76→Asn

| Mutant | pH | T | $\Delta G$ | $c_{1/2}$ | m | Approach/Remarks | | Ref |
|---|---|---|---|---|---|---|---|---|
| RNase T1 | 7.0 | 25 | 25.5 | 5.20 | 4.90 | urea | (1–3) | 99G7 |
| | 7.0 | 20 | 28.1 | 5.71 | 4.92 | urea | (1–3) | 99G7 |
| | 7.0 | 15 | 30.7 | 6.43 | 4.77 | urea | (1–3) | 99G7 |
| | 7.0 | 10 | 32.9 | 6.70 | 4.92 | urea | (1–3) | 99G7 |
| | 7.0 | 5 | 34.8 | 7.04 | 4.94 | urea | (1–3) | 99G7 |
| Asp76→Asn | 7.0 | 25 | 10.8 | 1.80 | 6.00 | urea | (1–3) | 99G7 |
| | 7.0 | 20 | 14.8 | 2.39 | 6.19 | urea | (1–3) | 99G7 |
| | 7.0 | 15 | 17.8 | 3.01 | 5.90 | urea | (1–3) | 99G7 |
| | 7.0 | 10 | 20.9 | 3.30 | 6.34 | urea | (1–3) | 99G7 |
| | 7.0 | 5 | 23.2 | 3.63 | 6.40 | urea | (1–3) | 99G7 |

Remarks:

(1) linear extrapolation to zero denaturant concentration by a method that includes the pre- and postdenaturational baselines for a nonlinear regression of the data

(2) measured by fluorescence intensity at 320 nm in 30 mM MOPS buffer, for the procedure see also Refs. 98H7 and 97P2

(3) average errors are ±0.17 for m, and 0.03 for $c_{1/2}$

Ribonuclease T1, mutants concerning His92

| Mutant | pH | T | ΔG | $c_{1/2}$ | m | Approach/Remarks | | Ref |
|---|---|---|---|---|---|---|---|---|
| wild type | 6.0 | 25 | 34.81 | 6.57 | 5.30 | urea | (1–4) | 98D12 |
| | 9.0 | 25 | 16.19 | 3.06 | | urea | (1–4) | 98D12 |
| His92→Ala | 6.0 | 25 | 30.46 | 5.75 | | urea | (1–4) | 98D12 |
| | 9.0 | 25 | 15.82 | 2.99 | | urea | (1–4) | 98D12 |
| His92→Asn | 6.0 | 25 | 27.53 | 5.20 | | urea | (1–4) | 98D12 |
| | 9.0 | 25 | 12.51 | 2.36 | | urea | (1–4) | 98D12 |
| His92→Gln | 6.0 | 25 | 32.43 | 6.12 | | urea | (1–4) | 98D12 |
| | 9.0 | 25 | 18.87 | 3.56 | | urea | (1–4) | 98D12 |

Remarks:
(1) linear extrapolation
(2) m = 5.30 kJ/mol/M was derived from the wild-type protein and used in all calculations
(3) measured in 30 mM MES (pH 6.0) and 30 mM glycine (pH 9.0)
(4) transition monitored by fluorescence intensity at 320 nm

Ribonuclease T1 with all four cysteines replaced by alanines, RNaseT1(4A)

| Protein | pH | T | ΔG | Approach/Remarks | Ref |
|---|---|---|---|---|---|
| RNaseT1(4A) | 7 | 25 | −8 | (1–3) | 98S13 |

Remarks:
(1) multiple mutant (Cys2→Ala, Cys6→Ala Cys10→Ala, and Cys103→Ala), i.e., the disulfide bridges 2–10 and 6–103 are removed
(2) RNaseT1(4A) is largely unfolded under the given conditions
(3) RNaseT1(4A) shows a NaCl-induced folding transition with $C_{1/2} \approx 0.8$ M NaCl in 100 mM Tris-HCl, pH 8.0, at 15°C

Ribonuclease T1 with improved stability obtained by a phage-based method termed Proside (*Pro*tein *S*tability *I*ncreased by *D*irect *E*volution)

| Mutant | pH | $T_{trs}$ | ΔT | Δ(ΔG) | Approach/Remarks | | Ref |
|---|---|---|---|---|---|---|---|
| wild type | 7.0 | 50.9 | 0.0 | 0.0 | heat | (1) | 98S13 |
| LAF | 7.0 | 57.1 | 6.2 | 9.5 | heat | (1–3) | 98S13 |
| AAF | 7.0 | 57.9 | 7.0 | 11.0 | heat | (1–3) | 98S13 |
| ALF | 7.0 | 58.7 | 7.8 | 12.4 | heat | (1–3) | 98S13 |

Remarks:
(1) measured in 100 mM cacodylate buffer, pH 7.0
(2) Δ(ΔG) refers to the wild-type protein at 55°C
(3) mutant LAF is the triple mutant (Ser17→Leu, Asp29→Ala, Tyr42→Phe)
(3) mutant AAF is the triple mutant (Ser17→Ala, Asp29→Ala, Tyr42→Phe)
(3) mutant ALF is the triple mutant (Ser17→Ala, Asp29→Leu, Tyr42→Phe)

Ribonuclease T1, see also helix propensity

*c) data from H/D exchange*

Ribonuclease T1, Lys25 variant, apparent free energies of hydrogen-deuterium exchange of amide protons monitored by NMR at pD = 5.6
($\Delta G_{op}$ = free energy of opening)

| Residue | $\Delta G(25°C)$ | $\Delta G(40°C)$ | $\Delta G(45°C)$ | $\Delta G(50°C)$ | Rem. | Ref |
|---------|------|------|------|------|------|------|
| Tyr4 | 30.8 | 28.6 | 28.1 | 23.8 | | 97M12 |
| Thr5 | 25.9 | 24.6 | 24.5 | 23.2 | | 97M12 |
| Cys6 | 31.4 | 29.3 | 29.9 | 29.7 | | 97M12 |
| Tyr11 | 43.6 | 34.3 | 31.8 | 27.3 | | 97M12 |
| Ser12 | 22.7 | 21.5 | 22.0 | | | 97M12 |
| Val16 | 27.1 | 24.5 | 25.4 | 22.0 | | 97M12 |
| Ser17 | 33.7 | 27.3 | 27.3 | | | 97M12 |
| Ala19 | 34.9 | 33.1 | 30.4 | 25.4 | | 97M12 |
| Gln20 | 34.9 | 34.6 | 30.8 | | | 97M12 |
| Ala21 | 35.8 | 32.3 | 31.8 | | | 97M12 |
| Ala22 | 34.5 | 32.4 | 30.4 | 25.4 | | 97M12 |
| Tyr24 | 33.1 | 34.7 | 29.4 | | | 97M12 |
| Lys25 | 29.2 | 25.2 | 25.9 | | | 97M12 |
| Leu26 | 33.1 | 26.9 | 26.7 | 23.5 | | 97M12 |
| His40 | 34.0 | 28.7 | 29.0 | 25.7 | | 97M12 |
| Tyr42 | 23.0 | | | | | 97M12 |
| Tyr57 | 45.0 | 34.4 | 31.8 | 26.1 | | 97M12 |
| Glu58 | 29.7 | 31.4 | 30.4 | | | 97M12 |
| Trp59 | 47.4 | 32.9 | 28.8 | 24.4 | | 97M12 |
| Ile61 | 48.8 | 32.6 | 29.2 | 23.1 | | 97M12 |
| Leu62 | 28.2 | 24.9 | 24.5 | 21.0 | | 97M12 |
| Asp76 | 23.1 | 20.2 | 20.6 | 19.7 | | 97M12 |
| Arg77 | 50.6 | 35.1 | 30.3 | 25.7 | | 97M12 |
| Val78 | 47.3 | 33.6 | 28.9 | 22.0 | | 97M12 |
| Val79 | 48.6 | 29.8 | 29.1 | 23.0 | | 97M12 |
| Phe80 | 49.0 | 33.5 | 30.1 | 24.1 | | 97M12 |
| Asn81 | 50.0 | 36.4 | 33.4 | 28.7 | | 97M12 |
| Asn84 | 35.1 | 33.3 | 33.1 | 29.6 | | 97M12 |
| Gln85 | 52.1 | 36.9 | 32.9 | 27.9 | | 97M12 |
| Ala87 | 47.4 | 33.2 | 31.1 | 25.1 | | 97M12 |
| Gly88 | | 36.3 | 32.0 | | | 97M12 |
| Val89 | 20.9 | | | | | 97M12 |
| Ile90 | 31.7 | 29.8 | 28.2 | | | 97M12 |
| Thr91 | 43.3 | 30.7 | 28.7 | 23.6 | | 97M12 |
| His92 | 33.1 | 26.7 | 26.6 | 22.2 | | 97M12 |
| Val101 | 41.3 | 30.2 | 28.0 | 23.8 | | 97M12 |
| Cys103 | 29.6 | 29.2 | 29.0 | 26.8 | | 97M12 |
| average values: | | | | | | |
| $\Delta G_{op}$ | 47.3 | 33.5 | 30.1 | 25.1 | (1,2) | 97M12 |
| $\Delta G_{unf}$ | 43.9 | 27.2 | 21.3 | 14.6 | (3) | 97P1 |
| $\Delta G_{unf}$ | 28.9 | 23.0 | 18.4 | | (4) | 97M12 |

Remarks:

(1) the free energy of opening for ten of the residues excedes the global free energy by 4–10 kJ/mol determined from either urea denaturation or calorimetry, see Refs. 92H8 and 94Y2

(2) the following residues are used to calculate the average $\Delta G_{op}$: Tyr11, Tyr57, Trp59, Ile61, Arg77, Val78, Val79, Phe80, Asn81, Gln85, Ala87, Thr91, and Val101

(3) $\Delta G_{unf}$ determined by urea denaturation in $H_2O$ (LEM-SB)

(4) $\Delta G_{unf}$ determined by urea denaturation in $D_2O$ (LEM-SB)

Ribonuclease T1, Lys25 variant, apparent free energies of hydrogen-deuterium exchange of amide protons monitored by NMR at pD = 5.6 ($\Delta G_{op}$ = free energy of opening)

| Residue | $\Delta G(40°C)$ | $\Delta G(45°C)$ | Remarks | Ref |
|---|---|---|---|---|
| Tyr4 | 29.5 | 25.3 | | 97M12 |
| Thr5 | 26.2 | 24.6 | | 97M12 |
| Cys6 | 30.0 | 33.2 | | 97M12 |
| Tyr11 | 33.0 | 28.3 | | 97M12 |
| Val16 | 26.9 | 23.5 | | 97M12 |
| Ala19 | 31.0 | 28.0 | | 97M12 |
| Ala22 | 30.9 | 26.3 | | 97M12 |
| Leu26 | 25.8 | 22.3 | | 97M12 |
| His40 | 29.8 | 27.0 | | 97M12 |
| Tyr57 | 31.8 | 27.9 | | 97M12 |
| Glu58 | 30.2 | 25.5 | | 97M12 |
| Trp59 | 31.0 | 26.2 | | 97M12 |
| Ile61 | 29.2 | 29.7 | | 97M12 |
| Leu62 | 24.9 | 22.5 | | 97M12 |
| Arg77 | 31.5 | 27.9 | | 97M12 |
| Val78 | 29.0 | 24.5 | | 97M12 |
| Val79 | 29.4 | 24.1 | | 97M12 |
| Phe80 | 30.3 | 25.8 | | 97M12 |
| Asn81 | 34.4 | 29.8 | | 97M12 |
| Asn84 | 33.8 | 29.8 | | 97M12 |
| Gln85 | 33.8 | 29.6 | | 97M12 |
| Ala87 | 31.2 | 27.1 | | 97M12 |
| Gly88 | 32.4 | | | 97M12 |
| Ile90 | 28.3 | 23.7 | | 97M12 |
| Thr91 | 29.5 | 24.7 | | 97M12 |
| His92 | 29.0 | | | 97M12 |
| Val101 | 29.0 | 24.7 | | 97M12 |
| Cys103 | 30.8 | 28.5 | | 97M12 |
| average values: | | | | |
| $\Delta G_{op}$ | 31.0 | 26.8 | (1,2) | 97M12 |

Remarks:

(1) the free energy of opening for ten of the residues excedes the global free energy by 4–10 kJ/mol determined from either urea denaturation or calorimetry, see Refs. 92H8 and 94Y2

(2) the following residues are used to calculate the average $\Delta G_{op}$: Tyr11, Tyr57, Trp59, Ile61, Arg77, Val78, Val79, Phe80, Asn81, Gln85, Ala87, Thr91, and Val101

Ribonuclease T1, Lys25 variant, comparison of free energies from hydrogen-deuterium exchange (HX) and free energies from urea denaturation in $D_2O$ at pD = 6.6, the importance of proline correction

| Res./Value | pD | $\Delta G(40°C)$ | $\Delta G(45°C)$ | $\Delta G(50°C)$ | Appr./Rem. | Ref |
|---|---|---|---|---|---|---|
| Arg77 | 5.6 | 35.1 | 30.5 | 25.9 | HX | 97M12 |
| Asn81 | 5.6 | 36.4 | 33.5 | 28.9 | HX | 97M12 |
| Gln85 | 5.6 | 36.8 | 33.1 | 28.0 | HX | 97M12 |
| average $\Delta G_{op}$ | 5.6 | 36.0±0.8 | 32.2±1.7 | 27.6±1.7 | HX | 97M12 |
| Pro correction | | 10.0 | 9.6 | 9.6 | (1) | 97M12 |
| $\Delta G_{unf}$ (HX) | 5.6 | 26.0 | 22.6 | 18.0 | (2) | 97M12 |
| $\Delta G_{unf}$ (urea) | 5.6 | 28.9 | 23.0 | 18.4 | (3) | 97M12 |

Remarks:

(1) correction for proline isomerization by $\Delta(\Delta G)_{corr} = \Delta G_{op} - \Delta G_{unf} = RT \ln(1 + K)$ with K the *cis/trans* isomerization equilibrium constant, i.e., $\Delta(\Delta G)_{corr} = RT \ln(1 + K_{Pro55}) + RT \ln(1 + K_{Pro39})$

(2) average $\Delta G_{unf}$ from hydrogen exchange with proline correction

(3) $\Delta G_{unf}$ from urea denaturation in $D_2O$ (LEM-SB)

Ribonuclease T1 (RNase T1), wild type and Ala→Gly substitutions, H/D exchange data

| Residue | $\Delta G_{w.t.}$ | $\Delta G_{Ala21\rightarrow Gly}$ | $\Delta G_{Gly23\rightarrow Ala}$ | $\Delta G_{D.M.}$ | Appr./Rem. | Ref |
|---|---|---|---|---|---|---|
| Tyr4 | 38.1 | 34.7 | 34.7 | 32.2 | H/D (1) | 99H16 |
| Thr5 | 28.5 | 26.8 | 27.2 | 24.7 | H/D (1) | 99H16 |
| Cys6 | 33.1 | 31.8 | 31.4 | 29.3 | H/D (1) | 99H16 |
| Tyr11 | 42.3 | 38.9 | 38.9 | 35.1 | H/D (1) | 99H16 |
| Ser12 | | 25.9 | | | H/D (1) | 99H16 |
| Val16 | 34.3 | 31.4 | 31.4 | 29.3 | H/D (1) | 99H16 |
| Ser17 | 31.4 | 29.3 | 29.7 | 27.6 | H/D (1) | 99H16 |
| Ala19 | 37.2 | 34.7 | 33.5 | 31.4 | H/D (1) | 99H16 |
| Gln20 | 43.5 | 38.5 | 38.1 | 34.7 | H/D (1) | 99H16 |
| Ala/Gly21 | 42.7 | 32.6 | 38.1 | 32.6 | H/D (1) | 99H16 |
| Ala22 | 38.5 | 32.2 | 33.9 | 28.9 | H/D (1) | 99H16 |
| Gly/Ala23 | 42.3 | 37.7 | 40.2 | 36.8 | H/D (1) | 99H16 |
| Tyr24 | 42.7 | 37.2 | 37.2 | 33.5 | H/D (1) | 99H16 |
| Lys25 | 28.9 | 25.5 | 23.4 | | H/D (1) | 99H16 |
| Leu26 | 31.0 | 28.5 | 25.5 | 23.4 | H/D (1) | 99H16 |
| His27 | 26.8 | 25.9 | | | H/D (1) | 99H16 |
| Glu28 | 25.5 | 24.7 | 24.3 | 23.8 | H/D (1) | 99H16 |
| His40 | 40.2 | 38.9 | 38.1 | 36.4 | H/D (1) | 99H16 |
| Tyr42 | | 23.0 | | | H/D (1) | 99H16 |
| Tyr57 | | | 38.1 | 34.7 | H/D (1) | 99H16 |
| Glu58 | | | 33.9 | 31.0 | H/D (1) | 99H16 |
| Trp59 | 43.1 | 39.7 | | | H/D (1) | 99H16 |
| Ile61 | 40.6 | 34.3 | 32.2 | 27.6 | H/D (1) | 99H16 |
| Leu62 | 30.5 | 28.9 | 27.6 | 26.4 | H/D (1) | 99H16 |
| Tyr68 | 24.3 | 22.6 | 23.0 | 23.0 | H/D (1) | 99H16 |
| Asp76 | | 22.6 | | | H/D (1) | 99H16 |
| Arg77 | 43.1 | 37.7 | 38.1 | 35.6 | H/D (1) | 99H16 |
| Val78 | 41.8 | 37.2 | 36.8 | 33.9 | H/D (1) | 99H16 |
| Phe80 | | | 37.7 | 34.3 | H/D (1) | 99H16 |
| Asn81 | 45.2 | 41.4 | 40.6 | 37.7 | H/D (1) | 99H16 |
| Asn84 | 38.1 | 36.8 | 34.3 | 33.1 | H/D (1) | 99H16 |
| Gln85 | 43.9 | 40.2 | 39.3 | 36.4 | H/D (1) | 99H16 |
| Ala87 | 43.5 | 40.2 | 37.2 | 34.3 | H/D (1) | 99H16 |
| Gly88 | 44.8 | 38.9 | 39.3 | 36.0 | H/D (1) | 99H16 |
| Val89 | 26.8 | 25.1 | 25.1 | 23.8 | H/D (1) | 99H16 |
| Ile90 | 40.2 | 36.0 | 34.7 | 30.5 | H/D (1) | 99H16 |
| Thr91 | 40.2 | 35.6 | 35.6 | 33.1 | H/D (1) | 99H16 |
| His92 | 41.0 | 39.3 | 38.5 | 35.1 | H/D (1) | 99H16 |
| Phe100 | 44.8 | 40.2 | 39.7 | 36.4 | H/D (1) | 99H16 |
| Val101 | 38.1 | 36.0 | 34.7 | 32.2 | H/D (1) | 99H16 |
| Glu102 | 22.6 | 21.3 | 20.9 | 19.7 | H/D (1) | 99H16 |
| Cys103 | 34.3 | 33.5 | 33.1 | 31.8 | H/D (1) | 99H16 |
| $\Delta G^*$ | 43.5 | 40.2 | 39.7 | 35.6 | H/D (1–3) | 99H16 |

Remarks:

(1) all data at 25°C, measured in 50 mM sodium acetate, $d3$ at pD 7.4 in $D_2O$, except for the wild-type protein, measured at pD 7.7

(2) $\Delta G^*$ corrected for proline isomerization (the correction for proline isomerization is 10.6 kJ/mol for RNase T1)

(3) the error in $\Delta G^*$ is estimated to be ±2.1 kJ/mol

*d) data obtained in the presence of osmolytes*

Ribonuclease T1 in the presence of osmolytes, difference in ΔG of transfer between native and unfolded protein as calculated from the increments of the amino acids

| Osmolyte Conc. | T | Δ(ΔG) | Approach/Remarks | Ref |
|---|---|---|---|---|
| 1 M TMAO | 25.1 | 17.6 | transfer (1–3) | 97W2 |
| 2 M urea | 25.1 | 31.0 | transfer (3) | 97W2 |
| 1 M TMAO in 2 M urea | 25.1 | 15.5 | transfer (1,3) | 97W2 |

Remarks:

(1) TMAO – trimethylamine N-oxide

(2) Δ(ΔG) means that the native to unfolded conversion is 17.6 kJ/mol less favourable than it is in water

(3) see also Ref. 94L8

## Ribosomal Protein L9

Ribosomal protein L9, synthesized N-terminal domain comprising residues 1–56

| pH | T | ΔG | $c_{1/2}$ | m | Appr./Rem. | Ref |
|---|---|---|---|---|---|---|
| 7.0 | 40 | 10.9±3.3 | 4.8 | 2.26±0.42 | urea (1–4) | 98K16 |
| 7.0 | 40 | 12.6±3.3 | 4.8 | 2.64±0.42 | urea (1–3,5) | 98K16 |

Remarks:

(1) linear extrapolation

(2) buffer: 10 mM MOPS, 100 mM NaCl, pH 7.0

(3) the midpoint temperatures at thermal denaturation obtained by NMR (77±2°C Tyr25, 77±2°C Thr40, 77±2°C Met1) coincide with 77±2°C by far-UV CD, and 78±2°C by near-UV CD

(4) transition monitored by CD at 222 nm

(5) transition monitored by fluorescence at 304 nm

Ribosomal protein L9 from *Bacillus stearothermophilus*, N-terminal domain (NTL9), dependence of the m value on temperature

| Protein/Solvent | | pH | T | m | Approach/Remarks | | Ref |
|---|---|---|---|---|---|---|---|
| NTL9 | $H_2O$ | 5.45 | 5.0 | 5.06±0.63 | GuHCl | (1,2) | 98K18 |
| NTL9 | $H_2O$ | 5.45 | 15.0 | 5.10±0.63 | GuHCl | (1,2) | 98K18 |
| NTL9 | $H_2O$ | 5.45 | 25.0 | 5.15±0.63 | GuHCl | (1,2) | 98K18 |
| NTL9 | $H_2O$ | 5.45 | 35.0 | 5.19±0.63 | GuHCl | (1,2) | 98K18 |
| NTL9 | $H_2O$ | 5.45 | 45.0 | 5.10±0.63 | GuHCl | (1,2) | 98K18 |

Remarks:

(1) measured in 20 mM sodium acetate, 100 mM NaCl, pH 5.45

(2) from global analysis of thermal and denaturant-induced unfolding

Ribosomal protein L9 from *Bacillus stearothermophilus*, N-terminal domain (NTL9), thermodynamic parameters in the presence of $H_2O$ and $D_2O$

| Protein/Solvent | | pH | T | $\Delta G$ | m | Approach/Remarks | | Ref |
|---|---|---|---|---|---|---|---|---|
| NTL9 | $H_2O$ | 5.45 | 25 | 16.5±1.9 | 5.52±0.60 | GuHCl | (1–3) | 98K18 |
| NTL9 | $H_2O$ | 5.45 | 25 | 16.9±1.6 | 2.76±0.25 | urea | (1–3) | 98K18 |
| NTL9 | $H_2O$ | 5.45 | 25 | 15.3±1.0 | 5.15±0.29 | glob. | (2–4) | 98K18 |
| NTL9 | $H_2O$ | 5.45 | 25 | 15.8±1.0 | | glob. | (2–4,6) | 98K18 |
| NTL9 | $D_2O$ | 5.45 | 25 | 19.3±2.1 | 6.32±0.65 | GuHCl | (1,3,5) | 98K18 |
| NTL9 | $D_2O$ | 5.45 | 25 | 19.7±2.2 | 6.28±0.65 | glob. | (3–5) | 98K18 |
| NTL9 | $D_2O$ | 5.45 | 17 | 20.0±2.2 | | glob. | (3–5,6) | 98K18 |

Remarks:

(1) linear extrapolation

(2) measured in 20 mM sodium acetate, 100 mM NaCl, pH 5.45

(3) the errors given here are rounded values, for detailed analysis of the confidence limits see Ref. 98K18

(4) from global analysis of thermal and denaturant-induced unfolding, assuming a linear dependence of $\Delta G$ on the denaturant concentration, and assuming $\Delta Cp$ and m to be temperature-independent

(5) measured in 20 mM sodium acetate, 100 mM NaCl, pD 5.45 (corrected value)

(6) conditions at which maximal stability is observed

Ribosomal protein L9 from *Bacillus stearothermophilus*, N-terminal domain (NTL9, residues 1–56)

| Protein | pH | T | $\Delta G$ | m | Approach/Remarks | | Ref |
|---|---|---|---|---|---|---|---|
| NTL9 | 5.45 | 19 | 17.15±2.09 | 5.61±0.67 | GuHCl | (1–3) | 98K17 |
| | 5.45 | 19 | 17.07±0.63 | 5.90±0.17 | GuHCl | (3–5) | 98K17 |
| | 5.45 | 25 | 19.70±2.50 | 6.28±0.67 | GuDCl | (6–8) | 98K17 |
| | 5.45 | 25 | 18.80±2.10 | 6.15±0.59 | GuDCl | (4–7) | 98K17 |

Remarks:

(1) data from equilibrium unfolding, linear extrapolation

(2) transition monitored by fluorescence

(3) buffer: $H_2O$, 20 mM sodium acetate, 100 mM NaCl, pH 5.45

(4) data from folding and unfolding kinetics, assuming a linear dependence of ln k on the denaturant concentration

(5) from stopped-flow fluorescence measurements

(6) data from global analysis of ln k depending on (deuterated) GuDCl concentration and temperature

(7) buffer: $D_2O$, 20 mM sodium acetate, 100 mM NaCl, pD 5.45

(8) data from equilibrium unfolding, transition monitored by far-UV CD

N-terminal domain of the ribosomal protein L9 from *Bacillus stearothermophilus* ($NTL9_{1-51}$ and $NTL9_{1-56}$)

| Protein | pD | T | $\Delta G$ | $c_{1/2}$ | m | Appr./Rem. | Ref |
|---|---|---|---|---|---|---|---|
| $NTL9_{1-51}$ | 5.4 | 25 | 10.9±1.7 | 2.00 | 5.36±0.63 | GuDCl (1–4) | 99L10 |
| $NTL9_{1-51}$ | 5.4 | 25 | 12.3±0.8 | | 6.19±0.33 | GuDCl (2,5,6) | 99L10 |
| $NTL9_{1-56}$ | 5.4 | 25 | 19.3±2.1 | 2.98 | 6.32±0.67 | GuDCl (1–3,7) | 99L10 |

Remarks:

(1) linear extrapolation

(2) measured in $D_2O$, 20 mM sodium acetate, 100 mM NaCl at $pD_{corr}$ 5.4

(3) transition monitored by far-UV CD

(4) thermal transition at $T_{trs}$ = 75°C

(5) from kinetics of folding and unfolding rates versus GuDCl conc.

(6) transition monitored by fluorescence

(7) thermal transition at $T_{trs}$ = 81°C

N-terminal domain of the ribosomal protein L9 from *Bacillus stearothermophilus* (NTL9, NTL9$_{1-51}$, and NTL9$_{1-56}$)

| Protein | H$_2$O/D$_2$O | pH/pD | T | ΔG | m | Appr./Rem. | Ref |
|---------|---------------|-------|------|-----------|------------|---------------|-------|
| NTL9 | H$_2$O | 5.45 | 25.2 | 18.95±1.4 | 6.61±0.35 | GuHCl (1–5) | 99S3 |
| NTL9$_{1-56}$ | H$_2$O | 5.45 | 25.2 | 16.99±1.8 | 6.11±0.46 | GuDCl (1–4,6) | 99S3 |
| NTL9$_{1-56}$ | D$_2$O | 5.05 | 25.2 | 18.70±2.1 | 6.15±0.59 | GuDCl (1–4,6) | 99S3 |
| NTL9$_{1-51}$ | D$_2$O | 5.05 | 25.2 | 12.34±0.8 | 6.23±0.33 | GuDCl (1–4,7) | 99S3 |

Remarks:

(1) from kinetics of folding and unfolding rates versus GuHCl or GuDCl conc.

(2) measured in H$_2$O or D$_2$O, 20 mM sodium acetate, 100 mM NaCl at pH 5.45 or pD 5.05 (apparent)

(3) using stopped-flow fluorescence

(4) for a more detailed error analysis see 99S3

(5) NTL9 is the N-terminal domain in the intact protein

(6) NTL9$_{1-56}$ is the isolated N-terminal domain, residues 1–56

(7) NTL9$_{1-51}$ is the isolated N-terminal domain, residues 1–51, see also Ref. 99L10

N-terminal domain of the ribosomal protein L9 from *Bacillus stearothermophilus* (residues 1–56)

| Protein | pH | T | ΔG | Approach/Remarks | | Ref |
|---------|------|------|----------|------|-------|-------|
| NTL9 in the presence of 100 mM NaCl: | | | | | | |
| NTL9 | 2.04 | 25 | 7.9±2.1 | heat | (1–3) | 99K19 |
| | 2.79 | 25 | 9.6±1.8 | heat | (1–3) | 99K19 |
| | 3.55 | 25 | 12.6±2.9 | heat | (1–3) | 99K19 |
| | 4.51 | 25 | 14.6±2.1 | urea | (1–4) | 99K19 |
| | 5.99 | 25 | 18.4±2.3 | urea | (1–4) | 99K19 |
| | 6.98 | 25 | 18.8±5.0 | urea | (1–4) | 99K19 |
| NTL9 in the presence of 750 mM NaCl: | | | | | | |
| NTL9 | 2.40 | 25 | 12.1±2.1 | urea | (1–4) | 99K19 |
| | 3.15 | 25 | 13.0±2.1 | urea | (1–4) | 99K19 |
| | 4.14 | 25 | 16.7±3.3 | urea | (1–4) | 99K19 |
| | 5.25 | 25 | 20.1±3.8 | urea | (1–4) | 99K19 |
| | 6.56 | 25 | 19.7±2.9 | urea | (1–4) | 99K19 |
| | 7.26 | 25 | 21.8±4.6 | urea | (1–4) | 99K19 |

Remarks:

(1) measured in 10 mM sodium phosphate with either 100 mM or 750 mM NaCl in 90% H$_2$O, 10% D$_2$O

(2) chemical denaturation was monitored with far-UV CD at 222 nm, and thermal denaturation was monitored with near-UV at 280 nm

(3) for a more detailed error analysis see Ref. 99K19

(4) nonlinear least-squares fit of the transition curve assuming a linear dependence of ΔG on the denaturant conc.

## Ribosomal Protein S6

Ribosomal protein S6 from *Thermus thermophilus* (*wt*S6), cloned in two vectors, and expressed in *Escherichia coli* (*cp*S6d and *cp*S6f)

| Protein | pH | T | ΔG | c$_{1/2}$ | m | Appr./Rem. | Ref |
|---------|-----|----|------------|-----------|-----------|-------------|-------|
| *wt*S6 | 7.4 | 25 | 48.33±0.21 | 8.68±0.04 | 5.56±0.21 | urea (1–4) | 99U2 |
| *cp*S6d | 7.4 | 25 | 27.41±0.21 | 5.18±0.04 | 5.31±0.21 | urea (1–4) | 99U2 |
| *cp*S6f | 7.4 | 25 | 24.31±0.25 | 4.69±0.04 | 5.19±0.21 | urea (1–4) | 99U2 |

Remarks:

(1) linear extrapolation

(2) transition monitored by fluorescence

(3) measured in 20 mM sodium phosphate buffer, pH 7.4

(4) thermal transition according to DSC at 99°C for *wt*S6, 74°C for *cp*S6d, and 69°C for *cp*S6f

Ribosomal protein S6 from *Thermus thermophilus*, wild type and mutants

| Protein | pH | T | $\Delta G$ | $c_{1/2}$ | m | Appr./Rem. | Ref |
|---|---|---|---|---|---|---|---|
| S6 | 6.25 | 25 | 30.29±2.68 | 3.32±0.03 | 6.69±0.59 | GuHCl (1–4) | 99O5 |
|  | 6.25 | 25 | 33.56±7.53 | 3.33±0.08 | 7.41±1.67 | GuHCl (1–3,5) | 99O5 |
|  | 6.25 | 25 | 34.39±0.75 | 3.46±0.05 | 7.32±0.08 | GuHCl (1–3,6) | 99O5 |
| Leu30→Ala | 6.25 | 25 |  |  | 8.54±0.50 | GuHCl (7) | 99O5 |
| Leu75→Ala | 6.25 | 25 |  |  | 8.16±0.71 | GuHCl (7) | 99O5 |
| Val85→Ala | 6.25 | 25 |  |  | 7.03±0.50 | GuHCl (7) | 99O5 |

Remarks:

(1) Ref. 99O5 deals with the interpretation of curved chevron plots, and contains a comparative study of the ribosomal protein S6 with the human spliceosomal protein U1A

(2) data from equilibrium measurements, nonlinear least-squares fit of the two-state transition curve assuming a linear dependence of $\Delta G$ on the denaturant conc.

(3) buffer: 50 mM MES, pH 6.25

(4) transition monitored by fluorescence

(5) transition monitored by CD spectra

(6) kinetic data from stopped-flow fluorescence measurements

(7) see chevron plots in Ref. 99O5

## ROP

ROP, four-α-helix-bundle protein, wild type

| Protein | pH | T | $\Delta G$ | $c_{1/2}$ | m | Appr./Rem. | Ref |
|---|---|---|---|---|---|---|---|
| ROP | 6.0 | 19 | 66.8±3.7 | 3.30 | 11.5±1.04 | GuHCl (1–4) | 99R17 |
|  | 6.0 | 25 | 70.8±4.2 | 3.23 | 12.8±1.15 | GuHCl (1–4) | 99R17 |
|  | 6.0 | 40 | 57.2±3.4 | 2.26 | 11.6±1.28 | GuHCl (1–4) | 99R17 |
|  | 6.0 | 25 | 71.1 |  |  | DSC  (5) | 99R17 |
|  | 6.0 | 25 | 69.1 |  |  | DSC  (6) | 98R10 |

Remarks:

(1) linear extrapolation, LEM-SB

(2) $\Delta G$ per mole of dimer

(3) transition monitored by far-UV CD at 222 nm

(4) buffer: 10 mM sodium phosphate, 10 mM sodium sulphate, 1 mM EDTA, pH 6.0

(5) reference data from Ref. 93S10

(6) reconsideration of data from Ref. 93S10 which yields $T_{trs}$ = 69.5°C, $\Delta H$ = 518 kJ/mol, $\Delta Cp$ = 2.9 kJ/mol/K, and a temperature increment of $\Delta Cp'$ = –0.25 kJ/mol/K²

ROP, four-α-helix bundle protein, mutant Ala31→Pro, comparison of various approaches

| Mutant | pH | T | $\Delta G$ | $c_{1/2}$ | m | Approach/Remarks | | Ref |
|---|---|---|---|---|---|---|---|---|
| Ala31→Pro | 6.0 | 6 | 46.2±1.1 | 2.3 | 8.3±0.5 | urea | (1–3) | 97P8 |
|  | 6.0 | 10 | 45.8±1.4 | 2.2 | 8.2±0.6 | urea | (1–3) | 97P8 |
|  | 6.0 | 20 | 46.6±0.8 | 1.9 | 9.6±0.3 | urea | (1–3) | 97P8 |
| Ala31→Pro | 6.0 | 6 | 46.5±2.2 | 2.3 | 8.6±0.8 | urea | (2–4) | 97P8 |
|  | 6.0 | 10 | 46.8±2.7 | 2.2 | 8.8±0.9 | urea | (2–4) | 97P8 |
|  | 6.0 | 20 | 43.9±1.7 | 1.9 | 8.6±0.6 | GuHCl | (2–4) | 97P8 |
| Ala31→Pro | 6.0 | 10 | 47.2±0.6 | 1.2 | 16.8±0.3 | GuHCl | (1–3) | 97P8 |
|  | 6.0 | 19 | 45.5±2.0 | 1.1 | 17.4±1.1 | GuHCl | (1–3) | 97P8 |
| Ala31→Pro | 6.0 | 15 | 47.8±1.0 | 2.0 | 9.4±0.5 | urea | (1,2,5) | 97P8 |
|  | 6.0 | 20 | 45.9±0.9 | 1.9 | 9.0±0.4 | urea | (1,2,5) | 97P8 |
|  | 6.0 | 25 | 43.9±0.5 | 1.7 | 8.5±0.2 | urea | (1,2,5) | 97P8 |

ROP, four-α-helix bundle protein, mutant Ala31→Pro, comparison of various approaches (continued)

| Mutant | pH | T | ΔG | $c_{1/2}$ | m | Approach/Remarks | | Ref |
|--------|----|----|----|-----------|---|------------------|---|-----|
| Ala31→Pro | 6.0 | 15 | 46.4±2.0 | 2.0 | 9.3±0.8 | urea | (2,4,5) | 97P8 |
| | 6.0 | 20 | 45.5±2.8 | 1.9 | 9.3±1.1 | urea | (2,4,5) | 97P8 |
| | 6.0 | 25 | 44.1±3.2 | 1.7 | 9.2±1.8 | urea | (2,4,5) | 97P8 |
| Ala31→Pro | 6.0 | 20 | 46.6 | 1.9 | 9.6 | urea | (1–3,6) | 97P8 |
| | 6.0 | 20 | 44.7 | 1.7 | 9.1 | urea | (1–3,7) | 97P8 |
| | 6.0 | 20 | 45.9 | 1.9 | 9.0 | urea | (1,2,5,6) | 97P8 |
| | 6.0 | 20 | 47.9 | 1.8 | 10.6 | urea | (1,2,5,7) | 97P8 |
| Ala31→Pro | 6.0 | 25 | 43.0 | | | DSC | (2) | 97P8 |
| wild type | 6.0 | 25 | 72.0 | | | DSC | (2) | 97P8 |

Remarks:

(1) linear extrapolation

(2) measured in 10 mM sodium phosphate, pH 6.0, 10 mM $Na_2SO_4$, 1 mM EDTA; the protein concentration was 0.05 mg/ml throughout

(3) transition monitored by CD at 222 nm

(4) linear extrapolation, LEM-SB method modified for the monomer → dimer equilibrium

(5) transition monitored by fluorescence intensity at 302 nm for urea-induced unfolding and 304 nm for GuHCl-induced unfolding, respectively

(6) denaturation experiment

(7) renaturation experiment

ROP, four-α-helix bundle protein, insertion of glycine linkers instead of the natural two-residue loop, ΔG from thermal denaturation

| Mutant | pH | $T_{trs}$ | ΔG(56.4°C) | Approach | Remarks | Ref |
|--------|----|-----------|------------|----------|---------|-----|
| $Gly_1$ | 7 | 72.7 | 17.2 | heat, v.H. | (1–5) | 97N1 |
| $Gly_2$ | 7 | 69.0 | 13.8 | heat, v.H. | (1–5) | 97N1 |
| $Gly_3$ | 7 | 61.2 | 6.7 | heat, v.H. | (1–5) | 97N1 |
| $Gly_4$ | 7 | 60.3 | 6.3 | heat, v.H. | (1–5) | 97N1 |
| $Gly_5$ | 7 | 58.4 | 4.6 | heat, v.H. | (1–5) | 97N1 |
| $Gly_6$ | 7 | 56.7 | 3.8 | heat, v.H. | (1–5) | 97N1 |
| $Gly_7$ | 7 | 56.3 | 2.9 | heat, v.H. | (1–5) | 97N1 |
| $Gly_8$ | 7 | 53.8 | 0.8 | heat, v.H. | (1–5) | 97N1 |
| $Gly_9$ | 7 | 52.9 | –0.8 | heat, v.H. | (1–5) | 97N1 |
| $Gly_{10}$ | 7 | 49.8 | –3.8 | heat, v.H. | (1–5) | 97N1 |

Remarks:

(1) the short loop region between helix 1 and helix 2 (Asp30, Ala31) was replaced by a series of nonnatural loops composed of one to ten Gly residues; all mutants retain wild-type RNA-binding activity

(2) given is the transition temperature $T_{trs}$ (±0.3°C) and ΔG at the reference temperature $T_{ref}$ = 56.4°C

(3) buffer: 100 mM sodium phosphate, 200 mM NaCl, 1 mM DTT, pH 7.0

(4) transition monitored by ellipticity at 222 nm

(5) the van't Hoff enthalpy amounts to about 335 kJ/mol for all proteins

ROP, four-α-helix bundle protein, insertion of glycine linkers instead of the natural two-residue loop, ΔG from GuHCl denaturation

| Mutant | pH | T | ΔG | $c_{1/2}$ | m | Approach/Remarks | | Ref |
|--------|----|----|----|-----------|---|------------------|---|-----|
| $Gly_1$ | 7 | 25 | 20.9 | 2.7 | 8.4 | GuHCl | (1–5) | 97N1 |
| $Gly_2$ | 7 | 25 | 26.8 | 2.4 | 11.3 | GuHCl | (1–5) | 97N1 |
| $Gly_3$ | 7 | 25 | 24.3 | 2.1 | 11.7 | GuHCl | (1–5) | 97N1 |
| $Gly_4$ | 7 | 25 | 21.8 | 2.0 | 10.5 | GuHCl | (1–5) | 97N1 |
| $Gly_5$ | 7 | 25 | 20.5 | 1.9 | 10.9 | GuHCl | (1–5) | 97N1 |

ROP, four-α-helix bundle protein, insertion of glycine linkers instead of the natural two-residue loop, ΔG from GuHCl denaturation (continued)

| Mutant | pH | T | ΔG | $c_{1/2}$ | m | Approach/Remarks | | Ref |
|---|---|---|---|---|---|---|---|---|
| Gly$_6$ | 7 | 25 | 19.2 | 1.6 | 10.9 | GuHCl | (1–5) | 97N1 |
| Gly$_7$ | 7 | 25 | 18.8 | 1.5 | 10.0 | GuHCl | (1–5) | 97N1 |
| Gly$_8$ | 7 | 25 | 18.8 | 1.5 | 13.0 | GuHCl | (1–5) | 97N1 |
| Gly$_9$ | 7 | 25 | 18.4 | 1.4 | 12.6 | GuHCl | (1–5) | 97N1 |
| Gly$_{10}$ | 7 | 25 | 16.3 | 1.3 | 12.6 | GuHCl | (1–5) | 97N1 |

Remarks:

(1) the short loop region between helix 1 and helix 2 (Asp30, Ala31) was replaced by a series of nonnatural loops composed of one to ten Gly residues; all mutants retain wild-type RNA-binding activity

(2) linear extrapolation, LEM-SB

(3) estimated error in ΔG ±1.7, and in m ±1.3

(4) buffer: 100 mM sodium phosphate, 200 mM NaCl, 1 mM DTT, pH 7.0

(5) transition monitored by ellipticity at 222 nm

ROP, four-α-helix bundle protein, wild type and insertion mutant

| Protein | pH | T | ΔG | Approach/Remarks | | Ref |
|---|---|---|---|---|---|---|
| wild type | 6.0 | 25 | 71.7 | DSC | (1–2) | 94V4 |
| <aa> | 6.0 | 25 | 71.2 | DSC | (1–3) | 94V4 |

Remarks:

(1) measured in 10 mM sodium phosphate, pH 6.0, 10 mM Na$_2$SO$_4$, 1 mM EDTA

(2) ΔG refers to the molecular mass of the dimer

(3) mutant <2aa> re-establishes the continuous heptad repeat pattern by inserting one Ala between ROP residues Leu29 and Asp30, and a second one between Asp30 and Ala31

ROP, four-α-helix bundle protein, wild type and loop excision that produces a hyperthermophilic variant

| Protein | pH | T | ΔG | Approach/Remarks | | Ref |
|---|---|---|---|---|---|---|
| wild type | 6.0 | 25 | 71.7 | DSC | (1–4) | 98L2 |
| RM7 | 6.0 | 25 | 71.2 | DSC | (1–3,5) | 98L2 |
| RM6 | 6.0 | 25 | 195.1 | DSC | (1–3,6) | 98L2 |
| RM6 | 6.0 | 25 | 195.1±5 | DSC | (1,3,6,7) | 98L1 |

Remarks:

(1) measured in 10 mM sodium phosphate, pH 6.0, 10 mM Na$_2$SO$_4$, 1 mM EDTA

(2) the data refer to a protein concentration of 0.5 mg/ml

(3) data per mole of cooperative unit, i.e., the dimer state for wild type and RM7, and to the tetrameric state of RM6

(4) data from Ref. 93S10

(5) data from Ref. 94V4, see also insertion mutant <aa>

(6) RM6 was obtained by removal of five amino acids (Asp30, Ala31, Asp32, Glu33, Gln34) from the loop which results in a reorganization of the protein, forming a homotetrameric four-α-helix structure instead of the homodimeric four-α-helix motif of the wild type

(7) the data refer to a protein concentration of 1.0 mg/ml

## Rubredoxin

Rubredoxin from the hyperthermophilic archaebacterium *Pyrococcus furiosus*

| pH | T | $\Delta G$ | Approach | Remarks | Ref |
|---|---|---|---|---|---|
| 5.5–9.5 | 100 | 63 | H/D in GuHCl | (1–3) | 97H5 |

Remarks:

(1) from hydrogen exchange in GuHCl-containing solutions

(2) $\Delta H$ amounts to 293 kJ/mol at 100°C, and m = 7.1 kJ/mol/M

(3) the extrapolated temperature at which $\Delta G$ approaches zero is in the range of 176° to 195°C

**Sac7d Protein**, see DNA-Binding Protein

## S-Adenosylhomocysteine Hydrolase

Human placental S-adenosylhomocysteine hydrolase, wild type and proteolytic fragments

| Mutant | pH | T | $\Delta G$ | $c_{1/2}$ | m | Approach/Remarks | | Ref |
|---|---|---|---|---|---|---|---|---|
| wild type | 7.2 | 25 | 12.6 | 0.82 | 15.3 | GuHCl | (1,2) | 97H8 |
| wild type | 7.2 | 25 | 21.7 | 1.12 | 14.6 | GuHCl | (1,3) | 97H8 |
| fragments | 7.2 | 25 | 12.1 | 0.86 | 19.4 | GuHCl | (1,2,4) | 97H8 |
| fragments | 7.2 | 25 | 15.4 | 0.96 | 14.6 | GuHCl | (1,3,4) | 97H8 |

Remarks:

(1) buffer: 50 mM phosphate containing 1 mM EDTA

(2) measured by CD at 220 nm

(3) measured by fluorescence from $\lambda_{max}$ data

(4) papain digested protein

**Selenomethionine Containing Proteins**, see Annexin V

**Serum Albumin**, see Human Serum Albumin

## SH3 Domain

The data entries are arranged as follows:

a) α-spectrin SH3 domain

b) src SH3 domain

c) SH3 domains of Btk, Itk, and Tec

d) SH3 domains of α-spectrin, Fyn, and Abl

e) SH3 domain of phosphatidylinositol 3'-kinase

*a) α-spectrin SH3 domain*

SH3 domain of α-spectrin in $D_2O$

| Protein | pH* | T | ΔG | Approach/Remarks | | Ref |
|---|---|---|---|---|---|---|
| Spectrin | 4.9 | 25 | 17.9 | DSC | (1–4) | 99S1 |
| | 4.0 | 25 | 16.7 | DSC | (1–4) | 99S1 |
| | 3.0 | 25 | 11.6 | DSC | (1–4) | 99S1 |
| | 2.5 | 25 | 7.7 | DSC | (1–4) | 99S1 |
| | 1.4 | 25 | 4.3 | DSC | (1–4) | 99S1 |

Remarks:
(1) measured in 20 mM glycine or 20 mM acetate in $D_2O$ at different pH*
(2) in Ref. 99S1 DSC results are compared with H/D exchange measured by NMR
(3) the Trp41, Trp42, and Arg49 exhibit at pH* 2.5 ΔG of 7.05, 6.85, and 7.09, respectively
(4) for all remaining residues are the individual ΔG lower than the global ΔG values by 1.6 kJ/mol or even lower

SH3 domain of α-spectrin

| Protein | pH | T | ΔG | RT m | Appr./Rem. | | Ref |
|---|---|---|---|---|---|---|---|
| wild type | 7.0 | 25 | 16.3±1.3 | 3.26 | urea | (1,2) | 99M10 |
| | 3.5 | 25 | 12.6±1.3 | 4.90 | urea | (1,2) | 99M10 |
| | 2.5 | 25 | 2.9±0.3 | −3.85 | urea | (1,2) | 99M10 |
| Asp48→Gly | 7.0 | 25 | 22.2±0.3 | 3.10 | urea | (1,2) | 99M10 |
| | 3.5 | 25 | 15.9±0.4 | 4.18 | urea | (1,2) | 99M10 |
| | 2.5 | 25 | 6.3±0.3 | −3.18 | urea | (1,2) | 99M10 |
| (Leu8→Ser and Asp48→Gly) | | | | | | | |
| | 7.0 | 25 | 18.0±0.1 | 3.10 | urea | (1,2) | 99M10 |
| | 3.5 | 25 | 12.6±0.4 | 4.18 | urea | (1,2) | 99M10 |
| | 2.5 | 25 | 3.1±0.3 | −3.47 | urea | (1,2) | 99M10 |
| (Ala11→Gly and Asp48→Gly) | | | | | | | |
| | 7.0 | 25 | 12.1±0.4 | 3.31 | urea | (1,2) | 99M10 |
| | 3.5 | 25 | 8.4±0.2 | 4.35 | urea | (1,2) | 99M10 |
| Asp14→Ser | 7.0 | 25 | 12.6±0.2 | 3.10 | urea | (1,2) | 99M10 |
| | 3.5 | 25 | 11.7±0.4 | 4.73 | urea | (1,2) | 99M10 |
| | 2.5 | 25 | 3.7±0.3 | −3.51 | urea | (1,2) | 99M10 |
| (Val23→Ala and Asp48→Gly) | | | | | | | |
| | 7.0 | 25 | 13.8±0.4 | 3.01 | urea | (1,2) | 99M10 |
| | 3.5 | 25 | 9.2±0.8 | 4.77 | urea | (1,2) | 99M10 |
| (Thr24→Ala and Asp48→Gly) | | | | | | | |
| | 7.0 | 25 | 17.6±0.3 | 3.10 | urea | (1,2) | 99M10 |
| | 3.5 | 25 | 11.7±0.4 | 4.06 | urea | (1,2) | 99M10 |
| | 2.5 | 25 | 3.3±0.2 | −3.22 | urea | (1,2) | 99M10 |
| (Asp29→Ala and Asp48→Gly) | | | | | | | |
| | 7.0 | 25 | 17.2±1.0 | 3.43 | urea | (1,2) | 99M10 |
| | 3.5 | 25 | 13.0±0.3 | 4.73 | urea | (1,2) | 99M10 |
| | 2.5 | 25 | 3.6±0.3 | −3.43 | urea | (1,2) | 99M10 |
| (Leu33→Val and Asp48→Gly) | | | | | | | |
| | 7.0 | 25 | 17.6±0.8 | 3.14 | urea | (1,2) | 99M10 |
| | 3.5 | 25 | 12.1±0.4 | 4.18 | urea | (1,2) | 99M10 |
| | 2.5 | 25 | 2.6±0.2 | −3.22 | urea | (1,2) | 99M10 |
| (Ser36→Asn and Asp48→Gly) | | | | | | | |
| | 7.0 | 25 | 17.6±2.1 | 3.18 | urea | (1,2) | 99M10 |
| | 3.5 | 25 | 13.4±0.4 | 4.18 | urea | (1,2) | 99M10 |
| | 2.5 | 25 | 3.6±0.5 | −3.18 | urea | (1,2) | 99M10 |

SH3 domain of α-spectrin (continued)

| Protein | pH | T | ΔG | RT m | Appr./Rem. | Ref |
|---|---|---|---|---|---|---|
| (Lys43→Ala and Asp48→Gly) | | | | | | |
| | 7.0 | 25 | 18.0±0.3 | 3.18 | urea (1,2) | 99M10 |
| | 3.5 | 25 | 13.4±0.4 | 4.10 | urea (1,2) | 99M10 |
| | 2.5 | 25 | 3.6±0.2 | −3.26 | urea (1,2) | 99M10 |
| (Val44→Ala and Asp48→Gly) | | | | | | |
| | 7.0 | 25 | 7.9±0.2 | 3.68 | urea (1,2) | 99M10 |
| | 3.5 | 25 | 2.5±0.4 | 4.23 | urea (1,2) | 99M10 |
| (Val46→Ala and Asp48→Gly) | | | | | | |
| | 7.0 | 25 | 18.4±0.2 | 3.26 | urea (1,2) | 99M10 |
| | 3.5 | 25 | 12.1±0.4 | 4.52 | urea (1,2) | 99M10 |
| | 2.5 | 25 | 2.5±0.2 | −3.47 | urea (1,2) | 99M10 |
| Asn48→Gly | 7.0 | 25 | 18.0±0.3 | 3.26 | urea (1,2) | 99M10 |
| | 3.5 | 25 | 15.1±0.4 | 4.52 | urea (1,2) | 99M10 |
| | 2.5 | 25 | 4.8±0.1 | −3.51 | urea (1,2) | 99M10 |
| (Phe52→Ala and Asp48→Gly) | | | | | | |
| | 7.0 | 25 | 10.9±0.3 | 3.31 | urea (1,2) | 99M10 |
| | 3.5 | 25 | 5.4±0.4 | 4.10 | urea (1,2) | 99M10 |
| (Val53→Ala and Asp48→Gly) | | | | | | |
| | 7.0 | 25 | 13.0±0.8 | 3.43 | urea (1,2) | 99M10 |
| | 3.5 | 25 | 7.5±0.8 | 4.23 | urea (1,2) | 99M10 |
| (Val55→Gly and Asp48→Gly) | | | | | | |
| | 7.0 | 25 | 13.8±0.8 | 3.43 | urea (1,2) | 99M10 |
| | 3.5 | 25 | 7.5±0.2 | 4.35 | urea (1,2) | 99M10 |
| (Val58→Ala and Asp48→Gly) | | | | | | |
| | 7.0 | 25 | 12.1±0.8 | 3.51 | urea (1,2) | 99M10 |
| | 3.5 | 25 | 7.1±0.4 | 4.54 | urea (1,2) | 99M10 |

Remarks:

(1) data from kinetic measurements followed by stopped-flow fluorimetry

(2) buffer: 50 mM sodium phosphate at pH 7.0, and 50 mM glycine-HCl at pH 3.5 and pH 2.5

SH3 domain of α-spectrin, wild type and Asp48→Gly mutant

| Mutant | pH | T | ΔG | Approach/Remarks | Ref |
|---|---|---|---|---|---|
| wild type | 2.0 | 25 | 2.3 | DSC (1) | 98M6 |
| | 2.5 | 25 | 6.9 | DSC (1) | 98M6 |
| | 3.0 | 25 | 11.6 | DSC (1) | 98M6 |
| | 3.5 | 25 | 13.9 | DSC (1) | 98M6 |
| | 4.0 | 25 | 15.6 | DSC (1) | 98M6 |
| Asp48→Gly | 2.0 | 25 | 6.4 | DSC (2,3) | 98M6 |
| | 2.5 | 25 | 9.8 | DSC (2,3) | 98M6 |
| | 3.0 | 25 | 14.0 | DSC (2,3) | 98M6 |
| | 3.5 | 25 | 18.3 | DSC (2,3) | 98M6 |
| | 4.0 | 25 | 20.5 | DSC (2,3) | 98M6 |

Remarks:

(1) data from Ref. 94V2, see Ref. 94V2 for additional data

(2) measured in 50 mM glycine/HCl or 50 mM acetic acid/sodium acetate

(3) for further thermodynamic quantitites see also Table 2

SH3 domain of α-spectrin, wild type and mutants, introduction of helical tendency in the all β-sheet

| Mutant | pH | T | ΔG | m | Approach/Remarks | | Ref |
|---|---|---|---|---|---|---|---|
| wild type | 7.0 | 25 | 15.0±0.1 | 3.01±0.04 | urea | (1,2) | 97P14 |
| wild type | 7.0 | 25 | 15.8±0.4 | 3.31±0.04 | urea | (1–3) | 97P14 |
| wild type | 7.0 | 25 | 15.9±2.9 | | heat | (1,2,4) | 97P14 |
| wild type | 2.5 | 25 | 7.9±1.7 | | heat | (1,2,4) | 97P14 |
| Asp14→Ser | 7.0 | 25 | 10.5±0.2 | 2.93±0.04 | urea | (1,2) | 97P14 |
| Asp14→Ser | 7.0 | 25 | 12.5±0.4 | 3.26±0.04 | urea | (1,2) | 97P14 |
| Asp14→Ser | 7.0 | 25 | 10.5±2.1 | | heat | (1,2,4) | 97P14 |
| Asp14→Ser | 2.5 | 25 | 7.9±1.7 | | heat | (1,2,4) | 97P14 |
| double mutant (Thr4→Ala and Gly5→Glu) | | | | | | | |
| | 7.0 | 25 | 13.8±0.3 | 2.97±0.04 | urea | (1,2) | 97P14 |
| | 7.0 | 25 | 14.9±0.4 | 3.22±0.04 | urea | (1,2) | 97P14 |
| | 7.0 | 25 | 13.0±2.5 | | heat | (1,2,4) | 97P14 |
| | 2.5 | 25 | 6.3±1.3 | | heat | (1,2,4) | 97P14 |
| triple mutant (Thr4→Ala, Gly5→Glu, and Asp14→Ser) | | | | | | | |
| | 7.0 | 25 | 9.4±0.2 | 2.89±0.04 | urea | (1,2) | 97P14 |
| | 7.0 | 25 | 11.0±0.4 | 3.01±0.04 | urea | (1,2) | 97P14 |
| | 7.0 | 25 | 7.5±1.7 | | heat | (1,2,4) | 97P14 |
| | 2.5 | 25 | 5.9±1.3 | | heat | (1,2,4) | 97P14 |
| triple mutant (Thr4→Ala, Gly5→Glu, and Asp14→Lys) | | | | | | | |
| | 7.0 | 25 | 1.3±0.6 | 2.68±0.04 | urea | (1,2) | 97P14 |
| | 7.0 | 25 | 3.2±1.7 | 2.85±0.04 | urea | (1,2) | 97P14 |

Remarks:
(1) linear extrapolation, LEM-SB
(2) buffer: 50 mM sodium phosphate, pH 7.0
(3) m and ΔG from kinetics
(4) ΔG was calculated using primary data contained in Table 2, and temperature dependent ΔCp from Ref. 94V2

SH3 domain of α-spectrin and circular permutants with different loop lengths

| Mutant | pH | T | ΔG | Approach/Remarks | | Ref |
|---|---|---|---|---|---|---|
| wild type | 2.0 | 25 | 2.3 | DSC | (1) | 99M11 |
| | 2.5 | 25 | 6.9 | DSC | (1) | 99M11 |
| | 3.5 | 25 | 13.9 | DSC | (1) | 99M11 |
| pseudo-w.t. | 2.0 | 25 | 4.0 | DSC | | 99M11 |
| | 2.5 | 25 | 7.6 | DSC | | 99M11 |
| | 3.5 | 25 | 16.9 | DSC | | 99M11 |
| | 7.0 | 25 | 18.5 | DSC | | 99M11 |
| Ser19-Pro20s | 2.5 | 25 | 1.9 | DSC | | 99M11 |
| | 3.5 | 25 | 9.7 | DSC | | 99M11 |
| | 7.0 | 25 | 11.1 | DSC | | 99M11 |
| Ser19-Pro20s1G | 2.5 | 25 | 1.5 | DSC | | 99M11 |
| | 3.5 | 25 | 10.9 | DSC | | 99M11 |
| | 7.0 | 25 | 13.6 | DSC | | 99M11 |
| Ser19-Pro20s3G | 3.5 | 25 | 9.5 | DSC | | 99M11 |
| | 7.0 | 25 | 11.8 | DSC | | 99M11 |
| Ser19-Pro20s5G | 3.5 | 25 | 9.3 | DSC | | 99M11 |
| | 7.0 | 25 | 11.3 | DSC | | 99M11 |
| Asn47-Asp48s | 2.0 | 25 | 1.4 | DSC | | 99M11 |
| | 2.5 | 25 | 6.0 | DSC | | 99M11 |
| | 3.5 | 25 | 13.5 | DSC | | 99M11 |
| | 7.0 | 25 | 11.2 | DSC | | 99M11 |

SH3 domain of α-spectrin and circular permutants with different loop lengths (continued)

| Mutant | pH | T | ΔG | Approach/Remarks | Ref |
|---|---|---|---|---|---|
| Asn47-Asp48s1G | 2.5 | 25 | 3.4 | DSC | 99M11 |
|  | 3.5 | 25 | 13.2 | DSC | 99M11 |
|  | 7.0 | 25 | 13.2 | DSC | 99M11 |
| Asn47-Asp48s3G | 3.5 | 25 | 12.4 | DSC | 99M11 |
|  | 7.0 | 25 | 11.8 | DSC | 99M11 |
| Asn47-Asp48s5G | 3.5 | 25 | 12.0 | DSC | 99M11 |
|  | 7.0 | 25 | 11.0 | DSC | 99M11 |

Explanations:

| | |
|---|---|
| wild type | for the sequence see Ref. 99M11 |
| pseudo-w.t. | the second and third residues were substituted by a Gly |
| Ser19-Pro20s | circular permutant, the pseudo-w.t. sequence was interrupted between Ser19 and Pro20, the previous N- and C-termini have been joined by the linker Lys60-Leu61-Asp62-Ser2-Gly3-Thr4-Gly5-Lys6, "s" refers to a Ser in the linker which was not present in shorter versions (see Ref. 95V5) |
| Asn47-Asp48s | circular permutant analogous to Ser19-Pro20s, the pseudo-w.t. sequence was interrupted between Asn47 andAsp48 |
| 1G, 3G, 5G | the connecting sequence was elongated with one, three, or five extra Gly residues |

Remark:
(1) data taken from Ref. 94V2

*b) src SH3 domain*

src SH3 domain, wild type and mutants

| Mutant | pH | T | Δ(ΔG) | Approach/Remarks | | Ref |
|---|---|---|---|---|---|---|
| wild type | 6.0 | 22 | 0.00 | GuHCl | (1,2) | 99R12 |
| Thr9→Ala | 6.0 | 22 | −2.68±0.33 | GuHCl | (1,2) | 99R12 |
| Phe10→Ala | 6.0 | 22 | −3.51±0.29 | GuHCl | (1,2) | 99R12 |
| Phe10→Ile | 6.0 | 22 | −6.90±0.71 | GuHCl | (1,2) | 99R12 |
| Val11→Ala | 6.0 | 22 | −6.86±0.50 | GuHCl | (1,2) | 99R12 |
| Ala12→Gly | 6.0 | 22 | −4.18±0.38 | GuHCl | (1,2) | 99R12 |
| Leu13→Ala | 6.0 | 22 | −6.23±0.54 | GuHCl | (1,2) | 99R12 |
| Tyr14→Ala | 6.0 | 22 | −1.30±0.25 | GuHCl | (1,2) | 99R12 |
| Asp15→Ala | 6.0 | 22 | −1.80±0.54 | GuHCl | (1,2) | 99R12 |
| Tyr16→Ala | 6.0 | 22 | −0.50±1.09 | GuHCl | (1,2) | 99R12 |
| Tyr16→Phe | 6.0 | 22 | −0.75±0.42 | GuHCl | (1,2) | 99R12 |
| Ser18→Ala | 6.0 | 22 | 2.18±0.42 | GuHCl | (1,2) | 99R12 |
| Arg19→Ala | 6.0 | 22 | −0.29±0.33 | GuHCl | (1,2) | 99R12 |
| Thr20→Ala | 6.0 | 22 | 0.27±0.29 | GuHCl | (1,2) | 99R12 |
| Thr22→Ala | 6.0 | 22 | −0.04±0.29 | GuHCl | (1,2) | 99R12 |
| Asp23→Ala | 6.0 | 22 | −2.34±0.54 | GuHCl | (1,2) | 99R12 |
| Leu24→Ala | 6.0 | 22 | −7.49±0.38 | GuHCl | (1,2) | 99R12 |
| Ser25→Ala | 6.0 | 22 | −3.43±0.33 | GuHCl | (1,2) | 99R12 |
| Phe26→Ala | 6.0 | 22 | −8.24±0.42 | GuHCl | (1,2) | 99R12 |
| Lys27→Ala | 6.0 | 22 | −1.84±0.46 | GuHCl | (1,2) | 99R12 |
| Lys28→Ala | 6.0 | 22 | −0.38±0.29 | GuHCl | (1,2) | 99R12 |
| Gly29→Ala | 6.0 | 22 | −6.95±0.50 | GuHCl | (1,2) | 99R12 |
| Glu30→Ala | 6.0 | 22 | −8.12±0.54 | GuHCl | (1,2) | 99R12 |
| Arg31→Ala | 6.0 | 22 | −1.34±0.50 | GuHCl | (1,2) | 99R12 |
| Leu32→Ala | 6.0 | 22 | −9.46±1.55 | GuHCl | (1,2) | 99R12 |
| Leu32→Val | 6.0 | 22 | −5.06±0.46 | GuHCl | (1,2) | 99R12 |
| Gln33→Ala | 6.0 | 22 | −0.88±0.38 | GuHCl | (1,2) | 99R12 |
| Ile34→Ala | 6.0 | 22 | −1.34±0.50 | GuHCl | (1,2) | 99R12 |

src SH3 domain, wild type and mutants (continued)

| Mutant | pH | T | $\Delta(\Delta G)$ | Approach/Remarks | | Ref |
|---|---|---|---|---|---|---|
| Ile34→Val | 6.0 | 22 | −0.38±0.50 | GuHCl | (1,2) | 99R12 |
| Val35→Ala | 6.0 | 22 | −3.22±0.50 | GuHCl | (1,2) | 99R12 |
| Asn36→Ala | 6.0 | 22 | −0.84±0.38 | GuHCl | (1,2) | 99R12 |
| Asn37→Ala | 6.0 | 22 | 0.29±0.25 | GuHCl | (1,2) | 99R12 |
| Gly40→Ala | 6.0 | 22 | −1.17±0.33 | GuHCl | (1,2) | 99R12 |
| Trp42→Ala | 6.0 | 22 | −5.40±0.42 | GuHCl | (1,2) | 99R12 |
| Trp43→Ala | 6.0 | 22 | −5.02±0.46 | GuHCl | (1,2) | 99R12 |
| Trp43→Ile | 6.0 | 22 | −3.22±0.54 | GuHCl | (1,2) | 99R12 |
| Leu44→Ala | 6.0 | 22 | −6.86±0.63 | GuHCl | (1,2) | 99R12 |
| Ala45→Gly | 6.0 | 22 | −3.85±0.63 | GuHCl | (1,2) | 99R12 |
| His46→Ala | 6.0 | 22 | −2.59±0.25 | GuHCl | (1,2) | 99R12 |
| Ser47→Ala | 6.0 | 22 | −6.11±0.38 | GuHCl | (1,2) | 99R12 |
| Leu48→Ala | 6.0 | 22 | −2.55±0.33 | GuHCl | (1,2) | 99R12 |
| Ser49→Ala | 6.0 | 22 | 0.75±0.46 | GuHCl | (1,2) | 99R12 |
| Thr50→Ala | 6.0 | 22 | −7.49±0.42 | GuHCl | (1,2) | 99R12 |
| Gly51→Ala | 6.0 | 22 | −5.06±0.59 | GuHCl | (1,2) | 99R12 |
| Gln52→Ala | 6.0 | 22 | −1.46±0.50 | GuHCl | (1,2) | 99R12 |
| Thr53→Ala | 6.0 | 22 | −4.64±0.46 | GuHCl | (1,2) | 99R12 |
| Gly54→Ala | 6.0 | 22 | −7.57±0.50 | GuHCl | (1,2) | 99R12 |
| Tyr55→Ala | 6.0 | 22 | −6.36±0.42 | GuHCl | (1,2) | 99R12 |
| Ile56→Ala | 6.0 | 22 | −7.70±0.42 | GuHCl | (1,2) | 99R12 |
| Pro57→Ala | 6.0 | 22 | −5.69±0.46 | GuHCl | (1,2) | 99R12 |
| Ser58→Ala | 6.0 | 22 | 1.00±0.33 | GuHCl | (1,2) | 99R12 |
| Asn59→Ala | 6.0 | 22 | −0.59±0.29 | GuHCl | (1,2) | 99R12 |
| Tyr60→Ala | 6.0 | 22 | 0.96±0.38 | GuHCl | (1,2) | 99R12 |
| Val61→Ala | 6.0 | 22 | −4.94±0.38 | GuHCl | (1,2) | 99R12 |
| Ala62→Gly | 6.0 | 22 | −2.22±0.33 | GuHCl | (1,2) | 99R12 |
| Pro63→Ala | 6.0 | 22 | −0.59±0.46 | GuHCl | (1,2) | 99R12 |
| Ser64→Ala | 6.0 | 22 | 1.84±0.25 | GuHCl | (1,2) | 99R12 |
| Asp65→Ala | 6.0 | 22 | 0.54±0.29 | GuHCl | (1,2) | 99R12 |

Remarks:
(1) for the method see also Ref. 98M17
(2) measured in 50 mM sodium phosphate, pH 6.0

src SH3 domain, wild type and mutants

| Mutant | pH | T | $\Delta G$ | Approach/Remarks | | Ref |
|---|---|---|---|---|---|---|
| wild type | 6.0 | 22 | 15.5 | GuHCl | (1,2) | 98G14 |
| Phe10→Ile | 6.0 | 22 | 11.3 | GuHCl | (1,2) | 98G14 |
| Asp15→Ala | 6.0 | 22 | 13.8 | GuHCl | (1,2) | 98G14 |
| Tyr16→Ala | 6.0 | 22 | 3.8 | GuHCl | (1,2) | 98G14 |
| Ser18→Ala | 6.0 | 22 | 17.2 | GuHCl | (1,2) | 98G14 |
| Leu24→Ala | 6.0 | 22 | 7.9 | GuHCl | (1,2) | 98G14 |
| Gly29→Ala | 6.0 | 22 | 8.8 | GuHCl | (1,2) | 98G14 |
| Glu30→Ala | 6.0 | 22 | 7.5 | GuHCl | (1,2) | 98G14 |
| Leu32→Ala | 6.0 | 22 | 0.4 | GuHCl | (1,2) | 98G14 |
| Ile34→Ala | 6.0 | 22 | 12.6 | GuHCl | (1,2) | 98G14 |
| Asn36→Ala | 6.0 | 22 | 16.7 | GuHCl | (1,2) | 98G14 |
| Gly40→Ala | 6.0 | 22 | 15.9 | GuHCl | (1,2) | 98G14 |
| Trp42→Ala | 6.0 | 22 | 10.5 | GuHCl | (1,2) | 98G14 |
| Leu44→Ala | 6.0 | 22 | 6.3 | GuHCl | (1,2) | 98G14 |
| Ala45→Ala | 6.0 | 22 | 9.2 | GuHCl | (1,2) | 98G14 |

src SH3 domain, wild type and mutants (continued)

| Mutant | pH | T | ΔG | Approach/Remarks | | Ref |
|---|---|---|---|---|---|---|
| Ser47→Ala | 6.0 | 22 | 5.0 | GuHCl | (1,2) | 98G14 |
| Thr50→Ala | 6.0 | 22 | 7.9 | GuHCl | (1,2) | 98G14 |
| Gly51→Ala | 6.0 | 22 | 8.4 | GuHCl | (1,2) | 98G14 |
| Tyr55→Ala | 6.0 | 22 | 7.9 | GuHCl | (1,2) | 98G14 |
| Ile56→Ala | 6.0 | 22 | 5.9 | GuHCl | (1,2) | 98G14 |
| Tyr60→Ala | 6.0 | 22 | 17.2 | GuHCl | (1,2) | 98G14 |
| Val61→Ala | 6.0 | 22 | 13.0 | GuHCl | (1,2) | 98G14 |

Remarks:

(1) linear extrapolation

(2) measured in 50 mM sodium phosphate

src SH3 domain, wild type and mutants

| Protein | pH | T | ΔG | Approach/Remarks | | Ref |
|---|---|---|---|---|---|---|
| wild type | | | 17.2 | GuHCl | (1) | 99G18 |
| Glu21→Ile | | | 19.7 | | | 99G18 |
| Thr22→Ile | | | 9.6 | | | 99G18 |
| Ser25→Ile | | | 10.9 | | | 99G18 |
| Lys27→Ile | | | 16.3 | | | 99G18 |
| double mutant (Arg19→Ile and Thr20→Ile) | | | | | | |
| | | | 15.1 | kinetics | | 99G18 |

Remark:

(1) see also Ref. 97G11

Chicken src SH3 domain in H$_2$O and D$_2$O

| Solvent | pH/pD | T | ΔG | m | Approach/Remarks | | Ref |
|---|---|---|---|---|---|---|---|
| H$_2$O | 6 | 22 | 17.2±0.4 | 6.7±0.1 | GuHCl | (1) | 97G11 |
| H$_2$O | 6 | 22 | 15.9±0.4 | 6.3±0.1 | GuHCl | (2,3) | 97G11 |
| H$_2$O | 6 | 22 | 15.5±0.3 | 5.9±0.1 | GuHCl | (4) | 97G11 |
| D$_2$O | 6 | 22 | 19.7±0.7 | 7.1±0.3 | GuHCl | (1) | 97G11 |
| D$_2$O | 6 | 22 | 19.7±0.9 | 7.1±0.3 | heat/GuHCl | (5) | 97G11 |

Remarks:

(1) transition monitored by fluorescence

(2) value obtained from the extrapolated rates of folding and unfolding

(3) transition monitored by CD at 235 nm

(4) data from kinetic experiments

(5) heat denaturation, global fit of the temperature-GuHCl denaturation surface

src SH3 domain, mutants produced by simplified amino acid sequences

| Mutant | pH | T | ΔG | Approach | Remarks | Ref |
|---|---|---|---|---|---|---|
| wild type | 7 | 22 | 15.5 | GuHCl | (1–3) | 97R3 |
| S1 | 7 | 22 | 7.5 | GuHCl | (1–3) | 97R3 |
| S2 | 7 | 22 | 14.2 | GuHCl | (1–3) | 97R3 |
| S3 | 7 | 22 | 15.5 | GuHCl | (1–3) | 97R3 |
| FP1 | 7 | 22 | 12.6 | GuHCl | (1–3) | 97R3 |
| FP2 | 7 | 22 | 7.1 | GuHCl | (1–3) | 97R3 |

Remarks:

(1) linear extrapolation

(2) for the sequences see Ref. 97R3

(3) for the approach see also Ref. 97S3

*c) SH3 domains of Btk, Itk, and Tec*

SH3 domains of the nonreceptor protein-tyrosin kinases Btk, Itk, and Tec

| Protein | pH | T | ΔG | Approach/Remarks | Ref |
|---|---|---|---|---|---|
| Btk-SH3 | 5.1–6.2 | 25 | 16.6 | heat (1) | 98K10 |
| Itk-SH3 | 4.1–6.0 | 25 | 12.7 | heat (1) | 98K10 |
| Tec-SH3 | 5.9–7.5 | 25 | 13.1 | heat (1) | 98K10 |
| Sso7d | 7.0 | 25 | 34.0 | DSC/heat(2) | 98K10 |

Remarks:

(1) pH range where the protein achieves maximal thermostability, the data were taken from Fig. 5 in Ref. 98K10

(2) Sso7d from the hyperthermophile *Sulfolobus solfataricus* is topologically identical to the eukaryotic SH3 domains, data from Ref. 96K5

*d) SH3 domains of α-spectrin, Fyn, and Abl*

SH3 domains of α-spectrin (Spc-SH3), Fyn (Fyn-SH3), and Abl protein (Abl-SH3)

| Protein | pH | T | ΔG | m | Approach/Remarks | Ref |
|---|---|---|---|---|---|---|
| Spc-SH3 | 7.0 | 25 | 13.6±0.8 | 2.6±0.1 | urea, LEM (1–3) | 99F6 |
| | 7.0 | 25 | 15.0±0.8 | 3.0±0.1 | urea, BEM (1,2,4) | 99F6 |
| | 7.0 | 25 | 15.4±0.8 | | urea, DBM (1,2,5) | 99F6 |
| | 7.0 | 25 | 15.6±1.2 | | DSC (6) | 99F6 |
| Spc-SH3 | 3.5 | 25 | 11.6±0.8 | 2.9±0.1 | urea, LEM (1–3) | 99F6 |
| | 3.5 | 25 | 13.0±0.8 | 3.3±0.1 | urea, BEM (1,2,4) | 99F6 |
| | 3.5 | 25 | 13.5±0.8 | | urea, DBM (1,2,5) | 99F6 |
| | 3.5 | 25 | 13.9±1.2 | | DSC (6) | 99F6 |
| Abl-SH3 (-DTT) | 7.0 | 25 | 14.7±0.8 | 2.3±0.1 | urea, LEM (1–3) | 99F6 |
| | 7.0 | 25 | 16.2±0.8 | 2.8±0.1 | urea, BEM (1,2,4) | 99F6 |
| | 7.0 | 25 | 16.9±0.8 | | urea, DBM (1,2,5) | 99F6 |
| | 7.0 | 25 | 15.0±1.5 | | DSC (6) | 99F6 |
| Abl-SH3 (+DTT) | 7.0 | 25 | 15.1±0.8 | 2.3±0.1 | urea, LEM (1–3) | 99F6 |
| | 7.0 | 25 | 17.6±0.8 | 2.8±0.1 | urea, BEM (1,2,4) | 99F6 |
| | 7.0 | 25 | 18.2±0.8 | | urea, DBM (1,2,5) | 99F6 |
| Fyn-SH3 | 7.0 | 25 | 21.1±1.5 | 2.7±0.2 | urea, LEM (1–3) | 99F6 |
| | 7.0 | 25 | 23.4±1.0 | 3.3±0.1 | urea, BEM (1,2,4) | 99F6 |
| | 7.0 | 25 | 24.0±1.5 | | urea, DBM (1,2,5) | 99F6 |
| | 7.0 | 25 | 20.5±2.0 | | DSC (6) | 99F6 |

Remarks:

(1) measured in 50 mM sodium phosphate

(2) transition monitored by fluorescence

(3) data treatment by LEM – linear extrapolation model

(4) data treatment by BEM – binominal extrapolation model, see also Refs. 86P1 and 94M9

(5) data treatment by DBM – denaturant binding model, with Δn ≈ 18

(6) from DSC measurements performed in 10 mM glycine, PIPES, or phosphate at pH 2–3.5, 7, or 10.5, see also Table 2

SH3 domain of human Fyn tyrosine kinase

| Protein | pH | T | ΔG | m | Approach/Remarks | | Ref |
|---------|-----|----|-----------|-----------|---------|-------|-------|
| Fyn-SH3 | 7.2 | 20 | 25.1±2.5 | 5.69±0.54 | GuHCl | (1–3) | 98P11 |
| Fyn-SH3 | 7.2 | 20 | 27.95±0.04 | 6.74±0.08 | GuHCl | (1,2,4) | 98P11 |

Remarks:

(1) linear extrapolation

(2) Fyn-SH3 consists of 67 amino acid residues without disulfide bonds and prosthetic groups

(3) equilibrium denaturation, transition monitored by fluorescence emission at 340±1.25 nm

(4) data derived from kinetics

SH3 domain of Fyn tyrosine kinase, mutants that affect a buried hydrogen bond between the side chains of Glu24 and Ser41

| Mutant | pH | T | ΔG | m | Approach/Remarks | | Ref |
|--------|-----|----|------------|-----------|---------|-------|------|
| wild type | 8.0 | 25 | 20.88±0.54 | 6.28±0.42 | GuHCl | (1–3) | 98M9 |
| Glu24→Ala | 8.0 | 25 | 15.56 | 6.40 | GuHCl | (1–3) | 98M9 |
| Glu24→Arg | 8.0 | 25 | 6.74 | 7.87 | GuHCl | (1–3) | 98M9 |
| Glu24→Asp | 8.0 | 25 | 23.72 | 6.57 | GuHCl | (1–3) | 98M9 |
| Glu24→Gln | 8.0 | 25 | 17.78 | 7.20 | GuHCl | (1–3) | 98M9 |
| Glu24→His | 8.0 | 25 | 16.74 | 6.82 | GuHCl | (1–3) | 98M9 |
| Glu24→Ile | 8.0 | 25 | 9.54 | 8.03 | GuHCl | (1–3) | 98M9 |
| Glu24→Leu | 8.0 | 25 | 8.24 | 7.66 | GuHCl | (1–3) | 98M9 |
| Glu24→Lys | 8.0 | 25 | 16.82 | 7.03 | GuHCl | (1–3) | 98M9 |
| Glu24→Phe | 8.0 | 25 | 14.06 | 7.57 | GuHCl | (1–3) | 98M9 |
| Glu24→Pro | 8.0 | 25 | 13.72 | 7.41 | GuHCl | (1–3) | 98M9 |
| Glu24→Ser | 8.0 | 25 | 14.02 | 6.86 | GuHCl | (1–3) | 98M9 |
| Glu24→Thr | 8.0 | 25 | 18.37 | 7.20 | GuHCl | (1–3) | 98M9 |
| Glu24→Tyr | 8.0 | 25 | 10.83 | 8.28 | GuHCl | (1–3) | 98M9 |
| Glu24→Val | 8.0 | 25 | 12.51 | 7.07 | GuHCl | (1–3) | 98M9 |
| Ser41→Ala | 8.0 | 25 | 17.91 | 6.69 | GuHCl | (1–3) | 98M9 |
| Ser41→Arg | 8.0 | 25 | 20.13 | 7.24 | GuHCl | (1–3) | 98M9 |
| Ser41→Asn | 8.0 | 25 | 17.82 | 6.49 | GuHCl | (1–3) | 98M9 |
| Ser41→Asp | 8.0 | 25 | 12.43 | 7.36 | GuHCl | (1–3) | 98M9 |
| Ser41→Gly | 8.0 | 25 | 13.85 | 9.04 | GuHCl | (1–3) | 98M9 |
| Ser41→His | 8.0 | 25 | 15.65 | 6.95 | GuHCl | (1–3) | 98M9 |
| Ser41→Ile | 8.0 | 25 | 14.98 | 6.15 | GuHCl | (1–3) | 98M9 |
| Ser41→Leu | 8.0 | 25 | 19.33 | 7.11 | GuHCl | (1–3) | 98M9 |
| Ser41→Lys | 8.0 | 25 | 21.33 | 7.45 | GuHCl | (1–3) | 98M9 |
| Ser41→Phe | 8.0 | 25 | 12.93 | 7.49 | GuHCl | (1–3) | 98M9 |
| Ser41→Thr | 8.0 | 25 | 13.97 | 7.20 | GuHCl | (1–3) | 98M9 |
| Ser41→Tyr | 8.0 | 25 | 11.38 | 8.03 | GuHCl | (1–3) | 98M9 |
| Ser41→Val | 8.0 | 25 | 16.28 | 6.69 | GuHCl | (1–3) | 98M9 |
| double mutants: | | | | | | | |
| (Glu24→Ala and Ser41→Ala) | | | | | | | |
| | 8.0 | 25 | 14.69 | 6.69 | GuHCl | (1–3) | 98M9 |
| (Glu24→Ala and Ser41→Leu) | | | | | | | |
| | 8.0 | 25 | 19.08 | 6.90 | GuHCl | (1–3) | 98M9 |
| (Glu24→Ala and Ser41→Lys) | | | | | | | |
| | 8.0 | 25 | 15.98 | 6.74 | GuHCl | (1–3) | 98M9 |
| (Glu24→Arg and Ser41→Asp) | | | | | | | |
| | 8.0 | 25 | 15.27 | 8.45 | GuHCl | (1–3) | 98M9 |
| (Glu24→Leu and Ser41→Ala) | | | | | | | |
| | 8.0 | 25 | 9.54 | 8.12 | GuHCl | (1–3) | 98M9 |

Remarks:

(1) linear extrapolation

(2) transition monitored by CD at 220 nm

(3) buffer: 10 mM Tris–HCl, pH 8.0, 0.2 mM EDTA, 250 mM KCl

*e) SH3 domain of phosphatidylinositol 3'-kinase*

SH3 domain of phosphatidylinositol 3'-kinase (PI3-SH3)

| | pH | T | ΔG | $c_{1/2}$ | m | Appr./Rem. | Ref |
|---|---|---|---|---|---|---|---|
| | 7.2 | 20 | 13.52±0.80 | 1.39 | 9.75±0.59 | GuHCl (1–3) | 98G20 |
| | 7.2 | 20 | 15.27±0.21 | | 9.83±0.21 | kinetics (4) | 98G20 |
| | 7.2 | 20 | 14.14±0.79 | | | kinetics (5) | 98G20 |

Remarks:

(1) linear extrapolation

(2) transition monitored by intrinsic fluorescence at 303 nm (excitation 268 nm)

(3) buffer: 20 mM sodium phosphate, pH 7.2

(4) from kinetics of folding and unfolding at different GuHCl conc. studied by fluorescence

(5) using relative amplitudes of fast and slow phase (corrected value)


## Sonic Hedgehog

N-terminal fragment of human Sonic hedgehog (ShhN), wild type and mutants

| Protein | pH | T | ΔG | $c_{1/2}$ | Approach/Remarks | | Ref |
|---|---|---|---|---|---|---|---|
| wild type | 7.5 | 25 | ~20 | 1.58 | GuHCl | (1–4) | 99D2 |
| wild type + EDTA | 7.5 | 25 | ~10 | 1.03 | GuHCl | (1–5) | 99D2 |
| His140→Ala | 7.5 | 25 | ~8 | 1.04 | GuHCl | (1–4) | 99D2 |
| His140→Ala + EDTA | 7.5 | 25 | ~6 | | GuHCl | (1–5) | 99D2 |
| His140→Ala + ZnCl₂ | 7.5 | 25 | ~10 | 1.56 | GuHCl | (1–4,6) | 99D2 |
| Asp147→Ala | 7.5 | 25 | ~6 | 1.09 | GuHCl | (1–4) | 99D2 |
| Asp147→Ala + EDTA | 7.5 | 25 | ~10 | | GuHCl | (1–5) | 99D2 |
| Asp147→Ala + ZnCl₂ | 7.5 | 25 | ~13 | 1.33 | GuHCl | (1–4,6) | 99D2 |
| Glu176→Ala | 7.5 | 25 | ~20 | 1.69 | GuHCl | (1–4) | 99D2 |
| Glu176→Ala + EDTA | 7.5 | 25 | ~10 | | GuHCl | (1–5) | 99D2 |

Remarks:

(1) linear extrapolation

(2) transiton monitored by fluorescence emission at 334 nm

(3) buffer: 100 mM HEPES, 150 mM NaCl, 0.5 mM DTT, pH 7.5

(4) ΔG values were taken from Fig. 4 in Ref. 99D2

(5) measured in the presence of a 10-fold molar excess of EDTA

(6) measured in the presence of a 16-fold molar excess of ZnCl₂


**α-Spectrin**, see SH3 Domain

## Spherulin

Spherulin 3a from slime mold *Physarum polycephalum* in the presence and absence of calcium

| Protein | pH | T | $\Delta G$ | Approach/Remarks | | Ref |
|---|---|---|---|---|---|---|
| 3a ($+Ca^{2+}$) | 7.0 | 25 | 135 | GuHCl | (1–4) | 99K17 |
| | 7.0 | 25 | 137±11 | GuHCl | (1–5) | 99K16 |
| 3a ($-Ca^{2+}$) | 7.0 | 25 | 77 | GuHCl | (1–4) | 99K17 |
| | 7.0 | 25 | 81±8 | GuHCl | (1–5) | 99K16 |
| | 7.0 | 15 | 77±6 | GuHCl | (1–5) | 99K16 |
| | 7.0 | 5 | 68±7 | GuHCl | (1–5) | 99K16 |

Remarks:

(1) two-state unfolding according to $N_2 \to 2U$ assuming a linear dependence of $\Delta G$ on the denaturant conc., data per mole of dimer

(2) transition monitored by fluorescence intensity at 360 nm

(3) buffer: 25 mM MOPS/NaOH, plus either 3 mM $CaCl_2$ or 1 mM EDTA

(4) thermal transition in the presence of 0.8 M GuHCl, at ~50°C for 3a ($-Ca^{2+}$), and ~70°C for 3a ($+Ca^{2+}$)

(5) thermal transition in the absence of GuHCl, at 75°C for 3a ($-Ca^{2+}$), and 92°C for 3a ($+Ca^{2+}$)

**Sso7d Protein**, see DNA-Binding Protein

## Stefin

Recombinant human stefin A and B, and stefin A dimer

| Protein | pH | T | $\Delta G$ | $c_{1/2}$ | Approach/Remarks | | Ref |
|---|---|---|---|---|---|---|---|
| stefin A | 8.0 | 25 | | 2.8±0.1 | GuHCl | (1) | 91Z3 |
| stefin B | 8.0 | 25 | | 1.6 | GuHCl | (2) | 92Z6 |
| | 8.0 | 25 | | 1.75 | GuHCl | (3) | 92Z6 |
| stefin A, dimer | 6.0 | 25 | | 2.7±0.1 | GuHCl | (4–6) | 99J9 |

Remarks:

(1) transition monitored by fluorescence, near- and far-UV CD

(2) transition monitored by fluorescence and near-UV CD

(3) transition monitored by far-UV CD

(4) the dimer is a structured protein (no more functional) produced under partly denaturing conditions

(5) transition monitored by far-UV CD and size exclusion chromatography

(6) the dimerization enthalpy amounts to –464±33 kJ/mol as determined from the temperature dependence of the dissociation constant

## Sterol Carrier Protein

Recombinant human sterol carrier protein 2, wild type and mutant

| Mutant | pH | T | $\Delta G$ | $c_{1/2}$ | m | Appr./Rem. | Ref |
|---|---|---|---|---|---|---|---|
| wild type | 6.8 | 20 | 15.5 | 0.82 | 19.4 | GuHCl (1–3) | 99J8 |
| Cys71→Ser | 6.8 | 20 | 8.4 | 0.55 | 13.6 | GuHCl (1–3) | 99J8 |

Remarks:

(1) linear extrapolation, LEM-SB

(2) transition monitored by far-UV CD at 222 nm

(3) buffer: 15 mM $K_2HPO_4$, 15 mM $KH_2PO_4$, 1 mM EDTA, pH 6.8

## Subtilisin BPN'

Subtilisin BPN' pro-domain stability, introduction of stabilizing mutations in the largely unfolded prodomain

| Mutant | pH | T | $\Delta(\Delta G)$ | Remarks | Ref |
|---|---|---|---|---|---|
| wild type | 5.0 | 20 | 0.0 | (1–4) | 97R8 |
| Gln40→Leu | 5.0 | 20 | 4.6 | (1–4) | 97R8 |
| Lys57→Glu | 5.0 | 20 | 4.2 | (1–4) | 97R8 |
| double mutant (Glu32→Gln and Lys57→Glu) | | | | | |
| | 5.0 | 20 | 5.4 | (1–4) | 97R8 |
| double mutant (Gln40→Leu and Lys57→Glu) | | | | | |
| | 5.0 | 20 | 8.8 | (1–4) | 97R8 |
| triple mutant (Glu32→Gln, Gln40→Leu, and Lys57→Glu) | | | | | |
| | 5.0 | 20 | 10.5 | (1–4) | 97R8 |

Remarks:

(1) the prodomain of subtilisin BPN' consists of 77 residues; the isolated prodomain is 97% unfolded under optimal conditions (pH 7, 20°C)

(2) $\Delta(\Delta G)$ refers to the intrinsic equilibrium for folding independent of dimerization, for details of the approach see Ref. 97R8

(3) buffer: 100 mM sodium acetate

(4) transition monitored by CD at 222 nm

Subtilisin BPN' pro domain, stabilization obtained by phage display selection

| Protein | pH | T | $\Delta G$ | Approach/Remarks | | Ref |
|---|---|---|---|---|---|---|
| wild-type pro domain | | | | | | |
| | 7.0 | 25 | ~–8.4 | EG | (1,2) | 95B12 |
| | | 25 | –8.8 | | (1,3) | 98R13 |
| pro-R1 | 5.0 | 20 | 16.7 | DSC | (1,4,5) | 98R13 |
| Ser31→Ile | 5.0 | | –10.9 | kinet. | (1,6,7) | 98R13 |
| Ser31→Leu | 5.0 | | –11.7 | kinet. | (1,6,8) | 98R13 |

Remarks:

(1) the pro domain consists of 77 amino acid residues compared with 275 of the mature protein

(2) from an ethylene glycol (EG)-induced folding transition monitored by CD at 220 nm in 0.1 M potassium phosphate, pH 7.0

(3) the isolated pro domain is to about 97% unfolded at 25°C which corresponds to $\Delta G = -8.8$ kJ/mol, based on data from Ref. 95B12

(4) pro R1 is a multiple mutant (Ala23→Cys, Lys27→Glu, Val37→Leu, and Gln40→Cys) obtained by phage-display selection

(5) measured in 100 mM sodium acetate buffer, pH 5.0

(6) data from the rate of subtilisin folding

(7) mutant Ser31→Ile is 2.1 kJ/mol less stable than pro-wild type

(8) mutant Ser31→Leu is 2.9 kJ/mol less stable than pro-wild type

Subtilisin pro domain (proR9)

| Protein | pH | T | $\Delta G$ | Approach/Remarks | | Ref |
|---|---|---|---|---|---|---|
| proR9 | 7.0 | 25 | 16.7 | DSC | (1–3) | 99R20 |

Remarks:

(1) in proR9 the residues 17–21 (Thr-Met-Ser-Thr-Met) were replaced by Ser-Gly-Ile-Lys, and His72 and His75 by Lys

(2) $\Delta G$ is the same as observed for proR1, see Ref. 98R13

(3) measured in 100 mM potassium phosphate, pH 7.0

## Subtilisin Inhibitor

Subtilisin inhibitor from *Streptomyces*, urea-induced changes in cold- and heat-denatured states

| State | pH | T | $\Delta G$ | m | Approach/Remarks | | Ref |
|-------|-----|----|------------|------|------------------|------|------|
| D' | 1.8 | 3 | 2.2 | 6.2 | urea | (1,2) | 97K9 |
| D' | 1.8 | 3 | 2.3 | 5.6 | urea | (1,3) | 97K9 |
| N | 3.0 | 20 | 8.3 | 10.3 | urea | (2–4) | 97K9 |

Remarks:

(1) D' = cold-denatured state

(2) transition monitored by CD at 292 nm

(3) transition monitored by NMR of the His106 C2 proton signal, given is the pH*

(4) Ref. 97K9 contains additionally small-angle X-ray scattering data

Subtilisin inhibitor from *Streptomyces* (SSI) and mutants

| Mutant | pH | T | $\Delta(\Delta G)$ | Approach/Remarks | | Ref |
|--------|-----|----|-----------|-----------------|------|------|
| wild type | | 25 | 0.0 | DSC | (1,3–6) | 98T2 |
| Val13→Ala | | 25 | −19.2 | DSC | (2–5) | 98T2 |
| Val13→Gly | | 25 | −53.5 | DSC | (2–5) | 98T2 |
| Val13→Phe | | 25 | −33.4 | DSC | (2–5) | 98T2 |
| Val13→Ile | | 25 | −6.7 | DSC | (2–5) | 98T2 |
| Val13→Leu | | 25 | −14.6 | DSC | (2–5) | 98T2 |
| Val13→Met | | 25 | −25.5 | DSC | (2–5) | 98T2 |
| Met73→Lys | | 25 | 1.3 | DSC | (2–5) | 98T2 |
| Met73→Asp | | 25 | 0.4 | DSC | (2–5) | 98T2 |
| Met73→Glu | | 25 | 1.7 | DSC | (2–5) | 98T2 |
| Met73→Gly | | 25 | 0.0 | DSC | (2–5) | 98T2 |
| Met73→Ala | | 25 | −1.7 | DSC | (2–5) | 98T2 |
| Met73→Val | | 25 | −0.4 | DSC | (2–5) | 98T2 |
| Met73→Leu | | 25 | 0.3 | DSC | (2–5) | 98T2 |
| Met73→Ile | | 25 | −0.8 | DSC | (2–5) | 98T2 |
| Met103→Gly | | 25 | −21.7 | DSC | (2–5) | 98T2 |
| Met103→Ala | | 25 | −3.3 | DSC | (2–5) | 98T2 |
| Met103→Val | | 25 | −6.3 | DSC | (2–5) | 98T2 |
| Met103→Ile | | 25 | −11.3 | DSC | (2–5) | 98T2 |
| Met103→Leu | | 25 | −1.3 | DSC | (2–5) | 98T2 |

Remarks:

(1) wild-type SSI was obtained by cultivating *Streptomyces albogriseolus* S-3253

(2) SSI mutant proteins were harvested from *Streptomyces lividans* 66

(3) the $\Delta(\Delta G)$ values are based on data from Refs. 94T3, 95T2, and 95T3

(4) Ref. 98T2 contains DSC data on heat and cold denaturation of SSI wild type and mutant proteins obtained at pH 2.51 and pH 2.99, see Table 2

(5) Ref. 98T2 contains additional $\Delta(\Delta G)$ values calculated at 100°C

(6) reference values for the wild-type protein at 25°C are $\Delta H$ = 98.2 kJ/mol, and $\Delta G$ = 89.9 kJ/mol, see Refs. 98T2 and 94T3

## Sunflower Albumin

Sunflower albumin 8 (SFA-8)

| Protein | pH | T | $\Delta G$ | m | Approach/Remarks | Ref |
|---------|-----|----|-----------|----------|-------------------|-------|
| SFA-8 | 7.2 | 10 | 28.0±0.4 | 19.2±0.4 | GuHSCN (1–4) | 99P3 |
|        | 7.2 | 25 | 35.1±2.9 |          | GuHSCN (1–4) | 99P3 |

Remarks:
(1) nonlinear least-squares fit of the transition curve assuming a linear dependence of $\Delta G$ on the denaturant conc.
(2) transition monitored by fluorescence emission
(3) buffer: 50 mM Tris-HCl
(4) Ref. 99P3 contains additional data on a compact misfolded state

## Superoxide Dismutase

Human Cu,Zn superoxide dismutase (SOD), wild type and circularly permuted variants

| Protein | Component | pH | $T_{trs}$ | $\Delta T$ | $\Delta(\Delta G)$ | Appr./Rem. | Ref |
|---------|-----------|-----|-----------|------------|---------------------|-------------|--------|
| native | minor | 7.8 | 81.2±0.3 | 0.0 | 0.0 | DSC (1,2) | 97B10 |
|        | major | 7.8 | 84.3±0.3 | 0.0 | 0.0 | DSC (1,2) | 97B10 |
| cp(–1+1) | minor | 7.8 | 83.5±0.5 | 2.3 | 2.9±0.4 | DSC (1,2) | 97B10 |
|          | major | 7.8 | 86.2±0.1 | 1.9 | 3.8±0.4 | DSC (1,2) | 97B10 |
| cp(–2+2) | minor | 7.8 | 80.9±1.0 | –0.3 | –0.4±1.3 | DSC (1,2) | 97B10 |
|          | major | 7.8 | 83.5±0.8 | –0.8 | –1.7±1.7 | DSC (1,2) | 97B10 |
| cp(–3+3) | minor | 7.8 | 77.5±0.3 | –3.7 | –4.2±1.3 | DSC (1,2) | 97B10 |
|          | major | 7.8 | 82.7±0.5 | –1.6 | –2.5±0.8 | DSC (1,2) | 97B10 |

Remarks:
(1) measured in 100 mM potassium phosphate, pH 7.8
(2) components resolved by deconvolution

Bovine Cu,Zn superoxide dismutase (Cu,Zn-SOD), wild type and mutant

| Mutant | | pH | $T_{trs}$ | $\Delta T$ | $\Delta(\Delta G)$ | Appr./Rem. | | Ref |
|--------|-----------|-----|-----------|------------|---------------------|------|-------|--------|
| wild type | trans. (1) | 7.8 | 82.8 | 0.0 | 0.0 | DSC | (1,2) | 90M6 |
|           | trans. (2) | 7.8 | 89.5 | 0.0 | 0.0 | DSC | (1,2) | 90M6 |
| Cys6→Ala | trans. (1) | 7.8 | 80.7 | –2.1 | –3.1 | DSC | (1,2) | 90M6 |
|          | trans. (2) | 7.8 | 85.8 | –3.7 | –5.4 | DSC | (1,2) | 90M6 |

Remarks:
(1) measured in 100 mM potassium phosphate buffer, pH 7.8
(2) $\Delta(\Delta G)$ refers to $T_{trs}$ of the corresponding transition of the wild-type protein

Cu,Zn superoxide dismutase from *Photobacterium leiognathi* (PISOD), holo- and Cu-free enzyme

| Protein/Transition | | pH | T | $\Delta G$ | m | Appr./Rem. | Ref |
|--------------------|-----------|-----|----|-----------|----------|-------------|--------|
| holo-PISOD | trans. (1) | 7.4 | 4 | 49.8±2.1 | 12.1±0.8 | GuHCl (1–3) | 99M5 |
|            | trans. (2) | 7.4 | 4 | 22.6±2.1 | 5.0±0.4 | GuHCl (1–3) | 99M5 |
| Cu-free PISOD | trans. (1) | 7.4 | 4 | 45.6±2.1 | 11.7±0.4 | GuHCl (1–3) | 99M5 |
|               | trans. (2) | 7.4 | 4 | 15.5±2.1 | 6.7±0.4 | GuHCl (1–3) | 99M5 |

Remarks:
(1) data were analyzed by the three-state model $N_2 \rightarrow 2X \rightarrow 2U$ with native dimer and monomeric intermediate assuming a linear dependence of $\Delta G$ on the denaturant conc.
(2) transition monitored by steady state fluorescence anisotropy and CD
(3) measured in 10 mM Tris-HCl, pH 7.4 at 4°C

Human manganese superoxide dismutase, mutants at Tyr34 and Glu143

| Protein | Component | pH | $T_{trs}$ | $\Delta(\Delta G)$ | Appr./Rem. | Ref |
|---|---|---|---|---|---|---|
| native | B | 7.8 | 70 | 0.0 | DSC (1–3) | 98G19 |
| | C | 7.8 | 88.9 | 0.0 | DSC (1–4) | 98G19 |
| Tyr34→Phe | B | 7.8 | 85.9 | 64.9 | DSC (1–5) | 98G19 |
| Tyr34→Phe | C | 7.8 | 95.6 | 14.6 | DSC (1–5) | 98G19 |
| Gln143→Asn | A | 7.8 | 57.4 | | DSC (1–4) | 98H10 |
| Gln143→Asn | B | 7.8 | 85.3 | 14.0 | DSC (1–4) | 98H10 |
| Gln143→Asn | C | 7.8 | 90.7 | 3.3 | DSC (1–4) | 98H10 |

Remarks:

(1) $\Delta(\Delta G)$ is given per mole of tetramer

(2) $\Delta(\Delta G)$ was calculated at the appropriate $T_{trs}$ of the native protein, assuming constant $\Delta H$ ($\Delta Cp = 0$), for further details see Refs. 98G19 and 98H10

(2) measured in 2 mM potassium phosphate buffer, pH 7.8

(3) for labeling of the transitions see also Ref. 96B9

(4) component C is the main transition

## Tailspike Protein

P22 tailspike protein, fragment: the isolated β-helix domain (residues 109–544)

| Protein | pH | T | $\Delta G$ | m | Approach/Remarks | | Ref |
|---|---|---|---|---|---|---|---|
| fragment | 7.0 | 10 | 32 | 12.7 | urea | (1,2) | 98M19 |

Remarks:

(1) linear extrapolation, LEM-SB

(2) transition monitored by fluorescence at 345 nm

P22 tailspike protein, stability of the isolated β-helix domain, wild type and mutants

| Mutant | pH | T | $\Delta G$ | $c_{1/2}$ | m | Approach/Remarks | | Ref |
|---|---|---|---|---|---|---|---|---|
| wild type | 7.0 | 10 | 32 | 2.52±0.06 | 12.6 | urea | (1–4) | 98S3 |
| Asp238→Ser | 7.0 | 10 | 28 | 2.20±0.02 | | urea | (1–4) | 98S3 |
| Gly244→Arg | 7.0 | 10 | 22 | 1.76±0.06 | | urea | (1–4) | 98S3 |
| Val331→Ala | 7.0 | 10 | 35 | 2.80±0.10 | | urea | (1–4) | 98S3 |
| Val331→Gly | 7.0 | 10 | 38 | 3.00±0.10 | | urea | (1–4) | 98S3 |
| Ala334→Ile | 7.0 | 10 | 37 | 2.91±0.07 | | urea | (1–4) | 98S3 |
| Ala334→Val | 7.0 | 10 | 39 | 3.09±0.04 | | urea | (1–4) | 98S3 |

Remarks:

(1) linear extrapolation, LEM-SB

(2) transition monitored by fluorescence at 338 nm

(3) buffer: 50 mM sodium phosphate

(4) m = 12.6 kJ/mol/M is an averaged value that was used for all mutant proteins

**Telluromethionine Containing Proteins**, see Annexin V

## Tendamistat

Tendamistat, $\Delta G$ from equilibrium equilibrium unfolding and kinetics

| pH | T | $\Delta G$ | m | Approach/Remarks | | Ref |
|---|---|---|---|---|---|---|
| 7.0 | 25 | 34.0±0.7 | 5.31±0.06 | GuHCl | (1,2) | 97S5 |
| 7.0 | 25 | 34.6±2.0 | 5.36±0.21 | GuHCl | (2,3) | 97S5 |

Remarks:
(1) equilibrium unfolding, linear extrapolation, LEM-SB
(2) buffer: 100 mM cacodylate, pH 7.0
(3) from kinetics

## Thermolysin

Thermolysin from *Bacillus thermoproteolyticus*, C-terminal fragment 255–316, and semisynthetic analogs obtained by replacement of alanines by $\alpha$-aminoisobutyric acid (Aib)

| Mutant | pH | $T_{trs}$ | $\Delta T$ | $\Delta(\Delta G)$ | Approach/Remarks | | Ref |
|---|---|---|---|---|---|---|---|
| wild type | 7.5 | 63.5 | 0.0 | 0.0 | heat | (1–4) | 98D10 |
| Ala304→Aib | 7.5 | 65.7 | 2.2 | 2.97 | heat | (1–5) | 98D10 |
| Ala309→Aib | 7.5 | 68.9 | 5.4 | 7.36 | heat | (1–5) | 98D10 |
| Ala312→Aib | 7.5 | 62.9 | −0.6 | −0.84 | heat | (1–5) | 98D10 |
| double mutant (Ala304→Aib and Ala309→Aib) | | | | | | | |
| | 7.5 | 71.5 | 8.0 | 10.17 | heat | (1–5) | 98D10 |

Remarks:
(1) wild type is the natural fragment 255–316
(2) data treatment by a two-state model assuming that the native dimer ($N_2$) unfolds to denatured monomers (U), see also Ref. 94C9
(3) transition monitored by CD at 222 nm
(4) buffer: 20 mM sodium phosphate, pH 7.5, 0.1 M NaCl
(5) $\Delta Cp$ used for the Aib analogs was $\Delta Cp = 2.5±0.4$ kJ/mol/K obtained from DSC studies of the natural fragment, see Ref. 94C9

**Tet Repressor**, see Repressor Proteins

## Thioredoxin

*Bacillus acidocaldarius* thioredoxin (BacTrx), wild type and mutants

| Mutant | pH | $T_{trs}$ | $\Delta T$ | $\Delta(\Delta G)$ | Appr./Rem. | | Ref |
|---|---|---|---|---|---|---|---|
| BacTrx | 7.0 | 102.6 | 0.0 | 0.00 | heat | (1,2) | 99P5 |
| Lys18→Gly | 7.0 | 90.6 | −12 | −9.42 | heat | (1,2) | 99P5 |
| Arg82→Glu | 7.0 | 90.6 | −12 | −9.57 | heat | (1,2) | 99P5 |
| double mutant (Lys18→Gly and Arg82→Glu) | | | | | | | |
| | 7.0 | 88.7 | −13.9 | −10.9 | heat | (1,2) | 99P5 |
| Asp102→X | 7.0 | 86.1 | −16.5 | −12.9 | heat | (1–3) | 99P5 |

Remarks:
(1) $\Delta(\Delta G)$ was calculated using $\Delta(\Delta G) = \Delta T_{mut} \times \Delta S_{mut}$
(2) see also Table 2
(3) mutant Asp102→X, the last four amino acids were deleted

*Escherichia coli* thioredoxin, wild type and mutants

| Protein | pH | T | ΔG | $c_{1/2}$ | m | Appr./Rem. | Ref |
|---------|-----|----|----------|-----------|-----------|------------|------|
| wild type* | 7.0 | 5 | 36.0±3.3 | 2.47 | 14.6±1.3 | GuHCl (1–3) | 99C4 |
| | 7.0 | 10 | 38.5±2.5 | 2.71 | 14.2±0.8 | GuHCl (1–3) | 99C4 |
| | 7.0 | 15 | 30.5±4.6 | 2.76 | 10.9±1.7 | GuHCl (1–3) | 99C4 |
| | 7.0 | 20 | 43.5±3.3 | 2.85 | 15.1±1.3 | GuHCl (1–3) | 99C4 |
| | 7.0 | 25 | 48.1±4.6 | 2.87 | 16.7±1.7 | GuHCl (1–3) | 99C4 |
| | 7.0 | 30 | 42.3±2.9 | 2.84 | 15.1±0.8 | GuHCl (1–3) | 99C4 |
| | 7.0 | 37 | 33.9±3.3 | 2.73 | 12.6±1.3 | GuHCl (1–3) | 99C4 |
| | 7.0 | 40 | 39.3±5.4 | 2.71 | 14.2±2.1 | GuHCl (1–3) | 99C4 |
| | 7.0 | 45 | 36.0±7.1 | 2.53 | 14.2±2.9 | GuHCl (1–3) | 99C4 |
| | 7.0 | 50 | 20.9±6.3 | 2.41 | 8.8±2.5 | GuHCl (1–3) | 99C4 |
| | 7.0 | 55 | 27.6±9.6 | 2.10 | 13.0±4.2 | GuHCl (1–3) | 99C4 |
| | 7.0 | 60 | 14.2±3.3 | 2.10 | 6.7±1.3 | GuHCl (1–3) | 99C4 |
| | 7.0 | 65 | 9.2±4.2 | 1.44 | 6.3±1.7 | GuHCl (1–3) | 99C4 |
| | 7.0 | 25 | 38.9±1.3 | | | DSC (4) | 99C4 |
| double m. | 7.0 | 4 | 33.5±3.8 | 2.09 | 15.9±1.7 | GuHCl (1–3) | 99C4 |
| | 7.0 | 6 | 33.5±5.4 | 2.11 | 16.3±2.5 | GuHCl (1–3) | 99C4 |
| | 7.0 | 8 | 25.9±4.6 | 2.15 | 12.1±2.1 | GuHCl (1–3) | 99C4 |
| | 7.0 | 10 | 24.7±2.1 | 2.47 | 10.0±0.8 | GuHCl (1–3) | 99C4 |
| | 7.0 | 15 | 33.5±3.8 | 2.34 | 14.2±1.7 | GuHCl (1–3) | 99C4 |
| | 7.0 | 25 | 38.1±5.4 | 2.58 | 15.5±2.9 | GuHCl (1–3) | 99C4 |
| | 7.0 | 40 | 25.5±5.0 | 2.74 | 9.2±1.7 | GuHCl (1–3) | 99C4 |
| | 7.0 | 45 | 31.8±4.6 | 2.45 | 13.0±2.1 | GuHCl (1–3) | 99C4 |
| | 7.0 | 55 | 22.2±3.3 | 2.29 | 8.8±1.7 | GuHCl (1–3) | 99C4 |
| | 7.0 | 60 | 12.1±3.3 | 1.40 | 8.4±2.1 | GuHCl (1–3) | 99C4 |
| | 7.0 | 25 | 32.6±0.8 | | | DSC (4) | 99C4 |

Explanations:

> wild type* – pseudo-wild type, mutant Met37→Leu
> double m.  – double mutant Met37→Leu and Pro40→Ser

Remarks:
(1) linear extrapolation
(2) transition monitored by fluorescence emission at 350 nm
(3) buffer: 50 mM phosphate, pH 7.0
(4) ΔG derived from joint analysis of GuHCl and DSC studies (see also Table 2)

Thioredoxin from *E. coli*, wild type and Pro40→Ala mutant

| Mutant | pH | T | ΔG | $c_{1/2}$ | m | Appr./Rem. | Ref |
|--------|-----|----|------|-----------|------|------------|------|
| wild type | 7.2 | 25 | 37.2 | 2.4 | 15.5 | GuHCl (1,2) | 97L2 |
| Pro40→Ala | 7.2 | 25 | 25.1 | 2.8 | 9.2 | GuHCl (1,2) | 97L2 |

Remarks:
(1) linear extrapolation
(2) measured by fluorescence at 340 nm

Stability of intermediates formed during refolding of denatured oxidized *E. coli* thioredoxin

| kinetic phase | pH | T | ΔG | $c_{1/2}$ | Approach/Remarks | Ref |
|---------------|-----|----|----------|-----------|------------------|------|
| burst phase | 7.0 | 20 | 14.6±7.5 | 0.6±0.3 | kinetics (1,2,5) | 98G5 |
| burst phase | 7.0 | 6 | 5.4±2.5 | | kinetics (1,2,5) | 98G5 |
| very rapid phase | 7.0 | 20 | 19.2±3.3 | 1.3±0.2 | kinetics (1,3,5) | 98G5 |
| very rapid phase | 7.0 | 6 | 16.3±4.6 | 1.0±0.2 | kinetics (1,3,5) | 98G5 |
| rapid phase | 7.0 | 20 | 21.3±4.6 | 1.7±0.2 | kinetics (1,4,5) | 98G5 |

Stability of intermediates formed during refolding of denatured oxidized *E. coli* thioredoxin (continued)

| kinetic phase | pH | T | $\Delta G$ | $c_{1/2}$ | Approach/Remarks | Ref |
|---|---|---|---|---|---|---|
| equilibrium N → D | 7.0 | 20 | 10.9±0.4 | 2.7±0.2 | GuHCl (5,7) | 98G5 |
| burst phase | 7.0 | 20 | 18.0±4.6 | 1.3±0.3 | kinetics (1,2,6) | 98G5 |
| burst phase | 7.0 | 6 | 7.1±4.2 | | kinetics (1,2,6) | 98G5 |
| very rapid phase | 7.0 | 20 | 26.4±5.0 | 1.5±0.3 | kinetics (1,3,6) | 98G5 |
| very rapid phase | 7.0 | 6 | 21.8±3.3 | 0.9±0.3 | kinetics (1,3,6) | 98G5 |
| rapid phase | 7.0 | 20 | 26.4±5.9 | 1.6±0.2 | kinetics (1,4,6) | 98G5 |
| rapid phase | 7.0 | 6 | 9.2±2.1 | 1.0±0.3 | kinetics (1,4,6) | 98G5 |
| equilibrium N → D | 7.0 | 20 | 39.7±4.2 | 9.6±1.0 | GuHCl (6,7) | 98G5 |

Remarks:

(1) kinetics of refolding was monitored by the following independent criteria: far-UV CD, binding of ANS, near-UV CD, and intrinsic fluorescence

(2) burst phase: < 4 ms

(3) very rapid phase: rate constant of about 33±3 $sec^{-1}$ at 0.087 M GuHCl

(4) rapid phase: relaxation time of 1.58±0.06 sec

(5) transition monitored by far-UV CD

(6) transition monitored by fluorescence

(7) linear extrapolation, LEM-SB

Oxidized *E. coli* thioredoxin (Trx), cleaved and uncleaved Trx

| Protein | pH | T | $\Delta G$ | $c_{1/2}$ | m | Appr./Rem. | Ref |
|---|---|---|---|---|---|---|---|
| Trx | 5.7 | 20 | 39.7±0.8 | 2.61±0.12 | 15.5±1.7 | GuHCl (1–3) | 99G5 |
| NC | 5.7 | 20 | 41.8±1.7 | | 16.3±1.3 | GuHCl (1–5) | 99G5 |
| | 5.7 | 20 | 41.0±0.8 | | | Titr. (6) | 99G5 |

Remarks:

(1) two-state unfolding of Trx, unfolding/dissociation of cleaved Trx, for the data analysis see Ref. 99G5

(2) transition monitored by fluorescence and CD

(3) measured in 10 mM potassium phosphate, pH 5.7

(4) NC = cleaved Trx (1–73, 74–108)

(5) from measurements varying protein conc.

(6) $\Delta G$ estimated from titration experiments

Thioredoxin from *E. coli*, construction of $Cys_2His_2$ zinc binding sites

| Mutant/Form | pH | T | $\Delta G$ | m | Approach/Remarks | Ref |
|---|---|---|---|---|---|---|
| 0 background (1) | | | | | | |
| apo | 7.5 | 25 | 49.4±0.8 | 15.9±0.4 | GuHCl (1–4) | 98W9 |
| Zn | 7.5 | 25 | 51.0±0.8 | 16.3±0.4 | GuHCl (1–5) | 98W9 |
| multiple mutant (Leu53→Cys, Leu103→Cys, Leu24→His, and Ile45→His) | | | | | | |
| apo | 7.5 | 25 | | | GuHCl (1,6) | 98W9 |
| Zn | 7.5 | 25 | 21.8±0.8 | 16.7±0.4 | GuHCl (1–5) | 98W9 |
| multiple mutant (Leu24→Cys, Val55→Cys, Leu26→His, and Ile45→His) | | | | | | |
| apo | 7.5 | 25 | | | GuHCl (1,6) | 98W9 |
| Zn | 7.5 | 25 | 15.5±0.8 | 11.7±0.4 | GuHCl (1–5) | 98W9 |
| multiple mutant (Leu26→Cys, Leu42→Cys, Val55→His, and Lys57→His) | | | | | | |
| apo | 7.5 | 25 | 18.4±0.4 | 14.6±0.4 | GuHCl (1–4) | 98W9 |
| Zn | 7.5 | 25 | 26.8±0.4 | 13.0±0.4 | GuHCl (1–5) | 98W9 |
| multiple mutant (Leu26→Cys, Leu42→Cys, Leu24→His, and Lys57→His) | | | | | | |
| apo | 7.5 | 25 | | | GuHCl (1,6) | 98W9 |
| Zn | 7.5 | 25 | | | GuHCl (1,6) | 98W9 |

Thioredoxin from *E. coli*, construction of Cys₂His₂ zinc binding sites (continued)

| Mutant/Form | pH | T | ΔG | m | Approach/Remarks | Ref |
|---|---|---|---|---|---|---|
| multiple mutant (Leu26→Cys, Leu42→Cys, Leu38→His, and Lys57→His) | | | | | | |
| apo | 7.5 | 25 | 25.1±0.4 | 15.1±0.4 | GuHCl  (1–4) | 98W9 |
| Zn | 7.5 | 25 | 18.8±0.4 | 10.5±0.4 | GuHCl  (1–5) | 98W9 |
| multiple mutant (Leu58→Cys, Thr66→Cys, Phe12→His, and Phe27→His) | | | | | | |
| apo | 7.5 | 25 | 17.2±0.8 | 17.6±0.8 | GuHCl  (1–4) | 98W9 |
| Zn | 7.5 | 25 | 27.2±0.4 | 10.5±0.4 | GuHCl  (1–5) | 98W9 |

Remarks:

(1) the thioredoxin 0 background is defined by the following mutations: Asp2→Ala, Asp26→Leu, Cys32→Ser, and Cys35 →Ser, relative to wild-type thioredoxin

(2) linear extrapolation

(3) transition monitored by molar ellipticity at 222 nm

(4) buffer: 20 mM Tris, 200 mM NaCl

(5) in the presence of 11 μM ZnCl₂

(6) the protein is partly unfolded, i.e., no pretransition baseline for the folded form observed, precluding accurate quantitation of ΔG

Thioredoxin from *E. coli* and variants mimicking the active site of other thiol/disulfide oxidoreductases

| Variant/Form | pH | T | ΔG | $c_{1/2}$ | m | Appr./Rem. | Ref |
|---|---|---|---|---|---|---|---|
| wild type oxidized | 7.0 | 25 | 41.4±1.8 | 2.50 | 16.6±0.7 | GuHCl  (1–3) | 98M22 |
| wild type reduced | 7.0 | 25 | 24.5±0.6 | 1.50 | 16.2±0.4 | GuHCl  (1–3) | 98M22 |
| TR-type   oxidized | 7.0 | 25 | 36.4±0.5 | 2.38 | 15.3±0.2 | GuHCl  (1–3) | 98M22 |
| TR-type   reduced | 7.0 | 25 | 32.7±0.9 | 2.06 | 16.0±0.4 | GuHCl  (1–3) | 98M22 |
| PDI-type oxidized | 7.0 | 25 | 29.7±1.1 | 1.85 | 16.1±0.6 | GuHCl  (1–3) | 98M22 |
| PDI-type reduced | 7.0 | 25 | 30.2±1.2 | 1.69 | 17.9±0.7 | GuHCl  (1–3) | 98M22 |
| DsbA-type oxidized | 7.0 | 25 | 30.1±0.7 | 1.94 | 15.6±0.3 | GuHCl  (1–3) | 98M22 |
| DsbA-type reduced | 7.0 | 25 | 30.0±0.9 | 1.96 | 15.3±0.4 | GuHCl  (1–3) | 98M22 |
| Grx-type  oxidized | 7.0 | 25 | 31.7±1.3 | 2.04 | 15.6±0.6 | GuHCl  (1–3) | 98M22 |
| Grx-type  reduced | 7.0 | 25 | 32.3±0.7 | 2.20 | 14.7±0.3 | GuHCl  (1–3) | 98M22 |

Remarks:

(1) the variants concerning the active site disulfide consensus sequence Cys-Xaa-Xaa-Cys:

– wild type = thioredoxin, Cys-Gly-Pro-Cys

– TR = thioredoxin reductase-type from *E. coli*, Cys-Ala-Thr-Cys

– PDI = protein disulfide isomerase-type, Cys-Gly-His-Cys

– DsbA-type, Cys-Pro-His-Cys

– Grx = glutaredoxin-type, Cys-Pro-Tyr-Cys

(2) linear extrapolation, LEM-SB

(3) transition monitored by CD at 220 nm

(4) buffer: 100 mM sodium phosphate, pH 7.0, 100 μM EDTA, and for reduced proteins 3 mM DTT

Thioredoxin from *E. coli*, nonconservative amino acid substitutions

| Mutant | pH | $\Delta T$ | $\Delta(\Delta G°)$ | Approach/Remarks | | Ref |
|---|---|---|---|---|---|---|
| wild type | 7.0 | 19.4 | 17.2 | DSC | (1–4) | 97O1 |
| double mutant (Cys32→Ser and Cys33→Ser) | | | | | | |
| | 7.0 | 9.4 | 9.6 | DSC | (1–4) | 97O1 |
| Leu78→Cys | 7.0 | 0.0 | 0.0 | DSC | (1–4) | 97O1 |
| methyl derivative | 7.0 | 2.9 | 3.3 | DSC | (1–5) | 97O1 |
| ethyl derivative | 7.0 | 3.0 | 3.3 | DSC | (1–5) | 97O1 |
| propyl derivative | 7.0 | 1.9 | 1.7 | DSC | (1–5) | 97O1 |
| butyl derivative | 7.0 | −1.3 | −0.8 | DSC | (1–5) | 97O1 |

Remarks:

(1) buffer: 50 mM sodium phosphate pH 7.0

(2) $\Delta(\Delta G°)$ was calculated using $\Delta Cp = 7.87$ kJ/mol/K from Ref. 93L1

(3) for the primary data see Table 2

(4) reference value at 120 μM as standard conc.

(5) side chain added to the Cys78 residue

## Toxins

Equinatoxin II (EqTxII) from the sea anemone *Actinia equina* L.

| pH | T | $\Delta G$ | Approach | Remarks | Ref |
|---|---|---|---|---|---|
| 2.0 | 20 | 3.4 | DSC | (1,2) | 97P13 |
| 3.0 | 20 | 21.5 | DSC | (1,2) | 97P13 |
| 3.5 | 20 | 30.6 | DSC | (1,2) | 97P13 |
| 5.5–6.0 | 20 | 41.0 | DSC | (1,2) | 97P13 |

Remarks:

(1) using $\Delta Cp = 5.8$ kJ/mol/K

(2) relative error in $\Delta G$ is estimated to be ±10–20%

Cardiotoxin analogue III (CTX III) and cobrotoxin (CBTX) from Taiwan cobra (*Naja naja astra*)

| Protein | pH | T | $\Delta G$ | $c_{1/2}$ | m | Appr./Rem. | Ref |
|---|---|---|---|---|---|---|---|
| CTX III | 7.0 | | 26.4 | 4.5 | 3.68 | GuHCl (1–3) | 99S12 |
| CBTX | 7.0 | | 19.7 | 2.6 | 4.35 | GuHCl (1–3) | 99S12 |

Remarks:

(1) linear extrapolation

(2) transition monitored by far-UV CD

(2) thermal transition at $T_{trs} = 84°C$ for CTX III, and $T_{trs} = 74°C$ for CBTX

## Trigger Factor

Trigger factor from *E. coli* having peptidyl-prolyl *cis/trans* isomerase (PPIase) activity, authentic protein and fragments

| Protein | pH | T | $\Delta G$ | $c_{1/2}$ | m | Appr./Rem. | Ref |
|---------|-----|----|-----------|-----------|---------|------------|------|
| authentic | 7.8 | 20 | 24.2±1.6 | 3.14 | 7.7±0.5 | urea (1–4) | 97Z2 |
|  | 7.8 | 20 | 23.9±1.7 | 3.03 | 7.9±0.5 | urea (1–3,5) | 97Z2 |
| N fragment | 7.8 | 20 | 12.3±1.1 | 2.28 | 5.4±0.4 | urea (1–3,6) | 97Z2 |
|  | 7.8 | 20 | 13.4±1.1 | 2.35 | 5.7±0.4 | urea (1–3,7) | 97Z2 |
| NM fragment | 7.8 | 20 | 14.9±1.2 | 2.57 | 5.8±0.4 | urea (1–4) | 97Z2 |
|  | 7.8 | 20 | 12.1±1.5 | 2.33 | 5.2±0.5 | urea (1–3,7) | 97Z2 |
| M fragment | 7.8 | 20 | 15.1±0.7 | 2.56 | 5.9±0.3 | urea (1–4) | 97Z2 |
|  | 7.8 | 20 | 14.5±0.8 | 2.46 | 5.9±0.3 | urea (1–3,5) | 97Z2 |
| MC fragment | 7.8 | 20 | 15.7±0.5 | 2.80 | 5.6±0.5 | urea (1–3,6) | 97Z2 |

Remarks:
(1) abbreviations:

   authentic = residues 1 to 432, catalytic core at residues 146 to 251
   N  = fragment, residues 1 to 145
   NM = fragment, residues 1 to 251
   M  = fragment, residues 145 to 251 with (His)$_6$ tag
   MC = fragment, residues 145 to 432

(2) linear extrapolation, LEM-SB
(3) buffer: 0.1 M Tris-HCl, pH 7.8
(4) transition monitored by fluorescence at 327 nm
(5) transition monitored by PPIase activity
(6) transition monitored by fluorescence at 306 nm
(7) transition monitored by CD at 222 nm

## Triose Phosphate Isomerase

Triose phosphate isomerase (TIM) from rabbit muscle, equilibrium subunit dissociation/unfolding

| Protein Conc. | pH | T | $\Delta G$ | m | Appr./Rem. | Ref |
|---------------|-----|---------|-----------|----|------------|------|
| 0.6 µM | 7.4 | 25±0.3 | 71 | 63 | GuHCl (1–3) | 98R6 |
| 6.0 µM | 7.4 | 25±0.3 | 69 | 63 | GuHCl (1–3) | 98R6 |

Remarks:
(1) linear extrapolation
(2) transition monitored by fluorescence
(3) buffer: 50 mM phosphate, pH 7.4, 1 mM DTT

Triose phosphate isomerase (TIM) and PGK-TIM fusion protein from the hyperthermophile *Thermotoga maritima*

| Mutant | pH | T | $\Delta G$ | $c_{1/2}$ | Approach | Remarks | Ref |
|--------|-----|----|-----------|-----------|-------------|---------|------|
| TIM | 8.0 | 40 | | 3.4 | activity | (1) | 97B5 |
| TIM | 8.0 | 40 | | 3.6 | fluorescence | (1) | 97B5 |
| PGK-TIM | 8.0 | 40 | | 3.8 | activity | (2,3) | 97B5 |
| PGK-TIM | 8.0 | 40 | | 3.9 | fluorescence | (2,3) | 97B5 |

Remarks:
(1) thermal transition at 82°C in the absence of GuHCl
(2) covalently linked phosphoglycerate kinase with TIM at the C-terminus
(3) thermal transition at 90°C in the absence of GuHCl

## Troponin C

N-terminal domain of chicken troponin C (TC, residues 1–90) and Phe29→Trp mutant

| Protein | pH | T | ΔG | Approach/Remarks | | Ref |
|---|---|---|---|---|---|---|
| TC 1–90 | 7.0 | | 19.2 | heat | (1,2) | 99Y3 |
| Phe29→Trp | 7.0 | | 14.2 | heat | (1,2) | 99Y3 |

Remarks:
(1) transition monitored by NMR chemical shifts versus temperature
(2) buffer: 20 mM Tris-*d11*, 5 mM DTT, 2 mM EGTA, 100 mM KCl, 1.5 mM TSP in D$_2$O, pH* 7.00±0.05

## Trypsin Inhibitor (BPTI)

Bovine pancreatic trypsin inhibitor (BPTI), aromatic to leucine substitutions

| Mutant | pH | T | Δ(ΔG) | Approach/Remarks | Ref |
|---|---|---|---|---|---|
| native form relative to the wild-type protein: | | | | | |
| Phe4→Leu | 8.7 | 25 | −9.2 | kinetics (1,2) | 97Z6 |
| Tyr10→Leu | 8.7 | 25 | −12.1 | kinetics (1,2) | 97Z6 |
| Tyr21→Leu | 8.7 | 25 | −14.2 | kinetics (1,2) | 97Z6 |
| Phe22→Leu | 8.7 | 25 | −23.4 | kinetics (1,2) | 97Z6 |
| Tyr23→Leu | 8.7 | 25 | −29.7 | kinetics (1,2) | 97Z6 |
| Phe33→Leu | 8.7 | 25 | −17.2 | kinetics (1,2) | 97Z6 |
| Tyr35→Leu | 8.7 | 25 | −27.2 | kinetics (1,2) | 97Z6 |
| Phe45→Leu | 8.7 | 25 | −26.4 | kinetics (1,2) | 97Z6 |
| Native-like two-disulfide intermediate containing the 30–51 and 5–55 disulfides: | | | | | |
| Phe4→Leu | 8.7 | 25 | −7.1 | kinetics (1,2) | 97Z6 |
| Tyr10→Leu | 8.7 | 25 | −5.0 | kinetics (1,2) | 97Z6 |
| Tyr21→Leu | 8.7 | 25 | −13.0 | kinetics (1,2) | 97Z6 |
| Phe22→Leu | 8.7 | 25 | −18.8 | kinetics (1,2) | 97Z6 |
| Tyr23→Leu | 8.7 | 25 | −25.5 | kinetics (1,2) | 97Z6 |
| Phe33→Leu | 8.7 | 25 | −8.8 | kinetics (1,2) | 97Z6 |
| Tyr35→Leu | 8.7 | 25 | −9.6 | kinetics (1,2) | 97Z6 |
| Phe45→Leu | 8.7 | 25 | −20.5 | kinetics (1,2) | 97Z6 |
| intermediate II: | | | | | |
| Phe4→Leu | 8.7 | 25 | −0.8 | kinetics (1,2) | 97Z6 |
| Tyr10→Leu | 8.7 | 25 | −0.8 | kinetics (1,2) | 97Z6 |
| Tyr21→Leu | 8.7 | 25 | −1.7 | kinetics (1,2) | 97Z6 |
| Phe22→Leu | 8.7 | 25 | −4.2 | kinetics (1,2) | 97Z6 |
| Tyr23→Leu | 8.7 | 25 | −6.3 | kinetics (1,2) | 97Z6 |
| Phe33→Leu | 8.7 | 25 | −1.3 | kinetics (1,2) | 97Z6 |
| Tyr35→Leu | 8.7 | 25 | −3.8 | kinetics (1,2) | 97Z6 |
| Phe45→Leu | 8.7 | 25 | −2.9 | kinetics (1,2) | 97Z6 |
| one-disulfide intermediate containing the 30–51 disulfide: | | | | | |
| Phe4→Leu | 8.7 | 25 | <−0.8 | kinetics (1,2) | 97Z6 |
| Tyr10→Leu | 8.7 | 25 | <−0.8 | kinetics (1,2) | 97Z6 |
| Tyr21→Leu | 8.7 | 25 | −6.3 | kinetics (1,2) | 97Z6 |
| Phe22→Leu | 8.7 | 25 | −3.3 | kinetics (1,2) | 97Z6 |
| Tyr23→Leu | 8.7 | 25 | >−6.3 | kinetics (1,2) | 97Z6 |
| Phe33→Leu | 8.7 | 25 | −3.3 | kinetics (1,2) | 97Z6 |
| Tyr35→Leu | 8.7 | 25 | <−0.8 | kinetics (1,2) | 97Z6 |
| Phe45→Leu | 8.7 | 25 | −6.3 | kinetics (1,2) | 97Z6 |

Bovine pancreatic trypsin inhibitor (BPTI), aromatic to leucine substitutions (continued)

| Mutant | pH | T | $\Delta(\Delta G)$ | Approach/Remarks | Ref |
|---|---|---|---|---|---|
| intermediate I: | | | | | |
| Phe4→Leu | 8.7 | 25 | −1.7 | kinetics (1,2) | 97Z6 |
| Tyr10→Leu | 8.7 | 25 | −0.8 | kinetics (1,2) | 97Z6 |
| Tyr21→Leu | 8.7 | 25 | −2.1 | kinetics (1,2) | 97Z6 |
| Phe22→Leu | 8.7 | 25 | −2.5 | kinetics (1,2) | 97Z6 |
| Tyr23→Leu | 8.7 | 25 | −4.2 | kinetics (1,2) | 97Z6 |
| Phe33→Leu | 8.7 | 25 | −0.8 | kinetics (1,2) | 97Z6 |
| Tyr35→Leu | 8.7 | 25 | −0.8 | kinetics (1,2) | 97Z6 |
| Phe45→Leu | 8.7 | 25 | −1.3 | kinetics (1,2) | 97Z6 |

Remarks:
(1) from refolding and unfolding kinetics of various selectively reduced forms
(2) buffer: 0.1 M Tris, pH 8.7, containing 0.2 M KCl and 1 mM EDTA

Bovine pancreatic trypsin inhibitor (BPTI), 30–51 cysteine substitutions

| Mutant | pH | T | $\Delta G$ | $c_{1/2}$ | m | Appr. | Remarks | Ref |
|---|---|---|---|---|---|---|---|---|
| double mutant (Cys30→Ala and Cys51→Ala) | | | | | | | | |
| | 7.0 | 25 | 23.0±1.3 | 4.25 | 5.52±0.13 | GuHCl | (1–4) | 97L11 |
| | 7.0 | 25 | 22.2 | | | heat | (2,4,6) | 97L11 |
| | 7.0 | 25 | 24.3±1.3 | | | DSC | (2,4,7) | 97L11 |
| double mutant (Cys30→Val and Cys51→Ala) | | | | | | | | |
| | 7.0 | 25 | 27.6±0.8 | 4.72 | 5.98±0.21 | GuHCl | (1–3,5) | 97L11 |
| | 7.0 | 25 | 23.8 | | | heat | (2,4,5) | 97L11 |
| | 7.0 | 25 | 25.5±1.3 | | | DSC | (2,4,7) | 97L11 |
| double mutant (Cys30→Gly and Cys51→Ala) | | | | | | | | |
| | 7.0 | 25 | 13.4±0.8 | 2.20 | 5.77±0.25 | GuHCl | (1–4) | 97L11 |
| | 7.0 | 25 | 16.3 | | | heat | (2,4,6) | 97L11 |
| | 7.0 | 25 | 19.2±0.4 | | | DSC | (2,4,7) | 97L11 |
| double mutant (Cys30→Thr and Cys51→Ala) | | | | | | | | |
| | 7.0 | 25 | 18.8±1.3 | 3.25 | 5.94±0.33 | GuHCl | (1–3,5) | 97L11 |
| | 7.0 | 25 | 18.4 | | | heat | (2,4,5) | 97L11 |
| | 7.0 | 25 | 21.3±0.4 | | | DSC | (2,4,5) | 97L11 |
| double mutant (Cys30→Ser and Cys51→Ala) | | | | | | | | |
| | 7.0 | 25 | 18.4±0.8 | 3.12 | 6.15±0.21 | GuHCl | (1–4) | 97L11 |
| | 7.0 | 25 | 17.2 | | | heat | (2,4,6) | 97L11 |
| | 7.0 | 25 | 20.1±0.8 | | | DSC | (2,4,7) | 97L11 |
| double mutant (Cys30→Ala and Cys51→Ser) | | | | | | | | |
| | 7.0 | 25 | 14.2±0.8 | 2.23 | 5.73±0.21 | GuHCl | (1–4) | 97L11 |
| | 7.0 | 25 | 12.6 | | | heat | (2,4,6) | 97L11 |
| | 7.0 | 25 | 17.2±0.4 | | | DSC | (2,4,7) | 97L11 |
| double mutant (Cys30→Ser and Cys51→Ser) | | | | | | | | |
| | 7.0 | 25 | 11.3±0.8 | 1.75 | 5.94±0.29 | GuHCl | (1–3,5) | 97L11 |
| | 7.0 | 25 | 10.9 | | | heat | (2,4,5) | 97L11 |
| double mutant (Cys30→Gly and Cys51→Met) | | | | | | | | |
| | 7.0 | 25 | 8.8±0.8 | 1.20 | 6.02±0.21 | GuHCl | (1–3,5) | 97L11 |
| | 7.0 | 25 | 7.1 | | | heat | (2,4,5) | 97L11 |

Remarks:
(1) GuHCl-induced equilibrium unfolding monitored by CD at 222 nm
(2) buffer: 10 mM potassium phosphate, 0.1 M NaCl, 0.2 mM EDTA, pH 7.0
(3) linear extrapolation, LEM-SB
(4) data from multiple measurements
(5) data from single experiments
(6) data from thermally induced unfolding monitored by CD at 222 and 205 nm, see also Table 2
(7) data from DSC, see also Table 2

Bovine pancreatic trypsin inhibitor (BPTI), truncated and modified protein

| Protein | pH | T | $\Delta(\Delta G)$ | Approach/Remarks | Ref |
|---|---|---|---|---|---|
| 3–58BPTI | 1 | 45 | –10.5 | HX, NMR (1,2) | 98H5 |
| | 6.5 | 58 | –13.8 | HX, NMR (1,2) | 98H5 |
| 3–58BPTI, mutant (Arg42→Ser) | | | | | |
| | 1 | 45 | –15.5 | HX, NMR (1,2) | 98H5 |
| | 6.5 | 58 | –18.0 | HX, NMR (1,2) | 98H5 |
| 3–58BPTI, double mutant (Arg17→Ala and Arg42→Ser) | | | | | |
| | 1 | 45 | –20.1 | HX, NMR (1,2) | 98H5 |
| 3–58BPTI, triple mutant (Lys15→Arg, Arg17→Ala and Arg42→Ser) | | | | | |
| | 1 | 45 | –20.5 | HX, NMR (1,2) | 98H5 |

Remarks:

(1) the truncated protein 3–58BPTI is des(Arg1, Pro2)BPTI

(2) $\Delta(\Delta G)$ is based on the ratio of exchange rates of Tyr23 NH of wild type and mutant protein

(3) the thermal transition temperature at pH 5.5 monitored by DSC is $T_{trs}$ >100, 84, 79, 75, and 71°C for wild-type BPTI, 3–58BPTI, single mutant, double mutant, and triple mutant, respectively

## Trypsin Inhibitor, Hen Egg White

Trypsin inhibitor from hen egg white, in aqueous polyol solution

| Cosolvent | Conc. | pH | $T_{trs}$ | $\Delta T$ | $\Delta(\Delta G)$ | Appr./Rem. | Ref |
|---|---|---|---|---|---|---|---|
| control | | 7.0 | 59.0 | 0.0 | 0.00 | heat (1,2) | 98K3 |
| mannitol | 1.00 M | 7.0 | 64.8 | 5.8 | 4.52 | heat (1,2) | 98K3 |
| inositol | 0.75 M | 7.0 | 64.5 | 5.5 | 4.44 | heat (1,2) | 98K3 |
| xylitol | 2.00 M | 7.0 | 67.3 | 8.3 | 6.40 | heat (1,2) | 98K3 |
| adonitol | 2.00 M | 7.0 | 67.3 | 8.3 | 6.44 | heat (1,2) | 98K3 |
| sorbitol | 2.00 M | 7.0 | 71.8 | 12.8 | 9.29 | heat (1,2) | 98K3 |

Remarks:

(1) transition monitored by optical absorption at 287 nm

(2) buffer: 20 mM phosphate or MOPS, pH 7.0, 1.5 M GuHCl

## Tryptophan Repressor, see Repressor Proteins

## Tryptophan Synthase

α-subunit of tryptophan synthase from *Escherichia coli*, transitions

| Transition | pH | T | $\Delta G$ | m | Approach/Remarks | Ref |
|---|---|---|---|---|---|---|
| N → I1 | 7.80 | 25 | 25.1±0.4 | 9.2±0.3 | urea, FLI (1–3) | 99G19 |
| | 7.80 | 25 | 23.4±0.4 | 8.8±0.2 | urea, FLA (1,2,4) | 99G19 |
| | 7.80 | 25 | 27.6±0.4 | 10.0±0.2 | urea, CD (1,5,7) | 99G19 |
| | 7.80 | 25 | 25.1±0.4 | 9.2±0.2 | urea, GLF (1,6) | 99G19 |
| | 7.80 | 25 | 23.8±1.7 | 8.4±0.4 | urea, SEC (1,2,8) | 99G20 |
| | 7.80 | 25 | 22.2±4.2 | 8.4±1.7 | urea, SAXS(1,9,10) | 99G20 |
| I1 → I2 | 7.80 | 25 | 18.8±1.3 | 4.6±0.3 | urea, FLI (1–3) | 99G19 |
| | 7.80 | 25 | 20.1±0.4 | 4.6±0.2 | urea, FLA (1,2,4) | 99G19 |
| | 7.80 | 25 | 19.2±1.7 | 4.6±0.4 | urea, CD (1,5,7) | 99G19 |
| | 7.80 | 25 | 18.8±0.8 | 4.6±0.2 | urea, GLF (1,6) | 99G19 |
| | 7.80 | 25 | 20.5±2.1 | 5.0±0.8 | urea, SEC (1,2,8) | 99G20 |
| | 7.80 | 25 | 13.0±3.8 | 3.8±0.8 | urea, SAXS(1,9,10) | 99G20 |

α-subunit of tryptophan synthase from *Escherichia coli*, transitions (continued)

| Transition | pH | T | ΔG | m | Approach/Remarks | Ref |
|---|---|---|---|---|---|---|
| I2 → U | 7.80 | 25 | 20.5±2.1 | 3.47±0.29 | urea, FLI (1–3) | 99G19 |
| | 7.80 | 25 | 18.4±2.1 | 2.76±0.38 | urea, FLA (1,2,4) | 99G19 |
| | 7.80 | 25 | 18.4±4.2 | 2.89±2.09 | urea, CD (1,5,7) | 99G19 |
| | 7.80 | 25 | 20.5±1.7 | 3.51±0.25 | urea, GLF (1,6) | 99G19 |
| | 7.80 | 25 | 20.9 | 3.6 | urea, SAXS(1,9,10) | 99G20 |

Remarks:

(1)  data fit to the four-state equilibrium unfolding model N → I1 → I2 → U assuming a linear dependence of ΔG on the denaturant conc., for further details (e.g., singular value decomposition) see Ref. 99G19

(2)  buffer: 100 mM potassium phosphate, 0.2 mM K$_2$EDTA, 1 mM DTE, pH 7.80

(3)  transition monitored by fluorescence intensity (FLI)

(4)  transition monitored by fluorescence anisotropy at 308 nm (FLA)

(5)  transition monitored by CD spectra (CD)

(6)  data from global fit (GLF)

(7)  buffer: 100 mM potassium phosphate, 0.2 mM K$_2$EDTA, 1 mM 2-mercaptoethanol, pH 7.80

(8)  transition monitored by size exclusion chromatography (SEC)

(9)  buffer: 10 mM potassium phosphate, 0.2 mM K$_2$EDTA, 1 mM DTE, pH 7.80

(10) transition monitored by small-angle X-ray scattering (SAXS)

α-subunit of tryptophan synthase from *Escherichia coli*, multiple intermediate forms

| Transition | pH | T | ΔG | $c_{1/2}$ | m | Appr./Rem. | Ref |
|---|---|---|---|---|---|---|---|
| N → I1 | 7.8 | 25 | 22.2±0.8 | 2.7 | 8.4±0.4 | urea (1–3) | 99B13 |
| I → I2/U | 7.8 | 25 | 20.1±0.8 | 4.2 | 5.0±0.1 | urea (1–3) | 99B13 |
| remark (4) | 7.8 | 25 | 24.7±0.8 | | 9.3±0.3 | urea (3,4) | 99B13 |
| N → I | 7.8 | 25 | 33.1±8.4 | | 12.1±2.1 | urea (3,5) | 99B13 |
| I → U | 7.8 | 25 | 22.6±4.2 | | 5.4±0.8 | urea (3,5) | 99B13 |

Remarks:

(1) linear extrapolation, for further details see Ref. 99B13

(2) data from equilibrium unfolding, transition monitored by CD at 222 nm

(3) buffer: 10 mM potassium phosphate, pH 7.8, 0.2 mM EDTA, 1 mM 2-mercaptoethanol

(4) transition observed from unfolding varying initial urea conc., using unfolding amplitudes of both kinetic phases, data treatment by a two-state model

(5) transitions observed from refolding to 0.6 M urea from varying initial urea concentrations, using amplitudes of the 10 sec, 40 sec, and 300 sec phases

α-subunit of tryptophan synthase from *Escherichia coli*, proline to alanine mutants, comparison of equilibrium unfolding and kinetics

| Mutant/Transition | | pH | T | ΔG | Approach/Remarks | | Ref |
|---|---|---|---|---|---|---|---|
| wild type | N → U | 7 | 25 | 34.3±0.1 | GuHCl | (1–3) | 97O2 |
| | N → I | 7 | 25 | 17.7±0.8 | GuHCl | (1–3) | 97O2 |
| | N → I | 7 | 25 | 15.8 | GuHCl | (3,4) | 97O2 |
| | I → U | 7 | 25 | 16.7±0.6 | GuHCl | (1–3) | 97O2 |
| Pro28→Ala | N → U | 7 | 25 | 26.3±0.1 | GuHCl | (1–3) | 97O2 |
| | N → I | 7 | 25 | 9.5±0.5 | GuHCl | (1–3) | 97O2 |
| | N → I | 7 | 25 | 9.3 | GuHCl | (3,4) | 97O2 |
| | I → U | 7 | 25 | 16.9±0.3 | GuHCl | (1–3) | 97O2 |
| Pro57→Ala | N → U | 7 | 25 | 32.3±0.1 | GuHCl | (1–3) | 97O2 |
| | N → I | 7 | 25 | 16.7±0.2 | GuHCl | (1–3) | 97O2 |
| | N → I | 7 | 25 | 15.9 | GuHCl | (3,4) | 97O2 |
| | I → U | 7 | 25 | 15.7±0.1 | GuHCl | (1–3) | 97O2 |

α-subunit of tryptophan synthase from *Escherichia coli*, proline to alanine mutants, comparison of equilibrium unfolding and kinetics (continued)

| Mutant/Transition | | pH | T | ΔG | Approach/Remarks | | Ref |
|---|---|---|---|---|---|---|---|
| Pro62→Ala | N → U | 7 | 25 | 33.0±0.3 | GuHCl | (1–3) | 97O2 |
| | N → I | 7 | 25 | 14.8±0.7 | GuHCl | (1–3) | 97O2 |
| | N → I | 7 | 25 | 14.1 | GuHCl | (3,4) | 97O2 |
| | I → U | 7 | 25 | 18.2±0.4 | GuHCl | (1–3) | 97O2 |
| Pro96→Ala | N → U | 7 | 25 | 25.4±0.6 | GuHCl | (1–3) | 97O2 |
| | N → I | 7 | 25 | 11.0±0.2 | GuHCl | (1–3) | 97O2 |
| | N → I | 7 | 25 | 11.0 | GuHCl | (3,4) | 97O2 |
| | I → U | 7 | 25 | 14.4±0.4 | GuHCl | (1–3) | 97O2 |
| Pro132→Ala | N → U | 7 | 25 | 29.9±0.2 | GuHCl | (1–3) | 97O2 |
| | N → I | 7 | 25 | 13.9±0.3 | GuHCl | (1–3) | 97O2 |
| | N → I | 7 | 25 | 13.2 | GuHCl | (3,4) | 97O2 |
| | I → U | 7 | 25 | 16.1±0.3 | GuHCl | (1–3) | 97O2 |
| Pro207→Ala | N → U | 7 | 25 | 29.4 | GuHCl | (1,5) | 97O2 |
| | N → I | 7 | 25 | 10.2 | GuHCl | (1,5) | 97O2 |
| | N → I | 7 | 25 | 10.3 | GuHCl | (3,4) | 97O2 |
| | I → U | 7 | 25 | 19.3 | GuHCl | (1,5) | 97O2 |

Remarks:
(1) equilibrium unfolding, three-state treatment
(2) transition monitored by CD at 222 nm
(3) buffer: 20 mM Tris-HCl, 0.1 mM EDTA, 0.1 mM DTT, pH 7
(4) data from kinetics, kinetic phases detected by CD at 222 nm and fluorescence intensity above 300 nm
(5) data from Ref. 91Y2

α-subunit of tryptophan synthase from *Escherichia coli*, folding unit comprising residues 1–188 (αTS[1–188])

| Protein Conc. | pH | T | ΔG | $c_{1/2}$ | m | Appr./Rem. | Ref |
|---|---|---|---|---|---|---|---|
| 13.9 µM | 7.8 | 25 | 17.41±1.84 | 3.73 | 4.69±0.46 | urea (1–3) | 99Z7 |
| 2.5 µM | 7.8 | 25 | 17.07±4.14 | 3.67 | 4.64±1.05 | urea (1,2,4) | 99Z7 |
| 13.9 µM | 7.8 | 25 | 16.61±1.17 | 3.72 | 4.48±0.29 | urea (1,2,4) | 99Z7 |
| 20.0 µM | 7.8 | 25 | 16.57±1.55 | 3.65 | 4.56±0.42 | urea (1,2,4) | 99Z7 |
| 5.0 µM | 7.8 | 25 | 18.03±3.72 | 3.94 | 4.56±0.46 | urea (1,2,5) | 99Z7 |
| 5.0 µM | 7.8 | 25 | 17.32±1.88 | 3.81 | 4.56±0.46 | urea (1,2,6) | 99Z7 |
| | 7.8 | 25 | 16.65±0.79 | 3.73 | 4.48±0.21 | urea (1,2,7) | 99Z7 |

Remarks:
(1) linear extrapolation
(2) buffer: 10 mM potassium phosphate, pH 7.8, 0.2 mM $K_2$EDTA, 1 mM 2-mercaptoethanol
(3) transition monitored by optical absorption at 287 nm
(4) transition monitored by CD at 222 nm
(5) transition monitored by fluorescence intensity at 305 nm
(6) transition monitored by fluorescence center-of-mass wavelength
(7) simultaneous fit of three data sets from absorption, CD, and fluorescence

α-subunit of tryptophan synthase from *Escherichia coli* (αTS), N-terminal fragments

| Fragment | pH | T | ΔG | $c_{1/2}$ | m | Appr./Rem. | Ref |
|---|---|---|---|---|---|---|---|
| data referring to the N → U transition [remark (1)]: | | | | | | | |
| αTS full length | | | | | | | |
| N → I1 | 7.8 | 25 | 22.05±0.33 | 2.62 | 8.41±0.17 | urea (1–3) | 99Z6 |
| I → I2 | 7.8 | 25 | 20.08±0.75 | 3.93 | 5.10±0.17 | urea (1–3) | 99Z6 |
| αTS(1–214) | 7.8 | 25 | 15.52±3.26 | 3.71 | 4.18±0.84 | urea (1–3) | 99Z6 |
| αTS(1–204) | 7.8 | 25 | 15.06±2.13 | 3.72 | 4.06±0.54 | urea (1–3) | 99Z6 |
| αTS(1–188) | 7.8 | 25 | 16.95±0.92 | 3.73 | 4.56±0.25 | urea (1–3) | 99Z6 |
| αTS(1–171) | 7.8 | 25 | 10.38±1.72 | 2.87 | 3.64±0.46 | urea (1–3) | 99Z6 |
| αTS(1–147) | 7.8 | 25 | 10.08±0.92 | 2.80 | 3.60±0.46 | urea (1–3) | 99Z6 |
| αTS(1–133) | 7.8 | 25 | 9.67±2.22 | 2.70 | 3.56±0.63 | urea (1–3) | 99Z6 |
| data referring to the I2 → U transition [remark (4)]: | | | | | | | |
| αTS full length | | | | | | | |
| | 7.8 | 25 | 32.2±3.8 | 5.9 | 5.4±0.4 | urea (3–5) | 99Z6 |
| αTS(1–214) | 7.8 | 25 | 18.0±4.2 | 6.1 | 2.9±0.8 | urea (3–5) | 99Z6 |
| αTS(1–204) | 7.8 | 25 | 20.1±1.7 | 5.2 | 3.8±0.4 | urea (3–5) | 99Z6 |
| αTS(1–188) | 7.8 | 25 | 17.6±2.1 | 5.1 | 3.3±0.8 | urea (3–5) | 99Z6 |
| αTS(1–171) | 7.8 | 25 | 17.6±1.3 | 5.4 | 3.3±0.4 | urea (3–5) | 99Z6 |
| αTS(1–157) | 7.8 | 25 | 14.6±2.1 | 5.5 | 2.5±0.4 | urea (3–5) | 99Z6 |
| αTS(1–147) | 7.8 | 25 | 8.8±0.8 | ~4.5 | 2.1±0.4 | urea (3–5) | 99Z6 |

Remarks:

(1) linear extrapolation, the data were fit to the two-state model N → U, with the exception of the full-length data (see Ref. 99B13), which were fit to the three-state model N → I1 → I2. The I2 → U transition is not apparent by CD under the conditions of these experiments

(2) data from equilibrium unfolding, transition monitored by CD at 222 nm

(3) buffer: 10 mM potassium phosphate, pH 7.8, 0.2 mM $K_2EDTA$, 1 mM 2-mercaptoethanol

(4) transition monitored by $^1H$-NMR, the data were fit to the two-state model I2 → U

(5) buffer: 58 mM potassium phosphate, pH 7.8, 0.2 mM $K_2EDTA$, 1 mM 2-mercaptoethanol, the pH is uncorrected for the effect of $D_2O$

## Tumour Suppressor

Tumour suppressor protein p16, wild type and variants

| Mutant | pH | T | ΔG | $c_{1/2}$ | m | Appr./Rem. | Ref |
|---|---|---|---|---|---|---|---|
| wild type | 7.5 | 25 | 13.0±0.4 | 1.87±0.01 | 7.1±0.4 | urea (1,2) | 99T6 |
| | 7.5 | 25 | | 1.98±0.07 | 7.1±1.3 | urea (1,3) | 99T6 |
| | 7.5 | 25 | 15.1±1.7 | 2.05±0.07 | 7.1±0.8 | urea (1,4) | 99T6 |
| | 7.5 | 25 | 22.2±0.8 | 2.90±0.02 | 7.5±0.4 | urea (1,5) | 99T6 |
| wild type | 7.5 | 25 | 15.5±0.4 | 1.9 ±0.04 | 8.4±0.4 | urea (1,6) | 99T6 |
| Asp74→Asn | 7.5 | 25 | 7.1±0.8 | 1.4 ±0.05 | 5.0±0.4 | urea (1,6) | 99T6 |
| Pro81→Leu | 7.5 | 25 | 15.5±0.8 | 3.4 ±0.2 | 4.6±1.3 | urea (1,6) | 99T6 |
| Asp84→Asn | 7.5 | 25 | 15.1±0.4 | 1.8 ±0.03 | 8.4±0.8 | urea (1,6) | 99T6 |

Remarks:

(1) linear extrapolation based on a two-state fit in which the signal of the native and denatured state is dependent on the denaturant conc., see also Ref. 93C7

(2) measured in 50 mM Tris, 1 mM DTT or DTE, transition monitored by ellipticity

(3) measured in 50 mM Tris, 1 mM DTT or DTE, transition monitored by fluorescence

(4) measured in 15 mM sodium phosphate, 100 mM NaCl, 1 mM DTT, transition monitored by size exclusion chromatography

(5) measured in 15 mM sodium phosphate, 350 mM sodium sulfate, 1 mM DTT, transition monitored by ellipticity

(6) measured in 15 mM sodium phosphate, 100 mM NaCl, 1 mM DTT, transition monitored by CD at 222 nm

Tumour suppressor INK4

| Protein | pH | T | ΔG | m | Approach/Remarks | | Ref |
|---|---|---|---|---|---|---|---|
| p16[INK4] | 7.5 | 20 | 8.12±0.42 | 13.51±0.71 | GuHCl | (1–3) | 99Y4 |
| p18[INK4] | 7.5 | 20 | 12.47±0.75 | 15.77±1.00 | GuHCl | (1–3) | 99Y4 |

Remarks:

(1) linear extrapolation to zero denaturant concentration by a method that includes the pre- and postdenaturational baselines for a nonlinear regression of the data

(2) transition monitored by far-UV CD at 222 nm

(3) buffer: 20 mM sodium borate, 40 μM DTE, pH 7.5

**β-Turn Propensity**, see also Immunoglobulin Variable Domain

**Tyrosine Kinase**, see SH3 Domains

# U1A

U1A, human spliceosomal protein, equilibrium and kinetic data

| Protein | pH | T | ΔG | $c_{1/2}$ | m | Appr./Rem. | Ref |
|---|---|---|---|---|---|---|---|
| U1A | 6.25 | 25 | 39.3±0.8 | 4.07±0.02 | 7.11±0.04 | GuHCl (1–5) | 99O5 |
| | 6.25 | 25 | 48.5±7.1 | 4.04±0.04 | 8.79±1.26 | GuHCl (1–3,6) | 99O5 |
| | 6.25 | 25 | 41.4±0.8 | 4.09±0.05 | 7.57±0.17 | GuHCl (1–3,6) | 99O5 |

Remarks:

(1) Ref. 99O5 deals with the interpretation of curved chevron plots, and contains a comparative study of the protein U1A and the ribosomal protein S6

(2) data from equilibrium measurements, nonlinear least-squares fit of the two-state transition curve assuming a linear dependence of ΔG on the denaturant conc.

(3) buffer: 50 mM MES, pH 6.25

(4) transition monitored by fluorescence

(5) U1A has no Trp residues, here the variant Phe56→Trp was used

(6) transition monitored by CD spectra

(7) kinetic data from stopped-flow fluorescence measurements, $m_{kin}$ was found to be dependent on the denaturant conc. $(7.57±0.17) − (0.040±0.025)×c_{GuHCl}$

U1A, human spliceosomal protein, equilibrium and kinetic data

| Protein | pH | T | ΔG | $c_{1/2}$ | m | Appr./Rem. | Ref |
|---|---|---|---|---|---|---|---|
| U1A  1 μM | 6.3 | 25 | 41.2±1.3 | 3.98±0.03 | 10.4±0.8 | GuHCl (1–3) | 99S10 |
| U1A 10 μM | 6.3 | 25 | 39.8±2.5 | 4.07±0.05 | 9.7±1.3 | GuHCl (1–3) | 99S10 |
| U1A | 6.3 | 25 | 38.9±2.5 | 4.13±0.04 | 9.5±1.3 | GuHCl (1,2,4) | 99S10 |
| | 6.3 | 25 | 48.9±3.4 | 4.03±0.03 | 12.1±1.7 | GuHCl (1,2,5) | 99S10 |
| | 6.3 | 25 | 38.1±0.8 | 4.08 | 9.2±0.8 | GuHCl (1,2,6) | 99S10 |

Remarks:

(1) data from equilibrium measurements, nonlinear least-squares fit of the two-state transition curve assuming a linear dependence of ΔG on the denaturant conc.

(2) buffer: 50 mM MES, pH 6.3

(3) transition monitored by fluorescence

(4) transition monitored by CD spectra

(5) transition monitored by fluorescence anisotropy

(6) kinetic data from stopped-flow fluorescence measurements

U1A, human spliceosomal protein, wild type and mutants, kinetic data

| Protein | pH | T | $\Delta(\Delta G)$ | $c_{1/2}$ | m | Appr./Rem. | Ref |
|---|---|---|---|---|---|---|---|
| wild type | 6.3 | 25 | 0.0 | 4.1 | 7.61 | GuHCl (1,2) | 99T9 |
| Ile12→Val | 6.3 | 25 | −6.8 | 3.4 | 7.78 | GuHCl (1,2) | 99T9 |
| Ile14→Ala | 6.3 | 25 | −18.8 | 2.2 | 7.57 | GuHCl (1,2) | 99T9 |
| Leu17→Ala | 6.3 | 25 | −15.4 | 2.6 | 7.53 | GuHCl (1,2) | 99T9 |
| Leu26→Ala | 6.3 | 25 | −9.7 | 3.1 | 7.07 | GuHCl (1,2) | 99T9 |
| Leu30→Ala | 6.3 | 25 | −21.7 | 2.0 | 7.57 | GuHCl (1,2) | 99T9 |
| Phe34→Ala | 6.3 | 25 | −18.8 | 2.1 | 6.49 | GuHCl (1,2) | 99T9 |
| Ile40→Ala | 6.3 | 25 | −21.7 | 2.0 | 7.28 | GuHCl (1,2) | 99T9 |
| Ile43→Val | 6.3 | 25 | −2.3 | 3.9 | 7.78 | GuHCl (1,2) | 99T9 |
| Val45→Ala | 6.3 | 25 | −9.7 | 3.1 | 7.15 | GuHCl (1,2) | 99T9 |
| Ile58→Ala | 6.3 | 25 | −5.7 | 3.5 | 6.49 | GuHCl (1,2) | 99T9 |
| Leu69→Ala | 6.3 | 25 | −9.7 | 3.1 | 6.53 | GuHCl (1,2) | 99T9 |
| Ile84→Ala | 6.3 | 25 | −12.5 | 2.7 | 5.86 | GuHCl (1,2) | 99T9 |
| Ile84→Val | 6.3 | 25 | −5.7 | 3.6 | 8.03 | GuHCl (1,2) | 99T9 |

Remarks:

(1) data from stopped-flow kinetics of folding and unfolding (chevron plot)

(2) buffer: 50 mM MES, pH 6.3

Human spliceosomal protein U1A, mutant protein

| Mutant | pH | T | $\Delta G$ | $c_{1/2}$ | m | Appr./Rem. | Ref |
|---|---|---|---|---|---|---|---|
| Phe56→Trp | 6.3 | 25 | 38.9±0.8 | 4.07±0.02 | 9.6±0.4 | GuHCl (1,2) | 97S12 |
| Phe56→Trp | 6.3 | 25 | 38.1±0.8 | 4.08 | 9.2±0.4 | GuHCl (1,3) | 97S12 |

Remarks:

(1) equilibrium unfolding, linear extrapolation

(2) kinetics

(3) the paper is aimed to exclude misinterpretations of folding intermediates caused by transient aggregates

N-terminal RNA binding domain of human U1A protein (RBD1), variants

| Mutant | pH | $T_{trs}$ | $\Delta T$ | $\Delta(\Delta G)$ | Approach/Remarks | | Ref |
|---|---|---|---|---|---|---|---|
| RBD1(102A) | 2.3 | 56.6 | 0.0 | 0.0 | DSC | (1,2) | 97L13 |
| RBD1(95A) | 2.3 | 57.0 | 0.4 | 0.0 | DSC | (1,3) | 97L13 |
| RBD1(8–99) | 2.3 | 47.0 | −9.6 | −5.9 | DSC | (1,4,5) | 97L13 |
| RBD1(Δloop3) | 2.3 | 65.4 | 8.8 | 5.9 | DSC | (1,6,7) | 97L13 |

Remarks:

(1) $\Delta(\Delta G)$ is given at 56°C, see also Table 2

(2) RBD1(102A), the construct containing 1–102 amino acids of the human U1A protein is referred to as wild type

(3) RBD1(95A), the construct containing residues 1–95

(4) RBD1(8–99), the construct containing residues 8–99 with Arg7→Met and Asn8→Ala

(5) calculated using $\Delta Cp(56°C) = 3.43$ kJ(mol/K

(6) RBD1(Δloop3) is RBD1(102A) with the six amino acids of loop3 (Ser46-Arg47-Ser48-Leu49-Lys50-Met51) replaced with four residues (Val-Pro-Gly-Arg)

(7) calculated using $\Delta Cp(56°C) = 3.93$ kJ/mol/K

## Ubiquitin

Ubiqitin from bovine red blood cells, wild type and mutant Phe45→Trp

| Mutant | pH | T | $\Delta G$ | $c_{1/2}$ | m | Approach/Remarks | | Ref |
|---|---|---|---|---|---|---|---|---|
| wild type | 3.5 | 25 | 20.9 | 3.7 | 5.65 | GuHCl | (1,4,7) | 92B6 |
| | 4.0 | 25 | | 3.86±0.10 | 8.2±0.4 | GuHCl | (1–3) | 99I2 |
| | 5.0 | 25 | 31.4 | 4.01 | 7.87 | GuHCl | (1,4,5) | 93K3 |
| | 5.0 | 25 | 28.0 | 3.85 | 7.32 | GuHCl | (1,4,6) | 93K3 |
| Phe45→Trp | 2.4 | 25 | 26.4 | 3.52 | 7.49 | GuHCl | (1,4,8) | 93K3 |
| | 5.0 | 25 | 29.7 | 3.82 | 7.82 | GuHCl | (1,4,5) | 93K3 |
| | 5.0 | 25 | 30.5 | 3.69 | 8.28 | GuHCl | (1,4,6) | 93K3 |
| | 5.0 | 25 | 29.3 | 3.65 | 8.03 | GuHCl | (1,4,8) | 93K3 |

Remarks:

(1) linear extrapolation

(2) measured in 10 mM acetate buffer, transition monitored by CD at 222 nm

(3) Ref. 99I2 contains additional DSC data for cold denaturation of ubiquitin in the presence of GuHCl

(4) reference data in Ref. 99I2

(5) transition monitored by NMR at pD 5

(6) measured in 25 mM acetate buffer, transition monitored by CD at 222 nm

(7) transition monitored by NMR at pD 3.5

(8) transition monitored by fluorescence emission at 353 nm

Yeast and bovine ubiquitin, results of thermal- versus GuHCl-induced unfolding

| Protein | pH | T | $\Delta G$ | $c_{1/2}$ | m | Appr./Rem. | Ref |
|---|---|---|---|---|---|---|---|
| bovine ubi. | 2.0 | 25 | 23 | 3.75 | 6.2 | GuHCl (1–6) | 99I1 |
| | 3.0 | 25 | 22 | 3.9 | 5.6 | GuHCl (1–6) | 99I1 |
| | 4.0 | 25 | 21 | 3.9 | 5.3 | GuHCl (1–6) | 99I1 |
| | 2.0 | 25 | 9 | | | DSC (3–6) | 99I1 |
| | 3.0 | 25 | 19 | | | DSC (3–6) | 99I1 |
| | 4.0 | 25 | 30 | | | DSC (3–6) | 99I1 |
| yeast ubi. | 2.0 | 25 | 19 | 2.75 | 7 | GuHCl (1–6) | 99I1 |
| | 3.0 | 25 | 18 | 3.3 | 5.4 | GuHCl (1–6) | 99I1 |
| | 4.0 | 25 | 16 | 3.2 | 5 | GuHCl (1–6) | 99I1 |
| | 5.5 | 25 | 14 | 3.2 | 4.4 | GuHCl (1–6) | 99I1 |
| | 2.0 | 25 | 7.5 | 2.1 | 3.6 | urea (1,2) | 99I1 |
| | 5.0 | 25 | 26.4 | 5.9 | 4.5 | urea (1,2) | 99I1 |
| | 2.0 | 25 | 4.5 | | | DSC (3–6) | 99I1 |
| | 3.0 | 25 | 12.5 | | | DSC (3–6) | 99I1 |
| | 4.0 | 25 | 22 | | | DSC (3–6) | 99I1 |

Remarks:

(1) linear extrapolation

(2) transition monitored by far UV-CD

(3) buffer: 10 mM sodium acetate from pH 3.5 to 4.5, and 10 mM glycine from pH 1.5 to 3.5

(4) selected values taken from Figs. 3 and 7 from Ref. 99I1

(5) urea denaturation is found to be in agreement with DSC data whereas the pH dependencies of $\Delta G$ from DSC and GuHCl-induced unfolding disagree

(6) Ref. 99I1 contains a variety of data obtained by DSC and GuHCl-induced unfolding (CD and fluorescence) and clarifies the discrepancy in the pH dependence of $\Delta G$ between GuHCl and thermal denaturation of ubiquitin (see also Refs. 94W5, 98M2, and 99I2) by an analysis of the contributions from charge-charge interactions

Bovine ubiquitin, wild type and mutant Phe45→Trp

| Mutant | pH | T | $\Delta G$ | $c_{1/2}$ | m | Approach/Remarks | Ref |
|--------|-----|-----|------|------|------|------------------|------|
| wild type | 5 | 25 | 31.4 | 4.01 | 7.87 | GuHCl, NMR (1–3) | 93K12 |
| | 5 | 25 | 28.0 | 3.85 | 7.32 | GuHCl, CD    (1,4,5) | 93K12 |
| | 3.5 | 25 | 20.9 | 3.7 | 5.65 | H/D, NMR | 93K12 |
| Phe45→Trp | 5 | 25 | 29.7 | 3.82 | 7.82 | GuHCl, NMR (1–3) | 93K12 |
| | 5 | 25 | 30.5 | 3.69 | 8.28 | GuHCl, CD    (1,4,5) | 93K12 |
| | 5 | 25 | 29.3 | 3.65 | 8.03 | GuHCl, Fl    (1,5,6) | 93K12 |
| | 2.4 | 25 | 26.4 | 3.52 | 7.49 | GuHCl, Fl    (1,5,7) | 93K12 |

Remarks:

(1) linear extrapolation

(2) transition monitored by NMR

(3) buffer: 25 mM deuterated sodium acetate, pD 5

(4) transition monitored by far-UV CD at 222 nm

(5) buffer: 25 mM sodium acetate, pH 5

(6) transition monitored by fluorescence at 353 nm

(7) buffer: 20 mM phosphate, pH 2.4

Ubiquitin, wild type and mutants

| Protein | pH | T | $\Delta G$ | $c_{1/2}$ | m | Appr./Rem. | Ref |
|---------|-----|-----|----------|------|-----------|------------|------|
| wild type | 5.0 | 25 | 25.2±1.0 | 5.99 | 4.21±0.16 | urea (1–4) | 99L8 |
| Lys6→Gln | 5.0 | 25 | 26.3±1.1 | 6.25 | | urea (1–3) | 99L8 |
| Lys6→Glu | 5.0 | 25 | 27.4±1.1 | 6.51 | | urea (1–3) | 99L8 |
| Lys27→Gln | 5.0 | 25 | 17.2±0.7 | 4.09 | | urea (1–3) | 99L8 |
| Lys29→Gln | 5.0 | 25 | 19.0±0.8 | 4.51 | | urea (1–3) | 99L8 |
| Lys29→Asn | 5.0 | 25 | 18.2±0.7 | 4.32 | | urea (1–3) | 99L8 |
| Arg42→Glu | 5.0 | 25 | 32.0±1.3 | 7.60 | | urea (1–3) | 99L8 |
| His68→Gln | 5.0 | 25 | 27.5±1.1 | 6.53 | | urea (1–3) | 99L8 |
| His68→Glu | 5.0 | 25 | 28.4±1.2 | 6.75 | | urea (1–3) | 99L8 |
| Arg72→Gln | 5.0 | 25 | 23.8±1.0 | 5.65 | | urea (1–3) | 99L8 |

Remarks:

(1) linear extrapolation

(2) transition monitored by ellipticity at 222 nm

(3) buffer: 50 mM sodium acetate, pH 5.0

(4) the same m value was used for the mutant proteins

Ubiquitin, de novo design of the hydrophobic core

| Mutant | pH | T | ΔG | $c_{1/2}$ | m | Approach/Remarks | | Ref |
|---|---|---|---|---|---|---|---|---|
| wild type | 7.0 | 25 | 30.1 | 3.8 | 7.9 | GuHCl | (1–3) | 97L4 |
| 1D7 (5) | 7.0 | 25 | 23.4 | 2.4 | 9.6 | GuHCl | (1–3) | 97L4 |
| 3D6 (6) | 7.0 | 25 | 19.7 | 2.3 | 8.4 | GuHCl | (1–3) | 97L4 |
| 3D3 (7) | 7.0 | 25 | 20.5 | 2.1 | 9.6 | GuHCl | (1–3) | 97L4 |
| 1D8 (8) | 7.0 | 25 | 19.2 | 2.2 | 8.8 | GuHCl | (1–3) | 97L4 |
| 3D4 (9) | 7.0 | 25 | 18.4 | 1.9 | 9.6 | GuHCl | (1–3) | 97L4 |
| 2D6 (10) | 7.0 | 25 | 16.7 | 1.8 | 9.2 | GuHCl | (1–3) | 97L4 |
| R6 (11) | 7.0 | 25 | 11.3 | 1.4 | 7.9 | GuHCl | (1–3) | 97L4 |
| 2D7 (12) | 7.0 | 25 | 11.7 | 1.3 | 9.2 | GuHCl | (1–3) | 97L4 |
| R7 (13) | 7.0 | 25 | 8.4 | 1.1 | 7.5 | GuHCl | (1–3) | 97L4 |

Remarks:

(1) linear extrapolation, LEM-SB
(2) buffer: 10 mM potassium phosphate pH 7.0
(3) the transition was monitored by CD at 222 nm
(5) 1D7 = multiple mutant (Ile3→Val, Val5→Leu, Ile13→Val, Leu15→Ile, Ile23→Val, Val26→Phe, and Leu67→Ile)
(6) 3D6 = multiple mutant (Ile3→Leu, Leu15→Ile, Val17→Ile, Ile23→Val, Ile61→Leu, and Leu67→Ile)
(7) 3D3 = multiple mutant (Ile3→Leu, Val17→Leu, and Ile23→Val)
(8) 1D8 = multiple mutant (Ile3→Leu, Ile13→Leu, Leu15→Val, Val17→Leu, Ile23→Val, Val26→Leu, Ile61→Leu, and Leu67→Ile)
(9) 3D4 = multiple mutant (Ile3→Leu, Val17→Leu, Ile23→Val, and Ile61→Val)
(10) 2D6 = multiple mutant (Ile3→Leu, Val5→Ile, Ile13→Val, Val26→Leu, Ile30→Leu, and Leu69→Ile)
(11) R6 = multiple mutant (Leu15→Ile, Val17→Ile, Ile30→Leu, Leu43→Ile, Leu50→Ile, and Leu56→Ile)
(12) 2D7 = multiple mutant (Ile3→Leu, Val5→Leu, Ile13→Val, Val26→Ile, Ile30→Leu, Leu43→Phe, and Leu69→Ile)
(13) R7 = multiple mutant (Val5→Ile, Ile13→Leu, Leu15→Val, Ile30→Leu, Leu43→Ile, Leu50→Ile, and Leu56→Ile)


## Villin

Recombinant villin 14T

| Protein | pH | T | ΔG | m | Approach/Remarks | | Ref |
|---|---|---|---|---|---|---|---|
| villin | 4.1 | 25 | 41.0±0.4 | 9.71±0.08 | urea eq. | (1–4) | 98C8 |
| | 4.1 | 25 | 39.7±2.5 | 8.8±0.4 | urea kin. | (4–7) | 98C8 |
| | 4.1 | 25 | 41.0±0.4 | 16.44±0.13 | GuHCl eq. | (1–4) | 98C8 |
| | 4.1 | 25 | 23.0±0.8 | 10.5±0.4 | GuHCl kin. | (4–6) | 98C8 |
| | 4.1 | 37 | 25.9±0.4 | 6.65±0.17 | urea eq. | (1–4) | 98C8 |
| | 4.1 | 37 | 24.7±0.8 | 6.3±0.4 | urea kin. | (4–6) | 98C8 |

Remarks:

(1) equilibrium data
(2) linear extrapolation
(3) transition monitored by fluorescence emission at 320 nm
(4) buffer: 50 mM acetic acid, 15 mM NaCl, 1 mM 2-mercaptoethanol, pH 4.1
(5) data from the kinetics of villin folding and refolding in the presence of denaturant
(6) transition measured by stopped-flow fluorescence
(7) at 25°C and low final urea conc. the refolding kinetics deviates from the two-state model. Data fit yields an I → U transition with ΔG = 17.2±1.3 kJ/mol and m = 6.3±0.4 kJ/mol/M

## WW Domain

Recombinant WW domain

| Protein | pH | T | $\Delta G$ | $c_{1/2}$ | m | Appr./Rem. | Ref |
|---------|----|----|-----------|-----------|----|------------|-----|
| WW domain | 6.0 | 20 | 9.37 | 3.86 | 2.43 | urea (1–4) | 99K12 |

Remarks:

(1) linear extrapolation, LEM-SB

(2) transition monitored by fluorescence at 340 nm

(3) buffer: 10 mM potassium phosphate, 100 mM NaCl, 0.1 mM DTT, 0.1 mM EDTA, 0.02% $NaN_3$, pH 6.0

(4) thermal transition at $T_{trs} = 48.9\pm0.6°C$ monitored by CD at 230 nm

## Xylanase

Xylanase from *Bacillus circulans*, wild type and disulfide-bridge containing mutant (Ser100→Cys and Asn148→Cys)

| Mutant | pH | $T_{trs}$ | $\Delta T$ | $\Delta(\Delta G)$ | Approach/Remarks | | Ref |
|--------|----|-----------|-----------|--------------------|------------------|----|-----|
| wild type | 6.0 | 51.2 | 0.0 | 0.0 | DSC | (1–3) | 98D9 |
| disulfide m. | 6.0 | 58.6 | 7.4 | 12.1±0.8 | DSC | (1–4) | 98D9 |

Remarks:

(1) measured in the presence of 2.5 M urea

(2) in the presence of low (nondenaturing) concentrations of urea, reversible calorimetric transitions (reheating) were observed whereas the transitions are irreversible in the absence of urea

(3) the data show scan rate dependence of the transition temperature in the case of reversible denaturation (see also J.R. Lepock et al., Biochemistry 31, 1992, 12706–12712)

(4) $\Delta(\Delta G)$ was calculated assuming $\Delta Cp = 10.5$ kJ/mol/K

Xylanase from *Streptomyces halstedii* JM8 and variants

| Variant/Transition | pH | T | $\Delta(\Delta G)$ | Approach/Remarks | | Ref |
|--------------------|----|----|--------------------|------------------|----|-----|
| xylanase type Xys1S (1): | | | | | | |
| trans. (1) | 7.5 | 60.6 | 0.0 | DSC | (1,4) | 98R14 |
| trans. (2) | 7.5 | 63.9 | 0.0 | DSC | (1,4) | 98R14 |
| xylanase variant Xys1Δ (2): | | | | | | |
| trans. (1) | 7.5 | 60.6 | −15.1 | DSC | (2,4) | 98R14 |
| trans. (2) | 7.5 | 63.9 | −23.4 | DSC | (2,4) | 98R14 |
| xylanase variant Xys1VW (3): | | | | | | |
| trans. (1) | 7.5 | 60.6 | − 7.5 | DSC | (3,4) | 98R14 |
| trans. (2) | 7.5 | 63.9 | −13.8 | DSC | (3,4) | 98R14 |
| xylanase type Xys1S (1): | | | | | | |
| trans. (1) | 4.1 | 43.1 | 0.0 | DSC | (1,4) | 98R14 |
| trans. (2) | 4.1 | 50.2 | 0.0 | DSC | (1,4) | 98R14 |
| xylanase variant Xys1Δ (2): | | | | | | |
| trans. (1) | 4.1 | 43.1 | − 0.8 | DSC | (2,4) | 98R14 |
| trans. (2) | 4.1 | 50.2 | − 7.9 | DSC | (2,4) | 98R14 |
| xylanase variant Xys1VW (3): | | | | | | |
| trans. (1) | 4.1 | 43.1 | − 1.3 | DSC | (3,4) | 98R14 |
| trans. (2) | 4.1 | 50.2 | − 4.2 | DSC | (3,4) | 98R14 |
| xylanase type Xys1S (1): | | | | | | |
| trans. (1) | 8.9 | 46.5 | 0.0 | DSC | (1,4) | 98R14 |
| trans. (2) | 8.9 | 53.7 | 0.0 | DSC | (1,4) | 98R14 |

Xylanase from *Streptomyces halstedii* JM8 and variants (continued)

| Variant/Transition | pH | T | Δ(ΔG) | Approach/Remarks | | Ref |
|---|---|---|---|---|---|---|
| xylanase variant Xys1Δ (2): | | | | | | |
| trans. (1) | 8.9 | 46.5 | − 5.9 | DSC | (2,4) | 98R14 |
| trans. (2) | 8.9 | 53.7 | −12.5 | DSC | (2,4) | 98R14 |
| xylanase variant Xys1VW (3): | | | | | | |
| trans. (1) | 8.9 | 46.5 | 4.6 | DSC | (3,4) | 98R14 |
| trans. (2) | 8.9 | 53.7 | − 7.9 | DSC | (3,4) | 98R14 |

Remarks:

(1) *Streptomyces halstedii* JM8 secrets two xylanase forms, Xys1L and Xys1S, see also Ref. 94R5

(2) xylanase variant Xys1Δ was obtained by deletion of a Gly-rich linker region at the C-terminus

(3) xylanase Xys1VW was obtained by Val399→Trp point mutation

(4) Δ(ΔG) refers to $T_{trs}$ of Xys1S

Xylanase A from *Streptomyces lividans*, wild type and mutants

| Mutant | pH | $T_{trs}$ | ΔT | Δ(ΔG) | Approach/Remarks | | Ref |
|---|---|---|---|---|---|---|---|
| wild type | 6.0 | 68.0 | 0.0 | 0.0 | heat | (1–3) | 99R13 |
| Trp85→Ala | 6.0 | 70.0 | 2.0 | 4.5 | heat | (1–3) | 99R13 |
| Trp85→Phe | 6.0 | 64.0 | −4.0 | −9.2 | heat | (1–3) | 99R13 |
| Trp85→His | 6.0 | 67.0 | −1.0 | −2.2 | heat | (1–3) | 99R13 |
| Tyr172→Ala | 6.0 | 63.0 | −5.0 | −11.6 | heat | (1–3) | 99R13 |
| Tyr172→Phe | 6.0 | 67.1 | −0.9 | −2.1 | heat | (1–3) | 99R13 |
| Tyr172→Ser | 6.0 | 69.6 | 1.6 | 3.6 | heat | (1–3) | 99R13 |
| Trp266→Ala | 6.0 | 63.3 | −4.7 | −10.8 | heat | (1–3) | 99R13 |
| Trp266→Phe | 6.0 | 64.7 | −3.3 | −7.6 | heat | (1–3) | 99R13 |
| Trp266→His | 6.0 | 60.7 | −7.3 | −16.7 | heat | (1–3) | 99R13 |
| Trp274→Ala | 6.0 | 72.1 | 4.1 | 9.5 | heat | (1–3) | 99R13 |
| Trp274→Phe | 6.0 | 68.8 | 0.8 | 1.8 | heat | (1–3) | 99R13 |
| Trp274→His | 6.0 | 68.2 | 0.2 | 0.6 | heat | (1–3) | 99R13 |

Remarks:

(1) Δ(ΔG) was calculated at $T_{trs}$ of the wild-type protein using $\Delta(\Delta G) = \Delta H_{w.t.} \times (T_{trs.w.t.} - T_{trs.mut})$

(2) transition monitored by far-UV CD

(3) measured in 10 mM sodium phosphate buffer

**Table 2.**
**Enthalpy and Heat Capacity Changes – Molar Values**

## Abrin

Abrin II, type II ribosome-inactivating protein from *Abrus precatorius*

| pH | $T_{trs1}$ | $\Delta H_{trs1}$ | $T_{trs2}$ | $\Delta H_{trs2}$ | Appr./Rem. | Ref |
|---|---|---|---|---|---|---|
| abrin: | | | | | | |
| 4.50 | 67.6 | 1356 | 70.1 | 946 | DSC (1–4) | 99K18 |
| 5.43 | 65.3 | 1230 | 66.9 | 854 | DSC (1–4) | 99K18 |
| 5.80 | 61.2 | 1125 | 63.3 | 816 | DSC (1–4) | 99K18 |
| 6.43 | 54.2 | 1067 | 57.9 | 766 | DSC (1–4) | 99K18 |
| 6.83 | 51.0 | 937 | 55.2 | 686 | DSC (1–4) | 99K18 |
| 7.20 | 46.0 | 854 | 51.4 | 573 | DSC (1–4) | 99K18 |
| 7.82 | 44.0 | 736 | 50.6 | 527 | DSC (1–4) | 99K18 |
| 8.30 | 41.6 | 586 | 47.8 | 519 | DSC (1–4) | 99K18 |
| 9.16 | 38.4 | 561 | 44.8 | 452 | DSC (1–4) | 99K18 |
| 10.16 | 36.6 | 456 | 43.2 | 410 | DSC (1–4) | 99K18 |
| abrin A–subunit: | | | | | | |
| 4.50 | | | 49.7 | 494 | DSC (1,2,5) | 99K18 |
| 7.20 | | | 49.0 | 506 | DSC (1,2,5) | 99K18 |

Remarks:

(1) results of deconvolution

(2) buffers: 50 mM sodium acetate (pH 4.5–5.8), 50 mM sodium phosphate (pH 6.4–7.9), 50 mM sodium borate (pH 8.3–9.2), 50 mM sodium carbonate (pH 10.16)

(3) $\Delta Cp$ from $\Delta H^{cal}$ versus $T_{trs}$ amounts to 27±2 and 20±1 kJ/mol/K for the lower- and higher-temperature transition, respectively

(4) Ref. 99K18 contains further data obtained in the presence of lactose and NaCl

(5) measured in the presence of 5 mM 2-mercaptoethanol

## Acylphosphatase

Recombinant muscle acylphosphatase, thermodynamic parameters obtained at different urea concentrations

| urea conc. | pH | $T_{trs}$ | $\Delta Cp$ | $\Delta H$ | Approach/Remarks | | Ref |
|---|---|---|---|---|---|---|---|
| 0.0 | 5.50 | 56.5 | 6.15 | 351 | heat | (1–3) | 98C5 |
| 0.4 | 5.50 | 54.6 | 6.07 | 335 | heat | (1–3) | 98C5 |
| 0.8 | 5.50 | 52.6 | 6.28 | 324 | heat | (1–3) | 98C5 |
| 1.2 | 5.50 | 50.7 | 6.19 | 305 | heat | (1–3) | 98C5 |
| 1.6 | 5.50 | 48.2 | 5.86 | 286 | heat | (1–3) | 98C5 |
| 2.0 | 5.50 | 45.4 | 6.19 | 276 | heat | (1–3) | 98C5 |
| 2.4 | 5.50 | 43.1 | 6.20 | 256 | heat | (1–3) | 98C5 |
| 2.8 | 5.50 | 40.0 | 5.77 | 238 | heat | (1–3) | 98C5 |
| 0.0 | 5.50 | | 6.15±0.60 | | heat | (4,5) | 98C5 |
| 0.0 | 5.50 | | 6.61±0.70 | | heat | (6) | 98C5 |

Remarks:

(1) transition monitored by CD at 222 nm

(2) measured in 50 mM acetate buffer with urea

(3) experimental error 0.5°C for $T_{trs}$, 5% for $\Delta H$, and 10% for $\Delta Cp$

(4) $\Delta Cp$ in the absence of urea, extrapolated value

(5) this value is considered as the best one, and was used in further calculations in Ref. 98C5

(6) $\Delta Cp$ from $\Delta H^{v.H.}$ versus $T_{trs}$ after taking into account the urea-protein interaction

Recombinant muscle acylphosphatase, wild type and lysine to glutamine mutants

| Mutant | pH | $T_{trs}$ | $\Delta H$ | Approach/Remarks | | Ref |
|---|---|---|---|---|---|---|
| wild type | 5.5 | 56.5 | 391 | heat | (1–3) | 98C6 |
| Lys32→Gln | 5.5 | 52.2 | 392 | heat | (1–3) | 98C6 |
| Lys57→Gln | 5.5 | 50.1 | 370 | heat | (1–3) | 98C6 |
| Lys67→Gln | 5.5 | 56.4 | 358 | heat | (1–3) | 98C6 |
| Lys84→Gln | 5.5 | 50.6 | 324 | heat | (1–3) | 98C6 |
| Lys88→Gln | 5.5 | 55.1 | 370 | heat | (1–3) | 98C6 |

Remarks:
(1) transition monitored by CD at 222 nm
(2) buffer: 50 mM acetate
(3) the mean errors in $T_{trs}$ are ±1°C, and in $\Delta H$ within ±10%

Common-type acylphosphatase (CT AcP) and muscle acylphosphatase (M AcP)

| Protein | pH | $T_{trs}$ | $\Delta Cp$ | $\Delta H$ | Appr./Rem. | | Ref |
|---|---|---|---|---|---|---|---|
| CT AcP | 5.5 | 53.9±0.5 | 6.10 | 290 | heat | (1,2) | 99T1 |
| M AcP | 5.5 | 56.6±0.5 | 6.35 | 350 | heat | (3) | 99T1 |

Remarks:
(1) data from fitting the stability curve obtained by urea denaturation from 5 to 40°C, see also Table 1
(2) experimental error 5% for $\Delta H$, and 10% for $\Delta Cp$
(3) data from Ref. 98C5

## Adenylate Kinase

Adenylate kinase (AK) from the archaeon *Sulfolobus acidocaldarius*

| Protein | pH | $T_{trs}$ | $\Delta Cp$ | $\Delta H$ | Appr./Rem. | Ref |
|---|---|---|---|---|---|---|
| AK | 7.0 | 89.0±0.4 | 11.97±0.21 | 808.7±7.4 | GuHCl (1,2,4) | 98B1 |
| | 7.0 | 89.0±1.6 | 16.16±1.12 | 926.8±34.2 | GuHCl (1,3,4) | 98B1 |
| | 7.0 | 95.5 | | 812 | DSC    (4,5) | 98B1 |

Remarks:
(1) data from GuHCl-induced unfolding of AK in the temperature range from 5 to 70°C, see also Table 1
(2) data treatment by linear extrapolation, LEM-SB
(3) data treatment by the denaturant binding model using k = 0.8
(4) data per trimer
(5) data obtained by dissociation-upon-unfolding fitting of the DSC data

Adenylate kinase from *E. coli* (AK$_e$), a 4-cysteine variant (AKC$_4$) that binds zinc ions, and further mutants

| Protein | Ligand | pH | $T_{trs}$ | $\Delta H$ | Approach/Remarks | | Ref |
|---|---|---|---|---|---|---|---|
| AK$_e$ | none | 7.2 | 52.5±0.2 | | heat | (1) | 98B12 |
| AK$_e$ | | 7.4 | 54 | 414 | heat | (1) | 87M2 |
| AK$_e$ | | 7.45 | 53.8 | 472 | heat | (1) | 90R5 |
| AK$_e$ | | 7.45 | 55.5 | 426 | heat | (3) | 90R5 |
| AK$_e$ | | 7.4 | 51.8 | | DSC | (4) | 98P8 |
| AK$_e$ | Ap$_5$A | 7.45 | 55.5 | 425 | heat | (1) | 90R5 |
| AK$_e$ | Ap$_5$A | 7.2 | 61.1±0.2 | | heat | (1) | 98B12 |
| apoAKC$_4$ | none | 7.2 | 58.1±0.2 | | heat | (1) | 98B12 |
| apoAKC$_4$ | Ap$_5$A | 7.2 | 65.5±0.2 | | heat | (1) | 98B12 |
| AKC$_4$ | none | 7.2 | 61.6±0.2 | | heat | (1) | 98B12 |

Adenylate kinase from *E. coli* (AK$_e$), a 4-cysteine variant (AKC$_4$) that binds zinc ions, and further mutants (continued)

| Protein   Ligand | pH | T$_{trs}$ | ΔH | Approach/Remarks | Ref |
|---|---|---|---|---|---|
| AKC$_4$ | 7.4 | 63 | | DSC    (4) | 98P8 |
| AKC$_4$          Ap$_5$A | 7.2 | 68.9±0.2 | | heat    (1) | 98B12 |
| AK$_e$ (Ser87→Pro) | 7.4 | 38 | 226 | heat    (1,2) | 87M2 |
| AK$_e$ (Gln28→His) | 7.45 | 52.5 | 460 | heat    (1) | 90R5 |
| AK$_e$ (Gln28→His) | 7.45 | 59.5 | 453 | heat    (3) | 90R5 |

Explanation:

Ap$_5$A = P$^1$,P$^5$-di(adenosin-5')-pentaphosphate

Remarks:

(1) transition monitored by CD at 222 nm

(2) thermosensitive mutant (Gln28→His) derived from AK$_e$

(3) transition monitored by fluorescence

(4) measured in 50 mM Tris-HCl, pH 7.4


## Adrenodoxin

Recombinant bovine adrenodoxin (Adx) and preadrenodoxin (Padx)

| Protein | NaCl | pH | T$_{trs}$ | ΔCp | ΔH | Appr./Rem. | Ref |
|---|---|---|---|---|---|---|---|
| Adx | 0 mM | 8.5 | 51.4±0.1 | 7.5±1.2 | 355±13 | DSC  (1,2) | 98G9 |
| | 50 mM | 8.5 | 55.6±0.3 | 8.5±2.0 | 319±23 | DSC  (1) | 98G9 |
| Padx | 0 mM | 8.5 | 47.8±0.3 | | 295±15 | DSC  (1) | 98G9 |
| | 50 mM | 8.5 | 48.4±0.4 | 4.6±2.2 | 280±12 | DSC  (1) | 98G9 |
| | 100 mM | 8.5 | 49.5±0.2 | | 323±15 | DSC  (1) | 98G9 |
| | 400 mM | 8.5 | 55.4 | | 320 | DSC  (1) | 98G9 |
| | 0 mM | 8.5 | 47.8±0.2 | | 313±16 | DSC  (3) | 98G9 |

Remarks:

(1) buffer: 40 mM glycine, pH 8.5, 20 mM sodium sulfide, 1 mM ascorbate, 10 mM 2-mercaptoethanol

(2) data from Ref. 95B9

(3) the above buffer with variation of sulfide concentration from 0 to 20 mM

Recombinant bovine adrenodoxin, wild type and mutants, reduced and oxidized

| Protein | Form | pH | T$_{trs}$ | ΔH | Appr./Rem. | Ref |
|---|---|---|---|---|---|---|
| wild type | ox. | 8.0 | 51.3 | 359 | heat  (1,2) | 98I1 |
| | red. | 8.0 | 49.3 | 448 | heat  (1–3) | 98I1 |
| Thr54→Ala | ox. | 8.0 | 48.3 | 279 | heat  (1,2) | 98I1 |
| | red. | 8.0 | 38.8 | 344 | heat  (1–3) | 98I1 |
| Thr54→Ser | ox. | 8.0 | 52.6 | 355 | heat  (1,2) | 98I1 |
| | red. | 8.0 | 45.7 | 389 | heat  (1–3) | 98I1 |
| His56→Arg | ox. | 8.0 | 45.3 | 233 | heat  (1,2) | 98I1 |
| | red. | 8.0 | 46.7 | 480 | heat  (1–3) | 98I1 |
| His56→Thr | ox. | 8.0 | 45.2 | 227 | heat  (1,2) | 98I1 |
| | red. | 8.0 | 44.2 | 363 | heat  (1–3) | 98I1 |
| His56→Gln | ox. | 8.0 | 44.1 | 245 | heat  (1,2) | 98I1 |
| | red. | 8.0 | 42.4 | 344 | heat  (1–3) | 98I1 |

Remarks:

(1) van't Hoff treatment, transition monitored by CD at 440 nm for the oxidized form and at 405 nm for the reduced form

(2) buffer: 0.1 M Tris/HCl, pH 8.0

(3) measured in the presence of 10 mM sodium dithionite

Recombinant bovine adrenodoxin (Adx), truncated mutant Met-Adx-(4–107), role of Pro108

| Mutant | pH | $T_{trs}$ | $\Delta H$ | Appr./Rem. | Ref |
|---|---|---|---|---|---|
| wild type (oxidized) | 8.5 | 51.3 | 359 | heat (1,2) | 97U1 |
| wild type (reduced) | 8.5 | 49.3 | 448 | heat (1,2) | 97U1 |
| Met-4–108 (oxidized) | 8.5 | 51.7 | 336 | heat (1,2) | 97U1 |
| Met-4–108 (reduced) | 8.5 | 48.2 | 432 | heat (1,2) | 97U1 |
| Met-4–107 (oxidized) | 8.5 | 44.8 | 305 | heat (1,2) | 97U1 |
| Met-4–107 (reduced) | 8.5 | 43.7 | 435 | heat (1,2) | 97U1 |

Remarks:
(1) measured in protectant buffer, see Refs. 95B9, 96B10
(2) transition monitored by CD at 440 nm for oxidized proteins, and at 405 nm for reduced proteins

Recombinant bovine adrenodoxin, pseudo-wild type and mutants at the Pro108-Arg14 hydrogen bond

| Protein | pH | $T_{trs}$ | $\Delta H$ | Approach/Remarks | Ref |
|---|---|---|---|---|---|
| pseudo-w.t. | 9.2 | 53.5±0.5 | 364±7 | heat (1–3) | 98G15 |
| Pro108→Ala | 9.2 | 49.2±0.2 | 306±5 | heat (1–3) | 98G15 |
| Arg14→Ala | 9.2 | 47.2±0.2 | 329±6 | heat (1–3) | 98G15 |
| Arg14→Glu | 9.2 | 40.0±0.2 | 288±7 | heat (1–3) | 98G15 |

Remarks:
(1) pseudo-wild type is the truncated protein Adx(4–108)
(2) measured in protectant buffer: 40 mM glycine, pH 9.2, 10 mM $Na_2S$, 1 mM ascorbate, 10 mM 2-mercaptoethanol
(3) transition monitored by CD at 440 nm

Recombinant bovine adrenodoxin, wild type, truncated form 4–108 with Pro108, and mutants derived from the truncated protein (Pro108→X)

| Protein | pH | $T_{trs}$ | $\Delta Cp$ | $\Delta H$ | Appr./Rem. | Ref |
|---|---|---|---|---|---|---|
| wild type | 8.5 | 48.9±0.6 | | 332±15 | heat (1,2) | 98G16 |
| 4–108Pro | 8.5 | 51.7±0.4 | | 336± 9 | heat (1–3) | 98G16 |
| 4–108Ala | 8.5 | 46.3±0.6 | | 280±11 | heat (1–3) | 98G16 |
| 4–108Lys | 8.5 | 38.6±0.6 | | 214±10 | heat (1–3) | 98G16 |
| 4–108Ser | 8.5 | 42.4±0.3 | | 256± 8 | heat (1–3) | 98G16 |
| 4–108Trp | 8.5 | 39.6±0.5 | | 209±11 | heat (1–3) | 98G16 |
| | 8.5 | 38–49 | 10.35±1.1 | 209–336 | heat (4) | 98G16 |

Remarks:
(1) transition monitored by CD at 440 nm
(2) measured in protectant buffer: 40 mM glycine, pH 8.5, 10 mM $Na_2S$, 1 mM ascorbate, 10 mM 2-mercaptoethanol
(3) mutant description: truncated form consisting of residues 4–108 with Pro or mutants with other residues in position 108
(4) $\Delta Cp$ from $\Delta H$ versus $T_{trs}$ of the wild type and mutant proteins

## Agglutinin

Peanut agglutinin (tetramer)

| Transition | pH | $T_{trs}$ | $\Delta Cp$ | $\Delta H$ | Approach/Remarks | Ref |
|---|---|---|---|---|---|---|
| trans. (1) | 7.4 | 56.12–60.67 | | 1038±93 | DSC (1–3) | 99R6 |
| trans. (2) | 7.4 | 63.6±0.5 | | 145±18 | DSC (1,3) | 99R6 |
| trans. (1) | 5.5 | 56.92 | | 888 | DSC (4) | 99R6 |
| trans. (2) | 5.5 | 61.20 | | 473 | DSC (4) | 99R6 |
| trans. (1) | 6.0 | 57.52 | | 913 | DSC (4) | 99R6 |

Peanut agglutinin (tetramer) (continued)

| Transition | pH | $T_{trs}$ | $\Delta C_p$ | $\Delta H$ | Approach/Remarks | | Ref |
|---|---|---|---|---|---|---|---|
| trans. (2) | 6.0 | 61.89 | | 482 | DSC | (4) | 99R6 |
| trans. (1) | 6.5 | 58.00 | | 943 | DSC | (4) | 99R6 |
| trans. (2) | 6.5 | 62.47 | | 504 | DSC | (4) | 99R6 |
| trans. (1) | 7.0 | 58.53 | | 968 | DSC | (4) | 99R6 |
| trans. (2) | 7.0 | 63.12 | | 524 | DSC | (4) | 99R6 |
| trans. (1) | 7.4 | 58.97 | | 989 | DSC | (4) | 99R6 |
| trans. (2) | 7.4 | 63.75 | | 540 | DSC | (4) | 99R6 |
| trans. (1) | 7.8 | 60.11 | | 1026 | DSC | (4) | 99R6 |
| trans. (2) | 7.8 | 64.42 | | 553 | DSC | (4) | 99R6 |
| trans. (1) | 8.2 | 61.18 | 38.9 | 1052 | DSC | (4,5) | 99R6 |
| trans. (2) | 8.2 | 64.96 | 25.4 | 565 | DSC | (4,5) | 99R6 |

Remarks:

(1) measured at 0.05–0.260 mM protein conc. in PBS buffer

(2) $T_{trs}$ of the first transition is scan rate dependent

(3) $\Delta H^{cal}/\Delta H^{v.H.}$ is about 3.3 for the first and 0.22 for the second transition

(4) pH dependent measurements at 0.100 mM peanut agglutinin tetramer

(5) $\Delta C_p$ from $\Delta H$ versus $T_{trs}$

Peanut agglutinin (tetramer) in the presence of ligands

| Ligand | | Transition | pH | $T_{trs}$ | $\Delta H$ | Appr./Rem. | | Ref |
|---|---|---|---|---|---|---|---|---|
| TAG | 0.5–5.0 mM | trans. (1) | 7.4 | 57.70–61.53 | 976±35 | DSC | (1,2) | 99R6 |
| | | trans. (2) | 7.4 | 63.25–66.01 | 137±12 | DSC | (1,2) | 99R6 |
| Lac | 1.0–20 mM | trans. (1) | 7.4 | 57.65–59.48 | 1046±82 | DSC | (1,2) | 99R6 |
| | | trans. (2) | 7.4 | 63.21–64.23 | 145±6 | DSC | (1,2) | 99R6 |
| MAG | 0.5–10 mM | trans. (1) | 7.4 | 56.75–59.14 | 1059±73 | DSC | (1,2) | 99R6 |
| | | trans. (2) | 7.4 | 62.29–63.73 | 138±9 | DSC | (1,2) | 99R6 |
| MBG | 5.0–30 mM | trans. (1) | 7.4 | 57.01–59.08 | 1012±93 | DSC | (1,2) | 99R6 |
| | | trans. (2) | 7.4 | 62.52–63.70 | 143±7 | DSC | (1,2) | 99R6 |
| LacNAc | 5.0–20 mM | trans. (1) | 7.4 | 56.80–57.29 | 1121±188 | DSC | (1,2) | 99R6 |
| | | trans. (2) | 7.4 | 62.59–63.13 | 150±7 | DSC | (1,2) | 99R6 |

Explanations:

| | | |
|---|---|---|
| | TAG | T-antigen |
| | Lac | lactose |
| | MAG | methyl α-galactopyranoside |
| | MBG | methyl β-galactopyranoside |
| | LacNAc | N-acetyllactosamine |

Remarks:

(1) measured at 0.080 mM peanut agglutinin tetramer in PBS buffer, pH 7.4

(2) $\Delta H^{cal}/\Delta H^{v.H.}$ is about 3 for the first and 0.22 for the second transition

## α-Amylase

α-Amylase from various species

| Protein/Transition | | pH | $T_{trs}$ | ΔH | Approach/Remarks | | Ref |
|---|---|---|---|---|---|---|---|
| AH-A | | 7.2 | 43.7 | 966 | DSC | (1,2) | 99F3 |
| PP-A | trans. (1) | 7.2 | 60.7 | 510 | DSC | (1,3) | 99F3 |
| | trans. (2) | 7.2 | 65.2 | 824 | DSC | (1,3) | 99F3 |
| TM-A | trans. (1) | 7.2 | 61.4 | 423 | DSC | (1,4) | 99F3 |
| | trans. (2) | 7.2 | 63.9 | 766 | DSC | (1,4) | 99F3 |
| | trans. (3) | 7.2 | 66.4 | 1033 | DSC | (1,4) | 99F3 |
| AO-A | trans. (1) | 7.2 | 64.2 | 364 | DSC | (1,5) | 99F3 |
| | trans. (2) | 7.2 | 67.7 | 615 | DSC | (1,5) | 99F3 |
| | trans. (3) | 7.2 | 70.1 | 1063 | DSC | (1,5) | 99F3 |
| HS-A | trans. (1) | 7.2 | 67.4 | 602 | DSC | (1,3) | 99F3 |
| | trans. (2) | 7.2 | 70.4 | 1042 | DSC | (1,3) | 99F3 |
| BA-A | trans. (1) | 7.2 | 79.6 | 552 | DSC | (1,4,6) | 99F3 |
| | trans. (2) | 7.2 | 84.2 | 766 | DSC | (1,4,6) | 99F3 |
| | trans. (3) | 7.2 | 87.7 | 720 | DSC | (1,4,6) | 99F3 |
| BL-A | trans. (1) | 7.2 | 96.0 | 690 | DSC | (1,4) | 99F3 |
| | trans. (2) | 7.2 | 100.7 | 975 | DSC | (1,4) | 99F3 |
| | trans. (3) | 7.2 | 104.3 | 1021 | DSC | (1,4) | 99F3 |

Explanations:

AH-A = *Alteromonas haloplanctis* α-amylase
PP-A = porcine pancreas α-amylase
TM-A = *Tenobrio molitor* (meal beetle) α-amylase
AO-A = *Aspergillus oryzae* (Taka amylase) α-amylase
HS-A = human salivary α-amylase
BA-A = *Bacillus amyloliquefaciens* α-amylase
BL-A = *Bacillus licheniformis* α-amylase

Remarks:

(1) measured in 30 mM MOPS, 50 mM NaCl, 1 mM $CaCl_2$, pH 7.2
(2) AH-A melting proceeds as a reversible two-state transition
(3) results of deconvolution, two subtransitions have been observed
(4) results of deconvolution, three subtransitions have been observed
(5) data from Ref. 87F
(6) ΔCp from ΔH versus $T_{trs}$, ΔCp amounts to 14.48, 9.08, and 12.47 kJ/mol/K for the first, second, and third transition, respectively

α-Amylase from various species, effect of calcium and chloride removal

| Protein/Conditions | | | pH | $\Delta T_{trs}$ | Δ(ΔH) | Appr./Rem. | | Ref |
|---|---|---|---|---|---|---|---|---|
| AH-A | no $Ca^{2+}$ | | 7.2 | −0.2 | −146 | DSC | (1,2) | 99F3 |
| | 5 mM EGTA | | 7.2 | 1.0 | −259 | DSC | (1) | 99F3 |
| PP-A | no $Ca^{2+}$ | | | | | | | |
| | | trans. (1) | 7.2 | −3.6 | −13 | DSC | (1,2) | 99F3 |
| | | trans. (2) | 7.2 | −2.8 | −54 | DSC | (1,2) | 99F3 |
| PP-A | 20 mM EGTA | | | | | | | |
| | | trans. (1) | 7.2 | −19.5 | −205 | DSC | (1) | 99F3 |
| | | trans. (2) | 7.2 | −20.2 | −172 | DSC | (1) | 99F3 |
| BA-A | no $Ca^{2+}$ | | | | | | | |
| | | trans. (1) | 7.2 | −8.3 | −88 | DSC | (1,2) | 99F3 |
| | | trans. (2) | 7.2 | −7.5 | −159 | DSC | (1,2) | 99F3 |
| | | trans. (3) | 7.2 | −6.0 | −184 | DSC | (1,2) | 99F3 |

α-Amylase from various species, effect of calcium and chloride removal (continued)

| Protein/Conditions | | pH | $\Delta T_{trs}$ | $\Delta(\Delta H)$ | Appr./Rem. | Ref |
|---|---|---|---|---|---|---|
| BA-A | 5 mM EGTA | | | | | |
| | trans. (1) | 7.2 | −41.6 | −280 | DSC (1) | 99F3 |
| | trans. (2) | 7.2 | −39.3 | −623 | DSC (1) | 99F3 |
| | trans. (3) | 7.2 | −39.1 | −439 | DSC (1) | 99F3 |
| AH-A | no Cl⁻, 1 mM Ca²⁺ | | | | | |
| | | 7.2 | −0.1 | −42 | DSC (1) | 99F3 |
| PP-A | no Cl⁻, 1 mM Ca²⁺ | | | | | |
| | trans. (1) | 7.2 | −4.7 | −100 | DSC (1) | 99F3 |
| | trans. (2) | 7.2 | −4.0 | −138 | DSC (1) | 99F3 |

Explanations:

AH-A = *Alteromonas haloplanctis* α-amylase
PP-A = porcine pancreas α-amylase
BA-A = *Bacillus amyloliquefaciens* α-amylase

Remarks:
(1) buffer: 30 mM MOPS, 50 mM NaCl, 1 mM CaCl₂, pH 7.2
(2) no calcium added, ~ 1 µM Ca²⁺ in buffer estimated by atomic absorption

α-amylase from psychrophilic *Alteromonas haloplanctis* (AH-A)

| Protein | pH | $T_{trs}$ | $\Delta H$ | $\Delta Cp$ | Appr./Rem. | Ref |
|---|---|---|---|---|---|---|
| AH-A | 5.7 | 30.6 | 577 | | DSC | 99F3 |
| | 5.8 | 32.0 | 582 | | DSC | 99F3 |
| | 5.9 | 34.4 | 695 | | DSC | 99F3 |
| | 6.0 | 34.7 | 665 | | DSC | 99F3 |
| | 6.1 | 37.7 | 749 | | DSC | 99F3 |
| | 6.2 | 37.9 | 816 | | DSC | 99F3 |
| | 6.3 | 39.1 | 824 | | DSC | 99F3 |
| | 6.4 | 40.7 | 933 | | DSC | 99F3 |
| | 7.2 | 43.7 | 996 | | DSC | 99F3 |
| | 5.7–6.4 | 30–41 | | 35.44±0.67 | DSC (1) | 99F3 |

Remark:
(1) $\Delta Cp$ from $\Delta H^{cal}$ versus $T_{trs}$

## Annexin

Annexin I, recombinant porcine protein, and annexin V, recombinant human protein

| Protein | pH | $T_{trs}$ | $\Delta Cp$ | $\Delta H$ | Appr./Rem. | Ref |
|---|---|---|---|---|---|---|
| annexin I | 8.0 | 58.5±0.1 | | 824±20 | DSC (1,2) | 99R18 |
| | 7.0 | 59.9±0.1 | | 836±20 | DSC (1,2) | 99R18 |
| | 6.0 | 61.8±0.1 | | 824±10 | DSC (1,2) | 99R18 |
| | 5.0 | 62.7±0.3 | | 849± 6 | DSC (1,2) | 99R18 |
| | 4.5 | 62.3±0.2 | | 802± 5 | DSC (1,2) | 99R18 |
| | 4.0 | 58.8 | | 678 | DSC (1) | 99R18 |
| | 3.8 | 56.4 | | 634 | DSC (1) | 99R18 |
| | 3.6 | 53.3 | | 613 | DSC (1) | 99R18 |
| | 3.4 | 48.5 | | 581 | DSC (1) | 99R18 |
| | 3.2 | 41.1 | | 427 | DSC (1) | 99R18 |
| | 3.2–8.0 | 41–63 | 19.0 | | DSC (1,3) | 99R18 |

Annexin I, recombinant porcine protein, and annexin V, recombinant human protein (continued)

| Protein | pH | $T_{trs}$ | $\Delta Cp$ | $\Delta H$ | Appr./Rem. | Ref |
|---------|-----|------|------|-----|-----------|-------|
| annexin I | 7.2 | 52 | | 674 | DSC (4) | 99R18 |
| annexin V | 6.9 | 58 | | 623 | DSC (4) | 99R18 |
| | 8.0 | 54.4 | | 679 | DSC (5) | 99R18 |
| | 6.0 | 50.5 | 10.3 | 619 | DSC (5) | 99R18 |

Remarks:

(1) measured in 50 mM sodium phosphate at a heating rate of 2 K/min

(2) Ref. 99R18 contains additional data for various models and scan rate dependence

(3) $\Delta Cp$ from $\Delta H^{cal}$ versus $T_{trs}$ at varying pH

(4) measured in 50 mM Tris-HCl

(5) measured in 50 mM sodium phosphate

Membrane-binding protein annexin V, wild type and mutant (Glu17→Gly)

| Mutant | pH | $T_{trs}$ | $\Delta Cp$ | $\Delta H$ | Approach/Remarks | | Ref |
|--------|-----|----------|----------|---------|-----------|--------|-------|
| wild type | 8.0 | 52.5±0.5 | | 682±18 | DSC | (1–3) | 97V5 |
| Glu17→Gly | 8.0 | 52.8±0.5 | | 686±25 | DSC | (1–3) | 97V5 |
| Glu17→Gly | 8.0 | 54.4±0.5 | | 679±31 | DSC | (1,4,5) | 97V5 |
| Glu17→Gly | 7.2 | 52.0±0.5 | | 674±25 | DSC | (1,4–6) | 97V5 |
| Glu17→Gly | 7.0 | 53.4±0.5 | | 680±9 | DSC | (1,4,5) | 97V5 |
| Glu17→Gly | 6.0 | 50.5±0.5 | | 616±2 | DSC | (1,4,5) | 97V5 |
| Glu17→Gly | 5.0 | 46.1±0.4 | | 619±10 | DSC | (1,4,5) | 97V5 |
| Glu17→Gly | 4.0 | 40.6±0.1 | | 537±18 | DSC | (1,4,5) | 97V5 |
| Glu17→Gly | 4–8 | 40–55 | 10.3±1.3 | | DSC | (1,4,5,7) | 97V5 |

Remarks:

(1) Ref. 97V5 is a case study of quasi-equilibrium treatment of thermodynamic quantities derived from irreversible denaturation, see also the table below

(2) heating rate 1.99 K/min

(3) buffer: 50 mM Tris, 100 mM KCl, pH 8.0 at 25°C and pH 7.2 at 52°C

(4) heating rate 1.19 K/min

(5) buffer: 20 mM sodium phosphate

(6) pH value at 52°C

(7) $\Delta Cp$ from $\Delta H$ versus $T_{trs}$

Membrane-binding protein annexin V, mutant (Glu17→Gly), variation of thermodynamic transition parameters with heating rate

| Mutant | Rate | pH | $T_{trs}$ | $\Delta H$ | Approach/Remarks | | Ref |
|--------|-------|-----|------|-----|-----------|--------|-------|
| Glu17→Gly | 0.125 | 7.0 | 51.1 | 695 | DSC | (1,2) | 97V5 |
| Glu17→Gly | 0.589 | 7.0 | 52.5 | 671 | DSC | (1,2) | 97V5 |
| Glu17→Gly | 1.189 | 7.0 | 53.4 | 671 | DSC | (1,2) | 97V5 |
| Glu17→Gly | 2.342 | 7.0 | 54.7 | 690 | DSC | (1,2) | 97V5 |

Remarks:

(1) heating rate in K/min

(2) buffer: 20 mM sodium phosphate, pH 7.0

Human recombinant annexin V, isomorphous replacement of Met with aminohexanoic acid, selenomethionine, and telluromethionine

| Protein | pH | $T_{trs}$ | $\Delta H$ | Approach/Remarks | | Ref |
|---|---|---|---|---|---|---|
| wild type | 7.5 | 51.79±0.03 | 165.6±13.3 | heat | (1–3) | 98B11 |
| Sem | 7.5 | 50.77±0.04 | 492.4±8.8 | heat | (1–3) | 98B11 |
| Ahx | 7.5 | 50.60±0.04 | 484.5±8.8 | heat | (1–3) | 98B11 |
| Tem | 7.5 | 46.00±0.09 | 155.0±24.5 | heat | (1–3) | 98B11 |

Remarks:
(1) abbreviations:
Sem – selenomethionine
Ahx – 2-aminohexanoic acid
Tem – telluromethionine
(2) $\Delta Cp$ was calculated as 10.5 kJ/mol/k
(3) transition monitored by CD at 222 nm

Human recombinant annexin, the single tryptophan residue (Trp187) replaced by fluorine analogs

| Protein | pH | $T_{trs}$ | $\Delta H$ | Approach/Remarks | | Ref |
|---|---|---|---|---|---|---|
| wild type | 8.0 | 51.74±0.03 | 693.3±13.2 | heat | (1) | 99M17 |
| 4-F-Trp | 8.0 | 48.41±0.07 | 502.8±18.1 | heat | (1,2a) | 99M17 |
| 5-F-Trp | 8.0 | 53.24±0.03 | 747.7±15.2 | heat | (1,2b) | 99M17 |
| 6-F-Trp | 8.0 | 48.50±0.06 | 482.2±13.9 | heat | (1,2c) | 99M17 |

Remarks:
(1) measured in Tris-HCl, pH 8.0, transition monitored by CD at 222 nm
(2a) 4-F-Trp = 4-fluorotryptophan
(2b) 5-F-Trp = 5-fluorotryptophan
(2c) 6-F-Trp = 6-fluorotryptophan

**Apocalmodulin**, see Calmodulin

## Apolipoprotein

Human plasma apolipoprotein C-1 (apoC-1)

| Protein | pH | $T_{trs}$ | $\Delta Cp$ | $\Delta H$ | Approach/Remarks | | Ref |
|---|---|---|---|---|---|---|---|
| plasma apoC-1 | 7.5 | 50±3 | ≤1.3 | 79±13 | heat | (1,2) | 98G21 |
| synthetic apoC-1 | | | | | | | |
| wild type | 7.5 | 50±2 | | 75±8 | heat | (1,3) | 99G22 |
| Gly15→Ala | 7.5 | 57±2 | | 79±8 | heat | (1,3) | 99G22 |

Remarks:
(1) transition monitored by CD at 222 nm
(2) measured in 1 mM sodium phosphate buffer, pH 7.5
(3) measured in 2-5 mM sodium phosphate buffer, pH 7.5

**Arc Repressor**, see Repressor Proteins

## Ascorbate Peroxidase

Ascorbate peroxidase, wild type and mutants

| Protein | pH | $T_{trs}$ | $\Delta Cp$ | $\Delta H$ | Appr./Rem. | Ref |
|---|---|---|---|---|---|---|
| wild type | 7.0 | 58.3±0.5 | 0.7 | 1025±121 | DSC (1,2) | 98M5 |
| Glu112→Ala | 7.0 | 56.0±0.8 | | 833±159 | DSC (1,2) | 98M5 |
| Glu112→Lys | 7.0 | 53.0±0.9 | | 711±105 | DSC (1,2) | 98M5 |

Remarks:
(1) unfolding of a native dimer into two unfolded monomers ($N_2 \leftrightarrow 2U$), values are calculated per monomer
(2) measured in 50 mM potassium phosphate, 0.15 M NaCl, pH 7.0, at a scan rate of 15 K/h

## Aspartate Transcarbamoylase

Aspartate transcarbamoylase (ATCase), catalytic (C), and regulatory (R) subunits in the presence of ligands

| Sample | pH | $T_{trs}$ | $\Delta H$ | Approach/Remarks | | Ref |
|---|---|---|---|---|---|---|
| intact ATCase: | | | | | | |
| ATCase | 7.0 | 82 | 6318±88 | DSC | (1) | 78V1 |
| ATCase + PALA | 7.0 | 89.5 | 6820±134 | DSC | (1,2) | 78V1 |
| ATCase + CTP | 7.0 | 79.5 | 6067 | DSC | (1) | 78V1 |
| ATCase + ATP | 7.0 | 77.5 | 6004±268 | DSC | (1) | 78V1 |
| catalytic subunit: | | | | | | |
| C, isolated | 7.0 | 80 | 1653±42 | DSC | (1) | 78V1 |
| C, isolated, + PALA | 7.0 | 85.5 | 2071±42 | DSC | (1,2) | 78V1 |
| C in ATCase | 7.0 | 82 | 2272±75 | DSC | (1,3) | 78V1 |
| C in ATCase + PALA | 7.0 | 89.5 | 2749±59 | DSC | (1–3) | 78V1 |
| regulatory subunit: | | | | | | |
| R, isolated | 7.0 | 55 | 272±4 | DSC | (1) | 78V1 |
| R, isolated, + CTP | 7.0 | 60 | 439±8 | DSC | (1) | 78V1 |
| R, isolated, + ATP | 7.0 | 59 | 439±17 | DSC | (1) | 78V1 |
| R in ATCase | 7.0 | 72.5 | 611±42 | DSC | (1,3) | 78V1 |
| R in ATCase + PALA | 7.0 | 74.5 | 452±13 | DSC | (1–3) | 78V1 |

Remarks:
(1) measured in 40 mM potassium phosphate, 2 mM 2-mercaptoethanol, 0.2 mM EDTA
(2) PALA = *N*-(phosphonacetyl)-L-aspartate
(3) data from curve analysis

## ATPase

Mitochondrial $F_1$-ATPase ($MF_1$)

| Ligand | conc. (mM) | pH | $T_{trs}$ | $\Delta H$ | Appr./Rem. | Ref |
|---|---|---|---|---|---|---|
| remark (1) | ~$10^{-3}$ | 7.0 | 52.8 | 4600 | DSC (1,3) | 98V5 |
| remark (2) | ~$7.0\times10^{-3}$ | 7.0 | 55.2 | 5400 | DSC (2,3) | 98V5 |
| ADP | $1.5\times10^{-2}$ | 7.0 | 55.4 | 5600 | DSC (3) | 98V5 |
| ADP | $5.0\times10^{-2}$ | 7.0 | 55.5 | 5500 | DSC (3) | 98V5 |
| ADP | 2.0 | 7.0 | 62.4 | 5400 | DSC (3) | 98V5 |
| ADP | 5.0 | 7.0 | 64.0 | 5800 | DSC (3) | 98V5 |
| ATP | 0.3 | 7.0 | 62.2 | 5200 | DSC (3) | 98V5 |

Mitochondrial F₁-ATPase (MF₁) (continued)

| Ligand | conc. (mM) | pH | $T_{trs}$ | $\Delta H$ | Appr./Rem. | Ref |
|---|---|---|---|---|---|---|
| ATP | 2.0 | 7.0 | 66.5 | 5600 | DSC (3) | 98V5 |
| ATP | 5.0 | 7.0 | 66.8 | 6800 | DSC (3) | 98V5 |
| $Mg^{2+}$ | $7.0 \times 10^{-3}$ | 7.0 | 59.5 | 6100 | DSC (3) | 98V5 |
| $ADP \cdot Mg^{2+}$ | 5.0 | 7.0 | 61.9 | ~4000 | DSC (3,4) | 98V5 |
| $ATP \cdot Mg^{2+}$ | 0.3 | 7.0 | 62.7 | ~5000 | DSC (3,4) | 98V5 |
| $ATP \cdot Mg^{2+}$ | 5.0 | 7.0 | 62.5 | ~6000 | DSC (3,4) | 98V5 |
| $AMPPNP \cdot Mg^{2+}$ | 0.25 | 7.0 | 68.4 | 6600 | DSC (3,4) | 98V5 |
| $AMPPNP \cdot Mg^{2+}$ | 5.0 | 7.0 | 70.1 | 6000 | DSC (3,4,5) | 98V5 |

Remarks:
(1) nucleotide depleted MF₁
(2) purified MF₁ containing three endogenous nucleotides
(3) buffer: 50 mM HEPES-NaOH, pH 7.0, 2 mM EDTA, 20 % glycerol
(4) magnesium: 7 mM $MgCl_2$
(5) additionally 3 mM sodium azide and 5 µM ADP


## ATP Synthase

$F_0F_1$ ATP synthase from thermophilic *Bacillus* PS3 in the presence of nucleotides

| Component added | pH | $T_{trs}$ | $\Delta H$ | Appr./Rem. | Ref |
|---|---|---|---|---|---|
| none | 8.0 | 81.7 | 4120 | DSC (1) | 97V4 |
| 0.6 mM ADP | 8.0 | 82.3 | 4100 | DSC (1) | 97V4 |
| 0.6 mM ATP | 8.0 | 82.5 | 4200 | DSC (1) | 97V4 |
| 1.0 mM $Mg^{2+}$ | 8.0 | 83.7 | | DSC (1,2) | 97V4 |
| 0.6 mM ADP + 1.0 mM $Mg^{2+}$ | 8.0 | 84.4 | | DSC (1) | 97V4 |
| 0.6 mM ATP + 1.0 mM $Mg^{2+}$ | 8.0 | 80.2 | | DSC (1,2) | 97V4 |
| 0.6 mM ADP, 0.1 mM ATP + 1.0 mM $Mg^{2+}$ | 8.0 | 82.5 | | DSC (1,2) | 97V4 |
| 0.6 mM ADP, 1.0 mM ATP + 1.0 mM $Mg^{2+}$ | 8.0 | 81.0 | | DSC (1,2) | 97V4 |

Remarks:
(1) buffer: 5 mM Tris-$H_2SO_4$
(2) overlapping with an exothermic transition, $T_{trs}$ might be underestimated by 0.4°C, $\Delta H$ was not determined for the same reason

$F_0F_1$ ATP synthase from thermophilic *Bacillus* PS3 in the presence of nucleotides and detergents

| Component added | pH | $T_{trs}$ | $\Delta H$ | Approach/Remarks | | Ref |
|---|---|---|---|---|---|---|
| none | 8.0 | 82.2 | 3470 | DSC | (1) | 97V4 |
| 1.0 mM $Mg^{2+}$ | 8.0 | 84.2 | | DSC | (1,2) | 97V4 |
| in the presence of 17 mg/ml cholate (chol.): | | | | | | |
| chol. | 8.0 | 75.2 | 3730 | DSC | (1) | 97V4 |
| chol. + 0.6 mM ADP | 8.0 | 76.3 | 3980 | DSC | (1) | 97V4 |
| chol. + 0.6 mM ATP | 8.0 | 77.8 | 3680 | DSC | (1) | 97V4 |
| chol. + 1.0 mM $Mg^{2+}$ | 8.0 | 76.4 | 3680 | DSC | (1) | 97V4 |
| in the presence of 9 mg/ml n-Octyl-β-D-glucopyranoside (OGP): | | | | | | |
| OGP | 8.0 | 75.0 | 2900 | DSC | (1) | 97V4 |
| OGP + 1.0 mM ATP | 8.0 | 76.0 | | DSC | (1,2) | 97V4 |
| OGP + 1.0 mM $Mg^{2+}$ | 8.0 | 77.5 | | DSC | (1) | 97V4 |

$F_0F_1$ ATP synthase from thermophilic *Bacillus* PS3 in the presence of nucleotides and detergents (continued)

| Component added | pH | $T_{trs}$ | $\Delta H$ | Approach/Remarks | | Ref |
|---|---|---|---|---|---|---|
| OGP + 0.6 mM ADP + 1.0 mM $Mg^{2+}$ | | | | | | |
| | 8.0 | 77.7 | | DSC | (1) | 97V4 |
| OGP + 1.0 mM ATP + 1.0 mM $Mg^{2+}$ | | | | | | |
| | 8.0 | 75.5 | 3270 | DSC | (1) | 97V4 |
| OGP + 0.1 mM ADP, 1.0 mM ATP + 1.0 mM $Mg^{2+}$ | | | | | | |
| | 8.0 | 73.0 | 2500 | DSC | (1) | 97V4 |

Remarks:
(1) buffer: 5 mM Tris-$H_2SO_4$ with 5 mg/ml Triton X-100
(2) overlapping with an exothermic transition, $T_{trs}$ might be underestimated by 0.4°C, $\Delta H$ was not determined for the same reason

## Avidin

Chicken egg avidin, glycosylated (native) and deglycosylated

| Protein | pH | $T_{trs}$ | $\Delta H$ | Approach/Remarks | | Ref |
|---|---|---|---|---|---|---|
| native | 8.0 | 74.4±0.1 | 1807±17 | DSC | (1) | 96W3 |
| deglycosylated | 8.0 | 74.2±0.1 | 1766±13 | DSC | (1) | 96W3 |

Remark:
(1) measured in 50 mM sodium phosphate in the presence of 1 M GuHCl

Avidin and avidin-biotin complex

| Protein | pH | $T_{trs}$ | $\Delta H$ | Approach/Remarks | Ref |
|---|---|---|---|---|---|
| avidin | 6.62 | 85 | 1289±50 | DSC | 73D1 |
| avidin | 6.84 | 85 | 1247±42 | DSC | 73D1 |
| avidin | 9.04 | 85 | 1109±84 | DSC | 73D1 |
| avidin | 9.14 | 85 | 1205±63 | DSC | 73D1 |
| avidin-biotin | 6.62 | 132 | 4636±230 | DSC | 73D1 |
| avidin-biotin | 6.84 | 131 | 4456±126 | DSC | 73D1 |
| avidin-biotin | 9.04 | 133 | 4243±209 | DSC | 73D1 |
| avidin-biotin | 9.14 | 132 | 3929±126 | DSC | 73D1 |

## Azurin

Azurin from *Pseudomonas aeruginosa*, wild type and double mutant (Cys3→Ala and Cys26→Ala)

| Protein | pH | $T_{trs}$ | $\Delta Cp$ | $\Delta H$ | Appr./Rem. | | Ref |
|---|---|---|---|---|---|---|---|
| wild type | 7.03 | 86.3 | 8.5 | 624 | DSC | (1,2,5) | 96M14 |
| | 7.03 | 86.3±0.5 | | 624.6±73 | DSC | (1,2) | 95L14 |
| | 7.03 | 20 | | 61.0±6.7 | DSC | (3,4) | 99G23 |
| double mutant (Cys3→Ala and Cys26→Ala) | | | | | | | |
| | 7.03 | 64.72±0.06 | 8.9±0.9 | 444±18 | DSC | (1,2,6) | 99G23 |
| | 7.03 | 20 | | 67.1±5.4 | DSC | (3) | 99G23 |

Remarks:
(1) DSC data extrapolated to infinite scan rate
(2) measured in 10 mM phosphate buffer, pH 7.03, with 0.1 M NaCl
(3) data extrapolated to the temperature of maximum stability (20°C)
(4) data from Ref. 96M14 cited in Ref. 99G23
(5) $\Delta Cp$ is a calculated value
(6) $\Delta Cp$ from $\Delta H$ versus $T_{trs}$ using extrapolated unfolding enthalpies at different pH values from 4.8 to 7.0

**Bacteriophage λ cI repressor**, see Repressor Proteins

## Barnase

Barnase, ribonuclease from *Bacillus amyloliquefaciens*, wild type and mutants, transition N → D

| Protein | pH | $T_{trs}$ | $\Delta Cp$ | $\Delta H$ | Appr./Rem. | | Ref |
|---|---|---|---|---|---|---|---|
| wild type | 6.3 | 54.4 | 7.11 | 591.6 | DSC | (1,2) | 98D2 |
| Ile4→Ala | 6.3 | 51.1 | | 537.6 | DSC | (1,2) | 98D2 |
| Thr6→Gly | 6.3 | 51.1 | | 551.9 | DSC | (1,2) | 98D2 |
| Asp8→Ala | 6.3 | 52.4 | | 586.2 | DSC | (1,2) | 98D2 |
| Asp8→Gly | 6.3 | 51.4 | | 575.7 | DSC | (1,2) | 98D2 |
| Asp12→Gly | 6.3 | 52.2 | | 565.7 | DSC | (1,2) | 98D2 |
| Tyr17→Ala | 6.3 | 49.3 | | 556.9 | DSC | (1,2) | 98D2 |
| His18→Asn | 6.3 | 50.3 | | 551.5 | DSC | (1,2) | 98D2 |
| Val36→Ala | 6.3 | 51.6 | | 551.8 | DSC | (1,2) | 98D2 |
| Asn41→Ala | 6.3 | 46.2 | | 477.8 | DSC | (1,2) | 98D2 |
| Asn58→Ala | 6.3 | 49.9 | | 536.8 | DSC | (1,2) | 98D2 |
| Ile88→Val | 6.3 | 51.3 | | 560.2 | DSC | (1,2) | 98D2 |
| Ser91→Ala | 6.3 | 49.7 | | 515.5 | DSC | (1,2) | 98D2 |
| Ser92→Ala | 6.3 | 47.7 | | 518.4 | DSC | (1,2) | 98D2 |
| Ile96→Val | 6.3 | 52.3 | | 580.3 | DSC | (1,2) | 98D2 |
| Thr105→Val | 6.3 | 49.3 | | 534.7 | DSC | (1,2) | 98D2 |
| double mutant (Ala43→Cys and Ser80→Cys) | | | | | | | |
| | 6.3 | 59.4 | | 585.8 | DSC | (1–3) | 98D1 |
| wild type in D$_2$O | | | | | | | |
| | 7.8 | 55.0 | | 569.4 | DSC | (1,4) | 98D1 |

Remarks:
(1)  $\Delta Cp = 7.11$ was taken from Ref. 95J1 and assumed to be the same for all mutants
(2) measured in 50 mM MES buffer/H$_2$O
(3) the stabilized disulfide mutant Cys43-Cys80, see also Ref. 93C7
(4) measured in 50 mM imidazole buffer/D$_2$O, pD 7.8

Barnase, ribonuclease from *Bacillus amyloliquefaciens*, wild type and mutants, transition I → D

| Protein | pH | $T_{ref}$ | $\Delta Cp$ | $\Delta H$ | Appr./Rem. | Ref |
|---|---|---|---|---|---|---|
| wild type | 6.3 | 25 | 3.01 | 113.8 | (1,2) | 98D2 |
| Ile4→Ala | 6.3 | 25 | 3.77 | 128.9 | (1,2) | 98D2 |
| Thr6→Gly | 6.3 | 25 | 3.85 | 127.6 | (1,2) | 98D2 |
| Asp8→Ala | 6.3 | 25 | 3.81 | 97.5 | (1,2) | 98D2 |
| Asp8→Gly | 6.3 | 25 | 3.18 | 109.6 | (1,2) | 98D2 |
| Asp12→Gly | 6.3 | 25 | 3.14 | 106.3 | (1,2) | 98D2 |
| Tyr17→Ala | 6.3 | 25 | 3.93 | 100.8 | (1,2) | 98D2 |
| His18→Asn | 6.3 | 25 | 3.18 | 101.7 | (1,2) | 98D2 |
| Val36→Ala | 6.3 | 25 | 3.47 | 77.8 | (1,2) | 98D2 |
| Asn41→Ala | 6.3 | 25 | 3.22 | 26.4 | (1,2) | 98D2 |
| Asn58→Ala | 6.3 | 25 | 3.93 | 67.4 | (1,2) | 98D2 |
| Ile88→Val | 6.3 | 25 | 5.15 | 98.3 | (1,2) | 98D2 |
| Ser91→Ala | 6.3 | 25 | 3.64 | 72.4 | (1,2) | 98D2 |
| Ser92→Ala | 6.3 | 25 | 7.61 | 44.4 | (1,2) | 98D2 |
| Ile96→Val | 6.3 | 25 | 4.94 | 125.1 | (1,2) | 98D2 |
| Thr105→Val | 6.3 | 25 | 3.93 | 117.6 | (1,2) | 98D2 |

Remarks:
(1) data from kinetics and equilibrium unfolding, for details see Ref. 98D2
(2) $\Delta Cp$ and $\Delta H$ from fitting $\Delta G(I \rightarrow D)$ versus T, for details see Ref. 98D2

Barnase, ribonuclease from *Bacillus amyloliquefaciens*, wild type and mutants in the presence and absence of chaperone

| Mutant | pH | $T_{trs}$ | $\Delta H$ | $\Delta T$(GroEL) | $\Delta T$(SecB) | Appr./Rem. | Ref |
|---|---|---|---|---|---|---|---|
| wild type | 6.3 | 53.9 | 590 | −9.2 | −7.1 | heat (1–4) | 97P7 |
| pro-barnase | 6.3 | 53.6 | 569 | −9.3 | −6.4 | heat (1–3) | 97P7 |
| Trp94→Tyr | 6.3 | 51.0 | 540 | −9.7 | −6.1 | heat (1–3) | 97P7 |
| Ser91→Ala | 6.3 | 49.6 | 552 | −9.2 | −7.3 | heat (1–4) | 97P7 |
| Asn58→Ala | 6.3 | 49.4 | 515 | −11 | −8.4 | heat (1–3) | 97P7 |
| Arg87→Ala | 6.3 | 47.6 | 494 | −8.2 | −5.8 | heat (1–3) | 97P7 |
| double mutant (Ile4→Ala and Ile51→Val) | | | | | | | |
| | 6.3 | 46.5 | 473 | −11 | | heat (1–3) | 97P7 |
| w.t., low-salt | 6.3 | 53.1 | 536 | −13 | −10 | heat (1–3,5) | 97P7 |
| w.t., high-salt | 6.3 | 55.1 | 598 | −4.0 | −1.3 | heat (1–3,6) | 97P7 |

Remarks:

(1) heat denaturation, van't Hoff treatment

(2) buffer: 50 mM MES, pH 6.3

(3) the transition temperature in the presence of GroEL and SecB is reduced; both chaperones bind to the denatured state of barnase, so lowering $T_{trs}$

(4) data from Zahn et al., Ref. 96Z

(5) buffer: 5 mM MES pH 6.3

(6) buffer: 50 mM MES pH 6.3 with 200 mM NaCl

Barnase, ribonuclease from *Bacillus amyloliquefaciens*, wild type and mutants with disulfide crosslinks

| Mutant | $T_{trs}$ | $\Delta Cp$ | Approach/Remarks | | Ref |
|---|---|---|---|---|---|
| wild type | 18–54 | 7.11±0.25 | DSC | (1) | 97J4 |
| CL(85–102) | 28–65 | 6.15±0.25 | DSC | (1,2) | 97J4 |
| CL(43–80) | 20–65 | 6.44±0.29 | DSC | (1,2) | 97J4 |
| CL(70–92) | 10–52 | 7.20±0.42 | DSC | (1,2) | 97J4 |

Remarks:

(1) $\Delta Cp$ from $\Delta H$ versus $T_{trs}$, $\Delta H$ from DSC and melting curves monitored by CD

(2) barnase mutants with crosslinks between residues 85 and 102, 43 and 80, and 70 and 92, respectively

Barnase (Ba) from *Bacillus amyloliquefaciens*, binase (Bi) from *Bacillus intermedius* 7P, and hybrides [remark (8)]

| Protein | pH | $T_{trs}$ | $\Delta H$ | Approach/Remarks | | Ref |
|---|---|---|---|---|---|---|
| barnase | 2.4 | 29.7 | 364 | DSC | (1,2) | 97R1 |
| | 5.5 | 54.3 | 527 | DSC | (1,2) | 97R1 |
| hybrid 73–110 | 2.4 | 32.9 | 356 | DSC | (1–4) | 97R1 |
| | 5.5 | 54.8 | 481 | DSC | (1–4) | 97R1 |
| hybrid 26–72 | 2.4 | 28.7 | 381 | DSC | (1–3,5) | 97R1 |
| | 5.5 | 53.7 | 540 | DSC | (1–3,5) | 97R1 |
| hybrid 26–110 | 2.4 | 31.0 | 351 | DSC | (1–3,6) | 97R1 |
| | 5.5 | 53.0 | 536 | DSC | (1–3,6) | 97R1 |

Barnase (Ba) from *Bacillus amyloliquefaciens*, binase (Bi) from *Bacillus intermedius* 7P, and hybrides [remark (8)] (continued)

| Protein | pH | $T_{trs}$ | $\Delta H$ | Approach/Remarks | | Ref |
|---|---|---|---|---|---|---|
| hybrid 1–72 | 2.4 | 29.2 | 343 | DSC | (1–3,7) | 97R1 |
| | 5.5 | 54.6 | 477 | DSC | (1–3,7) | 97R1 |
| binase | 2.4 | 33.6 | 389 | DSC | (1,2) | 97R1 |
| | 5.5 | 56.3 | 515 | DSC | (1,2) | 97R1 |

Remarks:

(1) buffers: 10 mM glycine at pH 2.4, and 10 mM sodium acetate at pH 5.5

(2) barnase and binase consist of 110 and 109 amino acid residues, respectively, and differ in 17 positions

(3) amino acid replacements are based on the barnase sequence

(4) hybrid 73–110 consists of (1–72)Ba/(73–110)Bi, and corresponds to barnase (Thr79→Val, Ser85→Ala, Ile88→Leu, Leu89→Val, Gln104→Ala, Lys108→Arg)

(5) hybrid 26–72 consists of (1–25)Ba/(26–72)Bi/(73–110)Ba, and corresponds to barnase (Glu29→Gln, Gln31→Ser, Asp44→Glu, Ile55→Val, Lys62→Arg, Gly65→Ser, Lys66→Ala)

(6) hybrid 26–110 consists of (1–25)Ba/(26–110)Bi, and corresponds to barnase (Glu29→Gln, Gln31→Ser, Asp44→Glu, Ile55→Val, Lys62→Arg, Gly65→Ser, Lys66→Ala, Thr79→Val, Ser85→Ala, Ile88→Leu, Leu89→Val, Gln104→Ala, Lys108→Arg)

(7) hybrid 1–72 consists of (1–72)Bi/(73–110)Ba, and corresponds to barnase (ΔGln2, Gln15→Ile, Thr16→Arg, His18→Lys, Lys19→Arg, Glu29→Gln, Gln31→Ser, Asp44→Glu, Ile55→Val, Lys62→Arg, Gly65→Ser, Lys66→Ala)

(8) see also Ref. 98S4

Barnase, ribonuclease from *Bacillus amyloliquefaciens*, complexes with barstar (BSR) and barstar mutants

| Sample | pH | $T_{trs}$ | $\Delta H$ | Appr./Rem. | Ref |
|---|---|---|---|---|---|
| barnase | 6.2 | 54.5 | 540 | DSC (1) | 99P12 |
| barnase + wild-type BSR | 6.2 | 78.4 | 1046 | DSC (1,3) | 99P12 |
| barnase + wild-type BSR | 6.2 | 78.3 | 1046 | DSC (1,2) | 99P12 |
| barnase + BSR Cys40→Ala | 6.2 | 75.3 | 778 | DSC (1) | 99P12 |
| barnase + BSR Cys82→Ala | 6.2 | 76.7 | 883 | DSC (1,4) | 99P12 |
| barnase + BSR double mutant (Cys40→Ala and Cys82→Ala) | | | | | |
| | 6.2 | 74.5 | 849 | DSC (1) | 99P12 |

Remarks:

(1) buffer: 10 mM Na-acetate, 0.05 M NaCl, pH 6.2

(2) measured in the presence of 1 mM DTT

(3) deconvolution yields $T_{trs1}$ = 72.8°C, $\Delta H_1$ = 364 kJ/mol, $T_{trs2}$ = 77.8°C, and $\Delta H_2$ = 664 kJ/mol

(4) deconvolution yields $T_{trs1}$ = 73.0°C, $\Delta H_1$ = 318 kJ/mol, $T_{trs2}$ = 75.0°C, and $\Delta H_2$ = 669 kJ/mol

## Barstar

Barstar, wild type and mutant (Cys82→Ala)

| Mutant | pH | $T_{trs}$ | $\Delta Cp$ | $\Delta H$ | Appr./Rem. | Ref |
|---|---|---|---|---|---|---|
| wild type | 6.4 | 74.1±0.3 | 4.9±1 | 320±15 | DSC (1) | 97S6 |
| wild type | 7.4 | 72.6±0.3 | 5.3±1 | 314±15 | DSC (1) | 97S6 |
| wild type | 8.0 | 71.9±0.3 | 4.7±1 | 308±15 | DSC (1) | 97S6 |
| wild type | 8.3 | 71.4±0.3 | 5.0±1 | 309±15 | DSC (1) | 97S6 |
| Cys82→Ala | 7.4 | 71.3±0.3 | 3.6±1 | 279±15 | DSC (1) | 97S6 |
| wild type | 7.4 | 73.0±0.3 | 5.3±1 | 296±15 | DSC (2) | 97S6 |
| Cys82→Ala | 7.4 | 71.2±0.3 | | 270±15 | DSC (2) | 97S6 |
| wild type | 6.4–8.0 | 71.4–74.1 | 4.3±2 | | DSC (3) | 97S6 |

Barstar, wild type and mutant (Cys82→Ala) (continued)

| Mutant | pH | $T_{trs}$ | $\Delta Cp$ | $\Delta H$ | Appr./Rem. | Ref |
|---|---|---|---|---|---|---|
| wild type | 7.4 | 73 | 5.3 | | DSC (4) | 97S6 |
| wild type | 7.4 | | 3.2 | | DSC (5,7) | 97S6 |
| wild type | 7.4 | 48.8 | | 160 | DSC (5,6) | 97S6 |

Remarks:
(1) measured in the presence of DTT, buffer: 50 mM sodium phosphate, 1 mM EDTA, 10 mM DTT
(2) measured in the absence of DTT, buffer: 50 mM sodium phosphate, 1 mM EDTA
(3) $\Delta Cp$ from $\Delta H$ versus $T_{trs}$
(4) average $\Delta Cp$ from individual transition curves, the value is regarded as the more reliable one in Ref. 97S6
(5) measured in buffer (2) containing additionally 1.87 M GuHCl
(6) heat denaturation peak in the presence of GuHCl, $\Delta H$ corrected for the barstar poulation taking part in the transition
(7) cold denaturation peak in the presence of GuHCl

Barstar mutant Cys40→Ala, Cys82→Ala and Pro27→Ala (b*C40A/C82A/P27A)

| Protein | pH | $T_{trs}$ | $\Delta H$ | Approach/Remarks | | Ref |
|---|---|---|---|---|---|---|
| b*C40A/C82A | 8.0 | 57.2 | 167 | heat | (1,2,4) | 99G9 |
| | 8.0 | 57.1 | 163 | heat | (1,3,4) | 99G9 |

Remarks:
(1) buffer: 50 mM sodium phosphate, pH 8.0
(2) transition monitored by far-UV CD at 222 nm
(3) transition monitored by near-UV CD at 280 nm
(4) Ref. 99G9 deals with the catalysis of the petidyl-prolyl cis/trans isomerization by human cytosolic cyclophilin

Barstar, wild type and Cys→Ala mutants

| Mutant | pH | $T_{trs}$ | $\Delta H$ | Approach/Remarks | | Ref |
|---|---|---|---|---|---|---|
| wild type | 6.2 | 73.3 | 222 | DSC | (1,2) | 99P12 |
| Cys40→Ala | 6.2 | 74.1 | 213 | DSC | (1,2) | 99P12 |
| Cys82→Ala | 6.2 | 71.9 | 205 | DSC | (1,2) | 99P12 |
| double mutant (Cys40→Ala and Cys82→Ala) | | | | | | |
| | 6.2 | 73.6 | 305 | DSC | (1) | 99P12 |

Remarks:
(1) buffer: 10 mM Na-acetate, 0.05 M NaCl, pH 6.2
(2) measured in the presence of 1 mM DTT

## Binase

Binase, ribonuclease from *Bacillus intermedius*, complexes with barstar (BSR) and barstar mutants

| Sample | pH | $T_{trs}$ | $\Delta H$ | Appr./Rem. | Ref |
|---|---|---|---|---|---|
| binase | 6.2 | 56.5 | 519 | DSC (1) | 99P12 |
| binase + wild-type BSR | 6.2 | 83.3 | 816 | DSC (1,2) | 99P12 |
| binase + BSR Cys40→Ala | 6.2 | 76.5 | 774 | DSC (1) | 99P12 |
| binase + BSR Cys82→Ala | 6.2 | 81.5 | 799 | DSC (1,3) | 99P12 |
| binase + BSR double mutant (Cys40→Ala and Cys82→Ala) | | | | | |
| | 6.2 | 76.6 | 799 | DSC (1) | 99P12 |

Remarks:
(1) buffer: 10 mM Na-acetate, 0.05 M NaCl, pH 6.2
(2) deconvolution yields $T_{trs1}$ = 75.8°C, $\Delta H_1$ = 343 kJ/mol, $T_{trs2}$ = 82.4°C, and $\Delta H_2$ = 519 kJ/mol
(3) deconvolution yields $T_{trs1}$ = 76.1°C, $\Delta H_1$ = 347 kJ/mol, $T_{trs2}$ = 81.2°C, and $\Delta H_2$ = 531 kJ/mol

**Binase,** see also Barnase

## Blood Coagulation Factor

Blood coagulation factor VIIa (fVIIa) and variants

| Protein | Form | pH | $T_{trs}$ | $\Delta H$ | Approach/Remarks | | Ref |
|---|---|---|---|---|---|---|---|
| fVIIa | $Ca^{2+}$ | 7.5 | 58.1 | 695 | DSC | (1–3) | 98F3 |
| des(1–44)-fVIIa | $Ca^{2+}$ | 7.5 | 58.7 | 427 | DSC | (2–4) | 98F3 |
| FFR-fVIIa | $Ca^{2+}$ | 7.5 | 56.6 | 778 | DSC | (2,3,5) | 98F3 |
| fVIIa | EDTA | 7.5 | 55.4 | 351 | DSC | (1,2,6) | 98F3 |
| des(1–44)-fVIIa | EDTA | 7.5 | 56.0 | 406 | DSC | (2,4,6) | 98F3 |
| FFR-fVIIa | EDTA | | | | | | |
| trans. (1) | | 7.5 | 53.2 | 402 | DSC | (2,5,6) | 98F3 |
| trans. (2) | | 7.5 | 44.5 | 247 | DSC | (2,5,6) | 98F3 |

Remarks:
(1) recombinant factor VIIa consisting of 406 amino acid residues
(2) buffer: 20 mM HEPES, 150 mM NaCl, pH 7.5
(3) measured in the presence of 5 mM $Ca^{2+}$
(4) des(1–44)-fVIIa = fragment lacking the 44 N-terminal residues
(5) FFR-fVIIa = fVIIa inhibited with Phe-Phe-Arg chloromethyl ketone
(6) measured in the presence of EDTA

## Bovine Serum Albumin (BSA)

Bovine serum albumin (BSA)

| | pH | $T_{trs}$ | $\Delta Cp$ | $\Delta H$ | Approach/Remarks | | Ref |
|---|---|---|---|---|---|---|---|
| | 6.3 | 63 | 38.1±1.8 | 683±16 | DSC | (1,2) | 92B10 |
| | 6.7 | 64 | 36.3±1.8 | 690±17 | DSC | (1,2) | 92B10 |
| | 7.0 | 64.5 | 32.4±1.8 | 720±3 | DSC | (1,2) | 92B10 |
| | 7.3 | 65 | 27.8±1.8 | 811±15 | DSC | (1,2) | 92B10 |
| | 7.5 | 63 | 26.1±1.8 | 742±6 | DSC | (1,2) | 92B10 |
| | 8.0 | 61 | 38.1±1.8 | 623±20 | DSC | (1,2) | 92B10 |
| | 8.5 | 57 | 22.3±1.8 | 505±6 | DSC | (1,2) | 92B10 |

Remarks:
(1) measured in 0.2 M Tris buffer
(2) BSA concentration $1 \times 10^{-4}$ M

Bovine serum albumin (BSA), concentration dependence

| Protein conc. | pH | $T_{trs}$ | $\Delta Cp$ | $\Delta H$ | Approach/Remarks | | Ref |
|---|---|---|---|---|---|---|---|
| $2.9 \times 10^{-5}$ | 5.0 | 62 | 49.1 | 524 | DSC | (1–3) | 92B10 |
| $2.9 \times 10^{-5}$ | 5.0 | 62 | 51.5 | 528 | DSC | (1–3) | 92B10 |
| $2.9 \times 10^{-5}$ | 5.0 | 62 | 49.8 | 518 | DSC | (1–3) | 92B10 |
| $2.9 \times 10^{-4}$ | 5.0 | 62 | 49.2 | 551 | DSC | (1–3) | 92B10 |
| $2.9 \times 10^{-4}$ | 5.0 | 62 | 49.2 | 540 | DSC | (1–3) | 92B10 |

Remarks:
(1) protein concentration in mol/l
(2) measured in 0.1 M acetate buffer, pH 5.0
(3) Ref. 92B10 contains a dissection of the heat effects of unfolding and aggregation

Bovine serum albumin (BSA), defatted

| Transition | pH | $T_{trs}$ | $\Delta Cp$ | $\Delta H$ | Approach/Remarks | | Ref |
|---|---|---|---|---|---|---|---|
| trans. (1) | 6.0 | 60.6 | | 254 | DSC | (1,2) | 97G6 |
| trans. (2) | 6.0 | 66.5 | | 681 | DSC | (1,2) | 97G6 |
| trans. (3) | 6.0 | 70.9 | | −73 | DSC | (1,2) | 97G6 |
| trans. (1) | 7.0 | 63.0 | | 228 | DSC | (1,2) | 97G6 |
| trans. (2) | 7.0 | 63.9 | | 739 | DSC | (1,2) | 97G6 |
| trans. (1) | 8.0 | 59.0 | | 194 | DSC | (1,2) | 97G6 |
| trans. (2) | 8.0 | 59.3 | | 589 | DSC | (1,2) | 97G6 |

Remarks:

(1) buffer: 0.01 M phosphate with 0.15 M NaCl

(2) results of deconvolution

Bovine serum albumin (BSA), defatted, dependence on ionic strength

| I | $T_{trs1}$ | $\Delta H_{trs1}$ | $T_{trs2}$ | $\Delta H_{trs2}$ | Appr./Rem. | Ref |
|---|---|---|---|---|---|---|
| 0 | 40.3 | 65 | 56.3 | 500 | DSC (1,2) | 97G6 |
| 0.15 | 63.0 | 238 | 63.9 | 739 | DSC (1,2) | 97G6 |
| 0.30 | 62.8 | 167 | 66.3 | 819 | DSC (1,2) | 97G6 |
| 0.50 | 62.8 | 100 | 68.1 | 834 | DSC (1,2 | 97G6 |
| 1.0 | | | | | DSC (1–3) | 97G6 |

Remarks:

(1) buffer: 0.01 M phosphate pH 7.0

(2) results of deconvolution

(3) peak maximum at $T_{trs}$ = 72.1°C with total heat $\Delta H_{trs}$ = 1100 kJ/mol

Bovine serum albumin (BSA), defatted, dependence on SDS concentration

| r=[SDS]/[BSA] | pH | $T_{max1}$ | $T_{max2}$ | $\Delta H_{total}$ | Appr./Rem. | Ref |
|---|---|---|---|---|---|---|
| 0.0 | 7.0 | 64 | | 938 | DSC (1,2) | 97G6 |
| 2.0 | 7.0 | 66.5 | 80.9 | 1271 | DSC (1,2) | 97G6 |
| 3.0 | 7.0 | 68.2 | 81.8 | 1402 | DSC (1,2) | 97G6 |
| 5.0 | 7.0 | 73.6 | 81.6 | 1497 | DSC (1,2) | 97G6 |
| 10.0 | 7.0 | | 84.0 | 1564 | DSC (1,2) | 97G6 |
| 20.0 | 7.0 | | 83.1 | 1455 | DSC (1,2) | 97G6 |
| 35.0 | 7.0 | | 78.2 | 649 | DSC (1,2) | 97G6 |

Remarks:

(1) buffer: 0.01 M phosphate with 0.15 M NaCl

(2) ratio r=[SDS]/[BSA]

Bovine serum albumin (BSA), fatty acid-free, in the presence of halothane

| Protein/Halothane | | pH | $T_{trs}$ | $\Delta Cp$ | $\Delta H$ | Appr./Rem. | Ref |
|---|---|---|---|---|---|---|---|
| BSA | 0 mM | 7 | 62.3 | 23.0±4.2 | 502 | DSC (1) | 99T7 |
| BSA | 10 mM | 7 | 71 | | 711 | DSC (1–4) | 99T7 |

Remarks:

(1) buffer: 10 mM potassium phosphate, 150 mM NaCl, pH 7

(2) the measurements performed in the presence of 0.6 to 10 mM halothane give a general expression

$\Delta H^{cal}$ = 520 + (T-62.3)×39.7 and $\Delta H^{v.H.}$ = 662 + (T-62.3)×13.6 ($\Delta H$ in kJ/mol, T in °C))

(3) $\Delta(\Delta G)$ due to halothane is estimated to be 16.7 kJ/mol

(4) for DSC of BSA in the presence of ethanol see Ref. 97Z3

Bovine serum albumin (BSA), in the presence of fatty acids and sodium dodecyl sulfate (SDS)

| Protein | pH | $T_{trs}$ | $\Delta H$ | Approach/Remarks | | Ref |
|---|---|---|---|---|---|---|
| BSA | 6.8±0.1 | 68 | 535 | DSC | (1,2) | 79G2 |
| BSA + 8.0 mol/mol lauric acid | | | | | | |
| | 6.8±0.1 | 92 | 1340 | DSC | (1,2) | 79G2 |
| BSA + 3.4 mol/mol stearic acid | | | | | | |
| | 6.8±0.1 | 83 | 535 | DSC | (1,2) | 79G2 |
| BSA + 10 mol/mol SDS | | | | | | |
| | 6.8±0.1 | 87 | 915 | DSC | (1,2) | 79G2 |

Remarks:
(1) Ref. 79G2 contains concentration dependent data
(2) the DSC measurements were performed at 5% (w/v) BSA and a scan rate of 10 K/min

*Bst*HU, see DANN-Binding Protein

## Cactaceae Proteins

Cactaceae proteins from *Machaerocereus gummosus* (Pitahaya agria, P.a.), *Lophocereu schottii* (Garambullo, G.), and *Cholla opuntia* (Cholla, C.)

| Protein | pH | $T_{trs}$ | $\Delta H$ | Approach/Remarks | | Ref |
|---|---|---|---|---|---|---|
| P.a. | | 61.2±0.8 | 918±15 | DSC | (1) | 99G10 |
| G. | | 52.0±1.3 | 315±31 | DSC | (1) | 99G10 |
| C. | | 59.0±1.1 | 508±19 | DSC | (1) | 99G10 |

Remark:
(1) Ref. 99G10 contains further DSC data of the cactaceae proteins obtained in the presence of urea

## Calmodulin

Calmodulin, synthetic protein hybrid of mammalian and plant calmodulin (SynCaM), apo-form, mutants

| Mutant | $T_{trs1}$ | $\Delta H_{trs1}$ | $T_{trs2}$ | $\Delta H_{trs2}$ | Appr./Rem. | Ref |
|---|---|---|---|---|---|---|
| in 10 mM cacodylate buffer, pH 7.5 | | | | | | |
| SynCaM | 47.7 | 275 | 61.2 | 184 | DSC (1) | 97P15 |
| SynCaM8 | 42.9 | 267 | 61.1 | 183 | DSC (1,2) | 97P15 |
| SynCaM12A | 43.4 | 243 | 58.5 | 170 | DSC (1,3) | 97P15 |
| SynCaM18A | 40.1 | 232 | 65.5 | 149 | DSC (1,4) | 97P15 |
| in 50 mM cacodylate buffer, pH 7.5 | | | | | | |
| SynCaM | 41.2 | 306 | 62.0 | 196 | DSC (1) | 97P15 |
| SynCaM8 | 50.1 | 320 | 64.7 | 200 | DSC (1,2) | 97P15 |
| SynCaM12A | 40.0 | 300 | 60.0 | 195 | DSC (1,3) | 97P15 |
| SynCaM18A | 34.1 | 221 | 62.8 | 123 | DSC (1,4) | 97P15 |

Remarks:
(1) results of deconvolution
(2) SynCaM8 = triple mutant Glu82,83,84→Lys
(3) SynCaM12A = triple mutant Asp118→Lys and Glu119,120→Lys
(4) SynCaM18A = mutations of SynCaM8 and SynCaM18A combined

Calmodulin from *Drosophila*, apo-form, wild type and mutants

| Mutant | domain | pH | $T_{trs}$ | $\Delta H$ | Approach/Remarks | | Ref |
|---|---|---|---|---|---|---|---|
| wild type | C-domain | 8.0 | 42 | 84 | heat | (1,2) | 97B12 |
| Ile27→Gly | N-domain | 8.0 | <20 | | heat | (1,2) | 97B12 |
| Ile27→Gly | C-domain | 8.0 | 42 | 84 | heat | (1,2) | 97B12 |
| Ile63→Gly | N-domain | 8.0 | <20 | | heat | (1,2) | 97B12 |
| Ile100→Gly | N-domain | 8.0 | 55 | 167 | heat | (1,2) | 97B12 |
| Ile100→Gly | C-domain | 8.0 | ≈0 | (80) | heat | (1,2) | 97B12 |
| Val136→Gly | N-domain | 8.0 | 55 | 167 | heat | (1,2) | 97B12 |
| Val136→Gly | C-domain | 8.0 | ≈0 | (80) | heat | (1,2) | 97B12 |

Remarks:
(1) buffer: 25 mM Tris, 100 mM KCl, 1 mM EGTA, pH 8.0
(2) Ref. 97B12 contains additional data on calcium binding and thermostability of the holo-protein

**Catabolic Activator Protein**, see CRP, cAMP Receptor Protein

**Catalase**

Catalase I from *Bacillus stearothermophilus*, high- and low-activity form

| Protein/Transition | pH | $T_{trs}$ | $\Delta Cp$ | $\Delta H$ | Appr./Rem. | | Ref |
|---|---|---|---|---|---|---|---|
| low-activity form | | | | | | | |
| trans. (1) | 7.0 | 55.8 | 4.2 | 91 | DSC | (1,2) | 97K6 |
| trans. (2) | 7.0 | 75.9 | 12.3 | 2145 | DSC | (1,2) | 97K6 |
| high-activity form | 7.0 | 75.8 | 13.7 | 2074 | DSC | (1,3,4) | 97K6 |

Remarks:
(1) measured in 50 mM potassium phosphate, pH 7.0
(2) unheated enzyme
(3) single melting peak
(4) the high-activity form was produced by preheating to 65°C

**Cell Surface Receptor Protein CD2**

Domain 1 of cell surface receptor protein CD2 from rat, transient intermediates and transition states (TS)

| Transition | pH | $T_{ref}$ | $\Delta Cp$ | $\Delta H$ | Approach/Remarks | | Ref |
|---|---|---|---|---|---|---|---|
| I → U | 4.5 | 25 | 3.60±0.59 | 19.5±3.7 | GuHCl | (1–3) | 98P5 |
| N → I | 4.5 | 25 | 4.35±0.84 | 78.7±5.9 | GuHCl | (1–3) | 98P5 |
| TS → I | 4.5 | 25 | 1.09±0.67 | 52.7±4.2 | GuHCl | (1–3) | 98P5 |
| N → TS | 4.5 | 25 | 3.26±1.13 | 131.4±7.5 | GuHCl | (1–3) | 98P5 |

Remarks:
(1) from folding and unfolding rates measured by stopped-flow kinetics
(2) transition monitored by fluorescence intensity
(3) buffer: 50 mM sodium acetate and GuHCl

## Cellulase

Isolated catalytic domain of cellulase E2 from thermophile *Thermomonospora fusca* (E2$_{cd}$) compared with the analogous domain of the cellulase from the mesophile *Cellulomonas fimi* (CenA$_{P30}$)

| Protein | pH | T$_{trs}$ | ΔH | Approach/Remarks | | Ref |
|---|---|---|---|---|---|---|
| E2$_{cd}$ | 6.8 | 72.3±0.3 | 933±92 | heat | (1,4) | 99B5 |
| | 6.8 | 71.4±1.5 | 1092±105 | heat | (2,4) | 99B5 |
| | 6.8 | 72.6±0.1 | 904±59 | heat | (3,4) | 99B5 |
| CenA$_{P30}$ | 6.8 | 49.9±0.6 | 352±67 | heat | (1,5) | 99B5 |
| | 6.8 | 50.1±0.4 | 377±23 | heat | (3,5) | 99B5 |
| E2$_{cd}$ | 6.8 | 72.2±0.2 | 795±59 | heat | (1,6) | 99B5 |
| CenA$_{P30}$ | 6.8 | 56.4±0.3 | 448±13 | heat | (1,6) | 99B5 |

Remarks:
(1) transition monitored by far-UV CD at 223 or 228 nm
(2) transition monitored by near-UV CD at 285 nm
(3) transition monitored by fluorescence
(4) buffer: 50 mM potassium phosphate, 100 mM KCl, pH 6.8
(5) buffer: 50 mM potassium phosphate, 800 mM KCl, 45% ethylene glycol, pH 6.8
(5) buffer: 50 mM potassium phosphate, 225 mM KCl, 11.25% ethylene glycol, pH 6.8

Cellulase from *Streptomyces halstedii* JM8

| pH | T$_{trs}$ | ΔH | Approach/Remarks | | Ref |
|---|---|---|---|---|---|
| 3.25 | 43.3 | 288 | DSC | (1–3) | 96G11 |
| 6.0 | 54.5 | 348 | DSC | (1–3) | 96G11 |

Remarks:
(1) the protein undergoes irreversible denaturation, the thermal transition temperature is scan rate dependent
(2) measured in 0.1 M phosphate buffer at a scan rate of 59.5 K/h
(3) for a detailed analysis of various models of irreversible protein denaturation, see Refs. 97K12 and 98L8

## Ceruloplasmin

Ceruloplasmin from various species

| Protein | Transition | pH | T$_{trs}$ | ΔH | Appr./Rem. | | Ref |
|---|---|---|---|---|---|---|---|
| sheep, native protein | | | | | | | |
| | overall | 6.8 | 71.2 | 2301 | DSC | (1,4) | 97C1 |
| | trans. (1) | 6.8 | 67.6 | (32%) | DSC | (2,3) | 97C1 |
| | trans. (2) | 6.8 | 71.3 | (39%) | DSC | (2,3) | 97C1 |
| | trans. (3) | 6.8 | 76.1 | (29%) | DSC | (2,3) | 97C1 |
| | trans. (3) | 6.8 | 73.5 | (29%) | DSC | (2,3) | 90B12 |
| sheep, apoceruloplasmin | | | | | | | |
| | overall | 6.8 | 57.2 | 1565 | DSC | (1,4) | 90B12 |
| | trans. (1) | 6.8 | 55.9 | (42%) | DSC | (2,3) | 90B12 |
| | trans. (2) | 6.8 | 58.0 | (58%) | DSC | (2,3) | 90B12 |
| chicken | | | | | | | |
| | overall | 7.4 | 82.1 | 1435 | DSC | (1,4) | 97C1 |
| | trans. (1) | 7.4 | 74.0 | (17%) | DSC | (2,3) | 97C1 |
| | trans. (2) | 7.4 | 78.0 | (27%) | DSC | (2,3) | 97C1 |
| | trans. (3) | 7.4 | 82.1 | (56%) | DSC | (2,3) | 97C1 |

Ceruloplasmin from various species (continued)

| Protein | Transition | pH | $T_{trs}$ | $\Delta H$ | Appr./Rem. | | Ref |
|---|---|---|---|---|---|---|---|
| turtle | | | | | | | |
| | overall | 7.4 | 57.8 | 1506 | DSC | (1,4) | 97C1 |
| | trans. (1) | 7.4 | 42.8 | (27%) | DSC | (2,3) | 97C1 |
| | trans. (2) | 7.4 | 57.8 | (47%) | DSC | (2,3) | 97C1 |
| | trans. (3) | 7.4 | 58.8 | (26%) | DSC | (2,3) | 97C1 |
| dolphin | | | | | | | |
| | overall | 7.4 | 70.6 | 2096 | DSC | (1,5) | 97C1 |
| | trans. (1) | 7.4 | 64.8 | (24%) | DSC | (2,3) | 97C1 |
| | trans. (2) | 7.4 | 69.8 | (37%) | DSC | (2,3) | 97C1 |
| | trans. (3) | 7.4 | 71.5 | (39%) | DSC | (2,3) | 97C1 |
| human | | | | | | | |
| | overall | 7.4 | 85.1 | 1435 | DSC | (1,6) | 97C1 |
| | trans. (1) | 7.4 | 79.5 | (19%) | DSC | (2,3) | 97C1 |
| | trans. (2) | 7.4 | 83.1 | (35%) | DSC | (2,3) | 97C1 |
| | trans. (3) | 7.4 | 85.2 | (46%) | DSC | (2,3) | 97C1 |

Remarks:
(1) given is the peak maximum temperature and the total calorimetric heat
(2) results of deconvolution with per cent of total calorimetric heat
(3) a further transition (omitted here) which was observed at 59.1°C may be due to a portion of apoceruloplasmin
(4) for further details including pretreated samples see Ref. 90B12
(5) for further details see Ref. 92B12
(6) for further details see Ref. 96M16

## Chaperone

GroEL from *E. coli*, influence of ligands on thermostability

| Protein Conditions | pH | $T_{trs}$ | $\Delta H$ | Approach/Remarks | | Ref |
|---|---|---|---|---|---|---|
| GroEL, buffer A (1) | 7.6 | 67±0.5 | 1140±215 | DSC | (3) | 97S15 |
| GroEL, buffer B (2) | 7.6 | 70±0.5 | 1140±215 | DSC | (3) | 97S15 |
| GroEL, 2 mM ADP, buffer B | 7.6 | 66±0.5 | 1290±215 | DSC | (2,3) | 97S15 |
| GroES, 2 mM ADP, buffer B | 7.6 | 76±0.5 | 170±40 | DSC | (2,3) | 97S15 |
| GroEL, GroES, 2 mM ADP, buffer B | | | | | | |
| | 7.6 | 69±0.5 | 1360±215 | DSC | (2,3,4) | 97S15 |

Remarks:
(1) buffer A: 50 mM HEPES, pH 7.6
(2) buffer B: 50 mM HEPES, 100 mM KCl, 10 mM magnesium acetate, pH 7.6
(3) $\Delta H$ per subunit
(4) GroEL and GroES molar ratio 1 : 1.5

Co-chaperonin GroES

| Potein conc. | pH | $T_{trs}$ | $\Delta Cp$ | $\Delta H$ | Approach/Remarks | | Ref |
|---|---|---|---|---|---|---|---|
| 11.5 µM | 7.0 | 68.7 | 4.18 | 310 | heat, v.H. | (1–5) | 97B11 |
| 15.0 µM | 7.0 | 69.6 | | 226 | heat, v.H. | (1–5) | 97B11 |
| 22.5 µM | 7.0 | 70.5 | | 285 | heat, v.H. | (1–5) | 97B11 |
| 30.0 µM | 7.0 | 71.9 | | 267 | heat, v.H. | (1–5) | 97B11 |
| 37.5 µM | 7.0 | 72.5 | | 267 | heat, v.H. | (1–5) | 97B11 |
| 45.0 µM | 7.0 | 73.0 | | 281 | heat, v.H. | (1–5) | 97B11 |
| 52.5 µM | 7.0 | 72.6 | | 251 | heat, v.H. | (1–5) | 97B11 |

Co-chaperonin GroES (continued)

| Potein conc. | pH | $T_{trs}$ | $\Delta Cp$ | $\Delta H$ | Approach/Remarks | | Ref |
|---|---|---|---|---|---|---|---|
| | 7.0 | | | 285 | heat, v.H. | (1–6) | 97B11 |
| 18.6 μM | 7.0 | 68.7 | 4.80 | 249 | DSC | (1–3,5) | 97B11 |
| 36.0 μM | 7.0 | 70.5 | 4.28 | 245 | DSC | (1–3,5) | 97B11 |
| 54.4 μM | 7.0 | 71.8 | 4.65 | 261 | DSC | (1–3,5) | 97B11 |
| 87.0 μM | 7.0 | 72.0 | 4.05 | 256 | DSC | (1–3,5) | 97B11 |
| 148.0 μM | 7.0 | 74.2 | 4.59 | 266 | DSC | (1–3,5) | 97B11 |
| 175.0 μM | 7.0 | 74.7 | 4.18 | 264 | DSC | (1–3,5) | 97B11 |
| | 7.0 | | 4.31 | 266 | DSC | (1–3,5,6) | 97B11 |
| | 7.0 | | 5.73 | | DSC/urea (1,7) | | 97B11 |

Remarks:

(1) $T_{trs}$ is defined here by $T_{1/2}$, and $\Delta H$ by $\Delta H(70°C)$

(2) transition monitored by CD at 222 nm

(3) buffer: 20 mM sodium phosphate

(4) $\Delta Cp$ was fixed at 4.18 kJ/mol/K to fit the CD data

(5) GroES melting is a spontaneous reversible process involving a highly cooperative transition between folded heptamers and unfolded monomers

(6) result of a global fit

(7) from $\Delta H$ versus $T_{trs}$ of measurements performed in the presence of 0 to 3 M urea at 31 μM protein

**Catabolic Activator Protein**, see also CRP, cAMP Receptor Protein

# Chitosanase

Chitosanase from *Streptomyces* sp. N174 and tryptophan to phenylalanine mutants

| Protein/Transition | pH | $T_{trs}$ | $\Delta T$ | $\Delta H$ | Approach/Remarks | | Ref |
|---|---|---|---|---|---|---|---|
| Data obtained by CD measurements: | | | | | | | |
| wild type | | | | | | | |
| trans. (1) | 7.0 | 42.7 | 0.0 | 502 | heat | (1–3) | 99H11 |
| trans. (2) | 7.0 | 52.2 | 0.0 | 544 | heat | (1–3) | 99H11 |
| Trp28→Phe | | | | | | | |
| trans. (1) | 7.0 | 35.6 | −7.1 | 649 | heat | (1–3) | 99H11 |
| trans. (2) | 7.0 | 45.2 | −7.0 | 335 | heat | (1–3) | 99H11 |
| Trp101→Phe | | | | | | | |
| trans. (1) | 7.0 | 35.6 | −7.1 | 711 | heat | (1–3) | 99H11 |
| trans. (2) | 7.0 | 44.7 | −7.5 | 335 | heat | (1–3) | 99H11 |
| Trp227→Phe | | | | | | | |
| trans. (1) | 7.0 | 35.9 | −6.8 | 565 | heat | (1–3) | 99H11 |
| trans. (2) | 7.0 | 45.8 | −6.4 | 272 | heat | (1–3) | 99H11 |
| double mutant (Trp28→Phe and Trp101→Phe) | | | | | | | |
| trans. (1) | 7.0 | 31.3 | −11.4 | 649 | heat | (1–3) | 99H11 |
| trans. (2) | 7.0 | 44.2 | −8.0 | 251 | heat | (1–3) | 99H11 |
| Data obtained by DSC measurements: | | | | | | | |
| wild type | | | | | | | |
| trans. (1) | 7.0 | 43.8 | | | DSC | | 99H11 |
| trans. (2) | 7.0 | 50–52 | | | DSC | (6) | 99H11 |
| double mutant (Trp28→Phe and Trp101→Phe) | | | | | | | |
| trans. (1) | 7.0 | 32.6 | | | DSC | | 99H11 |
| trans. (2) | 7.0 | 47.4 | | | DSC | | 99H11 |

Chitosanase from *Streptomyces* sp. N174 and tryptophan to phenylalanine mutants (continued)

| Protein/Transition | pH | $T_{trs}$ | $\Delta T$ | $\Delta H$ | Approach/Remarks | | Ref |
|---|---|---|---|---|---|---|---|
| Data obtained by fluorescence measurments: | | | | | | | |
| wild type | 7.0 | 43.2 | 0.0 | 672 | heat | (3–5) | 99H11 |
| Trp28→Phe | 7.0 | 36.2 | –7.0 | 822 | heat | (3–5) | 99H11 |
| Trp101→Phe | 7.0 | 36.6 | –6.6 | 838 | heat | (3–5) | 99H11 |
| Trp227→Phe | 7.0 | 37.3 | –5.9 | 552 | heat | (3–5) | 99H11 |
| double mutant (Trp28→Phe and Trp101→Phe) | | | | | | | |
| | 7.0 | 31.2 | –12.0 | 852 | heat | (3–5) | 99H11 |

Remarks:

(1) transition monitored by CD at 222 nm

(2) the transition profile characteristic of a three-state transition was treated by a model derived from $N \rightarrow I \rightarrow U$

(3) measured in 50 mM sodium phosphate buffer, pH 7.0

(4) transition monitored by fluorescence at 330 nm

(5) the transition profile was treated by a two-state transition

(6) a very small second peak was observed at 50–52°C

## Cholinesterase

Butyrylcholinesterase from human plasma (wtBuChE)

| Protein | pH | $T_{trs}$ | $\Delta H$ | Approach/Remarks | | Ref |
|---|---|---|---|---|---|---|
| native wtBuChE | 8.0 | 62.8 | 780 | DSC | (1,2) | 99M12 |

Remarks:

(1) native (non-phosphorylated) wtBuChE is compared with the soman-aged protein and mutants (Glu197→Asp) and (Asp70→Gly) in Ref. 99M12

(2) measured in 50 mM sodium phosphate, pH 8.0

## Chymotrypsin

Bovine pancreatic α-chymotrypsin (α-CT) in the absence and presence of maltosyl-β-cyclodextrin and maltose

| Protein | pH | $T_{trs}$ | $\Delta H$ | Approach/Remarks | | Ref |
|---|---|---|---|---|---|---|
| α-CT | 7.4 | 48.9±0.2 | 699± 8 | DSC | (1) | 97M3 |
| α-CT + G2-β-CyD | 7.4 | 48.7±0.1 | 527±25 | DSC | (1,2) | 97M3 |
| α-CT + maltose | 7.4 | 54.2±0.4 | 787±59 | DSC | (1,3) | 97M3 |

Remarks:

(1) measured in isotonic phosphate buffer, pH 7.4

(2) measured in the presence of 0.1 M 6-O-maltosyl-β-cyclodextrin (G2-β-CyD)

(3) measured in the presence of 1 M maltose

## Chymotrypsinogen

Chymotrypsinogen (CGN), α-chymotrypsin (α-CT) and derivatives

| Protein | pH | $T_{trs}$ | $\Delta Cp$ | $\Delta H$ | Appr./Rem. | | Ref |
|---------|-----|------|-----------|------|------|------|------|
| CGN | 2.0 | 41.3 | 13.4±1.7 | 389 | heat | | 71S1 |
| CGN | 4.5–1 | 50 | 12.6 | 515 | ITC | (3,4) | 71B1 |
| CGN | | 25 | | 295 | DSC | (5,6) | 96P5 |
| DMSCGN | 2.0 | 24.5 | 9.6±0.8 | 141 | heat | (1) | 71S1 |
| DMSCGN | 4.5–1 | 40 | 7.9 | 305 | ITC | (3,4) | 71B1 |
| DMSCGN | 4.5–1 | 25 | | 134 | ITC | (3,4) | 71B1 |
| DPC-α-CT | 2.0 | 30.9 | 17.6±1.3 | 331 | heat | (2) | 71S1 |
| α-CT | 4.5–1 | 40 | 13.8 | 460 | ITC | (3,4) | 71B1 |

Remarks:

(1) DMSCGN = dimethionine sulfoxide derivative of chymotrypsinogen

(2) DPC-α-CT = diphenylcarbamyl-α-chymotrypsin

(3) $\Delta H$ from isothermal calorimetric pH titrations

(4) $\Delta Cp$ from precision calorimetry, difference value of the folded state at pH 4 and the unfolded state at pH 2

(5) extrapolated to 25°C from DSC measurements

(6) Ref. 96P5 reports on interactions of α-chymotrypsinogen A with alkylureas

α-chymotrypsinogen A, heat-induced denatured states [remarks (1–3)]

| pH | $T_{trs}$ | $\Delta Cp$ | Approach/Remarks | | Ref |
|-----|------|-----------|------|------|------|
| 1.3–3.6 | 37–59 | 8.4 | heat, v.H.(4) | | 97C6 |
| 1.3–3.6 | 35–57 | 9.6±2.1 | DSC | (5) | 97C6 |

Remarks:

(1) Ref. 97C6 contains a phase diagram ($T_{trs}$ versus pH) obtained by a combination of ultrasonic velocimetry, densimetry, and differential scanning calorimetry, in conjunction with UV absorbance and CD spectroscopy

(2a) at neutral and alkaline pH an unfolded heat denatured state is found lacking both tertiary and secondary structure, being similar to the urea-unfolded state

(2b) below pH 3.1 the heat denatured protein is molten globule-like

(2c) at intermediate pH values both molten globule and unfolded properties are found

(3) at low pH increase in compressibility of the native protein, i.e. less tight packing, is shown by volumetric data

(4) $\Delta Cp$ from $\Delta H$ versus $T_{trs}$ using $\Delta H^{v.H}$ from UV melting profiles

(5) $\Delta Cp$ from $\Delta H$ versus $T_{trs}$ using $\Delta H^{cal}$ from DSC

α-Chymotrypsinogen in aqueous polyol solutions

| Cosolvent | Conc. | pH | $T_{trs}$ | $\Delta H$ | Appr./Rem. | Ref |
|-----------|-------|-----|------|------|------|------|
| control | | 2.5 | 44.9 | 402.9 | heat (1,2) | 98K3 |
| mannitol | 1.00 M | 2.5 | 50.5 | 393.3 | heat (1,2) | 98K3 |
| inositol | 0.75 M | 2.5 | 52.3 | 419.2 | heat (1,2) | 98K3 |
| xylitol | 2.00 M | 2.5 | 52.6 | 426.3 | heat (1,2) | 98K3 |
| adonitol | 2.00 M | 2.5 | 52.5 | 416.3 | heat (1,2) | 98K3 |
| sorbitol | 2.00 M | 2.5 | 54.7 | 410.9 | heat (1,2) | 98K3 |

Remarks:

(1) transition monitored by optical absorption at 287 nm

(2) buffer: 20 mM glycine, pH 2.5

Recombinant rat wild-type chymotrypsinogen and Δ-chymotrypsinogen

| Protein | pH | $T_{trs}$ | ΔH | Approach/Remarks | | Ref |
|---|---|---|---|---|---|---|
| chymotrypsinogen | 5.4 | 59.9 | 567 | DSC | (1,2) | 99K2 |
| Δ-chymotrypsinogen | 5.4 | 51.6 | 260 | DSC | (1–3) | 99K2 |

Remarks:
(1) buffer: 50 mM sodium acetate, 100 mM NaCl, 10 mM CaCl$_2$
(2) measured at 0.5 mg/ml protein measured at 60°C/h scan rate
(3) Δ-chymotrypsinogen, mutant of rat chymotrypsinogen containing the six-amino acid propeptide of rat trypsinogen instead of the 15-amino acid propeptide of rat chymotrypsinogen with a Cys122→Ser

c-Myb, see Protooncogene Product

## Coiled-Coil

Coiled-coil model peptides

| Protein | pH | $T_{trs}$ | ΔCp | ΔH | Approach/Remarks | | Ref |
|---|---|---|---|---|---|---|---|
| dKL27ox | 4.7 | 75.1±0.04 | 2.8±0.22 | 46.8±1.1 | heat | (1,2,4–6) | 97H7 |
| mKL27ox | 4.7 | 75.5±0.06 | 0.6±0.07 | 72.4±0.6 | heat | (1,3–6) | 97H7 |

Remarks:
(1) model peptide: Trp-Cys(SH)-Lys-Ala-Val-Lys-Lys-Leu-Ala-Lys-Ala-Val-Lys-Lys-Leu- Ala-Lys-Ala-Val-Lys-Lys-Leu-Ala-Lys-Ala-Gly-Cys(SH)-NH$_2$
(2) dKL27ox: parallel dimer connected by the Cys residues at the N- and C-termini
(3) mKL27ox: monomeric circular species in which a disulfide bond was formed intramolecularly between the N-terminal and C-terminal Cys residues
(4) buffer: 10 mM sodium acetate, pH 4.7, with 20 mM NaClO$_4$
(5) van't Hoff treatment of a two-state transition $U_2 \rightarrow H_2$, with U = unfolded and H = helical state
(6) the transition was monitored by the ellipticity at 222 nm

## Cold Shock Protein

Cold shock protein (CspB) from *Bacillus subtilis*, wild type and Phe to Ala mutants

| Mutant | pH | $T_{ref}$ | ΔCp | ΔH | Appr./Rem. | Ref |
|---|---|---|---|---|---|---|
| wild type | 7.0 | 45 | 4.7±0.6 | 161±2 | heat (1,2) | 99J3 |
| Phe15→Ala | 7.0 | 45 | 3.9±0.5 | 157±8 | heat (1,2) | 99J3 |
| Phe17→Ala | 7.0 | 45 | 4.1±0.3 | 153±3 | heat (1,2) | 99J3 |
| Phe27→Ala | 7.0 | 45 | 3.8±0.5 | 140±2 | heat (1,2) | 99J3 |

Remarks:
(1) transition monitored by CD at 223 nm
(2) buffer: 0.1 M sodium cacodylate/HCl, pH 7.0

Cold shock protein CspB from *Bacillus subtilis* in the presence of ethylene glycol

| Ethylene glycol | pH | $T_{trs}$ | ΔH | Approach/Remarks | | Ref |
|---|---|---|---|---|---|---|
| 0 g/100 ml | 7.0 | 47.0±0.2 | 167±4 | heat | (1–3) | 99J2 |
| 5 g/100 ml | 7.0 | 47.1±0.1 | 176±2 | heat | (1–3) | 99J2 |
| 10 g/100 ml | 7.0 | 48.0±0.1 | 184±3 | heat | (1–3) | 99J2 |
| 15 g/100 ml | 7.0 | 47.1±0.1 | 199±2 | heat | (1–3) | 99J2 |

Cold shock protein CspB from *Bacillus subtilis* in the presence of ethylene glycol (continued)

| Ethylene glycol | pH | $T_{trs}$ | $\Delta H$ | Approach/Remarks | | Ref |
|---|---|---|---|---|---|---|
| 20 g/100 ml | 7.0 | 48.5±0.1 | 209±3 | heat | (1–3) | 99J2 |
| 25 g/100 ml | 7.0 | 48.7±0.1 | 219±2 | heat | (1–3) | 99J2 |
| 30 g/100 ml | 7.0 | 48.9±0.1 | 236±1 | heat | (1–3) | 99J2 |
| 35 g/100 ml | 7.0 | 48.5±0.1 | 241±1 | heat | (1–3) | 99J2 |
| 40 g/100 ml | 7.0 | 48.9±0.1 | 254±2 | heat | (1–3) | 99J2 |

Remarks:
(1) Ref. 99J2 concerns the effect of protein stability on the folding kinetics
(2) transition monitored by CD at 223 nm
(3) buffer: 10 mM cacodylate-HCl, pH 7.0

Cold shock protein (Csp) from *Thermotoga maritima*

| Protein | pH | $T_{trs}$ | $\Delta Cp$ | $\Delta H$ | Appr./Rem. | | Ref |
|---|---|---|---|---|---|---|---|
| Csp | 4.0 | 87.9 | −1.51 | 209.6 | DSC | (1,2) | 99W6 |
| | 4.25 | 89.3 | 0.08 | 224.3 | DSC | (1,2) | 99W6 |
| | 4.5 | 91.6 | 0.21 | 235.1 | DSC | (1,2) | 99W6 |
| | 4.8 | 94.1 | −0.92 | 259.4 | DSC | (1,2) | 99W6 |
| | 5.7 | 93.3 | 0.33 | 300.8 | DSC | (1,2) | 99W6 |
| | 6.0 | 92.1 | −0.71 | 312.1 | DSC | (1,2) | 99W6 |
| | 6.1 | 91.2 | −2.30 | 290.8 | DSC | (1,2) | 99W6 |
| | 6.35 | 89.6 | −2.09 | 284.5 | DSC | (1,2) | 99W6 |
| | 6.5 | 88.0 | 0.00 | 295.0 | DSC | (1,2) | 99W6 |
| | 6.75 | 85.8 | −0.75 | 272.0 | DSC | (1,2) | 99W6 |
| | 6.9 | 83.4 | −0.13 | 274.5 | DSC | (1,2) | 99W6 |
| | 7.0 | 82.0 | 1.51 | 261.9 | DSC | (1,2) | 99W6 |
| | 7.2 | 83.1 | 0.92 | 273.2 | DSC | (1,2) | 99W6 |
| | 7.25 | 82.8 | −1.67 | 256.5 | DSC | (1,2) | 99W6 |
| | 7.5 | 81.1 | 0.46 | 227.6 | DSC | (1,2) | 99W6 |
| | 7.5 | 81.4 | −0.38 | 243.5 | DSC | (1,2) | 99W6 |
| | 7.6 | 80.8 | −1.59 | 254.4 | DSC | (1,2) | 99W6 |
| | 8.0 | 79.5 | 0.17 | 246.4 | DSC | (1,2) | 99W6 |
| | 8.1 | 81.0 | 0.04 | 228.9 | DSC | (1,2) | 99W6 |
| | 8.4 | 77.5 | −2.09 | 234.7 | DSC | (1,2) | 99W6 |
| | 8.5 | 77.6 | 0.25 | 225.5 | DSC | (1,2) | 99W6 |
| | 8.9 | 75.5 | −2.09 | 234.7 | DSC | (1,2) | 99W6 |
| | 5.7–8.5 | 93–77 | 4.6±0.4 | 301–225 | DSC | (1,3) | 99W6 |

Remarks:
(1) buffers: 20 mM sodium acetate (pH 4.0–4.8), 20 mM cacodylate/HCl (pH 5.7–7.4), 10 mM sodium phosphate (pH 7.0–7.8), 20 mM borate/NaOH (pH 8.0–8.9)
(2) error limits are ±0.2 K for $T_{trs}$, ±5% in $\Delta H^{cal}$ and ±10% in $\Delta Cp$, pH at 25°C (detailed pH correction at $T_{trs}$ is given in Ref. 99W6)
(3) $\Delta Cp$ from $\Delta H^{cal}$ versus $T_{trs}$

Cold shock protein from *Thermotoga maritima* (TmCsp), and mutant Asp9→Asn

| Protein | pH | $T_{trs}$ | $\Delta Cp$ | $\Delta H$ | Approach/Remarks | Ref |
|---|---|---|---|---|---|---|
| TmCsp | 7.0 | | 4.1±0.4 | | GuHCl (1–3) | 99F8 |
| Asp9→Asn | 7.0 | | 5.7±0.7 | | GuHCl (1–3) | 99F8 |

Remarks:
(1) data from GuHCl-induced equilibrium unfolding between 6 and 35°C
(2) buffer: 100 mM Na-cacodylate, pH 7.0
(3) transition monitored by fluorescence at 337 nm

## Colicin

Colicin E1 channel domain, soluble and membrane-bound state of the C-terminal 190-residue channel polypeptide (P190)

| Protein | pH | $T_{trs}$ | $\Delta Cp$ | $\Delta H$ | Appr./Rem. | | Ref |
|---|---|---|---|---|---|---|---|
| P190 soluble | 4.0 | 52 | 4.5 | | DSC | (1) | 98Z2 |
| P190 membrane-bound | 4.0 | 40 | ~0 | | DSC | (1) | 98Z2 |

Remark:
(1) measured in 20 mM sodium acetate, 100 mM NaCl, pH 4.0

## Collagen

Collagen III model peptides, coil to triple helix transition

| Peptide | $T_{trs}$ | $\Delta H^{v.H.}$(kJ/mol) | $\Delta H^{v.H.}$(kJ/triplet) | Appr./Rem. | Ref |
|---|---|---|---|---|---|
| P*PG(8)HBTU | 58.5 | –116 | –3.52 | DSC (1,2) | 99H6 |
| GPP*(8)HBTU | 62.5 | –122 | –3.39 | DSC (1,2) | 99H6 |
| PP*G(8)HATU | 50.7 | –123 | –3.73 | DSC (1,2) | 99H6 |
| PP*G(7)HATU | 44.4 | –102 | –3.4 | DSC (1,2) | 99H6 |
| PP*G(6)HATU | 39.4 | –158 | –5.85 | DSC (1,2) | 99H6 |

Remarks:
(1) for the solid-phase synthesis and the sequences of the peptides see Ref. 99H6
(2) measured in 0.1 M acetic acid, P*PG(8)HBTU in 0.1 M acetic acid, 85% methanol

Collagen-water interactions:
see Ref. 99R14 for the dependence of the enthalpy and temperature of denaturation on water content and sample preheating treatment

## Complement Receptor

Complement receptor type 1 (CR1), segment corresponding to the modules 15–17, and fragments

| Transition | pH | $T_{trs}$ | $\Delta H$ | Approach/Remarks | | Ref |
|---|---|---|---|---|---|---|
| segment CR1~15–17 | | | | | | |
| trans. (1) | 6.4 | 50.5 | 181 | DSC | (1,2) | 99K6 |
| trans. (2) | 6.4 | 57.5 | 239 | DSC | (1,2) | 99K6 |
| trans. (3) | 6.4 | 61.5 | 378 | DSC | (1,2) | 99K6 |
| fragment CR1~15–16 | | | | | | |
| trans. (2) | 6.4 | 54.0 | 185 | DSC | (1,2) | 99K6 |
| trans. (3) | 6.4 | 65.5 | 370 | DSC | (1,2) | 99K6 |
| fragment CR1~16 | | | | | | |
| trans. (3) | 6.4 | 69.0 | 252 | DSC | (1,2) | 99K6 |
| fragment CR1~16–17 | | | | | | |
| total | 6.5 | 54/64 | 502 | DSC | (3,4) | 99K5 |

Remarks:
(1) results of deconvolution
(2) measured in 25 mM phosphate buffer, pH 6.4
(3) total heat
(4) measured in 20 mM sodium phosphate buffer, pH 6.5

**Concanavalin A**, see Lectin

## Creatine Kinase

Creatine kinase from rabbit muscle

| Protein | pH | $T_{trs}$ | $\Delta H$ | Approach/Remarks | | Ref |
|---|---|---|---|---|---|---|
| | 8.0 | 55.85 | 1079±1 | DSC | (1–4) | 99L13 |

Remarks:

(1) the thermograms are independent of protein conc., but strongly scan rate-dependent

(2) the thermal denaturation of creatine kinase follows a two-state irreversible model

(3) the present data are fit parameters, the activation energy amounts to $E_a = 461.0\pm0.7$ kJ/mol

(4) for a mathematical analysis of the Lumry/Eyring model see also Ref. 99L12

## Cro Protein

Phage 434 Cro protein, thermal denaturation in the presence of urea

| Urea Conc. | pH | $T_{trs}$ | $\Delta Cp$ | $\Delta H$ | Appr./Rem. | Ref |
|---|---|---|---|---|---|---|
| 0.0 M | 6.0 | 57.0±0.1 | 3.81±0.13 | 195.0±2.1 | heat (1–4) | 99P2 |
| 1.0 M | 6.0 | 49.8±0.1 | 4.23±0.21 | 169.5±2.5 | heat (1–4) | 99P2 |
| 2.0 M | 6.0 | 41.6±0.1 | 4.31±0.21 | 128.4±1.3 | heat (1–4) | 99P2 |
| 2.5 M | 6.0 | 35.1±0.1 | 4.06±0.08 | 91.6±1.3 | heat (1–4) | 99P2 |
| 3.0 M | 6.0 | 29.7±0.1 | 4.18±0.08 | 72.4±0.8 | heat (1–4) | 99P2 |

Remarks:

(1) van't Hoff heat calculated using a two state model, for details of the data treatment see Ref. 99P2

(2) transition monitored by far-UV-CD at 222 nm

(3) buffer: 25 mM potassium phosphate, 100 mM KCl

(4) Ref. 99P2 contains additional cold denaturation values

Phage 434 Cro protein, DSC data at various pH values

| Protein | pH | $T_{trs}$ | $\Delta Cp$ | $\Delta H$ | Appr./Rem. | Ref |
|---|---|---|---|---|---|---|
| 434 Cro | 3.50 | 32.6±2.0 | | 115±13 | DSC (1,2) | 99P2 |
| | 3.75 | 39.5±1.0 | | 136±10 | DSC (1,2) | 99P2 |
| | 4.00 | 41.9±1.0 | | 136±10 | DSC (1,2) | 99P2 |
| | 4.50 | 47.4±0.5 | | 157±10 | DSC (1,2) | 99P2 |
| | 5.00 | 49.0±0.5 | | 167±10 | DSC (1,2) | 99P2 |
| | 6.00 | 57.2±0.5 | | 190±17 | DSC (1,2) | 99P2 |
| | 3.5–6.0 | | 3.97±0.13 | | DSC (3) | 99P2 |
| | 6.0 | | 4.10±0.21 | | urea (4) | 99P2 |

Remarks:

(1) data from individual curve fittings based on linear $\Delta H^{cal}/T_{trs}$ equations (also other expressions are tested in Ref. 99P2)

(2) buffers: 20 or 25 mM sodium glycine, acetate or phosphate, 100 mM KCl

(3) $\Delta Cp$ from $\Delta H^{cal}$ versus $T_{trs}$ at varying pH, for details see Ref. 99P2

(4) data from urea denaturation at varying temperature, for details see Ref. 99P2

Phage 434 cro protein

| | pH | $T_{trs}$ | $\Delta Cp$ | $\Delta H$ | Approach/Remarks | Ref |
|---|---|---|---|---|---|---|
| | | | 4.7±0.8 | | (1) | 97P3 |
| | | | 6.3±0.4 | | (1) | 97P3 |

Remark:

(1) estimated values from the solvent-accessible surface areas, for details see Ref. 97P3

## CRP, cAMP Receptor Protein

CRP, cAMP receptor protein from *E. coli*

| pH | $T_{trs}$ | $\Delta Cp$ | $\Delta H$ | Approach/Remarks | Ref |
|---|---|---|---|---|---|
| 7.9 | 63.1±1.5 | 10.7±0.7 | | heat/GuHCl (1,2) | 97M2 |

Remarks:

(1) CRP is a homodimer and each monomer is folded into two structural domains

(2) from temperature stability curve in the absence of denaturant

## Cytochrome $b_5$

Cytochrome $b_5$, hydrophilic domain, wild type and mutants, reduced and oxidized forms

| Mutant | pH | $T_{trs}$ | $\Delta H$ | Approach/Remarks | Ref |
|---|---|---|---|---|---|
| wild type, ox. | 7.0 | 66.2 | 298 | heat, v.H. (1,2) | 97Y2 |
| Phe35→His, ox. | 7.0 | 54.9 | 396 | heat, v.H. (1,2) | 97Y2 |
| Phe35→Leu, ox. | 7.0 | 59.1 | 434 | heat, v.H. (1,2) | 97Y2 |
| Phe35→Tyr, ox. | 7.0 | 69.6 | 361 | heat, v.H. (1,2) | 97Y2 |
| wild type, red. | 7.0 | 76.7 | 345 | heat, v.H. (1,2) | 97Y2 |
| Phe35→His, red. | 7.0 | 63.1 | 492 | heat, v.H. (1,2) | 97Y2 |
| Phe35→Leu, red. | 7.0 | 68.8 | 383 | heat, v.H. (1,2) | 97Y2 |
| Phe35→Tyr, red. | 7.0 | 74.4 | 358 | heat, v.H. (1,2) | 97Y2 |

Remarks:

(1) recombinant trypsin-solubilized bovine microsomal cytochrome $b_5$, 82 residues in length

(2) buffer: 0.1 M sodium phosphate, pH 7.0

Abridged rat cytochrome $b_5$ and parent proteins

| Protein | Form | pH | $T_{trs}$ | $\Delta Cp$ | $\Delta H$ | Appr./Rem. | Ref |
|---|---|---|---|---|---|---|---|
| rat cyt. $b_5$ | holo | 7.5 | 69.9±0.5 | 7±1 | 320±30 | heat (1,2) | 98C10 |
| rat cyt. $b_5$ | apo | 7.5 | 46.0±1.0 | 4±1 | 136±12 | heat (1,2) | 98C10 |
| double mutant (His80→Ala and Pro81→Ala) | | | | | | | |
| | holo | 7.5 | 62.7±0.5 | 7±1 | 320±30 | heat (1,2) | 98C10 |
| | apo | 7.5 | 27.0±2.0 | 4±1 | 90±12 | heat (1,2) | 98C10 |
| abridged $b_5$ | | 6.8 | 39.0±3.0 | 3±1 | 66±11 | heat (1–4) | 98C10 |

Remarks:

(1) transition monitored by CD at 220 nm

(2) buffer: 20 mM phosphate

(3) rat hepatic holocytochrome $b_5$ contains elements of secondary structure in the sequence β1-α1-β4-β3-α2-α3-β5-α4-α5-β2-α6. The abridged protein consist of the sequence β1-α1-β4-β3-Λ-β2-α6, where Λ is a seven-residue linker bypassing the heme binding site

(4) the abridged protein consist of a organizing scaffold without the heme binding site

Recombinant trypsin-solubilized bovine liver microsomal cytochrome $b_5$ and mutants

| Mutant | pH | $T_{trs}$ | $\Delta H$ | Approach/Remarks | Ref |
|---|---|---|---|---|---|
| wild type | 7.0 | 67.5 | 449.8 | heat, v.H. (1) | 98Q1 |
| Glu44→Ala | 7.0 | 67.1 | 446.0 | heat, v.H. (1) | 98Q1 |
| Glu56→Ala | 7.0 | 68.3 | 463.6 | heat, v.H. (1) | 98Q1 |
| double mutant (Glu44→Ala and Glu56→Ala) | | | | | |
| | 7.0 | 68.1 | 478.7 | heat, v.H. (1) | 98Q1 |

Remark:

(1) buffer: 100 mM phosphate

Bovine cytochrome $b_5$, tryptic fragment, wild type and mutants

| Mutant | pH | $T_{trs}$ | $\Delta H$ | Approach/Remarks | | Ref |
|---|---|---|---|---|---|---|
| wild type | 7.0 | 67.2 | 398 | heat | (1,2) | 99X1 |
| Val61→Glu | 7.0 | 63.2 | 390 | heat | (1,2) | 99X1 |
| Val61→His | 7.0 | 61.3 | 369 | heat | (1,2) | 99X1 |
| Val61→Lys | 7.0 | 58.5 | 364 | heat | (1,2) | 99X1 |
| Val61→Tyr | 7.0 | 62.5 | 382 | heat | (1,2) | 99X1 |

Remarks:

(1) the transition was monitored by optical absorption at 412 nm

(2) buffer: 100 mM phosphate buffer, pH 7.0

# Cytochrome $b_{562}$

Cytochrome $b_{562}$, reduced and oxidized forms

| Mutant | pH | $T_{trs}$ | $\Delta H$ | Approach/Remarks | Ref |
|---|---|---|---|---|---|
| oxidized | 7.0 | 66.7±0.5 | 460±25 | heat, v.H. (1) | 91F3 |
| oxidized | 7.0 | 68.4±3.1 | 523±54 | heat, v.H. (2) | 91F3 |
| reduced | 7.0 | 81.0±0.5 | 573±42 | heat, v.H. (3) | 91F3 |

Remarks:

(1) thermal transition monitored by heme absorption at 418 nm

(2) thermal transition monitored by second-derivative tyrosine solvation changes at 287 nm

(3) thermal transition monitored by heme absorption at 426 nm

Cytochrome $b_{562}$, oxidized

| Protein | pH | $T_{ref}$ | $\Delta H$ | Approach/Remarks | Ref |
|---|---|---|---|---|---|
| oxidized | | 20 | 155±5 | GuHCl/heat (1,2) | 99W13 |

Remarks:

(1) from heat denaturation in the presence of varying GuHCl conc. with extrapolation to zero denaturant conc. and 20°C

(2) transition monitored by far-UV CD at 220 nm

Cytochrome $b_{562}$ from the periplasm of *E. coli*, holo- and apo-protein

| Protein | pH | $T_{trs}$ | $\Delta Cp$ | $\Delta H$ | Approach/Remarks | | Ref |
|---|---|---|---|---|---|---|---|
| holo | 7.4 | 66.99±0.15 | 3.85±1.13 | 436±10 | DSC | (1,2) | 98R8 |
| holo | 5 and 6 | 73.81±0.32 | 4.35±0.84 | 436±8 | DSC | (3) | 98R8 |
| apo | 7.4 | 54.04±0.02 | 2.34±0.96 | 198±12 | DSC | (2) | 98R8 |
| apo | 8.5 | 56.26±0.02 | 0.67±0.59 | 188±27 | DSC | (4) | 98R8 |

Remarks:
(1) mean of 16 measurements varying protein conc. from 1.07 to 5.34 mg/ml
(2) buffer: 50 mM sodium phosphate, 100 mM sodium chloride, pH 7.4
(3) mean of 6 measurements in 50 mM acetate at pH 5.0 and 6.0
(4) mean of three measurements in 20 mM BICINE buffer

Apocytochrome $b_{562}$

| pH | $T_{trs}$ | $\Delta Cp$ | $\Delta H$ | Approach/Remarks | | Ref |
|---|---|---|---|---|---|---|
| 4.5 | 47–60 | 6.3 | | DSC/GuHCl | (1,2) | 98F6 |
| 4.5 | 47–60 | 2.1 | | DSC/GuHCl | (2,3) | 98F6 |

Remarks:
(1) $\Delta Cp$ from $\Delta H$ versus $T_{trs}$, measured in the presence of 0 to 0.4 M GuHCl
(2) measured in 10 mM acetate buffer, 90% $D_2O$, 0.65 M NaCl, pH* 4.5
(3) $\Delta Cp = 2.1$ kJ/mol after corrections for protein-denaturant interaction

# Cytochrome *c*

Equine cytochrome *c*, oxidized (ox.) and reduced (red.), $\Delta H$ for individual amino acid residues

| Position | pD | T | $\Delta H_{ox.}$ | $\Delta H_{red.}$ | Appr./Rem. | | Ref |
|---|---|---|---|---|---|---|---|
| a) N/C helix, global unfolding: | | | | | | | |
| Leu98 | 7.0 | 20 | 155 | | H/D | (1,2) | 99M16 |
| Ile9 | 7.0 | 20 | 59 | 67 | H/D | (1) | 99M16 |
| Phe10 | 7.0 | 20 | 176 | 108 | H/D | (1) | 99M16 |
| Val11 | 7.0 | 20 | 54 | 33 | H/D | (1) | 99M16 |
| Glu92 | 7.0 | 20 | 84 | 59 | H/D | (1) | 99M16 |
| Asp93 | 7.0 | 20 | 63 | 63 | H/D | (1) | 99M16 |
| Leu94 | 7.0 | 20 | 176 | 42 | H/D | (1) | 99M16 |
| Ala96 | 7.0 | 20 | 75 | 75 | H/D | (1) | 99M16 |
| Tyr97 | 7.0 | 20 | | 113 | H/D | (1) | 99M16 |
| Lys99 | 7.0 | 20 | 63 | | H/D | (1) | 99M16 |
| b) 60s helix, subglobal unfolding: | | | | | | | |
| Leu68 | 7.0 | 20 | 218 | | H/D | (1,2) | 99M16 |
| Leu32 | 7.0 | 20 | 100 | 63 | H/D | (1) | 99M16 |
| Met65 | 7.0 | 20 | 113 | 63 | H/D | (1) | 99M16 |
| Tyr67 | 7.0 | 20 | 75 | 38 | H/D | (1) | 99M16 |
| c) omega loop, subglobal unfolding: | | | | | | | |
| Leu64 | 7.0 | 20 | 84 | 105 | H/D | (1,2) | 99M16 |
| Lys60 | 7.0 | 20 | | 138 | H/D | (1,2) | 99M16 |
| Trp59 $N_1H$ | 7.0 | 20 | 105 | 130 | H/D | (1,2) | 99M16 |
| Phe36 | 7.0 | 20 | 63 | 46 | H/D | (1) | 99M16 |
| Gly37 | 7.0 | 20 | | 29 | H/D | (1) | 99M16 |

Equine cytochrome *c*, oxidized (ox.) and reduced (red.), ΔH for individual amino acid residues (continued)

| Position | pD | T | ΔH$_{ox.}$ | ΔH$_{red.}$ | Appr./Rem. | Ref |
|---|---|---|---|---|---|---|
| d) region connecting 60s helix and C helix, subglobal unfolding: | | | | | | |
| Ile75 | 7.0 | 20 | 75 | 84 | H/D  (1,2) | 99M16 |
| Tyr74 | 7.0 | 20 | | 33 | H/D  (1) | 99M16 |

Remarks:
(1) Ref. 99M16 contains data from H/D exchange at varying denaturant conc. and temperature
(2) marker proton

Cytochrome *c* from tuna

| | pH | T$_{trs}$ | ΔCp | ΔH | Approach/Remarks | Ref |
|---|---|---|---|---|---|---|
| | | 64.5 | 4.7 | 322 | DSC  (1,2) | 98Y2 |

Remarks:
(1) ΔCp from from ΔH versus T$_{trs}$
(2) T$_{trs}$ and ΔH were taken from Fig. 8 in Ref. 98Y2

Horse heart cytochrome *c*, interaction with polyelectrolytes

| Electrolyte | Conc. | pH | T$_{trs}$ | ΔH | Appr./Rem. | Ref |
|---|---|---|---|---|---|---|
| cytochrome *c* in the presence of NaCl: | | | | | | |
| NaCl | 0 mM | 7.0 | 84.1 | 410 | DSC (1,2) | 94B1 |
| | 50 mM | 7.0 | 83.0 | 388 | DSC (1,2) | 94B1 |
| | 100 mM | 7.0 | 82.0 | 402 | DSC (1,2) | 94B1 |
| | 200 mM | 7.0 | 80.9 | 386 | DSC (1,2) | 94B1 |
| | 400 mM | 7.0 | 80.1 | 384 | DSC (1,2) | 94B1 |
| | 600 mM | 7.0 | 79.8 | 392 | DSC (1,2) | 94B1 |
| | 1000 mM | 7.0 | 79.8 | 398 | DSC (1,2) | 94B1 |
| cytochrome *c* in the presence of heparin (units/ml): | | | | | | |
| heparin | 25 u/ml | 7.0 | 59.8 | 340 | DSC (1,3) | 94B1 |
| | 50 u/ml | 7.0 | 59.7 | 355 | DSC (1,3) | 94B1 |
| | 75 u/ml | 7.0 | 60.9 | 362 | DSC (1,3) | 94B1 |
| | 100 u/ml | 7.0 | 61.4 | 354 | DSC (1,3) | 94B1 |
| cytochrome *c* in the presence of heparin (50 units/ml) and NaCl: | | | | | | |
| NaCl | 0 mM | 7.0 | 59.7 | 355 | DSC (1,4) | 94B1 |
| | 50 mM | 7.0 | 63.8 | 350 | DSC (1,4) | 94B1 |
| | 100 mM | 7.0 | 65.9 | 342 | DSC (1,4) | 94B1 |
| | 200 mM | 7.0 | 71.8 | 360 | DSC (1,4) | 94B1 |
| | 400 mM | 7.0 | 78.6 | 377 | DSC (1,4) | 94B1 |
| | 600 mM | 7.0 | 79.6 | 372 | DSC (1,4) | 94B1 |
| | 1000 mM | 7.0 | 79.8 | 375 | DSC (1,4) | 94B1 |
| cytochrome *c* in the presence of polyglutamate: | | | | | | |
| polyglutamate | 1.0 mg/ml | 7.0 | 66.8±0.5 | 462±23 | heat (5,6) | 97B1 |
| | 0.0 mg/ml | 7.0 | 57.4±0.5 | | heat (5,6) | 97B1 |
| | 1.5 mg/ml | 7.0 | 68.7±0.2 | 354±16 | DSC (5,7) | 97B1 |
| | 0.0 mg/ml | 7.0 | 85.4±0.2 | 315±15 | DSC (5,8) | 97B1 |

Remarks:
(1) measured in 2 mM Tris/maleate buffer, pH 7.0
(2) the cooperative ratio is close to unity
(3) the cooperative ratio deviates from unity
(4) the cooperative ratio deviates from unity at low NaCl conc. and approaches unity at high salt conc.
(5) measured in 2 mM HEPES buffer, pH 7.0
(6) transition monitored by absorption difference ΔA = 695–800 nm and ΔA = 345–395 nm, respectively
(7) the van't Hoff heat amounts to 496±15 kJ/mol, ΔCp = 5.62±0.17 kJ/mol/K
(8) the van't Hoff heat amounts to 337±10 kJ/mol, ΔCp = 4.52±0.30 kJ/mol/K

Cytochrome $c$ in aqueous polyol solution

| Cosolvent | Conc. | pH | $T_{trs}$ | $\Delta H$ | Appr./Rem. | Ref |
|---|---|---|---|---|---|---|
| control | | 4.0 | 64.5 | 257.3 | heat (1,2) | 98K3 |
| mannitol | 1.00 M | 4.0 | 69.8 | 278.2 | heat (1,2) | 98K3 |
| inositol | 0.75 M | 4.0 | 70.9 | 226.8 | heat (1,2) | 98K3 |
| xylitol | 2.00 M | 4.0 | 72.5 | 259.8 | heat (1,2) | 98K3 |
| adonitol | 2.00 M | 4.0 | 72.6 | 282.0 | heat (1,2) | 98K3 |
| sorbitol | 2.00 M | 4.0 | 73.8 | 271.1 | heat (1,2) | 98K3 |
| control | | 7.0 | 48.0 | 160.7 | heat (1,3) | 98K3 |
| mannitol | 1.00 M | 7.0 | 52.7 | 160.7 | heat (1,3) | 98K3 |
| inositol | 0.75 M | 7.0 | 54.0 | 168.2 | heat (1,3) | 98K3 |
| xylitol | 2.00 M | 7.0 | 56.8 | 167.4 | heat (1,3) | 98K3 |
| adonitol | 2.00 M | 7.0 | 56.5 | 166.9 | heat (1,3) | 98K3 |
| sorbitol | 2.00 M | 7.0 | 61.1 | 190.4 | heat (1,3) | 98K3 |

Remarks:

(1) transition monitored by optical absorption at 287 nm

(2) buffer: 40 mM acetate, pH 4.0

(3) buffer: 20 mM phosphate or MOPS, pH 7.0, 1.5 M GuHCl

Horse heart cytochrome $c$, thermal unfolding of the molten globule in the presence of various solutes

| Solute | pH | $T_{trs}$ | $\Delta Cp$ | $\Delta H$ | Appr./Rem. | Ref |
|---|---|---|---|---|---|---|
| sorbitol 2.2 M | 2 | 32.5 | 5.2 | 226±1 | DSC | 99K1 |
| 2.5 M | 2 | 34.8 | 3.5 | 236±2 | DSC | 99K1 |
| 2.7 M | 2 | 36.8 | 3.1 | 243±1 | DSC | 99K1 |
| 3.0 M | 2 | 39.3 | 2.5 | 246±3 | DSC | 99K1 |
| 3.3 M | 2 | 41.7 | 3.4 | 259±3 | DSC | 99K1 |
| | | | 3.5±0.5 | | average | 99K1 |
| NaCl 300 mM | 2 | 47.4 | 1.7 | 116±9 | DSC | 99K1 |

Cytochrome $c$ (cyt. $c$) from horse heart in the presence of poly(vinylsulfate) (PVS)

| Protein | pH | $T_{trs}$ | $\Delta H$ | Approach/Remarks | | Ref |
|---|---|---|---|---|---|---|
| cyt. $c$ | 7.0 | 84.9 | 390 | DSC | | 99S6 |
| cyt. $c$ + PVS | 7.0 | 52.8 | 350 | DSC | (1,2) | 99S6 |

Remarks:

(1) measured in the presence of 3.3 mg/ml PVS

(2) see also Ref. 98S9

## Cytochrome *c₃*

Cytochrome *c₃* from *Desulfovibrio vulgaris* Hildenborough, wild type and mutants

| Protein | pH | T$_{trs}$ | ΔH | Approach/Remarks | | Ref |
|---|---|---|---|---|---|---|
| wild type | 7.6 | 81 | 1675 | DSC | | 95F2 |
| wild type | 7.5 | 121 | | heat | (1–3) | 95D7 |
| wild type | 7.6 | 121.6 | | DSC | (1,4) | 99D6 |
| His22→Met | 7.5 | 76 | | heat | (1–3) | 95D7 |
| His25→Met | 7.5 | 83 | | heat | (1–3) | 95D7 |
| His35→Met | 7.5 | 83 | | heat | (1–3) | 95D7 |
| His70→Met | 7.5 | 91 | | heat | (1–3) | 95D7 |
| Phe20→Leu | 7.6 | 114.3 | | DSC | (1,4) | 99D6 |

Remarks:

(1) tetraheme cytochrome *c₃* (M 13,000)
(2) transition monitored by CD in the far-UV and Soret region
(3) buffer: 0.01 M sodium phosphate, pH 7.5
(4) buffer: 100 mM Tris-HCl, pH 7.6

## Cytochrome *c₅₅₁*

Cytochrome *c₅₅₁* from *Pseudomonas aeruginosa*, wild type and mutants based on cytochrome *c₅₅₂* from thermophilic *Hydrogenobacter thermophilus*

| Protein | pH | T$_{trs}$ | ΔH | Approach/Remarks | | Ref |
|---|---|---|---|---|---|---|
| wild type | 5.0 | 50.4 | 157.3 | heat | (1,2) | 99H3 |
| Phe7→Ala | 5.0 | 59.9 | 201.7 | heat | (1,2) | 99H3 |
| Val13→Met | 5.0 | 53.6 | 183.3 | heat | (1,2) | 99H3 |
| Phe34→Tyr | 5.0 | 66.4 | 256.1 | heat | (1,2) | 99H3 |
| Gln37→Arg | 5.0 | 54.7 | 214.6 | heat | (1,2) | 99H3 |
| Glu43→Tyr | 5.0 | 55.5 | 240.2 | heat | (1,2) | 99H3 |
| Val78→Ile | 5.0 | 58.8 | 183.3 | heat | (1,2) | 99H3 |
| double mutant (Phe7→Ala and Val13→Met) | | | | | | |
| | 5.0 | 62.4 | 195.4 | heat | (1,2) | 99H3 |
| double mutant (Phe34→Tyr and Gln37→Arg) | | | | | | |
| | 5.0 | 62.9 | 228.0 | heat | (1,2) | 99H3 |
| double mutant (Phe34→Tyr and Glu43→Tyr) | | | | | | |
| | 5.0 | 70.7 | 266.5 | heat | (1,2) | 99H3 |
| double mutant (Gln37→Arg and Glu43→Tyr) | | | | | | |
| | 5.0 | 54.7 | 193.3 | heat | (1,2) | 99H3 |
| multiple mutant (Phe34→Tyr, Gln37→Arg and Glu43→Tyr) | | | | | | |
| | 5.0 | 57.9 | 236.0 | heat | (1,2) | 99H3 |

Remarks:
(1) transition monitored by far-UV CD at 222 nm
(2) measured in water, pH 5.0, adjusted with HCl

## Iso-1-Cytochrome *c*

The data entries are arranged as follows:

a) wild type and mutants
b) spin-labeled protein
c) transitions

*a) wild type and mutants*

Yeast iso-1-cytochrome *c*, pseudo-wild type (Cys102→Thr, w.t.*), and mutants derived from w.t.*

| Mutant/Form | | pH | $T_{trs}$ | $\Delta C_p$ | $\Delta H$ | Approach/Remarks | | Ref |
|---|---|---|---|---|---|---|---|---|
| wild type | ox. | 6.0 | 57.1 | | | DSC | | 99L3 |
| | red. | 6.0 | 81.4 | | | DSC | | 99L3 |
| | ox. | 4.7 | 51.4 | | | DSC | | 99L3 |
| | red. | 4.7 | 77.3 | | | DSC | | 99L3 |
| Cys102→Thr | ox. | 6.0 | 60.2 | | 360 | DSC | (1,2) | 99L3 |
| | red. | 6.0 | 84.7 | | 510 | DSC | (1,2) | 99L3 |
| | ox. | 4.7 | 53.9 | | 289 | DSC | (1,2) | 99L3 |
| | red. | 4.7 | 79.9 | | 485 | DSC | (1,2) | 99L3 |
| | | | | 7.1±0.4 | | DSC | (5) | 99L3 |
| Thr69→Glu | ox. | 6.0 | 62.6 | | 410 | DSC | (1–3) | 99L3 |
| | red. | 6.0 | 86.3 | | 565 | DSC | (1–3) | 99L3 |
| | ox. | 4.7 | 57.1 | | 372 | DSC | (1–3) | 99L3 |
| | red. | 4.7 | 81.6 | | 523 | DSC | (1–3) | 99L3 |
| | | | | 6.3±0.8 | | DSC | (3,5) | 99L3 |
| Thr96→Ala | ox. | 6.0 | 61.1 | | 381 | DSC | (1–3) | 99L3 |
| | red. | 6.0 | 85.9 | | 602 | DSC | (1–3) | 99L3 |
| | ox. | 4.7 | 54.8 | | 351 | DSC | (1–3) | 99L3 |
| | red. | 4.7 | 80.7 | | 506 | DSC | (1–3) | 99L3 |
| | | | | 7.5±1.3 | | DSC | (3,5) | 99L3 |
| double mutant (Thr69→Glu and Thr96→Ala) | | | | | | | | |
| | ox. | 6.0 | 62.6 | | 385 | DSC | (1–3) | 99L3 |
| | red. | 6.0 | 87.0 | | 598 | DSC | (1–3) | 99L3 |
| | ox. | 4.7 | 57.2 | | 365 | DSC | (1–3) | 99L3 |
| | red. | 4.7 | 82.5 | | 544 | DSC | (1–3) | 99L3 |
| | | | | 7.1±1.3 | | DSC | (3,5) | 99L3 |
| deletion mutant Δ(−5/−1) | | | | | | | | |
| | ox. | 6.0 | 57.1 | | 326 | DSC | (1–4) | 99L3 |
| | red. | 6.0 | 83.0 | | 410 | DSC | (1–4) | 99L3 |
| | ox. | 4.7 | 50.3 | | 167 | DSC | (1–4) | 99L3 |
| | red. | 4.7 | 77.3 | | 356 | DSC | (1–4) | 99L3 |
| | | | | 5.9±1.7 | | DSC | (3–5) | 99L3 |

Remarks:
(1) measured in 100 mM phosphate buffer at pH 6.0 or 50 mM sodium acetate, pH 4.7
(2) errors in $T_{trs}$ are estimated to be ±0.05°C
(3) yeast iso-1-cytochromes all have Cys102→Thr modification
(4) amino acids are referred to as residues −5 to −1, when using the eukaryotic numbering system for cytochromes *c*
(5) $\Delta C_p$ from $\Delta H$ versus $T_{trs}$

Yeast iso-1-cytochrome *c*, pseudo-wild type (Cys102→Thr, w.t.*), and mutants derived from w.t.*, data at reference temperature

| Protein | | pH | $T_{trs}$ | $\Delta H$ | Approach/Remarks | | Ref |
|---|---|---|---|---|---|---|---|
| Cys102→Thr | ox. | 6.0 | 57.1 | 339 | DSC | (1,2) | 99L3 |
| | red. | 6.0 | 57.1 | 314 | DSC | (1,2) | 99L3 |
| | ox. | 4.7 | 51.4 | 272 | DSC | (1,2) | 99L3 |
| | red. | 4.7 | 51.4 | 285 | DSC | (1,2) | 99L3 |
| Thr69→Glu | ox. | 6.0 | 57.1 | 377 | DSC | (1,2) | 99L3 |
| | red. | 6.0 | 57.1 | 381 | DSC | (1,2) | 99L3 |
| | ox. | 4.7 | 51.4 | 335 | DSC | (1,2) | 99L3 |
| | red. | 4.7 | 51.4 | 335 | DSC | (1,2) | 99L3 |

Yeast iso-1-cytochrome *c*, pseudo-wild type (Cys102→Thr, w.t.*), and mutants derived from w.t.*, data at reference temperature (continued)

| Protein | | pH | $T_{trs}$ | $\Delta H$ | Approach/Remarks | | Ref |
|---|---|---|---|---|---|---|---|
| Thr96→Ala | ox. | 6.0 | 57.1 | 351 | DSC | (1,2) | 99L3 |
| | red. | 6.0 | 57.1 | 385 | DSC | (1,2) | 99L3 |
| | ox. | 4.7 | 51.4 | 326 | DSC | (1,2) | 99L3 |
| | red. | 4.7 | 51.4 | 280 | DSC | (1,2) | 99L3 |
| double mutant (Thr69→Glu and Thr96→Ala) | | | | | | | |
| | ox. | 6.0 | 57.1 | 347 | DSC | (1,2) | 99L3 |
| | red. | 6.0 | 57.1 | 385 | DSC | (1,2) | 99L3 |
| | ox. | 4.7 | 51.4 | 314 | DSC | (1,2) | 99L3 |
| | red. | 4.7 | 51.4 | 322 | DSC | (1,2) | 99L3 |
| deletion mutant Δ(−5/−1) | | | | | | | |
| | ox. | 6.0 | 57.1 | 326 | DSC | (1–3) | 99L3 |
| | red. | 6.0 | 57.1 | 259 | DSC | (1–3) | 99L3 |
| | ox. | 4.7 | 51.4 | 176 | DSC | (1–3) | 99L3 |
| | red. | 4.7 | 51.4 | 205 | DSC | (1–3) | 99L3 |

Remarks:
(1) the reference temperature is defined by the wild-type protein (ox.), i.e., $T_{ref}$ = 57.1°C at pH 6.0 and $T_{ref}$ = 51.4°C at pH 4.7
(2) yeast iso-1-cytochromes all have Cys102→Thr modification
(3) amino acids are referred to as residues −5 to −1, when using the eukaryotic numbering system for cytochromes *c*

Iso-1-cytochrome *c* from *Saccharomyces cerevisiae*, Cys102→Thr mutant, stabilization upon reduction

| Mutant/Form | pH | $T_{trs}$ | $\Delta Cp$ | $\Delta H$ | Approach/Remarks | | Ref |
|---|---|---|---|---|---|---|---|
| Cys102→Thr, red. | 4.6 | 79.3 | | 522 | DSC | (1) | 95C7 |
| | 4.6 | 79.6±0.3 | | 496±5 | heat | (1,2) | 95C7 |
| | 4.6 | 27 | | 169±25 | DSC | (1,3) | 95C7 |
| | 3.2–5.5 | | 6.19±0.33 | | DSC | (4) | 95C7 |
| Cys102→Thr, ox. | 4.6 | 27 | | 197±19 | DSC | (3,5) | 95C7 |
| | 3.0–5.0 | | 5.73±0.25 | | DSC | (4,5) | 95C7 |

Remarks:
(1) measured in 100 mM sodium acetate, for further details see Ref. 95C7
(2) transition monitored by CD at 222, 332, 416, and 424 nm
(3) reference value at 300 K
(4) $\Delta Cp$ from $\Delta H$ versus $T_{trs}$
(5) data from Ref. 94C7

Yeast iso-1-cytochrome *c*, Cys102→Thr mutant

| Protein | pH | $T_{trs}$ | $\Delta H$ | Approach/Remarks | | Ref |
|---|---|---|---|---|---|---|
| Cys102→Thr | 4.6 | 52.8±0.2 | 334.7±15.9 | heat | (1,2) | 98A3 |

Remarks:
(1) Ref. 98A3 analyses the error propagation in $\Delta G$ and $\Delta H$ due to removed data points in the predenaturational baseline
(2) measured in 50 mM sodium acetate, pH 4.6, transition monitored by CD at 222 nm

**Table 2. Enthalpy and Heat Capacity Changes – Molar Values**

Yeast iso-1-cytochrome *c*, wild type and mutants at position 73

| Mutant | pH | $T_{trs}$ | $\Delta Cp$ | $\Delta H$ | Approach/Remarks | Ref |
|---|---|---|---|---|---|---|
| wild type | 3.00 | 21.5 | | 166±6 | heat, v.H. (1) | 97H4 |
| | 3.25 | 22.6 | | 176±18 | heat, v.H. (1) | 97H4 |
| | 3.50 | 30.7 | | 227±4 | heat, v.H. (1) | 97H4 |
| | 3.75 | 34.9 | | 261±5 | heat, v.H. (1) | 97H4 |
| | 4.00 | 40.7 | | 292±3 | heat, v.H. (1) | 97H4 |
| | 4.25 | 43.1 | | 308±17 | heat, v.H. (1) | 97H4 |
| | 4.50 | 45.0 | | 307±4 | heat, v.H. (1) | 97H4 |
| | 4.75 | 48.6 | | 330±3 | heat, v.H. (1) | 97H4 |
| | 5.00 | 52.7 | | 343±15 | heat, v.H. (1) | 97H4 |
| | 3–5 | 21–53 | 5.86±0.25 | | heat      (2) | 97H4 |
| Lys73→Trp | 3.00 | 22.8 | | 182±16 | heat, v.H. (1) | 97H4 |
| | 3.25 | 22.8 | | 208±20 | heat, v.H. (1) | 97H4 |
| | 3.50 | 27.5 | | 211±11 | heat, v.H. (1) | 97H4 |
| | 3.75 | 29.2 | | 245±31 | heat, v.H. (1) | 97H4 |
| | 4.00 | 30.4 | | 284±11 | heat, v.H. (1) | 97H4 |
| | 4.25 | 31.5 | | 287±12 | heat, v.H. (1) | 97H4 |
| | 4.50 | 36.1 | | 269±24 | heat, v.H. (1) | 97H4 |
| | 4.75 | 39.0 | | 299±25 | heat, v.H. (1) | 97H4 |
| | 5.00 | 47.6 | | 326±8 | heat, v.H. (1) | 97H4 |
| | 5.25 | 49.3 | | 320±13 | heat, v.H. (1) | 97H4 |
| | 5.50 | 47.4 | | 332±6 | heat, v.H. (1) | 97H4 |
| | 3–5.5 | 23–49 | 4.81±0.71 | | heat      (2) | 97H4 |
| Lys73→Val | 3.00 | 22.7 | | 155±23 | heat, v.H. (1) | 97H4 |
| | 3.25 | 28.7 | | 194±6 | heat, v.H. (1) | 97H4 |
| | 3.50 | 31.4 | | 198±3 | heat, v.H. (1) | 97H4 |
| | 3.75 | 36.6 | | 245±3 | heat, v.H. (1) | 97H4 |
| | 4.00 | 38.9 | | 252±9 | heat, v.H. (1) | 97H4 |
| | 4.25 | 42.9 | | 271±3 | heat, v.H. (1) | 97H4 |
| | 4.50 | 46.7 | | 303±8 | heat, v.H. (1) | 97H4 |
| | 4.75 | 47.9 | | 322±6 | heat, v.H. (1) | 97H4 |
| | 5.00 | 49.5 | | 316±6 | heat, v.H. (1) | 97H4 |
| | 3–5 | 23–50 | 6.28±0.25 | | heat      (2) | 97H4 |
| Lys73→Ile | 3.00 | 22.3 | | 138±21 | heat, v.H. (1) | 97H4 |
| | 3.25 | 26.7 | | 171±5 | heat, v.H. (1) | 97H4 |
| | 3.50 | 31.7 | | 215±4 | heat, v.H. (1) | 97H4 |
| | 3.75 | 35.5 | | 241±5 | heat, v.H. (1) | 97H4 |
| | 4.00 | 38.7 | | 256±3 | heat, v.H. (1) | 97H4 |
| | 4.25 | 41.3 | | 282±9 | heat, v.H. (1) | 97H4 |
| | 4.50 | 45.2 | | 289±7 | heat, v.H. (1) | 97H4 |
| | 4.75 | 46.9 | | 309±20 | heat, v.H. (1) | 97H4 |
| | 5.00 | 47.7 | | 324±3 | heat, v.H. (1) | 97H4 |
| | 3–5 | 22–48 | 6.90±0.29 | | heat      (2) | 97H4 |

Remarks:

(1) transition monitored by CD at 222 nm

(2) $\Delta Cp$ from $\Delta H$ versus $T_{trs}$

Yeast iso-1-cytochrome *c*, pseudo-wild type (w.t.*) and mutants in the hydrophobic heme pocket region

| Protein | pH | $T_{trs}$ | $\Delta Cp$ | $\Delta H$ | Appr./Rem. | Ref |
|---|---|---|---|---|---|---|
| w.t.* | 6.0 | 60.0 | | 339±17 | DSC (1,2) | 99L6 |
| | 6.0 | 25.0 | | 109±17 | DSC (1–3) | 99L6 |
| | | | 6.95 | | DSC (5) | 99L6 |
| | | | 6.61 | | DSC (6,7) | 99L6 |
| Phe82→Ala | 6.0 | 46.3 | | 238±13 | DSC (2,4) | 99L6 |
| | 6.0 | 25.0 | | 96±13 | DSC (2–4) | 99L6 |
| Phe82→Tyr | 6.0 | 55.8 | | 318±17 | DSC (2,4) | 99L6 |
| | 6.0 | 25.0 | | 113±17 | DSC (2–4) | 99L6 |
| double mutant (Phe82→Tyr and Leu85→Ala) | | | | | | |
| | 6.0 | 45.6 | | 238±13 | DSC (2,4) | 99L6 |
| | 6.0 | 25.0 | | 100±13 | DSC (2–4) | 99L6 |

Remarks:

(1) iso-1-cytochrome *c* pseudo-wild type (Cys102→Thr)

(2) measured in 0.1 M sodium phosphate, pH 6.0

(3) data at reference temperature $T_{ref}$ = 25.0°C

(4) mutants derived from the pseudo-wild type protein

(5) $\Delta Cp$ from $\Delta H^{cal}$ versus $T_{trs}$ at varying pH for w.t.*

(6) $\Delta Cp$ from $\Delta H^{cal}$ versus $T_{trs}$ for oxidized and reduced states of w.t.* and mutants

(7) this value is considered as the more reliable one in Ref. 99L6

*b) spin-labeled protein*

Yeast iso-1-cytochrome *c*, wild type (Cys102) and spin-labeled protein

| Mutant | pH | $T_{trs}$ | $\Delta H$ | Approach/Remarks | Ref |
|---|---|---|---|---|---|
| Cys102 | 5 | 54.3±0.6 | 384±74 | heat, v.H. (1–3) | 97Q1 |
| Cys102→Thr | 5 | 54.7±1.5 | 340±51 | heat, v.H. (1–3) | 97Q1 |
| Cys102-SL | 5 | 35.7±2.0 | 179±33 | heat, v.H. (1–4) | 97Q1 |

Remarks:

(1) buffer: 0.1 M acetate, pH 5

(2) transition monitored by CD at 222, 281.5, and 287.5 nm, optical absorption at 280, 287, 360, and 399 nm, and EPR (Cys102-SL)

(3) more detailed material see below

(4) Cys102-SL is at Cys102 spin-labeled iso-1-cytochrome *c*

Yeast iso-1-cytochrome *c*, wild type (Cys102) and spin-labeled protein, comparison of results obtained by various methods

| Mutant | pH | Method | Wavelength | $T_{trs}$ | $\Delta H$ | Rem. | Ref |
|---|---|---|---|---|---|---|---|
| Cys102 | 5.0 | CD | 222 nm | 53.7 | 416 | (1–3) | 97Q1 |
| | 5.0 | CD | 281.5 nm | 54.1 | 400 | (1–3) | 97Q1 |
| | 5.0 | CD | 287.5 nm | 54.2 | 393 | (1–3) | 97Q1 |
| | 5.0 | UV-Vis | 280 nm | 54.8 | 308 | (1–3) | 97Q1 |
| | 5.0 | UV-Vis | 287 nm | 53.9 | 428 | (1–3) | 97Q1 |
| | 5.0 | UV-Vis | 360 nm | 55.4 | 240 | (1–3) | 97Q1 |
| | 5.0 | UV-Vis | 399 nm | 53.7 | 510 | (1–3) | 97Q1 |
| | 5.0 | UV-Vis | 415 nm | 54.5 | 334 | (1–3) | 97Q1 |
| Cys102→Thr | 5.0 | CD | 222 nm | 56.7 | 343 | (1–3) | 97Q1 |
| | 5.0 | CD | 281.5 nm | 56.6 | 389 | (1–3) | 97Q1 |
| | 5.0 | CD | 287.5 nm | 55.9 | 400 | (1–3) | 97Q1 |
| | 5.0 | UV-Vis | 280 nm | 53.5 | 298 | (1–3) | 97Q1 |
| | 5.0 | UV-Vis | 287 nm | 53.4 | 320 | (1–3) | 97Q1 |

Yeast iso-1-cytochrome *c*, wild type (Cys102) and spin-labeled protein, comparison of results obtained by various methods (continued)

| Mutant | pH | Method | Wavelength | $T_{trs}$ | $\Delta H$ | Rem. | Ref |
|--------|-----|--------|------------|-----|-----|------|------|
|  | 5.0 | UV-Vis | 360 nm | 53.9 | 280 | (1–3) | 97Q1 |
|  | 5.0 | UV-Vis | 399 nm | 53.0 | 397 | (1–3) | 97Q1 |
|  | 5.0 | UV-Vis | 415 nm | 54.9 | 289 | (1–3) | 97Q1 |
| Cys102-SL | 5.0 | EPR |  | 36.5 | 169 | (1–4) | 97Q1 |
|  | 5.0 | CD | 222 nm | 36.6 | 202 | (1–4) | 97Q1 |
|  | 5.0 | CD | 281.5 nm | 38.2 | 220 | (1–4) | 97Q1 |
|  | 5.0 | CD | 287.5 nm | 38.1 | 222 | (1–4) | 97Q1 |
|  | 5.0 | UV-Vis | 280 nm | 34.0 | 145 | (1–4) | 97Q1 |
|  | 5.0 | UV-Vis | 287 nm | 35.5 | 183 | (1–4) | 97Q1 |
|  | 5.0 | UV-Vis | 360 nm | 32.8 | 136 | (1–4) | 97Q1 |
|  | 5.0 | UV-Vis | 399 nm | 33.7 | 154 | (1–4) | 97Q1 |

Remarks:
(1) data from Supporting Information to Ref. 97Q1 (page 3, table 1S)
(2) heat denaturation, van't Hoff treatment
(3) buffer: 0.1 M acetate, pH 5
(4) Cys102-SL is at Cys102 spin-labeled iso-1-cytochrome *c*

*c) transitions*

Yeast iso-1-cytochrome *c*, N → D transition of wild type and mutants of Ala7

| Mutant | pH | $T_{trs}$ | $\Delta Cp$ | $\Delta H$ | Approach/Remarks | Ref |
|--------|-----|-----|------|-----|------------------|-----|
| wild type | 4.6 | 54.7 | 5.77±0.96 | 355 | heat, v.H. (1–4) | 98M23 |
| Ala7→Gly | 4.6 | 50.7 | 6.11±0.29 | 346 | heat, v.H. (1–4) | 98M23 |
| Ala7→Leu | 4.6 | 53.5 | 6.32±1.05 | 352 | heat, v.H. (1–4) | 98M23 |
| Ala7→Phe | 4.6 | 54.4 | 6.23±0.50 | 374 | heat, v.H. (1–4) | 98M23 |
| Ala7→Tyr | 4.6 | 55.9 | 5.77±0.67 | 369 | heat, v.H. (1–4) | 98M23 |

Remarks:
(1) the Cys102→Thr variant is referred to as the wild-type protein
(2) measured in 50 mM sodium acetate, pH 4.6
(3) transition monitored by CD at 222 nm
(3) $\Delta Cp$ from $\Delta H$ versus $T_{trs}$

Yeast iso-1-cytochrome *c*, A → D transition of the interface variants in 0.33 M $Na_2SO_4$/$H_2SO_4$, pH 1.8–2.0

| Mutant | pH | $T_{trs}$ | $\Delta H$ | Approach/Remarks | Ref |
|--------|-----|-----|-----|------------------|-----|
| Phe10→Trp | 2.0 | 22.7 | 109 | heat, v.H. (1–3) | 98M23 |
|  | 1.9 | 22.3 | 118 | heat, v.H. (1–3) | 98M23 |
|  | 1.8 | 20.4 | 104 | heat, v.H. (1–3) | 98M23 |
| Phe10→Tyr | 2.0 | 25.5 | 139 | heat, v.H. (1–3) | 98M23 |
|  | 1.9 | 24.8 | 129 | heat, v.H. (1–3) | 98M23 |
|  | 1.8 | 23.2 | 135 | heat, v.H. (1–3) | 98M23 |
| Leu94→Ala | 2.0 | <–1 |  | heat, v.H. (1–3) | 98M23 |
| Leu94→Thr | 2.0 | <10 |  | heat, v.H. (1–3) | 98M23 |
|  | 1.8 | <–1 |  | heat, v.H. (1–3) | 98M23 |
| Leu94→Val | 2.0 | 30.9 | 164 | heat, v.H. (1–3) | 98M23 |
|  | 1.9 | 27.2 | 149 | heat, v.H. (1–3) | 98M23 |
|  | 1.8 | 25.8 | 128 | heat, v.H. (1–3) | 98M23 |

Yeast iso-1-cytochrome *c*, A → D transition of the interface variants in 0.33 M Na$_2$SO$_4$/H$_2$SO$_4$, pH 1.8–2.0 (continued)

| Mutant | pH | T$_{trs}$ | ΔH | Approach/Remarks | Ref |
|---|---|---|---|---|---|
| Tyr97→Phe | 2.0 | 30.0 | 130 | heat, v.H. (1–3) | 98M23 |
| | 1.9 | 31.3 | 127 | heat, v.H. (1–3) | 98M23 |
| | 1.8 | 27.4 | 118 | heat, v.H. (1–3) | 98M23 |

Remarks:
(1) mutants derived from the Cys102→Thr variant which is referred to as the wild-type protein
(2) the A form measured in 0.33 M Na$_2$SO$_4$/H$_2$SO$_4$, pH 1.8–2.0
(3) transition monitored by CD at 222 nm

Yeast iso-1-cytochrome *c*, A → D transition of the non-interface variants in 0.33 M Na$_2$SO$_4$/H$_2$SO$_4$, pH 1.8–2.0

| Mutant | pH | T$_{trs}$ | ΔH | Approach/Remarks | Ref |
|---|---|---|---|---|---|
| Ala7→Gly | 2.0 | 22.6 | 123 | heat, v.H. (1–3) | 98M23 |
| | 1.9 | 23.7 | 130 | heat, v.H. (1–3) | 98M23 |
| | 1.8 | 18.2 | 113 | heat, v.H. (1–3) | 98M23 |
| Ala7→Leu | 2.0 | 32.9 | 175 | heat, v.H. (1–3) | 98M23 |
| | 1.9 | 28.6 | 159 | heat, v.H. (1–3) | 98M23 |
| | 1.8 | 29.9 | 150 | heat, v.H. (1–3) | 98M23 |
| Ala7→Phe | 2.0 | 29.9 | 164 | heat, v.H. (1–3) | 98M23 |
| | 1.9 | 30.8 | 153 | heat, v.H. (1–3) | 98M23 |
| | 1.8 | 27.7 | 140 | heat, v.H. (1–3) | 98M23 |
| Ala7→Tyr | 2.0 | 33.5 | 155 | heat, v.H. (1–3) | 98M23 |
| | 1.9 | 32.0 | 163 | heat, v.H. (1–3) | 98M23 |
| | 1.8 | 31.5 | 154 | heat, v.H. (1–3) | 98M23 |

Remarks:
(1) mutants derived from the Cys102→Thr variant which is referred to as the wild-type protein
(2) the A form measured in 0.33 M Na$_2$SO$_4$/H$_2$SO$_4$, pH 1.8–2.0
(3) transition monitored by CD at 222 nm

Yeast iso-1-cytochrome *c*, A → D transition of the interface variants at enhanced sodium sulfate concentration (>0.33 M)

| Mutant | [SO$_4$]$^{2-}$ | pH | T$_{trs}$ | ΔH | Approach/Remarks | Ref |
|---|---|---|---|---|---|---|
| Phe10→Trp | 1.00 M | 1.9 | 44.9 | 168 | heat, v.H. (1–3) | 98M23 |
| | 0.68 M | 1.9 | 38.7 | 156 | heat, v.H. (1–3) | 98M23 |
| Leu94→Ala | 1.00 M | 1.9 | 25.4 | 144 | heat, v.H. (1–3) | 98M23 |
| Leu94→Thr | 1.00 M | 2.0 | 26.2 | 147 | heat, v.H. (1–3) | 98M23 |
| | 1.00 M | 1.8 | 21.9 | 126 | heat, v.H. (1–3) | 98M23 |
| | 0.67 M | 2.0 | 14.5 | 108 | heat, v.H. (1–3) | 98M23 |
| Tyr97→Ala | 1.00 M | 2.0 | ≤25 | | heat, v.H. (1–3) | 98M23 |
| | 1.00 M | 1.9 | ≤25 | | heat, v.H. (1–3) | 98M23 |

Remarks:
(1) mutants derived from the Cys102→Thr variant which is referred to as the wild-type protein
(2) the A form measured in Na$_2$SO$_4$/H$_2$SO$_4$, pH 1.8–2.0, at the indicated sulfate conc.
(3) transition monitored by CD at 222 nm

Yeast iso-1-cytochrome *c*, mutants in position 68 derived from the pseudo-wild type (Cys102→Thr), N → D, and N → A transitions

| Protein | pH | $T_{trs}$ | $\Delta Cp$ | $\Delta H$ | Appr./Rem. | Ref |
|---|---|---|---|---|---|---|
| N → D transition: | | | | | | |
| wild type | 4.6 | 53.8±0.8 | 6.15±0.50 | 367.4±43.5 | heat (1–4) | 99H14 |
| Leu68→Ile | 4.6 | 49.0±0.2 | 7.70±1.26 | 349.8±13.4 | heat (1–4) | 99H14 |
| Leu68→Val | 4.6 | 43.5±0.4 | 5.06±0.33 | 269.9±19.7 | heat (1–4) | 99H14 |
| Leu68→Met | 4.6 | 42.9±0.4 | 4.94±0.46 | 270.3± 6.3 | heat (1–4) | 99H14 |
| N → A transition: | | | | | | |
| wild type | 4.6 | 38.7±0.3 | | 218.0±13.4 | heat (2,3,5) | 99H14 |
| Leu68→Ile | 4.6 | 32.9±2.0 | | 159.8± 4.6 | heat (2,3,5) | 99H14 |
| Leu68→Val | 4.6 | 27.9±2.1 | | 144.3±17.2 | heat (2,3,5) | 99H14 |
| Leu68→Met | 4.6 | 30.4±1.5 | | 144.8±13.8 | heat (2,3,5) | 99H14 |

Remarks:
(1) measured in 0.05 M sodium acetate, pH 4.6
(2) transition monitored by ellipticity at 222 nm
(3) for the procedure see also Ref. 95P3
(4) $\Delta Cp$ from $\Delta H^{cal}$ versus $T_{trs}$ at varying pH
(5) measured in 0.5 M $Na_2SO_4/H_2SO_4$, pH 2.1

## Cytochrome Oxidase

$Cu_A$ domain of cytochrome oxidase from *Thermus thermophilus*

| Protein | GuHCl | pH | $T_{trs}$ | $\Delta H$ | Appr./Rem. | Ref |
|---|---|---|---|---|---|---|
| reduced | 0.0 M | 7.0 | 83 | 322±5 | heat (1,2) | 98W12 |
| reduced | 5.5 M | 7.0 | 40 | 188±5 | heat (1,2) | 98W12 |
| reduced | 7.0 M | 7.0 | <20 | | heat (1,2) | 98W12 |
| oxidized | 0.0 M | 7.0 | >100 | | heat (1,2) | 98W12 |
| oxidized | 5.5 M | 7.0 | 80 | 225±5 | heat (1,2) | 98W12 |
| oxidized | 7.0 M | 7.0 | 70 | 165±5 | heat (1,2) | 98W12 |

Remarks:
(1) van't Hoff heat
(2) transition monitored by CD at 218 nm

## Cytochrome P450

Cytochrome P450 from the acidothermophilic archae *Sulfolobus solfataricus* (CYP 119) compared with cytochrome P450 from *Pseudomonas putida* (CYP 101)

| Protein | pH | $T_{trs}$ | $\Delta H$ | Approach/Remarks | | Ref |
|---|---|---|---|---|---|---|
| CYP 119 | 7.0 | 91 | | DSC | (1,2) | 98M12 |
| CYP 101 | 7.0 | 54 | | DSC | (1–4) | 98M12 |

Remarks:
(1) buffer: 100 mM potassium phosphate, pH 7.0
(2) the temperature is the peak maximum temperature
(3) substrate free cytochrome P450
(4) for deconvolution of melting profiles of cytochrome P450 in the presence and in the absence of the substrate camphor see Refs. 93P4 and 94J4

## De Novo Designed Proteins/Peptides

Design of a native-like three-helix bundle

| Protein/GuHCl conc. | | pH | $T_{trs}$ | $\Delta Cp$ | $\Delta H$ | Approach/Remarks | | Ref |
|---|---|---|---|---|---|---|---|---|
| $\alpha_3$B | 3.0 M | 7.3 | 70 | 2.30 | 103 | heat | (1–4) | 98B9 |
| $\alpha_3$B | 3.5 M | 7.3 | 58 | 2.20 | 74 | heat | (1–4) | 98B9 |
| $\alpha_3$B | 4.0 M | 7.3 | 49 | 2.58 | 65 | heat | (1–4) | 98B9 |
| $\alpha_3$B | 4.5 M | 7.3 | 31 | 2.52 | 15 | heat | (1–4) | 98B9 |
| $\alpha_3$B | | 7.3 | | 2.40±0.18 | | heat | (5) | 98B9 |
| $\alpha_3$B | | 7.3 | | 2.46 | | heat | (6) | 98B9 |
| $\alpha_3$C | 1.0 M | 7.3 | 74 | 2.72 | 153 | heat | (1–4) | 98B9 |
| $\alpha_3$C | 1.5 M | 7.3 | 63 | 2.65 | 135 | heat | (1–4) | 98B9 |
| $\alpha_3$C | 2.0 M | 7.3 | 51 | 2.48 | 118 | heat | (1–4) | 98B9 |
| $\alpha_3$C | 2.5 M | 7.3 | 32 | 2.77 | 55 | heat | (1–4) | 98B9 |
| $\alpha_3$C | 3.0 M | 7.3 | 17 | 2.49 | 9 | heat | (1–4) | 98B9 |
| $\alpha_3$C | | 7.3 | | 2.62±0.13 | | heat | (5) | 98B9 |
| $\alpha_3$C | | 7.3 | | 2.88 | | heat | (6) | 98B9 |

Remarks:
(1) for details of the design see Ref. 98B9
(2) from thermal denaturation in the presence of GuHCl monitored by CD at 222 nm
(3) data treatment using the Gibbs-Helmholtz equation
(4) buffer: 25 mM MOPS, 100 mM NaCl, pH 7.3
(5) average $\Delta Cp$ value from data treatment using the Gibbs-Helmholtz equation
(6) $\Delta Cp$ from $\Delta H$ versus $T_{trs}$

De novo designed ROP-like four-helix bundles

| Form | pH | $T_{trs}$ | $\Delta Cp$ | $\Delta H^{v.H.}$ | Approach/Remarks | | Ref |
|---|---|---|---|---|---|---|---|
| protein concentration 75 µM: | | | | | | | |
| RLP–1 | 6.97 | 69.9 | 4.27 | 191 | heat, v.H. | (1,2) | 97B8 |
| RLP–1 | 6.97 | 71.9 | 4.31 | 193 | heat, GH | (1,3) | 97B8 |
| RLP–2 | 6.97 | 66.3 | 4.81 | 192 | heat, v.H. | (1,2) | 97B8 |
| RLP–2 | 6.97 | 67.2 | 4.56 | 188 | heat, GH | (1,3) | 97B8 |
| RLP–3 | 6.97 | 81.6 | 5.40 | 259 | heat, v.H. | (1,2) | 97B8 |
| RLP–3 | 6.97 | 83.4 | 5.06 | 263 | heat, GH | (1,3) | 97B8 |
| protein concentration 58 µM | | | | | | | |
| RLP–3 | 6.97 | 78.6 | | 199 | DSC | (1,4) | 97B8 |
| RLP–3 | 6.97 | 77.7 | | 196 | DSC | (1,5) | 97B8 |
| protein concentration 260 µM | | | | | | | |
| RLP–3 | 6.97 | 83.5 | | 230 | DSC | (1,6) | 97B8 |
| RLP–3 | 6.97 | 82.4 | | 210 | DSC | (1,5) | 97B8 |

Remarks:
(1) 51-residue helix-turn-helix peptides that form dimers in solution. RLP–1, –2, and –3 designate the number of Val residues in the second helix
(2) van't Hoff treatment, $\Delta Cp$ from $\Delta H$ versus $T_{trs}$
(3) GH = fit of entire melting curves by the Gibbs-Helmholtz equation
(4) DSC, $\Delta H^{v.H} = 199$ kJ/mol, $\Delta H^{cal} = 96$ kJ/mol
(5) van't Hoff heat obtained from melting curves monitored by CD at 222 nm
(6) DSC, $\Delta H^{v.H} = 230$ kJ/mol, $\Delta H^{cal} = 105$ kJ/mol

DHP1, 108 amino acid designed four helix bundle protein constructed from a reduced alphabet of seven amino acids

| pH | $T_{trs}$ | $\Delta C_p$ | $\Delta H$ | Approach/Remarks | Ref |
|---|---|---|---|---|---|
| 7 | 122 | 5.44 | 519±71 | heat/GuHCl (1–3) | 97S4 |

Remarks:
(1) from 17 thermal denaturation curves obtained varying GuHCl conc. from 0 to 5.76 M
(2) transition monitored by CD at 222 nm
(3) data extrapolated to zero GuHCl conc.

GCN4brNC, dicyclic 29-residue helix-forming model peptide

| Peptide | pH | $T_{trs}$ | $\Delta C_p$ | $\Delta H$ | Appr./Rem. | Ref |
|---|---|---|---|---|---|---|
| GCN4brNC | 7.4 | 34.1 | 0.0 | 65.3 | heat (1–5) | 99T8 |
| | 7.4 | 30.1 | 0.460 | 61.5 | heat (1–4,6) | 99T8 |
| | 7.4 | 32.3±0.6 | 0.0 | 71.1±2.9 | DSC (1–3,5) | 99T8 |
| | 7.4 | 28.0±1.6 | 0.460±0.084 | 68.6±3.8 | DSC (1–3) | 99T8 |

Remarks:
(1) unfolding of an α-helix studied on a 29-residue peptide
(2) the first amino acid residue (Lys) and the fifth residue (Asp) at the N-terminal end, and the 25th residue (Lys) and the 29th residue (Asp) at the C-terminal end are covalently linked
(3) measured in phosphate buffered 0.1 M NaCl solution, pH 7.4
(4) transition monitored by CD at 208 and 222 nm, mean value
(5) curve fit with $\Delta C_p$ = 0 kJ/mol/K fixed
(6) curve fit with $\Delta C_p$ = 0.460 kJ/mol/K fixed

β-hairpin peptide (YTV) that folds in water (folding data)

| Peptide | pH | $T_{ref}$ | $\Delta C_p$ | $\Delta H$ | Approach/Remarks | Ref |
|---|---|---|---|---|---|---|
| YTV | 5.5 | 25 | −1.130±0.075 | 3.7±0.4 | heat, NMR (1,2) | 99G14 |
| | 5.5 | 25 | −1.335±0.123 | 3.7±0.7 | heat, NMR (1,3) | 99G14 |
| YTV + urea | | 25 | −0.718±0.042 | 1.1±0.3 | heat, NMR (1,2,4) | 99G14 |
| YTV + MeOH | | 25 | −0.011±0.070 | −38.5±0.7 | heat, NMR (1,2,5) | 99G14 |

Remarks:
(1) β-hairpin peptide (YTV): Lys-Lys-Tyr-Thr-Val-Ser-Ile-Asn-Gly-Lys-Lys-Ile-Thr-Val-Ser-Ile
(2) transition monitored by chemical shift, 40% YTV folded at 25°C
(3) transition monitored by $H^\alpha$ splitting, 44% YTV folded at 25°C
(4) in 7 M urea, 28% YTV folded at 25°C
(5) in 50% aqueous methanol, 86% YTV folded at 25°C

Minimal peptides (Min-21, Min-23) that fold like a cysteine-stabilized β-sheet motif

| Protein | pH | $T_{trs}$ | $\Delta H$ | Approach | Rem. | Ref |
|---|---|---|---|---|---|---|
| Min-23 | 2.7 | 89 | 46 | heat, NMR, Cys[21]Hβ | (1,2) | 99H5 |
| | 2.7 | 85 | 54 | heat, NMR, Leu[6]Hδ | (1,2) | 99H5 |
| | 2.7 | 106 | 46 | heat, NMR, Gly[25]Hα2 | (1,2) | 99H5 |
| | 2.7 | 96 | 38 | heat, NMR, Phe[26]H2,H6 | (1,2) | 99H5 |
| | 2.7 | 102 | 63 | heat, NMR, Phe[26]Hα | (1,2) | 99H5 |
| Min-21 | 2.7 | 54 | 59 | heat, NMR, Leu[6]Hδ | (1,2) | 99H5 |
| | 2.7 | 82 | 33 | heat, NMR, Gly[25]Hα2 | (1,2) | 99H5 |
| | 2.7 | 62 | 38 | heat, NMR, Phe[26]H2,H6 | (1,2) | 99H5 |
| | 2.7 | 62 | 38 | heat, NMR, Phe[26]Hα | (1,2) | 99H5 |
| EETI II | 2.7 | 127 | 42 | heat, NMR, Cys[21]Hβ | (1,3) | 99H5 |

Remarks:
(1) data from thermal denaturation followed by 1D NMR
(2) Min-23 and Min-21 = 23 and 21 residue peptide, respectively, that corresponds to the cysteine-stabilized β-sheet motif
(3) EETI II = *Ecballium elaterium* trypsin inhibitor II

## Dihydrofolate Reductase

Dihydrofolate reductase from *E. coli*, mutants at Gly121

| Protein | pH | $T_{trs}$ | $\Delta H$ | Approach/Remarks | | Ref |
|---|---|---|---|---|---|---|
| wild type | 7.0 | 49.3 | 393 | DSC | (1) | 94G8 |
| Gly121→Ala | 7.0 | 46.6 | 344 | DSC | (1) | 94G8 |
| Gly121→Asp | 7.0 | 43.8 | 324 | DSC | (1) | 94G8 |
| Gly121→Cys | 7.0 | 48.1 | 215 | DSC | (1) | 94G8 |
| Gly121→His | 7.0 | 46.4 | 211 | DSC | (1) | 94G8 |
| Gly121→Leu | 7.0 | 44.1 | 325 | DSC | (1) | 94G8 |
| Gly121→Ser | 7.0 | 46.9 | 379 | DSC | (1) | 94G8 |
| Gly121→Tyr | 7.0 | 43.8 | 370 | DSC | (1) | 94G8 |
| Gly121→Val | 7.0 | 46.9 | 265 | DSC | (1) | 94G8 |

Remark:
(1) buffer: 10 mM potassium phosphate, 0.1 mM EDTA, 0.1 mM DTT

Dihydrofolate reductase from the hyperthermophilic bacterium *Thermotoga maritima* (TmDHFR)

| Protein | pH | $T_{trs}$ | $\Delta Cp$ | $\Delta H$ | Approach/Remarks | Ref |
|---|---|---|---|---|---|---|
| TmDHFR | 7.8 | 60 | 22.4 | 466 | heat/GuHCl (1–3) | 99D1 |

Remarks:
(1) data from the temperature profile of $\Delta G$ at 2.9 M GuHCl, see also Table 1
(2) transition monitored by far-UV CD at 222 nm, fluorescence emission at 330 nm, and optical absorption at 280 nm
(3) measured at 0.52 μM TmDHFR conc. in 10 mM potassium phosphate buffer, 0.2 mM EDTA, pH 7.8

## DNA Binding Protein

The data entries are arranged as follows:

a) HU from *B. stearothermophilus* and *B. subtilis,* wild type and mutants
b) DNA-binding proteins: Sac7d, Sso7d

*a) BstHU from B. stearothermophilus and B. subtilis, wild type and mutants*

DNA-binding protein HU from *Bacillus stearothermophilus* (*Bst*HU), *Bacillus subtilis* (*Bsu*HU), and mutant proteins

| Protein | pH | $T_{trs}$ | $\Delta H$ | Approach/Remarks | | Ref |
|---|---|---|---|---|---|---|
| *Bst*Hu wild type | 7.0 | 63.9 | 304.2 | heat | (1) | 96K12 |
| | 7.0 | 65.8 | | DSC | (4) | 98K4 |
| *Bst*Hu-Thr13→Ala | 7.0 | 67.0 | 276.8 | heat | (1) | 96K12 |
| *Bst*Hu-Gly15→Glu | 7.0 | 54.0 | 271.2 | heat | (1) | 96K12 |
| | 7.0 | 55.7 | | DSC | | 98K4 |
| *Bst*Hu-Ala27→Ser | 7.0 | 58.4 | 319.6 | heat | (1) | 98K4 |
| | 7.0 | 59.6 | | DSC | | 98K4 |
| *Bst*Hu-Ser31→Thr | 7.0 | 65.8 | 338.8 | heat | (1) | 98K4 |
| *Bst*Hu-Thr33→Leu | 7.0 | 65.6 | 245.0 | heat | (1) | 96K12 |
| *Bst*Hu-Glu34→Asp | 7.0 | 61.6 | | heat | (1) | 98K4 |
| | 7.0 | 63.5 | | DSC | | 98K4 |
| *Bst*Hu-Lys38→Asn | 7.0 | 61.5 | | heat | (1) | 98K4 |
| | 7.0 | 61.7 | | DSC | | 98K4 |
| *Bst*Hu-Val42→Ile | 7.0 | 60.1 | 228.8 | heat | (1) | 98K4 |
| | 7.0 | 59.6 | | DSC | | 98K4 |

DNA-binding protein HU from *Bacillus stearothermophilus* (*Bst*HU), *Bacillus subtilis* (*Bsu*HU), and mutant proteins (continued)

| Protein | pH | $T_{trs}$ | $\Delta H$ | Approach/Remarks | | Ref |
|---|---|---|---|---|---|---|
| *Bst*Hu-Ala56→Ser | 7.0 | 63.3 | 347.0 | heat | (1) | 98K4 |
| *Bst*Hu-Met69→Ile | 7.0 | 63.9 | 331.2 | heat | (1) | 98K4 |
| *Bst*Hu-Lys90→ext | 7.0 | 65.7 | 296.6 | heat | (1,2) | 98K4 |
| *Bst*Hu double mutant (Thr13→Ala and Gly15→Glu) | | | | | | |
| | 7.0 | 57.2 | 262.1 | heat | (1) | 96K12 |
| *Bsu*Hu    wild type | 7.0 | 48.6 | 176.2 | heat | (1,4) | 96K12 |
| *Bsu*Hu-Glu15→Gly | 7.0 | 60.4 | 220.0 | heat | (1) | 96K12 |
| *Bsu*Hu-Ser27→Ala | 7.0 | 54.2 | 186.9 | heat | (1) | 98K4 |
| *Bsu*Hu-Ile42→Val | 7.0 | 52.6 | 256.1 | heat | (1) | 98K4 |
| *Bsu*Hu double mutant (Asp34→Glu and Asn38→Lys) | | | | | | |
| | 7.0 | 52.1 | | heat | (1) | 98K4 |
| *Bst*Hu multiple mutant (Glu15→Gly, Ser27→Ala, Asp34→Glu, Asn38→Lys, Ile42→Val) | | | | | | |
| | 7.0 | 71.3 | | heat | (1,3) | 98K4 |

Remark:

(1) measured in 0.05 M phosphate buffer, pH 7.0, transition monitored by CD at 222 nm

(2) *Bst*Hu-Lys90→ext, protein extended by -Ala90-Gly91-Lys92

(3) the five thermostabilizing mutations were simultaneously introduced into the mesophilic protein *Bsu*Hu

(4) HU protein from *Thermotoga maritima* exhibits $T_{trs}$ = 96°C (pH 7.5), see Ref. 99E2

*Bsu*HU, DNA-binding protein from *Bacillus subtilis*, mesophilic protein, wild type and mutants

| Mutant | pH | $T_{trs}$ | $\Delta H$ | Approach/Remarks | | Ref |
|---|---|---|---|---|---|---|
| wild type | 7.0 | 48.6 | 176.2 | heat | (1,2) | 97K3 |
| Glu15→Gly | 7.0 | 60.4 | 220.0 | heat | (1,2) | 97K3 |
| double mutant (Asp34→Glu and Asn38→Lys) | | | | | | |
| | 7.0 | 52.1 | 265.4 | heat | (1,2) | 97K3 |
| triple mutant (Glu15→Gly, Asp34→Glu, and Asn38→Lys) | | | | | | |
| | 7.0 | 64.5 | 203.6 | heat | (1,2) | 97K3 |

Remarks:

(1) buffer: 5 mM sodium phosphate, 0.2 M NaCl

(2) the transition was monitored by CD at 222 nm

*Bst*HU, DNA-binding protein from *Bacillus stearothermophilus*, wild type and mutants

| Mutant | pH | $T_{trs}$ | $\Delta H$ | Approach/Remarks | | Ref |
|---|---|---|---|---|---|---|
| wild type | 7.0 | 63.9 | 304.2 | heat | (1,2) | 97K3 |
| Glu34→Asp | 7.0 | 61.6 | 317.4 | heat | (1,2) | 97K3 |
| Glu34→Gln | 7.0 | 60.9 | 351.7 | heat | (1,2) | 97K3 |
| Arg37→Lys | 7.0 | 63.9 | 336.8 | heat | (1,2) | 97K3 |
| Lys38→Asn | 7.0 | 61.5 | 309.7 | heat | (1,2) | 97K3 |
| double mutant (Glu34→Asp and Lys38→Asn) | | | | | | |
| | 7.0 | 60.5 | 324.6 | heat | (1,2) | 97K3 |

Remarks:

(1) buffer: 5 mM sodium phosphate, 0.2 M NaCl

(2) the transition was monitored by CD at 222 nm

*Bst*HU, DNA-binding protein from *Bacillus stearothermophilus*, wild type and mutants, pH dependence

| Mutant | pH | $T_{trs}$ | $\Delta H$ | Approach/Remarks | | Ref |
|---|---|---|---|---|---|---|
| wild type | 2.0 | 48.4 | 229.3 | heat | (1,2) | 97K3 |
| | 3.0 | 61.3 | 301.1 | heat | (1,2) | 97K3 |
| | 4.0 | 64.1 | 265.2 | heat | (1,2) | 97K3 |
| | 5.0 | 64.3 | 354.8 | heat | (1,2) | 97K3 |
| | 7.0 | 63.9 | 304.2 | heat | (1,2) | 97K3 |
| Glu34→Asp | 2.0 | 46.7 | 239.7 | heat | (1,2) | 97K3 |
| | 3.0 | 59.0 | 328.1 | heat | (1,2) | 97K3 |
| | 4.0 | 62.2 | 321.4 | heat | (1,2) | 97K3 |
| | 5.0 | 62.3 | 282.3 | heat | (1,2) | 97K3 |
| | 7.0 | 61.6 | 317.4 | heat | (1,2) | 97K3 |
| Glu34→Gln | 2.0 | 47.1 | 222.6 | heat | (1,2) | 97K3 |
| | 7.0 | 60.9 | 351.7 | heat | (1,2) | 97K3 |
| Lys38→Asn | 2.0 | 48.1 | 234.0 | heat | (1,2) | 97K3 |
| | 3.0 | 58.6 | 222.2 | heat | (1,2) | 97K3 |
| | 4.0 | 61.9 | 267.7 | heat | (1,2) | 97K3 |
| | 5.0 | 61.7 | 291.6 | heat | (1,2) | 97K3 |
| | 7.0 | 61.5 | 309.7 | heat | (1,2) | 97K3 |
| double mutant (Glu34→Asp and Lys38→Asn) | | | | | | |
| | 2.0 | 46.5 | 233.4 | heat | (1,2) | 97K3 |
| | 7.0 | 60.5 | 324.6 | heat | (1,2) | 97K3 |

Remarks:
(1) buffer: 5 mM phosphate and 0.2 M NaCl below pH 4, and 5 mM sodium phosphate and 0.2 M NaCl above pH 5
(2) the transition was monitored by CD at 222 nm

*Bst*HU, DNA-binding protein from *Bacillus stearothermophilus*, wild type and mutants, salt dependence

| Mutant | NaCl | pH | $T_{trs}$ | $\Delta H$ | Approach/Remarks | | Ref |
|---|---|---|---|---|---|---|---|
| wild type | 0.0 M | 7.0 | 52.9 | 216.5 | heat | (1,2) | 97K3 |
| | 0.1 M | 7.0 | 57.8 | 268.7 | heat | (1,2) | 97K3 |
| | 0.2 M | 7.0 | 63.9 | 304.2 | heat | (1,2) | 97K3 |
| | 0.5 M | 7.0 | 65.4 | 265.9 | heat | (1,2) | 97K3 |
| Glu34→Asp | 0.0 M | 7.0 | 50.7 | 255.5 | heat | (1,2) | 97K3 |
| | 0.1 M | 7.0 | 56.3 | 333.9 | heat | (1,2) | 97K3 |
| | 0.2 M | 7.0 | 61.6 | 317.4 | heat | (1,2) | 97K3 |
| | 0.5 M | 7.0 | 65.0 | 332.7 | heat | (1,2) | 97K3 |
| Lys38→Asn | 0.0 M | 7.0 | 49.6 | 226.1 | heat | (1,2) | 97K3 |
| | 0.1 M | 7.0 | 54.9 | 323.9 | heat | (1,2) | 97K3 |
| | 0.2 M | 7.0 | 61.5 | 309.7 | heat | (1,2) | 97K3 |
| | 0.5 M | 7.0 | 65.0 | 310.0 | heat | (1,2) | 97K3 |

Remarks:
(1) buffer: 5 mM sodium phosphate with the indicated NaCl concentration
(2) the transition was monitored by CD at 222 nm

*b) DNA-binding proteins: Sac7d, Sso7d*

Sac7d, 7 kDa DNA-binding protein from *Sulfolobus acidocaldarius*, linkage of protonation and anion binding

| pH | $T_{trs}$ | $\Delta Cp$ | $\Delta H$ | Approach/Remarks | | Ref |
|---|---|---|---|---|---|---|
| 5.2 | 93 | | 238 | DSC | (1–3) | 98M10 |
| 4.0 | 91 | | 243 | DSC | (1–3) | 98M10 |
| 3.7 | 86 | | 222 | DSC | (1–3) | 98M10 |
| 3.4 | 87 | | 167 | DSC | (1–3) | 98M10 |

Sac7d, 7 kDa DNA-binding protein from *Sulfolobus acidocaldarius*, linkage of protonation and anion binding (continued)

| pH | $T_{trs}$ | $\Delta Cp$ | $\Delta H$ | Approach/Remarks | | Ref |
|---|---|---|---|---|---|---|
| 3.1 | 70 | | 142 | DSC | (1–3) | 98M10 |
| 2.5 | 50 | | 109 | DSC | (1–3) | 98M10 |
| 2.3 | 47 | | 113 | DSC | (1–3) | 98M10 |
| 2.1 | 33 | | 23 | DSC | (1–3) | 98M10 |
| 1.1 | 56 | | 92 | DSC | (1–3) | 98M10 |
| 0.5 | 61 | | 113 | DSC | (1–3) | 98M10 |
| | | 2.09 | | | (4,5) | 98M10 |
| | | 2.08 | | | (4,6) | 98M10 |
| | | 3.03 | | | (4,7) | 98M10 |

Remarks:

(1) measured in 0.001 M buffer (glycine from pH 0.5 to 3.7, potassium acetate from pH 4 to 5.2)

(2) for data measured in the presence of 0.3 M KCl, see Ref. 96M2

(3) estimated error in $\Delta H$ ±20 kJ/mol

(4) the data obtained by multidimensional DSC can be explained by the linkage of at least two protonation reactions and two anion binding sites to a two-state unfolding process

(5) from $\Delta H$ versus $T_{trs}$, data at 0.3 M KCl from Ref. 96M2, simulated data

(6) from $\Delta H$ versus $T_{trs}$, experimental data at 0.3 M KCl from Ref. 96M2

(7) $\Delta Cp$ is the intrinsic $\Delta Cp°$ value obtained from the global fit to the linkage model, i.e., the slope from $\Delta H$ versus $T_{trs}$ is not the intrinsic $\Delta Cp$

Sso7d, 7 kDa DNA-binding protein Sso7d from thermoacidophilic *Sulfolobus solfataricus*, wild type and mutants

| Protein | pH | $T_{trs}$ | $\Delta Cp$ | $\Delta H$ | Approach/Remarks | | Ref |
|---|---|---|---|---|---|---|---|
| Sso7d | 2.3 | 64.0 | 2.0 | 170 | DSC | (1–3) | 98C2 |
| | 2.5 | 65.5 | 2.4 | 180 | DSC | (1–3) | 98C2 |
| | 2.7 | 71.1 | 2.4 | 195 | DSC | (1–3) | 98C2 |
| | 2.8 | 74.0 | 1.9 | 205 | DSC | (1–3) | 98C2 |
| | 2.9 | 76.3 | 2.7 | 210 | DSC | (1–3) | 98C2 |
| | 3.0 | 78.8 | 3.1 | 215 | DSC | (1–3) | 98C2 |
| | 3.2 | 84.2 | 3.0 | 230 | DSC | (1–3) | 98C2 |
| | 3.6 | 90.3 | 2.5 | 245 | DSC | (1–3) | 98C2 |
| | 3.8 | 92.4 | 2.9 | 250 | DSC | (1–3) | 98C2 |
| | 4.0 | 95.1 | 2.3 | 265 | DSC | (1–3) | 98C2 |
| | | 64–95 | 2.8±0.2 | | DSC | (4) | 98C2 |
| | 4.5–7.0 | 98.2±0.4 | | 267±5 | DSC | (5,6) | 98C2 |
| Trp23→Ala | 2.1 | 57.9 | 2.4 | 155 | DSC | (1,2) | 98C2 |
| | 2.5 | 61.8 | 2.0 | 165 | DSC | (1,2) | 98C2 |
| | 2.8 | 63.9 | 2.6 | 170 | DSC | (1,2) | 98C2 |
| | 3.0 | 67.1 | 1.9 | 175 | DSC | (1,2) | 98C2 |
| | 3.3 | 74.2 | 2.2 | 195 | DSC | (1,2) | 98C2 |
| | 3.5 | 79.3 | 3.1 | 210 | DSC | (1,2) | 98C2 |
| | 3.7 | 83.8 | 2.8 | 220 | DSC | (1,2) | 98C2 |
| | 3.8 | 87.2 | 2.7 | 230 | DSC | (1,2) | 98C2 |
| | 4.6 | 90.1 | 2.0 | 240 | DSC | (1,2) | 98C2 |
| | 5.0 | 91.6 | 3.2 | 245 | DSC | (1,2) | 98C2 |
| | 5.5 | 92.5 | 3.0 | 250 | DSC | (1,2) | 98C2 |
| | 6.0 | 92.3 | 2.5 | 245 | DSC | (1,2) | 98C2 |
| | 6.5 | 92.7 | 2.2 | 250 | DSC | (1,2) | 98C2 |
| | 7.0 | 92.1 | 2.7 | 245 | DSC | (1,2) | 98C2 |
| | | 58–93 | 2.7±0.2 | | DSC | (4) | 98C2 |
| | 4.9–7.2 | 92.2±0.4 | | 247±2 | DSC | (5,6) | 98C2 |
| Phe31→Ala | 3.0 | 53.0 | 1.9 | 135 | DSC | (1,2) | 98C2 |
| | 3.2 | 56.0 | 2.0 | 140 | DSC | (1,2) | 98C2 |

Sso7d, 7 kDa DNA-binding protein Sso7d from thermoacidophilic *Sulfolobus solfataricus*, wild type and mutants (continued)

| Protein | pH | $T_{trs}$ | $\Delta Cp$ | $\Delta H$ | Approach/Remarks | | Ref |
|---|---|---|---|---|---|---|---|
| | 3.5 | 60.5 | 2.2 | 150 | DSC | (1,2) | 98C2 |
| | 4.0 | 71.0 | 2.3 | 175 | DSC | (1,2) | 98C2 |
| | 4.5 | 73.0 | 3.0 | 180 | DSC | (1,2) | 98C2 |
| | 4.9 | 74.7 | 1.9 | 190 | DSC | (1,2) | 98C2 |
| | 5.0 | 74.3 | 2.0 | 185 | DSC | (1,2) | 98C2 |
| | 5.5 | 74.7 | 2.1 | 190 | DSC | (1,2) | 98C2 |
| | 5.6 | 74.5 | 1.9 | 185 | DSC | (1,2) | 98C2 |
| | 5.8 | 73.7 | 2.4 | 180 | DSC | (1,2) | 98C2 |
| | 6.0 | 75.1 | 2.0 | 190 | DSC | (1,2) | 98C2 |
| | 6.9 | 74.1 | 2.4 | 185 | DSC | (1,2) | 98C2 |
| | 7.2 | 73.9 | 2.1 | 180 | DSC | (1,2) | 98C2 |
| | | 53–75 | 2.5±0.2 | | DSC | (4) | 98C2 |
| | 4.9–7.2 | 74.4±0.5 | | 186±5 | DSC | (5) | 98C2 |
| Phe31→Tyr | 2.6 | 63.0 | 2.3 | 160 | DSC | (1,2) | 98C2 |
| | 3.0 | 74.2 | 2.8 | 185 | DSC | (1,2) | 98C2 |
| | 3.6 | 78.4 | 2.6 | 195 | DSC | (1,2) | 98C2 |
| | 3.8 | 80.1 | 1.9 | 200 | DSC | (1,2) | 98C2 |
| | 4.0 | 83.6 | 2.5 | 210 | DSC | (1,2) | 98C2 |
| | 4.2 | 84.8 | 3.0 | 210 | DSC | (1,2) | 98C2 |
| | 4.6 | 86.0 | 2.1 | 215 | DSC | (1,2) | 98C2 |
| | 4.9 | 88.0 | 2.0 | 225 | DSC | (1,2) | 98C2 |
| | 5.2 | 87.6 | 2.4 | 220 | DSC | (1,2) | 98C2 |
| | 5.5 | 88.5 | 2.8 | 230 | DSC | (1,2) | 98C2 |
| | 6.0 | 88.2 | 3.0 | 225 | DSC | (1,2) | 98C2 |
| | 6.5 | 88.0 | 2.7 | 220 | DSC | (1,2) | 98C2 |
| | 7.0 | 87.8 | 2.4 | 220 | DSC | (1,2) | 98C2 |
| | | 63–88 | 2.6±0.3 | | DSC | (4) | 98C2 |
| | 4.9–7.2 | 88.0±0.3 | | 223±4 | DSC | (5,6) | 98C2 |

Remarks:
(1) measured in the presence of 100 mM NaCl
(2) each figure is the average of three measurements, uncertainty in $T_{trs}$ ≤0.3°C, in $\Delta H$ 5%, and in $\Delta Cp$ 10%
(3) for further data on Sso7d see also Ref. 96K5
(4) $\Delta Cp$ from $\Delta H$ versus $T_{trs}$
(5) conditions of maximal thermostability
(6) based on data from Ref. 96K5

**DNA-binding protein Sso7d**, see also Ribonuclease P2

## DsbA

DsbA, disulfide-bond forming enzyme from *E. coli*, variants

| Mutant/Form | | pH | $T_{trs}$ | $\Delta Cp$ | $\Delta H$ | Appr./Rem. | | Ref |
|---|---|---|---|---|---|---|---|---|
| DsbA | oxidized | 7.0 | 68.40±0.03 | 9.28±2.09 | 618.0±25.0 | DSC | (1) | 99M22 |
| | reduced | 7.0 | 76.77±0.16 | 8.15±2.18 | 720.2±31.2 | DSC | (1) | 99M22 |
| DsbA Pro151→Ala | | | | | | | | |
| | oxidized | 7.0 | 57.40±0.80 | 10.77±4.88 | 431.7± 5.1 | DSC | (1) | 99M22 |
| | reduced | 7.0 | 66.06±0.23 | 7.72±4.10 | 526.9±40.0 | DSC | (1) | 99M22 |
| DsbA | oxidized | 8.0 | 68.04 | 9.25 | 592.3 | DSC | (2) | 99M22 |
| DsbA | oxidized | 8.5 | 66.8 | 9.33 | 565.6 | DSC | (2) | 99M22 |

Remarks:
(1) measured in 100 mM phosphate, 1 mM EDTA, pH 7.0
(2) measured in 40 mM glycine-NaOH, 1 mM EDTA

## Eglin

Recombinant *Hirudo medicinalis* eglin c

| Protein | pH | $T_{trs}$ | $\Delta Cp$ | $\Delta H$ | Approach/Remarks | | Ref |
|---|---|---|---|---|---|---|---|
| eglin | 1.5 | 45.1 | | 189.5 | heat | (1–3) | 99W3 |
| | 2.0 | 47.3 | | 183.3 | heat | (1–3) | 99W3 |
| | 2.5 | 54.6 | | 214.2 | heat | (1–3) | 99W3 |
| | 3.0 | 62.2 | | 240.2 | heat | (1–3) | 99W3 |
| | 3.3 | 69.0 | | 269.0 | heat | (1–3) | 99W3 |
| | 1.5–3.3 | 45–69 | 3.51±0.29 | | heat | (1–4) | 99W3 |
| | 1.5 | 45.4 | | 190.8 | DSC | (1,5) | 99W3 |
| | 2.0 | 48.1 | | 243.9 | DSC | (1,5) | 99W3 |
| | 2.5 | 56.8 | | 256.5 | DSC | (1,5) | 99W3 |
| | 3.0 | 66.5 | | 297.1 | DSC | (1,5) | 99W3 |
| | 3.3 | 75.0 | | 305.0 | DSC | (1,5) | 99W3 |
| | 1.5–3.3 | 45–75 | 3.43±0.79 | | DSC | (1,6) | 99W3 |

Remarks:

(1) buffer: 50 mM glycine-HCl

(2) transition monitored by CD at 227 nm

(3) uncertainties in $T_{trs}$ ±1.6°C and in $\Delta H^{v.H.}$ ±26.4 kJ/mol

(4) $\Delta Cp$ from $\Delta H^{v.H.}$ versus $T_{trs}$ at varying pH

(5) uncertainties in $T_{trs}$ ±1.6°C and in $\Delta H^{cal}$ ±20.9 kJ/mol

(6) $\Delta Cp$ from $\Delta H^{cal}$ versus $T_{trs}$ at varying pH

Recombinant eglin c, effect of a polyhistidine terminal extension (His tag) on the protein stability [remark (5)]

| Protein | pH | $T_{trs}$ | $\Delta Cp$ | $\Delta H$ | Approach/Remarks | | Ref |
|---|---|---|---|---|---|---|---|
| wild-type eglin c | | | | | | | |
| | 1.5 | 45.1 | 3.51±0.29 | 190 | heat | (1–3) | 98W1 |
| | 2.0 | 47.3 | | 183 | heat | (1,2) | 98W1 |
| | 2.5 | 54.6 | | 214 | heat | (1,2) | 98W1 |
| | 3.0 | 62.2 | | 240 | heat | (1,2) | 98W1 |
| | 3.3 | 69.0 | | 269 | heat | (1,2) | 98W1 |
| His-tagged eglin c | | | | | | | |
| | 1.5 | 45.9 | 4.73±0.54 | 171 | heat | (1–4) | 98W1 |
| | 2.0 | 48.4 | | 192 | heat | (1,2,4) | 98W1 |
| | 2.5 | 56.4 | | 235 | heat | (1,2,4) | 98W1 |
| | 3.0 | 63.4 | | 264 | heat | (1,2,4) | 98W1 |
| | 3.3 | 69.0 | | 290 | heat | (1,2,4) | 98W1 |

Remarks:

(1) measured in 50 mM glycine-HCl buffer

(2) transition monitored by ellipticity at 227 nm, error in $T_{trs}$ ±1.6°C, in $\Delta H$ ±26 kJ/mol (mean from four repetitions)

(3) $\Delta Cp$ from $\Delta H$ versus $T_{trs}$

(4) eglin c with a six residue N-terminal His tag

(5) for similar results obtained on β-lactamase from a thermophilic *Bacillus licheniformis* strain, see Ref. 97L7

## Elastin

Elastin-like peptides, temperature-induced structural transitions

| Peptide | pH | $T_{trs}$ | $\Delta H$ | Approach/Remarks | Ref |
|---|---|---|---|---|---|
| A | 7.0 | 41 | 46.0±11.3 | heat, v.H. (1) | 98R3 |
| B | 9.5 | 35 | 62.3±20.5 | heat, v.H. (1) | 98R3 |
| C | 4.0 | 45 | 83.7±12.1 | heat, v.H. (1) | 98R3 |
| D | 9.5 | 40 | 54.4±20.1 | heat, v.H. (1) | 98R3 |
| G | 9.5 | 35 | 57.3±11.7 | heat, v.H. (1) | 98R3 |
| H | 9.5 | 47 | 60.2±9.6 | heat, v.H. (1) | 98R3 |
| I | 7.0 | 20 | 108.8±44.4 | heat, v.H. (1) | 98R3 |
| J | 7.0 | 18 | 105.9±10.5 | heat, v.H. (1) | 98R3 |
| L | 7.0 | 37 | 79.1±19.7 | heat, v.H. (1) | 98R3 |
| M | 7.0 | 14 | 66.5±44.8 | heat, v.H. (1) | 98R3 |
| N | 7.0 | 17 | 50.6±10.0 | heat, v.H. (1) | 98R3 |

Remark:
(1) for the sequences of the peptides see Ref. 98R3

## Erythropoietin

Erythropoietin (EPO), glycosylated and unglycosylated protein

| Protein | pH | $T_{trs}$ | $\Delta H$ | Approach/Remarks | | Ref |
|---|---|---|---|---|---|---|
| EPO, CHO | 7.4 | 48 | | heat | (1,2) | 97N3 |
| EPO, E. coli | 7.4 | 56 | | heat | (2,3) | 97N3 |

Remarks:
(1) EPO, CHO = glycosylated protein derived from Chinese hamster ovary cells
(2) measured in PBS buffer, transition monitored by far-UV CD at 215, 222, and 231 nm
(3) EPO, E. coli = unglycosylated protein derived from E. coli

## Esterase

Bacillus subtilis p-nitrobenzyl esterase, mutants obtained by in vitro evolution

| Generation | pH | $T_{trs}$ | $\Delta H$ | Approach/Remarks | | Ref |
|---|---|---|---|---|---|---|
| wild type | 7.0 | 52.5 | | DSC | (1) | 98G6 |
| 1A5D1 | 7.0 | 57.3 | | DSC | (1,2) | 98G6 |
| 2A12 | 7.0 | 58.2 | | DSC | (1,2) | 98G6 |
| 3H5 | 7.0 | 62.1 | | DSC | (1–3) | 98G6 |
| 4G4 | 7.0 | 64.2 | | DSC | (1–3) | 98G6 |
| 5H3 | 7.0 | 65.0 | | DSC | (1–3) | 98G6 |
| 6sF9 | 7.0 | 66.5 | | DSC | (1,2) | 98G6 |

Remarks:
(1) buffer: 0.1 M PIPES, pH 7.0
(2) for a detailed mutant description see Ref. 98G6
(3) data taken from Fig. 1 in Ref. 98G6

## Ferredoxin

[4Fe-4S] ferredoxin from hyperthermophile *Thermotoga maritima*, DSC in the presence of guanidinium salts

| Denaturant | pH | $T_{trs}$ | $\Delta Cp$ | $\Delta H$ | Approach/Remarks | | Ref |
|---|---|---|---|---|---|---|---|
| 2.0 M GuHCl | 8.5 | 99.9 | | 263.7 | DSC | (1) | 97P9 |
| 2.0 M GuHCl | 8.5 | 99.7 | | 274.5 | DSC | (1) | 97P9 |
| 1.75 M GuHSCN | 8.5 | 86.0 | | 230.8 | DSC | (1) | 97P9 |
| 1.75 M GuHSCN | 8.5 | 82.1 | | 204.2 | DSC | (2) | 97P9 |
| 1.91 M GuHSCN | 7.0 | 81.1 | | 191.1 | heat, v.H. | (3) | 97P9 |
| 2.0 M GuHSCN | 8.5 | 78.5 | | 159.7 | DSC | (1) | 97P9 |
| 2.38 M GuHSCN | 7.0 | 73.0 | | 175.6 | heat, v.H. | (3) | 97P9 |
| 2.38 M GuHSCN | 7.0 | 73.9 | | 156.7 | heat, v.H. | (3) | 97P9 |
| 2.49 M GuHSCN | 8.5 | 77.0 | | 193.6 | heat, v.H. | (2) | 97P9 |
| 2.62 M GuHSCN | 7.0 | 64.4 | | 140.4 | heat, v.H.( | 3) | 97P9 |
| 2.62 M GuHSCN | 7.0 | 65.1 | | 125.0 | heat, v.H.( | 3) | 97P9 |
| 2.82 M GuHSCN | 7.0 | 55.3 | | 133.3 | heat, v.H. | (3) | 97P9 |
| | | 55–100 | 3.5±0.4 | | DSC, heat (4) | | 97P9 |
| | | 78 | 3.6±0.1 | | DSC | (5) | 97P9 |
| 0.0 M | | 124.5 | | 350 | DSC, heat (7) | | 97P9 |

Remarks:
(1) buffer: 40 mM glycine, pH 8.5, 10 mM sodium sulfide, 10 mM 2-mercaptoethanol, 1 mM ascorbic acid
(2) buffer: as buffer (1), however, 40 mM HEPPS instead of glycine
(3) buffer: 50 mM potassium phosphate, pH 7.0, 10 mM 2-mercaptoethanol
(4) $\Delta Cp$ from $\Delta H$ versus $T_{trs}$ (assuming linear dependence)
(5) from Cp of native and unfolded protein
(6) $\Delta H$ and $T_{trs}$ extrapolated to zero denaturant concentration

## Ferritin

Recombinant L (rL) and H (rH) subunits of human ferritin

| Protein | pH | $T_{trs}$ | $\Delta H$ | Approach/Remarks | | Ref |
|---|---|---|---|---|---|---|
| rL ferritin subunit: | | | | | | |
| | 2.0 | 37.2 | 293.9 | DSC | (1,2) | 98M8 |
| | 2.2 | 53.1 | 360.9 | DSC | | 98M8 |
| | 2.4 | 65.1 | 431.3 | DSC | | 98M8 |
| | 2.8 | 79.8 | 440.9 | DSC | | 98M8 |
| rH ferritin subunit: | | | | | | |
| | 2.0 | 50.0 | 168.7 | DSC | | 98M8 |
| | 2.4 | 50.3 | 169.6 | DSC | | 98M8 |
| | 2.8 | 50.4 | 171.7 | DSC | | 98M8 |

Remarks:
(1) deconvolution yields $T_{trs1}$ = 35°C with $\Delta H_1$ = 150.7 kJ/mol, and $T_{trs2}$ = 40°C with $\Delta H_2$ = 142.3 kJ/mol
(2) specific heat capacity change $\Delta cp$ = 0.23 J/g/K from $\Delta h$ versus $T_{trs}$

Horse spleen apoferritin and human recombinant apoferritin

| Protein | pH | $T_{trs}$ | $\Delta H$ | Approach/Remarks | | Ref |
|---|---|---|---|---|---|---|
| horse spleen | 3.5 | 42.1 | 351 | DSC | (1–3) | 96S17 |
| human recombinant | 3.5 | 50.0 | 376 | DSC | (1–3) | 96S17 |
| human recombinant | 7.0 | 77.3 | 8899 | DSC | (4) | 96S17 |
| human recombinant | 7.0 | 59.8 | 9912 | DSC/GuHCl | (4,5) | 96S17 |

Remarks:

(1) dimeric subunit obtained by treatment at pH 1.8 followed by dialysis versus 40 mM glycine-HCl pH 3.5

(2) data calculated for molar mass = 2×19.5 and 2×21.0 kDa for horse spleen and human recombinant protein, respectively

(3) for $\Delta$cp see Table 3

(4) data per 24-mer, molar mass = 504 kDa

(5) measured in the presence of 2 M GuHCl

## Fetuin

Bovine serum fetuin (BSF)

| Transition | pH | $T_{trs}$ | $\Delta$Cp | $\Delta H$ | Approach/Remarks | | Ref |
|---|---|---|---|---|---|---|---|
| N → I | 6.0 | 59.5±0.5 | 11.13±0.75 | 367±18 | DSC | (1) | 98W4 |
| N → I | 6.0 | 58.9±1.2 | 11.30±0.92 | 356±34 | heat | (1,2) | 98W4 |
| I → U | 6.0 | 91.5±2.1 | 3.26±1.05 | 191±15 | DSC | (1) | 98W4 |

Remarks:

(1) buffer: 50 mM sodium phosphate, pH 6.0

(2) van't Hoff heat, transition monitored by CD at 285 nm

Bovine serum fetuin, glycosylated (native) and deglycosylated protein

| Protein | pH | $T_{trs}$ | $\Delta H$ | Approach/Remarks | | Ref |
|---|---|---|---|---|---|---|
| native | 5.5 | 60.3 | 375 | DSC | (1) | 96W3 |
| native | 6.0 | 59.6±0.2 | 367±1 | DSC | (2) | 96W3 |
| native | 7.0 | 54.2±0.4 | 325±11 | DSC | (2) | 96W3 |
| native | 8.0 | 48.4±0.3 | 297±8 | DSC | (2) | 96W3 |
| native | 8.8 | 46.8 | 284 | DSC | (3) | 96W3 |
| deglycosylated | 6.0 | 58.2±0.1 | 354±1 | DSC | (2) | 96W3 |
| deglycosylated | 7.0 | 53.0±0.4 | 314±11 | DSC | (2) | 96W3 |
| deglycosylated | 8.0 | 46.4±0.2 | 288±5 | DSC | (3) | 96W3 |
| deglycosylated | 8.8 | 45.5 | 263 | DSC | (3) | 96W3 |

Remarks:

(1) measured in 50 mM sodium acetate buffer

(2) measured in 50 mM sodium phosphate buffer

(3) measured in 50 mM glycine buffer

**Fibronectin**, see also Tenascin

## Flagellin

Flagellin from *Salmonella typhimurium* and its proteolytic fragments

| Protein/Transition | | pH | T | $\Delta$Cp | $\Delta$H | Appr./Rem. | Ref |
|---|---|---|---|---|---|---|---|
| flagellin | trans. (1) | 7.0 | 25 | 5 | 330 | DSC (1,2) | 99H13 |
| | trans. (2) | 7.0 | 25 | 17 | 880 | DSC (1,2) | 99H13 |
| | trans. (3) | 7.0 | 25 | 20 | 1187 | DSC (1,2) | 99H13 |
| F40$^T$ | trans. (1) | 7.0 | 25 | 9 | 329 | DSC (1,2) | 99H13 |
| | trans. (2) | 7.0 | 25 | 15 | 854 | DSC (1,2) | 99H13 |
| | trans. (3) | 7.0 | 25 | 20 | 1125 | DSC (1,2) | 99H13 |
| F27$^S$ | trans. (1) | 7.0 | 25 | 8 | 466 | DSC (1,2) | 99H13 |
| | trans. (2) | 7.0 | 25 | 11 | 844 | DSC (1,2) | 99H13 |

Explanations:

flagellin  residues 1–494
F40$^T$    residues 66–450
F27$^S$    residues 173–422

Remark:
(1) Ref. 99H13 contains an approach which enables better correlation of the results of deconvolution with molecular masses and morphological domains
(2) Ref. 99H13 contains additional thermodynamic data from CD melting measurements

## Flavodoxin

Apoflavodoxin from *Anabaena* PCC7119, fragment comprising residues 1–149 (Fld1–149)

| Protein | pH | $T_{trs}$ | $\Delta$Cp | $\Delta$H | Appr./Rem. | Ref |
|---|---|---|---|---|---|---|
| Fld1–149 | 7.0 | 46.0±2.5 | | 51.5±2.5 | heat (1,2,4) | 98M4 |
| | 7.0 | 46.7 | | 54.0 | heat (1,2,5) | 98M4 |
| | 7.0 | 48.3±1.3 | 3.05±0.21 | 71.1±7.5 | heat (1,3,4) | 98M4 |

Remarks:
(1) measured in 5 mM sodium phosphate, 0.5 M NaCl
(2) transition monitored by fluorescence
(3) transition monitored by CD
(4) measured at 2.0 µM Fld1–149
(5) measured at 20.0 µM Fld1–149

## GABA Receptor

N-terminal part of the recombinant metabotropic GABA receptor (GABABR1a)

| Protein | pH | $T_{trs}$ | $\Delta$H | Approach/Remarks | Ref |
|---|---|---|---|---|---|
| GABABR1a | 8.0 | 64 | 377 | DSC | 98H6 |

## Galectin-1

Bovine spleen galectin-1

| Protein conc. | pH | $T_{trs}$ | $\Delta H^{cal}$ | $\Delta H^{v.H.}$ | Appr./Rem. | Ref |
|---|---|---|---|---|---|---|
| 0.043–0.179 mM | 7.6 | 71.2±0.8 | 215±9 | 956±19 | DSC (1–4) | 98S6 |
| 0.043–0.089 mM | 7.6 | 72.3±0.5 | 248±25 | 958±80 | DSC (1–3,5) | 98S6 |
| 0.044 mM | 7.6 | 72.3±0.2 | 342±57 | 1013±73 | DSC (1–4) | 98S6 |

Remarks:

(1) $\Delta H^{cal}$ is in kJ per mole of galectin dimer (M = 28,000)

(2) Ref. 98S6 contains additional data on galectin-1 melting in the presence of disaccharides measured by DSC, and disaccharide binding measured by ITC

(3) measured in PBS buffer + 2 mM DTT

(4) scan rate 15 K/h

(5) scan rate 25 K/h

## β-Galactosidase

β-Galactosidase from *Aspergillus oryzae*

| Transition | pH | $T_{trs}$ | $\Delta H$ | Approach/Remarks | | Ref |
|---|---|---|---|---|---|---|
| trans. (1) | 6.5 | 60.3 | 720 | DSC | (1,2) | 99B18 |
| trans. (2) | 6.5 | 66.4 | 640 | DSC | (1,2) | 99B18 |
| trans. (3) | 6.5 | 68.6 | 803 | DSC | (1,2) | 99B18 |

Remarks:

(1) results of deconvolution

(2) Ref. 99B18 contains further data obtained on spray-dried β-galactosidase

## Glucanase

Endoglucanase CenC, *Cellulomonas fimi* β-1,4-glucanase CenC, N1 cellulose-binding domain (CBDN1)

| pH | $T_{trs}$ | $\Delta Cp$ | $\Delta H$ | Appr./Rem. | Ref |
|---|---|---|---|---|---|
| 5.14 | 50.8±0.4 | | 363.6±18 | DSC (1) | 98C12 |
| 5.50 | 50.0 | | 372.1 | DSC | 98C12 |
| 6.10 | 51.0 | | 410.4 | DSC | 98C12 |
| 7.09 | 49.3 | | 391.4 | DSC | 98C12 |
| 7.36 | 48.5 | | 370.4 | DSC | 98C12 |
| 9.06 | 45.0 | | 361.8 | DSC | 98C12 |
| 10.58 | 43.5 | | 322.2 | DSC (2) | 98C12 |
| 10.86 | 43.3 | | 319.4 | DSC (2) | 98C12 |
| 11.06 | 36.2 | | 295.4 | DSC (2) | 98C12 |
| 5.5–11 | 36–51 | 7.5±1.3 | | DSC (3–5) | 98C12 |

Remarks:

(1) thermal unfolding was irreversible at pH 5.14 and below

(2) thermal unfolding was partly reversible at pH 10.58 and above

(3) $\Delta Cp$ from $\Delta H$ versus $T_{trs}$

(4) $\Delta Cp$ was found to be temperature independent in the transition regions (33–55°C)

(5) Ref. 98C12 contains additional thermograms for unfolding of CBDN1 in the presence of cellopentaose

## Glucoamylase

Glucoamylase 1 (GA1) from *Aspergillus niger*, and modified forms

| Protein/Transition | | pH | $T_{trs}$ | $\Delta H$ | Appr./Rem. | Ref |
|---|---|---|---|---|---|---|
| GA1 | trans. (1) | 6.0 | 73.0±0.1 | 2122±15 | DSC (1,2) | 99C10 |
| GA2 | trans. (1) | 6.0 | 73.0±0.2 | 1160±24 | DSC (1,2) | 99C10 |
| | trans. (2) | 6.0 | 79.3±0.2 | 387±24 | DSC (1,2) | 99C10 |
| GA1 | trans. (3) | 7.5 | 64.6±0.5 | 1759±279 | DSC (3) | 99C10 |
| GA1 + acarbose | trans. (1) | 7.5 | 61.0±0.6 | 382±57 | DSC (3) | 99C10 |
| | trans. (3) | 7.5 | 74.4±0.4 | 1503±56 | DSC (3) | 99C10 |
| GA1 + β-CD | trans. (3) | 7.5 | 64.9±0.4 | 1896±56 | DSC (3) | 99C10 |
| GA2 | trans. (1) | 7.5 | 45.0±0.4 | 183±15 | DSC (3) | 99C10 |
| | trans. (2) | 7.5 | 56.1±1.5 | 147±54 | DSC (3) | 99C10 |
| | trans. (3) | 7.5 | 66.0±0.3 | 1148±50 | DSC (3) | 99C10 |
| GA2 + acarbose | trans. (1) | 7.5 | 45.9±0.5 | 243±24 | DSC (3) | 99C10 |
| | trans. (3) | 7.5 | 77.3±0.2 | 1935±23 | DSC (3) | 99C10 |
| GACD | trans. (3) | 7.5 | 64.4±0.4 | 1460±64 | DSC (3) | 99C10 |
| GACD + acarbose | trans. (3) | 7.5 | 74.9±0.1 | 1984±21 | DSC (3) | 99C10 |
| SBD | | 7.5 | 56.7±0.2 | 311±12 | DSC (3) | 99C10 |
| SBD + β-CD | | 7.5 | 59.8±0.2 | 336±5 | DSC (3) | 99C10 |
| SBD + acarbose | | 7.5 | 56.7±0.2 | 202±7 | DSC (3) | 99C10 |

Abbreviations:
- GA1     glucoamylase 1 from *Aspergillus niger*, consisting of a catalytic domain, a starch-binding domain, and a O-glycosylated linker region
- GA2     glucoamylase 2, a form lacking the starch-binding domain
- GACD     a proteolytically cleaved form of glucoamylase containing the catalytic domain and the first 30 amino acid residues of the linker region
- SBD     starch-binding domain
- β-CD     β-cyclodextrin

Remarks:
(1) average values from scan rate dependent measurements
(2) buffer: 50 mM MES, pH 6.0
(3) buffer: 50 mM phosphate buffer, pH 7.5

Glucoamylase from *Aspergillus niger*, glycosylated (native) and deglycosylated

| Protein | pH | $T_{trs}$ | $\Delta H$ | Approach/Remarks | | Ref |
|---|---|---|---|---|---|---|
| native | 4.5 | 67.8 | | DSC | (1,3) | 96W3 |
| native | 5.0 | 69.7 | | DSC | (1,3) | 96W3 |
| naive | 5.5 | 69.1±0.1 | 2209±17 | DSC | (1) | 96W3 |
| native | 6.0 | 64.0 | 2017 | DSC | (2) | 96W3 |
| native | 7.0 | 62.4±0 | 2050±25 | DSC | (2) | 96W3 |
| native | 8.0 | 57.8±0.1 | 1833±25 | DSC | (2) | 96W3 |
| deglycosylated | 4.5 | 67.5 | | DSC | (1,3) | 96W3 |
| deglycosylated | 5.0 | 68.2 | | DSC | (1,3) | 96W3 |
| deglycosylated | 5.5 | 67.7±0.1 | 2000±4 | DSC | (1) | 96W3 |
| deglycosylated | 7.0 | 60.0±0.1 | 1837±25 | DSC | (2) | 96W3 |
| deglycosylated | 8.0 | 55.9±0.4 | 1674±29 | DSC | (2) | 96W3 |

Remarks:
(1) measured in 50 mM sodium acetate buffer
(2) measured in 50 mM sodium phosphate buffer
(3) $\Delta H$ is uncertain due to precipitation upon thermal unfolding

## Glucose Transporter

Glucose transporter (GLUT 1) from human erythrocytes

| D-glucose | scan rate | pH | $T_{trs}$ | $\Delta H$ | Appr./Rem. | Ref |
|---|---|---|---|---|---|---|
| 0 mM | 0.25 | 7.4 | 68.5±0.2 | 502 | DSC  (1–3) | 99E1 |
| 0 mM | 0.50 | 7.4 | 70 | 540 | DSC  (1,2) | 99E1 |
| 0 mM | 1.00 | 7.4 | 71 | 787 | DSC  (1,2) | 99E1 |
| 0 mM | 1.50 | 7.4 | 73 | 669 | DSC  (1,2) | 99E1 |
| 0 mM | | 7.4 | | 623±130 | DSC  (5) | 99E1 |
| 500 mM | 0.25 | 7.4 | 72.6±0.2 | 657 | DSC  (1–4) | 99E1 |
| 500 mM | 0.50 | 7.4 | 74 | 519 | DSC  (1,2) | 99E1 |
| 500 mM | 1.00 | 7.4 | 77 | 674 | DSC  (1,2) | 99E1 |
| 500 mM | 1.50 | 7.4 | 78 | 724 | DSC  (1,2) | 99E1 |
| 500 mM | | 7.4 | | 644±88 | DSC  (5) | 99E1 |

Remarks:
(1) measured in 10 mM Tris, pH 7.4
(2) scan rate in K/min
(3) $T_{trs}$ from multiple measurements
(4) $T_{trs}$ = 69.4±0.3°C was observed in the presence of 500 mM L-glucose
(5) average from all scans

## Glutathione S-Transferase

26 kDa glutathione S-transferase from *Schistosoma japonicum*

| Protein conc. | pH | $T_{trs}$ | $\Delta Cp$ | $\Delta H$ | Appr./Rem. | Ref |
|---|---|---|---|---|---|---|
| 50  μM | 6.5 | 61.2 | 5.17 | 921.7 | DSC  (1) | 97K1 |
| 25  μM | 6.5 | 60.2 | 10.97 | 933.3 | DSC  (1) | 97K1 |
| 10.6 μM | 6.5 | 59.4 | 6.20 | 995.9 | DSC  (1) | 97K1 |
| average | 6.5 | | 7.4±3.1 | 950±40 | DSC  (1) | 97K1 |

Remark:
(1) buffer: 20 mM sodium phosphate, pH 6.5

## Glyceraldehyde-3-Phosphate Dehydrogenase

Phosphorylating D-glyceraldehyde-3-phosphate dehydrogenase (GAPDH) from rabbit muscle (*R.m.*), *Escherichia coli* (*E.c.*), and *Bacillus stearothermophilus* (*B.s.*)

| Protein | NAD$^+$ | pH | $T_{trs}$ | $\Delta H$ | Appr./Rem. | Ref |
|---|---|---|---|---|---|---|
| *holo-R.m.*-GAPDH | 1 mM | 6.0 | 68.9 | 2900±400 | DSC  (1) | 99L4 |
| | 1 mM | 8.0 | 67.2 | 3600±400 | DSC  (1) | 99L4 |
| apo-*R.m.*-GAPDH | 0 μM | 8.0 | 58.4 | 4300±200 | DSC  (1) | 99L4 |
| | 7 μM | 8.0 | 59.0 | 4100±200 | DSC  (1) | 99L4 |
| | 21 μM | 8.0 | 59.5 | 4500±200 | DSC  (1) | 99L4 |
| | 100 μM | 8.0 | 60.8 | 4800±200 | DSC  (1) | 99L4 |
| | 1000 μM | 8.0 | 64.5 | 4800±200 | DSC  (1) | 99L4 |
| carboxymethylated apo-*R.m.*-GAPDH | | | | | | |
| | 0.0 mM | 8.0 | 56.3 | 2400±100 | DSC  (1) | 99L4 |
| | 0.3 mM | 8.0 | 56.9 | 2300±100 | DSC  (1) | 99L4 |
| | 1.0 mM | 8.0 | 57.7 | 2600±100 | DSC  (1) | 99L4 |
| | 3.0 mM | 8.0 | 58.7 | 2700±100 | DSC  (1) | 99L4 |

Phosphorylating D-glyceraldehyde-3-phosphate dehydrogenase (GAPDH) from rabbit muscle (*R.m.*), *Escherichia coli* (*E.c.*), and *Bacillus stearothermophilus* (*B.s.*) (continued)

| Protein | NAD$^+$ | pH | $T_{trs}$ | $\Delta H$ | Appr./Rem. | Ref |
|---|---|---|---|---|---|---|
| *holo-E.c.*-GAPDH | 1 mM | 6.0 | 75.3 | 3800±400 | DSC (1) | 99L4 |
| | 1 mM | 8.0 | 72.6 | 4500±400 | DSC (1) | 99L4 |
| *holo-B.s.*-GAPDH | 1 mM | 6.0 | 98.8 | 5600±400 | DSC (1) | 99L4 |
| | 1 mM | 8.0 | 92.1 | 7800±400 | DSC (1) | 99L4 |
| apo-*B.s.*-GAPDH | 0.0 mM | 8.0 | 78.3 | 4415±300 | DSC (1) | 99L4 |
| | 0.1 mM | 8.0 | 87.4 | 5780±300 | DSC (1) | 99L4 |
| | 0.3 mM | 8.0 | 89.7 | 6360±300 | DSC (1) | 99L4 |
| | 1.0 mM | 8.0 | 91.9 | 6780±300 | DSC (1) | 99L4 |
| isolated NAD$^+$-binding *B.s.*-GAPDH domain | | | | | | |
| | 0 mM | 6.0 | 54.3 | 46±2 | DSC (1) | 99L4 |
| | 10 mM | 6.0 | 55.1 | 180±50 | DSC (1) | 99L4 |
| | 0 mM | 8.0 | 61.9 | 437±50 | DSC (1) | 99L4 |
| | 10 mM | 8.0 | 64.7 | 546±50 | DSC (1) | 99L4 |
| apo-*B.s.*-GAPDH, mutant Cys149→Ser | | | | | | |
| | 0.0 mM | 8.0 | 80.8 | 5796±300 | DSC (1) | 99L4 |
| | 0.1 mM | 8.0 | 89.8 | 7601±300 | DSC (1) | 99L4 |
| | 0.3 mM | 8.0 | 91.6 | 7968±300 | DSC (1) | 99L4 |
| | 1.0 mM | 8.0 | 93.7 | 7680±300 | DSC (1) | 99L4 |
| apo-*B.s.*-GAPDH, mutant His176→Asn | | | | | | |
| | 0.0 mM | 8.0 | 65.8 | 3340±200 | DSC (1) | 99L4 |
| | 0.1 mM | 8.0 | 66.3 | 3610±200 | DSC (1) | 99L4 |
| | 0.3 mM | 8.0 | 68.2 | 3950±200 | DSC (1) | 99L4 |
| | 1.0 mM | 8.0 | 72.0 | 4090±200 | DSC (1) | 99L4 |
| | 3.0 mM | 8.0 | 76.6 | 4240±200 | DSC (1) | 99L4 |

Remark:
(1) measured in 100 mM KH$_2$PO$_4$-KOH buffer containing the indicated NAD$^+$ conc.

## Glycosidase

β-Glycosidase from the thermophilic archaeon *Sulfolobus solfataricus*, wild type and recombinant protein (Sβgly expressed in *E. coli*)

| Protein | Conc. | pH | $T_{trs}$ | $\Delta Cp$ | $\Delta H$ | Approach/Remarks | | Ref |
|---|---|---|---|---|---|---|---|---|
| wild type | 2.4 µM | 10.1 | 86.5 | 200 | 9800 | DSC | (1) | 98D7 |
| | 3.2 µM | 10.1 | 87.6 | 220 | 9500 | DSC | (1) | 98D7 |
| | 3.8 µM | 10.1 | 88.2 | 200 | 9700 | DSC | (1) | 98D7 |
| | 4.7 µM | 10.1 | 88.9 | 190 | 9500 | DSC | (1) | 98D7 |
| wild type | 1.9 µM | 10.6 | 85.8 | 180 | 7700 | DSC | (2) | 98D7 |
| | 4.6 µM | 10.6 | 84.2 | 190 | 7600 | DSC | (2) | 98D7 |
| reco. Sβgly | 2.4 µM | 10.1 | 85.8 | 180 | 9400 | DSC | (1) | 98D7 |
| | 3.3 µM | 10.1 | 86.4 | 170 | 9000 | DSC | (1) | 98D7 |
| | 4.8 µM | 10.1 | 87.9 | 200 | 9200 | DSC | (1) | 98D7 |
| reco. Sβgly | 3.0 µM | 10.6 | 86.0 | 200 | 8200 | DSC | (1) | 98D7 |
| | 4.8 µM | 10.6 | 84.3 | 190 | 8600 | DSC | (1) | 98D7 |

Remarks:
(1) buffer: 10 mM Capso, pH 10.1
(2) buffer: 50 mM sodium glycine, pH 10.6

**GroEL**, see Chaperone

**GroES**, see Chaperone

## Growth Factor

Human acidic fibroblast growth factor (haFGF), calorimetric results for the unfolding of haFGF in the presence of GuHCl

| GuHCl conc. | pH | $T_{trs}$ | $\Delta H$ | Approach/Remarks | | Ref |
|---|---|---|---|---|---|---|
| 0.0 M | 7.3 | 46.2 | 339 | DSC | (1–3) | 98A1 |
| 0.3 M | 7.3 | 45.3 | 251 | DSC | (1,3) | 98A1 |
| 0.6 M | 7.3 | 44.9 | 289 | DSC | (1,3) | 98A1 |
| 0.6 M | 7.3 | 46.0 | 276 | DSC | (1,4) | 98A1 |

Remarks:
(1) increasing reversibility was observed with increasing GuHCl conc.
(2) Ref. 98A1 contains additional data for haFGF unfolding (and precipitation) in HEPES buffer in the presence of EDTA + DTT, NaCl, sulfate and phosphate (3) scan rate 60 K/h
(4) scan rate 120 K/h

Human acidic fibroblast growth factor (FGF-1) in the presence of GuHCl

| Protein | GuHCl | pH | $T_{trs}$ | $\Delta Cp$ | $\Delta H$ | Appr./Rem. | Ref |
|---|---|---|---|---|---|---|---|
| FGF-1 | 0.6 M | 6.60 | 42.5±0.1 | 0.160±0.012 | 305.6±3.4 | DSC (1–3) | 99B15 |
| | 0.7 M | 6.60 | 39.4±0.1 | 0.409±0.033 | 257.3±3.1 | DSC (1) | 99B15 |
| | 0.8 M | 6.60 | 36.9±0.2 | 0.525±0.043 | 234.4±3.7 | DSC (1) | 99B15 |
| | 0.9 M | 6.60 | 33.8±0.2 | 0.621±0.025 | 197.0±2.9 | DSC (1) | 99B15 |
| | 1.0 M | 6.60 | 31.6±0.2 | 0.577±0.025 | 171.9±2.5 | DSC (1) | 99B15 |
| | 1.1 M | 6.60 | 26.7±0.8 | 0.639±0.043 | 123.1±11.6 | DSC (1) | 99B15 |
| | 0.6–1.1 | 6.60 | | 0.587±0.036 | | DSC (1,4) | 99B15 |

Remarks:
(1) measured in 20 mM ADA buffer, 0.1 M NaCl, pH 6.60, in the presence of GuHCl
(2) $T_{trs}$ was determined to 42°C by DSC and CD, and 35°C by fluorescence in 50 mM sodium phosphate pH 6.5 in the absence of GuHCl in Ref. 91C8
(3) for data on the interaction of FGF-1 with heparin see Ref. 91C8
(4) average value

Recombinant human Flt3 ligand (rhFlt3L)

| Protein | pH | $T_{trs}$ | $\Delta H$ | Approach/Remarks | | Ref |
|---|---|---|---|---|---|---|
| RhFlt3L | 3.0 | 50.2 | 237.7 | DSC | (1) | 99R9 |
| | 4.0 | 66.6 | 323.4 | DSC | (1) | 99R9 |
| | 4.3 | 66.8 | 322.2 | DSC | (1) | 99R9 |
| | 5.3 | 76.8 | 375.3 | DSC | (1) | 99R9 |
| | 6.1 | 78.6 | 413.8 | DSC | (1) | 99R9 |
| | 7.0 | 78.5 | 484.1 | DSC | (1) | 99R9 |
| | 7.2 | 80.0 | 447.3 | DSC | (1) | 99R9 |
| | 7.4 | 79.4 | 480.7 | DSC | (1) | 99R9 |

Remark:
(1) Flt3 protein is a member of a family of growth factors that stimulate growth of hematopoietic cells

## Growth Hormone

Human growth hormone

| | pH | $T_{trs}$ | $\Delta Cp$ | $\Delta H$ | Appr./Rem. | Ref |
|---|---|---|---|---|---|---|
| | | 60 | | 220 | DSC (1) | 98G12 |
| | 2.0–3.5 | 59–81 | 4.73±0.30 | 210–313 | DSC (1–3) | 98G12 |
| | 6.0 | 79.2 | | | DSC (4) | 98B2 |
| | 6.0 | 79.1 | | | heat, CD (4) | 98B2 |

Remarks:

(1) thermal two-state transition into an intermediate state

(2) $\Delta Cp$ from $\Delta H$ versus $T_{trs}$

(3) Ref. 98G12 contains additional data on the influence of alcohols on thermal unfolding of human growth hormone

(4) Ref. 98B2 contains further $T_{trs}$ values obtained in the presence of Tween and at varying protein conc.

Human growth hormone

| Protein | pH | $T_{trs}$ | $\Delta Cp$ | $\Delta H$ | Appr./Rem. | Ref |
|---|---|---|---|---|---|---|
| data from heat capacity function: | | | | | | |
| 1.25 mg/ml | 3.0 | ~76 | ~8 | 318±13 | DSC (1,2) | 98K2 |
| 9.6 mg/ml | 3.0 | ~72,93 | ~8 | 318±13 | DSC (1,2) | 98K2 |
| data from a thermodynamic treatment (3): | | | | | | |
| 1st process | 3.0 | 51.5±0.8 | 3.70±0.25 | 139±6 | DSC (3–5) | 98K2 |
| 2nd process | 3.0 | 106.4±4 | 4.25±0.42 | 241±4 | DSC (3–5) | 98K2 |

Remarks:

(1) peak maximum temperatures and $\Delta Cp$ were taken from the Figs. in Ref. 98K2

(2) at high protein conc. the transition is characterized by the presence of two separate peaks, and these peaks merge into one at lower protein conc.

(3) the thermodynamic treatment implies a partially folded dimeric intermediate according to $2N \rightarrow I_2 \rightarrow 2D$

(4) for details see Ref. 98K2

(5) the reference temperature $T_0$, at which $\Delta G(T_0) = 0$, does not coincide with the peak maximum temperature for a denaturation transition coupled to an association/dissociation process

## Heat Shock Protein

Heat shock protein 90 (hsp90), effect of calcium and magnesium

| Added comp. | pH | $T_{trs1}$ | $T_{trs2}$ | $\Delta H$(total) | Approach/Remarks | Ref |
|---|---|---|---|---|---|---|
| none | 7 | 53.8 | 63.1 | 2050 | DSC (1–3) | 98G3 |
| | 7 | 53.6 | 64.6 | 1925 | DSC (1,4) | 98G3 |
| geldanamycin | 7 | 57.4 | 63.1 | 2470 | DSC (1,3) | 98G3 |
| 5 mM $Mg^{2+}$ | 7 | 49.9 | 60.9 | 1675 | DSC (1–3) | 98G3 |
| 5 mM $Ca^{2+}$ | 7 | 45.7 | 56.3 | 1505 | DSC (1–3) | 98G3 |

Remarks:

(1) $T_{trs1}$ and $T_{trs2}$ as determined by deconvolution

(2) $T_{trs1}$ agrees with the denaturation temperature obtained by CD

(3) scan rate 1 K/h

(4) scan rate 2 K/h

## Hexokinase

Hexokinase from yeast, in the absence and presence of D-glucose

| Glucose conc. | pH | $T_{trs}$ | $\Delta Cp$ | $\Delta H$ | Appr./Rem. | Ref |
|---|---|---|---|---|---|---|
| 0 mM | 8.0 | 46.7 | 30.0 | 700 | DSC (1) | 97C3 |
| 10 mM | 8.0 | 48.4 | 32.0 | 750 | DSC | 97C3 |
| 30 mM | 8.0 | 49.4 | 29.0 | 780 | DSC | 97C3 |
| 50 mM | 8.0 | 50.0 | 31.0 | 800 | DSC | 97C3 |
| 80 mM | 8.0 | 50.5 | 28.0 | 820 | DSC | 97C3 |
| 100 mM | 8.0 | 52.0 | 30.0 | 870 | DSC | 97C3 |
| 300 mM | 8.0 | 54.0 | 33.0 | 930 | DSC | 97C3 |
| 500 mM | 8.0 | 56.0 | 29.0 | 1000 | DSC | 97C3 |
| 800 mM | 8.0 | 58.0 | 30.0 | 1060 | DSC | 97C3 |
| 1000 mM | 8.0 | 59.7 | 33.0 | 1120 | DSC | 97C3 |
| 1500 mM | 8.0 | 62.2 | 35.0 | 1200 | DSC | 97C3 |
| 0 mM | 8.5 | 48.2 | 30.0 | 750 | DSC (1) | 97C3 |
| 10 mM | 8.5 | 49.5 | 30.0 | 790 | DSC | 97C3 |
| 30 mM | 8.5 | 50.4 | 31.0 | 820 | DSC | 97C3 |
| 80 mM | 8.5 | 51.7 | 32.0 | 860 | DSC | 97C3 |

Remark:

(1) mean of three measurements

Hexokinase from yeast, in the absence and presence of D-glucose, results of deconvolution

| Transition | pH | $T_{trs}$ | $\Delta Cp$ | $\Delta H$ | Appr./Rem. | Ref |
|---|---|---|---|---|---|---|
| hexokinase in the absence of D-glucose | | | | | | |
| trans. (1) | 8.0 | 38.8±0.5 | | 320±20 | DSC (1,2) | 97C3 |
| trans. (1) | 8.0 | 46.6±0.5 | | 380±25 | DSC (1,2) | 97C3 |
| hexokinase in the presence of D-glucose | | | | | | |
| trans. (1) | 8.0 | 38.9±0.6 | 13±2 | 310±30 | DSC (1,2,4) | 97C3 |
| trans. (1) | 8.0 | 46.9±0.6 | 15±3 | 380±25 | DSC (1,2,4) | 97C3 |
| hexokinase in the absence of D-glucose: | | | | | | |
| trans. (1) | 8.0 | 38.7±0.4 | | 325±20 | DSC (1,3) | 97C3 |
| trans. (1) | 8.0 | 46.7±0.5 | | 390±30 | DSC (1,3) | 97C3 |
| hexokinase in the absence of D-glucose: | | | | | | |
| trans. (1) | 8.5 | 40.2±0.5 | | 350±35 | DSC (1,2) | 97C3 |
| trans. (1) | 8.5 | 48.2±0.5 | | 400±30 | DSC (1,2) | 97C3 |
| hexokinase in the presence of D-glucose | | | | | | |
| trans. (1) | 8.5 | 40.6±0.4 | 13±2 | 360±35 | DSC (1,2,4) | 97C3 |
| trans. (1) | 8.5 | 48.7±0.4 | 15±3 | 400±25 | DSC (1,2,4) | 97C3 |
| hexokinase in the absence of D-glucose | | | | | | |
| trans. (1) | 8.5 | 40.4±0.5 | | 355±40 | DSC (1,3) | 97C3 |
| trans. (1) | 8.5 | 48.6±0.5 | | 405±30 | DSC (1,3) | 97C3 |

Remarks:

(1) deconvolution by two-dimensional nonlinear regression of excess heat capacity

(2) sequential transition model

(3) independent transition model

(4) the substrate binding enthalpy was found to be zero

## High Mobility Group 1 Protein

High mobility group 1 protein (HMG1), isolated domains A and B, and truncated protein A-B

| Protein | | pH | $T_{trs}$ | $\Delta H$ | $\Delta Cp$ | Appr./Rem. | | Ref |
|---|---|---|---|---|---|---|---|---|
| domain A: | | | | | | | | |
| | | 5.1 | 43 | 194.1 | 2.1 | DSC | (1,2) | 99R4 |
| | | 5.1 | 44 | 198.4 | | heat | (4,5) | 99R4 |
| domain B: | | | | | | | | |
| | | 5.1 | 41 | 193.3 | 1.9 | DSC | (1,2) | 99R4 |
| | | 5.1 | 42 | 167.4 | | heat | (4,5) | 99R4 |
| protein A-B: | | | | | | | | |
| | trans. (1) | 5.1 | 40 | 184.1 | 1.3 | DSC | (1–3) | 99R4 |
| | trans. (2) | 5.1 | 41 | 192.5 | 1.0 | DSC | (1–3) | 99R4 |
| protein HMG1: | | | | | | | | |
| | trans. (1) | 5.1 | 38 | 167.4 | 1.2 | DSC | (1,2) | 99R4 |
| | trans. (2) | 5.1 | 55 | 173.6 | –4.1 | DSC | (1,2) | 99R4 |

Remarks:

(1) measured in 3 mM sodium acetate, pH 5.1, 0.2 mM dithiothreitol

(2) with temperature dependence $\Delta Cp = (a + b) \times T$ in J/mol

   $4.184 \times (12.000 – 32 \times T)$ for domain A,

   $4.184 \times (13.850 – 42 \times T)$ for domain B,

   $4.184 \times (17.200 – 54 \times T)$ for protein A-B,

   $4.184 \times (23.200 – 74 \times T)$ for HMG1

(3) the protein A-B corresponds to HMG1 without the acidic tail C

(4) van't Hoff treatment

(5) transition monitored by CD

## Histidine-Containing Phosphocarrier Protein HPr

Histidine-containing phosphocarrier protein HPr from *E. coli*

| GuHCl conc. | pH | $T_{trs}$ | $\Delta H^{cal}$ | $\Delta H^{v.H.}$ | $\Delta Cp$ | Appr./Rem. | Ref |
|---|---|---|---|---|---|---|---|
| 0.0 M | 7.0 | 64.3 | 268.4 | 284.3 | | DSC (1) | 98V3 |
| 0.0 M | 7.0 | 65.5±0.1 | | 287.9±4.5 | | heat (1,2) | 98V3 |
| 0.0 M | 7.6 | 60.5 | 306.0 | 325.2 | | DSC (1) | 98V3 |
| 0.5 M | 7.6 | 55.0 | 244.1 | 255.8 | | DSC (1) | 98V3 |
| 1.0 M | 7.6 | 49.0 | 219.0 | 224.0 | | DSC (1) | 98V3 |
| 1.5 M | 7.6 | 40.4 | 170.1 | 167.6 | | DSC (1) | 98V3 |
| 0–2 M | 7.0 | 29–68 | | | 6.29±0.43 | heat (1–3) | 98V3 |
| 0–1.5 M | 7.6 | 40–61 | | | 6.43±0.93 | DSC (1,3) | 98V3 |

Remarks:

(1) measured in 10 mM sodium phosphate

(2) transition monitored by CD at 222 nm

(3) $\Delta Cp$ from $\Delta H$ versus $T_{trs}$ with $T_{trs}$ varied by the addition of GuHCl

Histidine-containing phosphocarrier protein HPr, wild type and mutant Lys49→Glu

| Protein | pH | T$_{trs}$ | ΔH | Approach/Remarks | | Ref |
|---|---|---|---|---|---|---|
| wild type | 7.0 | 63.4±0.1 | 317±5 | heat | (1,2) | 99P7 |
| Lys49→Glu | 7.0 | 72.2±0.1 | 351±4 | heat | (1,2) | 99P7 |

Remarks:
(1) transition monitored by far-UV CD at 222 nm
(2) measured in 10 mM potassium phosphate


**His-Tagged Proteins**, see Arc Repressor, Eglin, and Mannitol Permease


## Histone

Recombinant archaeal histones from the mesophile *Methanobacterium formicicum* (rHFoB), hyperthermophile *Methanothermus fervidus* (rHMfA, rHMfB), and hyperthermophile *Pyrococcus* strain GB-3a (rHPyA1)

| Protein/KCl | | pH | T* | ΔH$^{v.H.}$ | CR | ΔCp | Appr./Rem. | Ref |
|---|---|---|---|---|---|---|---|---|
| rHMfA | 0.2 M | 3.0 | 96.1±0.1 | 583.2±2.5 | 0.101±0.001 | 8.79±0.07 | DSC (1–4) | 98L4 |
| | | 3.0 | 84.6±0.1 | 561.1±2.5 | | | heat (1–3,5) | 98L4 |
| | | 4.0 | 97.6±0.1 | 662.3±2.5 | 0.261±0.003 | 9.31±0.08 | DSC (1–4) | 98L4 |
| | | 4.0 | 100.2±0.2 | 674.5±3.3 | | | heat (1–3,5) | 98L4 |
| | | 5.0 | 104.5±0.0 | 687.4±1.3 | 0.227±0.001 | 9.03±0.15 | DSC (1–4) | 98L4 |
| | | 5.0 | 104.1±0.3 | 686.6±5.4 | | | heat (1–3,5) | 98L4 |
| | | 6.0 | 101.0±0.0 | 662.7±0.8 | 0.281±0.001 | 9.11±0.08 | DSC (1–4) | 98L4 |
| | | 6.0 | 101.1±0.2 | 658.1±2.9 | | | heat (1–3,5) | 98L4 |
| | 1.0 M | 2.0 | 94.0±0.3 | 570.3±5.0 | | 7.81±0.15 | heat (1–3,5) | 98L4 |
| | | 3.0 | 100.7±0.2 | 640.6±4.2 | | 8.84±0.11 | heat (1–3,5) | 98L4 |
| | | 4.0 | 111.6±0.2 | 720.1±4.2 | | 8.99±0.11 | heat (1–3,5) | 98L4 |
| | | 3–6 | | | | 9.06±0.21 | DSC (6) | 98L4 |
| | | 3–6 | 91–112 | | | 8.16±0.14 | DSC (7) | 98L4 |
| rHFoB | 0.2 M | 4.0 | 75.6±0.0 | 381.2±1.3 | 0.501±0.009 | | DSC (1–4) | 98L4 |
| | | 5.0 | 76.9±0.0 | 474.9±0.4 | 0.325±0.001 | 10.68±0.02 | DSC (1–4) | 98L4 |
| | | 5.0 | 74.8±0.2 | 484.9±0.4 | | | heat (1–3,5) | 98L4 |
| | | 6.0 | 76.5±0.0 | 397.1±0.4 | 0.673±0.005 | 12.38±0.38 | DSC (1–4) | 98L4 |
| | | 6.0 | 70.9±0.6 | 505.0±10.0 | | | heat (1–3,5) | 98L4 |
| | 1.0 M | 2.0 | 64.7±0.1 | 480.7±1.7 | 0.091±0.009 | 8.10±0.07 | DSC (1–4) | 98L4 |
| | | 2.0 | 60.3±0.5 | 442.2±4.2 | | | heat (1–3,5) | 98L4 |
| | | 3.0 | 77.0±0.0 | 455.6±0.0 | 0.325±0.001 | 9.76±0.03 | DSC (1–4) | 98L4 |
| | | 3.0 | 73.0±0.2 | 485.3±0.4 | | | heat (1–3,5) | 98L4 |
| | | 4.0 | 90.3±0.0 | 538.1±0.8 | 0.366±0.001 | 8.00±0.05 | DSC (1–4) | 98L4 |
| | | 4.0 | 91.5±0.1 | 525.9±1.7 | | | heat (1–3,5) | 98L4 |
| | | 5.0 | 90.9±0.0 | 520.1±0.8 | 0.330±0.002 | 8.13±0.10 | DSC (1–4) | 98L4 |
| | | 5.0 | 92.0±0.1 | 530.1±3.8 | | | heat (1–3,5) | 98L4 |
| | | 6.0 | 90.5±0.2 | 512.5±0.4 | 0.411±0.001 | 9.96±0.09 | DSC (1–4) | 98L4 |
| | | 6.0 | 87.5±0.2 | 567.8±2.9 | | | heat (1–3,5) | 98L4 |
| | | 2–6 | | | | 8.79±0.98 | DSC (6) | 98L4 |
| rHMfB | 0.2 M | 2.5 | 78.2±0.1 | 409.6±0.4 | | | heat (1–3,5) | 98L4 |
| | | 2.7 | 84.9±0.1 | 423.0±0.8 | | 7.09±0.02 | heat (1–3,5) | 98L4 |
| | | 3.0 | 89.3±0.2 | 470.7±2.1 | | 6.68±0.02 | heat (1–3,5) | 98L4 |
| | | 3.3 | 94.1±0.1 | 524.3±2.1 | | 7.17±0.06 | heat (1–3,5) | 98L4 |
| | | 3.7 | 102.9±0.1 | 618.4±2.1 | | 8.46±0.05 | heat (1–3,5) | 98L4 |
| | | 4.0 | 103.4±0.1 | 607.9±2.1 | | 8.26±0.05 | heat (1–3,5) | 98L4 |
| | | 5.0 | 112.8±0.3 | 628.9±2.1 | | 7.82±0.09 | heat (1–3,5) | 98L4 |
| | | 6.0 | 108.9±0.2 | 659.4±3.8 | | 9.60±0.09 | heat (1–3,5) | 98L4 |

Recombinant archaeal histones from the mesophile *Methanobacterium formicicum* (rHFoB), hyperthermophile *Methanothermus fervidus* (rHMfA, rHMfB), and hyperthermophile *Pyrococcus* strain GB-3a (rHPyA1) (continued)

| Protein/KCl | pH | T* | $\Delta H^{v.H.}$ | CR | $\Delta C_p$ | Appr./Rem. | Ref |
|---|---|---|---|---|---|---|---|
| | 2–6 | | | | 8.31±0.81 | heat (6) | 98L4 |
| | 2–6 | 89–113 | | | 6.67±0.73 | heat (7) | 98L4 |
| rHPyA1  0.2 M | 4.0 | 107.2±0.3 | 720.5±8.8 | | 8.88±0.38 | heat (1–3,5) | 98L4 |
| | 5.0 | 114.1±0.6 | 771.5±12.6 | | 10.00±0.64 | heat (1–3,5) | 98L4 |
| | 7.5 | 109.8±0.3 | 720.9±5.4 | | 9.23±0.14 | heat (1–3,5) | 98L4 |
| | 4–7 | | | | 9.37±0.57 | heat (6) | 98L4 |

Remarks:
(1) the molar unit for the van't Hoff enthalpy and the change in heat capacity upon unfolding is the cooperative unit, i.e., a dimer for the archaeal histones
(2) according to the thermodynamic treatment is T* the temperature at which the free energy of unfolding of a 1 M protein solution is zero
(3) T* does not coincide with $T_{trs}$:
   (a) for a 153 µM protein solution (total monomer) at pH 4, $\Delta H^{cal}$ = 172.8 kJ/mol and $T_{trs}$ = 83°C according to DSC
   (b) for a 2.3 µM protein solution (total monomer) at pH 4, $\Delta H^{v.H.}$ = 674.5 kJ/mol and $T_{trs}$ = 74°C according to heat denaturation monitored by CD (see also Figs. 3A and 3B in Ref. 98L4)
(4) CR is the cooperative ratio, i.e., $\Delta H^{cal}/\Delta H^{v.H.}$
(5) transition monitored by CD at 222 nm
(6) average $\Delta C_p$ from the tabulated data
(7) $\Delta C_p$ from $\Delta H^{v.H.}$ versus $T_{trs}$

## HIV-1 Protease

Plasmid-encoded mutant HIV-1 protease (Gln7→Lys, Leu33→Ile, Leu63→Ile)

| Protein | pH | $T_{trs}$ | $\Delta C_p$ | $\Delta H$ | Appr./Rem. | Ref |
|---|---|---|---|---|---|---|
| | 3.4 | 59 | 13.4 | 420±42 | DSC  (1–3) | 98T9 |

Remarks:
(1) the data refer to a two state transition of folded dimer into two unfolded monomers
(2) mean of three measurements
(3) measured in 10 mM Na-formate at 25 µM protease conc.

## Homeodomain

MATα2 homeodomain from yeast, DNA-bindng protein domain

| pH | [NaCl] | [MATα2] | $T_{trs}$ | $\Delta C_p$ | $\Delta H$ | Approach/Remarks | Ref |
|---|---|---|---|---|---|---|---|
| 3.5 | 100 mM | 0.84 | 47.8 | | 169 | DSC (1–5) | 97C2 |
| 4.0 | 0 mM | 2.66 | 49.0 | | 158 | DSC (1–5) | 97C2 |
| 4.0 | 100 mM | 0.30 | 55.2 | | 155 | heat, v.H. (1,6) | 97C2 |
| 4.0 | 100 mM | 3.07 | 52.3 | | 170 | DSC (1–5) | 97C2 |
| 4.0 | 200 mM | 2.17 | 54.2 | | 174 | DSC (1–5) | 97C2 |
| 4.0 | 300 mM | 2.19 | 56.0 | | 171 | DSC (1–5) | 97C2 |
| 4.0 | 400 mM | 2.25 | 57.0 | | 170 | DSC (1–5) | 97C2 |
| 5.0 | 100 mM | 0.92 | 56.4 | | 164 | DSC (1–5) | 97C2 |
| 6.0 | 100 mM | 0.94 | 58.2 | | 158 | DSC (1–5) | 97C2 |
| 7.0 | 0 mM | 1.24 | 56.3 | | 136 | DSC (1–5) | 97C2 |
| 7.0 | 0 mM | 0.26 | 56.8 | | 178 | heat, v.H. (1,6) | 97C2 |
| 7.0 | 50 mM | 0.57 | 57.3 | | 159 | DSC (1–5) | 97C2 |
| 7.0 | 100 mM | 0.59 | 56.4 | | 149 | DSC (1–5) | 97C2 |
| 7.0 | 100 mM | 0.59 | 55.8 | | 144 | DSC (1–5) | 97C2 |
| 7.0 | 100 mM | 0.80 | 60.1 | | 161 | DSC (1–5) | 97C2 |

MATα2 homeodomain from yeast, DNA-binding protein domain (continued)

| pH | [NaCl] | [MATα2] | $T_{trs}$ | ΔCp | ΔH | Approach/Remarks | | Ref |
|----|--------|---------|-----------|-----|-----|------------------|---|-----|
| 7.0 | 100 mM | 1.55 | 57.4 | | 165 | DSC | (1–5) | 97C2 |
| 7.0 | 100 mM | 1.61 | 59.3 | | 163 | DSC | (1–5) | 97C2 |
| 7.0 | 100 mM | 3.23 | 59.5 | | 161 | DSC | (1–5) | 97C2 |
| 7.0 | 200 mM | 0.57 | 60.1 | | 170 | DSC | (1–5) | 97C2 |
| 7.0 | 300 mM | 0.57 | 61.4 | | 174 | DSC | (1–5) | 97C2 |
| | | | | 1±1 | | DSC | (1,7) | 97C2 |

Remarks:
(1) the 128–210 amino acid fragment of the MATα2 protein containing the homeodomain expressed in *E. coli*
(2) NaCl conc. in mM
(3) protein conc. in mg/ml
(4) error in $T_{trs}$ ±1 K
(5) ΔH obtained by integrating the area under the curve, errors are ±10%
(6) transition monitored by CD at 222 nm
(7) ΔCp from extrapolated heat capacity between unfolded and folded states

## Hook Protein

Flagellar hook protein from *Salmonella typhimurium*, proteolytic fragments

| Fragment/Transition | | pH | $T_{trs}$ | ΔCp | ΔH | Appr./Rem. | | Ref |
|---------------------|---|----|-----------|-----|-----|-----------|---|-----|
| H32 | overall | 7.0 | 52.2 | 20 | 1032 | DSC | (1,3) | 99U1 |
| | Trans. (1) | 7.0 | 43.7±0.4 | 7 | 164±7 | DSC | (1,3,4) | 99U1 |
| | trans. (2) | 7.0 | 48.5±0.1 | 23 | 627±7 | DSC | (1,3,4) | 99U1 |
| | trans. (3) | 7.0 | 51.3±0.1 | 23 | 1024±8 | DSC | (1–3) | 99U1 |
| H22 | overall | 7.0 | 54.8 | 5 | 476 | DSC | (2,3) | 99U1 |
| | trans. (1) | 7.0 | 55.8±1.7 | 0 | 112±8 | DSC | (2–4) | 99U1 |
| | trans. (2) | 7.0 | 54.8±0.2 | 5.1±0.1 | 432±4 | DSC | (2–4) | 99U1 |

Remarks:
(1) fragment H32 contains the residues 72–370 of hook protein
(2) fragment H22 contains the residues 148–355 of hook protein
(3) measured in 10 mM sodium phosphate, pH 7.0
(4) results of deconvolution

**HU (*Bsu*HU), see DNA-Binding Protein**

## Human Serum Albumin

Human serum albumin (HSA)

| pH | $T_{trs}$ | ΔCp | ΔH | Approach/Remarks | | Ref |
|----|-----------|-----|-----|------------------|---|-----|
| 6.5 | 60 | 60.6±2 | 562±5 | DSC | (1,2) | 92B10 |
| 7.0 | 65 | 52.5±1 | 688±4 | DSC | (1,2) | 92B10 |
| 7.3 | 64 | 46.3±4 | 595±3 | DSC | (1,2) | 92B10 |
| 7.7 | 62.5 | 37.5±1 | 571±10 | DSC | (1,2) | 92B10 |
| 8.0 | 61 | 34.5±2 | 480±4 | DSC | (1,2) | 92B10 |
| 8.5 | 60 | 26.9±3 | 446±5 | DSC | (1,2) | 92B10 |

Remarks:
(1) measured in 0.2 M Tris buffer
(2) BSA concentration $1 \times 10^{-4}$ M

Human serum albumin, fatty acid free

| pH | $T_{trs}$ | $\Delta C_p$ | $\Delta H$ | Appr./Rem. | Ref |
|---|---|---|---|---|---|
| 9.99 | 65.6±0.5 | 19.2±1.7 | 384±15 | DSC (1) | 97P10 |
| 8.88 | 65.5±0.4 | 17.2±2.5 | 427±19 | DSC (1) | 97P10 |
| 7.4 | 63.2±0.4 | 16.3±1.7 | 372±19 | DSC (1) | 97P10 |
| 6.4 | 62.8±0.2 | 18.9±1.3 | 386±9 | DSC (1) | 97P10 |
| 5.4 | 65.4±0.4 | 21.4±1.3 | 418±14 | DSC (1) | 97P10 |
| 4.3 | 46.2±0.1 | 9.6±0.4 | 233±14 | DSC (1,2) | 97P10 |
| | 60.2±0.7 | 16.0±1.3 | 250±6 | DSC (1,2) | 97P10 |

Remarks:
(1) heating rate 1 K/min
(2) occurrence of a bimodal melting pattern at pH 4.3, i.e., a second peak at 46.2°C

Human serum albumin (HSA), fatty acid free

| Protein | pH | $T_{trs}$ | $\Delta C_p$ | $\Delta H$ | Appr./Rem. | Ref |
|---|---|---|---|---|---|---|
| HSA | 9.9 | 65.6±0.5 | 19.2±1.7 | 384±15 | DSC | 99F2 |
| | 7.4 | 63.2±0.4 | 16.3±1.7 | 372±19 | DSC (1) | 99F2 |
| | 5.4 | 65.4±0.4 | 21.4±1.3 | 418±14 | DSC | 99F2 |

Remark:
(1) for serum albumin in the presence of polyethylene glycols see Ref. 99F1

Human serum albumin (HSA), defatted

| Protein | pH | $T_{trs}$ | $\Delta H$ | Approach/Remarks | Ref |
|---|---|---|---|---|---|
| HSA | 7.4 | 60.1±0.1 | 594±8 | DSC (1,2) | 98R11 |

Remarks:
(1) measured in 100 mM phosphate buffer, pH 7.4
(2) Ref. 98R11 contains additional data for HSA in the presence of the herbicide 2,4-dichlorophenoxyacetic acid

Human serum albumin, fatty acid free, dependence of thermodynamic quantities on scan rate

| Scan rate | pH | $T_{trs}$ | $\Delta C_p$ | $\Delta H$ | Appr./Rem. | Ref |
|---|---|---|---|---|---|---|
| 0.25 K/min | 7.4 | 64.1±0.3 | 16.3±2.9 | 451±8 | DSC (1) | 97P10 |
| 0.50 K/min | 7.4 | 62.7±0.4 | 24.6±2.9 | 430±4 | DSC (1) | 97P10 |
| 1.00 K/min | 7.4 | 63.2±0.4 | 16.7±4.2 | 376±19 | DSC (1) | 97P10 |
| 2.00 K/min | 7.4 | 64.8±0.1 | 15.5±0.4 | 370±19 | DSC (1) | 97P10 |
| 0.7 K/min | 7.4 | 62.1±0.3 | | 377±17 | heat (2) | 97P10 |

Remarks:
(1) buffer: 100 mM sodium phosphate
(2) transition monitored by fluorescence, excitation 280 nm, emission 345 nm

## Immunoglobulin

The data entries are arranged as follows:

a) antibodies
b) domains

*a) antibodies*

Rabbit immunoglobulin (IgG), conformers studied by DSC changing pH from 7.0 to 2.0 and from 2.0 to 7.0, results of deconvolution

| Transition | pH | $T_{trs}$ | $\Delta Cp$ | $\Delta H$ | Approach/Remarks | | Ref |
|---|---|---|---|---|---|---|---|
| measurements at pH decrease from 7.0 to 2.0: | | | | | | | |
| trans. (5) | 7.0 | 69.1 | | 356 | DSC | (1) | 96V6 |
| trans. (4) | 7.0 | 73.6 | | 720 | DSC | (1) | 96V6 |
| trans. (3) | 7.0 | 76.3 | | 893 | DSC | (1) | 96V6 |
| trans. (2) | 7.0 | 79.0 | | 795 | DSC | (1) | 96V6 |
| trans. (1) | 7.0 | 83.4 | | 628 | DSC | (1) | 96V6 |
| trans. (5) | 6.5 | 71.9 | | 437 | DSC | (1) | 96V6 |
| trans. (4) | 6.5 | 75.2 | | 687 | DSC | (1) | 96V6 |
| trans. (3) | 6.5 | 77.8 | | 840 | DSC | (1) | 96V6 |
| trans. (2) | 6.5 | 80.7 | | 789 | DSC | (1) | 96V6 |
| trans. (1) | 6.5 | 85.1 | | 632 | DSC | (1) | 96V6 |
| trans. (5) | 5.5 | 68.6 | | 397 | DSC | (1) | 96V6 |
| trans. (4) | 5.5 | 72.0 | | 582 | DSC | (1) | 96V6 |
| trans. (3) | 5.5 | 77.0 | | 784 | DSC | (1) | 96V6 |
| trans. (2) | 5.5 | 80.3 | | 840 | DSC | (1) | 96V6 |
| trans. (1) | 5.5 | 83.8 | | 718 | DSC | (1) | 96V6 |
| trans. (5) | 5.0 | 66.7 | | 500 | DSC | (1) | 96V6 |
| trans. (4) | 5.0 | 72.5 | | 563 | DSC | (1) | 96V6 |
| trans. (3) | 5.0 | 77.0 | | 732 | DSC | (1) | 96V6 |
| trans. (2) | 5.0 | 80.5 | | 757 | DSC | (1) | 96V6 |
| trans. (1) | 5.0 | 84.0 | | 672 | DSC | (1) | 96V6 |
| trans. (6) | 4.5 | 59.2 | | 338 | DSC | (1) | 96V6 |
| trans. (5) | 4.5 | 63.6 | | 328 | DSC | (1) | 96V6 |
| trans. (4) | 4.5 | 70.7 | | 524 | DSC | (1) | 96V6 |
| trans. (3) | 4.5 | 75.2 | | 626 | DSC | (1) | 96V6 |
| trans. (2) | 4.5 | 78.9 | | 677 | DSC | (1) | 96V6 |
| trans. (1) | 4.5 | 82.7 | | 648 | DSC | (1) | 96V6 |
| trans. (6) | 4.0 | 53.1 | | 261 | DSC | (1) | 96V6 |
| trans. (5) | 4.0 | 55.9 | | 277 | DSC | (1) | 96V6 |
| trans. (4) | 4.0 | 65.4 | | 454 | DSC | (1) | 96V6 |
| trans. (3) | 4.0 | 71.3 | | 562 | DSC | (1) | 96V6 |
| trans. (2) | 4.0 | 75.9 | | 650 | DSC | (1) | 96V6 |
| trans. (1) | 4.0 | 80.0 | | 650 | DSC | (1) | 96V6 |
| trans. (6) | 3.0 | 37.7 | | 248 | DSC | (1) | 96V6 |
| trans. (5) | 3.0 | 46.2 | | 297 | DSC | (1) | 96V6 |
| trans. (4) | 3.0 | 54.4 | | 276 | DSC | (1) | 96V6 |
| trans. (3) | 3.0 | 62.1 | | 451 | DSC | (1) | 96V6 |
| trans. (2) | 3.0 | 66.9 | | 547 | DSC | (1) | 96V6 |
| trans. (1) | 3.0 | 70.9 | | 549 | DSC | (1) | 96V6 |
| trans. (4) | 2.0 | 51.2 | | 314 | DSC | (1) | 96V6 |
| trans. (3) | 2.0 | 56.6 | | 405 | DSC | (1) | 96V6 |
| trans. (2) | 2.0 | 60.4 | | 430 | DSC | (1) | 96V6 |
| trans. (1) | 2.0 | 63.6 | | 441 | DSC | (1) | 96V6 |
| measurements at pH increase from 2.0 to 7.0: | | | | | | | |
| trans. (5) | 7.0 | 69.7 | | 450 | DSC | (1) | 96V6 |
| trans. (4) | 7.0 | 74.4 | | 617 | DSC | (1) | 96V6 |
| trans. (3) | 7.0 | 77.9 | | 746 | DSC | (1) | 96V6 |
| trans. (2) | 7.0 | 81.2 | | 662 | DSC | (1) | 96V6 |
| trans. (1) | 7.0 | 85.5 | | 562 | DSC | (1) | 96V6 |
| trans. (5) | 6.5 | 68.3 | | 371 | DSC | (1) | 96V6 |
| trans. (4) | 6.5 | 73.4 | | 516 | DSC | (1) | 96V6 |

Rabbit immunoglobulin (IgG), conformers studied by DSC changing pH from 7.0 to 2.0 and from 2.0 to 7.0, results of deconvolution (continued)

| Transition | pH | $T_{trs}$ | $\Delta Cp$ | $\Delta H$ | Approach/Remarks | | Ref |
|---|---|---|---|---|---|---|---|
| trans. (3) | 6.5 | 77.0 | | 705 | DSC | (1) | 96V6 |
| trans. (2) | 6.5 | 80.9 | | 700 | DSC | (1) | 96V6 |
| trans. (1) | 6.5 | 84.1 | | 621 | DSC | (1) | 96V6 |
| trans. (5) | 5.5 | 67.3 | | 348 | DSC | (1) | 96V6 |
| trans. (4) | 5.5 | 73.8 | | 519 | DSC | (1) | 96V6 |
| trans. (3) | 5.5 | 78.0 | | 632 | DSC | (1) | 96V6 |
| trans. (2) | 5.5 | 81.2 | | 576 | DSC | (1) | 96V6 |
| trans. (1) | 5.5 | 84.0 | | 531 | DSC | (1) | 96V6 |
| trans. (5) | 5.0 | – | | – | DSC | (1) | 96V6 |
| trans. (4) | 5.0 | 68.8 | | 417 | DSC | (1) | 96V6 |
| trans. (3) | 5.0 | 75.4 | | 609 | DSC | (1) | 96V6 |
| trans. (2) | 5.0 | 79.6 | | 718 | DSC | (1) | 96V6 |
| trans. (1) | 5.0 | 83.2 | | 680 | DSC | (1) | 96V6 |
| trans. (6) | 4.5 | – | | – | DSC | (1) | 96V6 |
| trans. (5) | 4.5 | – | | – | DSC | (1) | 96V6 |
| trans. (4) | 4.5 | 68.2 | | 355 | DSC | (1) | 96V6 |
| trans. (3) | 4.5 | 74.1 | | 584 | DSC | (1) | 96V6 |
| trans. (2) | 4.5 | 78.4 | | 702 | DSC | (1) | 96V6 |
| trans. (1) | 4.5 | 82.1 | | 668 | DSC | (1) | 96V6 |
| trans. (6) | 4.0 | – | | – | DSC | (1) | 96V6 |
| trans. (5) | 4.0 | – | | – | DSC | (1) | 96V6 |
| trans. (4) | 4.0 | 67.3 | | 385 | DSC | (1) | 96V6 |
| trans. (3) | 4.0 | 73.0 | | 544 | DSC | (1) | 96V6 |
| trans. (2) | 4.0 | 76.9 | | 645 | DSC | (1) | 96V6 |
| trans. (1) | 4.0 | 80.4 | | 603 | DSC | (1) | 96V6 |
| trans. (6) | 3.0 | – | | – | DSC | (1) | 96V6 |
| trans. (5) | 3.0 | – | | – | DSC | (1) | 96V6 |
| trans. (4) | 3.0 | 59.3 | | 320 | DSC | (1) | 96V6 |
| trans. (3) | 3.0 | 63.3 | | 441 | DSC | (1) | 96V6 |
| trans. (2) | 3.0 | 67.2 | | 554 | DSC | (1) | 96V6 |
| trans. (1) | 3.0 | 71.0 | | 590 | DSC | (1) | 96V6 |
| N → D | | | 50±4 | | DSC | (2,4) | 96V6 |
| I → D | | | 42±3 | | DSC | (3,4) | 96V6 |

Remarks:

(1) the results are interpreted in Ref. 96V6 by four conformers: N, $N_A$, I, and $N_I$, differing in the amount of structure in the $C_H2$ domain as well as in its interaction with the neighboring domains

(2) data from the heat capacity increase of the calorimetric recording

(3) data from the heat capacity increase of the calorimetric recording at pH 2–2.5

(4) for $\Delta G$ values see Table 1

Monoclonal immunoglobulin M (IgM) compared with rheumatoid immunoglobulin M (IgM-RF) and fragments

| Protein | pH | $T_{trs}$ | $\Delta H$ | Approach/Remarks | | Ref |
|---|---|---|---|---|---|---|
| IgM | 7.0 | 67.1 | 3582 | DSC | (1) | 97P16 |
| IgM-RF | 7.0 | 71.1 | 3574 | DSC | (1) | 97P16 |
| Fab(IgM) | 7.0 | 73.7 | 1272 | DSC | (1) | 97P16 |
| Fab(IgM-RF) | 7.0 | 73.3 | 1356 | DSC | (1) | 97P16 |
| Fc(IgM) | 7.0 | 71.2 | 1214 | DSC | (1) | 97P16 |
| FC(IgM-RF) | 7.0 | 70.7 | 1541 | DSC | (1) | 97P16 |

Remark:

(1) measured in 0.01 M potassium phosphate buffer containing 0.15 M NaCl

Monoclonal immunoglobulin M (IgM) compared with rheumatoid immunoglobulin M (IgM-RF) and fragments, results of deconvolution

| Protein | Transition | pH | $T_{trs}$ | $\Delta H$ | Appr./Rem. | Ref |
|---|---|---|---|---|---|---|
| IgM | trans. (1) | 7.0 | 63.2 | 476 | DSC (1) | 97P16 |
|  | trans. (2) | 7.0 | 64.3 | 689 | DSC (1) | 97P16 |
|  | trans. (3) | 7.0 | 66.1 | 863 | DSC (1) | 97P16 |
|  | trans. (4) | 7.0 | 68.1 | 878 | DSC (1) | 97P16 |
|  | trans. (5) | 7.0 | 72.1 | 688 | DSC (1) | 97P16 |
| Fab(IgM) | trans. (1) | 7.0 | 70.1 | 566 | DSC (1) | 97P16 |
|  | trans. (2) | 7.0 | 74.0 | 710 | DSC (1) | 97P16 |
| IgM-RF | trans. (1) | 7.0 | 64.5 | 572 | DSC (1) | 97P16 |
|  | trans. (2) | 7.0 | 66.8 | 830 | DSC (1) | 97P16 |
|  | trans. (3) | 7.0 | 69.7 | 1034 | DSC (1) | 97P16 |
|  | trans. (4) | 7.0 | 71.6 | 1157 | DSC (1) | 97P16 |
| Fab(IgM-RF) | trans. (1) | 7.0 | 70.1 | 493 | DSC (1) | 97P16 |
|  | trans. (2) | 7.0 | 73.3 | 885 | DSC (1) | 97P16 |

Remark:

(1) measured in 0.01 M potassium phosphate buffer containing 0.15 M NaCl

Murine monoclonal IgG2a, κ anti-p24-antibody CB4-1 and its Fab and Fc fragments

| Protein | pH | $T_{trs}$ | $\Delta H$ | Approach/Remarks | | Ref |
|---|---|---|---|---|---|---|
| intact CB4-1 | 7.5 | 62–75 | 3498 | DSC | (1) | 99W7 |
| Fab CB4-1 | 7.5 | 64–67 | 1071 | DSC | (1) | 99W7 |
| Fc CB4-1 | 7.5 | 69–72 | 1013 | DSC | (1) | 99W7 |

Remark:

(1) Ref. 99W7 contains results of deconvolution for the irreversible transitions obtained by varying pH

Monoclonal anti-ferritin antibodies

| Protein/Subclass | pH | $T_{trs}$ | $\Delta H$ | Approach/Remarks | | Ref |
|---|---|---|---|---|---|---|
| G10, IgG2a | 7.0 | 66–80 | 4420 | DSC | (1,2) | 98C9 |
| F11, IgG2a | 7.0 | 66–80 | 3720 | DSC | (1–3) | 98C9 |
| C5, IgG1 | 7.0 | 66–80 | 4710 | DSC | (1,2) | 98C9 |

Remarks:

(1) measured in 50 mM sodium phosphate, pH 7.0
(2) the transition profile resolved by deconvolution shows individual transitions from 66 to 80°C
(3) in the transition profile the subtransition at 74°C is lacking

*b) domains*

Fc fragments from myeloma immunoglobulin G (for the deconvolution of the first peak see the data entries below)

| Fragment | pH | $T_{peak1}$ | $\Delta H_{peak1}$ | $T_{peak2}$ | $\Delta H_{peak2}$ | Appr./Rem. | Ref |
|---|---|---|---|---|---|---|---|
| Fc (donor 1) | 3.8 | 63.0 | 470 | 80.0 | 718 | DSC | 98T7 |
|  | 3.8 | 63.8 |  | 80.9 |  | heat (1) | 98T7 |
| Fc (donor 1) | 4.2 | 54.6 | 668 | 73.7 | 525 | DSC | 98T7 |
|  | 4.2 | 54.1 |  | 74.2 |  | heat (1) | 98T7 |
| Fc (donor 1) | 4.2 | 55.0 | 661 | 73.6 | 519 | DSC | 98T7 |
| Fc (donor 2) | 4.2 | 54.9 | 675 | 74.1 | 531 | DSC | 98T7 |
| Fc (donor 3) | 4.2 | 54.2 | 663 | 73.4 | 519 | DSC | 98T7 |

Fc fragments from myeloma immunoglobulin G (for the deconvolution of the first peak see the data entries below) (continued)

| Fragment | pH | $T_{peak1}$ | $\Delta H_{peak1}$ | $T_{peak2}$ | $\Delta H_{peak2}$ | Appr./Rem. | Ref |
|---|---|---|---|---|---|---|---|
| Fc (donor 1) | 4.6 | 63.5 | 811 | 81.1 | 596 | DSC | 98T7 |
|  | 4.6 | 64.0 |  | 81.4 |  | heat (1) | 98T7 |
| Fc (donor 1) | 5.5 | 66.1 | 861 | 82.2 | 638 | DSC | 98T7 |
|  | 5.5 | 66.5 |  | 81.5 |  | heat (1) | 98T7 |
| Fc (donor 1) | 8.0 | 70.8 | 899 | 83.3 | 693 | DSC | 98T7 |
|  | 8.0 | 72.1 |  | 82.9 |  | heat (1) | 98T7 |
| pFc (donor 1) | 4.6 |  |  | 81.3 | 542 | DSC (2) | 98T7 |
|  | 4.6 |  |  | 74.6 |  | heat (1,2) | 98T7 |
| Fc' (donor 1) | 4.6 |  |  | 80.7 | 537 | DSC (2) | 98T7 |
|  | 4.6 |  |  | 74.0 |  | heat (1,2) | 98T7 |

Remarks:
(1) transition monitored by fluorescence using FITC-labeled FC fragments
(2) the dependence of $T_{peak2}$ on the protein concentration is analyzed in detail in Ref. 98T7 (see also Ref. 82T2)

Melting of $C_H2$ and $C_H3$ domains with Fc and pFc' fragments from different myeloma immunoglobulin G, results of the deconvolution of the first calorimetric peak (see also data above)

| Fragment | pH | $\Delta H_{trs1}$ | $\Delta H_{trs2}$ | Approach/Remarks | | Ref |
|---|---|---|---|---|---|---|
| Fc (donor 1) | 3.8 | 255 | 213 | DSC | (1,2) | 98T7 |
| Fc (donor 1) | 4.2 | 357 | 309 | DSC | (1,2) | 98T7 |
| Fc (donor 1) | 4.6 | 452 | 370 | DSC | (1,2) | 98T7 |
| Fc (donor 1) | 5.5 | 476 | 391 | DSC | (1,2) | 98T7 |
| Fc (donor 1) | 8.0 | 489 | 420 | DSC | (1,2) | 98T7 |
| Fc (donor 2) | 8.0 | 496 | 426 | DSC | (1,2) | 98T7 |
| Fc (donor 3) | 8.0 | 487 | 423 | DSC | (1,2) | 98T7 |

Remarks:
(1) for the conformational stability of the $C_H2$ and $C_H3$ domains see Table 1
(2) the heat capacity change is $\Delta Cp = 11$ and 14 kJ/mol/K for $C_H2$ at pH 3.8 and 8.0, respectively, and $\Delta Cp = 12.2$ kJ/mol/K at pH 3.8 and 8.0 for $C_H3$ domains

Glycosylated and aglycosylated mouse IgG2b domains

| Domain | Glycos. | pH | $T_{trs}$ | $\Delta H$ | Approach/Remarks | | Ref |
|---|---|---|---|---|---|---|---|
| $C_H2$ | (+) | 4.1 | 47.3 | 283 | DSC | (1,2) | 98T8 |
| $C_H2$ | (–) | 4.1 | 49.5 | 185 | DSC | (1,2) | 98T8 |
| $C_H3$ | (+) | 4.1 | 66.1 | 645 | DSC | (1–3) | 98T8 |
| $C_H3$ | (–) | 4.1 | 63.5 | 624 | DSC | (1–3) | 98T8 |
| pFc' | (+) | 4.1 | 65.9 | 623 | DSC | (1,2) | 98T8 |
| pFc | (–) | 4.1 | 65.2 | 629 | DSC | (1,2) | 98T8 |

Remarks:
(1) measured in 10 mM acetate buffer
(2) (+) = glycosylated, (–) = aglycosylated
(3) $\Delta H$ refers to a dimer of $C_H3$ domains

$C_H2$ domains in glycosylated and aglycosylated Fc fragments from mouse IgG2b

| Domain | Glycos. | pH | $T_{trs}$ | $\Delta H$ | Approach/Remarks | | Ref |
|---|---|---|---|---|---|---|---|
| $C_H2$ | (+) | 3.4 | 32.2 | 117 | DSC | (1–3) | 98T8 |
| $C_H2$ | (−) | 3.4 | 34.8 | 73 | DSC | (1–3) | 98T8 |
| $C_H2$ | (+) | 3.6 | 40.4 | 179 | DSC | (1–3) | 98T8 |
| $C_H2$ | (−) | 3.6 | 44.0 | 117 | DSC | (1–3) | 98T8 |
| $C_H2$ | (+) | 3.8 | 42.1 | 217 | DSC | (1–3) | 98T8 |
| $C_H2$ | (−) | 3.8 | 44.9 | 128 | DSC | (1–3) | 98T8 |
| $C_H2$ | (+) | 4.0 | 46.1 | 254 | DSC | (1–3) | 98T8 |
| $C_H2$ | (−) | 4.0 | 47.9 | 145 | DSC | (1–3) | 98T8 |
| $C_H2$ | (+) | 4.1 | 47.3 | 283 | DSC | (1–3) | 98T8 |
| $C_H2$ | (−) | 4.1 | 49.5 | 185 | DSC | (1–3) | 98T8 |

Remarks:
(1) measured in 10 mM acetate buffer
(2) (+) = glycosylated, (−) = aglycosylated
(3) Ref. 98T8 contains a plot of $\Delta H$ versus $T_{trs}$ which yields a heat capacity change of 11.3 kJ/mol/K and 7.8 kJ/mol/K for glycoslated and aglycosylated $C_H2$ domains, respectively

Glycosylation of the Fc region of IgG

| Fragment | Transition | pH | $T_{trs}$ | $\Delta H$ | Appr./Rem. | | Ref |
|---|---|---|---|---|---|---|---|
| IgG1-Fc | trans. (1) | 7.4 | 65.2±0.6 | 238±50 | DSC | (1,2) | 99G24 |
| | trans. (2) | 7.4 | 81.9±0.2 | 151±8 | DSC | (1,2) | 99G24 |
| IgG4-Fc (G0) | | 7.4 | 64.8±0.2 | 347±8 | DSC | (1,3,4) | 99G24 |
| IgG4-Fc (G2) | | 7.4 | 65.1±0.02 | 393±67 | DSC | (1,3,5) | 99G24 |
| IgG4-Fc, deglycosylated | | | | | | | |
| | trans. (1) | 7.4 | 60.9±0.1 | 351±33 | DSC | (1,2) | 99G24 |
| | trans. (2) | 7.4 | 68.1±0.04 | 109±33 | DSC | (1,2) | 99G24 |

Remarks:
(1) measured in phosphate buffered saline (PBS) at pH 7.4
(2) modelled as two sequential two-state transitions
(3) modelled in terms of a single two-state transition
(4) deglycosylated IgG4-Fc
(5) fully galactoslyated (G2) IgG4-Fc

Recombinant human light chain variable domains (rV$_\lambda$6)

| Protein | pH | $T_{trs}$ | $\Delta Cp$ | $\Delta H$ | Approach/Remarks | | Ref |
|---|---|---|---|---|---|---|---|
| rV$_\lambda$6Wil | 7.5 | 38.3 | 5.9 | 335±4 | heat | (1–4) | 99W4 |
| rV$_\lambda$6Jto | 7.5 | 45.2 | 6.3 | 364±3 | heat | (1–4) | 99W4 |

Remarks:
(1) Ref. 99W4 compares proteins differing in thermodynamic properties and fibrillogenic potential, rV$_\lambda$6 of two patients
(2) transition monitored by intrinsic tryptophan fluorescence
(3) buffer: 10 mM $Na_2HPO_4$, 10 mM $NaH_2PO_4$, 150 mM NaCl, pH 7.5
(4) data from heat denaturation, for details see Ref. 99W4

Rabbit IgG, increase in stability of the $C_H2$ domain after acid treatment

| Protein | pH | $T_{trs}$ | $\Delta H$ | Approach/Remarks | | Ref |
|---|---|---|---|---|---|---|
| native IgG | 3.5 | 75 | 2790±150 | DSC | (1,2) | 94M24 |
| | 7.0 | 78 | 3350±207 | DSC | (1,2) | 94K14 |
| renat. IgG | 3.5 | 78 | 2860±140 | DSC | (3,4) | 94M24 |
| | 7.0 | 78 | 3040±150 | DSC | (3,4) | 94K14 |
| IgG | 2.0 | 58 | 1470±110 | DSC | (1,2) | 94K14 |

Remarks:

(1) measured in 50 mM glycine/HCl at pH 2 and 3.5, and in 50 mM sodium phosphate at pH 7

(2) integral heat, see also Refs. 94K14 and 95M6

(3) renatured after exposure of the protein to pH 2

(4) the 58°C transition of native IgG is not visible

# Interferon

Recombinant human γ-interferon

| Buffer conc. | pH | $T_{trs}$ | $\Delta H$ | Approach/Remarks | | Ref |
|---|---|---|---|---|---|---|
| 5 mM | 4.1 | 52.5 | 523 | DSC | (1) | 99B7 |
| | 4.3 | 53.9 | 530 | DSC | (1) | 99B7 |
| | 4.9 | 57.5 | 600 | DSC | (1) | 99B7 |
| | 6.6 | 60.6 | 649 | DSC | (1) | 99B7 |
| 10 mM | 3.0 | 36.0 | 165 | DSC | (1) | 99B7 |
| | 3.5 | 40.3 | 310 | DSC | (1) | 99B7 |
| | 3.8 | 44.1 | 375 | DSC | (1) | 99B7 |
| | 3.9 | 45.9 | 401 | DSC | (1) | 99B7 |
| | 4.1 | 50.1 | 460 | DSC | (1) | 99B7 |
| | 4.3 | 51.3 | 486 | DSC | (1) | 99B7 |
| | 4.5 | 52.5 | 503 | DSC | (1) | 99B7 |
| | 4.7 | 53.8 | 525 | DSC | (1) | 99B7 |
| | 4.9 | 54.9 | 547 | DSC | (1) | 99B7 |
| | 5.1 | 56.0 | 560 | DSC | (1) | 99B7 |
| | 5.3 | 56.9 | 578 | DSC | (1) | 99B7 |
| | 5.8 | 58.3 | 597 | DSC | (1) | 99B7 |
| | 6.6 | 59.5 | 606 | DSC | (1) | 99B7 |
| | 7.8 | 60.2 | 600 | DSC | (1) | 99B7 |
| 20 mM | 3.8 | 44.6 | 323 | DSC | (1) | 99B7 |
| | 3.9 | 47.2 | 392 | DSC | (1) | 99B7 |
| | 4.1 | 50.8 | 438 | DSC | (1) | 99B7 |
| | 4.3 | 51.8 | 460 | DSC | (1) | 99B7 |
| | 4.7 | 54.3 | 500 | DSC | (1) | 99B7 |
| | 4.9 | 56.8 | 518 | DSC | (1) | 99B7 |
| | 5.8 | 59.0 | 548 | DSC | (1) | 99B7 |
| | 6.6 | 60.1 | 560 | DSC | (1) | 99B7 |
| | 7.8 | 60.2 | 562 | DSC | (1) | 99B7 |
| | 8.8 | 60.2 | 554 | DSC | (1) | 99B7 |
| 50 mM | 3.9 | 45.2 | 300 | DSC | (1) | 99B7 |
| | 4.1 | 46.7 | 332 | DSC | (1) | 99B7 |
| | 4.3 | 48.4 | 348 | DSC | (1) | 99B7 |
| | 4.9 | 54.5 | 420 | DSC | (1) | 99B7 |
| | 6.6 | 60.0 | 509 | DSC | (1) | 99B7 |

Recombinant human γ-interferon (continued)

| Buffer conc. | pH | $T_{trs}$ | ΔH | Approach/Remarks | | Ref |
|---|---|---|---|---|---|---|
| 100 mM | 4.0 | 44.3 | 252 | DSC | (1) | 99B7 |
| | 4.3 | 47.7 | 294 | DSC | (1) | 99B7 |
| | 5.7 | 58.3 | 401 | DSC | (1) | 99B7 |
| | 6.6 | 59.7 | 438 | DSC | (1) | 99B7 |
| | 7.6 | 59.1 | 448 | DSC | (1) | 99B7 |

Remarks:

(1) measured in acetic acid/sodium acetate below pH 6.0 and in sodium phosphate at and above pH 6.0 (measured at 25°C)

(2) ΔCp measured directly from the calorimetric traces was 15.3±3.0 kJ/mol/K in agreement with ΔCp from ΔH versus $T_{trs}$

## Interleukin-1 Receptor

Recombinant human interleukin-1 receptor (IL-1R), formulation development using DSC

| Preservative | Transition | pH | $T_{trs}$ | ΔH | Appr./Rem. | | Ref |
|---|---|---|---|---|---|---|---|
| none (control) | | | | | | | |
| | trans. (1) | 6 | 50.8 | 232 | DSC | (1–3) | 98R4 |
| | trans. (2) | 6 | 53.7 | 198 | DSC | (1–3) | 98R4 |
| | trans. (3) | 6 | 66.3 | 146 | DSC | (1–3) | 98R4 |
| 0.9% benzyl alcohol | | | | | | | |
| | trans. (1) | 6 | 45.2 | 250 | DSC | (1–3) | 98R4 |
| | trans. (2) | 6 | 48.5 | 152 | DSC | (1–3) | 98R4 |
| | trans. (3) | 6 | 63.6 | 215 | DSC | (1–3) | 98R4 |
| 0.1% m-cresol | | | | | | | |
| | trans. (1) | 6 | 48.4 | 232 | DSC | (1–3) | 98R4 |
| | trans. (2) | 6 | 51.9 | 180 | DSC | (1–3) | 98R4 |
| | trans. (3) | 6 | 65.8 | 133 | DSC | (1–3) | 98R4 |
| 0.065% phenol | | | | | | | |
| | trans. (1) | 6 | 50.3 | 253 | DSC | (1–3) | 98R4 |
| | trans. (2) | 6 | 53.4 | 216 | DSC | (1–3) | 98R4 |
| | trans. (3) | 6 | 66.5 | 155 | DSC | (1–3) | 98R4 |

Remarks:

(1) measured in 20 mM sodium citrate, pH 6, 100 mM NaCl

(2) results of deconvolution

(3) the study is aimed to elucidate solution conditions that confer stability of aqueous IL-1R solutions for pharmaceutical purposes

## Invertase

Yeast external invertase, glycosylated (native) and deglycosylated

| Protein | pH | $T_{trs}$ | ΔH | Approach/Remarks | | Ref |
|---|---|---|---|---|---|---|
| native | 5.5 | 64.7 | 4008±25 | DSC | (1) | 96W3 |
| native | 6.0 | 62.9 | 3833 | DSC | (2) | 96W3 |
| native | 7.0 | 51.7±0.3 | 2870±46 | DSC | (2) | 96W3 |
| native | 8.0 | 47.9±0.2 | 2326±46 | DSC | (2) | 96W3 |
| native | 8.8 | 46.2±0.1 | 1929±17 | DSC | (3) | 96W3 |
| deglycosylated | 5.5 | 62.4±0.05 | | DSC | (2,4) | 96W3 |

Yeast external invertase, glycosylated (native) and deglycosylated (continued)

| Protein | pH | $T_{trs}$ | $\Delta H$ | Approach/Remarks | | Ref |
|---|---|---|---|---|---|---|
| deglycosylated | 6.0 | 60.9 | | DSC | (2,4) | 96W3 |
| deglycosylated | 7.0 | 50.0±0.5 | 2301±33 | DSC | (2) | 96W3 |
| deglycosylated | 8.0 | 44.4±0.3 | 1757±25 | DSC | (2) | 96W3 |
| deglycosylated | 8.8 | 43.8±0.1 | 1674±59 | DSC | (3) | 96W3 |

Remarks:
(1) measured in 50 mM sodium acetate buffer
(2) measured in 50 mM sodium phosphate buffer
(3) measured in 50 mM glycine buffer
(4) $\Delta H$ is uncertain due to precipitation upon thermal unfolding

**Iso-1-Cytochrome *c*, see Cytochrome**

## 3-Isopropylmalate Dehydrogenase

3-Isopropylmalate dehydrogenase (IPMDH) from thermophilic bacterium *Thermus thermophilus* HB8 and from the mesophilic *E. coli*, in $H_2O$ and $D_2O$

| Protein | | pH/pD | $T_{trs}$ | Approach/Remarks | | Ref |
|---|---|---|---|---|---|---|
| IPMDH *Thermus th.* | $H_2O$ | 7.6 | 90.0 | DSC | (1) | 98Z3 |
| IPMDH *Thermus th.* | $D_2O$ | 7.15 | 90.9 | DSC | (1) | 98Z3 |
| IPMDH *Thermus th.* | $D_2O$ | 8.15 | 92.2 | DSC | (1) | 98Z3 |
| IPMDH *E. coli* | $H_2O$ | 7.6 | 73.5 | DSC | (2) | 98Z3 |
| IPMDH *E. coli* | $D_2O$ | 7.15 | 77.4 | DSC | (2) | 98Z3 |
| IPMDH *E. coli* | $D_2O$ | 8.15 | 74.5 | DSC | (2) | 98Z3 |

Remarks:
(1) buffer: 20 mM potassium phosphate, 0.3 M KCl (optimal working conditions of IPMDH from *Thermus thermophilus*)
(2) buffer: 20 mM potassium phosphate, 1.0 M KCl (optimal working conditions of IPMDH from *E. coli*)

## Kinase Related Protein

Kinase related protein (KRP) from chicken gizzard

| Protein | pH | $T_{trs}$ | $\Delta H$ | Approach/Remarks | | Ref |
|---|---|---|---|---|---|---|
| KRP | 7 | 54.2±0.2 | 160 | DSC | (1) | 99B21 |

Remark:
(1) measured in 20 mM MOPS, pH 7, 50 mM NaCl, 0.1 mM EGTA, 0.1 mM PMSF, 2 mM DTT, 1 mM $MgCl_2$

## Kinetoplastid Membrane Protein

Kinetoplastid membrane protein-11 from *Leishmania infantum* (KMP-11)

| Protein | pH | $T_{trs}$ | $\Delta H$ | Approach/Remarks | | Ref |
|---|---|---|---|---|---|---|
| KMP-11 | 7.5 | 65 | 183.9 | heat | (1,2) | 99F10 |

Remarks:
(1) transition monitored by far-UV CD at 222 nm
(2) buffer: 0.1 M phosphate, 0.15 M NaCl, pH 7.5

## α-Lactalbumin

Bovine α-lactalbumin

| Protein | pH | $T_{trs}$ | $\Delta H$ | Approach/Remarks | | Ref |
|---|---|---|---|---|---|---|
| Ca$^{2+}$ containing α-lactalbumin: | | | | | | |
| | 7.0 | 66.0±0.5 | 299±11 | DSC | (1,2) | 98P9 |
| | 7.0 | 64.7±0.1 | 267±5 | heat | (1,3) | 98P9 |
| | 7.0 | 64.5±0.1 | 235±10 | heat | (1,4) | 98P9 |
| | 7.0 | 66.1±0.3 | | heat | (1,5a) | 98P9 |
| | 7.0 | 65.4±0.3 | | heat | (1,5b) | 98P9 |
| | 7.0 | 64.4±0.5 | | heat | (1,5c) | 98P9 |
| Ca$^{2+}$ depleted α-lactalbumin: | | | | | | |
| | 7.0 | 42.0±0.4 | 227±5 | DSC | (2,6) | 98P9 |
| | 7.0 | 41.9±0.1 | 221±4 | heat | (3,6) | 98P9 |
| | 7.0 | 41.6±0.5 | 217±10 | heat | (7,8) | 98G4 |
| | 7.0 | 41.5±0.1 | 216±10 | heat | (4,6) | 98P9 |
| | 7.0 | 42.0±1.7 | | heat | (5a,6) | 98P9 |
| | 7.0 | 41.6±1.1 | | heat | (5b,6) | 98P9 |
| | 7.0 | 41.4±1.7 | | heat | (5c,6) | 98P9 |

Remarks:

(1) buffer: 10 mM HEPES, 0.1 mM CaCl$_2$

(2) calorimetric heat, scan rate 1 K/min, mean of 8 and 9 measurements for the holo- and apo-protein, respectively. Ref. 98P9 contains additional heat capacity data

(3) van't Hoff heat, upscan at a scan rate of 0.5 K/min, mean of 4 measurements, transition monitored by difference absorption at 295 nm

(4) van't Hoff heat, downscan at a scan rate of –0.5 K/min, mean of 2 measurements, transition monitored by difference absorption at 295 nm

(5) $T_{trs}$ derived from second-derivative spectroscopy with spectral bands related to (a) Phe, (b) Tyr, and (c) Trp

(6) buffer: 10 mM HEPES, 100 mM NaCl, 4 mM EGTA

(7) buffer: 50 mM sodium cacodylate, 100 mM NaCl, 2 mM EGTA

(8) transition monitored by CD at 269 nm

Bovine holo- and apo-α-lactalbumin (α-LA) under varying conditions

| Buffer | pH | $T_{trs}$ | $\Delta Cp$ | $\Delta H$ | Appr./Rem. | Ref |
|---|---|---|---|---|---|---|
| apo-α-LA in Tris-HCl | | | | | | |
| 5 mM Tris-HCl | 8.0 | 28.2 | 6.18±0.6 | 147±26 | DSC | 99G16 |
| 300 mM Tris-HCl | 8.0 | 32.7 | 5.34±0.5 | 172±18 | DSC | 99G16 |
| 500 mM Tris-HCl | 8.0 | 39.0 | 8.91±0.4 | 200±10 | DSC | 99G16 |
| holo-α-LA in citrate | | | | | | |
| 10 mM citrate | 6.2 | 51.2 | 6.71±0.7 | 247±12 | DSC | 99G16 |
| 50 mM citrate | 6.2 | 50.8 | 6.45±0.7 | 243±12 | DSC | 99G16 |
| 100 mM citrate | 6.2 | 46.3 | 8.52±0.7 | 222±11 | DSC | 99G16 |
| 200 mM citrate | 6.2 | 42.5 | 5.92±0.6 | 220±18 | DSC | 99G16 |
| holo-α-LA in Tris-HCl | | | | | | |
| 5 mM Tris-HCl | 8.0 | 64.3 | 7.10±0.7 | 295±15 | DSC | 99G16 |
| 500 mM Tris-HCl | 8.0 | 65.8 | 7.90±0.8 | 315±16 | DSC | 99G16 |

Bovine α-lactalbumin, denaturation versus unfolding, heat capacity changes

| Transition | pH | $T_{trs}$ | ΔCp | ΔH | Approach/Remarks | | Ref |
|---|---|---|---|---|---|---|---|
| N → U | 8.0 | 70 | 7.725* | | DSC | (1,2) | 99G15 |
| N → X | 8.0 | 70 | 5.455* | | DSC | (1,3) | 99G15 |

Remarks:

(1) data from precision heat capacity measurements

(2) general expression for unfolding (N → U): $\Delta Cp = A \times T^4 + B \times T^3 + C \times T^2 + D \times T + E$ with A = 0, B = $-1.75576028 \times 10^{-6}$, C = 0.00118435778, D = $-0.227692892$, E = 17.4343415

(3) general expression for denaturation (N → X): $\Delta Cp = A \times T^4 + B \times T^3 + C \times T^2 + D \times T + E$ with A = 0, B = $-1.0329069 \times 10^{-5}$, C = 0.0100323126, D = $-3.2039497$, E = 340.927949

Bovine α-lactalbumin (Bα-LA)

| Protein | pH | $T_{trs}$ | ΔH | Approach/Remarks | | Ref |
|---|---|---|---|---|---|---|
| Bα-LA | 6.8 | 66–69 | 230 | heat | (1–3) | 99Z4 |

Remarks:

(1) transition monitored by FTIR spectroscopy

(2) the approach suggests the presence of two additional transitions between 5 and 45°C, and between 45 and 60°C

(3) measured in $D_2O$ with 4 mM sodium cacodylate (pH* 6.8) containing 10 mM KCl

Bovine holo-α-lactalbumin, ethanol-induced transitions

| Ethanol | pH | $T_{trs}$ | ΔCp | ΔH | Appr./Rem. | | Ref |
|---|---|---|---|---|---|---|---|
| 0.0 | 8.0 | 65.4±0.10 | 5.3±0.1 | 310.5±0.5 | DSC | (1,2) | 98G17 |
| 5.0 | 8.0 | 61.4±0.20 | 5.3±0.3 | 312.5±1.0 | DSC | (1,2) | 98G17 |
| 10.0 | 8.0 | 57.2±0.10 | 5.6±0.1 | 288.3±0.4 | DSC | (1,2) | 98G17 |
| 20.0 | 8.0 | 48.3±0.10 | 6.8±0.1 | 269.7±1.3 | DSC | (1,2) | 98G17 |
| 30.0 | 8.0 | 36.6±0.02 | 8.3±0.1 | 237.4±0.4 | DSC | (1,2) | 98G17 |

Remarks:

(1) ethanol conc. in % (v/v)

(2) the heat capacity function of holo-α-lactalbumin in the presence of ethanol follows a linear dependence according to
$$\Delta Cp(T) = \Delta Cp(T_{trs}) + \Delta Cp'(T - T_{trs})$$
with

| Ethanol | $T_{trs}$ | ΔCp | ΔCp' (kJ/mol/K²) |
|---|---|---|---|
| 0.0 | 65.4 | 5.3±0.1 | −0.050±0.004 |
| 5.0 | 61.4 | 5.3±0.3 | −0.036±0.009 |
| 10.0 | 57.2 | 5.6±0.1 | −0.085±0.002 |
| 20.0 | 48.3 | 6.8±0.1 | −0.020±0.010 |
| 30.0 | 36.6 | 8.3±0.1 | −0.022±0.003 |

Recombinant bovine α-lactalbumin (BLA), modified by an extraneous N-terminal Met residue

| Mutant/Form | | pH | $T_{trs}$ | ΔCp | ΔH | Approach/Remarks | | Ref |
|---|---|---|---|---|---|---|---|---|
| BLA | apo | 8.0 | 33.9±0.1 | 8.5±0.1 | 189±3 | heat | (1,2) | 98I3 |
| Met-BLA | apo | 8.0 | 16.5±0.9 | 6.6±0.3 | 93±11 | heat | (1,2) | 98I3 |
| BLA | Ca²⁺ | 8.0 | 64.2±0.1 | 7.6±1.2 | 288±3 | heat | (1,3) | 98I3 |
| Met-BLA | Ca²⁺ | 8.0 | 55.9±0.1 | 7.3±0.3 | 269±3 | heat | (1,3) | 98I3 |

Remarks:

(1) van't Hoff heat, transition monitored by CD at 270 nm

(2) buffer: 10 mM borate, 50 mM NaCl, 1 mM $CaCl_2$

(3) buffer: 10 mM borate, 50 mM NaCl, 1 mM EDTA

Recombinant bovine α-lactalbumin (mLA), wild type and mutants

| Mutant | pH | $T_{trs}$ | $\Delta Cp$ | $\Delta H$ | Appr./Rem. | Ref |
|---|---|---|---|---|---|---|
| mLA | 7.4 | 56.2±0.05 | 5.7±0.6 | 270±4 | heat (1–4) | 99G13 |
| functional site: | | | | | | |
| His32→Ala | 7.4 | 44.6±0.07 | | 211±4 | heat (1–3) | 99G13 |
| His32→Tyr | 7.4 | 56.5±0.04 | | 308±5 | heat (1–3) | 99G13 |
| Leu110→Arg | 7.4 | 58.1±0.03 | | 320±4 | heat (1–3) | 99G13 |
| Leu110→Glu | 7.4 | 55.1±0.10 | | 245±2 | heat (1–3) | 99G13 |
| Leu110→His | 7.4 | 62.5±0.04 | | 324±4 | heat (1–3) | 99G13 |
| Gln117→Ala | 7.4 | 52.1±0.04 | | 307±3 | heat (1–3) | 99G13 |
| Trp118→His | 7.4 | 52.9±0.07 | | 235±6 | heat (1–3) | 99G13 |
| Trp118→Tyr | 7.4 | 50.8±0.04 | | 281±2 | heat (1–3) | 99G13 |
| adjacent to site: | | | | | | |
| Ala106→Ser | 7.4 | 50.1±0.04 | | 216±2 | heat (1–3) | 99G13 |
| His107→Ala | 7.4 | 52.1±0.05 | | 253±3 | heat (1–3) | 99G13 |
| His107→Trp | 7.4 | 47.5±0.04 | | 242±2 | heat (1–3) | 99G13 |
| His107→Tyr | 7.4 | 55.2±0.03 | | 263±2 | heat (1–3) | 99G13 |
| Lys114→Asn | 7.4 | 66.9±0.05 | | 382±7 | heat (1–3) | 99G13 |
| Lys114→Gln | 7.4 | 53.5±0.07 | | 291±6 | heat (1–3) | 99G13 |
| Lys114→Glu | 7.4 | 53.0±0.05 | | 263±4 | heat (1–3) | 99G13 |
| other residues: | | | | | | |
| Val42→Ala | 7.4 | 51.7±0.04 | | 269±3 | heat (1–3) | 99G13 |
| Val42→Asn | 7.4 | 55.1±0.03 | | 281±3 | heat (1–3) | 99G13 |
| Val42→Gly | 7.4 | 51.0±0.07 | | 287±6 | heat (1–3) | 99G13 |
| Gln54→Ala | 7.4 | 54.3±0.08 | | 283±3 | heat (1–3) | 99G13 |
| Ile59→Trp | 7.4 | 51.7±0.05 | | 272±5 | heat (1–3) | 99G13 |
| Tyr103→Ala | 7.4 | 45.5±0.05 | | 269±4 | heat (1–3) | 99G13 |
| Tyr103→Pro | 7.4 | 55.1±0.02 | | 260±2 | heat (1–3) | 99G13 |
| Trp104→Tyr | 7.4 | 43.7±0.03 | | 225±4 | heat (1–3) | 99G13 |
| Reference proteins: | | | | | | |
| bovine LA | 7.4 | 62.7±0.09 | | 309±5 | heat (1–3) | 99G13 |
| goat LA | 7.4 | 62.8±0.04 | | 307±4 | heat (1–3) | 99G13 |

Remarks:

(1) mLA differs from bovine LA in having Val substituted for Met90 and the N-terminal Met extension

(2) transition monitored by CD at 270 and 275 nm, in some cases additionally at 222 nm

(3) buffer: 20 mM Tris-HCl, 1 mM $CaCl_2$, pH 7.4

(4) $\Delta Cp$ from $\Delta H^{cal}$ versus $T_{trs}$ including all mutants from Ref. 99G13

# β-Lactamase

β-lactamase from *Staphylococcus aureus* PC1, nonproline *cis* peptide bond

| Mutant | pH | $T_{trs}$ | $\Delta H$ | Approach/Remarks | | Ref |
|---|---|---|---|---|---|---|
| wild type | 6.8 | 77.0±1.0 | 561±4 | DSC | (1,2) | 97B3 |
| wild type | 6.8 | 74.0±1.0 | | heat | (1,3,4) | 97B3 |
| Asn136→Ala | 6.8 | 64.0±1.0 | | DSC | (2,5) | 97B3 |
| Asn136→Ala | 6.8 | 68.0±1.0 | | heat | (1,3,4) | 97B3 |

Remarks:

(1) the nonproline *cis* peptide bond is present between Glu166 and Ile167

(2) buffer: 0.1 M potassium phosphate, 2.5 M ammonium sulfate, pH 6.8

(3) buffer: 0.1 M potassium phosphate, 0.8 M ammonium sulfate, pH 6.8

(4) transition monitored by CD at 222 nm

(5) mutant Asn136→Ala to examine the interaction between side chain of Asn136 and main chain of Glu166

Metallo-β-lactamase from *Aeromonas hydrophilia*, various forms

| Form | pH | $T_{trs}$ | $\Delta Cp$ | $\Delta H$ | Approach/Remarks | | Ref |
|---|---|---|---|---|---|---|---|
| apoprotein | 6.5 | 46.2 | | | DSC | (1) | 97V1 |
| mono-$Zn^{2+}$ | 6.5 | 60 | | | DSC | (1) | 97V1 |
| di-$Zn^{2+}$ | 6.5 | 64 | | | DSC | (1) | 97V1 |

Remark:
(1) irreversible transition

Recombinant β-lactamase AmpC from *Escherichia coli* and mutant Tyr150→Phe

| Protein | pH | $T_{trs}$ | $\Delta H$ | Approach/Remarks | | Ref |
|---|---|---|---|---|---|---|
| AmpC | 6.8 | 54.6±0.2 | 761±39 | heat | (1,2) | 99B6 |
| | 6.8 | 54.2±0.5 | 728±163 | heat | (1,3) | 99B6 |
| | 6.8 | 54.1±0.5 | 766±163 | heat | (1,4) | 99B6 |
| Tyr150→Phe | 6.8 | 53.3±0.1 | 900±63 | heat | (1,2) | 99B6 |

Remarks:
(1) buffer: 50 mM potassium phosphate, 200 mM KCl, 38% ethylene glycol, pH 6.8
(2) thermal denaturation monitored by far-UV CD
(3) thermal denaturation monitored by near-UV CD
(4) thermal denaturation monitored by fluorescence

Recombinant β-lactamase AmpC from *Escherichia coli* and mutant Tyr150→Phe in the presence of inhibitors

| Inhibitor | pH | $T_{trs}$ | $\Delta T$ | $\Delta H$ | Appr./Rem. | Ref |
|---|---|---|---|---|---|---|
| wild-type AmpC + inhibitor: | | | | | | |
| none | 6.8 | 54.6 | 0.0 | 761 | heat (1–3) | 99B6 |
| Cloxacillin | 6.8 | 60.4±0.1 | 5.8±0.1 | 741±46 | heat (2,3) | 99B6 |
| Azteonam | 6.8 | 59.5±0.2 | 4.9±0.2 | 757±13 | heat (2,3) | 99B6 |
| Moxalactam | 6.8 | 51.4±0.1 | −3.2±0.1 | 837±63 | heat (2,3) | 99B6 |
| Imipenem | 6.8 | 53.4±0.1 | −1.2±0.1 | 653±25 | heat (2,3) | 99B6 |
| PNPP | 6.8 | 57.1±0.1 | 2.5±0.1 | 782±46 | heat (2–4) | 99B6 |
| BZBTH2B | 6.8 | 57.6±0.2 | 3.0±0.2 | 757±38 | heat (2–4) | 99B6 |
| mutant Tyr150→Phe + inhibitor: | | | | | | |
| none | 6.8 | 53.3 | 0.0 | 900 | heat (1–3) | 99B6 |
| Cloxacillin | 6.8 | 51.5±0.1 | −1.8±0.1 | 770±88 | heat (2,3) | 99B6 |
| Moxalactam | 6.8 | 49.3±0.1 | −4.0±0.1 | 866±163 | heat (2,3) | 99B6 |
| PNPP | 6.8 | 53.2±0.2 | −0.1±0.2 | 812±54 | heat (2–4) | 99B6 |
| BZBTH2B | 6.8 | 53.2±0.3 | −0.1±0.3 | 770±234 | heat (2–4) | 99B6 |

Remarks:
(1) reference value
(2) buffer: 50 mM potassium phosphate, 200 mM KCl, 38% ethylene glycol, pH 6.8
(3) transition monitored by far-UV CD at 223 nm
(4) PNPP and BZBTH2B are transition state analogs, see Ref. 99B6

## Lactate Dehydrogenase

Porcine muscle lactate dehydrogenase (LDH), isoenzyme M4

| Protein | pH | $T_{trs}$ | $\Delta H$ | Approach/Remarks | | Ref |
|---------|----|-----------|------------|------------------|---|-----|
| LDH | 7.0 | 62 | 1763 | DSC | (1,2) | 96S15 |

Remarks:
(1) measured in 0.05 M phosphate buffer
(2) DSC scans in the presence of glycerol show two peaks at 40–50° and 60–65°

## β-Lactoglobulin

β-lactoglobulin

| pH | $T_{ref}$ | $\Delta Cp$ | $\Delta H$ | Approach/Remarks | | Ref |
|------|-----------|-------------|------------|------------------|---|-----|
| 2.5–3.5 | 25 | 8.8 | –84 | urea/heat | (1) | 68P1 |
| 2.5–3.5 | 25 | 8.8 | –88 | GuHCl/heat | (1) | 68P1 |

Remark:
(1) van't Hoff treatment

Bovine β-lactoglobulin B (β-LG)

| Protein | pH | $T_{trs}$ | $\Delta Cp$ | $\Delta H$ | Appr./Rem. | | Ref |
|---------|----|-----------|-------------|------------|------------|---|-----|
| β-LG | 7.2 | 71.0±1.0 | 7.23 | 350.4±0.5 | EL | (1,2) | 99R15 |

Remarks:
(1) measured by a novel approach which is based on capillary electrophoresis (EL)
(2) $\Delta Cp$ from $\Delta H^{v.H.}$ versus $T_{trs}$

Bovine β-lactoglobulin (BLG) and mixed disulfide derivatives

| Protein | pH | $T_{trs}$ | $\Delta Cp$ | $\Delta H$ | Appr./Rem. | | Ref |
|---------|----|-----------|-------------|------------|------------|---|-----|
| BLG | 2.05 | 77.72±0.27 | 9.79±1.07 | 383.1±10.2 | DSC | (1,2) | 98B14 |
| | | | 10.25±1.06 | | DSC | (3) | 98B14 |
| BLG-COOH | 2.05 | 72.59±0.31 | 8.58±1.48 | 276.9± 5.6 | DSC | (2,4) | 98B14 |
| BLG-OH | 2.05 | 70.84±0.07 | 9.03±1.14 | 277.1±10.6 | DSC | (2,5) | 98B14 |

Explanations:
    BLG-COOH:    mixed disulfide derivative obtained by reaction with mercaptopropionic acid
    BLG-OH:    mixed disulfide derivative obtained by reaction with 2-mercaptoethanol

Remarks:
(1) buffer: 40 mM glycine, pH 2.05
(2) data estimated for n = 11 with 95% confidence level
(3) $\Delta Cp$ from $\Delta H$ versus $T_{trs}$ obtained varying pH
(4) data estimated for n = 7 with 95% confidence level
(5) standard deviation for n = 2

**λ Repressor**, see Repressor Proteins

**Lectin**, see also Galectin-1

## Lectin

Legume lectins, thermal transitions

| Protein/conc. | (mM) | pH | $T_{trs}$ | $\Delta H$ | Approach/Remarks | | Ref |
|---|---|---|---|---|---|---|---|
| concanavalin A | | | | | | | |
| | 0.071–0.712 | 6.90±0.05 | 87.1–90.7 | 2376±212 | DSC | (1) | 93S13 |
| | 0.844 | 6.90±0.05 | 91.4±0.2 | 2312±40 | DSC | (2) | 93S13 |
| | 0.680 | 5.30±0.05 | 92.3±0.1 | 2228±48 | DSC | (1) | 93S13 |
| | 0.680 | 5.00±0.05 | 91.0±0.1 | 2120±100 | DSC | (1) | 93S13 |
| pea lectin | | | | | | | |
| | 0.085–0.268 | 7.40±0.05 | 70.7–71.9 | 1064±160 | DSC | (1) | 93S13 |
| | 0.176–0.209 | 6.90±0.05 | 73.8–74.0 | 1124±20 | DSC | (1) | 93S13 |
| | 0.281 | 6.90±0.05 | 77.0 | 1200±100 | DSC | (2) | 93S13 |
| | 0.046–0.180 | 6.00±0.05 | 71.0–72.3 | 840±92 | DSC | (1) | 93S13 |
| | 0.190 | 5.00±0.05 | 72.5±0.6 | 802±140 | DSC | (1) | 93S13 |
| lentil lectin | | | | | | | |
| | 0.120–0.635 | 6.90±0.05 | 72.7–74.6 | 1070±164 | DSC | (1) | 93S13 |
| | 0.348 | 6.90±0.05 | 75.7 | 1360±160 | DSC | (2) | 93S13 |
| | 0.147–0.415 | 5.00±0.05 | 67.1–68.3 | 686±78 | DSC | (1) | 93S13 |

Remarks:
(1) scan rate 13 K/h
(2) scan rate 55 K/h

Legume lectins in the presence of ligands, thermal transitions

| Protein | Ligand/Lectin conc. | $T_{trs}$ | $\Delta H$ | Approach/Remarks | | Ref |
|---|---|---|---|---|---|---|
| concanavalin A | | | | | | |
| MeαMan | 22–65/0.721 | 94.2–95.8 | 2882±68 | DSC | (1,2) | 93S13 |
| Man | 11–44/0.712 | 91.6–93.5 | 2784±200 | DSC | (1,2) | 93S13 |
| MeαGlu | 18–55/0.712 | 93.3–94.6 | 2744±120 | DSC | (1,2) | 93S13 |
| Glu | 21–66/0.712 | 91.9–93.0 | 2392±176 | DSC | (1,2) | 93S13 |
| pea lectin | | | | | | |
| MeαMan | 19–74/0.165 | 77.7–79.6 | 1696±40 | DSC | (1,2) | 93S13 |
| Man | 18–76/0.154 | 77.0–78.6 | 1742±139 | DSC | (1,2) | 93S13 |
| MeαGlu | 21–69/0.165 | 77.5–79.0 | 1494±130 | DSC | (1,2) | 93S13 |
| Glu | 18–76/0.150 | 76.6–78.2 | 1670±34 | DSC | (1,2) | 93S13 |
| lentil lectin | | | | | | |
| MeαMan | 24–73/0.635 | 76.8–78.0 | 1318±128 | DSC | (1,2) | 93S13 |
| Man | 24–78/0.635 | 76.0–77.9 | 1200±60 | DSC | (1,2) | 93S13 |
| MeαGlu | 20–75/0.635 | 76.0–78.2 | 1226±54 | DSC | (1,2) | 93S13 |
| Glu | 21–80/0.635 | 75.4–77.3 | 1240±80 | DSC | (1,2) | 93S13 |

Explanations:

MeαMan = α-methyl-D-mannopyranoside
Man = D-mannopyranoside
MeαGlu = α-methyl-D-glucopyranoside
Glu = D-glucopyranoside

Remarks:
(1) ligand concentration/lectin concentration in mM/mM
(2) the data were analyzed in terms of ligand binding and compared with ITC data

Winged bean acidic lectin (WBAII) from the seeds of *Psophocarpus tetrogonolobus*, dependence on the protein concentration

| Conc. | pH | $T_{trs}$ | $\Delta H$ | Approach/Remarks | | Ref |
|---|---|---|---|---|---|---|
| 0.0178 mM | 7.4 | 58.4 | 1092±39.41 | DSC | (1,2) | 98S18 |
| 0.0294 mM | 7.4 | 58.5 | 1010±36.16 | DSC | (1,2) | 98S18 |
| 0.0437 mM | 7.4 | 57.3 | 1012±36.64 | DSC | (1,2) | 98S18 |
| 0.0870 mM | 7.4 | 59.4 | 1006±39.13 | DSC | (1,2) | 98S18 |
| 0.116 mM | 7.4 | 59.5 | 1013±36.47 | DSC | (1,2) | 98S18 |

Remarks:

(1) measured at a fixed scan rate of 30 K/h

(2) a single calorimetric peak with a cooperative ratio CR = $\Delta H^{cal}/\Delta H^{v.H.}$ = 2.0 was observed, the process is a $N_2 \rightarrow 2U$ two-state transition

Winged bean acidic lectin (WBAII) from the seeds of *Psophocarpus tetrogonolobus*, dependence on the scan rate

| Scan rate | pH | $T_{trs}$ | $\Delta H$ | Approach/Remarks | | Ref |
|---|---|---|---|---|---|---|
| 10 K/h | 7.4 | 53.7 | 893±32 | DSC | (1,2) | 98S18 |
| 20 K/h | 7.4 | 56.1 | 806±29 | DSC | (1,2) | 98S18 |
| 30 K/h | 7.4 | 58.5 | 1010±36.16 | DSC | (1,2) | 98S18 |
| 60 K/h | 7.4 | 61.4 | 867±31 | DSC | (1,2) | 98S18 |

Remarks:

(1) measured at a fixed WBAII conc. of 0.0294 mM

(2) a single calorimetric peak with a cooperative ratio CR = $\Delta H^{cal}/\Delta H^{v.H.}$ = 2.0 was observed, the process is a $N_2 \rightarrow 2U$ two-state transition

Winged bean acidic lectin (WBAII) from the seeds of *Psophocarpus tetrogonolobus* in the presence of various carbohydrate ligands

| Ligand | Conc. (mM) | pH | $T_{trs}$ | $\Delta H$ | Approach/Remarks | Ref |
|---|---|---|---|---|---|---|
| Gal | 50–100 | 7.4 | 61–68.7 | 1008±98 | DSC | 98S18 |
| MeαGal | 5.0–500 | 7.4 | 58.8–68.7 | 971±81 | DSC | 98S18 |
| MeβGal | 5.0–91.8 | 7.4 | 59–64.5 | 932±92 | DSC | 98S18 |
| Lac | 5.0–200 | 7.4 | 59.1–68.1 | 887±27 | DSC | 98S18 |

Explanations:

Gal      galactose

MeαGal   methly-α-galactose

Lac      lactose

Pea lectin, thermal denaturation in the presence of urea

| urea conc. | pH | $T_{trs}$ | $\Delta Cp$ | $\Delta H$ | Approach/Remarks | | Ref |
|---|---|---|---|---|---|---|---|
| 0.0 M | 7.2 | 75.0 | 22.3 | 1131 | heat | (1,2) | 98A2 |
| 1.0 M | 7.2 | 70.5 | 23.9 | 1110 | heat | (1,2) | 98A2 |
| 3.8 M | 7.2 | 66.8 | 26.1 | 1065 | heat | (1,2) | 98A2 |
| 5.5 M | 7.2 | 64.3 | 26.4 | 1028 | heat | (1,2) | 98A2 |
| 5.7 M | 7.2 | 63.1 | 26.8 | 1018 | heat | (1,2) | 98A2 |
| 6.0 M | 7.2 | 61.2 | 27.4 | 1000 | heat | (1,2) | 98A2 |

Remarks:

(1) from thermal denaturation profiles monitored by far-UV CD, treatment by an equilibrium between folded dimer and unfolded monomers ($N_2 \rightarrow 2U$)

(2) measured in 50 mM phosphate buffer, pH 7.2

Pea lectin, cold denaturation in the presence of urea

| urea conc. | pH | $T_{trs}$ | $\Delta Cp$ | $\Delta H$ | Approach/Remarks | | Ref |
|---|---|---|---|---|---|---|---|
| 0.0 M | 7.2 | –17.5 | | –1123 | heat | (1–3) | 98A2 |
| 1.1 M | 7.2 | –12.9 | | –1102 | heat | (1–3) | 98A2 |
| 3.8 M | 7.2 | –7.0 | | –1063 | heat | (1–3) | 98A2 |
| 5.5 M | 7.2 | –2.94 | | –1028 | heat | (1–3) | 98A2 |
| 5.7 M | 7.2 | –1.99 | | –1015 | heat | (1–3) | 98A2 |
| 6.0 M | 7.2 | 0.0 | | –996 | heat | (1–3) | 98A2 |

Remarks:

(1) from thermal denaturation profiles monitored by far-UV CD, treatment by an equilibrium between folded dimer and unfolded monomers ($N_2 \rightarrow 2U$)

(2) measured in 50 mM phosphate buffer, pH 7.2

(3) $\Delta Cp$ is within experimental error identical with values from heat denaturation

Pea lectin, pH dependence

| Protein | pH | $T_{trs}$ | $\Delta Cp$ | $\Delta H$ | Appr./Rem. | Ref |
|---|---|---|---|---|---|---|
| pea lectin | 4.3 | 63.9 | | 837.2 | DSC | 98A2 |
| | 4.8 | 65.7 | | 887.6 | DSC | 98A2 |
| | 5.2 | 70.2 | | 926.0 | DSC | 98A2 |
| | 5.8 | 73.3 | | 985.9 | DSC | 98A2 |
| | 6.2 | 73.5 | | 1018.6 | DSC | 98A2 |
| | 6.8 | 74.1 | | 1068.8 | DSC | 98A2 |
| | 7.2 | 75.4 | | 1123.3 | DSC | 98A2 |
| | 4.3–7.2 | 64–75 | 21.42 | | DSC (1) | 98A2 |
| | 7.2 | 4–55 | 22.26 | | urea (2,3) | 98A2 |
| | 7.2 | 61–75 | 22.6 | | heat (4) | 98A2 |

Remarks:

(1) $\Delta Cp$ from $\Delta H$ versus $T_{trs}$

(2) from urea denaturation varying temperature, see also Table 1

(3) this $\Delta Cp$ value is regarded as the most reliable one in Ref. 98A2

(4) from thermal denaturation in the presence of urea, values extrapolated to zero denaturant concentration

Lentil lectin, Arrhenius parameters for the two-state irreversible model

| Protein | pH | $T^*$ | $Q_t$ | $E_A$ | Appr./Rem. | Ref |
|---|---|---|---|---|---|---|
| lectin | 5.0 | 79.5±0.07 | 414±8 | 267±2 | DSC (1–3) | 99M8 |
| | 7.4 | 80.8±0.08 | 527±33 | 365±3 | DSC (1–3) | 99M8 |
| | 8.5 | 78.2±0.07 | 435±8 | 291±1 | DSC (1–3) | 99M8 |
| | 10.0 | 73.3±0.04 | 351±4 | 251±1 | DSC (1–3) | 99M8 |

Remarks:

(1) the thermal transition is strongly scan rate-dependent, Ref. 99M8 contains further experimental data

(2) the present data refer to a scan rate of 1 K/min

(3) the irreversible thermal transition is under kinetic control and can be interpreted by a kinetic scheme $N \rightarrow D$

## Leghemoglobin

Leghemoglobin a from soybean, cyanide complex (Lba.CN)

| Protein | pH | $T_{trs}$ | $\Delta Cp$ | $\Delta H$ | Approach/Remarks | | Ref |
|---|---|---|---|---|---|---|---|
| Lba.CN | 5.00 | 69.0 | | 456 | DSC | | 98T12 |
| | 5.90 | 77.5 | | 544 | DSC | | 98T12 |
| | 7.25 | 79.0 | | 569 | DSC | | 98T12 |
| | 8.50 | 79.5 | | 552 | DSC | | 98T12 |
| | 9.20 | 81.0 | | 556 | DSC | | 98T12 |
| | 9.90 | 69.5 | | 456 | DSC | | 98T12 |
| | 11.15 | 65.5 | | 435 | DSC | | 98T12 |
| | 5–11 | 65–81 | 9.16 | | DSC | (1) | 98T12 |

Remark:

(1) $\Delta Cp$ from $\Delta H^{v.H.}$ versus $T_{trs}$

## Leucine Zipper

Designed leucine zipper by including a structure that cannot form interhelical electrostatic bonds – results from thermal unfolding

| Conc. ($\mu$M) | pH | $T_{trs}$ | $\Delta H$ | Approach/Remarks | | Ref |
|---|---|---|---|---|---|---|
| 253 | 5.1 | 45.0 | 201 | heat | (1–3) | 99D7 |
| 355 | 5.0 | 53.9 | 212 | heat | (1–3) | 99D7 |
| 12 | 4.7 | 43.9 | 216 | heat | (1–3) | 99D7 |
| 25 | 4.7 | 48.1 | 217 | heat | (1–4) | 99D7 |
| 48 | 4.7 | 50.4 | 239 | heat | (1–3) | 99D7 |
| 96 | 4.7 | 52.6 | 239 | heat | (1–3) | 99D7 |
| 188 | 4.7 | 57.7 | 244 | heat | (1–3) | 99D7 |
| 277 | 4.7 | 61.7 | 245 | heat | (1–3) | 99D7 |
| 336 | 4.7 | 62.2 | 234 | heat | (1–3) | 99D7 |
| 340 | 4.7 | 61.2 | 225 | heat | (1–3) | 99D7 |
| 378 | 4.7 | 62.4 | 241 | heat | (1–3) | 99D7 |
| 3.12 | 4.4 | 49.8 | 231 | heat | (1–3) | 99D7 |
| 6.25 | 4.4 | 53.3 | 228 | heat | (1–4) | 99D7 |
| 12.5 | 4.4 | 55.4 | 226 | heat | (1–3) | 99D7 |
| 25 | 4.4 | 59.2 | 228 | heat | (1–3) | 99D7 |
| 348 | 4.4 | 69.4 | 254 | heat | (1–3) | 99D7 |
| 25 | 4.0 | 66.0 | 249 | heat | (1–3) | 99D7 |
| 97 | 4.0 | 73.6 | 288 | heat | (1–3) | 99D7 |
| 160 | 4.0 | 76.0 | 292 | heat | (1–3) | 99D7 |
| 260 | 4.0 | 79.0 | 291 | heat | (1–3) | 99D7 |
| 310 | 3.5 | 84.2 | 274 | heat | (1–3) | 99D7 |

Remarks:

(1) van't Hoff heat derived from a two-state transition between folded dimer and unfolded monomers

(2) transition monitored by far-UV CD at 222 nm

(3) buffer: composed of 7.5 mM each of phosphoric, citric, and boric acid, adjusted to the desired pH with KOH or HCl, and to 0.1 M ionic strength by KCl

(4) mean of multiple measurements

Designed leucine zipper by including a structure that cannot form interhelical electrostatic bonds – results from DSC

| Conc. (μM) | pH | $T_{trs}$ | ΔH | ΔCp | Appr./Rem. | Ref |
|---|---|---|---|---|---|---|
| 105 | 4.7 | 60.9 | 230 | | heat (1,2) | 99D7 |
| 135 | 4.7 | 62.4 | 231 | | heat (1,2) | 99D7 |
| 188 | 4.7 | 59.8 | 209 | | heat (1,2) | 99D7 |
| 277 | 4.7 | 62.5 | 215 | | heat (1,2) | 99D7 |
| 336 | 4.7 | 63.9 | 225 | | heat (1,2) | 99D7 |
| 353 | 4.7 | 65.5 | 235 | | heat (1,2) | 99D7 |
| 378 | 4.7 | 64.4 | 203 | | heat (1,2) | 99D7 |
| 399 | 4.7 | 69.6 | 228 | | heat (1,2) | 99D7 |
| 150 | 4.4 | 68.8 | 223 | | heat (1,2) | 99D7 |
| 213 | 4.4 | 70.2 | 231 | | heat (1,2) | 99D7 |
| 335 | 4.4 | 74.8 | 247 | | heat (1,2) | 99D7 |
| 585 | 4.4 | 78.2 | 245 | | heat (1,2) | 99D7 |
| 97 | 4.0 | 74.4 | 237 | | heat (1,2) | 99D7 |
| 160 | 4.0 | 77.1 | 239 | | heat (1,2) | 99D7 |
| 260 | 4.0 | 74.7 | 242 | | heat (1,2) | 99D7 |
| 96 | 3.5 | 80.0 | 252 | | heat (1,2) | 99D7 |
| 190 | 3.5 | 83.1 | 265 | | heat (1,2) | 99D7 |
| 310 | 3.5 | 84.7 | 269 | | heat (1,2) | 99D7 |
| 394 | 3.5 | 85.9 | 256 | | heat (1,2) | 99D7 |
| | | | | 0.80±0.60 | DSC (3) | 99D7 |
| | | | | 1.86±0.45 | DSC (4) | 99D7 |
| | | | | 2.1 | heat (5) | 99D7 |
| | | | | 1.83 | DSC (6) | 99D7 |
| | | | | 1.83 | DSC (7) | 99D7 |
| | | | | 2.14 | DSC (8) | 99D7 |
| | | | | 1.98±0.60 | (9) | 99D7 |

Remarks:

(1) errors in $T_{trs}$ are approximately ±0.25 K and of $\Delta H^{cal}$ approximately ±10%

(2) buffer: composed of 7.5 mM each of phosphoric, citric, and boric acid, adjusted to the desired pH with KOH or HCl, and to 0.1 M ionic strength by KCl

(3) ΔCp from linear extrapolation of pre- and post-transition traces (mean of 22 measurements)

(4) ΔCp from the difference between the measured Cp of the folded state and the calculated Cp of the completely unfolded state (mean of 22 measurements)

(5) ΔCp from $\Delta H^{v.H.}$ versus $T_{trs}$, $\Delta H^{v.H.}$ from CD unfolding experiments

(6) ΔCp from $\Delta H^{cal}$ versus $T_{trs}$, $\Delta H^{cal}$ from DSC unfolding experiments

(7) ΔCp from $\Delta H^{fit}$ versus $T_{trs}$, $\Delta H^{fit}$ from two-state fit of DSC recordings

(8) ΔCp from $\Delta H^{v.H.}$ versus $T_{trs}$, $\Delta H^{v.H.}$ from DSC unfolding experiments

(9) mean of ΔCp from all experiments, regarded as the most reliable value in Ref. 99D7

## Lipase

Lipase from *Pseudomonas cepacia* in the absence and presence of alcohols

| Alcohol conc. | pH | $T_{trs}$ | $\Delta H$ | Approach/Remarks | | Ref |
|---|---|---|---|---|---|---|
| without alcohol | 7.0 | 75.8–80.9 | 849±25 | DSC | (1–4) | 98T4 |
| MeOH 0.24–1.12 M | 7.0 | 76.9–74.6 | 836±15 | DSC | (3–5) | 98T4 |
| EtOH 0.17–0.86 M | 7.0 | 76.7–73.9 | 837±30 | DSC | (3–5) | 98T4 |
| PrOH 0.13–0.67 M | 7.0 | 76.9–70.8 | 784±60 | DSC | (3–5) | 98T4 |
| BuOH 0.11,0.22 M | 7.0 | 75.6–74.0 | 828,745 | DSC | (3–5) | 98T4 |

Remarks:

(1) measured at protein concentration from 29.3 to 236 μM in the absence of alcohol, $T_{trs}$ varies from 75.8 to 80.9°C without significant change in $\Delta H$, see also specific heat capacity change $\Delta cp = 0.055 \pm 0.075$ J/g/K in Table 3

(2) analysis of the DSC tracings is consistent with the mechanism $N.Ca^{2+} \rightarrow D + Ca^{2+}$

(3) abbreviations: MeOH, methanol; EtOH, ethanol; PrOH, propanol; BuOH, butanol

(4) buffer: 20 mM phosphate, pH 7.0

(5) measured at protein conc. of 98 μM

## Luciferase

Firefly luciferase, irreversible thermal denaturation at varying scan rate

| Scan rate | pH | $T_{trs}$ | $\Delta H$ | Approach/Remarks | | Ref |
|---|---|---|---|---|---|---|
| 10 K/h | 7.8 | 38.5 | 385 | DSC | (1,2) | 98U1 |
| 10 K/h | 7.8 | 41.7 | 415 | DSC | (1,3) | 94C13 |
| 120 K/h | 7.8 | 43.3 | | heat | (1,4) | 98U1 |
| 60 K/h | 7.8 | 42.5 | | heat | (1,4) | 98U1 |
| 30 K/h | 7.8 | 39.3 | | heat | (1,4) | 98U1 |
| 6 K/h | 7.8 | 37.1 | | heat | (1,4) | 98U1 |
| 60 K/h | 7.8 | 38.0 | | heat | (1,5) | 98U1 |

Remarks:

(1) measured in 100 mM glycylglycine, pH 7.8

(2) reference value

(3) see also Table 3 and Ref. 98U1

(4) transition from the fraction of irreversibly denatured protein at 0.25 mg/ml lucifease conc.

(5) transition from the fraction of irreversibly denatured protein at 3.0 mg/ml lucifease conc.

## Lysozyme, Canine

Recombinant canine milk lysozyme

| Protein | Transition | pH | $T_{trs}$ | $\Delta H$ | Approach/Remarks | | Ref |
|---|---|---|---|---|---|---|---|
| Holo-protein | trans. (1) | 4.5 | 68 | | DSC | (1–5) | 99K13 |
| | trans. (2) | 4.5 | 90 | | DSC | (1–5) | 99K13 |
| apo-protein | trans. (1) | 4.5 | 45 | | DSC | (1,2) | 99K13 |
| | trans. (2) | 4.5 | 90 | | DSC | (1,2) | 99K13 |

Remarks:

(1) canine milk lysozyme belongs to the calcium–binding proteins

(2) measured in 50 mM acetate buffer, pH 4.5

(3) in the presence of 10 mM $CaCl_2$

(4) the recombinant protein shows lower thermostability than the authentic protein, $T_{trs}$ in the presence of 3.0 M GuHCl and 10 mM $CaCl_2$ amounts to 30.2°C for the recombinant protein and 32.4°C for the authentic protein

(5) see also Ref. 99K20

**Lysozyme, HEW** (Hen Egg White Lysozyme)

Hen egg white lysozyme in the presence of dimethylsulfoxide (DMSO)

| DMSO % (v/v) | pH | $T_{trs}$ | $\Delta H$ | Approach/Remarks | | Ref |
|---|---|---|---|---|---|---|
| 0.0 | 2.5 | 65 | 535 | DSC | (1,2) | 96K13 |
| 8.0 | 2.5 | 62 | 575 | DSC | (1,2) | 96K13 |
| 25.5 | 2.5 | 64 | 660 | DSC | (1,2) | 96K13 |
| 34.9 | 2.5 | 55.5 | 630 | DSC | (1,2) | 96K13 |
| 50.0 | 2.5 | 47 | 500 | DSC | (1,2) | 96K13 |
| 56.7 | 2.5 | 38 | 370 | DSC | (1–3) | 96K13 |
| 57.5 | 2.5 | 36.5 | 350 | DSC | (1–3) | 96K13 |
| 58.0 | 2.5 | 33 | 140 | DSC | (1–3) | 96K13 |
| 59.0 | 2.5 | 32 | 100 | DSC | (1–3) | 96K13 |

Remarks:
(1) Ref. 96K13 contains further data for lysozyme in the presence of DMSO from 0 to 90% (v/v) at pH from 2.5 to 9.0
(2) the data were taken from the figures 1, 2, and 4 in Ref. 96K13
(3) at 55–60% (v/v) DMSO a reversible thermal transition of lysozyme proceeds accompanied by a sharp decrease in $\Delta H^{cal}$ whereas $\Delta H^{v.H.}$ remains almost unchanged

Hen egg white lysozyme in the presence of polyols

| Comp. | Conc. | pH | $T_{trs}$ | $\Delta Cp$ | $\Delta H$ | Appr./Rem. | | Ref |
|---|---|---|---|---|---|---|---|---|
| none | | 2.50 | 63.0 | 6.7±1.1 | 437±11 | DSC | | 97J2 |
| glycerol | 2.0 | 2.50 | 65.3 | 6.8±0.4 | 466±9 | DSC | (1) | 97J2 |
| i-erythritol | 2.0 | 2.50 | 66.6 | 6.3±0.5 | 490±6 | DSC | (1) | 97J2 |
| adonitol | 2.0 | 2.50 | 68.5 | 7.3±0.9 | 510±10 | DSC | (1) | 97J2 |
| arabitol | 2.0 | 2.50 | 68.0 | 6.0±1.1 | 492±8 | DSC | (1) | 97J2 |
| glycerol | 0.0 | 2.50 | 63.0 | 6.7±1.1 | 437±11 | DSC | | 97J2 |
| | 2.0 | 2.50 | 65.3 | 6.8±0.4 | 466±9 | DSC | (1) | 97J2 |
| | 6.0 | 2.50 | 67.3 | 6.5±1.2 | 495±7 | DSC | (1) | 97J2 |
| | 10.0 | 2.50 | 69.5 | 6.8±1.9 | 560±4 | DSC | (1) | 97J2 |
| glycerol | 0.0 | 6.00 | 75.1 | 6.4±1.1 | 495±10 | DSC | | 97J2 |
| | 2.0 | 6.00 | 76.3 | 5.7±0.7 | 519±6 | DSC | (1) | 97J2 |
| | 6.0 | 6.00 | 78.8 | 5.8±0.2 | 558±13 | DSC | (1) | 97J2 |
| | 10.0 | 6.00 | 81.0 | 4.8±0.6 | 588±6 | DSC | (1) | 97J2 |

Remark:
(1) polyol concentration is the molal conc.

Hen egg white lysozym in aqueous polyol solution

| Cosolvent | Conc. | pH | $T_{trs}$ | $\Delta H$ | Appr./Rem. | Ref |
|---|---|---|---|---|---|---|
| control | | 2.5 | 61.1 | 410.0 | heat (1,2) | 98K3 |
| mannitol | 1.00 M | 2.5 | 67.1 | 421.7 | heat (1,2) | 98K3 |
| inositol | 0.75 M | 2.5 | 66.8 | 410.0 | heat (1,2) | 98K3 |
| xylitol | 2.00 M | 2.5 | 68.7 | 433.9 | heat (1,2) | 98K3 |
| adonitol | 2.00 M | 2.5 | 69.1 | 423.0 | heat (1,2) | 98K3 |
| sorbitol | 2.00 M | 2.5 | 71.0 | 431.4 | heat (1,2) | 98K3 |
| control | | 4.0 | 73.1 | 412.1 | heat (1,3) | 98K3 |
| mannitol | 1.00 M | 4.0 | 77.9 | 421.7 | heat (1,3) | 98K3 |
| inositol | 0.75 M | 4.0 | 77.7 | 422.6 | heat (1,3) | 98K3 |
| xylitol | 2.00 M | 4.0 | 80.1 | 422.6 | heat (1,3) | 98K3 |
| adonitol | 2.00 M | 4.0 | 79.6 | 426.8 | heat (1,3) | 98K3 |
| sorbitol | 2.00 M | 4.0 | 82.5 | 438.9 | heat (1,3) | 98K3 |

Hen egg white lysozym in aqueous polyol solution (continued)

| Cosolvent | Conc. | pH | $T_{trs}$ | $\Delta H$ | Appr./Rem. | Ref |
|---|---|---|---|---|---|---|
| control | | 7.0 | 58.5 | 397.1 | heat (1,4) | 98K3 |
| mannitol | 1.00 M | 7.0 | 63.5 | 404.2 | heat (1,4) | 98K3 |
| inositol | 0.75 M | 7.0 | 64.5 | 402.1 | heat (1,4) | 98K3 |
| xylitol | 2.00 M | 7.0 | 66.5 | 422.2 | heat (1,4) | 98K3 |
| adonitol | 2.00 M | 7.0 | 67.0 | 426.8 | heat (1,4) | 98K3 |
| sorbitol | 2.00 M | 7.0 | 70.1 | 441.4 | heat (1,4) | 98K3 |

Remarks:

(1) transition monitored by optical absorption at 287 nm

(2) buffer: 20 mM glycine, pH 2.5

(3) buffer: 40 mM acetate, pH 4.0

(4) buffer: 20 mM phosphate or MOPS, pH 7.0, 1.5 M GuHCl

Hen egg white lysozyme in the presence of tetrabutylammonium bromide ($Bu_4NBr$) and polyols

| Components | $T_{trs}$ | $\Delta Cp$ | $\Delta H$ | Appr./Rem. | Ref |
|---|---|---|---|---|---|
| measurements at pH 2.50: | | | | | |
| 0.0 m $Bu_4NBr$ + 0.0 m glycerol | 63.0 | 6.7±1.1 | 437±11 | DSC | 97J2 |
| 0.5 m $Bu_4NBr$ + 0.0 m glycerol | 47.3 | 3.5±1.5 | 348±6 | DSC (1) | 97J2 |
| 0.5 m $Bu_4NBr$ + 2.0 m glycerol | 49.7 | 4.0±1.3 | 375±10 | DSC (1) | 97J2 |
| 0.5 m $Bu_4NBr$ + 2.0 m erythritol | 51.7 | 2.7±0.3 | 389±6 | DSC (1) | 97J2 |
| 0.5 m $Bu_4NBr$ + 2.0 m arabitol | 53.0 | 3.9±0.2 | 394±8 | DSC (1) | 97J2 |
| 0.5 m $Bu_4NBr$ + 2.0 m adonitol | 52.9 | 5.9±1.5 | 398±12 | DSC (1) | 97J2 |
| 0.0 m $Bu_4NBr$ + 0.0 m glycerol | 63.0 | 6.7±1.1 | 437±11 | DSC | 97J2 |
| 0.5 m $Bu_4NBr$ + 0.0 m glycerol | 47.3 | 3.5±1.5 | 348±6 | DSC (1) | 97J2 |
| 0.5 m $Bu_4NBr$ + 2.0 m glycerol | 49.7 | 4.0±1.3 | 375±10 | DSC (1) | 97J2 |
| 0.5 m $Bu_4NBr$ + 6.0 m glycerol | 52.3 | 6.7±0.1 | 403±6 | DSC (1) | 97J2 |
| 0.5 m $Bu_4NBr$ +10.0 m glycerol | 53.3 | 6.9±1.8 | 435±5 | DSC (1) | 97J2 |
| measurements at pH 2.50: | | | | | |
| 0.0 m $Bu_4NBr$ + 0.0 m glycerol | 75.1 | 6.4±1.1 | 495±10 | DSC | 97J2 |
| 0.5 m $Bu_4NBr$ + 0.0 m glycerol | 55.8 | 3.4±1.5 | 391±9 | DSC (1) | 97J2 |
| 0.5 m $Bu_4NBr$ + 2.0 m glycerol | 58.6 | 5.3±2.1 | 437±7 | DSC (1) | 97J2 |
| 0.5 m $Bu_4NBr$ + 6.0 m glycerol | 65.6 | 5.6±1.9 | 462±8 | DSC (1) | 97J2 |
| 0.5 m $Bu_4NBr$ +10.0 m glycerol | 72.2 | 5.4±0.9 | 481±10 | DSC (1) | 97J2 |

Remark:

(1) concentrations are the molal conc.

Hen egg white lysozyme in the presence of urea and glycerol at pH 6.0

| Components | $T_{trs}$ | $\Delta Cp$ | $\Delta H$ | Appr./Rem. | Ref |
|---|---|---|---|---|---|
| 0.0 m urea + 0.0 m glycerol | 75.1 | 6.4±1.1 | 495±10 | DSC | 97J2 |
| 0.5 m urea + 0.0 m glycerol | 72.6 | 5.9±0.1 | 478±9 | DSC (1) | 97J2 |
| 0.5 m urea + 2.0 m glycerol | 74.3 | 7.5±0.6 | 514±30 | DSC (1) | 97J2 |
| 0.5 m urea + 6.0 m glycerol | 76.6 | 5.4±0.9 | 540±20 | DSC (1) | 97J2 |

Remark:

(1) concentrations are the molal conc.

## Lysozyme, Human

The data entries are arranged as follows:

a) wild type and mutants
b) mutants that create a calcium binding site

*a) wild type and mutants*

Human lysozyme, Val to Ala mutants

| Mutant | pH | $T_{trs}$ | $\Delta Cp$ | $\Delta H$ | Appr./Rem. | Ref |
|---|---|---|---|---|---|---|
| Val2→Ala | 3.17 | 68.7 | 7.0 | 490 | DSC (1) | 97T2 |
| | 2.88 | 63.7 | 6.7 | 464 | DSC (1) | 97T2 |
| | 2.71 | 60.3 | 6.4 | 439 | DSC (1) | 97T2 |
| | 2.55 | 57.6 | 4.5 | 418 | DSC (1) | 97T2 |
| | | | 6.2±1.4 | | DSC (2) | 97T2 |
| Val77→Ala | 2.88 | 67.1 | 6.6 | 490 | DSC (1) | 97T2 |
| | 2.71 | 64.1 | 5.6 | 469 | DSC (1) | 97T2 |
| | 2.55 | 61.0 | 6.4 | 452 | DSC (1) | 97T2 |
| | | | 6.2±0.7 | | DSC (2) | 97T2 |
| Val93→Ala | 3.17 | 71.1 | 7.4 | 502 | DSC (1) | 97T2 |
| | 3.02 | 68.3 | 4.4 | 490 | DSC (1) | 97T2 |
| | 3.02 | 68.6 | 3.8 | 494 | DSC (1) | 97T2 |
| | 2.88 | 66.1 | 4.4 | 473 | DSC (1) | 97T2 |
| | 2.71 | 62.6 | 4.6 | 448 | DSC (1) | 97T2 |
| | 2.55 | 59.8 | 4.9 | 435 | DSC (1) | 97T2 |
| | | | 4.9±1.4 | | DSC (2) | 97T2 |
| Val99→Ala | 3.02 | 67.8 | 6.4 | 481 | DSC (1) | 97T2 |
| | 2.88 | 65.2 | 4.6 | 464 | DSC (1) | 97T2 |
| | 2.71 | 62.1 | 6.3 | 448 | DSC (1) | 97T2 |
| | 2.55 | 59.2 | 6.1 | 423 | DSC (1) | 97T2 |
| | | | 5.9±1.0 | | DSC (2) | 97T2 |
| Val100→Ala | 3.17 | 72.1 | 4.7 | 515 | DSC (1) | 97T2 |
| | 3.02 | 69.9 | 3.3 | 510 | DSC (1) | 97T2 |
| | 2.71 | 64.3 | 4.5 | 494 | DSC (1) | 97T2 |
| | 2.55 | 61.5 | 4.3 | 477 | DSC (1) | 97T2 |
| | | | 4.2±0.8 | | DSC (2) | 97T2 |
| Val110→Ala | 3.02 | 71.2 | 5.8 | 490 | DSC (1) | 97T2 |
| | 2.88 | 69.7 | 4.6 | 464 | DSC (1) | 97T2 |
| | 2.71 | 66.5 | 4.1 | 439 | DSC (1) | 97T2 |
| | 2.55 | 63.8 | 4.6 | 418 | DSC (1) | 97T2 |
| | | | 4.8±0.9 | | DSC (2) | 97T2 |
| Val121→Ala | 3.17 | 69.9 | 5.8 | 490 | DSC (1) | 97T2 |
| | 2.88 | 64.2 | 6.7 | 456 | DSC (1) | 97T2 |
| | 2.71 | 60.5 | 4.9 | 435 | DSC (1) | 97T2 |
| | 2.55 | 57.4 | 6.6 | 410 | DSC (1) | 97T2 |
| | | | 6.0±1.0 | | DSC (2) | 97T2 |
| Val125→Ala | 3.17 | 69.6 | 4.6 | 498 | DSC (1) | 97T2 |
| | 2.88 | 64.3 | 4.8 | 469 | DSC (1) | 97T2 |
| | 2.71 | 61.0 | 6.3 | 452 | DSC (1) | 97T2 |
| | 2.55 | 58.1 | 4.3 | 435 | DSC (1) | 97T2 |
| | 2.55 | 58.0 | 4.8 | 435 | DSC (1) | 97T2 |
| | | | 5.0±0.9 | | DSC (2) | 97T2 |

Human lysozyme, Val to Ala mutants (continued)

| Mutant | pH | $T_{trs}$ | $\Delta Cp$ | $\Delta H$ | Appr./Rem. | Ref |
|---|---|---|---|---|---|---|
| Val130→Ala | 3.17 | 70.6 | 3.6 | 502 | DSC (1) | 97T2 |
| | 3.02 | 68.0 | 3.2 | 485 | DSC (1) | 97T2 |
| | 2.88 | 65.7 | 4.6 | 469 | DSC (1) | 97T2 |
| | 2.71 | 62.4 | 6.7 | 452 | DSC (1) | 97T2 |
| | 2.55 | 59.6 | 7.0 | 435 | DSC (1) | 97T2 |
| | | | 5.0±2.0 | | DSC (2) | 97T2 |

Remarks:
(1) $\Delta Cp$ was obtained from each calorimetric recording
(2) average value

Human lysozyme, Val to Ala mutants, thermodynamic parameters at the denaturation temperature (64.9°C) of the wild-type protein at pH 2.7

| Mutant | pH | $T_{trs}$ | $\Delta Cp$ | $\Delta H$ | Appr./Rem. | Ref |
|---|---|---|---|---|---|---|
| wild type | 2.7 | 64.9±0.5 | 6.6±0.5 | 477±4 | DSC (1,2) | 97T2 |
| Val2→Ala | 2.7 | 60.3±0.2 | 6.5±0.5 | 468±4 | DSC (1) | 97T2 |
| Val74→Ala | 2.7 | 63.8±0.1 | 6.2±0.4 | 476±2 | DSC (1) | 97T2 |
| Val93→Ala | 2.7 | 62.6±0.2 | 6.4±0.5 | 466±4 | DSC (1) | 97T2 |
| Val99→Ala | 2.7 | 61.9±0.02 | 6.6±0.5 | 463±3 | DSC (1) | 97T2 |
| Val100→Ala | 2.7 | 64.1±0.2 | 4.2±0.4 | 461±3 | DSC (1) | 97T2 |
| Val110→Ala | 2.7 | 66.4±0.4 | 5.2±0.4 | 484±2 | DSC (1) | 97T2 |
| Val121→Ala | 2.7 | 60.4±0.2 | 6.4±0.3 | 460±3 | DSC (1) | 97T2 |
| Val125→Ala | 2.7 | 60.9±0.1 | 5.4±0.1 | 472±0.5 | DSC (1) | 97T2 |
| Val130→Ala | 2.7 | 62.3±0.1 | 6.3±0.6 | 464±5 | DSC (1) | 97T2 |

Remarks:
(1) $\Delta Cp$ from $\Delta H$ versus $T_{trs}$
(2) data from Ref. 95T1

Human lysozyme, Tyr to Phe mutants

| Mutant | pH | $T_{trs}$ | $\Delta Cp$ | $\Delta H$ | Approach/Remarks | Ref |
|---|---|---|---|---|---|---|
| Tyr20→Phe | 3.18 | 72.0 | 2.5 | 510 | DSC (1) | 98Y1 |
| | 3.06 | 69.8 | 5.2 | 498 | DSC (1) | 98Y1 |
| | 2.89 | 66.8 | 7.1 | 481 | DSC (1) | 98Y1 |
| | 2.74 | 64.3 | 5.3 | 456 | DSC (1) | 98Y1 |
| | 2.58 | 61.1 | 7.0 | 444 | DSC (1) | 98Y1 |
| | | | 6.5±1.4 | | DSC (2) | 98Y1 |
| | | | 6.6±0.5 | | DSC (3) | 98Y1 |
| Tyr38→Phe | 3.25 | 72.4 | 4.7 | 515 | DSC (1) | 98Y1 |
| | 3.10 | 70.3 | 4.0 | 506 | DSC (1) | 98Y1 |
| | 2.92 | 67.2 | 6.1 | 490 | DSC (1) | 98Y1 |
| | 2.71 | 64.6 | 4.3 | 473 | DSC (1) | 98Y1 |
| | | | 4.9±1.1 | | DSC (2) | 98Y1 |
| | | | 5.4±0.1 | | DSC (3) | 98Y1 |
| Tyr45→Phe | 3.13 | 72.5 | 6.8 | 506 | DSC (1) | 98Y1 |
| | 2.89 | 68.7 | 2.7 | 485 | DSC (1) | 98Y1 |
| | 2.71 | 64.9 | 4.8 | 464 | DSC (1) | 98Y1 |
| | 2.67 | 64.6 | 5.6 | 469 | DSC (1) | 98Y1 |
| | 2.48 | 61.4 | 3.7 | 456 | DSC (1) | 98Y1 |
| | | | 4.5±1.7 | | DSC (2) | 98Y1 |
| | | | 4.6±0.4 | | DSC (3) | 98Y1 |

Human lysozyme, Tyr to Phe mutants (continued)

| Mutant | pH | $T_{trs}$ | $\Delta Cp$ | $\Delta H$ | Approach/Remarks | | Ref |
|---|---|---|---|---|---|---|---|
| Tyr54→Phe | 3.13 | 69.1 | 4.7 | 490 | DSC | (1) | 98Y1 |
| | 2.99 | 67.0 | 5.6 | 477 | DSC | (1) | 98Y1 |
| | 2.82 | 64.0 | 8.0 | 452 | DSC | (1) | 98Y1 |
| | 2.69 | 61.7 | 3.7 | 439 | DSC | (1) | 98Y1 |
| | | | 5.5±2.3 | | DSC | (2) | 98Y1 |
| | | | 7.0±0.4 | | DSC | (3) | 98Y1 |
| Tyr63→Phe | 3.17 | 72.7 | 2.8 | 515 | DSC | (1) | 98Y1 |
| | 3.00 | 69.5 | 2.5 | 490 | DSC | (1) | 98Y1 |
| | 2.73 | 65.0 | 4.9 | 464 | DSC | (1) | 98Y1 |
| | 2.58 | 61.9 | 6.4 | 444 | DSC | (1) | 98Y1 |
| | | | 5.3±0.8 | | DSC | (2) | 98Y1 |
| | | | 6.4±0.3 | | DSC | (3) | 98Y1 |
| Tyr124→Phe | 3.06 | 69.7 | 7.6 | 506 | DSC | (1) | 98Y1 |
| | 2.70 | 64.0 | 6.4 | 473 | DSC | (1) | 98Y1 |
| | 2.68 | 63.4 | 4.5 | 469 | DSC | (1) | 98Y1 |
| | 2.53 | 60.8 | 6.2 | 452 | DSC | (1) | 98Y1 |
| | | | 6.0±1.2 | | DSC | (2) | 98Y1 |
| | | | 6.1±0.1 | | DSC | (3) | 98Y1 |

Remarks:
(1) $\Delta Cp$ from each calorimeric curve
(2) average value of $\Delta Cp$ from each calorimeric curve
(3) $\Delta Cp$ from $\Delta H$ versus $T_{trs}$

Human lysozyme, mutants derived from the 3SS variant (Cys77→Ala and Cys95→Ala), Ile to Val and Val to Ala mutants at different pH values

| Protein | pH | $T_{trs}$ | $\Delta H$ | Approach/Remarks | Ref |
|---|---|---|---|---|---|
| 3SS | 3.14 | 57.8 | 403 | DSC | 98T1 |
| | 3.04 | 55.4 | 383 | DSC | 98T1 |
| | 2.81 | 52.0 | 359 | DSC | 98T1 |
| | 2.70 | 49.2 | 345 | DSC | 98T1 |
| | 2.53 | 45.5 | 315 | DSC | 98T1 |
| Ile23→Val-3SS | 3.09 | 55.0 | 381 | DSC | 98T1 |
| | 2.96 | 52.3 | 357 | DSC | 98T1 |
| | 2.78 | 49.6 | 349 | DSC | 98T1 |
| | 2.66 | 46.7 | 319 | DSC | 98T1 |
| | 2.49 | 43.5 | 292 | DSC | 98T1 |
| Ile56→Val-3SS | 3.10 | 50.5 | 311 | DSC | 98T1 |
| | 3.00 | 48.2 | 290 | DSC | 98T1 |
| | 2.80 | 45.3 | 279 | DSC | 98T1 |
| | 2.70 | 42.6 | 266 | DSC | 98T1 |
| Ile59→Val-3SS | 3.10 | 52.0 | 315 | DSC | 98T1 |
| | 3.00 | 50.3 | 310 | DSC | 98T1 |
| | 2.81 | 46.9 | 282 | DSC | 98T1 |
| | 2.68 | 44.3 | 263 | DSC | 98T1 |
| Ile89→Val-3SS | 3.10 | 53.8 | 365 | DSC | 98T1 |
| | 2.96 | 51.8 | 344 | DSC | 98T1 |
| | 2.78 | 48.4 | 323 | DSC | 98T1 |
| | 2.66 | 45.5 | 300 | DSC | 98T1 |
| | 2.49 | 42.5 | 276 | DSC | 98T1 |

Human lysozyme, mutants derived from the 3SS variant (Cys77→Ala and Cys95→Ala), Ile to Val and Val to Ala mutants at different pH values (continued)

| Protein | pH | T$_{trs}$ | ΔH | Approach/Remarks | Ref |
|---|---|---|---|---|---|
| Ile106→Val-3SS | 3.10 | 51.3 | 308 | DSC | 98T1 |
| | 3.00 | 49.5 | 295 | DSC | 98T1 |
| | 2.79 | 46.0 | 276 | DSC | 98T1 |
| | 2.70 | 43.6 | 250 | DSC | 98T1 |
| Val2→Ala-3SS | 3.10 | 48.3 | 272 | DSC | 98T1 |
| | 3.02 | 47.3 | 259 | DSC | 98T1 |
| | 2.71 | 41.4 | 228 | DSC | 98T1 |
| Val74→Ala-3SS | 3.16 | 54.8 | 345 | DSC | 98T1 |
| | 3.03 | 52.8 | 331 | DSC | 98T1 |
| | 2.85 | 50.1 | 311 | DSC | 98T1 |
| | 2.69 | 47.5 | 299 | DSC | 98T1 |
| | 2.51 | 43.9 | 274 | DSC | 98T1 |
| Val93→Ala-3SS | 3.16 | 52.5 | 342 | DSC | 98T1 |
| | 3.02 | 50.1 | 324 | DSC | 98T1 |
| | 2.88 | 47.7 | 306 | DSC | 98T1 |
| | 2.70 | 45.2 | 297 | DSC | 98T1 |
| | 2.50 | 41.2 | 268 | DSC | 98T1 |
| Val99→Ala-3SS | 3.10 | 52.6 | 295 | DSC | 98T1 |
| | 3.02 | 51.0 | 288 | DSC | 98T1 |
| | 2.85 | 48.2 | 281 | DSC | 98T1 |
| | 2.72 | 45.3 | 260 | DSC | 98T1 |
| | 2.54 | 41.4 | 234 | DSC | 98T1 |
| Val100→Ala-3SS | 3.15 | 54.4 | 342 | DSC | 98T1 |
| | 3.02 | 52.0 | 325 | DSC | 98T1 |
| | 2.90 | 49.9 | 303 | DSC | 98T1 |
| | 2.72 | 46.8 | 288 | DSC | 98T1 |
| | 2.51 | 43.5 | 257 | DSC | 98T1 |
| Val110→Ala-3SS | 3.13 | 57.2 | 390 | DSC | 98T1 |
| | 3.00 | 55.0 | 372 | DSC | 98T1 |
| | 2.87 | 52.5 | 359 | DSC | 98T1 |
| | 2.73 | 50.0 | 347 | DSC | 98T1 |
| | 2.55 | 46.8 | 321 | DSC | 98T1 |
| Val121→Ala-3SS | 3.14 | 50.1 | 281 | DSC | 98T1 |
| | 3.00 | 47.0 | 257 | DSC | 98T1 |
| | 2.86 | 43.6 | 242 | DSC | 98T1 |
| | 2.72 | 40.3 | 224 | DSC | 98T1 |
| Val125→Ala-3SS | 3.17 | 50.8 | 339 | DSC | 98T1 |
| | 3.02 | 47.9 | 319 | DSC | 98T1 |
| | 2.87 | 45.9 | 302 | DSC | 98T1 |
| | 2.72 | 42.8 | 281 | DSC | 98T1 |
| | 2.53 | 38.7 | 261 | DSC | 98T1 |
| Val130→Ala-3SS | 3.09 | 52.5 | 352 | DSC | 98T1 |
| | 2.97 | 49.9 | 335 | DSC | 98T1 |
| | 2.82 | 47.0 | 305 | DSC | 98T1 |
| | 2.69 | 43.8 | 292 | DSC | 98T1 |
| | 2.50 | 40.4 | 255 | DSC | 98T1 |

Human lysozyme, mutants derived from the 3SS variant (Cys77→Ala and Cys95→Ala), Ile to Val and Val to Ala mutants, values normalized to $T_{trs}$ (49.2°C) of the 3SS protein at pH 2.7

| Protein | $T_{trs}$ | $\Delta T$ | $\Delta Cp$ | $\Delta H$ | Appr./Rem. | | Ref |
|---|---|---|---|---|---|---|---|
| 3SS | 49.2±0.5 | 0.0 | 7.0±0.3 | 342±3 | DSC | (1) | 98T1 |
| Ile23→Val-3SS | 47.6±0.3 | −1.6 | 7.6±0.6 | 338±6 | DSC | (1) | 98T1 |
| Ile56→Val-3SS | 42.9±0.5 | −6.3 | 5.4±0.7 | 300±4 | DSC | (1) | 98T1 |
| Ile59→Val-3SS | 44.8±0.1 | −4.4 | 7.0±0.6 | 298±4 | DSC | (1) | 98T1 |
| Ile89→Val-3SS | 46.5±0.4 | −2.7 | 7.7±0.3 | 328±3 | DSC | (1) | 98T1 |
| Ile106→Val-3SS | 44.5±0.1 | −4.7 | 7.2±0.8 | 294±5 | DSC | (1) | 98T1 |
| Val2→Ala-3SS | 41.3±0.3 | −7.9 | 6.0±0.9 | 274±5 | DSC | (1) | 98T1 |
| Val74→Ala-3SS | 47.3±0.3 | −1.9 | 6.4±0.2 | 308±2 | DSC | (1) | 98T1 |
| Val93→Ala-3SS | 44.8±0.3 | −4.4 | 6.4±0.4 | 319±3 | DSC | (1) | 98T1 |
| Val99→Ala-3SS | 44.8±0.3 | −4.4 | 5.3±0.5 | 280±5 | DSC | (1) | 98T1 |
| Val100→Ala-3SS | 46.6±0.2 | −2.6 | 7.7±0.4 | 302±4 | DSC | (1) | 98T1 |
| Val110→Ala-3SS | 49.5±0.1 | +0.3 | 6.3±0.4 | 338±3 | DSC | (1) | 98T1 |
| Val121→Ala-3SS | 39.9±0.1 | −9.3 | 5.7±0.5 | 273±3 | DSC | (1) | 98T1 |
| Val125→Ala-3SS | 42.2±0.4 | −7.0 | 6.5±0.4 | 326±4 | DSC | (1) | 98T1 |
| Val130→Ala-3SS | 44.4±0.3 | −4.8 | 7.8±0.6 | 327±6 | DSC | (1) | 98T1 |

Remark:
(1) $\Delta Cp$ from $\Delta H$ versus $T_{trs}$

Human lysozyme, Ser to Ala mutants

| Mutant | pH | $T_{trs}$ | $\Delta Cp$ | $\Delta H$ | Appr./Rem. | Ref |
|---|---|---|---|---|---|---|
| Ser24→Ala | 3.09 | 70.3 | 5.9 | 494 | DSC | 99T5 |
| | 2.82 | 65.7 | 6.5 | 469 | DSC | 99T5 |
| | 2.66 | 62.5 | 7.0 | 444 | DSC | 99T5 |
| | 2.46 | 58.8 | 6.1 | 423 | DSC | 99T5 |
| Ser36→Ala | 3.09 | 68.7 | 5.1 | 477 | DSC | 99T5 |
| | 2.84 | 64.0 | 5.2 | 448 | DSC | 99T5 |
| | 2.67 | 60.8 | 5.5 | 435 | DSC | 99T5 |
| | 2.47 | 57.1 | 5.5 | 412 | DSC | 99T5 |
| Ser51→Ala | 3.11 | 70.8 | 5.7 | 502 | DSC | 99T5 |
| | 2.86 | 66.7 | 6.5 | 481 | DSC | 99T5 |
| | 2.69 | 64.2 | 6.0 | 464 | DSC | 99T5 |
| | 2.51 | 61.0 | 6.2 | 444 | DSC | 99T5 |
| Ser61→Ala | 3.10 | 68.3 | 5.5 | 473 | DSC | 99T5 |
| | 2.84 | 63.6 | 4.8 | 448 | DSC | 99T5 |
| | 2.71 | 60.7 | 5.6 | 427 | DSC | 99T5 |
| | 2.51 | 57.0 | 4.6 | 414 | DSC | 99T5 |
| Ser80→Ala | 3.11 | 73.6 | 5.8 | 540 | DSC | 99T5 |
| | 2.85 | 69.1 | 5.6 | 510 | DSC | 99T5 |
| | 2.69 | 66.2 | 5.6 | 498 | DSC | 99T5 |
| | 2.50 | 62.7 | 5.9 | 473 | DSC | 99T5 |
| Ser82→Ala | 3.12 | 73.5 | 5.9 | 523 | DSC | 99T5 |
| | 2.85 | 68.9 | 5.0 | 502 | DSC | 99T5 |
| | 2.70 | 66.0 | 5.3 | 485 | DSC | 99T5 |
| | 2.51 | 62.5 | 5.9 | 464 | DSC | 99T5 |

Human lysozyme, Ser to Ala mutants, thermodynamic parameters at the denaturation temperature (64.9°C) of the wild-type protein at pH 2.7

| Mutant | $T_{trs}$ | $\Delta T_{trs}$ | $\Delta Cp$ | $\Delta H$ | Appr./Rem. | Ref |
|---|---|---|---|---|---|---|
| wild type | 64.9±0.5 | 0.0 | 6.6±0.5 | 477±4 | DSC (1,2) | 99T5 |
| Ser24→Ala | 63.3±0.2 | −1.6 | 6.3±0.3 | 461±2 | DSC (1) | 99T5 |
| Ser36→Ala | 61.4±0.1 | −3.5 | 5.5±0.3 | 455±3 | DSC (1) | 99T5 |
| Ser51→Ala | 64.2±0.2 | −0.7 | 6.0±0.3 | 468±2 | DSC (1) | 99T5 |
| Ser61→Ala | 60.6±0.2 | −4.3 | 5.3±0.5 | 454±4 | DSC (1) | 99T5 |
| Ser80→Ala | 66.3±0.1 | 1.4 | 6.0±0.4 | 487±3 | DSC (1) | 99T5 |
| Ser82→Ala | 66.0±0.2 | 1.1 | 5.3±0.3 | 479±2 | DSC (1) | 99T5 |

Remarks:
(1) $\Delta Cp$ from $\Delta H^{cal}$ versus $T_{trs}$ at varying pH
(2) data from Ref. 95T1

Human lysozyme mutants, thermodynamic parameters at different pH values

| Mutant | pH | $T_{trs}$ | $\Delta H$ | Approach/Remarks | Ref |
|---|---|---|---|---|---|
| Asn27→Leu | 2.70 | 65.3 | | DSC | 99T2 |
| Ala32→Leu | 2.61 | 63.1 | 397 | DSC | 99T2 |
| | 2.83 | 66.9 | 427 | DSC | 99T2 |
| | 3.23 | 73.2 | 452 | DSC | 99T2 |
| Glu35→Leu | 2.56 | 61.2 | 397 | DSC | 99T2 |
| | 3.25 | 70.8 | 439 | DSC | 99T2 |
| Gly37→Gln | 2.48 | 60.7 | 427 | DSC | 99T2 |
| | 2.73 | 64.5 | 452 | DSC | 99T2 |
| | 3.09 | 70.1 | 473 | DSC | 99T2 |
| Arg50→Gly | 2.58 | 63.7 | 418 | DSC | 99T2 |
| | 2.78 | 67.3 | 427 | DSC | 99T2 |
| | 3.11 | 73.0 | 473 | DSC | 99T2 |
| Gln58→Gly | 2.59 | 68.3 | 473 | DSC | 99T2 |
| | 2.78 | 72.5 | 498 | DSC | 99T2 |
| | 3.15 | 78.7 | 531 | DSC | 99T2 |
| Val74→Ser | 2.61 | 62.1 | 439 | DSC | 99T2 |
| | 2.83 | 66.0 | 456 | DSC | 99T2 |
| | 3.28 | 72.2 | 490 | DSC | 99T2 |
| His78→Gly | 2.53 | 61.5 | 431 | DSC | 99T2 |
| | 2.75 | 65.2 | 460 | DSC | 99T2 |
| | 3.10 | 71.0 | 494 | DSC | 99T2 |
| Ala96→Met | 2.56 | 62.5 | 398 | DSC | 99T2 |
| | 2.77 | 66.5 | 423 | DSC | 99T2 |
| | 3.19 | 72.9 | 456 | DSC | 99T2 |
| Val100→Phe | 2.62 | 56.8 | 331 | DSC | 99T2 |
| | 2.83 | 61.0 | 354 | DSC | 99T2 |
| | 3.26 | 67.6 | 374 | DSC | 99T2 |

Human lysozyme mutants, thermodynamic parameters at $T_{trs}$ (64.9°C) of the wild-type protein at pH 2.7

| Mutant | pH | $T_{trs}$ | $T_{ref}$ | $\Delta Cp$ | $\Delta H$ | Appr./Rem. | Ref |
|---|---|---|---|---|---|---|---|
| wild type | 2.70 | 64.9±0.5 | 64.9 | 6.6±0.5 | 477±4 | DSC (1–3) | 99T2 |
| Asn27→Leu | 2.70 | 65.3 | 64.9 | | | DSC | 99T2 |
| Ala32→Leu | 2.70 | 64.6±0.2 | 64.9 | 5.3±1.0 | 411±8 | DSC (1,2) | 99T2 |
| Glu35→Leu | 2.70 | 63.1 | 64.9 | 4.4 | 414 | DSC (1,2) | 99T2 |
| Gly37→Gln | 2.70 | 64.1±0.0 | 64.9 | 4.8±0.8 | 450±5 | DSC (1,2) | 99T2 |

Human lysozyme mutants, thermodynamic parameters at $T_{trs}$ (64.9°C) of the wild-type protein at pH 2.7 (continued)

| Mutant | pH | $T_{trs}$ | $T_{ref}$ | $\Delta Cp$ | $\Delta H$ | Appr./Rem. | Ref |
|---|---|---|---|---|---|---|---|
| Arg50→Gly | 2.70 | 65.8±0.1 | 64.9 | 6.1±1.5 | 420±10 | DSC (1,2) | 99T2 |
| Gln58→Gly | 2.70 | 70.6±0.5 | 64.9 | 5.6±0.2 | 455±1 | DSC (1,2) | 99T2 |
| Val74→Ser | 2.70 | 63.7±0.5 | 64.9 | 5.1±0.3 | 452±2 | DSC (1,2) | 99T2 |
| His78→Gly | 2.70 | 64.4±0.0 | 64.9 | 6.6±0.5 | 455±4 | DSC (1,2) | 99T2 |
| Ala96→Met | 2.70 | 65.0±0.4 | 64.9 | 5.5±0.3 | 413±2 | DSC (1,2) | 99T2 |
| Val100→Phe | 2.70 | 58.4±0.5 | 64.9 | 3.9±0.7 | 365±5 | DSC (1,2) | 99T2 |

Remarks:
(1) normalized data at $T_{trs}$ of the wild-type protein
(2) $\Delta Cp$ from $\Delta H^{cal}$ versus $T_{trs}$
(3) data from Ref. 95T1

Human lysozyme, wild type and Thr to Ala and Thr to Val mutants

| Mutant | pH | $T_{trs}$ | $\Delta H$ | Approach/Remarks | | Ref |
|---|---|---|---|---|---|---|
| Thr11→Ala | 3.22 | 73.4 | 499 | DSC | (1,2) | 99T4 |
| | 3.02 | 71.4 | 485 | DSC | (1,2) | 99T4 |
| | 2.82 | 68.5 | 481 | DSC | (1,2) | 99T4 |
| | 2.74 | 66.8 | 469 | DSC | (1,2) | 99T4 |
| | 2.58 | 63.7 | 439 | DSC | (1,2) | 99T4 |
| Thr40→Ala | 3.10 | 68.6 | 490 | DSC | (1,2) | 99T4 |
| | 3.00 | 66.5 | 477 | DSC | (1,2) | 99T4 |
| | 2.85 | 63.3 | 448 | DSC | (1,2) | 99T4 |
| | 2.70 | 59.4 | 418 | DSC | (1,2) | 99T4 |
| | 2.50 | 56.6 | 408 | DSC | (1,2) | 99T4 |
| Thr43→Ala | 3.08 | 69.7 | 477 | DSC | (1,2) | 99T4 |
| | 2.90 | 67.1 | 464 | DSC | (1,2) | 99T4 |
| | 2.74 | 64.5 | 452 | DSC | (1,2) | 99T4 |
| | 2.54 | 61.1 | 435 | DSC | (1,2) | 99T4 |
| Thr52→Ala | 3.10 | 68.9 | 456 | DSC | (1,2) | 99T4 |
| | 3.00 | 66.9 | 452 | DSC | (1,2) | 99T4 |
| | 2.85 | 64.4 | 439 | DSC | (1,2) | 99T4 |
| | 2.70 | 61.9 | 431 | DSC | (1,2) | 99T4 |
| | 2.50 | 58.8 | 413 | DSC | (1,2) | 99T4 |
| Thr70→Ala | 3.10 | 67.1 | 460 | DSC | (1,2) | 99T4 |
| | 3.00 | 64.7 | 448 | DSC | (1,2) | 99T4 |
| | 2.85 | 62.2 | 431 | DSC | (1,2) | 99T4 |
| | 2.70 | 60.0 | 423 | DSC | (1,2) | 99T4 |
| | 2.50 | 57.3 | 412 | DSC | (1,2) | 99T4 |
| Thr11→Val | 3.19 | 73.0 | 490 | DSC | (1,2) | 99T4 |
| | 2.81 | 68.7 | 477 | DSC | (1,2) | 99T4 |
| | 2.71 | 66.2 | 460 | DSC | (1,2) | 99T4 |
| | 2.51 | 62.9 | 435 | DSC | (1,2) | 99T4 |
| Thr40→Val | 3.10 | 68.7 | 481 | DSC | (1,2) | 99T4 |
| | 3.00 | 66.3 | 469 | DSC | (1,2) | 99T4 |
| | 2.85 | 63.2 | 460 | DSC | (1,2) | 99T4 |
| | 2.70 | 59.2 | 389 | DSC | (1,2) | 99T4 |
| | 2.50 | 57.0 | 404 | DSC | (1,2) | 99T4 |
| Thr43→Val | 3.20 | 75.4 | 473 | DSC | (1,2) | 99T4 |
| | 3.02 | 73.0 | 452 | DSC | (1,2) | 99T4 |
| | 2.71 | 68.4 | 439 | DSC | (1,2) | 99T4 |
| | 2.51 | 65.0 | 418 | DSC | (1,2) | 99T4 |

Human lysozyme, wild type and Thr to Ala and Thr to Val mutants (continued)

| Mutant | pH | $T_{trs}$ | $\Delta H$ | Approach/Remarks | | Ref |
|--------|------|------|------|------|------|------|
| Thr52→Val | 3.10 | 68.7 | 464 | DSC | (1,2) | 99T4 |
| | 3.00 | 66.6 | 448 | DSC | (1,2) | 99T4 |
| | 2.85 | 64.2 | 439 | DSC | (1,2) | 99T4 |
| | 2.70 | 62.1 | 427 | DSC | (1,2) | 99T4 |
| | 2.50 | 58.8 | 413 | DSC | (1,2) | 99T4 |
| Thr70→Val | 3.10 | 70.3 | 498 | DSC | (1,2) | 99T4 |
| | 3.00 | 67.6 | 481 | DSC | (1,2) | 99T4 |
| | 2.85 | 65.3 | 469 | DSC | (1,2) | 99T4 |
| | 2.70 | 62.6 | 448 | DSC | (1,2) | 99T4 |
| | 2.50 | 59.5 | 427 | DSC | (1,2) | 99T4 |

Remarks:
(1) measured in 0.05 M glycine-HCl, pH 2.5–3.3
(2) for details see also Ref. 95T1

Human lysozyme, wild type and Thr to Ala and Thr to Val mutants at the denaturation temperature (64.9°C) of the wild-type protein at pH 2.7

| Mutant | pH | $T_{trs}$ | $\Delta Cp$ | $\Delta H$ | Appr./Rem. | | Ref |
|--------|------|------|------|------|------|------|------|
| wild type | 2.7 | 64.9±0.5 | 6.6±0.5 | 477±4 | DSC | (1,2) | 99T4 |
| Thr11→Ala | 2.7 | 66.1±0.6 | 5.7±1.0 | 453±7 | DSC | (1) | 99T4 |
| Thr11→Val | 2.7 | 65.9±0.3 | 5.4±0.9 | 450±6 | DSC | (1) | 99T4 |
| Thr40→Ala | 2.7 | 60.2±0.6 | 7.2±0.5 | 462±5 | DSC | (1) | 99T4 |
| Thr40→Val | 2.7 | 60.7±0.3 | 6.7±0.9 | 459±8 | DSC | (1) | 99T4 |
| Thr43→Ala | 2.7 | 63.8±0.2 | 4.9±0.1 | 453±0 | DSC | (1) | 99T4 |
| Thr43→Val | 2.7 | 68.1±0.3 | 4.9±0.7 | 419±5 | DSC | (1) | 99T4 |
| Thr52→Ala | 2.7 | 62.0±0.2 | 3.5±0.5 | 445±5 | DSC | (1) | 99T4 |
| Thr52→Val | 2.7 | 62.1±0.3 | 5.0±0.3 | 442±3 | DSC | (1) | 99T4 |
| Thr70→Ala | 2.7 | 60.2±0.5 | 5.0±0.3 | 448±2 | DSC | (1) | 99T4 |
| Thr70→Val | 2.7 | 62.8±0.4 | 6.6±0.2 | 463±2 | DSC | (1) | 99T4 |

Remarks:
(1) $\Delta Cp$ from $\Delta H^{cal}$ versus $T_{trs}$
(2) wild-type data from Ref. 95T1

Human lysozyme, mutants at positions 56 and 59, pH dependence

| Mutant | pH | $T_{trs}$ | $\Delta Cp$ | $\Delta H$ | Approach/Remarks | | Ref |
|--------|------|------|------|------|------|------|------|
| Ile56→Leu | 2.48 | 61.2 | 5.9 | 431 | DSC | (1) | 99F13 |
| | 2.65 | 64.0 | 6.5 | 439 | DSC | (1) | 99F13 |
| | 2.96 | 68.7 | 5.3 | 469 | DSC | (1) | 99F13 |
| | 3.10 | 70.3 | 4.8 | 485 | DSC | (1) | 99F13 |
| | 2.5–3.1 | | 5.6±0.9 | | DSC | (2) | 99F13 |
| Ile56→Met | 2.54 | 56.3 | 5.8 | 387 | DSC | (1) | 99F13 |
| | 2.70 | 59.2 | 6.8 | 407 | DSC | (1) | 99F13 |
| | 2.83 | 61.3 | 7.2 | 423 | DSC | (1) | 99F13 |
| | 3.02 | 64.2 | 6.0 | 439 | DSC | (1) | 99F13 |
| | 3.11 | 65.9 | 6.4 | 448 | DSC | (1) | 99F13 |
| | 2.5–3.1 | | 6.4±0.8 | | DSC | (2) | 99F13 |
| Ile56→Phe | 2.75 | 50.8 | 7.9 | 318 | DSC | (1) | 99F13 |
| | 3.06 | 56.3 | 7.4 | 360 | DSC | (1) | 99F13 |
| | 2.7–3.1 | | 7.7±0.3 | | DSC | (2) | 99F13 |

**Table 2. Enthalpy and Heat Capacity Changes – Molar Values**

Human lysozyme, mutants at positions 56 and 59, pH dependence (continued)

| Mutant | pH | $T_{trs}$ | $\Delta Cp$ | $\Delta H$ | Approach/Remarks | | Ref |
|---|---|---|---|---|---|---|---|
| Ile59→Leu | 2.51 | 61.4 | 4.7 | 456 | DSC | (1) | 99F13 |
| | 2.66 | 64.5 | 4.9 | 473 | DSC | (1) | 99F13 |
| | 2.83 | 67.0 | 4.9 | 485 | DSC | (1) | 99F13 |
| | 2.96 | 69.8 | 5.5 | 498 | DSC | (1) | 99F13 |
| | 3.17 | 72.1 | 5.2 | 515 | DSC | (1) | 99F13 |
| | 2.5–3.2 | | 5.0±0.5 | | DSC | (2) | 99F13 |
| Ile59→Met | 2.47 | 57.4 | 5.5 | 402 | DSC | (1) | 99F13 |
| | 2.65 | 59.9 | 4.2 | 412 | DSC | (1) | 99F13 |
| | 2.80 | 62.0 | 5.0 | 431 | DSC | (1) | 99F13 |
| | 2.95 | 64.8 | 4.6 | 444 | DSC | (1) | 99F13 |
| | 3.13 | 66.5 | 4.8 | 448 | DSC | (1) | 99F13 |
| | 2.5–3.1 | | 4.8±0.7 | | DSC | (2) | 99F13 |
| Ile59→Phe | 2.54 | 59.8 | 5.9 | 418 | DSC | (1) | 99F13 |
| | 2.73 | 62.8 | 6.6 | 435 | DSC | (1) | 99F13 |
| | 2.90 | 65.4 | 5.0 | 452 | DSC | (1) | 99F13 |
| | 3.05 | 68.0 | 5.4 | 464 | DSC | (1) | 99F13 |
| | 3.15 | 69.8 | 5.7 | 477 | DSC | (1) | 99F13 |
| | 2.5–3.2 | | 5.7±0.9 | | DSC | (2) | 99F13 |
| Ile59→Ser | 2.51 | 50.2 | 5.3 | 364 | DSC | (1) | 99F13 |
| | 2.70 | 53.1 | 5.0 | 380 | DSC | (1) | 99F13 |
| | 2.84 | 55.3 | 6.5 | 396 | DSC | (1) | 99F13 |
| | 3.04 | 58.2 | 6.2 | 411 | DSC | (1) | 99F13 |
| | 3.17 | 60.0 | 5.2 | 418 | DSC | (1) | 99F13 |
| | 2.5–3.2 | | 5.6±0.9 | | DSC | (2) | 99F13 |
| Ile59→Thr | 2.40 | 52.9 | 6.9 | 378 | DSC | (1) | 99F13 |
| | 2.62 | 56.9 | 7.0 | 435 | DSC | (1) | 99F13 |
| | 2.79 | 59.5 | 5.3 | 444 | DSC | (1) | 99F13 |
| | 2.80 | 59.5 | 4.9 | 444 | DSC | (1) | 99F13 |
| | 2.99 | 62.9 | 5.8 | 448 | DSC | (1) | 99F13 |
| | 3.15 | 64.9 | 5.0 | 464 | DSC | (1) | 99F13 |
| | 2.4–3.2 | | 5.8±1.2 | | DSC | (2) | 99F13 |
| Ile59→Tyr | 2.71 | 51.4 | 6.5 | 332 | DSC | (1) | 99F13 |
| | 2.89 | 54.2 | 4.9 | 346 | DSC | (1) | 99F13 |
| | 3.18 | 59.5 | 5.3 | 379 | DSC | (1) | 99F13 |
| | 2.5–3.1 | | 5.6±0.9 | | DSC | (2) | 99F13 |

Remarks:
(1) $\Delta Cp$ from each calorimetric curve
(2) average $\Delta Cp$ value

Human lysozyme, mutants at positions 56 and 59, parameters at the denaturation temperature (64.9°C) of the wild-type human lysozyme at pH 2.7

| Mutant | pH | $T_{trs}$ | $\Delta Cp$ | $\Delta H$ | Appr./Rem. | | Ref |
|---|---|---|---|---|---|---|---|
| wild type | 2.7 | 64.9±0.5 | 6.6±0.5 | 477±4 | DSC | (1,2) | 99F13 |
| Ile56→Ala | | 52.4±0.4 | 5.6±0.2 | 440±2 | DSC | (1,3) | 99F13 |
| Ile56→Gly | | 62.2 | | | DSC | (3) | 99F13 |
| Ile56→Leu | | 64.6±0.3 | 5.9±0.8 | 449±1 | DSC | (1) | 99F13 |
| Ile56→Met | | 59.1±0.1 | 6.4±0.3 | 443±2 | DSC | (1) | 99F13 |
| Ile56→Phe | | 49.9 | 7.6 | 425 | DSC | (1) | 99F13 |
| Ile56→Thr | | 52.4 | 4.5±0.7 | 425±5 | DSC | (1,4) | 99F13 |
| Ile56→Val | | 61.3±0.3 | 5.6±1.7 | 475±13 | DSC | (1,2) | 99F13 |
| Ile59→Ala | | 59.7±0.2 | 6.1±0.4 | 480±4 | DSC | (1,3) | 99F13 |
| Ile59→Gly | | 52.2±0.6 | 5.2±0.5 | 444±5 | DSC | (1,3) | 99F13 |

Human lysozyme, mutants at positions 56 and 59, parameters at the denaturation temperature (64.9°C) of the wild-type human lysozyme at pH 2.7 (continued)

| Mutant | pH | $T_{trs}$ | $\Delta Cp$ | $\Delta H$ | Appr./Rem. | Ref |
|---|---|---|---|---|---|---|
| Ile59→Leu | | 64.9±0.5 | 5.3±0.2 | 475±1 | DSC (1) | 99F13 |
| Ile59→Met | | 60.7±0.4 | 5.3±0.5 | 442±2 | DSC (1) | 99F13 |
| Ile59→Phe | | 62.3±0.1 | 5.8±0.2 | 448±1 | DSC (1) | 99F13 |
| Ile59→Ser | | 53.1±0.1 | 5.6±0.3 | 447±4 | DSC (1) | 99F13 |
| Ile59→Thr | | 58.0±0.3 | 6.4±1.4 | 470±10 | DSC (1) | 99F13 |
| Ile59→Tyr | | 51.1±0.3 | 5.6±0.2 | 409±3 | DSC (1) | 99F13 |
| Ile59→Val | | 61.5±0.4 | 5.0±1.0 | 461±7 | DSC (1,2) | 99F13 |

Remarks:

(1) $\Delta Cp$ from $\Delta H^{cal}$ versus $T_{trs}$

(2) data from Ref. 95T1

(3) data from Ref. 97T1

(4) data from Ref. 96F6

Recombinant human lysozyme, modification of the N-terminus, pH dependence

| Protein | pH | $T_{trs}$ | $\Delta H$ | Approach/Remarks | Ref |
|---|---|---|---|---|---|
| Lys1→Ala | 2.58 | 60.9 | 379 | DSC | 99T3 |
| | 2.77 | 63.7 | 398 | DSC | 99T3 |
| | 2.97 | 66.3 | 415 | DSC | 99T3 |
| Lys1→Met | 2.50 | 61.7 | 370 | DSC | 99T3 |
| | 2.78 | 64.7 | 386 | DSC | 99T3 |
| | 2.95 | 69.1 | 399 | DSC | 99T3 |
| | 3.22 | 70.1 | 412 | DSC | 99T3 |
| Gly(−1) | 2.58 | 53.0 | 355 | DSC | 99T3 |
| | 2.81 | 56.8 | 384 | DSC | 99T3 |
| | 2.91 | 59.0 | 395 | DSC | 99T3 |
| | 3.09 | 60.8 | 413 | DSC | 99T3 |
| Met(−1) | 2.65 | 56.0 | 349 | DSC | 99T3 |
| | 2.93 | 61.0 | 380 | DSC | 99T3 |
| | 3.03 | 63.2 | 390 | DSC | 99T3 |
| Pro(−1) | 2.59 | 54.0 | 341 | DSC | 99T3 |
| | 2.74 | 56.8 | 354 | DSC | 99T3 |
| | 3.01 | 61.1 | 377 | DSC | 99T3 |
| | 3.25 | 63.5 | 397 | DSC | 99T3 |

Explanations:

the N-terminus of authentic human lysozyme is Lys1-Val2-Phe3-Xaa(−1) = additional N-terminal residue

Recombinant human lysozyme, modification of the N-terminus

| Mutant | pH | $T_{trs}$ | $\Delta Cp$ | $\Delta H$ | Approach/Remarks | Ref |
|---|---|---|---|---|---|---|
| wild type | 2.7 | 64.9 | 6.5 | 477 | DSC (1,3) | 99T3 |
| Lys1→Ala | 2.7 | 62.7 | 6.7 | 405 | DSC (2,3) | 99T3 |
| Lys1→Met | 2.7 | 64.4 | 4.3 | 384 | DSC (2,3) | 99T3 |
| Gly(−1) | 2.7 | 55.1 | 7.2 | 440 | DSC (2,3) | 99T3 |
| Met(−1) | 2.7 | 56.9 | 8.9 | 400 | DSC (2,3) | 99T3 |
| Pro(−1) | 2.7 | 55.9 | 5.7 | 402 | DSC (2,3) | 99T3 |

Explanations:

the N-terminus of authentic human lysozyme is Lys1-Val2-Phe3-Xaa(−1) = additional N-terminal residue

Remarks:

(1) data from Ref. 95T1

(2) data at the denaturation temperature of the wild-type protein at pH 2.7, see also Table 1

(3) $\Delta Cp$ from $\Delta H^{v.H.}$ versus $T_{trs}$ at varying pH

*b) mutants that create a calcium binding site*

Human lysozyme, mutants with partially introduced $Ca^{2+}$ binding site

| pH | $T_{trs}$ | $\Delta Cp$ | $\Delta H$ | Approach/Remarks | | Ref |
|---|---|---|---|---|---|---|
| 4.51 | 73.31 | 7.15 | 531 | DSC | (1–3) | 98K13 |
| 3.91 | 72.12 | | 527 | DSC | (4) | 98K13 |
| 3.22 | 70.01 | | 523 | DSC | (4) | 98K13 |
| 2.94 | 64.66 | | 473 | DSC | (4) | 98K13 |
| 2.75 | 62.87 | | 477 | DSC | (4) | 98K13 |
| 2.46 | 56.89 | | 431 | DSC | (4) | 98K13 |
| 2.26 | 53.31 | | 409 | DSC | (4) | 98K13 |
| 2.11 | 49.76 | | 344 | DSC | (4) | 98K13 |
| 1.84 | 46.65 | | 340 | DSC | (4) | 98K13 |

Remarks:

(1) measured in 50 mM sodium acetate buffer

(2) $\Delta Cp$ from from $\Delta H$ versus $T_{trs}$

(3) Ref. 98K13 contains additional thermodynamic data for unfolding of the wild-type and mutant lysozymes in the presence of $Ca^{2+}$ at pH 4.5

(4) measured in 50 mM glycine/HCl

Human lysozyme, mutants that create a calcium binding site

| Protein | $Ca^{2+}$ | pH | $T_{trs}$ | $\Delta Cp$ | $\Delta H$ | Appr./Rem. | | Ref |
|---|---|---|---|---|---|---|---|---|
| wild type | 0.0 mM | 4.5 | 80.3 | 7.15 | 579.1 | DSC | (1,2) | 98K19 |
| Gln86→Asp | 0.0 mM | 4.5 | 80.3 | 4.64 | 566.1 | DSC | (1) | 98K19 |
| Ala92→Asp | 0.0 mM | 4.5 | 77.0 | 5.44 | 561.1 | DSC | (1) | 98K19 |
| double mutant (Gln86→Asp and Ala92→Asp) | | | | | | | | |
| | 0.0 mM | 4.5 | 76.5 | 4.69 | 546.4 | DSC | (1,2) | 98K19 |
| triple mutant (Ala83→Lys, Gln86→Asp, and Ala92→Asp) | | | | | | | | |
| | 0.0 mM | 4.5 | 77.0 | 5.82 | 552.7 | DSC | (1) | 98K19 |
| wild type | 10.0 mM | 4.5 | 80.0 | 4.27 | 569.0 | DSC | (1,2) | 98K19 |
| Gln86→Asp | 10.0 mM | 4.5 | 80.3 | 7.32 | 560.7 | DSC | (1) | 98K19 |
| Ala92→Asp | 10.0 mM | 4.5 | 77.2 | 4.81 | 550.2 | DSC | (1) | 98K19 |
| double mutant (Gln86→Asp and Ala92→Asp) | | | | | | | | |
| | 10.0 mM | 4.5 | 89.2 | 3.35 | 595.0 | DSC | (1,2) | 98K19 |
| triple mutant (Ala83→Lys, Gln86→Asp, and Ala92→Asp) | | | | | | | | |
| | 10.0 mM | 4.5 | 89.3 | 5.23 | 607.1 | DSC | (1) | 98K19 |

Remarks:

(1) measured in 0.05 M sodium acetate buffer, pH 4.5, with and without 10 mM $CaCl_2$

(2) data from Ref. 92K12

## Lysozyme, Phage Lambda

Lysozyme phage lambda, wild type and histidine mutants

| Protein | pH | $T_{trs}$ | $\Delta H$ | Approach/Remarks | | Ref |
|---|---|---|---|---|---|---|
| wild type | 7.0 | 52.3 | 474.0 | heat | (1–3) | 99E4 |
| His31→Asp | 7.0 | 46.1 | 417.6 | heat | (1–3) | 99E4 |
| His48→Asn | 7.0 | 30.1 | 254.4 | heat | (1–3) | 99E4 |
| double mutant (His31→Asn and His48→Asn) | | | | | | |
| | 7.0 | 54.3 | 447.7 | heat | (1–3) | 99E4 |

Remarks:
(1) for the procedure see also Ref. 98S15
(2) transition monitored by fluorescence
(3) buffer: 100 mM phosphate, pH 7.0

Lysozyme phage lambda (λL), double mutant (His31→Asn and His137→Asn) with the remaining His48 replaced by 1,2,4-triazole-3-alanine (Taz)

| Protein | pH | $T_{trs}$ | $\Delta Cp$ | $\Delta H$ | Approach/Remarks | | Ref |
|---|---|---|---|---|---|---|---|
| His48-λL | 4 | 24.5 | 8.4±0.8 | 121 | heat | (1,2) | 98S15 |
| | 4 | 26 | | 148 | heat | (1,2) | 98S15 |
| | 4.5 | 39.3 | | 297 | heat | (1,2) | 98S15 |
| | 5 | 46.7 | | 393 | heat | (1,2) | 98S15 |
| | 5.5 | 51.1 | | 448 | heat | (1,2) | 98S15 |
| | 6 | 52.7 | | 389 | heat | (1,2) | 98S15 |
| | 6.5 | 53.6 | | 464 | heat | (1,2) | 98S15 |
| | 6.57 | 50.8 | | 418 | heat | (1,2) | 98S15 |
| | 7 | 54.3 | | 448 | heat | (1,2) | 98S15 |
| | 7.07 | 51.1 | | 439 | heat | (1,2) | 98S15 |
| | 7.57 | 51 | | 423 | heat | (1,2) | 98S15 |
| | 8.07 | 50.1 | | 381 | heat | (1,2) | 98S15 |
| Taz48-λL | 4 | 25.7 | | 119 | heat | (1,2) | 98S15 |
| | 4.5 | 37.8 | | 234 | heat | (1,2) | 98S15 |
| | 5 | 43.3 | | 281 | heat | (1,2) | 98S15 |
| | 5 | 44.3 | | 314 | heat | (1,2) | 98S15 |
| | 5.5 | 45 | | 310 | heat | (1,2) | 98S15 |
| | 6 | 48.4 | | 360 | heat | (1,2) | 98S15 |
| | 6.5 | 47.8 | | 322 | heat | (1,2) | 98S15 |
| | 6.79 | 44.4 | | 280 | heat | (1,2) | 98S15 |
| | 7 | 47.7 | | 326 | heat | (1,2) | 98S15 |
| | 7.29 | 43.7 | | 301 | heat | (1,2) | 98S15 |
| | 7.79 | 43.6 | | 276 | heat | (1,2) | 98S15 |
| | 8.29 | 42.5 | | 255 | heat | (1,2) | 98S15 |

Remarks:
(1) heat denaturation monitored by fluorescence
(2) for the procedure and $\Delta Cp$ see P. Soumillion et al., Protein Engng. 8 (1995) 451–456

## Lysozyme, Phage T4

Lysozyme phage T4, alanine mutants derived from pseudo-wild type w.t.*

| Protein | pH | $T_{trs}$ | $\Delta H$ | Approach/Remarks | | Ref |
|---|---|---|---|---|---|---|
| Met6→Ala | 5.4 | 60.8 | 506 | heat | (1–3) | 99G3 |
| Leu7→Ala | 5.4 | 59.0 | 469 | heat | (1–3) | 99G3 |
| Ile17→Ala | 5.4 | 58.9 | 389 | heat | (1–3) | 99G3 |
| Leu33→Ala | 5.4 | 56.5 | 347 | heat | (1–3) | 99G3 |
| Ile50→Ala | 5.4 | 61.1 | 485 | heat | (1–3) | 99G3 |
| Leu66→Ala | 5.4 | 55.2 | 351 | heat | (1–3) | 99G3 |
| Ile78→Ala | 5.4 | 61.8 | 531 | heat | (1–3) | 99G3 |
| Leu84→Ala | 5.4 | 54.8 | 423 | heat | (1–3) | 99G3 |
| Val87→Ala | 5.4 | 61.0 | 531 | heat | (1–3) | 99G3 |
| Leu91→Ala | 5.4 | 57.9 | 473 | heat | (1–3) | 99G3 |
| Leu99→Ala | 5.4 | 53.6 | 435 | heat | (1–3) | 99G3 |
| Met102→Ala | 5.4 | 57.1 | 452 | heat | (1–3) | 99G3 |
| Ile100→Ala | 5.4 | 58.2 | 494 | heat | (1–3) | 99G3 |
| Met102→Ala/Met106→Ala | | | | | | |
| | 5.4 | 54.5 | 423 | heat | (1–3) | 99G3 |
| Val103→Ala | 5.4 | 60.9 | 510 | heat | (1–3) | 99G3 |
| Phe104→Ala | 5.4 | 57.8 | 456 | heat | (1–3) | 99G3 |
| Met106→Ala | 5.4 | 60.1 | 477 | heat | (1–3) | 99G3 |
| Val111→Ala | 5.4 | 62.4 | 531 | heat | (1–3) | 99G3 |
| Leu118→Ala | 5.4 | 56.3 | 456 | heat | (1–3) | 99G3 |
| Leu121→Ala | 5.4 | 59.1 | 452 | heat | (1–3) | 99G3 |
| w.t.* (Ala129) | 5.4 | 65.3 | 544 | heat | (1–4) | 99G3 |
| Val149→Ala | 5.4 | 56.3 | 444 | heat | (1–3) | 99G3 |
| Phe153→Ala | 5.4 | 55.6 | 452 | heat | (1–3) | 99G3 |

Remarks:

(1) w.t.* is the cysteine-free pseudo-wild-type protein

(2) measured in 0.10 M sodium chloride, 1.4 mM acetic acid, 8.6 mM sodium acetate, pH 5.4, for the procedure see
    Ref. 93E2

(3) uncertainty in $T_{trs}$ ±0.14°C

(4) data from Ref. 96G2

Lysozyme phage T4, alanine mutants derived from pseudo-wild type w.t.*

| Protein | pH | $T_{trs}$ | $\Delta H$ | Approach/Remarks | | Ref |
|---|---|---|---|---|---|---|
| w.t.*(Met6) | 5.4 | 65.3 | 544 | heat | (1–4) | 99G3 |
| Ile17→Met | 5.4 | 59.4 | 427 | heat | (1–3) | 99G3 |
| Ile27→Met | 5.4 | 55.2 | 259 | heat | (1–3) | 99G3 |
| Ile17→Met/Leu33→Met | | | | | | |
| | 5.4 | 55.0 | 255 | heat | (1–3) | 99G3 |
| Leu33→Met | 5.4 | 60.0 | 452 | heat | (1–3) | 99G3 |
| Ile50→Met | 5.4 | 64.7 | 502 | heat | (1–3) | 99G3 |
| Leu66→Met | 5.4 | 62.6 | 510 | heat | (1–3) | 99G3 |
| Ile78→Met | 5.4 | 61.6 | 490 | heat | (1–4) | 99G3 |
| Leu84→Met | 5.4 | 60.4 | 460 | heat | (1–4) | 99G3 |
| Val87→Met | 5.4 | 59.0 | 473 | heat | (1–3) | 99G3 |
| Leu91→Met | 5.4 | 63.3 | 523 | heat | (1–4) | 99G3 |
| Leu99→Met | 5.4 | 64.0 | 561 | heat | (1–3,5) | 99G3 |
| WT*(Met102) | 5.4 | 65.3 | 544 | heat | (1–4) | 99G3 |
| Ile100→Met | 5.4 | 60.8 | 523 | heat | (1–3) | 99G3 |

Lysozyme phage T4, alanine mutants derived from pseudo-wild type w.t.* (continued)

| Protein | pH | $T_{trs}$ | $\Delta H$ | Approach/Remarks | | Ref |
|---|---|---|---|---|---|---|
| Val103→Met | 5.4 | 62.2 | 490 | heat | (1–4) | 99G3 |
| Phe104→Met | 5.4 | 64.5 | 506 | heat | (1–3) | 99G3 |
| w.t.*(Met106) | 5.4 | 65.3 | 544 | heat | (1–3) | 99G3 |
| Val111→Met | 5.4 | 63.3 | 531 | heat | (1–3) | 99G3 |
| Leu118→Met | 5.4 | 63.5 | 544 | heat | (1–4) | 99G3 |
| Leu121→Met | 5.4 | 63.2 | 540 | heat | (1–4) | 99G3 |
| Ala129→Met | 5.4 | 60.1 | 456 | heat | (1–4) | 99G3 |
| Val149→Met | 5.4 | 57.7 | 464 | heat | (1–3) | 99G3 |
| Phe153→Met | 5.4 | 63.7 | 536 | heat | (1–3,5) | 99G3 |

Remarks:

(1) w.t.* is the cysteine-free pseudo-wild-type protein

(2) measured in 0.10 M sodium chloride, 1.4 mM acetic acid, 8.6 mM sodium acetate, pH 5.4, for the procedure see Ref. 93E2

(3) uncertainty in $T_{trs}$ ±0.14°C

(4) data from Ref. 96G2

(5) data from Ref. 93E2

Lysozyme phage T4, large-to-small amino acid substitutions within the core

| Mutant | pH | $T_{trs}$ | $\Delta H$ | Approach/Remarks | Ref |
|---|---|---|---|---|---|
| wild type* | 3.0 | 51.65 | 473 | heat, v.H. (1–4) | 98X1 |
| Ile17→Ala | 3.0 | 43.25 | 364 | heat, v.H. (1–4) | 98X1 |
| Ile27→Ala | 3.0 | 41.55 | 318 | heat, v.H. (1–4) | 98X1 |
| Ile29→Ala | 3.0 | 43.45 | 356 | heat, v.H. (1–4) | 98X1 |
| Ile50→Ala | 3.0 | 45.85 | 393 | heat, v.H. (1–4) | 98X1 |
| Ile58→Ala | 3.0 | 41.25 | 335 | heat, v.H. (1–4) | 98X1 |
| Ile78→Ala | 3.0 | 46.95 | 439 | heat, v.H. (1–4) | 98X1 |
| Ile100→Ala | 3.0 | 40.95 | 356 | heat, v.H. (1–4) | 98X1 |
| Val71→Ala | 3.0 | 46.95 | 452 | heat, v.H. (1–4) | 98X1 |
| Val87→Ala | 3.0 | 46.75 | 427 | heat, v.H. (1–4) | 98X1 |
| Val94→Ala | 3.0 | 46.65 | 393 | heat, v.H. (1–4) | 98X1 |
| Val103→Ala | 3.0 | 45.05 | 393 | heat, v.H. (1–4) | 98X1 |
| Val111→Ala | 3.0 | 47.95 | 418 | heat, v.H. (1–4) | 98X1 |
| Val149→Ala | 3.0 | 40.65 | 276 | heat, v.H. (1–4) | 98X1 |
| Met6→Ala | 3.0 | 45.95 | 397 | heat, v.H. (1–4) | 98X1 |
| Met106→Ala | 3.0 | 44.55 | 372 | heat, v.H. (1–4) | 98X1 |
| Phe67→Ala | 3.0 | 45.95 | 423 | heat, v.H. (1–4) | 98X1 |
| Phe104→Ala | 3.0 | 41.95 | 343 | heat, v.H. (1–4) | 98X1 |
| Val7→Ala | 3.0 | 43.55 | 377 | heat, v.H. (1–4) | 98X1 |
| Val33→Ala | 3.0 | 39.35 | 280 | heat, v.H. (1–4) | 98X1 |
| Val66→Ala | 3.0 | 38.25 | 289 | heat, v.H. (1–4) | 98X1 |
| Val84→Ala | 3.0 | 38.25 | 280 | heat, v.H. (1–4) | 98X1 |
| Val91→Ala | 3.0 | 51.95 | 356 | heat, v.H. (1–4) | 98X1 |

Remarks:

(1) all mutant were constructed in the cysteine-free pseudo-wild-type lysozyme, wild type*

(2) measured in 25 mM KCl, 20 mM potassium phosphate, pH 3.0, see also Ref. 92E3

(3) thermodynamic calculations in Ref. 98X1 are based on $\Delta Cp$ = 7.5 kJ/mol/K

(4) the estimated error in $\Delta T$ is about 0.2°C

Lysozyme phage T4, pseudo-wild type (w.t.*), methionine mutants

| Mutant | pH | $T_{trs}$ | $\Delta T$ | $\Delta H$ | $\Delta Cp$ | Appr./Rem. | Ref |
|---|---|---|---|---|---|---|---|
| WT* | 3.0 | 51.7 | 0.0 | 447 | 7.53 | heat (1–3) | 98L5 |
| Met6→Leu | 3.0 | 41.1 | −10.6 | 255 | | heat (2,3) | 98L5 |
| Met102→Leu | 3.0 | 49.3 | − 2.4 | 402 | | heat (2–4) | 98L5 |
| Met106→Leu | 3.0 | 53.4 | 1.7 | 464 | | heat (2,3) | 98L5 |
| Met120→Leu | 3.0 | 53.4 | 1.7 | 485 | | heat (2,3) | 98L5 |
| Met102→Lys | 3.0 | 16.7 | −35 | 50 | | heat (2,3,5) | 98L5 |
| Met106→Lys | 3.0 | 41.2 | −10.5 | 377 | | heat (2,3) | 98L5 |
| Met120→Lys | 3.0 | 46.9 | − 4.8 | 414 | | heat (2,3) | 98L5 |

Remarks:
(1) w.t.* = cysteine-free pseudo-wild type (Cys54→Thr and Cys97→Ala)
(2) buffer: 25 mM potassium chloride, 3 mM phosphoric acid, 17 mM mono-potassium phosphate, pH 3.0
(3) $\Delta Cp$ = 7.53 kJ/mol/K at 47°C was assumed to be identical for all mutants
(4) data from Ref. 92H10
(5) data from Ref. 91D3

Lysozyme phage T4 substituted with methionine or selenomethionine

| Protein | Met residues | Met-$T_{trs}$ | SeMet-$T_{trs}$ | Appr./Rem. | Ref |
|---|---|---|---|---|---|
| w.t.* | 5 | 65.3 | 66.3 | heat (1–4) | 99G2 |
| 8M | 8 | 56.3 | 58.7 | heat (2–4) | 99G2 |
| 10M | 8 | 53.0 | 56.8 | heat (2–4) | 99G2 |
| 12Ma | 8 | 50.7 | 56.5 | heat (2–4) | 99G2 |
| 12Mb | 8 | 53.4 | 57.9 | heat (2–4) | 99G2 |
| 14M | 8 | 47.6 | 55.0 | heat (2–4) | 99G2 |

Remarks:
(1) w.t.* is the cysteine-free pseudo-wild type which contains five Met residues
(2) given is the total number of Met residues, the thermal transition temperature for the methionine variant (SeMet-$T_{trs}$) and for the selenomethionine variant (SeMet-$T_{trs}$)
(3) for the mutant description see Ref. 99G2
(4) buffer: 8.6 mM sodium acetate, 1.4 mM acetic acid, 0.10 M NaCl, pH 5.42

## Maltose-Binding Protein

Maltose-binding protein from *E. coli*, recombinant

| pH | $T_{trs}$ | $\Delta Cp$ | $\Delta H$ | Appr./Rem. | Ref |
|---|---|---|---|---|---|
| 4.5 | 62.3 | | 842 | DSC | 97G1 |
| 5.0 | 64.3 | | 1000 | DSC | 97G1 |
| 6.0 | 64.9 | | 1087 | DSC | 97G1 |
| 6.5 | 63.0 | | 1040 | DSC | 97G1 |
| 7.0 | 64.5 | | 1002 | DSC | 97G1 |
| 7.4 | 63.0 | | 1010 | DSC | 97G1 |
| 8.0 | 61.5 | | 849 | DSC | 97G1 |
| 8.5 | 60.0 | | 726 | DSC | 97G1 |
| 9.0 | 59.0 | | 842 | DSC | 97G1 |
| 9.5 | 57.0 | | 749 | DSC | 97G1 |
| 9.9 | 54.9 | | 651 | DSC | 97G1 |
| 10.4 | 51.1 | | 632 | DSC | 97G1 |
| 2.9–6.6 | | 26 | | DSC (1) | 97G1 |
| 4.5–10.4 | 51–65 | 33±5 | | DSC (2) | 97G1 |

Remarks:
(1) average $\Delta Cp$ from single calorimetric recordings
(2) $\Delta Cp$ from $\Delta H$ versus $T_{trs}$, the value is regarded as the more reliable one

Maltose-binding protein from *E. coli*, in the presence and absence of maltose

| Maltose conc. | pH | $T_{trs}$ | $\Delta Cp$ | $\Delta H$ | Appr./Rem. | Ref |
|---|---|---|---|---|---|---|
| 0 mM | 8.3 | 62.6 | | 862 | DSC (1) | 97N13 |
| 50 mM | 8.3 | 72.1 | | 946 | DSC (1) | 97N13 |
| 0 mM | 8.2 | 62.3 | | 816 | DSC (1) | 97N13 |
| 50 mM | 8.2 | 70.5 | | 1000 | DSC (1) | 97N13 |
| 0 mM | 3.4 | 56.1 | | 803 | DSC (1) | 97N13 |
| 50 mM | 3.4 | 68.4 | | 1176 | DSC (1) | 97N13 |
| 0 mM | 3.1 | 52.6 | | 720 | DSC (1) | 97N13 |
| 50 mM | 3.1 | 65.0 | | 879 | DSC (1) | 97N13 |
| 0 mM | 3.0 | 50.7 | | 674 | DSC (1) | 97N13 |
| 50 mM | 3.0 | 62.5 | | 841 | DSC (1) | 97N13 |
| 0 mM | 2.8 | 45.8 | | 586 | DSC (1) | 97N13 |
| 50 mM | 2.8 | 59.3 | | 912 | DSC (1) | 97N13 |
| 0–50 mM | 2.5–8.3 | 40–72 | 27.2±3.4 | | DSC (2) | 97N13 |
| 0 mM | 2.5–3.4 | 70 | 25.5 | | DSC (3) | 97N13 |

Remarks:
(1) error in $T_{trs}$ ±0.2°C, and in $\Delta H^{cal}$ ±10%
(2) $\Delta Cp$ from $\Delta H$ versus $T_{trs}$
(3) $\Delta Cp$ from extrapolated heat capacity between unfolded and folded states

Maltose binding protein (MBP), native and molten-globule state, data at standard temperature

| Protein | pH | $T_{ref}$ | $\Delta Cp$ | $\Delta H$ | Appr./Rem. | Ref |
|---|---|---|---|---|---|---|
| MBP | 3.0 | 25 | 24.7±3.3 | 184±33 | urea (1) | 99S9 |
| | 3.0 | 25 | 25.9±2.1 | | urea (2) | 99S9 |
| MBP | 7.1 | 25 | 34.7±2.9 | 50±25 | urea (1) | 99S9 |
| | 7.1 | 25 | 36.4±0.8 | | urea (2) | 99S9 |
| MBP | 7.1 | 25 | 36.0±2.1 | 151±17 | GuHCl (1) | 99S9 |
| | 7.1 | 25 | 35.9±0.1 | | GuHCl (2) | 99S9 |

Remarks:
(1) data from $\Delta G$ versus T, see Table 1
(2) data from $\Delta Cp$ versus denaturant conc.

## Mannitol Permease

Mannitol permease of *E. coli*, enzyme II and isolated domains

| Protein | Transition | pH | $T_{trs}$ | $\Delta H$ | Appr./Rem. | Ref |
|---|---|---|---|---|---|---|
| $EII^{mtl}$ | reconstituted and measured in the presence of 100 µM mannitol | | | | | |
| | trans. (1) | 7.5 | 61.3±2.9 | 274±101 | DSC (1–4) | 98M14 |
| | trans. (2) | 7.5 | 62.4±0.3 | 474±67 | DSC (1–4) | 98M14 |
| | trans. (3) | 7.5 | 81.0±1.6 | 802±80 | DSC (1–4) | 98M14 |
| $EII^{mtl}$ | reconstituted in the presence of mannitol, measured without | | | | | |
| | trans. (1) | 7.5 | 62.0±1.2 | 217±38 | DSC (1–3,5) | 98M14 |
| | trans. (2) | 7.5 | 62.8±0.4 | 420±12 | DSC (1–3,5) | 98M14 |
| | trans. (3) | 7.5 | 73.0±0.3 | 554±37 | DSC (1–3,5) | 98M14 |
| $EII^{mtl}$ | reconstituted in the absence mannitol | | | | | |
| | trans. (1) | 7.5 | 62.5±1.9 | 312±52 | DSC (1–3,6) | 98M14 |
| | trans. (2) | 7.5 | 63.1±0.6 | 552±18 | DSC (1–3,6) | 98M14 |
| | trans. (3) | 7.5 | 71.2±0.8 | 705±58 | DSC (1–3,6) | 98M14 |

Mannitol permease of *E. coli*, enzyme II and isolated domains (continued)

| Protein | Transition | pH | $T_{trs}$ | $\Delta H$ | Appr./Rem. | Ref |
|---|---|---|---|---|---|---|
| IIA$^{mtl}$ | trans. (2) | 7.5 | 65.1 | 396 | DSC (1) | 98M14 |
| | | 7.6 | 64.7±0.3 | 340±8 | DSC (8) | 96M13 |
| IIB$^{mtl}$ | trans. (1) | 7.5 | 65.3 | 220 | DSC (1) | 98M14 |
| | | 7.6 | 62.7±0.04 | 268±4 | DSC (8) | 96M13 |
| IIBA$^{mtl}$ | trans. (1) | 7.5 | 59.1 | 231 | DSC (1,2) | 98M14 |
| | trans. (2) | 7.5 | 64 | 473 | DSC (1,2) | 98M14 |
| IIBA$^{mtl}$ | trans. (1) | 7.6 | 59.3 | 156±12 | DSC (8) | 96M13 |
| | trans. (2) | 7.6 | 64.3 | 378±15 | DSC (8) | 96M13 |
| IICB$^{mtl}$ | trans. (3) | 7.5 | 80.6 | 849 | DSC (1,7) | 98M14 |
| His-IIC$^{mtl}$ | trans. (3) | 7.5 | 76.2 | 464 | DSC (1,7) | 98M14 |
| Tryp-IIC$^{mtl}$ | trans. (3) | 7.5 | 72.0 | 383 | DSC (1,7) | 98M14 |

Explanation:

| | |
|---|---|
| EII$^{mtl}$ | mannitol-specific enzyme II |
| IIA$^{mtl}$ | A domain of the mannitol-specific enzyme II |
| IIB$^{mtl}$ | B domain of the mannitol-specific enzyme II |
| IIBA$^{mtl}$ | BA domain of the mannitol-specific enzyme II |
| IICB$^{mtl}$ | CB domain of the mannitol-specific enzyme II |
| His-IIC$^{mtl}$ | C domain with a His tag of 6 histidines |
| Tryp-IIC$^{mtl}$ | C domain obtained by trypsin treatment of the wild-type enzyme |

Remarks:
(1) measured in 50 mM sodium phosphate, pH 7.5, 10 mM 2-mercaptoethanol, 1 mM NaN₃
(2) results of deconvolution
(3) data from Table 1 in Ref. 98M14, mean of results obtained using various models
(4) reconstituted and measured in the presence of 100 μM mannitol
(5) reconstituted in the presence of 100 μM mannitol, after which the mannitol was removed
(6) reconstituted in the absence of mannitol
(7) transition modeled by an irreversible two-state transition
(8) measured in 50 mM HEPES, pH 7.6, 10 mM 2-mercaptoethanol

## Meromyosin

Light meromyosin from carp, acclimated to various temperatures

| Form/Transition | | pH | $T_{trs}$ | $\Delta Cp$ | $\Delta H$ | Approach/Remarks | | Ref |
|---|---|---|---|---|---|---|---|---|
| 10°C-acclimated carp, type 10-LMM: | | | | | | | | |
| | trans. (1) | 8.0 | 32.5 | | 1125 | DSC | (1–3) | 97N2 |
| | trans. (2) | 8.0 | 39.5 | | 218 | DSC | (1–3) | 97N2 |
| 20°C-acclimated carp, type 20-LMM: | | | | | | | | |
| | trans. (1) | 8.0 | 34.5 | | 628 | DSC | (1,2,4) | 97N2 |
| | trans. (2) | 8.0 | 40.2 | | 84 | DSC | (1,2,4) | 97N2 |
| | trans. (3) | 8.0 | 46.9 | | 42 | DSC | (1,2,4) | 97N2 |
| 30°C-acclimated carp, type 30-LMM: | | | | | | | | |
| | trans. (1) | 8.0 | 39.2 | | 967 | DSC | (1,2,5) | 97N2 |
| | trans. (2) | 8.0 | 47.3 | | 163 | DSC | (1,2,5) | 97N2 |
| 30°C-acclimated carp, type 30-LMM': | | | | | | | | |
| | trans. (1) | 8.0 | 34.4 | | 490 | DSC | (1,2,6) | 97N2 |
| | trans. (2) | 8.0 | 39.5 | | 515 | DSC | (1,2,6) | 97N2 |
| | trans. (3) | 8.0 | 47.5 | | 117 | DSC | (1,2,6) | 97N2 |

Light meromyosin from carp, acclimated to various temperatures (continued)

| Form/Transition | pH | $T_{trs}$ | $\Delta Cp$ | $\Delta H$ | Approach/Remarks | | Ref |
|---|---|---|---|---|---|---|---|
| reference sample from rabbit: | | | | | | | |
| trans. (1) | 8.0 | 42.5 | | 770 | DSC | (1,2) | 97N2 |
| trans. (2) | 8.0 | 53.0 | | 318 | DSC | (1,2) | 97N2 |

Remarks:

(1) buffer: 50 mM Tris, pH 8.0, containing 0.6 M KCl, 5 mM MgCl$_2$, 0.1 mM DTT

(2) the transitions were resolved by deconvolution

(3) N-terminal sequence with Ala(2) and Ser(18)

(4) N-terminal sequence with Ala(2) and Ala(18)

(5) N-terminal sequence with Thr(2) and Ala(18)

(6) LMM' contains a second component with Ala(2) and Ala(18) which is predominant in 20-LMM

Light meromyosin (LMM), acclimation temperature-associated isoforms of carp light meromyosin, recombinant

| Protein | Transition | pH | $T_{trs}$ | $\Delta H$ | Appr./Rem. | | Ref |
|---|---|---|---|---|---|---|---|
| 10°C type | trans. (1) | 8.0 | 35.1 | 477.8 | DSC | (1) | 98K1 |
| | trans. (2) | 8.0 | 53.4 | 86.6 | DSC | (1) | 98K1 |
| intermediate type | | | | | | | |
| | trans. (1) | 8.0 | 34.9 | 122.6 | DSC | (1) | 98K1 |
| | trans. (2) | 8.0 | 40.6 | 167.8 | DSC | (1) | 98K1 |
| | trans. (3) | 8.0 | 44.9 | 11.3 | DSC | (1) | 98K1 |
| 30°C type | trans. (1) | 8.0 | 35.4 | 365.3 | DSC | (1) | 98K1 |
| | trans. (2) | 8.0 | 39.5 | 246.9 | DSC | (1) | 98K1 |
| 10N-30C chimera | | | | | | | |
| | trans. (1) | 8.0 | 36.4 | 285.8 | DSC | (1,2) | 98K1 |
| | trans. (2) | 8.0 | 39.5 | 401.2 | DSC | (1,2) | 98K1 |
| 30N-10C chimera | | | | | | | |
| | trans. (1) | 8.0 | 35.0 | 389.1 | DSC | (1,3) | 98K1 |
| | trans. (2) | 8.0 | 38.8 | 88.7 | DSC | (1,3) | 98K1 |

Remarks:

(1) buffer: 50 mM Tris-HCl (pH 8.0), 0.6 M KCl, 5 mM MgCl$_2$, 1 mM DTT

(2) 10N-30C chimera = LMM composed of th N-terminal half of the 10°C type and the C-terminal half of the 30°C type

(3) 30N-10C chimera = LMM composed of th N-terminal half of the 30°C type and the C-terminal half of the 10°C type

## Methionine Aminopeptidase

Methionine aminopeptidase from hyperthermophile *Pyrococcus furiosus*

| pH | $T_{trs}$ | $\Delta H$ | Approach/Remarks | | Ref |
|---|---|---|---|---|---|
| 10.2 | 106.2 | | DSC | (1,2) | 98O2 |
| 4.1 | 100 | 1730 | DSC | (1–4) | 98O2 |
| 2.8 | 73 | 970 | DSC | (1–4) | 98O2 |

Remarks:

(1) measured in 20 mM glycine-HCl buffer

(2) the thermal transition is irreversible

(3) the data were taken from Figs. 3 and 4 in Ref. 98O2

(4) $\Delta cp$ from from $\Delta H^{v.H.}$ versus $T_{trs}$ amounts to 3.43 J/g/K

## MHC, Major Histocompatibility Complex Protein

Recombinant human class II major histocompatibility complex protein HLA-DR1 (DR1), empty and peptide loaded DR1

| Protein | pH | $T_{trs}$ | $\Delta Cp$ | $\Delta H$ | Approach/Remarks | | Ref |
|---|---|---|---|---|---|---|---|
| DR1 | 7.0 | 67 | 3.3 | 146 | heat | (1) | 99F9 |
| DR1-Ha | 7.0 | 81 | 8.4 | 377 | heat | (2) | 99F9 |

Remarks:
(1) transition monitored by far-UV CD at 204 nm
(2) Ha = peptide 306–318 from influenza haemagglutinin

Human major histocompatibility complex protein HLA-DR1 (DR1) in the presence of peptides

| Complex | pH | $T_{trs}$ | $\Delta Cp$ | $\Delta H$ | Appr./Rem. | Ref |
|---|---|---|---|---|---|---|
| DR1 alone | 7.0 | 70 | 0.8±0.8 | 184±4 | heat (1,2) | 99Z2 |
| DR1-Ha | 7.0 | 85 | 6.3±0.4 | 310±13 | heat (1,2) | 99Z2 |
| DR1-Clip | 7.0 | ≥90 | 7.5±1.3 | 351±29 | heat (1,2) | 99Z2 |
| DR1-A2 | 7.0 | ≥90 | 16.3±1.3 | 602±29 | heat (1,2) | 99Z2 |
| DR1-Yak | 7.0 | 89 | 7.1±0.4 | 381±13 | heat (1,2) | 99Z2 |
| DR1-Min4 | 7.0 | 73 | 9.6±1.3 | 402±25 | heat (1,2) | 99Z2 |

Remarks:
(1) measured in 10 mM phosphate buffer, pH 7.0
(2) transition monitored by far-UV CD at 204 or 222 nm
(3) for the peptide sequences see Ref. 99Z2

## Myoglobin

Myoglobin from horse heart, recombinant, wild type and mutant, holo- and apo-protein

| Mutant | Form | pH | $T_{trs}$ | $\Delta H$ | Approach/Remarks | Ref |
|---|---|---|---|---|---|---|
| wild type | holo | 7.0 | 81.3±1 | | heat, CD | 97M4 |
| wild type | holo | 8.0 | 78.2±1 | | heat, CD | 97M4 |
| wild type | apo | 7.0 | 60.4±1 | | heat, CD | 97M4 |
| wild type | apo | 8.0 | 57.9±1 | | heat, CD | 97M4 |
| Leu104→Asn | holo | 7.0 | 71.8±1 | | heat, CD | 97M4 |
| Leu104→Asn | holo | 8.0 | 67.8±1 | | heat, CD | 97M4 |
| Leu104→Asn | apo | 7.0 | 52.0±1 | | heat, CD | 97M4 |
| Leu104→Asn | apo | 8.0 | 51.3±1 | | heat, CD | 97M4 |

**Myosin,** see also Meromyosin

## Myosin

Myosin II rod from *Acanthamoeba castellanii*

| Mutant | pH | Rod conc. | KCl conc. | $T_{trs}$ | $\Delta H$ | Remarks | Ref |
|---|---|---|---|---|---|---|---|
| wild type | | | | | | | |
| | 7.5 | 0.58 μM | 0.6 M | 39.2 | 1841 | (1,3) | 97Z7 |
| | 7.5 | 5.75 μM | 0.6 M | 39.9 | 1590 | (1,3) | 97Z7 |
| | 7.5 | 5.75 μM | 0.6 M | 40.5 | 2678 | (1,4) | 97Z7 |
| | 7.5 | 5.75 μM | 0.6 M | 40.5 | 2761 | (1,6) | 97Z7 |

Myosin II rod from *Acanthamoeba castellanii* (continued)

| Mutant | pH | Rod conc. | KCl conc. | $T_{trs}$ | $\Delta H$ | Remarks | Ref |
|---|---|---|---|---|---|---|---|
| | 7.5 | 1.1 µM | 0.6 M | 40.5 | 2218 | (2,6) | 97Z7 |
| | 7.5 | 1.9 µM | 2.2 M | 41.9 | 1590 | (1,5,7) | 97Z7 |
| Pro398→Ala | | | | | | | |
| | 7.5 | 4.0 µM | 0.6 M | 39.9 | 1715 | (1,3) | 97Z7 |
| | 7.5 | 4.0 µM | 0.6 M | 40.4 | 2134 | (1,5) | 97Z7 |
| | 7.5 | 4.0 µM | 0.6 M | 40.4 | 2803 | (1,6) | 97Z7 |
| | 7.5 | 1.2 µM | 0.6 M | 40.7 | 2259 | (2,6) | 97Z7 |
| | 7.5 | 1.3 µM | 2.2 M | 42.1 | 1674 | (1,3) | 97Z7 |
| hinge region, residues 384–408 deleted | | | | | | | |
| | 7.5 | 1.6 µM | 0.6 M | 41.3 | 1548 | (1,3) | 97Z7 |
| | 7.5 | 1.6 µM | 0.6 M | 41.9 | 1883 | (1,5) | 97Z7 |
| | 7.5 | 1.6 µM | 0.6 M | 41.7 | 2259 | (1,6) | 97Z7 |
| | 7.5 | 1.0 µM | 0.6 M | 42.0 | 1966 | (2,6) | 97Z7 |
| | 7.5 | 0.5 µM | 2.2 M | 43.6 | 1213 | (1,6) | 97Z7 |

Remarks:
(1) buffer: 10 mM imidazole, KCl as indicated, pH 7.5
(2) buffer: 20 mM potassium phosphate, KCl as indicated, pH 7.5
(3) heat, CD (equilibrium), two-state fit
(4) heat, CD (scan rate 10°C/h), two-state fit
(5) heat, CD (scan rate 7°C/h), two-state fit
(6) DSC (scan rate 7.5°C/h), calorimetric heat
(7) for a comparison of myosin from *Acanthamoeba castellanii* with other myosins see Ref. 96Z2

Motor domain fragments of *Dictyostelium discoideum* myosin II compared with skeletal muscle myosin subfragment 1

| Fragment or Complex | pH | $T_{trs}$ | $\Delta H$ | Approach/Remarks | Ref |
|---|---|---|---|---|---|
| fragment M761: | | | | | |
| M761 alone | 7.3 | 45.6 | 1417 | DSC (1,2) | 98L3 |
| M761.ADP | 7.3 | 49.1 | 1400 | DSC (1,2) | 98L3 |
| M761.ADP.BeF$_x$ | 7.3 | 52.7 | 1572 | DSC (1–3) | 98L3 |
| M761.ADP.AlF$_{4-}$ | 7.3 | 54.9 | 1772 | DSC (1,2,4) | 98L3 |
| M761.ADP.V$_i$ | 7.3 | 55.4 | 1588 | DSC (1,2,5) | 98L3 |
| fragment M754: | | | | | |
| M754 alone | 7.3 | 41.7 | 677 | DSC (1,2) | 98L3 |
| M754.ADP | 7.3 | 43.9 | 740 | DSC (1,2) | 98L3 |
| M754.ADP.BeF$_x$ | 7.3 | 47.1 | 949 | DSC (1–3) | 98L3 |
| skeletal S1: | | | | | |
| S1 alone | 7.3 | 49.5 | 1505 | DSC (1,2) | 98L3 |
| S1.ADP | 7.3 | 50.6 | 1618 | DSC (1,2) | 98L3 |
| S1.ADP.BeF$_x$ | 7.3 | 57.3 | 1726 | DSC (1–3) | 98L3 |
| S1.ADP.AlF$_{4-}$ | 7.3 | 58.4 | 2002 | DSC (1,2,4) | 98L3 |
| S1.ADP.V$_i$ | 7.3 | 58.9 | 1805 | DSC (1,2,5) | 98L3 |

Remarks:
(1) M761 and M754 are myosin head fragments of *Dictyostelium discoideum* myosin II containing 761 and 754 amino acid residues, respectively, S1 is skeletal muscle myosin subfragment 1
(2) measured in 30 mM HEPES, pH 7.3, 1 mM MgCl$_2$
(3) BeF$_x$ beryllium fluoride
(4) AlF$_{4-}$ aluminium fluoride
(5) V$_i$ orthovanadate

Motor domain fragment M761 of *Dictyostelium discoideum* myosin II compared with skeletal muscle myosin subfragment 1, results of deconvolution

| Fragment/Transition | | pH | $T_{trs}$ | $\Delta H$ | Appr./Rem. | Ref |
|---|---|---|---|---|---|---|
| S1 alone | (1–3) | 7.3 | 44/49/51 | 578/639/295 | DSC | 98L3 |
| S1.ADP | (1–3) | 7.3 | 44.7/48.6/51.1 | 342/403/871 | DSC | 98L3 |
| M761 alone | (1,2) | 7.3 | 43.5/45.8 | 636/795 | DSC | 98L3 |
| M761.ADP | (1,2) | 7.3 | 46.5/49.1 | 588/822 | DSC | 98L3 |

Myosin subfragment 1 (S1) and its interaction with F-actin from rabbit skeletal muscle

| Protein | pH | $T_{trs}$ | $\Delta H$ | Approach/Remarks | | Ref |
|---|---|---|---|---|---|---|
| unbound S1 | 7.3 | 50.1 | 1060±80 | DSC | (1) | 96N3 |
| actin-bound S1 | 7.3 | 55.1 | 1220±100 | DSC | (1) | 96N3 |

Remark:
(1) buffer: 15 mM HEPES/KOH, pH 7.3, 0.2 mM ADP, 2 mM $MgCl_2$

Turkey gizzard smooth muscle myosin subfragment 1 (S1) and heavy meromyosin (HMM) in the absence and presence of nucleotides

| Fragment or Complex | pH | $T_{trs}$ | $\Delta H$ | Approach/Remarks | | Ref |
|---|---|---|---|---|---|---|
| S1 and dephosphorylated RLC: | | | | | | |
| S1 | 7.3 | 51.5 | 639 | DSC | (1) | 98P6 |
| S1-ADP | 7.3 | 50.6 | 651 | DSC | (1) | 98P6 |
| S1-ADP-$V_i$ | 7.3 | 59.6 | 886 | DSC | (1,2) | 98P6 |
| S1-ADP-$BeF_x$ | 7.3 | 55.8 | 913 | DSC | (1,3) | 98P6 |
| S1-ADP-$AlF_{4-}$ | 7.3 | 59.6 | 959 | DSC | (1,4) | 98P6 |
| S1 and thiophosphorylated RLC: | | | | | | |
| S1($P_s$) | 7.3 | 50.4 | 658 | DSC | (1) | 98P6 |
| S1($P_s$)-ADP | 7.3 | 50.6 | 700 | DSC | (1) | 98P6 |
| S1($P_s$)-ADP-$V_i$ | 7.3 | 57.9 | 955 | DSC | (1,2) | 98P6 |
| S1($P_s$)-ADP-$BeF_x$ | 7.3 | 56.9 | 1056 | DSC | (1,3) | 98P6 |
| S1($P_s$)-ADP-$AlF_{4-}$ | 7.3 | 58.9 | 909 | DSC | (1,4) | 98P6 |
| HMM and dephosphorylated RLC: | | | | | | |
| HMM | 7.3 | 50.0 | 2277 | DSC | (1) | 98P6 |
| HMM-ADP | 7.3 | 50.0 | 2557 | DSC | (1) | 98P6 |
| HMM-ADP-$V_i$ | 7.3 | 56.4 | 2787 | DSC | (1,2) | 98P6 |
| HMM-ADP-$BeF_x$ | 7.3 | 53.6 | 2350 | DSC | (1,3) | 98P6 |
| HMM-ADP-$AlF_{4-}$ | 7.3 | 58.3 | 2435 | DSC | (1,4) | 98P6 |
| HMM and phosphorylated RLC: | | | | | | |
| HMM(P) | 7.3 | 49.2 | 2378 | DSC | (1) | 98P6 |
| HMM(P)-ADP | 7.3 | 49.3 | 2340 | DSC | (1) | 98P6 |
| HMM(P)-ADP-$V_i$ | 7.3 | 56.5 | 1975 | DSC | (1,2) | 98P6 |
| HMM(P)-ADP-$BeF_x$ | 7.3 | 52.9 | 1631 | DSC | (1,3) | 98P6 |

Turkey gizzard smooth muscle myosin subfragment 1 (S1) and heavy meromyosin (HMM) in the absence and presence of nucleotides (continued)

| Fragment or Complex | pH | $T_{trs}$ | $\Delta H$ | Approach/Remarks | | Ref |
|---|---|---|---|---|---|---|
| HMM and thiophosphorylated RLC: | | | | | | |
| HMM($P_s$) | 7.3 | 50.6 | 1849 | DSC | (1) | 98P6 |
| HMM($P_s$)-ADP | 7.3 | 49.7 | 2355 | DSC | (1) | 98P6 |
| HMM($P_s$)-ADP-$V_i$ | 7.3 | 56.5 | 2090 | DSC | (1,2) | 98P6 |
| HMM($P_s$)-ADP-BeF$_x$ | 7.3 | 53.4 | 1781 | DSC | (1,3) | 98P6 |
| HMM($P_s$)-ADP-AlF$_{4-}$ | 7.3 | 57.3 | 2163 | DSC | (1,4) | 98P6 |

Explanations:

RLC        – regulatory subunit
S1($P_s$)    – S1 and thiophosphorylated RLC
HMM(P) – HMM and phosphorylated RLC
HMM($P_s$)– HMM and thiophosphorylated RLC

Remarks:

(1) measured in 30 mM HEPES, pH 7.3, 1 mM $MgCl_2$

(2) $V_i$ orthovanadate

(3) BeF$_x$ beryllium fluoride

(4) AlF$_{4-}$ aluminium fluoride

## Nuclease from *Staphylococcus aureus* (Staphylococcal nuclease)

Staphylococcal nuclease (SN), wild type, GuHCl-induced unfolding

| Protein | pH | $T_{trs}$ | $\Delta Cp$ | $\Delta H$ | Appr./Rem. | | Ref |
|---|---|---|---|---|---|---|---|
| SN | 7.0 | 25.0 | | 100.8±4.2 | ITC | (1–3) | 99Y1 |
| | 7.0 | 25.0 | 7.5±4.2 | 110.5±11.7 | GuHCl | (1,3,4) | 99Y1 |

Remarks:

(1) Ref. 99Y1 is aimed to determine the ratio $\Delta H^{v.H.}/\Delta H^{cal}$ for SN in GuHCl

(2) data from heat of mixing of SN with GuHCl measured by ITC, using linear extrapolation to zero denaturant conc.

(3) measured in 25 mM phosphate, 0.1 M NaCl, pH 7.0

(4) data from the temperature dependence of GuHCl-induced unfolding of SN, see also Table 1

Staphylococcal nuclease, wild type and unstable mutant

| Mutant | pH | $T_{trs}$ | $\Delta Cp$ | $\Delta H$ | Approach/Remarks | | Ref |
|---|---|---|---|---|---|---|---|
| wild type | 7.0 | 51.1 | 9.29 | 305 | heat | (1,2) | 97E2 |
| wild type | 7.0 | 51.5 | 9.41 | 313 | heat | (1,3) | 97E2 |
| wild type | 7.0 | 51.3 | 9.37 | 312 | heat | (1,4) | 97E2 |
| wild type | 7.0 | 51.3 | 9.50 | 310 | heat | (1,5) | 97E2 |
| NCA (7) | 7.0 | 30.8 | 7.24 | 169 | heat | (1,2) | 97E2 |
| NCA (7) | 7.0 | 31.0 | 7.95 | 180 | heat | (1,3) | 97E2 |
| NCA (7) | 7.0 | 31.2 | 7.91 | 179 | heat | (1,4) | 97E2 |
| NCA (7) | 7.0 | 31.0 | 7.57 | 173 | heat | (1,5) | 97E2 |
| NCA-s. (8) | 7.0 | 25.9 | 5.61 | 116 | heat | (1,2) | 97E2 |
| NCA-s. (8) | 7.0 | 25.6 | 4.73 | 113 | heat | (1,3) | 97E2 |
| NCA-s. (8) | 7.0 | 25.7 | 4.10 | 118 | heat | (1,4) | 97E2 |
| NCA-s. (8) | 7.0 | 25.9 | 5.61 | 116 | heat | (1,5) | 97E2 |

Remarks:

(1) thermally induced transitions were simultaneously monitored by fluorescence and CD at various wavelengths, buffer: 0.01 M Tris-HCl, 0.1 M NaCl, pH 7.0

(2) transition monitored by fluorescence, excitation and emission wavelengths of 295 and 340 nm

(3) transition monitored by CD at 222 nm

(4) transition monitored by CD at 235 nm

(5) global fit of the three curves, two-state fit

(6) for comparison, previous thermodynamic data are available in Refs. 85C1, 88G5, 88S8, 91E1, 93T1, and 94C1, see also Ref. 97E2

(7) NCA = hybrid of nuclease having the hexapeptide (Ser-Ser-Asn-Gly-Ser-Pro) at positions 27–31 (a type I β-turn)

(8) NCA-s. = hybrid nuclease with additional substitution Ser28→Gly

## Odorant Binding Protein

Odorant binding protein from pig nasal mucosa (pOBP) in the presence of ligands

| Ligand/Conc. | | pH | $T_{trs}$ | $\Delta Cp$ | $\Delta H$ | Appr./Rem. | Ref |
|---|---|---|---|---|---|---|---|
| none | | 6.6 | 69.23±0.29 | 3.98±1.39 | 391.1±17.2 | DSC (1,2) | 99B20 |
| IBMP | 0.017 mM | 6.6 | 73.72 | 6.82 | 579.6 | DSC (2) | 99B20 |
| IBMP | 0.125 mM | 6.6 | 74.85 | 6.62 | 593.6 | DSC (2) | 99B20 |
| DMO | 0.139 mM | 6.6 | 77.95 | 6.72 | 581.6 | DSC (2) | 99B20 |
| ALDSon | 0.094 mM | 6.6 | 68.58 | 6.25 | 443.9 | DSC (2) | 99B20 |
| ALDSan | 0.091 mM | 6.6 | 68.83 | | 392.9 | DSC (2) | 99B20 |
| ALDSen | 0.066 mM | 6.6 | 68.84 | 5.68 | 393.7 | DSC (2) | 99B20 |
| LPDC | 0.126 mM | 6.6 | 70.07 | 5.23 | 484.1 | DSC (2) | 99B20 |
| CAP | 0.100 mM | 6.6 | 68.88 | 6.93 | 454.6 | DSC (2) | 99B20 |

Ligands:

| IBMP | 2-isobutyl-3-methoxypyrazine |
|---|---|
| DMO | 3,7-dimethyloctan-1-ol |
| ALDSon | Aldosterone |
| ALDSan | 5α-Androstan-17β-ol-3-one |
| ALDSen | 5-Androsten-3β-ol-17-one |
| LPDC | Lysophosphatidylcholine ($C_{16}$) |
| CAP | Capsaicin |

Remarks:

(1) mean of eight calorimetric scans at varying protein conc.

(2) measured in 10 mM phosphate, 1 mM EDTA, pH 6.6

## Oncomodulin

Rat oncomodulin mutants

| Protein | pH | $T_{trs}$ | $\Delta H$ | Approach/Remarks | | Ref |
|---|---|---|---|---|---|---|
| Ca$^{2+}$-free form: | | | | | | |
| Tyr57→Trp | 6.5 | 55.9±0.5 | 117±18 | heat | (1,2) | 98Z6 |
| Tyr65→Trp | 6.5 | 52.8±0.5 | 81±8 | heat | (1,2) | 98Z6 |
| Phe102→Trp | 6.5 | 50.9±0.5 | 226±26 | heat | (1,2) | 98Z6 |
| CDOM33 | 6.5 | 70.1±0.5 | 121±11 | heat | (1–3) | 98Z6 |
| Ca$^{2+}$-loaded form: | | | | | | |
| Tyr57→Trp | 6.5 | 70.0±0.5 | 119±13 | heat | (1,2) | 98Z6 |
| Tyr65→Trp | 6.5 | 58.0±0.4 | 100±11 | heat | (1,2) | 98Z6 |
| Phe102→Trp | 6.5 | 73.8±0.3 | 375±38 | heat | (1,2) | 98Z6 |
| CDOM33 | 6.5 | 79.1±0.3 | 322±33 | heat | (1–3) | 98Z6 |
| Tb$^{3+}$-loaded form: | | | | | | |
| Tyr57→Trp | 6.5 | 81.7±0.4 | 282±31 | heat | (1,4) | 98Z6 |
| Tyr65→Trp | 6.5 | 61.6±0.4 | 258±28 | heat | (1,2) | 98Z6 |
| Phe102→Trp | 6.5 | 86.4±0.4 | 311±35 | heat | (1,2) | 98Z6 |
| CDOM33 | 6.5 | 87.7±0.4 | 381±40 | heat | (1,3,4) | 98Z6 |

Remarks:

(1) buffer: 10 mM PIPES, 100 mM KCl, pH 6.5, and EGTA for apo proteins

(2) transition monitored by fluorescence intensity changes

(3) CDOM33 represents a CD loop prepared by insertion of a 12 amino acid sequence which has a significantly higher affinity for Ca$^{2+}$

(4) transition monitored by Tb$^{3+}$ luminescence

Recombinant rat oncomodulin and mutants (rOM), apo-, Mg$^{2+}$-, and Ca$^{2+}$-bound proteins

| Protein | pH | $T_{trs}$ | $\Delta H$ | Approach/Remarks | | Ref |
|---|---|---|---|---|---|---|
| apo-proteins: | | | | | | |
| rOM | 7.4 | 51.8 | 276±17 | DSC | (1) | 96H8 |
| Ser55→Asp | 7.4 | 48.9 | 213±13 | DSC | (1) | 96H8 |
| Gly98→Asp | 7.4 | 51.1 | 243±17 | DSC | (1) | 96H8 |
| double mutant (Ser55→Asp and Gly98→Asp) | | | | | | |
| | 7.4 | 47.9 | 205±13 | DSC | (1) | 96H8 |
| Ca$^{2+}$-bound proteins: | | | | | | |
| rOM | 7.4 | 92.8 | 452±21 | DSC | (2) | 96H8 |
| Ser55→Asp | 7.4 | 95.1 | 460±21 | DSC | (2) | 96H8 |
| Gly98→Asp | 7.4 | 95.3 | 502±25 | DSC | (2) | 96H8 |
| double mutant (Ser55→Asp and Gly98→Asp) | | | | | | |
| | 7.4 | 97.4 | 519±25 | DSC | (2) | 96H8 |
| Mg$^{2+}$-bound proteins | | | | | | |
| rOM | 7.4 | 68.5 | 368±17 | DSC | (3) | 96H8 |
| Ser55→Asp | 7.4 | 79.0 | 389±21 | DSC | (3) | 96H8 |
| Gly98→Asp | 7.4 | 69.0 | 381±21 | DSC | (3) | 96H8 |
| double mutant (Ser55→Asp and Gly98→Asp) | | | | | | |
| | 7.4 | 79.4 | 381±21 | DSC | (3) | 96H8 |

Remarks:

(1) buffer: 25 mM HEPES, 0.15 M NaCl, 5 mM EDTA

(2) buffer: 25 mM HEPES, 0.15 M NaCl, 5 mM CaCl$_2$

(3) buffer: 25 mM HEPES, 0.15 M NaCl, 20 mM MgCl$_2$, 1.0 mM EGTA

## Oncoprotein

BTB/POZ domain from the PLZF oncoprotein

| Protein | pH | $T_{trs}$ | $\Delta H$ | Approach/Remarks | | Ref |
|---|---|---|---|---|---|---|
| BTB/POZ domain | 8.5 | 70.4 | 201 | DSC | (1) | 97L8 |

Remark:
(1) measured in 40 mM boric acid, pH 8.5, 100 mM NaCl

## Ovomucoid

Chicken ovomucoid first domain (OMCHII), stabilization by glycosylation

| Form | pH | $T_{trs}$ | $\Delta Cp$ | $\Delta H$ | Appr./Rem. | | Ref |
|---|---|---|---|---|---|---|---|
| naturally occurring glycosylated OMCHI1 (gOMCHI1) | | | | | | | |
| | 5.50 | 69.3 | | 159±1 | DSC | (1) | 97D1 |
| | 4.75 | 69.8 | | 162±1 | DSC | (1) | 97D1 |
| | 4.50 | 68.9 | | 155±1 | DSC | (1) | 97D1 |
| | 4.30 | 67.9 | | 148±1 | DSC | (1) | 97D1 |
| | 4.00 | 64.7 | | 147±2 | DSC | (1) | 97D1 |
| | 3.60 | 61.3 | | 131±2 | DSC | (1) | 97D1 |
| | 3.30 | 59.0 | | 135±1 | DSC | (1) | 97D1 |
| | 3.3–5.5 | 59–69 | 2.59±0.75 | | DSC | (2) | 97D1 |
| recombinant nonglycosylated OMCHI1 (rOMCHI1) | | | | | | | |
| | 5.50 | 67.4 | | 164±2 | DSC | (1) | 97D1 |
| | 4.75 | 66.9 | | 159±2 | DSC | (1) | 97D1 |
| | 4.00 | 61.6 | | 148±2 | DSC | (1) | 97D1 |
| | 4.00 | 62.4 | | 146±1 | DSC | (1) | 97D1 |
| | 3.75 | 60.8 | | 146±2 | DSC | (1) | 97D1 |
| | 3.50 | 56.7 | | 138±1 | DSC | (1) | 97D1 |
| | 3.50 | 57.7 | | 137±1 | DSC | (1) | 97D1 |
| | 3.25 | 54.2 | | 131±1 | DSC | (1) | 97D1 |
| | 3.3–5.5 | 59–67 | 2.44±0.17 | | DSC | (2) | 97D1 |

Remarks:
(1) buffer: 10 mM potassium acetate, 10 mM potassium phosphate
(2) $\Delta Cp$ from $\Delta H$ versus $T_{trs}$

## Ovotransferrin

Chicken egg white ovotransferrin, glycosylated (native) and deglycosylated

| Protein | pH | $T_{trs}$ | $\Delta H$ | Approach/Remarks | | Ref |
|---|---|---|---|---|---|---|
| native | 8.0 | 63.5±0.1 | 1766±25 | DSC | (1) | 96W3 |
| deglycosylated | 8.0 | 63.5±0.1 | 1749±13 | DSC | (1) | 96W3 |

Remark:
(1) measured in 500 mM HEPES buffer

## Parvalbumin

Rat α- and β-parvalbumin (α-PV and β-PV), wild type and mutants

| Protein | pH | $T_{trs}$ | ΔH | Approach/Remarks | | Ref |
|---|---|---|---|---|---|---|
| α-PV | 7.4 | 45.8±0.5 | 293±8 | DSC | (1) | 99H8 |
| α-PV Δ108Glu | 7.4 | 41.8±0.5 | 259±8 | DSC | (1,2) | 99H8 |
| β-PV | 7.4 | 53.6±0.5 | 314±13 | DSC | (1) | 99H8 |
| β-PV 109Ser | 7.4 | 54.7±0.5 | 305±13 | DSC | (1,3) | 99H8 |

Remarks:

(1) measured in the presence of 0.20 M NaCl, 10 mM EDTA

(2) Glu-108 deleted, truncation of the F-helix in parvalbumin

(3) Ser-109 inserted, extension of the F-helix in parvalbumin

## Peripheral Subunit-Binding Protein

Peripheral subunit-binding domain (psbd) from the dihydrolipoamide acetyltransferase component of the *Bacillus stearothermophilus* pyruvate dehydrogenase multienzyme complex, truncation mutant of a protein minidomain

| Protein | pH | $T_{trs}$ | ΔCp | ΔH | Appr./Rem. | | Ref |
|---|---|---|---|---|---|---|---|
| psbd41 | 8.0 | 53±0.4 | 1.80±1.05 | 132.6± 7.1 | heat | (1–4) | 99S13 |
| psbd36 | 8.0 | 48±1.2 | 1.84±0.79 | 126.8±18.0 | heat | (1–4) | 99S13 |
| psbd33 | 8.0 | 39±3.6 | 1.84±1.17 | 109.2±32.2 | heat | (1–4) | 99S13 |

Remarks:

(1) $T_{trs}$ and ΔH were obtained from thermal denaturation monitored by CD at 222 nm and 280 nm

(2) ΔCp was obtained from the global analysis of thermal denaturation at various urea concentrations

(3) psbd41 corresponds to residues 3–43, psbd36 to residues 6–41, and psbd33 to residues 7–39

(4) buffer: 2 mM phosphate, 2 mM borate, 2 mM citrate, 50 mM NaCl, pH 8.0

## Phage P22 Scaffolding Protein

Phage P22 scaffolding protein, wild type and mutants

| Mutant | pH | $T_{trs}$ | ΔH | Approach/Remarks | | Ref |
|---|---|---|---|---|---|---|
| wild type | 7.6 | 58±2 | | heat | (1,3) | 99G12 |
| | 7.6 | 46 | | heat | (2–4) | 99G11 |
| Tyr214→Trp | 7.6 | 54±3 | | heat | (1,3) | 99G12 |
| Ser242→Phe | 7.6 | 54±2 | | heat | (1,3) | 99G12 |
| double mutant (Arg74→Cys and Gln149→Trp) | | | | | | |
| | 7.6 | 54±1 | | heat | (1,3) | 99G12 |
| double mutant (Arg74→Cys and Leu177→Ile) | | | | | | |
| | 7.6 | 54±1 | | heat | (1,3) | 99G12 |

Remarks:

(1) transition monitored by fluorescence emission at 330 nm

(2) transition monitored by CD at 222 nm

(3) buffer: 20 mM potassium phosphate, 25 mM NaCl, pH 7.6

(4) for the noncoincidence of $T_{trs}$ from fluorescence and CD, see also Ref. 99G11

**434-Phage Repressor**, see Repressor Proteins

## Phosphatidylethanolamine-Binding Protein

Bovine brain phosphatidylethanolamine-binding protein (PEBP)

| Protein | pH | $T_{trs}$ | $\Delta H$ | Approach/Remarks | | Ref |
|---------|------|------|-----|------|-------|------|
| PEBP | 4.0 | 52 | 523 | heat | (1,2) | 99V1 |
|  | 4.0 | 53 | 490 | heat | (1,3) | 99V1 |
|  | 4.0 | 53 | 494 | heat | (1,4) | 99V1 |
|  | 4.0 | 56 | 477 | DSC | (1) | 99V1 |
|  | 7.4 | 53 |  | heat | (2–5) | 99V1 |

Remarks:

(1) buffer: 10 mM sodium acetate, 2 mM 2-mercaptoethanol, pH 4.0

(2) transition monitored by far-UV CD at 215 nm

(3) transition monitored by near-UV CD at 283 nm

(4) transition monitored by fluorescence at 330 nm

(5) buffer: 10 mM Tris-HCl, 2 mM 2-mercaptoethanol, pH 7.4

## Phosphoenolpyruvate:Sugar Phosphotransferase

N-terminal domain of *E. coli* phosphoenolpyruvate:sugar phosphotransferase

| Residues | pH | $T_{trs}$ | $\Delta Cp$ | $\Delta H$ | Approach/Remarks | | Ref |
|----------|-----|-----------|-------------|------------|------|-------|------|
| (1–258 + Arg) | 7.5 | 56.9±0.4 | 11.3 | 590±13 | DSC | (1,2) | 98N3 |
| (1–258 + Arg) | 7.5 | 56.8 |  | 561 | heat | (1,3) | 98N3 |
| (1–268 + Cys) | 7.5 | 56.8 |  | 674 | DSC | (4) | 98N3 |

Remarks:

(1) the N-terminal domain of phosphoenolpyruvate:sugar phosphotransferase (EIN) expressed and purified from a $\Delta pts$ *E. coli* strain shows a single, reversible two-state transition. In contrast, monomeric EIN expressed in a wild-type strain (*pts*+) has two endotherms with $T_{trs} \cong 50$ and 57°C and overall $\Delta H = 586$ kJ/mol, and is completely converted into the more stable form on dephosphorylation

(2) $\Delta Cp = 11.3$ kJ/mol/K from curve analysis, average $\Delta Cp = 9.2±2.1$ kJ/mol/K from pre- and posttranslational baseline extrapolation

(3) transition monitored by CD at 222 nm, two-state fit

(4) data from Ref. 96C19

Phosphoenolpyruvate:sugar phosphotransferase, dephospho-enzyme I and its N-terminal domain from *Salmonella typhimurium* and *E. coli*

| pH | $T_{trs1}$ | $\Delta H_{trs1}$ | $T_{trs2}$ | $\Delta H_{trs2}$ | Approach/Remarks | | Ref |
|-----|-----------|-------------------|-----------|-------------------|------|-------|------|
| enzyme I | | | | | | | |
| 6.5 | 54.0 | 1213 |  |  | DSC | | 91L10 |
| 7.0 | 51.7 | 757 | 55.2 | 251 | DSC | | 91L10 |
| 7.5 | 47.6 | 456±42 | 54.9 | 452±13 | DSC | | 91L10 |
| 8.5 | 38.4 | 238 | 55.3 | 548 | DSC | | 91L10 |
| N-terminal domain of enzyme I | | | | | | | |
| 7.5 | 58.2 | 360 | n.t. | n.t. | DSC | (1,2) | 91L10 |
| 7.5 | 58.2 | 665 | n.t. | n.t. | DSC | (2–4) | 96C19 |

Remarks:

(1) 30 kDa fragment from *Salmonella typhimurium* obtained by proteolytic cleavage containing the N-terminus and the active site, His189

(2) n.t. = no second transition was observed

(3) recombinant N-terminal fragment from *E. coli* corresponding to the wild type coding sequence of enzyme I to Gln268

(4) $\Delta Cp = 20.4$ kJ/mol/K

*E. coli* phosphoenolpyruvate:sugar phosphotransferase, dephospho-enzyme I

| pH | $T_{trs1}$ | $\Delta H_{trs1}$ | $T_{trs2}$ | $\Delta H_{trs2}$ | Approach/Remarks | | Ref |
|---|---|---|---|---|---|---|---|
| 7.5 | 41.4±1.1 | 246±50 | 54.6±0.5 | 608±26 | DSC | (1,2) | 98N3 |
| 7.5 | 40.6 | 247 | 53.9 | 517 | heat | (1–3) | 98N3 |
| 7.5 | 40.6±0.6 | 240±13 | n.t. | n.t. | heat | (1,4,5) | 98N3 |
| 7.5 | 39.2 | ~335 | n.t. | n.t. | heat | (1,5,6) | 98N3 |
| 7.5 | 41.3 | 247 | n.t. | n.t. | heat | (1,4,7) | 98N3 |

Remarks:
(1) enzyme I (575 amino acid residues) was isolated from the Δ*pts E. coli* strain and than converted into the dephospho-enzyme I
(2) fit by two independent two-state transitions corresponding to the C- and N-terminal domains
(3) transition monitored by CD at 222 nm
(4) transition monitored by Trp fluorescence, fit by a single two-state transition
(5) n.t. = no transition was observed
(6) data from light scattering
(7) partially phosphorylated enzyme I, as isolated from the Δ*pts* strain

## Phosphoglycerate Kinase

N-terminal domain of phophoglycerate kinase from *Bacillus stearothermophilus*, transient intermediates and transition states (TS)

| Transition | pH | $T_{ref}$ | $\Delta Cp$ | $\Delta H$ | Approach/Remarks | | Ref |
|---|---|---|---|---|---|---|---|
| I → U | 7.5 | 25 | 5.44±0.84 | 19.3±2.3 | GuHCl | (1–3) | 98P5 |
| N → I | 7.5 | 25 | 3.05±0.25 | 0.8±0.8 | GuHCl | (1–3) | 98P5 |
| TS → I | 7.5 | 25 | 1.17±0.33 | 76.1±0.8 | GuHCl | (1–3) | 98P5 |
| N → TS | 7.5 | 25 | 1.88±0.25 | 77.0±0.8 | GuHCl | (1–3) | 98P5 |

Remarks:
(1) from folding and unfolding rates measured by stopped-flow kinetics
(2) transition monitored by fluorescence intensity
(3) buffer: 50 mM triethanolamine, 2 mM DTT, and GuHCl

Phosphoglycerate kinase from *Thermotoga maritima*, deconvolution based on a three-state model

| Transition | pH | $T_{trs}$ | $\Delta Cp$ | $\Delta H$ | Approach/Remarks | | Ref |
|---|---|---|---|---|---|---|---|
| trans. (1) | 3.0 | 61.1 | 3 | 215 | DSC | (1) | 98Z1 |
| trans. (2) | 3.0 | 67.5 | 14 | 290 | DSC | (1) | 98Z1 |
| trans. (1) | 3.2 | 67.7 | 3 | 203 | DSC | (1) | 98Z1 |
| trans. (2) | 3.2 | 73.1 | 14 | 340 | DSC | (1) | 98Z1 |
| trans. (1) | 3.4 | 70.0 | 8 | 221 | DSC | (1) | 98Z1 |
| trans. (2) | 3.4 | 78.4 | 16 | 437 | DSC | (1) | 98Z1 |
| trans. (1) | 3.6 | 75.7 | 6 | 242 | DSC | (1) | 98Z1 |
| trans. (2) | 3.6 | 81.6 | 20 | 491 | DSC | (1) | 98Z1 |
| trans. (1) | 3.8 | 80.0 | 2 | 267 | DSC | (1) | 98Z1 |
| trans. (2) | 3.8 | 82.5 | 17 | 467 | DSC | (1) | 98Z1 |
| trans. (1) | 4.0 | 84.4 | 10 | 773 | DSC | (2) | 98Z1 |
| trans. (1) | 3–3.8 | 61–80 | 3.0 | | DSC | (3) | 98Z1 |
| trans. (2) | 3–3.8 | 67–83 | 13.4 | | DSC | (3) | 98Z1 |

Remarks:
(1) results of deconvolution using a three-state model
(2) treatment by a two-state model
(3) $\Delta Cp$ from $\Delta H$ versus $T_{trs}$

Phosphoglycerate kinase from *Thermotoga maritima* (TmPGK), isolated N-terminal domain, recombinant protein (residues Met1-Ile175)

| Transition | pH | $T_{trs}$ | $\Delta Cp$ | $\Delta H$ | Approach/Remarks | | Ref |
|---|---|---|---|---|---|---|---|
| trans. (1) | 3.1 | 59.1 | 2.5 | 86.6 | DSC | (1–4) | 99Z1 |
| trans. (2) | 3.1 | 59.2 | 5.9 | 247.7 | DSC | (1–4) | 99Z1 |
| trans. (1) | 3.2 | 60.2 | 1.7 | 92.0 | DSC | (1–4) | 99Z1 |
| trans. (2) | 3.2 | 68.5 | 7.1 | 252.7 | DSC | (1–4) | 99Z1 |
| trans. (1) | 3.2 | 60.2 | 1.7 | 82.4 | DSC | (1–5) | 99Z1 |
| trans. (2) | 3.2 | 67.2 | 10.0 | 238.9 | DSC | (1–5) | 99Z1 |
| trans. (1) | 3.2 | 60.5 | 2.1 | 90.4 | DSC | (1–5) | 99Z1 |
| trans. (2) | 3.2 | 68.6 | 7.1 | 251.9 | DSC | (1–5) | 99Z1 |
| trans. (1) | 3.2 | 61.7 | 1.3 | 95.8 | DSC | (1–4) | 99Z1 |
| trans. (2) | 3.2 | 71.5 | 1.7 | 260.7 | DSC | (1–4) | 99Z1 |
| trans. (1) | 3.3 | 65.9 | 3.3 | 100.4 | DSC | (1–4) | 99Z1 |
| trans. (2) | 3.3 | 77.7 | 2.9 | 275.8 | DSC | (1–4) | 99Z1 |
| trans. (1) | 3.3 | 64.1 | 1.7 | 86.2 | DSC | (1–4) | 99Z1 |
| trans. (2) | 3.3 | 74.5 | 7.5 | 278.7 | DSC | (1–4) | 99Z1 |
| trans. (1) | 3.4 | 65.0 | 1.3 | 99.2 | DSC | (1–4) | 99Z1 |
| trans. (2) | 3.4 | 79.6 | 7.5 | 298.7 | DSC | (1–4) | 99Z1 |
| trans. (1) | 3.5 | 67.0 | 1.7 | 95.4 | DSC | (1–4) | 99Z1 |
| trans. (2) | 3.5 | 85.9 | 10.0 | 334.7 | DSC | (1–4) | 99Z1 |
| trans. (1) | 3.5 | 66.6 | 2.1 | 98.7 | DSC | (1–4) | 99Z1 |
| trans. (2) | 3.5 | 84.7 | 7.1 | 324.3 | DSC | (1–4) | 99Z1 |
| trans. (1) | 3.7 | 69.7 | 2.5 | 91.6 | DSC | (1–4) | 99Z1 |
| trans. (2) | 3.7 | 88.0 | 9.2 | 337.2 | DSC | (1–4) | 99Z1 |
| trans. (1) | 3.7 | 70.0 | 2.1 | 102.5 | DSC | (1–4) | 99Z1 |
| trans. (2) | 3.7 | 91.0 | 7.9 | 364.4 | DSC | (1–4) | 99Z1 |
| trans. (1) | 4.0 | 75.0 | 1.7 | 120.9 | DSC | (1–4) | 99Z1 |
| trans. (2) | 4.0 | 93.4 | 8.4 | 385.8 | DSC | (1–4) | 99Z1 |
| trans. (1) | 4.0 | 74.8 | 2.1 | 113.8 | DSC | (1–4) | 99Z1 |
| trans. (2) | 4.0 | 92.8 | 7.5 | 354.0 | DSC | (1–4) | 99Z1 |
| trans. (1) | 3.1–4.0 | | 1.7±0.8 | | DSC | (2,6) | 99Z1 |
| trans. (2) | 3.1–4.0 | | 5.0±0.8 | | DSC | (2,6) | 99Z1 |
| trans. (1) | 7.0 | 87.2 | 3.8 | 310.9 | DSC | (1,2,7,8) | 99Z1 |
| trans. (2) | 7.0 | 105.2 | 7.5 | 488.3 | DSC | (1,2,7,8) | 99Z1 |

Remarks:
(1) results of deconvolution, $\Delta Cp$ from calorimetric recordings
(2) buffers: 20 mM glycine-HCl (pH 2.5–3.6), 20 mM sodium acetate (pH 3.6–5.2), 20 mM Tris/HCl (pH 7.0–8.6)
(3) error limits are given to be about 1% in $T_{trs}$ and 5% in $\Delta H$
(4) scan rate of 1 K/min
(5) varying scan rate
(6) $\Delta Cp$ from $\Delta H^{cal}$ versus $T_{trs}$
(7) at pH 7.0 the calorimetric recording shows two well separated peaks, the transition is about 70% reversible (intact TmPGK and C-terminal domain melt irreversibly at 101 and 97°C)
(8) the data yield $\Delta G_1 = 31.8$ and $\Delta G_2 = 34.3$ kJ/mol at pH 7 and 25°C

Phosphoglycerate kinase from *Thermotoga maritima* (TmPGK), isolated C-terminal domain, recombinant protein (residues Met + Ile175-Lys399)

| Transition | pH | $T_{trs}$ | $\Delta Cp$ | $\Delta H$ | Approach/Remarks | | Ref |
|---|---|---|---|---|---|---|---|
| trans. (1) | 3.2 | 70.4 | 4.2 | 296.2 | DSC | (1–4) | 99Z1 |
| trans. (2) | 3.2 | 71.3 | 4.2 | 210.5 | DSC | (1–4) | 99Z1 |
| trans. (1) | 3.3 | 73.6 | 4.2 | 310.5 | DSC | (1–4) | 99Z1 |
| trans. (2) | 3.3 | 74.4 | 4.2 | 241.4 | DSC | (1–4) | 99Z1 |
| trans. (1) | 3.3 | 73.9 | 4.2 | 309.2 | DSC | (1–5) | 99Z1 |

Phosphoglycerate kinase from *Thermotoga maritima* (TmPGK), isolated C-terminal domain, recombinant protein (residues Met + Ile175-Lys399) (continued)

| Transition | pH | $T_{trs}$ | $\Delta Cp$ | $\Delta H$ | Approach/Remarks | | Ref |
|---|---|---|---|---|---|---|---|
| trans. (2) | 3.3 | 75.0 | 4.6 | 245.6 | DSC | (1–5) | 99Z1 |
| trans. (1) | 3.3 | 73.7 | 4.2 | 297.9 | DSC | (1–5) | 99Z1 |
| trans. (2) | 3.3 | 74.1 | 10.9 | 231.0 | DSC | (1–5) | 99Z1 |
| trans. (1) | 3.4 | 76.3 | 4.2 | 310.9 | DSC | (1–4) | 99Z1 |
| trans. (2) | 3.4 | 77.3 | 4.2 | 258.2 | DSC | (1–4) | 99Z1 |
| trans. (1) | 3.5 | 80.0 | 4.2 | 337.6 | DSC | (1–4) | 99Z1 |
| trans. (2) | 3.5 | 81.5 | 3.3 | 308.4 | DSC | (1–4) | 99Z1 |
| trans. (1) | 3.2–3.5 | | 4.2±0.8 | | DSC | (2,6) | 99Z1 |
| trans. (2) | 3.2–3.5 | | 9.6±0.8 | | DSC | (2,6) | 99Z1 |

Remarks:
(1) results of deconvolution, $\Delta Cp$ from calorimetric recordings
(2) buffer: 20 mM glycine-HCl (pH 2.5–3.5)
(3) error limits are given to be about 1% in $T_{trs}$ and 5% in $\Delta H$
(4) scan rate of 1 K/min
(5) varying scan rate
(6) $\Delta Cp$ from $\Delta H^{cal}$ versus $T_{trs}$

Phosphoglycerate kinase from *Thermotoga maritima* (TmPGK), intact recombinant protein

| Transition | pH | $T_{trs}$ | $\Delta Cp$ | $\Delta H$ | Approach/Remarks | | Ref |
|---|---|---|---|---|---|---|---|
| trans. (2) | 3.1 | 60.5 | 8.8 | 248.1 | DSC | (1–4) | 99Z1 |
| trans. (3) | 3.1 | 66.3 | 8.8 | 292.5 | DSC | (1–4) | 99Z1 |
| trans. (2) | 3.2 | 63.4 | 7.5 | 264.8 | DSC | (1–4) | 99Z1 |
| trans. (3) | 3.2 | 69.7 | 7.1 | 341.0 | DSC | (1–4) | 99Z1 |
| trans. (1) | 3.3 | 55.9 | 6.3 | 102.5 | DSC | (1–4) | 99Z1 |
| trans. (2) | 3.3 | 69.0 | 7.1 | 295.8 | DSC | (1–4) | 99Z1 |
| trans. (3) | 3.3 | 72.4 | 11.3 | 351.0 | DSC | (1–4) | 99Z1 |
| trans. (1) | 3.4 | 57.0 | 9.2 | 149.4 | DSC | (1–4) | 99Z1 |
| trans. (2) | 3.4 | 71.0 | 7.1 | 297.1 | DSC | (1–4) | 99Z1 |
| trans. (3) | 3.4 | 75.1 | 11.3 | 390.0 | DSC | (1–4) | 99Z1 |
| trans. (1) | 3.4 | 64.8 | 8.8 | 216.3 | DSC | (1–4) | 99Z1 |
| trans. (2) | 3.4 | 74.6 | 7.1 | 316.7 | DSC | (1–4) | 99Z1 |
| trans. (3) | 3.4 | 77.7 | 8.8 | 393.3 | DSC | (1–4) | 99Z1 |
| trans. (1) | 3.5 | 65.9 | 7.9 | 183.3 | DSC | (1–4) | 99Z1 |
| trans. (2) | 3.5 | 75.6 | 6.3 | 324.3 | DSC | (1–4) | 99Z1 |
| trans. (3) | 3.5 | 77.8 | 9.6 | 373.6 | DSC | (1–4) | 99Z1 |
| trans. (1) | 3.5 | 68.8 | 6.3 | 210.9 | DSC | (1–4) | 99Z1 |
| trans. (2) | 3.5 | 75.2 | 8.4 | 329.7 | DSC | (1–4) | 99Z1 |
| trans. (3) | 3.5 | 79.0 | 11.7 | 419.2 | DSC | (1–4) | 99Z1 |
| trans. (1) | 3.7 | 79.5 | 2.1 | 264.0 | DSC | (1–4) | 99Z1 |
| trans. (2) | 3.7 | 79.7 | 7.1 | 347.3 | DSC | (1–4) | 99Z1 |
| trans. (3) | 3.7 | 82.0 | 6.7 | 468.6 | DSC | (1–4) | 99Z1 |
| trans. (1) | 3.85 | 80.0 | 8.4 | 334.3 | DSC | (1–4) | 99Z1 |
| trans. (2) | 3.85 | 82.9 | 6.7 | 377.4 | DSC | (1–4) | 99Z1 |
| trans. (3) | 3.85 | 82.9 | 3.8 | 506.3 | DSC | (1–4) | 99Z1 |
| trans. (1) | 3.1–3.85 | | 7.5±2.5 | | DSC | (2,5) | 99Z1 |
| trans. (2) | 3.1–3.85 | | 5.4±0.8 | | DSC | (2,5) | 99Z1 |
| trans. (3) | 3.1–3.85 | | 11.3±2.1 | | DSC | (2,5) | 99Z1 |

Remarks:
(1) results of deconvolution, $\Delta Cp$ from calorimetric recordings
(2) buffers: 20 mM glycine-HCl (pH 2.5-3.6), 20 mM sodium acetate (pH 3.6–3.85)
(3) error limits are given to be about 1% in $T_{trs}$ and 5% in $\Delta H$
(4) scan rate of 1 K/min
(5) $\Delta Cp$ from $\Delta H^{cal}$ versus $T_{trs}$

## Phosphorylase

Phosphorylase *b* from rabbit skeletal muscle in the presence of ligands

| Ligand | pH | $T_{trs}$ | $\Delta H$ | Approach/Remarks | | Ref |
|---|---|---|---|---|---|---|
| without | 7.0 | 57.4 | 1290 | DSC | (1,2) | 97K10 |
| without | 6.8 | 56.8 | | DSC | (3) | 97K11 |
| AMP | 6.8 | 58.1 | | DSC | (3,4) | 97K11 |
| FMN | 6.8 | 59.6 | | DSC | (3,4) | 97K11 |
| glucose | 6.8 | 58.3 | | DSC | (3,4) | 97K11 |
| glucose-1-phosphat | 6.8 | 57.4 | | DSC | (3,4) | 97K11 |
| glucose-6-phosphat | 6.8 | 59.1 | | DSC | (3,4) | 97K11 |

Remarks:

(1) the protein undergoes irreversible denaturation, the transition temperature varies with protein concentration, increase of concentration from 0.5 to 6 mg/ml shifts the peak maximum temperature from 56.6 to 57.4°C

(2) measured in 30 mM HEPES-KOH, pH 7.0

(3) measured in 80 mM HEPES-NaOH, pH 6.8

(4) ligand concentrations: 3 mM AMP, 1 mM FMN, 100 mM glucose, 20 mM glucose-1-phosphat, 10 mM glucose-6-phosphat

## Plasminogen Activator

Kringle-2 domain of tissue-type plasminogen activator, recombinant, wild type and tryptophan mutants

| Mutant | pH | $T_{trs}$ | $\Delta H$ | Approach/Remarks | | Ref |
|---|---|---|---|---|---|---|
| wild type | 7.4 | 75.6 | 285 | DSC | (1,2) | 97C8 |
| wild type | 8.0 | 75.6 | | DSC | (1,3) | 94D4 |
| Trp25→Phe | 8.0 | 50.8 | | DSC | (1,3) | 94D4 |
| Trp25→Tyr | 8.0 | 58.0 | | DSC | (1,3) | 94D4 |
| Trp63→Phe | 7.4 | 58.5 | 167 | DSC | (1,2) | 97C8 |
| Trp63→His | 7.4 | 56.6 | 234 | DSC | (1,2) | 97C8 |
| Trp63→Ser | 7.4 | 56.6 | 238 | DSC | (1,2) | 97C8 |
| Trp63→Tyr | 7.4 | 63.3 | 188 | DSC | (1,2) | 97C8 |

Remarks:

(1) kringle 2 region of tissue-type plasminogen activator, residues 180 to 261

(2) buffer: 100 mM sodium phosphate, pH 7.4 (in the absence of EACA)

(3) buffer: Tris-acetate, pH 8.0

Kringle-2 domain of tissue-type plasminogen activator, recombinant, wild type in the presence and absence of EACA

| Protein | pH | $T_{trs}$ | $\Delta H$ | Approach/Remarks | | Ref |
|---|---|---|---|---|---|---|
| wild type | 7.4 | 75.6 | | DSC | (1) | 99C6 |
| wild type + EACA | 7.4 | 86.1 | | DSC | (1,2) | 99C6 |

Remarks:

(1) Ref. 99C6 contains further $T_{trs}$ values for mutants of the kringle-2 domain

(2) EACA = ε-amino caproic acid

## Plastocyanin

Plastocyanin from spinach chloroplasts

| Protein | pH | $T_{trs}$ | $\Delta Cp$ | $\Delta H$ | Appr./Rem. | | Ref |
|---|---|---|---|---|---|---|---|
| 0.3 | 7.03 | 68.03±0.05 | | 98±20 | DSC | | 98M17 |
| 0.3 | 7.03 | 58.34±0.03 | | | heat | (1) | 98M17 |
| 0.5 | 7.03 | 68.44±0.04 | | 140±22 | DSC | | 98M17 |
| 0.5 | 7.03 | 61.00±0.08 | | | heat | (1) | 98M17 |
| 0.7 | 7.03 | 70.65±0.03 | | 179±29 | DSC | | 98M17 |
| 0.7 | 7.03 | 61.47±0.06 | | | heat | (1) | 98M17 |
| 1.0 | 7.03 | 70.54±0.05 | | 194±21 | DSC | | 98M17 |
| 1.0 | 7.03 | 63.50±0.03 | | | heat | (1) | 98M17 |
| infinite | 7.03 | 69.0 | | 254±37 | DSC | (2) | 98M17 |
| infinite | 7.03 | 63.5 | | 284±41 | heat | (2) | 98M17 |
| | | | 5.7 | | | (3) | 98M17 |

Remarks:

(1) transition monitored by changes in optical density at 595 nm

(2) extrapolated to infinite scan rate

(3) estimated value

## Polyadenylation Factor

MEARA sequence repeat of human CstF-64 polyadenylation factor

| Protein | pH | $T_{trs}$ | $\Delta H$ | Approach/Remarks | | Ref |
|---|---|---|---|---|---|---|
| MEARA$_6$ | 2.0 | ~20 | 107 | DSC | (1–3) | 99R11 |

Remarks:

(1) MEARA$_6$ peptide analog of the CstF-64 polyadenylation factor which contains in position 409–469 12 nearly identical repeats of a consensus motif of Met-Glu-Ala-Arg-(Ala/Gly)

(2) the van't Hoff heat of the transition amounts to 38 kJ/mol

(3) $\Delta H$ with correction for 100% helix content

## Prion Protein

Recombinant human prion protein PrP(90–231), wild type and mutants

| Mutant | pH | $T_{trs}$ | $\Delta H$ | Approach/Remarks | Ref |
|---|---|---|---|---|---|
| wild type | 7.2 | 70.2±0.1 | 294±6 | heat, v.H. (1) | 98S22 |
| Pro102→Leu | 7.2 | 70.3±0.1 | 306±6 | heat, v.H. (1) | 98S22 |
| Glu200→Lys | 7.2 | 67.0±0.9 | 300±8 | heat, v.H. (1) | 98S22 |

Remark:

(1) transition monitored by CD at 222 nm

## Protease

Ca$^{2+}$-Zn$^{2+}$ proteases from psychrophilic and mesophilic organisms

| Form | pH | T$_{trs}$ | Q | Approach/Remarks | | Ref |
|------|----|-----------|---|------------------|--|-----|
| psychrophile | 8 | 44.4 | 3.7×10$^{-3}$ | DSC | (1–4) | 97G5 |
| mesophile | 8 | 64.3 | 5.6×10$^{-3}$ | DSC | (1–4) | 97G5 |

Remarks:

(1) Q is a normalized enthalpy value in arbitrary units

(2) measured in 20 mM Tris, 10 mM CaCl$_2$ at pH 8

(3) protein from the Antarctic strain *Pseudomonas aeruginosa* compared to that of the homologous enzyme from *P. aeruginosa* IF03455

(4) psychrophilic enzymes are reviewed in Ref. 97G5

## Protein A

Protein A from *Staphylococcus aureus*, domains

| Protein | pH | T$_{trs}$ | ΔH | Approach/Remarks | | Ref |
|---------|----|-----------|----|------------------|--|-----|
| E-domain | 5.0 | 43.0 | 138 | DSC | (1) | 99S15 |
| D-D fragments | 5.0 | 51.5 | 203 | DSC | (1) | 99S15 |

Remark:

(1) measured in 50 mM sodium acetate buffer

## Protein G (Immunoglobulin-Binding Protein)

Immunoglobulin G (IgG) binding protein G (β1), variants with charged residues at the parallel cross-strand site (positions 6 and 53)

| Mutant | pH | T$_{trs}$ | ΔH | Approach/Remarks | | Ref |
|--------|----|-----------|----|------------------|--|-----|
| Ala6, Ala53 | 7.2 | 38.19±0.09 | 151±4 | heat | (1,2) | 99M15 |
| Ala6, Glu53 | 7.2 | 40.3 ±0.1 | 157±2 | heat | (1,2) | 99M15 |
| Ala6, Arg53 | 7.2 | 46.62±0.08 | 163±4 | heat | (1,2) | 99M15 |
| Glu6, Ala53 | 7.2 | 39.55±0.06 | 155±3 | heat | (1,2) | 99M15 |
| Glu6, Glu53 | 7.2 | 32.55±0.06 | 142±3 | heat | (1,2) | 99M15 |
| Glu6, Arg53 | 7.2 | 52.25±0.09 | 187±1 | heat | (1,2) | 99M15 |
| Lys6, Ala53 | 7.2 | 50.9 ±0.2 | 184±4 | heat | (1,2) | 99M15 |
| Lys6, Glu53 | 7.2 | 55.19±0.01 | 200±3 | heat | (1,2) | 99M15 |
| Lys6, Arg53 | 7.2 | 52.9 ±0.1 | 201±8 | heat | (1,2) | 99M15 |

Remarks:

(1) transition monitored by far-UV CD at 222 nm

(2) buffer: 50 mM phosphate buffer, pH 7.2

Immunoglobulin G (IgG) binding protein G (β1), variants with polar and noncharged residues at the parallel cross-strand site (positions 6 and 53)

| Mutant | pH | T$_{trs}$ | ΔH | Approach/Remarks | | Ref |
|--------|----|-----------|----|------------------|--|-----|
| Ile6, Tyr53 | 5.2 | 79.3 ±0.2 | 238±8 | heat | (1,2) | 99M15 |
| Ile6, Phe53 | 5.2 | 77.78±0.03 | 239±1.3 | heat | (1,2) | 99M15 |
| Ile6, Ile53 | 5.2 | 77.2 ±0.2 | 218±4 | heat | (1,2) | 99M15 |
| Val6, Phe53 | 5.2 | 75.96±0.03 | 226±3 | heat | (1,2) | 99M15 |

Immunoglobulin G (IgG) binding protein G (β1), variants with polar and noncharged residues at the parallel cross-strand site (positions 6 and 53) (continued)

| Mutant | pH | $T_{trs}$ | $\Delta H$ | Approach/Remarks | | Ref |
|---|---|---|---|---|---|---|
| Ile6, Thr53 | 5.2 | 75.9 ±0.3 | 228±3 | heat | (1,2) | 99M15 |
| Ile6, Val53 | 5.2 | 74.3 ±0.2 | 209±4 | heat | (1,2) | 99M15 |
| Leu6, Ile53 | 5.2 | 73.96±0.06 | 209±4 | heat | (1,2) | 99M15 |
| Val6, Ile53 | 5.2 | 73.8 ±0.3 | 212±0.4 | heat | (1,2) | 99M15 |
| Phe6, Tyr53 | 5.2 | 72.2 ±0.2 | 213±8 | heat | (1,2) | 99M15 |
| Val6, Val53 | 5.2 | 71.5 ±0.1 | 201±4 | heat | (1,2) | 99M15 |
| Phe6, Phe53 | 5.2 | 70.7 ±0.3 | 219±2 | heat | (1,2) | 99M15 |
| Phe6, Ile53 | 5.2 | 69.42±0.06 | 197±4 | heat | (1,2) | 99M15 |
| Phe6, Val53 | 5.2 | 67.1 ±0.1 | 197±4 | heat | (1,2) | 99M15 |
| Thr6, Tyr53 | 5.2 | 62.75±0.05 | 192±8 | heat | (1,2) | 99M15 |
| Asn6, Thr53 | 5.2 | 62.7 ±0.1 | 206±1.3 | heat | (1,2) | 99M15 |
| Thr6, Thr53 | 5.2 | 62.3 ±0.1 | 205±4 | heat | (1,2) | 99M15 |
| Thr6, Phe53 | 5.2 | 61.62±0.01 | 184±8 | heat | (1,2) | 99M15 |
| Thr6, Ile53 | 5.2 | 60.9 ±0.1 | 180±0.8 | heat | (1,2) | 99M15 |
| Thr6, Val53 | 5.2 | 58.61±0.03 | 179±0.4 | heat | (1,2) | 99M15 |

Remarks:
(1) transition monitored by far-UV CD at 222 nm
(2) buffer: 50 mM phosphate buffer, pH 7.2

Immunoglobulin-binding domain GB1 of Streptococcal protein G, replacement of an α-helix 23–36 by a β-hairpin

| Mutant | pH | $T_{trs}$ | $\Delta H$ | Approach/Remarks | | Ref |
|---|---|---|---|---|---|---|
| $G_{B1}$ | 7.0 | 78.8±0.2 | 268±4 | heat | (1,2) | 99C14 |
| $G_{In}$ | 7.0 | 65.3±0.2 | 234±4 | heat | (1–3) | 99C14 |
| $G_{Out}$ | 7.0 | 49.8±0.2 | 172±4 | heat | (1–3) | 99C14 |
| $G_{Y}$ | 7.0 | 44.3±0.2 | 172±4 | heat | (1–3) | 99C14 |
| $G_{Hel}$ | 7.0 | 24.3±0.2 | 100±4 | heat | (1–3) | 99C14 |

Remarks:
(1) from heat denaturation monitored by far-UV CD (see also Table 1)
(2) measured in 10 mM sodium phosphate, pH 7
(3) mutant with the α-helix 23–36 replaced by a β-hairpin derived from residues 43–55 (with point mutations described in Ref. 99C14)

B1 domain of Streptococcal immunolglobulin-binding protein G, mutants in the antiparallel β-sheet, positions 44 and 53

| Mutant | pH | $T_{trs}$ | $\Delta H$ | Approach/Remarks | | Ref |
|---|---|---|---|---|---|---|
| Ala44, Ala53 | 5.2 | 62.27±0.09 | 190.4±3.3 | heat | (1,2) | 98M16 |
| Ala44, Gly53 | 5.2 | 51.41±0.04 | 164.4±7.5 | heat | (1,2) | 98M16 |
| Phe44, Ala53 | 5.2 | 64.1 ±0.1 | 185.4±7.5 | heat | (1,2) | 98M16 |
| Phe44, Gly53 | 5.2 | 57.2 ±0.2 | 166.9±3.8 | heat | (1,2) | 98M16 |

Remarks:
(1) transition monitored by CD at 222 nm
(2) measured in 50 mM sodium acetate buffer, pH 5.2

B1 domain of streptococcal protein G (B1G) and modified domain (DBF-B1G)

| Protein | pH | $T_{trs}$ | $\Delta H$ | Approach/Remarks | | Ref |
|---------|-----|-----------|------------|------------------|---|-----|
| B1G | 3.5 | 54.7±0.1 | 159.8±3.4 | heat | (1,2) | 99O1 |
| DBF-B1G | 3.5 | 52.0±0.2 | 157.7±4.3 | heat | (1–3) | 99O1 |

Remarks:

(1) transition monitored by the change in ellipticity at 218 nm

(2) buffer: 50 mM sodium acetate, pH 3.5

(3) DBF-B1G = the residues 10 and 11 of B1G are replaced by dibenzofuran (DBF)

Reconstituted immunoglobulin G-binding domains B1 of streptococcal protein G consisting of PGB1(1–40) and PGB1 (41–56) and disulfide mutant of PGB1(41–56)

| Protein | pH | $T_{trs}$ | $\Delta Cp$ | $\Delta H$ | Appr./Rem. | | Ref |
|---------|-----|-----------|-------------|------------|------------|---|-----|
| PGB1(1–40) + PGB1(41–56): | | | | | | | |
| | 7.0 | 40 | 2.6 | 121 | heat | (1,2) | 99K11 |
| PGB1(1–40) + disulfide mutant of PGB1(41–56), oxidized: | | | | | | | |
| | 7.0 | 54 | 1.9 | 126 | heat | (1–3) | 99K11 |
| PGB1(1–40) + disulfide mutant of PGB1(41–56), reduced: | | | | | | | |
| | 7.0 | 40 | 2.6 | 135 | heat | (1,3,4) | 99K11 |

Remarks:

(1) transition monitored by CD at 222 nm

(2) measured in 5 mM sodium phosphate

(3) double mutant Gly41→Cys and Glu56→Cys

(4) measured in 5 mM sodium phosphate with 30 mM dithiothreitol

Immunoglobulin G-binding domain B1 of streptococcal protein G, PGB1(1–56), and its reconstituted domain consisting of PGB1(1–40) and PGB1(41–56)

| Protein | pH | $T_{trs}$ | $\Delta Cp$ | $\Delta H$ | Appr./Rem. | | Ref |
|---------|-----|-----------|-------------|------------|------------|---|-----|
| reconstituted domain PGB1(1–40) + PGB1(41–56): | | | | | | | |
| term (1) | 4.9 | 42±0.8 | 3.2±0.3 | 114±5 | DSC | (1,2) | 99H12 |
| term (2) | | 28 | 0.0 | 20.4 | DSC | (1,2) | 99H12 |
| term (1) | 5.9 | 40±0.9 | 2.1±0.4 | 131±5 | DSC | (1,2) | 99H12 |
| term (2) | | 28 | 0.0 | 20.4 | DSC | (1,2) | 99H12 |
| term (1) | 4.9 | 46 | 2.9 | 161 | heat | (1–3) | 99H12 |
| term (2) | | 28 | 0.0 | 20.4 | heat | (1–3) | 99H12 |
| term (1) | 5.9 | 42 | 2.8 | 150 | heat | (1–3) | 99H12 |
| term (2) | | 28 | 0.0 | 20.4 | heat | (1–3) | 99H12 |
| term (1) | 7.0 | 41 | 2.6 | 119 | heat | (1–3) | 99H12 |
| term (2) | | 28 | 0.0 | 20.4 | heat | (1–3) | 99H12 |
| uncleaved domain PGB1(1–56): | | | | | | | |
| | 6.0 | 81 | 2.7±0.3 | 270±0.5 | DSC | (1,2) | 99H12 |
| [Met1]PGB1(1–56): | | | | | | | |
| | 5.4 | 88 | 2.9 | 253 | DSC | (4) | 99H12 |

Remarks:

(1) term (1) and term (2) refer to complex dissociation/folding equilibria

(2) measured in either 50 mM phosphate or 100 mM β,β'-dimethylglutaric acid

(3) transition monitored by CD at 222 nm

(4) data for Met replacement mutant from Ref. 92A2

## Protein L

IgG-binding domain of peptostreptococcal protein L, comparison of various approaches

| | pH | $T_{ref}$ | $\Delta C_p$ | $\Delta H$ | Appr./Rem. | Ref |
|---|---|---|---|---|---|---|
| | 7.0 | 22 | 3.22±0.08 | 84.1±1.7 | heat (1–3) | 97Y3 |

Remarks:
(1) data from global fit of protein melting carried out at different GuHCl concentration, reference temperature $T_{ref} = 22°C$
(2) buffer: 100 mM sodium phosphate pH 7.0, 0.5 M KCl, varying GuHCl conc. from 0 to 3.0 M
(3) transition monitored by CD at 220 nm

IgG-binding domain of peptostreptococcal protein L, mutant (Tyr43→Trp)

| Mutant | pH | $T_{ref}$ | $\Delta C_p$ | $\Delta H$ | Approach/Remarks | Ref |
|---|---|---|---|---|---|---|
| Tyr43→Trp | 7.0 | 22 | 3.22±0.08 | 84.1±1.7 | GuHCl (1,2) | 97S2 |

Remarks:
(1) for the procedure see Ref. 97S3
(2) m = 10.0±0.4 kJ/mol/M

## Protein S

Recombinant protein S, spore coat protein from *Myxococcus xanthus*, wild type and isolated domains

| Protein/Transition | | pH | $T_{trs}$ | $\Delta H$ | Approach/Remarks | Ref |
|---|---|---|---|---|---|---|
| protein S | | | | | | |
| | trans. (1) | 7.0 | 65 | 26 | DSC (1–3) | 98W7 |
| | trans. (2) | 7.0 | 66 | 19 | DSC (1–3) | 98W7 |
| N domain | | 7.0 | 70 | 20 | DSC (1,2) | 98W7 |
| C domain | | 7.0 | 50 | 18 | DSC (1,2) | 98W7 |

Remarks:
(1) protein S is a two-domain protein consisting of 172 amino acid residues with a 88 residues N-terminal domain and a 91 residues C-terminal domain
(2) measured in 20 mM sodium cacodylate, 3 mM $CaCl_2$, pH 7.0
(3) result of deconvolution

Protein S (PS) from *Myxococcus xanthus*, a βγ-crystallin homolog

| Prot. | pH | $T_{trs1}$ | $\Delta H_{trs1}$ | $T_{trs2}$ | $\Delta H_{trs2}$ | Appr./Rem. | Ref |
|---|---|---|---|---|---|---|---|
| protein S in the absence of $Ca^{2+}$: | | | | | | | |
| PS | 1.5 | 26.4 | 204 | 42.9 | 179 | DSC (1,2) | 99W10 |
| | 1.75 | 32.1 | 252 | 47.2 | 222 | DSC (1,2) | 99W10 |
| | 2.0 | 35.4 | 269 | 47.9 | 231 | DSC (1,2) | 99W10 |
| | 2.25 | 38.3 | 275 | 49.3 | 232 | DSC (1,2) | 99W10 |
| | 3.0 | 53.7 | 373 | 54.6 | 291 | DSC (1,2) | 99W10 |
| | 7.0 | 52.8 | 410 | 63.8 | 274 | DSC (1–3) | 99W10 |
| | 7.0 | 52.4 | 399 | 64.0 | 266 | DSC (1–3) | 99W10 |
| | 7.0 | 52.6 | 398 | 64.7 | 259 | DSC (1–3) | 99W10 |
| | 9.0 | 53.0 | 350 | 64.9 | 278 | DSC (1,2) | 99W10 |

Protein S (PS) from *Myxococcus xanthus*, a βγ-crystallin homolog (continued)

| Prot. | pH | $T_{trs1}$ | $\Delta H_{trs1}$ | $T_{trs2}$ | $\Delta H_{trs2}$ | Appr./Rem. | Ref |
|---|---|---|---|---|---|---|---|
| protein S in the presence of Ca²⁺: | | | | | | | |
| PS | 1.5 | 25.5 | 217 | 44.2 | 186 | DSC (1,4) | 99W10 |
| | 1.75 | 28.9 | 213 | 46.3 | 196 | DSC (1,4) | 99W10 |
| | 2.0 | 33.4 | 245 | 47.0 | 206 | DSC (1,4) | 99W10 |
| | 2.25 | 37.6 | 276 | 50.1 | 231 | DSC (1,4) | 99W10 |
| | 2.5 | 43.6 | 322 | 51.9 | 247 | DSC (1,4) | 99W10 |
| | 3.0 | 55.0 | 394 | 57.3 | 323 | DSC (1,4) | 99W10 |
| | 7.0 | 64.0 | 490 | 67.1 | 308 | DSC (1,3,4) | 99W10 |
| | 7.0 | 64.5 | 454 | 65.8 | 343 | DSC (1,3,4) | 99W10 |
| | 7.0 | 64.0 | 454 | 65.2 | 322 | DSC (1,3,4) | 99W10 |
| | 9.0 | 60.3 | 428 | 65.8 | 312 | DSC (1,4) | 99W10 |

Remarks:

(1) results of deconvoltion

(2) measured in the presence of 1 mM EDTA

(3) measured at varying protein conc. and scan rates

(4) measured in the presence of 3 mM $CaCl_2$

N-terminal fragment of protein S (NPS) from *Myxococcus xanthus*

| Protein | pH | $T_{trs}$ | $\Delta H$ | Approach/Remarks | | Ref |
|---|---|---|---|---|---|---|
| NPS in the absence of Ca²⁺: | | | | | | |
| | 6.0 | 68.9 | 284 | DSC | (1) | 99W10 |
| | 7.0 | 67.6 | 291 | DSC | (1,2) | 99W10 |
| | 7.0 | 68.2 | 279 | DSC | (1,2) | 99W10 |
| | 7.0 | 67.7 | 276 | DSC | (1,2) | 99W10 |
| | 9.0 | 66.7 | 272 | DSC | (1) | 99W10 |
| NPS in the presence of Ca²⁺: | | | | | | |
| | 1.75 | 50.1 | 211 | DSC | (3) | 99W10 |
| | 2.0 | 49.9 | 210 | DSC | (3) | 99W10 |
| | 2.25 | 52.7 | 228 | DSC | (3) | 99W10 |
| | 2.5 | 54.7 | 253 | DSC | (3) | 99W10 |
| | 3.0 | 59.0 | 260 | DSC | (3) | 99W10 |
| | 6.0 | 70.1 | 331 | DSC | (3) | 99W10 |
| | 7.0 | 69.4 | 342 | DSC | (2,3) | 99W10 |
| | 7.0 | 70.0 | 330 | DSC | (2,3) | 99W10 |
| | 7.0 | 69.7 | 330 | DSC | (2,3) | 99W10 |
| | 9.0 | 70.1 | 334 | DSC | (3) | 99W10 |

Remarks:

(1) measured in the presence of 1 mM EDTA

(2) measured at varying protein conc. and scan rates

(3) measured in the presence of 3 mM $CaCl_2$

C-terminal fragment of protein S (NPS) from *Myxococcus xanthus*

| Protein | pH | $T_{trs}$ | $\Delta H$ | Approach/Remarks | | Ref |
|---|---|---|---|---|---|---|
| CPS in the absence of $Ca^{2+}$: | | | | | | |
| | 7.0 | 45.2 | 287 | DSC | (1,2) | 99W10 |
| | 7.0 | 44.8 | 281 | DSC | (1,2) | 99W10 |
| | 7.0 | 44.7 | 281 | DSC | (1,2) | 99W10 |
| CPS in the presence of $Ca^{2+}$: | | | | | | |
| | 7.0 | 49.9 | 317 | DSC | (1,2) | 99W10 |
| | 7.0 | 50.2 | 314 | DSC | (1,2) | 99W10 |
| | 7.0 | 50.0 | 319 | DSC | (1,2) | 99W10 |
| | 9.0 | 48.4 | 294 | DSC | (1,2) | 99W10 |

Remarks:
(1) measured in the presence of 1 mM EDTA
(2) measured at varying protein conc. and scan rates
(3) measured in the presence of 3 mM $CaCl_2$

Protein S (PS) from *Myxococcus xanthus*, the N-terminal fragment (NPS) and the C-terminal fragment (CPS)

| Protein | pH | $T_{trs}$ | $\Delta Cp$ | $\Delta H$ | Appr./Rem. | | Ref |
|---|---|---|---|---|---|---|---|
| data in the absence of $Ca^{2+}$: | | | | | | | |
| PS trans. (1) | 7.0 | 52.5±1.1 | 6.7±0.7 | 399±20 | DSC | (1,2) | 99W10 |
| trans. (2) | 7.0 | 64.4±1.3 | 6.1±0.6 | 263±13 | DSC | (1,2) | 99W10 |
| NPS | 7.0 | 67.8±1.4 | 6.1±0.6 | 282±14 | DSC | (1,2) | 99W10 |
| CPS | 7.0 | 44.9±0.9 | 6.3±1.3 | 283±14 | DSC | (1,2) | 99W10 |
| data in the presence of $Ca^{2+}$: | | | | | | | |
| PS trans. (1) | 7.0 | 64.3±1.3 | 6.7±0.7 | 454±23 | DSC | (1,3) | 99W10 |
| trans. (2) | 7.0 | 65.5±1.3 | 6.1±0.6 | 332±17 | DSC | (1,3) | 99W10 |
| NPS | 7.0 | 69.7±1.4 | 6.1±0.6 | 334±17 | DSC | (1,3) | 99W10 |
| CPS | 7.0 | 50.0±1.0 | 6.3±1.3 | 317±16 | DSC | (1,3) | 99W10 |

Remarks:
(1) $\Delta Cp$ from $\Delta H^{cal}$ versus $T_{trs}$
(1) measured in the presence of 1 mM EDTA
(2) measured in the presence of 3 mM $CaCl_2$

## Proteins of Newcastle Disease Virus

Newcastle disease virus, constitutive proteins

| Component | pH | $T_{trs}$ | $\Delta H$ | Approach/Remarks | Ref |
|---|---|---|---|---|---|
| F protein | 7.4 | 51.6±0.5 | 280±29 | (1) | 97S11 |
| HN protein | 7.4 | 61.6±0.4 | 163±21 | (1) | 97S11 |
| P protein | 7.4 | 66.4±0.6 | 276±21 | (1) | 97S11 |
| NP protein | 7.4 | 68.5±0.5 | 213±25 | (1) | 97S11 |
| M protein | 7.4 | 71.9±0.5 | 293±29 | (1) | 97S11 |

Remark:
(1) data from combined use of various approaches including DSC, fluorescence, thermal aggregation and kinetics

**Protooncogene Product** (c-Myb)

c-myb protooncogene product, R2 subdomain of mouse c-Myb with natural and nonnatural amino acid replacements at position 103

| Protein | pH | $T_{trs}$ | $\Delta Cp$ | $\Delta H$ | Appr./Rem. | Ref |
|---------|-----|-----------|-------------|------------|------------|------|
| Ala-R2 | 7.5 | 2.8±0.3 | | 34.3±1.4 | heat (1–3) | 99M20 |
| Abu-R2 | 7.5 | 39.6±0.2 | | 138.2±2.3 | heat (1–3) | 99M20 |
| | 7.5 | 41.6 | | 129.0 | DSC (4,5) | 99M20 |
| Val-R2 | 7.5 | 39.6±0.1 | | 133.3±0.8 | heat (1–3,7) | 99M20 |
| | 7.5 | 39.0 | | 120.1 | DSC (4,5,7) | 99M20 |
| Cha-R2 | 7.5 | 39.6±0.3 | | 91.6±2.2 | heat (1–3) | 99M20 |
| | 7.5 | 40.0 | | 94.2 | DSC (4,5) | 99M20 |
| Ail-R2 | 7.5 | 48.0±0.1 | | 141.1±0.8 | heat (1–3) | 99M20 |
| | 7.5 | 45.9 | | 127.9 | DSC (4,5) | 99M20 |
| Ile-R2 | 7.5 | 46.6±0.2 | | 143.7±3.3 | heat (1–3) | 99M20 |
| | 7.5 | 47.5 | | 134.4 | DSC (4,5) | 99M20 |
| Nle-R2 | 7.5 | 53.4±0.1 | | 142.6±0.9 | heat (1–3) | 99M20 |
| | 7.5 | 50.9 | | 134.3 | DSC (4,5) | 99M20 |
| Chg-R2 | 7.5 | 54.4±0.1 | | 135.3±0.8 | heat (1–3) | 99M20 |
| | 7.5 | 51.9 | | 130.6 | DSC (4,5) | 99M20 |
| Nva-R2 | 7.5 | 55.6±0.1 | | 162.0±1.8 | heat (1–3) | 99M20 |
| | 7.5 | 55.5 | | 158.9 | DSC (4,5) | 99M20 |
| Leu-R2 | 7.5 | 62.0±0.2 | | 171.1±2.5 | heat (1–3) | 99M20 |
| | 7.5 | 62.0 | | 175.0 | DSC (4,5) | 99M20 |
| Leu-R2 | 5.5–7.0 | 41–61 | 1.34±0.31 | | heat (6) | 99M20 |
| Val-R2 | 5.5–7.0 | 21–40 | 1.35±0.34 | | heat (6–9) | 99M20 |

Explanations:

[Xaa103]-Myb(90–141) changing Val to Ala, Leu, Ile, 2-aminobutyric acid (Abu), norvaline (Nva), norleucine (Nle), *allo*-isoleucine (Ail), cyclohexylglycine (Chg), and cyclohexylalanine (Cha).

Remarks:

(1) measured in 50 mM potassium phosphate, 50 mM potassium chloride, 10 mM DTT, pH 7.5
(2) transition monitored by CD at 222 nm
(3) van't Hoff heat
(4) measured in 50 mM potassium phosphate, 50 mM potassium chloride, 1 mM DTT, pH 7.5
(5) calorimetric heat after correction for oxidation of free thiol groups
(6) $\Delta Cp$ from $\Delta H^{v.H.}$ versus $T_{trs}$ at varying pH (CD melting curves)
(7) Val-R2 = wild type
(8) $\Delta Cp$ from $\Delta H^{v.H.}$ versus $T_{trs}$ for the mutants Val-R2, Abu-R2, Ile-R2, Ail-R2, and Nva-R2 amounts to 1.62 kJ/mol
(9) mutant Cys130→Abu yields $T_{trs}$ = 40.8°C and $\Delta H$ = 109.8 kJ/mol

# RAP

Recombinant human 39-kDa receptor-associated protein (RAP), domain structure

| Fragment/Transition | | pH | $T_{trs}$ | $\Delta H$ | Appr./Rem. | Ref |
|---------------------|-----------|-----|-----------|------------|------------|------|
| RAP-(1–323) | trans. (1) | 8.7 | 41.0 | 197 | DSC (1,2) | 99M13 |
| | trans. (2) | 8.7 | 51.2 | 163 | DSC (1,2) | 99M13 |
| | trans. (3) | 8.7 | 60.4 | 172 | DSC (1,2) | 99M13 |
| | trans. (4) | 8.7 | 65.9 | 180 | DSC (1,2) | 99M13 |
| fragments: | | | | | | |
| (206–323) | trans. (1) | 8.7 | 43.1 | 188 | DSC (1,2) | 99M13 |
| (1–216) | trans. (2) | 8.7 | 54.1 | 176 | DSC (1,2) | 99M13 |
| | trans. (3) | 8.7 | 61.1 | 151 | DSC (1,2) | 99M13 |

Recombinant human 39-kDa receptor-associated protein (RAP), domain structure (continued)

| Fragment/Transition | | pH | $T_{trs}$ | $\Delta H$ | Appr./Rem. | Ref |
|---|---|---|---|---|---|---|
| | trans. (4) | 8.7 | 66.1 | 172 | DSC (1,2) | 99M13 |
| (1–164) | trans. (3) | 8.7 | 60.8 | 163 | DSC (1,2) | 99M13 |
| | trans. (4) | 8.7 | 64.6 | 167 | DSC (1,2) | 99M13 |
| (1–175) | trans. (3) | 8.7 | 60.6 | 172 | DSC (1,2) | 99M13 |
| | trans. (4) | 8.7 | 63.6 | 172 | DSC (1,2) | 99M13 |
| (1–99) | trans. (4) | 8.7 | 65.8 | 234 | DSC (1,2) | 99M13 |
| (1–92) | trans. (4) | 8.7 | 67.9 | 247 | DSC (1,2) | 99M13 |
| (89–216) | trans. (2) | 8.7 | 54.0 | 172 | DSC (1,2) | 99M13 |
| (82–323) | trans. (1) | 8.7 | 41.1 | 188 | DSC (1,2) | 99M13 |
| | trans. (2) | 8.7 | 53.6 | 155 | DSC (1,2) | 99M13 |

Remarks:

(1) results of deconvolution

(2) measured in 20 mM glycine, pH 8.7, in the presence of 0.25 M GuHCl

## Repressor Proteins

The data entries are arranged as follows:

a) Arc repressor
b) $\lambda$ repressor
c) Bacteriophage $\lambda$ cI repressor
d) 434-phage repressor
e) tryptophan repressor

*a) Arc repressor*

Arc *st11* repressor dimer, wild type and MYL mutant, transitions

| Transitions | pH | $T_{ref}$ | $\Delta Cp$ | $\Delta H$ | Approach/Remarks | | Ref |
|---|---|---|---|---|---|---|---|
| wild type, two-state: | | | | | | | |
| $N_2 \rightarrow 2D$ | 7.5 | 25 | 5.4 | 33 | heat | (1–4) | 97R5 |
| wild type, three-state: | | | | | | | |
| $N_2 \rightarrow D_2$ | 7.5 | 25 | 4.2 | 38 | heat | (1,2) | 97R5 |
| $D_2 \rightarrow 2D$ | 7.5 | 25 | 1.3 | –4 | heat | (1,2) | 97R5 |
| MYL mutant, two-state: | | | | | | | |
| $N_2 \rightarrow 2D$ | 7.5 | 25 | 5.4 | 42 | heat | (1,2) | 97R5 |
| MYL mutant, three-state: | | | | | | | |
| $N_2 \rightarrow D_2$ | 7.5 | 25 | 4.2 | 50 | heat | (1,2) | 97R5 |
| $D_2 \rightarrow 2D$ | 7.5 | 25 | 1.3 | –8 | heat | (1,2) | 97R5 |

Remarks:

(1) from thermal denaturation measured varying protein concentration from 5 to 334 μM, data at reference temperature $T_{ref} = 25°C$

(2) buffer: 50 mM Tris-HCl, 250 mM KCl, 0.1 mM EDTA, pH 7.5

(3) poor fit at protein concentration above 160 μM

(4) formation of a denatured dimer limits the thermal stability of Arc repressor

(5) MYL mutant = triple mutant (Arg31→Met, Glu36→Tyr, and Arg40→Leu) of Arc *st11* which has a C-terminal extension $(His)_6$-Lys-Asn-Gln-His-Glu

*b) λ repressor*

λ repressor, wild type and fragments

| Fragment/Trans. | | pH | $T_{trs}$ | $\Delta H$ | Approach/Remarks | | Ref |
|---|---|---|---|---|---|---|---|
| intact repressor (residues 1–236) | | | | | | | |
| | trans. (1) | 5.5 | 49 | 264 | DSC | (1) | 79P7 |
| | trans. (2) | 5.5 | 73 | 586 | DSC | (1) | 79P7 |
| fragments (residues): | | | | | | | |
| (1–92) | | 5.5 | 51 | 335 | DSC | (1) | 79P7 |
| (93–236) | | 5.5 | 72 | 481 | DSC | (1) | 79P7 |
| (122–236) | | 5.5 | 68 | 544 | DSC | (1) | 79P7 |
| (132–236) | | 5.5 | 67 | 523 | DSC | (1) | 79P7 |

Remark:

(1) measured in 50 mM sodium citrate

*c) Bacteriophage λ cI repressor*

Bacteriophage λ cI repressor, N-terminal domains of wild type and mutants

| Mutant | pH | $T_{trs}$ | $\Delta H$ | Approach/Remarks | | Ref |
|---|---|---|---|---|---|---|
| wild type | 7.0 | 50.7±0.1 | 224.3±16.7 | DSC | (1,2) | 98M15 |
| Glu102→Lys | 7.0 | 51.8±0.1 | 169.9±12.6 | DSC | (1,2) | 98M15 |
| Gly147→Asp | 7.0 | 56.4±0.1 | 220.1±12.6 | DSC | (1,2) | 98M15 |
| Pro158→Thr | 7.0 | 50.8±0.1 | 179.1± 7.5 | DSC | (1,2) | 98M15 |
| Ser228→Arg | 7.0 | 55.3±0.1 | 352.7±18.8 | DSC | (1,2) | 98M15 |
| Ser228→Asn | 7.0 | 54.6±0.1 | 261.9±20.9 | DSC | (1,2) | 98M15 |
| Glu188→Lys | 7.0 | 54.6±0.1 | 311.3±16.7 | DSC | (1,2) | 98M15 |
| Lys192→Asn | 7.0 | 56.0±0.1 | 241.4± 8.4 | DSC | (1,2) | 98M15 |
| Tyr210→His | 7.0 | 55.8±0.1 | 236.4±12.6 | DSC | (1,2) | 98M15 |

Remarks:

(1) buffer: 10 mM Bis-Tris, 200 mM KCl, 2.5 mM $MgCl_2$, 1.0 mM $CaCl_2$, pH 7.00

(2) for previous calorimetric data on the domain structure of λ repressor (fragments obtained by papain digestion) see C.O. Pabo et al., Proc. Natl. Acad. Sci. USA 76 (1979) 1608–1612

Bacteriophage λ cI repressor, C-terminal domains of wild type and mutants at pH 7.0

| Mutant | $T_{trs1}$ | $\Delta H_{trs1}$ | $T_{trs2}$ | $\Delta H_{trs2}$ | Appr./Rem. | | Ref |
|---|---|---|---|---|---|---|---|
| wild type | 69.1±1.2 | 131.8±12.6 | 72.4±0.5 | 341.0±14.2 | DSC | (1,2) | 98M15 |
| Glu102→Lys | 69.4±0.1 | 286.2±12.6 | 73.7±0.2 | 277.8±10.5 | DSC | (1,2) | 98M15 |
| Gly147→Asp | 65.3±0.2 | 113.0±37.7 | 67.2±0.2 | 221.8±12.6 | DSC | (1,2) | 98M15 |
| Pro158→Thr | 60.3±0.1 | 228.0±14.6 | 63.2±0.2 | 209.6±12.6 | DSC | (1,2) | 98M15 |
| Ser228→Arg | 58.6±0.4 | 80.8±25.1 | 59.7±0.3 | 188.7±12.6 | DSC | (1,2) | 98M15 |
| Ser228→Asn | 59.0±0.5 | 154.4±24.3 | 61.1±0.3 | 196.6± 8.4 | DSC | (1,2) | 98M15 |
| Glu188→Lys | 66.2±0.2 | 103.3±12.6 | 68.0±0.2 | 240.6±10.5 | DSC | (1,2) | 98M15 |
| Lys192→Asn | 60.7±0.2 | 74.5± 9.2 | 63.9±0.1 | 269.4± 5.0 | DSC | (1,2) | 98M15 |
| Tyr210→His | 64.1±0.2 | 121.8± 4.2 | 65.3±0.3 | 239.7± 8.4 | DSC | (1,2) | 98M15 |

Remarks:

(1) buffer: 10 mM Bis-Tris, 200 mM KCl, 2.5 mM $MgCl_2$, 1.0 mM $CaCl_2$, pH 7.00

(2) for previous calorimetric data on the domain structure of λ repressor (fragments obtained by papain digestion) see C.O. Pabo et al., Proc. Natl. Acad. Sci. USA 76 (1979) 1608–1612

Bacteriophage λ cI repressor, N-terminal domains of wild type and mutant repressors in the presence of OR1 DNA

| Mutant | pH | $T_{trs}$ | ΔH | Approach/Remarks | | Ref |
|---|---|---|---|---|---|---|
| wild type | 7.0 | 59.4 | 502 | DSC | (1,2) | 98M15 |
| Glu102→Lys | 7.0 | 57.3 | 251 | DSC | (1,2) | 98M15 |
| Gly147→Asp | 7.0 | 61.8 | 427 | DSC | (1,2) | 98M15 |
| Pro158→Thr | 7.0 | 55.4 | 351 | DSC | (1,2) | 98M15 |
| Ser228→Asn | 7.0 | 58.0 | 335 | DSC | (1,2) | 98M15 |

Remarks:

(1) buffer: 10 mM Bis-Tris, 200 mM KCl, 2.5 mM $MgCl_2$, 1.0 mM $CaCl_2$, pH 7.00

(2) repressor mixed with a stoichiometric excess of a 30-bp oligomer containing the right operator site 1 (OR1)

*d) 434-phage repressor*

434-phage repressor, N-terminal domain (R69) and the covalently linked dimers (RR69 and [R69C]$_2$)

| Protein/Salt | | pH | $T_{trs}$ | ΔCp | ΔH | Appr./Rem. | | Ref |
|---|---|---|---|---|---|---|---|---|
| R69 | L | 7.0 | 70.7±0.5 | | 247±10 | DSC | (1,2) | 99R21 |
| | H | 7.0 | 70.7±0.5 | | 245±10 | DSC | (1,2) | 99R21 |
| | L | 4.5 | 64.1±0.5 | | 222±10 | DSC | (1,2) | 99R21 |
| | H | 4.5 | 63.7±0.5 | | 226±10 | DSC | (1,2) | 99R21 |
| | L | 4.0 | 58.2±0.5 | | 209±8 | DSC | (1,2) | 99R21 |
| | H | 4.0 | 59.8±0.5 | | 214±10 | DSC | (1,2) | 99R21 |
| | L | 3.0 | 50.5±0.8 | | 183±8 | DSC | (1,2) | 99R21 |
| | H | 3.0 | 48.2±0.8 | | 176±8 | DSC | (1,2) | 99R21 |
| | L | 2.0 | 27.7±2.0 | | 101±8 | DSC | (1,2) | 99R21 |
| | H | 2.0 | 38.3±2.0 | | 140±10 | DSC | (1,2) | 99R21 |
| | H/L | 2–7 | | 3.33±0.07 | | DSC | (1–3) | 99R21 |
| RR69 | L | 7.0 | 64.8±0.7 | | 210±15 | DSC | (1–3) | 99R21 |
| | H | 7.0 | 66.6±0.6 | | 241±10 | DSC | (1–3) | 99R21 |
| | L | 4.5 | 60.1±0.7 | | 194±8 | DSC | (1,2) | 99R21 |
| | H | 4.5 | 63.5±0.5 | | 229±10 | DSC | (1,2) | 99R21 |
| | L | 4.0 | 58.2±0.5 | | 191±8 | DSC | (1,2) | 99R21 |
| | H | 4.0 | 57.8±0.7 | | 209±8 | DSC | (1,2) | 99R21 |
| | L | 3.0 | 42.3±1.0 | | 144±8 | DSC | (1,2) | 99R21 |
| | H | 3.0 | 46.9±1.0 | | 169±8 | DSC | (1,2) | 99R21 |
| | L | 2.0 | 32.2±1.5 | | 115±8 | DSC | (1,2) | 99R21 |

434-phage repressor, N-terminal domain (R69) and the covalently linked dimers (RR69 and [R69C]$_2$) (continued)

| Protein/Salt | | pH | $T_{trs}$ | $\Delta Cp$ | $\Delta H$ | Appr./Rem. | | Ref |
|---|---|---|---|---|---|---|---|---|
| [R69C]$_2$ | H | 2.0 | 39.4±1.2 | | 142±10 | DSC | (1,2) | 99R21 |
| | L | 7.0 | 67.2±0.5 | | 226±10 | DSC | (1–3) | 99R21 |
| | H | 7.0 | 67.6±0.6 | | 243±10 | DSC | (1–3) | 99R21 |
| | L | 4.5 | 58.8±0.5 | | 194±8 | DSC | (1,2) | 99R21 |
| | H | 4.5 | 60.2±0.6 | | 218±10 | DSC | (1,2) | 99R21 |
| | L | 4.0 | 53.6±0.5 | | 185±8 | DSC | (1,2) | 99R21 |
| | H | 4.0 | 57.2±0.5 | | 208±8 | DSC | (1,2) | 99R21 |
| | L | 3.0 | 37.2±1.3 | | 134±8 | DSC | (1,2) | 99R21 |
| | H | 3.0 | 42.8±1.0 | | 163±8 | DSC | (1,2) | 99R21 |
| | L | 2.0 | 24.4±2.0 | | 93±10 | DSC | (1,2) | 99R21 |
| | H | 2.0 | 35.6±1.5 | | 143±8 | DSC | (1,2) | 99R21 |

Explanations:

R69 = isolated 1–69 N-terminal fragment

RR69 = single-chain, heat-to-tail tandem of R69 linked by a 20-amino acid natural linker

[R69C]$_2$ = tail-to-tail cross-linked symmetric tandem of R69, linked via a disulfide bridge between Cys residues added to the termini

Remarks:

(1) the parameters are calculated per mole of R69 domain

(2) measured in 20 mM buffer: PIPES (pH 7.0), sodium acetate (pH 4.5 and 4.0), and glycine (pH 3.0 and 2.0), L = low salt, H = high salt – the same buffer with 200 mM NaCl

(3) $\Delta Cp$ from $\Delta H^{cal}$ versus $T_{trs}$ (DSC data), $\Delta Cp$ has the same value for RR69 and [R69C]$_2$

*e) tryptophan repressor*

Tryptophan repressor, wild type and temperature-sensitive mutant Leu75→Phe

| Protein | pH | $T_{trs}$ | $\Delta Cp$ | $\Delta H$ | Appr./Rem. | Ref |
|---|---|---|---|---|---|---|
| wild type | 7.5 | 91.5 | 5.0±0.8 | 377±21 | DSC (1) | 99J10 |
| Leu75→Phe | 7.5 | 91.5 | 4.6±0.8 | 381±21 | DSC (1) | 99J10 |

Remark:

(1) measured at 73 μM dimer concentration in 10 mM sodium phosphate, 0.1 M NaCl, pH 7.5

## Ribonuclease A

The data entries are arranged as follows:

a) comparison of various approaches
b) measurements in the presence of urea, alkylurea, and GuHCl
c) wild type and mutants
d) effect of selective deamidation
e) data obtained in the presence of cosolutes

*a) comparison of various approaches*

Bovine pancreatic ribonuclease A (RNase A)

| Protein | pH | $T_{trs}$ | $\Delta Cp$ | $\Delta H$ | Approach/Remarks | Ref |
|---|---|---|---|---|---|---|
| RNase A | 2.8 | 44.9 | | 332.2 | DSC (1) | 99P1 |
| | 3.0 | 49.1 | | 346.0 | DSC (1) | 99P1 |
| | 3.6 | 54.5 | | 382.8 | DSC (1) | 99P1 |
| | 4.0 | 56.1 | | 394.1 | DSC (1) | 99P1 |

Bovine pancreatic ribonuclease A (RNase A) (continued)

| Protein | pH | $T_{trs}$ | $\Delta Cp$ | $\Delta H$ | Approach/Remarks | | Ref |
|---------|-----|-----------|-------------|------------|------------------|---|-----|
| | 5.0 | 58.6 | | 414.6 | DSC | (1) | 99P1 |
| | 6.0 | 60.3 | | 421.3 | DSC | (1) | 99P1 |
| | 7.0 | 61.8 | | 428.0 | DSC | (1) | 99P1 |
| | 2–5 | 30–60 | 4.56±0.29 | | DSC | (2) | 99P1 |
| | 2–5 | 30–60 | 4.48±0.33 | | DSC | (3) | 99P1 |
| | 2–5 | 30–60 | 4.56±0.42 | | DSC | (4) | 99P1 |
| | 2–5 | 30–60 | 4.52±1.76 | | DSC | (5) | 99P1 |
| | 2–5 | 30–60 | 4.31±1.34 | | DSC | (6) | 99P1 |
| | | 25–60 | 4.73±0.33 | | DSC | (7) | 99P1 |
| | | | 5.36±0.84 | | DSC | (8) | 99P1 |
| | | | 4.69±0.67 | | DSC | (9) | 99P1 |
| | | | 4.81±0.38 | | DSC | (10) | 99P1 |

Remarks:

(1) selected values from Ref. 99P1, the data were used to calculate $\Delta G$, see also Table 1 (reconsideration of data from Refs. 89P2 and 90P1)

(2) $\Delta Cp$ from $\Delta H^{cal}$ versus $T_{trs}$ (DSC data)

(3) $\Delta Cp$ from $\Delta H^{v.H.}$ versus $T_{trs}$ (DSC data)

(4) $\Delta Cp$ from $\Delta H_{fit}$ versus $T_{trs}$ (DSC data)

(5) $\Delta Cp$ from global fit (DSC data)

(6) $\Delta Cp$ from Cp(D) – Cp(N)(DSC data)

(7) $\Delta Cp$ from $\Delta H^{v.H.}$ versus $T_{trs}$ (optically monitored thermal unfolding)

(8) data from urea-induced unfolding from Ref. 89P2 and values from Ref. 99P1

(9) data from urea-induced unfolding from Ref. 90P1 and values from Ref. 99P1

(10) average, recommended value from various approaches

Bovine pancreatic ribonuclease A

| pH | $T_{trs}$ | $\Delta Cp$ | $\Delta H$ | Appr./Rem. | | Ref |
|-----|-----------|-------------|------------|------------|---|-----|
| 2.0 | 26.7 | 8.4±0.8 | 276 | heat | (1) | 71S1 |
| 2.8 | 42.6 | 7.5±2.1 | 397±21 | DSC | (2) | 78F2 |

Remarks:

(1) measured without using buffer substances, transition monitored by optical absorption at 287 nm

(2) measured in 0.2 M glycine, 1 mM EDTA, pH 2.8, using 8 mg/ml protein

Bovine pancreatic ribonuclease A

| Protein | pH | $T_{trs}$ | $\Delta Cp$ | $\Delta H$ | Approach/Remarks | | Ref |
|---------|-----|-----------|-------------|------------|------------------|---|-----|
| RNase A | 2.0 | 35.2 | 4.8 | 274 | DSC | (1) | 92B11 |
| | 3.7 | 56.8 | 6.4 | 439 | DSC | (1) | 92B11 |
| | 5.0 | 62.9 | 5.3 | 492 | DSC | (1) | 92B11 |
| | 6.0 | 63.1 | 5.6 | 501 | DSC | (1) | 92B11 |
| | 7.0 | 63.9 | 5.3 | 519 | DSC | (1) | 92B11 |
| | 8.0 | 63.9 | 5.2 | 524 | DSC | (1) | 92B11 |

Remark:

(1) buffers used: 0.1 M formate for pH 3.7, 0.1 M acetate for pH 4.5 and 5.0, 0.1 M MES for pH 5.7 and 6.3, 0.1 MOPS for pH 7.0, and HEPPS for pH 8.0

Ribonuclease A, unfolding transition monitored by FTIR spectroscopy

| | pH | $T_{trs}$ | $\Delta H$ | Approach/Remarks | Ref |
|---|---|---|---|---|---|
| | 3.8 | 63.0 | 430 | heat, v.H.(1,3) | 96R4 |
| | 3.8 | 62.9 | 450 | heat, v.H.(2,3) | 96R4 |

Remarks:

(1) β-sheet specific transition monitored by the peak intensity of the amide I' band at 1632 cm$^{-1}$
(2) tyrosine-specific transition monitored by the frequency shift of the tyrosine band near 1515 cm$^{-1}$
(3) measured in 20 mM sodium citrate/$D_2O$ buffer , pH* 3.8, using 25 mg/ml RNase A

Bovine pancreatic ribonuclease A (RNase A) in $H_2O$ and $D_2O$

| Protein | Solvent | pH | $T_{trs}$ | $\Delta Cp$ | $\Delta H$ | Appr./Rem. | | Ref |
|---|---|---|---|---|---|---|---|---|
| RNase A | $H_2O$ | 2.5 | 40.0±0.2 | 4.8±0.2 | 356±8 | DSC | (1) | 99N3 |
| | $D_2O$ | 2.5 | 44.8±0.2 | 6.3±0.8 | 305±8 | DSC | (1) | 99N3 |
| | $H_2O$ | 6.0 | 63.7±0.2 | 5.4±0.4 | 477±33 | DSC | (1) | 99N3 |
| | $D_2O$ | 6.0 | 66.6±0.2 | 6.3±0.8 | 439±17 | DSC | (1) | 99N3 |

Remark:

(1) see also data for ribonuclease S in $H_2O$ and $D_2O$ in Ref. 99N3

Bovine pancreatic ribonuclease A, wild type and mutants, thermal denaturation compared with pressure denaturation (see Table 1)

| Protein | pH | $T_{trs}$ | $\Delta H$ | Approach/Remarks | | Ref |
|---|---|---|---|---|---|---|
| wild type | 5.0 | 58.9±0.0 | 487±9 | heat | (1–3) | 99T12 |
| rec. w.t. | 5.0 | 58.0±0.1 | 437±12 | heat | (1–4) | 99T12 |
| Tyr115→Trp | 5.0 | 54.2±0.1 | 416±19 | heat | (1–3) | 99T12 |
| Tyr115→Trp | 5.0 | 55.3±0.2 | 378±17 | heat | (1,3,5) | 99T12 |
| Ile106→Ala | 5.0 | 43.1±0.2 | 301±14 | heat | (1–3) | 99T12 |
| Lle106→Leu | 5.0 | 51.7±0.1 | 410±15 | heat | (1–3) | 99T12 |
| Ile106→Val | 5.0 | 55.0±0.1 | 431±21 | heat | (1–3) | 99T12 |
| Ile107→Ala | 5.0 | 47.7±0.1 | 405±13 | heat | (1–3) | 99T12 |
| Lle107→Leu | 5.0 | 49.4±0.1 | 435±19 | heat | (1–3) | 99T12 |
| Ile107→Val | 5.0 | 58.6±0.1 | 422±17 | heat | (1–3) | 99T12 |
| Val108→Ala | 5.0 | 43.3±0.1 | 350±10 | heat | (1–3) | 99T12 |
| Val108→Gly | 5.0 | 29.4±0.2 | 237±9 | heat | (1–3) | 99T12 |
| Ala109→Gly | 5.0 | 55.4±0.1 | 455±21 | heat | (1–3) | 99T12 |
| Val116→Ala | 5.0 | 54.8±0.2 | 447±27 | heat | (1–3) | 99T12 |
| Val116→Gly | 5.0 | 53.6±0.1 | 468±52 | heat | (1–3) | 99T12 |
| Val118→Ala | 5.0 | 52.5±0.1 | 454±89 | heat | (1–3) | 99T12 |
| Val118→Gly | 5.0 | 47.9±0.1 | 451±58 | heat | (1–3) | 99T12 |

Remarks:

(1) data from thermal denaturation, see also Table 1
(2) transition monitored by absorbance, 4$^{th}$ derivative spectroscopy
(3) buffer: 50 mM MES, pH 5.0
(4) rec. w.t. = recombinant wild-type protein
(5) transition monitored by fluorescence spectroscopy

*b) measurements in the presence of urea, substituted urea, and GuHCl*

Bovine pancreatic ribonuclease A in the presence of urea and GuHCl

| Denaturant conc. | | pH | $T_{trs}$ | $\Delta H$ | Approach/Remarks | Ref |
|---|---|---|---|---|---|---|
| | 0.00 M | 5.0 | 57.9±0.1 | 402±9 | heat, v.H. (1–3) | 97D2 |
| | 0.00 M | 5.0 | 57.7±0.1 | 410±10 | heat, v.H. (1–3) | 97D2 |
| urea | 0.25 M | 5.0 | 56.7±0.1 | 401±8 | heat, v.H. (1–3) | 97D2 |
| | 0.50 M | 5.0 | 55.3±0.1 | 393±19 | heat, v.H. (1–3) | 97D2 |
| | 0.75 M | 5.0 | 54.3±0.1 | 400±8 | heat, v.H. (1–3) | 97D2 |
| | 1.00 M | 5.0 | 53.0±0.1 | 394±20 | heat, v.H. (1–3) | 97D2 |
| GuHCl | 0.10 M | 5.0 | 58.4±0.1 | 410±8 | heat, v.H. (1–3) | 97D2 |
| | 0.25 M | 5.0 | 56.8±0.1 | 410±8 | heat, v.H. (1–3) | 97D2 |
| | 0.25 M | 5.0 | 57.1±0.1 | 428±9 | heat, v.H. (1–3) | 97D2 |
| | 0.50 M | 5.0 | 54.4±0.1 | 391±8 | heat, v.H. (1–3) | 97D2 |
| | 0.75 M | 5.0 | 52.1±0.1 | 379±8 | heat, v.H. (1–3) | 97D2 |
| | 1.00 M | 5.0 | 49.5±0.1 | 369±8 | heat, v.H. (1–3) | 97D2 |

Remarks:

(1) measured in 10 mM potassium acetate, pH 5.0

(2) transition monitored by CD at 222 nm

(3) in Ref. 97D2 it is shown that calorimetrically-derived parameters for protein-denaturant interaction are not consistent with denaturant m values

Bovine pancreatic ribonuclease A (RNase A) alone and in the presence of GuHCl, urea and alkylurea

| Conc. (M) | pH | $T_{trs}$ | $\Delta Cp$ | $\Delta H$ | Approach/Remarks | | Ref |
|---|---|---|---|---|---|---|---|
| RNase A alone, pH dependence: | | | | | | | |
| | 1.1 | 28.2 | | 181 | DSC | (1) | 99P9 |
| | 1.1 | 25.9 | | 258 | heat | (2) | 99P9 |
| | 1.5 | 32.3 | | 199 | DSC | (1) | 99P9 |
| | 1.5 | 28.7 | | 257 | heat | (2) | 99P9 |
| | 2.0 | 37.0 | | 288 | DSC | (1) | 99P9 |
| | 2.0 | 33.0 | | 274 | heat | (2) | 99P9 |
| | 3.0 | 50.2 | | 400 | DSC | (1) | 99P9 |
| | 3.0 | 46.1 | | 364 | heat | (2) | 99P9 |
| | 3.5 | 59.4 | | 458 | DSC | (1) | 99P9 |
| | 3.5 | 52.7 | | 394 | heat | (2) | 99P9 |
| | 7.0–7.4 | 64.4 | | 515 | DSC | (1) | 99P9 |
| | 7.0–7.4 | 62.8 | | 452 | heat | (2) | 99P9 |
| RNase A in the presence of GuHCl: | | | | | | | |
| 1.0 | 7.0–7.4 | 54.3 | | 424 | DSC | (1) | 99P9 |
| | 7.0–7.4 | 51.6 | | 407 | heat | (2) | 99P9 |
| 1.5 | 7.0–7.4 | 48.4 | | 336 | DSC | (1) | 99P9 |
| | 7.0–7.4 | 46.7 | | 365 | heat | (2) | 99P9 |
| 2.0 | 7.0–7.4 | 43.0 | | 270 | DSC | (1) | 99P9 |
| | 7.0–7.4 | 41.7 | | 332 | heat | (2) | 99P9 |
| 2.5 | 7.0–7.4 | 37.1 | | 217 | DSC | (1) | 99P9 |
| | 7.0–7.4 | 33.4 | | 268 | heat | (2) | 99P9 |
| 3.0 | 7.0–7.4 | 30.5 | | 171 | DSC | (1) | 99P9 |
| | | | 10.6 | | DSC | (3) | 99P9 |
| RNase A in the presence of urea: | | | | | | | |
| 2.0 | 7.0–7.4 | 57.4 | | | DSC | (1) | 99P9 |
| | 7.0–7.4 | 55.4 | | 401 | heat | (2) | 99P9 |
| 3.0 | 7.0–7.4 | 53.6 | | 364 | DSC | (1) | 99P9 |
| | 7.0–7.4 | 51.8 | | 351 | heat | (2) | 99P9 |

Bovine pancreatic ribonuclease A (RNase A) alone and in the presence of GuHCl, urea and alkylurea (continued)

| Conc. (M) | pH | $T_{trs}$ | $\Delta Cp$ | $\Delta H$ | Approach/Remarks | | Ref |
|---|---|---|---|---|---|---|---|
| 4.0 | 7.0–7.4 | 49.6 | | 372 | DSC | (1) | 99P9 |
| | 7.0–7.4 | 47.4 | | 319 | heat | (2) | 99P9 |
| 5.0 | 7.0–7.4 | 45.6 | | 307 | DSC | (1) | 99P9 |
| | 7.0–7.4 | 42.7 | | 306 | heat | (2) | 99P9 |
| 6.0 | 7.0–7.4 | 41.2 | | 250 | DSC | (1) | 99P9 |
| | 7.0–7.4 | 37.6 | | 278 | heat | (2) | 99P9 |
| 7.0 | 7.0–7.4 | 35.2 | | 184 | DSC | (1) | 99P9 |
| | 7.0–7.4 | 32.6 | | | heat | (2) | 99P9 |
| 8.0 | 7.0–7.4 | 32.6 | | 96 | DSC | (1) | 99P9 |
| | 7.0–7.4 | 28.6 | | | heat | (2) | 99P9 |
| | | | 9.6 | | DSC | (3) | 99P9 |

RNase A in the presence of methylurea:

| Conc. (M) | pH | $T_{trs}$ | $\Delta Cp$ | $\Delta H$ | Approach/Remarks | | Ref |
|---|---|---|---|---|---|---|---|
| 2.0 | 7.0–7.4 | 55.1 | | 483 | DSC | (1) | 99P9 |
| | 7.0–7.4 | 53.2 | | 473 | heat | (2) | 99P9 |
| 3.0 | 7.0–7.4 | 51.1 | | 455 | DSC | (1) | 99P9 |
| | 7.0–7.4 | 49.5 | | 381 | heat | (2) | 99P9 |
| 4.0 | 7.0–7.4 | 47.8 | | 420 | DSC | (1) | 99P9 |
| | 7.0–7.4 | 45.5 | | 389 | heat | (2) | 99P9 |
| 5.0 | 7.0–7.4 | 44.5 | | 372 | DSC | (1) | 99P9 |
| | 7.0–7.4 | 41.8 | | 357 | heat | (2) | 99P9 |
| 6.0 | 7.0–7.4 | 41.5 | | 364 | DSC | (1) | 99P9 |
| | 7.0–7.4 | 38.2 | | 355 | heat | (2) | 99P9 |
| 7.0 | 7.0–7.4 | 37.9 | | 349 | DSC | (1) | 99P9 |
| | 7.0–7.4 | 34.6 | | 348 | heat | (2) | 99P9 |
| 8.0 | 7.0–7.4 | 35.0 | | 335 | DSC | (1) | 99P9 |
| | | | 7.6 | | DSC | (3) | 99P9 |

RNase A in the presence of N,N'-dimethylurea:

| Conc. (M) | pH | $T_{trs}$ | $\Delta Cp$ | $\Delta H$ | Approach/Remarks | | Ref |
|---|---|---|---|---|---|---|---|
| 1.0 | 7.0–7.4 | 58.0 | | 466 | DSC | (1) | 99P9 |
| | 7.0–7.4 | 56.1 | | 423 | heat | (2) | 99P9 |
| 2.0 | 7.0–7.4 | 53.4 | | 465 | DSC | (1) | 99P9 |
| | 7.0–7.4 | 51.3 | | 421 | heat | (2) | 99P9 |
| 3.0 | 7.0–7.4 | 49.8 | | 469 | DSC | (1) | 99P9 |
| | 7.0–7.4 | 48.3 | | 420 | heat | (2) | 99P9 |
| 4.0 | 7.0–7.4 | 46.8 | | 454 | DSC | (1) | 99P9 |
| | 7.0–7.4 | 44.7 | | 419 | heat | (2) | 99P9 |
| 5.0 | 7.0–7.4 | 44.3 | | 420 | DSC | (1) | 99P9 |
| | 7.0–7.4 | 42.2 | | 427 | heat | (2) | 99P9 |
| 6.0 | 7.0–7.4 | 41.7 | | 431 | DSC | (1) | 99P9 |
| | 7.0–7.4 | 39.2 | | 410 | heat | (2) | 99P9 |
| 7.0 | 7.0–7.4 | 39.6 | | 415 | DSC | (1) | 99P9 |
| | 7.0–7.4 | 36.0 | | 420 | heat | (2) | 99P9 |
| 8.0 | 7.0–7.4 | 36.5 | | 399 | DSC | (1) | 99P9 |
| | 7.0–7.4 | 32.8 | | 405 | heat | (2) | 99P9 |
| | | | 4.8 | | DSC | (3) | 99P9 |

RNase A in the presence of ethylurea:

| Conc. (M) | pH | $T_{trs}$ | $\Delta Cp$ | $\Delta H$ | Approach/Remarks | | Ref |
|---|---|---|---|---|---|---|---|
| 1.0 | 7.0–7.4 | 57.3 | | 454 | DSC | (1) | 99P9 |
| | 7.0–7.4 | 55.7 | | 422 | heat | (2) | 99P9 |
| 2.0 | 7.0–7.4 | 51.8 | | 401 | DSC | (1) | 99P9 |
| | 7.0–7.4 | 49.7 | | 410 | heat | (2) | 99P9 |
| 3.0 | 7.0–7.4 | 46.6 | | 392 | DSC | (1) | 99P9 |
| | 7.0–7.4 | 44.7 | | 408 | heat | (2) | 99P9 |

Bovine pancreatic ribonuclease A (RNase A) alone and in the presence of GuHCl, urea and alkylurea (continued)

| Conc. (M) | pH | $T_{trs}$ | $\Delta Cp$ | $\Delta H$ | Approach/Remarks | | Ref |
|---|---|---|---|---|---|---|---|
| 4.0 | 7.0–7.4 | 42.1 | | 382 | DSC | (1) | 99P9 |
| | 7.0–7.4 | 39.3 | | 415 | heat | (2) | 99P9 |
| 5.0 | 7.0–7.4 | 37.8 | | 370 | DSC | (1) | 99P9 |
| | 7.0–7.4 | 34.6 | | 369 | heat | (2) | 99P9 |
| 6.0 | 7.0–7.4 | 33.9 | | 337 | DSC | (1) | 99P9 |
| | 7.0–7.4 | 30.5 | | 360 | heat | (2) | 99P9 |
| 7.0 | 7.0–7.4 | 30.6 | | 292 | DSC | (1) | 99P9 |
| | 7.0–7.4 | 27.6 | | 343 | heat | (2) | 99P9 |
| | | | 5.0 | | DSC | (3) | 99P9 |
| RNase A in the presence of butylurea: | | | | | | | |
| 0.2 | 7.0–7.4 | 60.4 | | 488 | DSC | (1) | 99P9 |
| | 7.0–7.4 | 58.9 | | 414 | heat | (2) | 99P9 |
| 0.4 | 7.0–7.4 | 56.8 | | 449 | DSC | (1) | 99P9 |
| | 7.0–7.4 | 55.2 | | 418 | heat | (2) | 99P9 |
| 0.6 | 7.0–7.4 | 53.2 | | 444 | DSC | (1) | 99P9 |
| | 7.0–7.4 | 52.1 | | 423 | heat | (2) | 99P9 |
| 0.8 | 7.0–7.4 | 49.7 | | 378 | DSC | (1) | 99P9 |
| | 7.0–7.4 | 48.3 | | 376 | heat | (2) | 99P9 |
| | | | 10.2 | | DSC | (3) | 99P9 |

Remarks:
(1) $\Delta H^{cal}$ determined by DSC in the following buffer: 0.1 M glycine, 0.1 M NaCl/0.1 M HCl with appropriate pH 1.1, 1.5, 2.0, 3.0, and 3.5; pH 7–7.4 distilled water
(2) $\Delta H^{v.H.}$ obtained by heat denaturation monitored by UV absorption at 286 nm in distilled water containing the appropriate conc. of alkylurea
(3) $\Delta Cp$ from $\Delta H^{cal}$ versus $T_{trs}$ (DSC data) in the presence of the denaturant

*c) wild type and mutants*

Bovine pancreatic ribonuclease A, Tyr to Phe mutants

| Mutant | pH | $T_{trs}$ | $\Delta H$ | Approach/Remarks | Ref |
|---|---|---|---|---|---|
| wild type | 4.0 | 52.6±0.2 | 384±21 | heat, v.H. | 97J5 |
| Tyr25→Phe | 4.0 | 46.1±0.4 | 343±38 | heat, v.H. | 97J5 |
| Tyr92→Phe | 4.0 | 54.4±0.4 | 309±33 | heat, v.H. | 97J5 |
| Tyr97→Phe | 4.0 | 47.7±0.6 | 295±47 | heat, v.H. | 97J5 |

Bovine pancreatic ribonuclease A, wild type and mutants at positions 106–118

| Protein | pH | $T_{trs}$ | $\Delta H$ | Approach/Remarks | | Ref |
|---|---|---|---|---|---|---|
| wild type | 5.0 | 60.8 | 482 | DSC | (1) | 99C13 |
| Ile106→Ala | 5.0 | 46.6 | 332 | DSC | (1) | 99C13 |
| Ile106→Leu | 5.0 | 54.0 | 404 | DSC | (1) | 99C13 |
| Ile106→Val | 5.0 | 56.5 | 400 | DSC | (1) | 99C13 |
| Ile107→Ala | 5.0 | 50.6 | 331 | DSC | (1) | 99C13 |
| Ile107→Leu | 5.0 | 51.8 | 351 | DSC | (1) | 99C13 |
| Ile107→Val | 5.0 | 60.4 | 365 | DSC | (1) | 99C13 |
| Val108→Ala | 5.0 | 46.8 | 299 | DSC | (1) | 99C13 |
| Val108→Gly | 5.0 | 33.8 | 177 | DSC | (1) | 99C13 |
| Ala109→Gly | 5.0 | 58.1 | 415 | DSC | (1) | 99C13 |

Bovine pancreatic ribonuclease A, wild type and mutants at positions 106–118 (continued)

| Protein | pH | $T_{trs}$ | $\Delta H$ | Approach/Remarks | | Ref |
|---|---|---|---|---|---|---|
| Val116→Ala | 5.0 | 57.5 | 387 | DSC | (1) | 99C13 |
| Val116→Gly | 5.0 | 55.5 | 396 | DSC | (1) | 99C13 |
| Val118→Ala | 5.0 | 54.3 | 404 | DSC | (1) | 99C13 |
| Val118→Gly | 5.0 | 50.0 | 327 | DSC | (1) | 99C13 |

Remark:
(1) $\Delta H$ values are accurate within ±6%, and $T_{trs}$ values are accurate within ±0.2°C

Bovine pancratic ribonuclease and variants

| Protein | pH | $T_{ref}$ | $\Delta Cp$ | $\Delta H$ | Appr./Rem. | Ref |
|---|---|---|---|---|---|---|
| wild type | 1.2–6 | 25 | 6.15±0.17 | 229.3±4.2 | heat (1) | 98Q2 |
| Asp121→Asn | 1.2–6 | 25 | 6.44±0.21 | 224.7±3.3 | heat (1) | 98Q2 |
| Asp121→Ala | 1.2–6 | 25 | 6.57±0.21 | 192.5±3.8 | heat (1) | 98Q2 |
| His119→Ala | 1.2–6 | 25 | 6.57±0.38 | 213 ±13 | heat (1) | 98Q2 |

Remark:
(1) $\Delta Cp$ and $\Delta H$ were calculated at 25°C from a linear fit of data obtained varying pH from 1.2 to 6 with $T_{trs}$ varying from about 26 to 65°C

Ribonuclease A, wild type and disulfide lacking mutants

| Mutant | pH | $T_{trs}$ | $\Delta H$ | Approach | Ref |
|---|---|---|---|---|---|
| wild type | 4.60 | 55.5±0.2 | 469±21 | heat, v.H. | 97L1 |
| double mutant (Cys40→Ala and Cys95→Ala) | | | | | |
| | 4.60 | 33.7±0.2 | 305±15 | heat, v.H. | 97L1 |
| double mutant (Cys40→Ser and Cys95→Ser) | | | | | |
| | 4.60 | 33.6±0.2 | 299±8 | heat, v.H. | 97L1 |

Ribonuclease A, recombinant, wild type and cysteine mutants

| Mutant | pH | $T_{trs}$ | $\Delta H$ | Approach/Remarks | Ref |
|---|---|---|---|---|---|
| wild type | 4.6±0.05 | 55.5±0.2 | 469±21 | heat, v.H. (1,2) | 97S10 |
| double mutant (Cys65→Ser and Cys72→Ser) | | | | | |
| | 4.6±0.05 | 38.5±0.2 | 350±8 | heat, v.H. (1,2) | 97S10 |
| des[65–72]RNase A | 4.6±0.05 | 38.4±0.2 | 336±33 | heat, v.H. (1–3) | 97S10 |

Remarks:
(1) the thermal transition was monitored by optical absorption at 287 nm
(2) buffer: 100 mM sodium acetate
(3) data from Ref. 94T2

Bovine pancreatc ribonuclease A (RNase A) and mutants containing the minimal structural requirements for dimerization and the N-terminal swapping of bovine seminal ribonuclease (BS-RNase)

| Protein | pH | $T_{trs}$ | $\Delta H$ | Approach/Remarks | | Ref |
|---|---|---|---|---|---|---|
| RNase A | 5.0 | 61.2 | 460±14 | heat | (1–4) | 98C1 |
| | 5.0 | 61.1 | 455±15 | heat | (1–3,5) | 98C1 |
| P-RNase A | 5.0 | 59.4 | 415±15 | heat | (1–4) | 98C1 |
| | 5.0 | 59.5 | 415±16 | heat | (1–3,5) | 98C1 |
| PL-RNase A | 5.0 | 58.8 | 400±14 | heat | (1–4) | 98C1 |
| | 5.0 | 59.0 | 410±18 | heat | (1–3,5) | 98C1 |

Bovine pancreatc ribonuclease A (RNase A) and mutants containing the minimal structural requirements for dimerization and the N-terminal swapping of bovine seminal ribonuclease (BS-RNase) (continued)

| Protein | pH | $T_{trs}$ | $\Delta H$ | Approach/Remarks | | Ref |
|---|---|---|---|---|---|---|
| MCAM-PLCC-RNase A | 5.0 | 56.1 | 385±15 | heat | (1–4) | 98C1 |
| | 5.0 | 56.3 | 395±19 | heat | (1–3,5) | 98C1 |
| MCAM-BS-RNase A | 5.0 | 55.1 | 375±15 | heat | (1–4) | 98C1 |
| | 5.0 | 55.0 | 370±16 | heat | (1–3,5) | 98C1 |

Explanations:

RNase A                  – bovine pancreatic ribonuclease A
P-RNase A                – monomeric mutant (Ala19→Pro) of RNase A
PL-RNase A               – monomeric double mutant (Ala19→Pro and Gln28→Leu) of RNase A
PLCC-RNase A             – dimeric multiple mutant (Ala19→Pro, Gln28→Leu, Lys31→Cys, and Ser32→Cys) of RNase A
MCAM-PLCC-RNase A
                         – monomeric and carboxyamidomethylated form of PLCC-RNase A
BS-RNase A               – bovine seminal ribonuclease
MCAM-BS-Rnase            – monomeric and carboxyamidomethylated form of BS-RNase

Remarks:
(1) van't Hoff treatment by nonlinear regression without considering $\Delta Cp$
(2) buffer: 10 mM acetate buffer, pH 5.0
(3) error in $T_{trs}$ does not exceed 0.2°C, error in $\Delta H$ from multiple determinations
(4) transition monitored by CD at 222 nm
(5) transition monitored by CD at 275 nm

*d) effect of selective deamidation*

Ribonuclease A, effect of selective monodeamidation at Asp76

| Form | pH | $T_{trs}$ | $\Delta Cp$ | $\Delta H$ | Appr./Rem. | | Ref |
|---|---|---|---|---|---|---|---|
| RNase A | | | | | | | |
| | 3.0 | 52.4 | 5.0±0.6 | 415±13 | DSC | (1) | 97C5 |
| | 3.5 | 54.7 | 6.0±0.7 | 430±12 | DSC | (1) | 97C5 |
| | 4.0 | 57.0 | 5.4±0.5 | 440±12 | DSC | (1) | 97C5 |
| | 4.5 | 59.2 | 4.9±0.5 | 455±14 | DSC | (1) | 97C5 |
| | 5.0 | 61.3 | 5.5±0.6 | 465±13 | DSC | (1) | 97C5 |
| | 5.0 | 61.2 | | 460±25 | heat | (2) | 97C5 |
| | 6.0 | 62.4 | 5.8±0.6 | 470±14 | DSC | (1) | 97C5 |
| | 8.5 | 61.3 | 5.9±0.6 | 530±16 | DSC | (1) | 97C5 |
| | 3–6 | 52–63 | 5.50±0.20 | | DSC | (3) | 97C5 |
| (Asn76→Asp)RNase A | | | | | | | |
| | 3.0 | 49.0 | 5.8±0.7 | 390±12 | DSC | (1) | 97C5 |
| | 3.5 | 52.1 | 4.9±0.6 | 410±13 | DSC | (1) | 97C5 |
| | 4.0 | 55.3 | 5.4±0.6 | 430±14 | DSC | (1) | 97C5 |
| | 4.5 | 58.5 | 6.0±0.5 | 445±15 | DSC | (1) | 97C5 |
| | 5.0 | 61.5 | 5.5±0.6 | 460±14 | DSC | (1) | 97C5 |
| | 5.0 | 61.6 | | 455±30 | heat | (2) | 97C5 |
| | 6.0 | 62.2 | 5.2±0.5 | 465±13 | DSC | (1) | 97C5 |
| | 8.5 | 61.6 | 5.7±0.6 | 470±15 | DSC | (1) | 97C5 |
| | 3–6 | 49–62 | 5.55±0.20 | | DSC | (3) | 97C5 |
| (Asn76→isoAsp)RNase A | | | | | | | |
| | 3.0 | 43.2 | 5.6±0.6 | 335±10 | DSC | (1) | 97C5 |
| | 3.5 | 46.3 | 4.7±0.5 | 355±12 | DSC | (1) | 97C5 |
| | 4.0 | 49.5 | 5.1±0.5 | 370±12 | DSC | (1) | 97C5 |
| | 4.5 | 52.4 | 4.8±0.6 | 385±11 | DSC | (1) | 97C5 |
| | 5.0 | 55.0 | 5.4±0.6 | 400±10 | DSC | (1) | 97C5 |

Ribonuclease A, effect of selective monodeamidation at Asp76 (continued)

| Form | pH | $T_{trs}$ | $\Delta Cp$ | $\Delta H$ | Appr./Rem. | | Ref |
|---|---|---|---|---|---|---|---|
| | 5.0 | 55.3 | | 390±20 | heat | (2) | 97C5 |
| | 6.0 | 55.8 | 5.8±0.6 | 405±11 | DSC | (1) | 97C5 |
| | 8.5 | 55.8 | 5.8±0.6 | 400±12 | DSC | (1) | 97C5 |
| | 3–6 | 43–56 | 5.45±0.25 | | DSC | (3) | 97C5 |

Remarks:

(1) each figure is the mean of at least four measurements; buffers: pH 3–3.5 glycine-HCl, pH 4–5 acetate, pH 6 MES, pH 8.5 HEPPS, each 0.1 M

(2) van't Hoff heat derived from transitions monitored by CD at 222 nm

(3) $\Delta Cp$ from $\Delta H$ versus $T_{trs}$

Ribonuclease A, effect of selective monodeamidation at Asp 76, measurements in the presence of 2'CMP (Ratio = molar ratio 2'CMP/protein)

| Form | Ratio | pH | $T_{trs}$ | $\Delta Cp$ | $\Delta H$ | Appr./Rem. | | Ref |
|---|---|---|---|---|---|---|---|---|
| RNase A | | | | | | | | |
| | 0 | 5.0 | 61.3 | 5.5±0.6 | 465±13 | DSC | (1) | 97C5 |
| | 1 | 5.0 | 63.5 | 5.0±0.6 | 535±16 | DSC | (1) | 97C5 |
| | 5 | 5.0 | 65.9 | 6.2±0.7 | 560±18 | DSC | (1) | 97C5 |
| | 10 | 5.0 | 67.1 | 5.5±0.5 | 570±17 | DSC | (1) | 97C5 |
| | 30 | 5.0 | 69.0 | 6.0±0.7 | 580±18 | DSC | (1) | 97C5 |
| (Asn76→Asp)RNase A | | | | | | | | |
| | 0 | 5.0 | 61.5 | 5.5±0.6 | 460±14 | DSC | (1) | 97C5 |
| | 1 | 5.0 | 63.3 | 5.9±0.7 | 530±17 | DSC | (1) | 97C5 |
| | 5 | 5.0 | 65.6 | 6.2±0.7 | 555±16 | DSC | (1) | 97C5 |
| | 10 | 5.0 | 66.7 | 5.0±0.7 | 560±18 | DSC | (1) | 97C5 |
| | 30 | 5.0 | 68.6 | 5.8±0.6 | 570±17 | DSC | (1) | 97C5 |
| (Asn76→isoAsp)RNase A | | | | | | | | |
| | 0 | 5.0 | 55.0 | 5.4±0.6 | 400±10 | DSC | (1) | 97C5 |
| | 1 | 5.0 | 58.1 | 4.7±0.5 | 475±15 | DSC | (1) | 97C5 |
| | 5 | 5.0 | 60.9 | 6.1±0.6 | 500±16 | DSC | (1) | 97C5 |
| | 10 | 5.0 | 62.2 | 5.3±0.7 | 510±16 | DSC | (1) | 97C5 |
| | 30 | 5.0 | 64.2 | 5.8±0.6 | 520±15 | DSC | (1) | 97C5 |

Remark:

(1) each figure is the mean of at least four measurements; buffer: 0.1 M acetate

*e) data obtained in the presence of cosolutes*

Bovine pancreatic ribonuclease A in aqueous polyol solutions

| Cosolvent | Conc. | pH | $T_{trs}$ | $\Delta H$ | Appr./Rem. | Ref |
|---|---|---|---|---|---|---|
| control | | 2.5 | 38.3 | 339.3 | heat (1,2) | 98K3 |
| mannitol | 1.00 M | 2.5 | 46.6 | 366.1 | heat (1,2,5) | 98K3 |
| inositol | 0.75 M | 2.5 | 47.0 | 370.7 | heat (1,2,5) | 98K3 |
| xylitol | 2.00 M | 2.5 | 48.5 | 371.5 | heat (1,2,5) | 98K3 |
| adonitol | 2.00 M | 2.5 | 50.2 | 372.4 | heat (1,2,5) | 98K3 |
| sorbitol | 2.00 M | 2.5 | 51.5 | 383.3 | heat (1,2,5) | 98K3 |
| control | | 4.0 | 54.2 | 393.3 | heat (1,3) | 98K3 |
| mannitol | 1.00 M | 4.0 | 59.0 | 406.7 | heat (1,3,5) | 98K3 |
| inositol | 0.75 M | 4.0 | 59.8 | 410.4 | heat (1,3,5) | 98K3 |
| xylitol | 2.00 M | 4.0 | 61.4 | 413.4 | heat (1,3,5) | 98K3 |
| adonitol | 2.00 M | 4.0 | 63.3 | 432.2 | heat (1,3,5) | 98K3 |

Bovine pancreatic ribonuclease A in aqueous polyol solutions (continued)

| Cosolvent | Conc. | pH | $T_{trs}$ | $\Delta H$ | Appr./Rem. | Ref |
|---|---|---|---|---|---|---|
| sorbitol | 2.00 M | 4.0 | 66.0 | 440.6 | heat (1,3,5) | 98K3 |
| control | | 7.0 | 46.0 | 384.9 | heat (1,4) | 98K3 |
| mannitol | 1.00 M | 7.0 | 50.2 | 408.4 | heat (1,4) | 98K3 |
| inositol | 0.75 M | 7.0 | 50.7 | 412.1 | heat (1,4) | 98K3 |
| xylitol | 2.00 M | 7.0 | 52.9 | 408.8 | heat (1,4) | 98K3 |
| adonitol | 2.00 M | 7.0 | 53.4 | 409.6 | heat (1,4) | 98K3 |
| sorbitol | 2.00 M | 7.0 | 56.1 | 416.3 | heat (1,4) | 98K3 |

Remarks:

(1) transition monitored by optical absorption at 287 nm

(2) buffer: 20 mM glycine, pH 2.5

(3) buffer: 40 mM acetate, pH 4.0

(4) buffer: 20 mM phosphate or MOPS, pH 7.0, 1.5 M GuHCl

(5) from $\Delta H$ obtained in the presence of polyols at pH 2.5 and 4.0 versus $T_{trs}$ it follows $\Delta Cp = 3.72 \pm 0.21$ kJ/mol/K, see Ref. 98K3

Ribonuclease A, preferential interactions in aqueous cosolvent systems, the sorbitol-water system

| Co-solvent | pH | $T_{trs}$ | $\Delta H$ | Approach/Remarks | Ref |
|---|---|---|---|---|---|
| 0 % sorbitol (w/v) | 2.0 | 30.1±0.3 | 305±4 | heat, upscan (1) | 97X2 |
| | 2.0 | 29.1±0.4 | | heat, downscan | 97X2 |
| 10 % sorbitol (w/v) | 2.0 | 33.1 | 292 | heat, upscan (1 | 97X2 |
| | 2.0 | 32.2 | | heat, downscan | 97X2 |
| 20 % sorbitol (w/v) | 2.0 | 36.3 | 308 | heat, upscan (1) | 97X2 |
| | 2.0 | 35.5 | | heat, downscan | 97X2 |
| 30 % sorbitol (w/v) | 2.0 | 39.6±0.3 | 316±21 | heat, upscan (1) | 97X2 |
| | 2.0 | 39.0±0.4 | | heat, downscan | 97X2 |
| 40 % sorbitol (w/v) | 2.0 | 43.1 | 328 | heat, upscan (1) | 97X2 |
| | 2.0 | 42.5 | | heat, downscan | 97X2 |
| 0 % sorbitol (w/v) | 5.5 | 60.4±0.2 | 453±25 | heat, upscan (1) | 97X2 |
| | 5.5 | 59.8±0.4 | | heat, downscan | 97X2 |
| 10 % sorbitol (w/v) | 5.5 | 62.0 | 471 | heat, upscan (1) | 97X2 |
| | 5.5 | 61.6 | | heat, downscan | 97X2 |
| 20 % sorbitol (w/v) | 5.5 | 64.0 | 492 | heat, upscan (1) | 97X2 |
| | 5.5 | 63.5 | | heat, downscan | 97X2 |
| 30 % sorbitol (w/v) | 5.5 | 66.6±0.4 | 523±29 | heat, upscan (1) | 97X2 |
| | 5.5 | 65.6±0.5 | | heat, downscan | 97X2 |
| 40 % sorbitol (w/v) | 5.5 | 68.6 | 542 | heat, upscan (1) | 97X2 |
| | 5.5 | 68.2 | | heat, downscan | 97X2 |
| 30 % sorbitol (w/v) | 1.5 | 36.4±0.2 | 299±17 | heat, upscan (1) | 97X3 |
| 30 % sorbitol (w/v) | 2.0 | 39.6±0.3 | 316±21 | heat, upscan (1) | 97X3 |
| 30 % sorbitol (w/v) | 3.0 | 51.6 | 356 | heat, upscan (1) | 97X3 |
| 30 % sorbitol (w/v) | 5.5 | 66.6±0.4 | 523±29 | heat, upscan (1) | 97X3 |

Remark:

(1) data treatment by means of a truncated form of the integrated van't Hoff equation with inherent $\Delta Cp$

Bovine pancreatic ribonuclease A, stabilization by trehalose

| Trehalose | pH | $T_{trs,unf}$ | $T_{trs,ref}$ | $\Delta H$ | Appr./Rem. | Ref |
|---|---|---|---|---|---|---|
| 0.0 M | 2.8 | 40.9±0.3 | 40.5±0.4 | 346.0±20.9 | heat (1–3) | 97X1 |
| 0.1 M | 2.8 | 41.8 | 41.6 | 346.4 | heat (1–3) | 97X1 |
| 0.2 M | 2.8 | 42.2 | 42.0 | 349.8 | heat (1–3) | 97X1 |
| 0.3 M | 2.8 | 43.4 | 43.1 | 351.0 | heat (1–3) | 97X1 |
| 0.4 M | 2.8 | 44.1 | 43.8 | 353.5 | heat (1–3) | 97X1 |
| 0.5 M | 2.8 | 45.0 | 45.0 | 356.9 | heat (1–3) | 97X1 |
| 0.6 M | 2.8 | 45.9 | 45.9 | 363.2 | heat (1–3) | 97X1 |
| 0.7 M | 2.8 | 47.0 | 46.8 | 368.6 | heat (1–3) | 97X1 |
| 0.8 M | 2.8 | 47.7 | 47.6 | 375.7 | heat (1–3) | 97X1 |
| 0.9 M | 2.8 | 48.7 | 48.5 | 378.2 | heat (1–3) | 97X1 |
| 0.0 M | 5.5 | 60.7±0.3 | 59.4±0.4 | 453.1±16.7 | heat (1,2,4) | 97X1 |
| 0.1 M | 5.5 | 61.1 | 60.1 | 447.7 | heat (1,2,4) | 97X1 |
| 0.2 M | 5.5 | 61.7 | 61.1 | 453.5 | heat (1,2,4) | 97X1 |
| 0.3 M | 5.5 | 62.3 | 61.8 | 450.6 | heat (1,2,4) | 97X1 |
| 0.4 M | 5.5 | 62.9 | 62.1 | 443.5 | heat (1,2,4) | 97X1 |
| 0.5 M | 5.5 | 63.3 | 62.6 | 441.8 | heat (1,2,4) | 97X1 |
| 0.6 M | 5.5 | 64.0 | 62.9 | 438.1 | heat (1,2,4) | 97X1 |
| 0.7 M | 5.5 | 64.8 | 64.2 | 447.7 | heat (1,2,4) | 97X1 |
| 0.8 M | 5.5 | 65.4 | 64.8 | 447.3 | heat (1,2,4) | 97X1 |
| 0.9 M | 5.5 | 66.2 | 65.8 | 461.9 | heat (1,2,4) | 97X1 |

Remarks:
(1) $T_{trs,unf}$ = temperature of the unfolding transition at heating, $T_{trs,ref}$ = temperature of the refolding transition at cooling
(2) transition monitored by optical absorption at 287 nm
(3) measured in 0.04 M glycine at pH 2.8
(4) measured in 0.04 M sodium acetate at pH 5.5

Bovine pancreatic ribonuclease A in the presence of β-hydroxyectoine and betaine

| Cosolvent | Conc. | pH | $T_{trs}$ | $\Delta H$ | $\Delta Cp$ | Appr./Rem. | Ref |
|---|---|---|---|---|---|---|---|
| β-hydroxyectoine: | | | | | | | |
| | 0.0 M | 5.5 | 59.8 | 364±14 | 4.4±0.6 | DSC (1,2) | 99K9 |
| | 1.5 M | 5.5 | 66.2 | 406±20 | 5.0±0.5 | DSC (1,2) | 99K9 |
| | 3.0 M | 5.5 | 71.8 | 426±23 | 5.6±0.6 | DSC (1,2) | 99K9 |
| betaine: | | | | | | | |
| | 0.0 M | 5.5 | 59.8 | 364±14 | 4.4±0.6 | DSC (1,2) | 99K9 |
| | 2.0 M | 5.5 | 66.3 | 413±17 | 4.7±0.5 | DSC (1,2) | 99K9 |
| | 3.0 M | 5.5 | 67.8 | 424±13 | 5.2±0.3 | DSC (1,2) | 99K9 |
| | 4.0 M | 5.5 | 68.6 | 427±25 | 5.1±0.5 | DSC (1,2) | 99K9 |
| | 5.0 M | 5.5 | 67.9 | 414±14 | 5.1±0.3 | DSC (1,2) | 99K9 |

Remarks:
(1) measured in 50 mM phosphate buffer, 200 mM sodium chloride, and the indicated osmolyte conc.
(2) Ref. 99K9 contains further data and general expressions for the concentration dependence of the osmolytes at varying pH

Bovine pancreatic ribonuclease A in the presence of osmolytes

| Osmolyte | Conc. | pH | $T_{trs}$ | $\Delta H$ | $\Delta Cp$ | Appr./Rem. | Ref |
|---|---|---|---|---|---|---|---|
| arginine | 0.0 M | 1.5–6 | 30–65 | | 4.97±0.21 | heat (1–3) | 98R7 |
| | 0.2 M | 1.5–6 | 35–65 | | 5.39±0.25 | heat (1–3) | 98R7 |
| | 0.5 M | 1.5–6 | 38–65 | | 5.43±0.29 | heat (1–3) | 98R7 |
| | 0.7 M | 1.5–6 | 40–62 | | 5.14±0.50 | heat (1–3) | 98R7 |
| | 1.0 M | 1.5–6 | 40–62 | | 5.52±0.63 | heat (1–3) | 98R7 |
| histidine | 0.2 M | 5.0 | 60.7 | 458 | | heat (1,3,4) | 98R7 |
| none | 0.0 M | 5.0 | 65.0 | 481 | | heat (1,3,4) | 98R7 |

Remarks:
(1) van't Hoff treatment, transition monitored by changes in optical absorption at 287 nm
(2) $\Delta Cp$ derived from $\Delta H^{v.H.}$ versus $T_{trs}$ with $T_{trs}$ varied by pH from 1.5 to 6.0
(3) measured in 0.05 M citrate, 0.1 M KCl
(4) reference values, measured at pH 5.0

Ribonuclease A, preferential interactions in aqueous cosolvent systems, magnesium salts containing aqueous solutions

| Salt conc. | pH | $T_{trs}$ | $\Delta H$ | Approach/Remarks | | Ref |
|---|---|---|---|---|---|---|
| buffer | 1.5 | 26.5±0.9 | 279±21 | heat | (1,2) | 97X3 |
| | 2.0 | 30.1±0.3 | 305±4 | heat | (1,2) | 97X3 |
| | 2.8 | 41.1±0.2 | 326±21 | heat | (1,2) | 97X3 |
| | 3.0 | 44.8 | 339 | heat | (1,2) | 97X3 |
| | 3.2 | 46.6 | 345 | heat | (1,2) | 97X3 |
| | 5.5 | 60.4±0.3 | 454±17 | heat | (1,2) | 97X3 |
| | 5.8 | 61.4 | 470 | heat | (1,2) | 97X3 |
| 0.6 M MgCl$_2$ | 1.5 | 43.5±0.1 | 298±8 | heat | (1,2) | 97X3 |
| | 2.0 | 45.5 | 345 | heat | (1,2) | 97X3 |
| | 3.0 | 52.8 | 375 | heat | (1,2) | 97X3 |
| | 5.5 | 60.5±0.2 | 453±21 | heat | (1,2) | 97X3 |
| 0.6 M MgSO$_4$ | 1.5 | 51.1±0.7 | 323±21 | heat | (1,2) | 97X3 |
| | 2.0 | 53.1 | 369 | heat | (1,2) | 97X3 |
| | 3.0 | 60.7 | 425 | heat | (1,2) | 97X3 |
| | 5.5 | 67.4±0.3 | 465±25 | heat | (1,2) | 97X3 |

Remarks:
(1) data treatment by means of a truncated form of the integrated van't Hoff equation with inherent $\Delta Cp$
(2) buffer: 0.04 M glycine at pH 1.5 and 2.0, 0.04 M acetate at pH 5.5

Bovine pancreatic ribonuclease A (RNase A)

| Protein | pH | $T_{trs}$ | $\Delta Cp$ | $\Delta H$ | Appr./Rem. | Ref |
|---|---|---|---|---|---|---|
| RNase A | 7.0 | 47–63 | 8.95±0.29 | 385–525 | heat (1) | 99K3 |
| | 7.0 | 52–70 | 6.65±0.71 | 400–510 | heat (1,2,3) | 99K3 |
| | 5.0 | 52–66 | 4.27±0.29 | 370–455 | heat (1,2,4) | 99K3 |

Remarks:
(1) $\Delta Cp$ from $\Delta H$ versus $T_{trs}$ in the presence of 0–1.5 M GuHCl at pH 7.0
(2) $\Delta Cp$ from $\Delta H$ versus $T_{trs}$ in the presence of carboxylic acids
(3) 0–1.5 M GuHCl was present along with various salts at pH 7.0
(4) 1.5 M GuHCl was always present at pH 5.0

## Ribonuclease B

Ribonuclease B, comparison with ribonuclease A
Ribonuclease B and ribonuclease A possess identical protein structures, but differ by the presence of a carbohydrate chain attached to Asn34

| Protein | pH | $T_{trs}$ | $\Delta Cp$ | $\Delta H$ | Approach/Remarks | Ref |
|---------|-----|-----------|-------------|------------|------------------|-----|
| RNase A | 8.0 | 60.4±0.1 | 9.40±0.06 | 574±4 | heat, v.H. (1,2) | 97A5 |
| RNase B | 8.0 | 61.9±0.1 | 8.78±0.11 | 576±8 | heat, v.H. (1,2) | 97A5 |

Remarks:
(1) buffer: 50 mM Tris-HCl, pH 8.0
(2) transition monitored by optical absorbance at 287 nm

## Ribonuclease H

Comparison of two ribonucleases H, one from mesophile *Escherichia coli* and one from thermophile *Thermus thermophilus*

| Protein | pH | $T_{trs}$ | $\Delta Cp$ | $\Delta H$ | Approach/Remarks | Ref |
|---------|-----|-----------|-------------|------------|------------------|-----|
| ribonuclease H from *E. coli*: | | | | | | |
| | 5.5 | 66 | 11.3±0.8 | 502±17 | GuHCl/heat (1–3) | 99H10 |
| ribonuclease H from *T. thermophilus*: | | | | | | |
| | 5.5 | 86 | 7.5±0.4 | 548±21 | GuHCl/heat (1–3) | 99H10 |

Remarks:
(1) data from GuHCl denaturation at 5, 10, 15, 20, 25, 30, 40, 50, and 60°C; thermal melts were fit using a two-state model and the Gibbs-Helmholtz relationship
(2) measured in 5 mM sodium acetate, 50 mM KCl
(3) transition monitored by CD at 225 nm

Recombinant ribonuclease H (RNase H) from *Thermus thermophilus* HB8 compared with RNase HI from *E. coli*

| Protein | pH | $T_{trs}$ | $\Delta H$ | Approach/Remarks | Ref |
|---------|-----|-----------|------------|------------------|-----|
| RNase HI from *E. coli*: | | | | | |
| | 5.5 | 48.2±0.6 | 349±6 | heat, v.H. (1–3) | 92K13 |
| RNase H from *Thermus thermophilus*: | | | | | |
| | 5.5 | 82.1±0.4 | 632±26 | heat, v.H. (1–3) | 92K13 |

Remarks:
(1) buffer: 20 mM sodium acetate, pH 5.5, 0.1 M NaCl
(2) measured in the presence of 1.2 M GuHCl
(3) transition monitored by CD at 220 nm

## Ribonuclease HI

Ribonuclease HI, wild type and mutants at position 52

| Mutant | pH | $T_{trs}$ | $\Delta T$ | $\Delta H$ | Approach/Remarks | Ref |
|--------|-----|-----------|------------|------------|------------------|-----|
| wild type | 3.2 | 53.0 | 0.0 | 415 | heat       (1) | 97A2 |
| Ala52→Asn | 3.2 | 47.1 | −5.9 | 400 | heat, v.H. (1) | 97A2 |
| Ala52→Asp | 3.2 | 46.9 | −6.1 | 433 | heat, v.H. (1) | 97A2 |
| Ala52→Cys | 3.2 | 55.5 | 2.5 | 537 | heat, v.H. (1) | 97A2 |
| Ala52→Gln | 3.2 | 49.1 | −3.9 | 378 | heat, v.H. (1) | 97A2 |
| Ala52→Glu | 3.2 | 48.0 | −5.0 | 421 | heat, v.H. (1) | 97A2 |

Ribonuclease HI, wild type and mutants at position 52 (continued)

| Mutant | pH | $T_{trs}$ | $\Delta T$ | $\Delta H$ | Approach/Remarks | Ref |
|---|---|---|---|---|---|---|
| Ala52→Gly | 3.2 | 44.1 | −8.9 | 381 | heat, v.H. (1) | 97A2 |
| Ala52→His | 3.2 | 41.2 | −11.8 | 354 | heat, v.H. (1) | 97A2 |
| Ala52→Ile | 3.2 | 59.2 | 6.2 | 355 | heat, v.H. (1) | 97A2 |
| Ala52→Leu | 3.2 | 57.3 | 4.3 | 428 | heat, v.H. (1) | 97A2 |
| Ala52→Lys | 3.2 | 33.5 | −19.5 | 296 | heat, v.H. (1) | 97A2 |
| Ala52→Met | 3.2 | 54.6 | 1.6 | 468 | heat, v.H. (1) | 97A2 |
| Ala52→Phe | 3.2 | 51.5 | −1.5 | 372 | heat, v.H. (1) | 97A2 |
| Ala52→Pro | 3.2 | 47.6 | −5.4 | 389 | heat, v.H. (1) | 97A2 |
| Ala52→Ser | 3.2 | 47.2 | −5.8 | 492 | heat, v.H. (1) | 97A2 |
| Ala52→Thr | 3.2 | 50.3 | −2.7 | 451 | heat, v.H. (1) | 97A2 |
| Ala52→Tyr | 3.2 | 45.4 | −7.6 | 296 | heat, v.H. (1) | 97A2 |
| Ala52→Val | 3.2 | 58.5 | 5.5 | 536 | heat, v.H. (1) | 97A2 |

Remark:

(1) heat denaturation, transition monitored by CD at 220 nm

Ribonuclease HI, wild type and double mutants at positions 52 and 74

| Mutant | pH | $T_{trs}$ | $\Delta T$ | $\Delta H$ | Approach/Remarks | Ref |
|---|---|---|---|---|---|---|
| wild type | 3.2 | 53.0 | 0.0 | 415 | heat, v.H. (1) | 97A2 |
| Val74→Ala | 3.2 | 45.4 | −7.6 | 423 | heat, v.H. (1,2) | 97A2 |
| Val74→Leu | 3.2 | 56.7 | 3.7 | 496 | heat, v.H. (1,2) | 97A2 |
| double mutant (Ala52→Ile and Val74→Ala ) | | | | | | |
| | 3.2 | 56.9 | 3.9 | 426 | heat, v.H. (1) | 97A2 |
| double mutant (Ala52→Leu and Val74→Ala) | | | | | | |
| | 3.2 | 54.2 | 1.2 | 440 | heat, v.H. (1) | 97A2 |
| double mutant (Ala52→Phe and Val74→Ala) | | | | | | |
| | 3.2 | 48.4 | −4.6 | 359 | heat, v.H. (1) | 97A2 |
| double mutant (Ala52→Val and Val74→Ala) | | | | | | |
| | 3.2 | 51.3 | −1.7 | 447 | heat, v.H. (1) | 97A2 |
| double mutant (Ala52→Gly and Val74→Leu) | | | | | | |
| | 3.2 | 48.4 | −4.6 | 409 | heat, v.H. (1) | 97A2 |
| double mutant (Ala52→Ile and Val74→Leu) | | | | | | |
| | 3.2 | 56.9 | 3.9 | 512 | heat, v.H. (1) | 97A2 |
| double mutant (Ala52→Leu and Val74→Leu) | | | | | | |
| | 3.2 | 57.9 | 4.9 | 494 | heat, v.H. (1) | 97A2 |
| double mutant (Ala52→Phe and Val74→Leu) | | | | | | |
| | 3.2 | 53.1 | 0.1 | 418 | heat, v.H. (1) | 97A2 |
| double mutant (Ala52→Val and Val74→Leu) | | | | | | |
| | 3.2 | 61.0 | 8.0 | 525 | heat, v.H. (1) | 97A2 |

Remarks:

(1) heat denaturation, transition monitored by CD at 220 nm
(2) reference protein, data from Ref. 93I2

Ribonuclease HI from *Thermus thermophilus* HB8, proteins with C-terminal truncations

| Protein | pH | $T_{trs}$ | $\Delta H$ | Approach/Remarks | Ref |
|---|---|---|---|---|---|
| wild type | 5.5 | 83.7 | 650 | heat, v.H. (1,2) | 98H9 |
| TRNH[Δ156–161] | 5.5 | 83.3 | 395 | heat, v.H. (1,2) | 98H9 |
| TRNH[Δ154–161] | 5.5 | 82.0 | 440 | heat, v.H. (1,2) | 98H9 |
| TRNH[Δ152–161] | 5.5 | 80.0 | 375 | heat, v.H. (1,2) | 98H9 |
| TRNH[Δ150–161] | 5.5 | 80.9 | 456 | heat, v.H. (1,2) | 98H9 |

Ribonuclease HI from *Thermus thermophilus* HB8, proteins with C-terminal truncations (continued)

| Protein | pH | $T_{trs}$ | $\Delta H$ | Approach/Remarks | Ref |
|---|---|---|---|---|---|
| TRNH[Δ149–161] | 5.5 | 72.0 | 565 | heat, v.H. (1,2) | 98H9 |
| TRNH[Δ148–161] | 5.5 | 68.1 | 366 | heat, v.H. (1,2) | 98H9 |
| TRNH[Δ146–161] | 5.5 | 69.1 | 437 | heat, v.H. (1,2) | 98H9 |
| TRNH[Δ144–161] | 5.5 | 66.7 | 361 | heat, v.H. (1,2) | 98H9 |

Remarks:
(1) measured in 20 mM sodium acetate (pH 5.5) in the presence of 1 M GuHCl
(2) transition monitored by CD at 220 nm

Ribonuclease HI from *Thermus thermophilus* HB8, proteins with C-terminal truncations and cysteine residues replaced

| Protein | pH | $T_{trs}$ | $\Delta H$ | Approach/Remarks | Ref |
|---|---|---|---|---|---|
| wild type | 5.5 | 83.7 | 650 | heat, v.H. (1,2) | 98H9 |
| wild type, reduced | 5.5 | 77.4 | 395 | heat, v.H. (1,2) | 98H9 |
| Cys149→Ala RNase HI | 5.5 | 74.7 | 490 | heat, v.H. (1–3) | 98H9 |
| TRNH[Δ150–161] | 5.5 | 80.9 | 456 | heat, v.H. (1–3) | 98H9 |
| Cys13→Ala TRNH[Δ150–161] | 5.5 | 80.1 | 447 | heat, v.H. (1–3) | 98H9 |
| Cys41→Ala TRNH[Δ150–161] | 5.5 | 71.7 | 600 | heat, v.H. (1–3) | 98H9 |
| Cys63→Ala TRNH[Δ150–161] | 5.5 | 78.8 | 470 | heat, v.H. (1–3) | 98H9 |
| Cys149→Ala TRNH[Δ150–161] | 5.5 | 72.5 | 567 | heat, v.H. (1–3) | 98H9 |
| Cys149→Ile TRNH[Δ150–161] | 5.5 | 74.6 | 523 | heat, v.H. (1–3) | 98H9 |
| Cys149→Ser TRNH[Δ150–161] | 5.5 | 73.1 | 507 | heat, v.H. (1–3) | 98H9 |
| Cys149→Thr TRNH[Δ150–161] | 5.5 | 75.0 | 529 | heat, v.H. (1–3) | 98H9 |
| Cys149→Val TRNH[Δ150–161] | 5.5 | 75.8 | 582 | heat, v.H. (1–3) | 98H9 |
| double mutant (Cys13→Ala and Cys63→Ala) TRNH[Δ150–161] | | | | | |
| | 5.5 | 78.5 | 461 | heat, v.H. (1–3) | 98H9 |
| oxidized form | 5.5 | 78.5 | 470 | heat, v.H. (1–3) | 98H9 |
| reduced form | 5.5 | 71.6 | 391 | heat, v.H. (1–3) | 98H9 |
| multiple mutant (Cys13→Ala, Cys41→Ala, Cys63→Ala, and Cys149→Ala) TRNH[Δ150–161] | | | | | |
| | 5.5 | 68.9 | 521 | heat, v.H. (1–3) | 98H9 |

Remarks:
(1) measured in 20 mM sodium acetate (pH 5.5) in the presence of 1 M GuHCl
(2) transition monitored by CD at 220 nm
(3) RNase HI from *Thermus thermophilus* HB8 contains four cysteine residues: Cys13, Cys41, Cys63, and Cys149

## Ribonuclease P2

Ribonuclease P2 from the hyperthermophile *Sulfolobus solfataricus*, heat and cold denaturation

| Mutant | pH | $T_{trs}$ | $\Delta H$ | Approach/Remarks | Ref |
|---|---|---|---|---|---|
| wild type, cold | 7.4 | −20 | −102±13 | heat, v.H. (1,4) | 97M8 |
| Phe31→Ala, heat | 7.4 | +38 | 170±32 | heat, v.H. (2,4) | 97M8 |
| Phe31→Ala, cold | 7.4 | −15 | −83±7 | heat, v.H. (2,4) | 97M8 |
| Phe31→Tyr, heat | 3 | +65 | 102±8 | heat, v.H. (3,4) | 97M8 |
| Phe31→Tyr, heat + p | 3 | +84 | 61±5 | heat, v.H. (3–5) | 97M8 |
| Phe31→Tyr, cold | 3 | −10 | −96±8 | heat, v.H. (3,4) | 97M8 |

Remarks:
(1) buffer: 50 mM phosphate with 4 M urea, pH 7.4
(2) buffer: 50 mM phosphate, pH 7.4
(3) buffer: 50 mM pyrophosphate, pH 3
(4) the transition was monitored by forth derivative absorption spectroscopy in the Tyr region
(5) heat denaturation at 400 mPa pressure

**Ribonuclease P2**, see also DNA-binding protein Sso7d

## Ribonuclease S

Bovine pancreatic ribonuclease S (RNase S) in $H_2O$ and $D_2O$

| Protein | Solvent | pH | $T_{trs}$ | $\Delta Cp$ | $\Delta H$ | Appr./Rem. | | Ref |
|---|---|---|---|---|---|---|---|---|
| RNase S | $H_2O$ | 6.0 | 47.7±0.2 | 8.4±1.3 | 448±8 | DSC | (1) | 99N3 |
| | $D_2O$ | 6.0 | 50.7±0.1 | 7.9±1.3 | 469±17 | DSC | (1) | 99N3 |

Remark:
(1) see also data for ribonuclease A in $H_2O$ and $D_2O$ in Ref. 99N3

## Ribonucleases Sa, Sa2, and Sa3

Ribonucleases (RNases) Sa, Sa2, and Sa3 from *Streptomyces aureofaciens* compared with selected ribonucleases at pH 7.0

| Protein | pH | $T_{trs}$ | $\Delta Cp$ | $\Delta H$ | Appr./Rem. | Ref |
|---|---|---|---|---|---|---|
| RNase Sa | 7.0 | 48.4±0.3 | 6.36±0.38 | 407.5 | heat (1,2) | 98P1 |
| RNase Sa2 | 7.0 | 41.1±0.3 | 5.31±0.25 | 286.2 | heat (1,2) | 98P1 |
| RNase Sa3 | 7.0 | 47.2±0.3 | 6.57±0.46 | 391.6 | heat (1,2) | 98P1 |
| RNase T1 | 7.0 | 51.6 | 6.90 | 442.2 | heat (3) | 98P1 |
| Barnase | 7.0 | 53.2 | 7.49 | 529.7 | heat (4) | 98P1 |
| RNase A | 7.0 | 62.8 | 7.95 | 499.6 | heat (5) | 98P1 |
| RNase A | 7.0 | 62.8 | 5.48 | 499.6 | heat (6) | 98P1 |

Remarks:
(1) $T_{trs}$, $\Delta H$, and $\Delta Cp$ are from Ref. 98P1
(2) $\Delta Cp$ from $\Delta H$ versus $T_{trs}$, varying pH from 2 to 10, and $T_{trs}$ over more than 30°C
(3) $T_{trs}$ and $\Delta H$ are averaged values from Refs. 89S8, 92S11, and 94Y2, and $\Delta Cp$ is from Ref. 89P2
(4) $T_{trs}$ and $\Delta H$ are from Ref. 98P1, $\Delta Cp$ is the average from Refs. 94O1 and 95J1
(5) $T_{trs}$ and $\Delta H$ are from Ref. 96C6, and $\Delta Cp$ is the average from 13 literature values: 67B, 67D, 70S, 70T2, Ref. 71S1, Ref. 78F2, 88S5, 89P2, Ref. 92B11, 95M2, 96L3, and Ref. 97C5
(6) $T_{trs}$, $\Delta H$ and $\Delta Cp$ from Ref. 96C6

Ribonucleases (RNases) Sa, Sa2, and Sa3 from *Streptomyces aureofaciens*, contribution of the disulfide bond to the stability of the RNases

| Protein | S-S bond | pH | $T_{trs}$ | $\Delta H$ | Approach/Remarks | Ref |
|---|---|---|---|---|---|---|
| RNase Sa | intact | 7.0 | 48.4 | 397 | heat, v.H. (1–3) | 98P1 |
| | reduced | 7.0 | 28.4 | 364 | heat, v.H. (1–3) | 98P1 |
| RNase Sa2 | intact | 5.0 | 48.9 | 326 | heat, v.H. (1–3) | 98P1 |
| | reduced | 5.0 | 17.4 | 184 | heat, v.H. (1–3) | 98P1 |
| RNase Sa3 | intact | 3.0 | 50.6 | 464 | heat, v.H. (1–3) | 98P1 |
| | reduced | 3.0 | 23.6 | 301 | heat, v.H. (1–3) | 98P1 |

Remarks:
(1) RNase Sa was measured in 30 mM PIPES, pH 7.0
(2) RNase Sa2 was measured in 30 mM acetate, pH 5.0
(3) RNase Sa3 was measured in 30 mM glycylglycine, pH 3.0

Ribonuclease Sa from *Streptomyces aureofaciens* (RNase Sa), mutants concerning Asn39

| Mutant | pH | $T_{trs}$ | $\Delta H$ | Approach/Remarks | | Ref |
|---|---|---|---|---|---|---|
| wild type | 7.0 | 48.4 | 386.6 | heat | (1–3) | 98H7 |
| Asn39→Ala | 7.0 | 40.8 | 294.6 | heat | (1–3) | 98H7 |
| Asn39→Asp | 7.0 | 43.2 | 349.4 | heat | (1–3) | 98H7 |
| Asn39→Ser | 7.0 | 40.4 | 307.5 | heat | (1–3) | 98H7 |
| wild type | 2.0 | 28.0 | 246.0 | heat | (1,2,4) | 98H7 |
| Asn39→Asp | 2.0 | 23.3 | 210.5 | heat | (1,2,4) | 98H7 |

Remarks:

(1) RNase Sa was expressed in *E. coli*

(2) the conserved Asn39 corresponds to position 44 in RNase T1 and 58 in barnase, see also RNase T1, Ref. 98H7

(3) measured by CD at 234 nm in 30 mM MOPS, pH 7.0

(4) measured by CD at 234 nm in 30 mM glycine, pH 2.0

Ribonuclease Sa, wild type and mutants

| Protein | pH | $T_{trs}$ | $\Delta H$ | Approach/Remarks | | Ref |
|---|---|---|---|---|---|---|
| wild type | 7.0 | 47.2±0.3 | 381±29 | heat | (1) | 99G17 |
| Asp25→Lys | 7.0 | 50.2±0.3 | 389±29 | heat | (1) | 99G17 |
| Glu74→Lys | 7.0 | 51.1±0.3 | 393±29 | heat | (1) | 99G17 |

Remark:

(1) buffer: 30 mM MOPS, pH 7.0

## Ribonuclease, Bovine Seminal

Ribonuclease A, step-wise mutation toward the dimeric bovine seminal ribonuclease

| Form | pH | $T_{trs}$ | $\Delta Cp$ | $\Delta H$ | Approach/Remarks | | Ref |
|---|---|---|---|---|---|---|---|
| RNase A | 5.0 | 61.3 | 5.5±0.6 | 465±13 | DSC | (1) | 97C4 |
| | 5.0 | 61.2 | | 460±14 | heat, v.H. | (1) | 97C4 |
| P-RNase A | 5.0 | 59.6 | 5.2±0.6 | 415±13 | DSC | (1) | 97C4 |
| | 5.0 | 59.4 | | 415±15 | heat, v.H. | (1) | 97C4 |
| PL-RNase A | 5.0 | 59.0 | 5.0±0.6 | 410±12 | DSC | (1) | 97C4 |
| | 5.0 | 58.8 | | 400±14 | heat, v.H. | (1) | 97C4 |
| MCAM-PLCC-RNase A | | | | | | | |
| | 5.0 | 56.4 | 4.8±0.5 | 390±12 | DSC | (1) | 97C4 |
| | 5.0 | 56.1 | | 385±15 | heat, v.H. | (1) | 97C4 |
| MCAM-BS-RNase | | | | | | | |
| | 5.0 | 55.2 | 4.7±0.5 | 380±12 | DSC | (1) | 97C4 |
| | 5.0 | 55.1 | | 375±15 | heat, v.H. | (1) | 97C4 |

Explanations:

| | |
|---|---|
| RNase A: | bovine pancreatic ribonuclease |
| P-RNase A: | monomeric RNase A, mutant Ala19→Pro |
| PL-RNase A: | monomeric RNase A, double mutant (Ala19→Pro and Gln28→Leu) |
| PLCC-RNase A: | monomeric RNase, multiple mutant (Ala19→Pro, Gln28→Leu, Lys31→Cys, and Ser32→Cys) |
| MCAM-PLCC-RNase A: | monomeric PLCC-RNase A, carboxyamidomethylated at Cys31 and Cys32 |
| MCAM-BS-RNase: | monomeric bovine seminal ribonuclease, carboxyamidomethylated at Cys31 and Cys32 |

Remark:

(1) buffer: 100 mM acetate

Bovine pancreatc ribonuclease A (RNase A) and mutants containing the minimal structural requirements for dimerization and the N-terminal swapping of bovine seminal ribonuclease (BS-RNase)

| Protein | pH | $T_{trs}$ | $\Delta H$ | Approach/Remarks | | Ref |
|---|---|---|---|---|---|---|
| RNase A | 5.0 | 61.2 | 460±14 | heat | (1–4) | 98C1 |
| | 5.0 | 61.1 | 455±15 | heat | (1–3,5) | 98C1 |
| P-RNase A | 5.0 | 59.4 | 415±15 | heat | (1–4) | 98C1 |
| | 5.0 | 59.5 | 415±16 | heat | (1–3,5) | 98C1 |
| PL-RNase A | 5.0 | 58.8 | 400±14 | heat | (1–4) | 98C1 |
| | 5.0 | 59.0 | 410±18 | heat | (1–3,5) | 98C1 |
| MCAM-PLCC-RNase A | 5.0 | 56.1 | 385±15 | heat | (1–4) | 98C1 |
| | 5.0 | 56.3 | 395±19 | heat | (1–3,5) | 98C1 |
| MCAM-BS-RNase A | 5.0 | 55.1 | 375±15 | heat | (1–4) | 98C1 |
| | 5.0 | 55.0 | 370±16 | heat | (1–3,5) | 98C1 |

Explanations:

RNase A – bovine pancreatic ribonuclease A
P-RNase A – monomeric mutant (Ala19→Pro) of RNase A
PL-RNase A – monomeric double mutant (Ala19→Pro and Gln28→Leu) of RNase A
PLCC-RNase A – dimeric mutliple mutant (Ala19→Pro, Gln28→Leu, Lys31→Cys, and Ser32→Cys) of RNase A
MCAM-PLCC-RNase A
– monomeric and carboxyamidomethylated form of PLCC-RNase A
BS-RNase A – bovine seminal ribonuclease
MCAM-BS-RNase – monomeric and carboxyamidomethylated form of BS-RNase

Remarks:
(1) van't Hoff treatment by nonlinear regression without considering $\Delta Cp$
(2) buffer: 10 mM acetate buffer, pH 5.0
(3) error in $T_{trs}$ does not exceed 0.2°C, error in $\Delta H$ from multiple determinations
(4) transition monitored by CD at 222 nm
(5) transition monitored by CD at 275 nm

## Ribonuclease T1

Ribonuclease T1 from *Aspergillus oryzae*, Lys25 isoform

| Protein | pH | $T_{trs}$ | $\Delta Cp$ | $\Delta H$ | Approach/Remarks | | Ref |
|---|---|---|---|---|---|---|---|
| RNase T1 | 3.7 | 59.0 | 4.1 | 403 | DSC | (1) | 92B11 |
| | 4.5 | 61.3 | 7.0 | 440 | DSC | (1) | 92B11 |
| | 5.0 | 61.9 | 5.9 | 452 | DSC | (1,2) | 92B11 |
| | 5.7 | 59.3 | 5.2 | 436 | DSC | (1) | 92B11 |
| | 6.3 | 57.3 | 4.0 | 402 | DSC | (1) | 92B11 |
| | 7.0 | 55.0 | 6.6 | 387 | DSC | (1) | 92B11 |
| | 8.0 | 50.2 | 6.7 | 375 | DSC | (1) | 92B11 |
| | 3.7–8.0 | | 6.54 | | DSC | (3) | 92B11 |

Remarks:
(1) buffers used: 0.1 M formate for pH 3.7, 0.1 M acetate for pH 4.5 and 5.0, 0.1 M MES for pH 5.7 and 6.3, 0.1 MOPS for pH 7.0, and HEPPS for pH 8.0
(2) $T_{trs}$ = 54°C and $\Delta H$ = 377 kJ/mol in the presence of 2 M urea, and $T_{trs}$ = 45°C and $\Delta H$ = 269 kJ/mol in the presence of 4 M urea
(3) $\Delta Cp$ from $\Delta H$ versus $T_{trs}$

Ribonuclease T1 from *Aspergillus oryzae*

| pH | $T_{trs}$ | $\Delta C_p$ | $\Delta H$ | Approach/Remarks | Ref |
|----|-----------|--------------|------------|------------------|-----|
| 7  | 51.0±0.5  |              | 497±120    | heat, v.H. (1,2) | 97G3 |

Remarks:
(1) $\Delta H$ from the temperature dependence of Stokes radius
(2) buffer: 10 mM sodium cacodylate, 1 mM EDTA, pH 7

Ribonuclease T1, mutants concerning Asn44

| Mutant | pH | $T_{trs}$ | $\Delta H$ | Approach/Remarks | | Ref |
|--------|----|-----------|------------|------------------|---|-----|
| wild type | 7.0 | 50.8 | 404.2 | heat | (1–3) | 98H7 |
| Asn44→Ala | 7.0 | 45.6 | 363.2 | heat | (1–3) | 98H7 |
| Asn44→Asp | 7.0 | 45.3 | 352.3 | heat | (1–3) | 98H7 |
| Asn44→Ser | 7.0 | 45.8 | 374.9 | heat | (1–3) | 98H7 |

Remarks:
(1) the conserved Asn44 corresponds to position 39 in RNase Sa and 58 in barnase, see also RNase Sa, Ref. 98H7
(2) measured by CD at 244 nm in 30 mM MOPS, pH 7.0

Ribonuclease T1, wild type, Asp49→Ala, and Asp49→His mutants

| Protein | pH | $T_{trs}$ | $\Delta H$ | Approach/Remarks | | Ref |
|---------|----|-----------|------------|------------------|---|-----|
| wild type | 7.0 | 52.3±0.3 | 435±29 | DSC | (1) | 99G17 |
| Asp49→Ala | 7.0 | 54.0±0.3 | 456±29 | DSC | (1,2) | 99G17 |
| wild type | 6.0 | 55.5±0.3 | 452±29 | heat | (1) | 99G17 |
| Asp49→His | 6.0 | 58.9±0.3 | 456±29 | heat | (1,2) | 99G17 |

Remark:
(1) buffer: 30 mM MOPS (pH 7.0), 30 mM MES (pH 6.0)

Ribonuclease T1, Asp49 mutants

| Protein | pH | $T_{trs}$ | $\Delta H$ | Approach/Remarks | | Ref |
|---------|----|-----------|------------|------------------|---|-----|
| Asp49→Ala | 7.0 | 54.0±0.3 | 456±29 | DSC | (1) | 99G17 |
| Asp49→Phe | 7.0 | 52.7±0.3 | 464±29 | DSC | (1) | 99G17 |
| Asp49→Trp | 7.0 | 51.2±0.3 | 435±29 | DSC | (1) | 99G17 |
| Asp49→Tyr | 7.0 | 52.4±0.3 | 452±29 | DSC | (1) | 99G17 |

Remark:
(1) buffer: 30 mM MOPS, pH 7.0

Ribonuclease T1, wild type and mutant Trp59→Tyr

| Mutant | pH | $T_{trs}$ | $\Delta H$ | Approach/Remarks | | Ref |
|--------|----|-----------|------------|------------------|---|-----|
| wild type | 7.0 | 55.3±0.2 | 430±40 | heat | (1,2) | 99R8 |
| Trp59→Tyr | 7.0 | 53.2±0.2 | 430±40 | heat | (1,2) | 99R8 |

Remarks:
(1) transition monitored by infrared spectroscopy (FTIR)
(2) buffer: 100 mM Na-cacodylate/$D_2O$ buffer, pH* 7.0

Ribonuclease T1, Asp76 mutants

| Protein | pH | $T_{trs}$ | $\Delta H$ | Approach/Remarks | | Ref |
|---|---|---|---|---|---|---|
| RNase Sa | 7.0 | 50.8 | 404 | heat | (1,2) | 99G7 |
| Asp76→Ala | 7.0 | 35.6 | 297 | heat | (1,2) | 99G7 |
| Asp76→Asn | 7.0 | 37.0 | 318 | heat | (1,2) | 99G7 |
| Asp76→Ser | 7.0 | 37.2 | 325 | heat | (1,2) | 99G7 |

Remarks:

(1) thermal denaturation measured in 30 mM MOPS, for the procedure see also Refs. 98H7 and 97P2

(2) average errors are ±29 for $\Delta H$ and ±0.3 for $T_{trs}$

Ribonuclease T1 and mutant Asp76→Asn

| Protein | pH | T | $\Delta Cp$ | Approach/Remarks | Ref |
|---|---|---|---|---|---|
| RNase T1 | 7.0 | 25 | 6.36 | urea/heat (1,2) | 99G7 |
| | 7.0 | 20 | 6.82 | urea/heat (1,2) | 99G7 |
| | 7.0 | 15 | 6.78 | urea/heat (1,2) | 99G7 |
| | 7.0 | 10 | 6.69 | urea/heat (1,2) | 99G7 |
| | 7.0 | 5 | 6.57 | urea/heat (1,2) | 99G7 |
| | 7.0 | 5–25 | 6.65±0.17 | average | 99G7 |
| Asp76→Asn | 7.0 | 25 | 6.40 | urea/heat (1,2) | 99G7 |
| | 7.0 | 20 | 5.61 | urea/heat (1,2) | 99G7 |
| | 7.0 | 15 | 5.98 | urea/heat (1,2) | 99G7 |
| | 7.0 | 10 | 5.56 | urea/heat (1,2) | 99G7 |
| | 7.0 | 5 | 5.61 | urea/heat (1,2) | 99G7 |
| | 7.0 | 5–25 | 5.82±0.38 | average | 99G7 |

Remarks:

(1) data from urea denaturation curves measured at different temperatures (see Table 1) and $\Delta H$ and $T_{trs}$, for details see Refs. 89P2 and 99G7

(2) measurements in 30 mM MOPS, pH 7.0

Ribonuclease T1 (RNase T1) and reduced and carboxyamidated RNase T1 (RTCAM)

| Protein | pH | $T_{trs}$ | $\Delta Cp$ | $\Delta H$ | Appr./Rem. | Ref |
|---|---|---|---|---|---|---|
| RNase T1 | 7.0 | 54.0±0.5 | 6.53±0.63 | 429.7±4.2 | urea (1–3) | 99B3 |
| | | 60 | | 470.3±4.2 | urea (1–4) | 99B3 |
| RTCAM | 7.0 | 12.4±0.2 | 6.65±0.42 | 137.7±2.1 | TMAO (1–3,5) | 99B3 |
| | | 60 | | 483.7±4.2 | TMAO (1–5) | 99B3 |

Remarks:

(1) data from solvent induced unfolding versus temperature, see also Table 1

(2) transition monitored by intrinsic fluorescence at 319 nm emission with excitation at 278 nm and 295 nm

(3) buffer: 30 mM MOPS, pH 7.0, 0.1 M NaCl, 2 mM EDTA

(4) data calculated at reference temperature $T_{ref}$ = 60°C

(5) TMAO – trimethylamine-N-oxide

### Ribosomal Protein

Ribosomal protein L9 from *Bacillus stearothermophilus*, N-terminal domain (NTL9), thermodynamic parameters in the presence of $H_2O$ and $D_2O$

| Protein/Solvent | | pH | $T_{trs}$ | $\Delta Cp$ | $\Delta H$ | Appr./Rem. | Ref |
|---|---|---|---|---|---|---|---|
| NTL9 | $H_2O$ | 5.45 | 77.5±1.3 | | | heat (1,2) | 98K18 |
| NTL9 | $H_2O$ | 5.45 | 79.8±1.6 | | | glob. (1–3) | 98K18 |
| NTL9 | $H_2O$ | 5.45 | 25 | 2.22±0.25 | 41.2±2.9 | glob. (1–4) | 98K18 |
| NTL9 | $D_2O$ | 5.45 | 81.0±1.5 | | | heat (2,5) | 98K18 |
| NTL9 | $D_2O$ | 5.45 | 85.6±3.0 | | | glob. (2,3,5) | 98K18 |
| NTL9 | $D_2O$ | 5.45 | 25 | 2.68±0.45 | 40.5±5.5 | glob. (2–5) | 98K18 |

Remarks:
(1) measured in 20 mM sodium acetate, 100 mM NaCl, pH 5.45
(2) the errors given here are rounded values, for detailed analysis of the confidence limits see Ref. 98K18
(3) from global analysis of thermal and denaturant-induced unfolding, assuming linear dependence of $\Delta G$ on the denaturant concentration, and assuming $\Delta Cp$ and m to be temperature independent
(4) data at reference temperature $T° = 25°C$
(5) measured in 20 mM sodium acetate, 100 mM NaCl, pD 5.45 (corrected value)

Ribosomal protein L9 from *Bacillus stearothermophilus*, N-terminal domain (NTL9, residues 1–56)

| Protein | pH | $T_{ref}$ | $\Delta Cp$ | $\Delta H$ | Approach/Remarks | | Ref |
|---|---|---|---|---|---|---|---|
| NTL9 | 5.45 | 25 | 2.68±0.50 | 40.6±5.9 | GuDCl | (1–4) | 98K17 |
| | 5.45 | 25 | 2.72±0.67 | 38.9±9.2 | GuDCl | (1,4–6) | 98K17 |

Remarks:
(1) data from global analysis of ln k in dependence of (deuterated) GuDCl concentration and temperature
(2) data from equilibrium unfolding, linear extrapolation
(3) transition monitored by far-UV CD
(4) buffer: $D_2O$, 20 mM sodium acetate, 100 mM NaCl, pD 5.45
(5) data from folding and unfolding kinetics, assuming linear dependence of ln k on the denaturant concentration
(6) from stopped-flow fluorescence measurements

Ribosomal protein L9 from *Bacillus stearothermophilus*, N-terminal domain (NTL9), $\Delta Cp$ in the presence of GuHCl

| Protein/Solvent | | GuHCl | pH | $\Delta Cp$ | Appr./Rem. | Ref |
|---|---|---|---|---|---|---|
| NTL9 | $H_2O$ | 0.00 M | 5.45 | 2.47±0.33 | glob. (1–3) | 98K18 |
| NTL9 | $H_2O$ | 0.74 M | 5.45 | 2.38±0.33 | glob. (1–3) | 98K18 |
| NTL9 | $H_2O$ | 1.18 M | 5.45 | 2.38±0.33 | glob. (1–3) | 98K18 |
| NTL9 | $H_2O$ | 2.02 M | 5.45 | 2.43±0.33 | glob. (1–3) | 98K18 |
| NTL9 | $H_2O$ | 2.69 M | 5.45 | 2.30±0.33 | glob. (1–3) | 98K18 |

Remarks:
(1) measured in 20 mM sodium acetate, 100 mM NaCl, pH 5.45
(2) $\Delta Cp$ from global analysis of thermal and denaturant-induced unfolding
(3) data at reference temperature $T° = 25°C$

Ribosomal protein L7/L12 from *E. coli*, wild type and mutants

| Protein | pH | $T_{trs}$ | $\Delta H$ | Approach/Remarks | | Ref |
|---|---|---|---|---|---|---|
| wild type | 7.3 | 67 | 293±11 | DSC | (1) | 97T4 |
| wild type | 7.3–9.2 | 68–70 | 293±11 | DSC | (1) | 96T5 |
| wild type | 7.3 | 63 | | heat | (2,3) | 97T4 |
| wild type | 5.88 | 63 | 134 | heat | (9,10) | 80L1 |

Ribosomal protein L7/L12 from *E. coli*, wild type and mutants (continued)

| Protein | pH | $T_{trs}$ | $\Delta H$ | Approach/Remarks | | Ref |
|---|---|---|---|---|---|---|
| Ser1→Tyr | 7.3 | 67 | 293±11 | DSC | (1,8) | 97T4 |
| Met14→Tyr | 7.3 | 67 | 293±11 | DSC | (1,8) | 97T4 |
| Met26→Tyr | 7.3 | 67 | 293±11 | DSC | (1) | 96T5 |
| L7 (27–120) | 7.6 | 72 | 244 | DSC | (4,5) | 78G1 |
| L7 oxidized | 7.6 | 75 | 260 | DSC | (5,6) | 78G1 |
| L7 oxidized | 9.2 | 71.5 | 239 | DSC | (6,7) | 78G1 |
| L7 dimer | 9.2 | 70 | 286 | DSC | (6,7) | 78G1 |

Remarks:
(1)  measured in 20 mM sodium phosphate buffer, pH 7.3
(2)  measured in 20 mM sodium acetate buffer, pH 7.3
(3)  transition monitored by CD at 220 nm
(4)  L7 fragment consisting of residues 27–120
(5)  buffer: 0.02 M Tris/HCl, pH 7.6, 0.3 M KCl, 8 mM $MgCl_2$
(6)  L7 oxidized monomer, with Met14, Met17, and Met26 oxidized
(7)  buffer: 0.02 M glycine/KOH, pH 9.2, 0.3 M KCl
(8)  for preliminary data see also Ref. 96T5
(9)  measured in a nonbuffered salt solution 20 mM $MgCl_2$, 360 mM KCl
(10) van't Hoff heat, transition monitored by CD at 222 nm

# ROP

ROP, four-helix-bundle protein, wild type

| Protein | pH | $T_{trs}$ | $\Delta Cp$ | $\Delta H$ | Approach/Remarks | | Ref |
|---|---|---|---|---|---|---|---|
| ROP | 6.0 | 69.5 | 2.9 | 518 | DSC | (1,2) | 98R10 |

Remarks:
(1) reconsideration of data from Ref. 93S10
(2) the temperature increment of $\Delta Cp$ amounts to $\Delta Cp' = -0.25$ kJ/mol/K²

ROP, four-α-helix bundle protein, mutant Ala31→Pro

| Protein | pH | $T_{trs}$ | $\Delta Cp$ | $\Delta H$ | Appr./Rem. | | Ref |
|---|---|---|---|---|---|---|---|
| wild type | 6.0 | 64.5 | 10.3±1.3 | 542±20 | DSC | (1–3) | 97P8 |
| Ala31→Pro | 6.0 | 43.0 | 7.2±0.8 | 272±12 | DSC | (1,2) | 97P8 |

Remarks:
(1) measured in 10 mM sodium phosphate, pH 6.0, 10 mM $Na_2SO_4$, 1 mM EDTA
(2) $T_{trs}$ refers to a protein concentration of 0.05 mg/ml
(3) data from Ref. 93S10

ROP, four-α-helix bundle protein, wild type and insertion mutant

| Protein | pH | $T_{trs}$ | $\Delta Cp$ | $\Delta H$ | Appr./Rem. | | Ref |
|---|---|---|---|---|---|---|---|
| wild type | 6.0 | 71.0±0.5 | 10.3±1.3 | 580±20 | DSC | (1–3) | 94V4 |
| <aa> | 6.0 | 69.5±0.3 | 10.2±1.5 | 575±25 | DSC | (1–5) | 94V4 |

Remarks:
(1) measured in 10 mM sodium phosphate, pH 6.0, 10 mM $Na_2SO_4$, 1 mM EDTA
(2) $T_{trs}$ refers to a protein concentration of 0.5 mg/ml
(3) the thermodynamic parameters refer to the molecular mass of the dimer
(4) $\Delta Cp$ from from $\Delta H^{v.H.}$ versus $T_{trs}$
(5) mutant <2aa> re-establishes the continuous heptad repeat pattern by inserting one Ala between ROP residues Leu29 and Asp30, and a second one between Asp30 and Ala31

ROP, four-$\alpha$-helix bundle protein, wild type and loop excision that produces a hyperthermophilic variant

| Protein | pH | $T_{trs}$ | $\Delta Cp$ | $\Delta H$ | Appr./Rem. | Ref |
|---|---|---|---|---|---|---|
| wild type | 6.0 | 71 | 10.3 | 580 | DSC (1–4) | 98L2 |
| RM7 | 6.0 | 69.5 | 10.2 | 575 | DSC (1–3,5) | 98L2 |
| RM6 | 6.0 | 98 | 14.9($\pm$3%) | 1099 | DSC (1–3,6) | 98L2 |
| RM6 | 6.0 | 101 | | 1073$\pm$30 | DSC (1,3,6,7) | 98L1 |

Remarks:

(1) measured in 10 mM sodium phosphate, pH 6.0, 10 mM $Na_2SO_4$, 1 mM EDTA

(2) $T_{trs}$ refers to a protein concentration of 0.5 mg/ml

(3) data per mole of cooperative unit, i.e., the dimer state for wild type and RM7, and to the tetrameric state of RM6

(4) data from Ref. 93S10

(5) data from Ref. 94V4, see insertion mutant <aa>

(6) RM6 was obtained by removal of five amino acids (Asp30, Ala31, Asp32, Glu33, Gln34) from the loop which results in a reorganization of the protein, forming a homotetrameric four-$\alpha$-helix structure instead of the homodimeric four-$\alpha$-helix motif of the wild type

(7) the data refer to a protein concentration of 1.0 mg/ml

**Sac7d Protein**, see DNA-Binding Protein

## SecA

SecA, peripheral subunit of the preprotein translocase from *Bacillus subtilis* in the absence and in the presence of ligands

| pH | Conc. | $T_{trs1}$ | $\Delta H_{trs1}$ | $T_{trs2}$ | $\Delta H_{trs2}$ | Appr./Rem. | Ref |
|---|---|---|---|---|---|---|---|
| 6.5 | | 38.3$\pm$0.5 | 548$\pm$82 | 45.8$\pm$1.0 | 448$\pm$133 | DSC (1,2) | 98D11 |
| 7.5 | | 40.5$\pm$0.5 | 644$\pm$96 | 48.6$\pm$0.5 | 732$\pm$109 | DSC (1,2) | 98D11 |
| 8.0 | | 39.6$\pm$0.5 | 628$\pm$94 | 47.3$\pm$0.5 | 669$\pm$100 | DSC (1,3) | 98D11 |
| SecA fragment N-SecA | | | | | | | |
| 8.0 | | | | 50.3$\pm$0.5 | 699$\pm$105 | DSC (1,3,4) | 98D11 |
| SecA in the presence of ADP: | | | | | | | |
| 7.5 | 0.0 | 40.5$\pm$0.5 | 644$\pm$96 | 48.6$\pm$0.5 | 732$\pm$109 | DSC (1,2) | 98D11 |
| 7.5 | 0.02 | 43.8$\pm$0.5 | 736$\pm$109 | 48.6$\pm$0.5 | 720$\pm$109 | DSC (1,2,5) | 98D11 |
| 7.5 | 0.2 | 46.3$\pm$0.2 | 824$\pm$121 | 49.1$\pm$0.2 | 469$\pm$71 | DSC (1,2,5) | 98D11 |
| 7.5 | 2 | 49.3$\pm$0.5 | 2950$\pm$444 | | | DSC (1,2,5) | 98D11 |
| SecA in the presence of AMP-PNP: | | | | | | | |
| 7.5 | 0.02 | 39.7$\pm$0.5 | 594$\pm$88 | 48.5$\pm$0.5 | 778$\pm$234 | DSC (1,2,6) | 98D11 |

Remarks:

(1) the homodimeric protein shows a bimodal DSC profile

(2) measured in phosphate buffer

(3) measured in Tris buffer, $\Delta Cp$ amounts to 8.4 kJ/mol/K per mole of dimer

(4) N-terminal fragment of SecA corresponding to amino acids 1–443

(5) ADP concentration in mM

(6) AMP-PNP is the nonhydrolysable ATP analogue adenyl-imido-diphosphate, AMP-PNP concentration in mM

**Selenomethionine Containing Proteins**, see Annexin V, and Lysozyme Phage T4

## Seminal Protein

Stallion seminal plasma protein HSP-3

| Protein | pH | $T_{trs}$ | $\Delta H$ | Approach/Remarks | | Ref |
|---------|-----|-----------|------------|------------------|---|-----|
| HSP-3 | 7.0 | 64 | 314 | DSC | (1–3) | 97M1 |

Remarks:
(1) measured in 20 mM HEPES, pH 7.0, at 0.33 K/min scan rate
(2) deconvolution of the single calorimetric peak reveals a non-two-state transition
(3) $\Delta H^{v.H.}$ amounts to 770 kJ/mol

Bovine seminal plasma protein PDC-10

| Ligand | $T_{trs1}$ | $\Delta H_{trs1}$ | $T_{trs2}$ | $\Delta H_{trs2}$ | Appr./Rem. | | Ref |
|--------|-----------|-------------------|-----------|-------------------|-----------|---|-----|
| none | 36.4 | 42 | 55.4 | 217 | DSC | (1,2) | 97G2 |
| 1 mM PC | 44.7 | 22 | 57.2 | 284 | DSC | (1–3) | 97G2 |
| 5 mM PC | | | 61.4 | 405 | DSC | (1–3) | 97G2 |
| 10 mM PC | | | 65.7 | 418 | DSC | (1–3) | 97G2 |
| 20 mM PC | | | 67.9 | 410 | DSC | (1–3) | 97G2 |
| 50 mM EDTA | | | 54.8 | 213 | DSC | (1–3) | 97G2 |
| 0.1 M NaCl | 34.8 | 29 | 55.2 | 221 | DSC | (1–3) | 97G2 |
| 0.5 M NaCl | | | 54.9 | 226 | DSC | (1–3) | 97G2 |
| 5 mM PC + 50 mM EDTA | | | 61.8 | 284 | DSC | (1–3) | 97G2 |
| 10 mM PC + 50 mM EDTA | | | 66.5 | 303 | DSC | (1–3) | 97G2 |
| 20 mM PC + 50 mM EDTA | | | 68.9 | 295 | DSC | (1–3) | 97G2 |

Remarks:
(1) measured in 10 mM MOPS, pH 7.0 at a heating rate of 0.33 K/min
(2) results of deconvolution, $T_{trs}$ was reproducible within 0.2°C, error in $\Delta H^{cal}$ about 5%
(3) PC = o-phosphorylcholine

**Serum Albumin**, see Bovine Serum Albumin, and Human Serum Albumin

## SH3 Domain

**The data entries are arranged as follows:**

a) α-spectrin SH3 domain
b) src SH3 domain
c) SH3 domains of Btk, Itk, and Tec
d) SH3 domains of α-spectrin, Fyn, and Abl

*a) α-spectrin SH3 domain*

α-spectrin SH3 domain in $D_2O$

| Protein | pH* | $T_{trs}$ | $\Delta Cp$ | $\Delta H$ | Approach/Remarks | | Ref |
|---------|-----|-----------|-------------|------------|------------------|---|-----|
| spectrin | 4.5 | 70.9 | | 227 | DSC | (1) | 99S1 |
| | 4.0 | 68.7 | | 219 | DSC | (1) | 99S1 |
| | 3.5 | 64.8 | | 211 | DSC | (1) | 99S1 |
| | 3.25 | 60.2 | | 193 | DSC | (1) | 99S1 |
| | 3.0 | 57.8 | | 183 | DSC | (1) | 99S1 |
| | 2.75 | 52.6 | | 167 | DSC | (1) | 99S1 |

α-spectrin SH3 domain in D$_2$O (continued)

| Protein | pH* | T$_{trs}$ | ΔCp | ΔH | Approach/Remarks | | Ref |
|---|---|---|---|---|---|---|---|
| | 2.5 | 49.6 | | 150 | DSC | (1) | 99S1 |
| | 2.25 | 45.3 | | 129 | DSC | (1) | 99S1 |
| | 2.0 | 42.4 | | 119 | DSC | (1) | 99S1 |
| | 1.0 | 41.4 | | 114 | DSC | (1) | 99S1 |
| | 1–4.5 | 41–71 | 3.9±0.3 | | DSC | (1,2) | 99S1 |

Remarks:
(1) measured in 20 mM glycine or 20 mM acetate in D$_2$O at different pH*
(2) ΔCp from ΔH$^{cal}$ versus T$_{trs}$

α-spectrin SH3 domain, wild type and mutants, introduction of helical tendency in the all β-sheet

| Mutant | pH | T$_{trs}$ | ΔH | Approach/Remarks | | Ref |
|---|---|---|---|---|---|---|
| wild type | 7.0 | 65.9±0.1 | 205±2 | heat | (1) | 97P14 |
| wild type | 2.5 | 51.6±0.5 | 144±3 | heat | (1) | 97P14 |
| Asp14→Ser | 7.0 | 57.9±0.2 | 164±2 | heat | (1) | 97P14 |
| Asp14→Ser | 2.5 | 49.4±0.3 | 145±2 | heat | (1) | 97P14 |
| double mutant (Thr4→Ala and Gly5→Glu) | | | | | | |
| | 7.0 | 63.2±0.1 | 180±2 | heat | (1) | 97P14 |
| | 2.5 | 44.7±0.5 | 134±2 | heat | (1) | 97P14 |
| triple mutant (Thr4→Ala, Gly5→Glu and Asp14→Ser) | | | | | | |
| | 7.0 | 52.0±0.3 | 136±1 | heat | (1) | 97P14 |
| | 2.5 | 44.0±0.4 | 132±3 | heat | (1) | 97P14 |

Remark:
(1) for temperature dependent ΔCp, see Ref. 94V2

α-spectrin SH3 domain, wild type and Asp48→Gly mutant

| Protein | pH | T$_{trs}$ | ΔCp | ΔH | Appr./Rem. | | Ref |
|---|---|---|---|---|---|---|---|
| wild type | 2.0 | 34.0 | 3.7 | 93 | DSC | (1) | 98M6 |
| | 2.5 | 47.0 | 3.4 | 139 | DSC | (1) | 98M6 |
| | 3.0 | 58.0 | 3.0 | 174 | DSC | (1) | 98M6 |
| | 3.5 | 63.0 | 2.9 | 188 | DSC | (1) | 98M6 |
| | 4.0 | 66.0 | 2.8 | 197 | DSC | (1) | 98M6 |
| Asp48→Gly | 2.0 | 45.3 | 3.6 | 140 | DSC | (2,3) | 98M6 |
| | 2.5 | 51.6 | 3.4 | 170 | DSC | (2,3) | 98M6 |
| | 3.0 | 60.4 | 3.0 | 196 | DSC | (2,3) | 98M6 |
| | 3.5 | 70.9 | 2.5 | 214 | DSC | (2,3) | 98M6 |
| | 4.0 | 74.2 | 2.4 | 225 | DSC | (2,3) | 98M6 |

Remarks:
(1) data from Ref. 94V2, see Ref. 94V2 also for additional data
(2) measured in 50 mM glycine/HCl or 50 mM acetic acid/sodium acetate
(3) ΔCp may be approximated by ΔCp(T) = −2.39+0.076×T−(1.795×10−4)×T$^2$ with T in K

α-spectrin SH3 domain and circular permutants with different loop lengths

| Mutant | pH | $T_{trs}$ | ΔCp | ΔH | Appr./Rem. | Ref |
|---|---|---|---|---|---|---|
| wild type | 2.0 | 33.8 | 3.7 | 93 | DSC (1) | 99M11 |
| | 2.5 | 46.8 | 3.4 | 139 | DSC (1) | 99M11 |
| | 3.5 | 62.8 | 2.9 | 188 | DSC (1) | 99M11 |
| pseudo-w.t. | 2.0 | 37.3±0.5 | 4.1 | 124 | DSC | 99M11 |
| | 2.5 | 45.5±0.4 | 3.4 | 150 | DSC | 99M11 |
| | 3.5 | 63.2±0.3 | 2.8 | 216 | DSC | 99M11 |
| | 7.0 | 66.0±0.8 | 2.7 | 225 | DSC | 99M11 |
| | | 25.0 | | 61.6 | DSC (2) | 99M11 |
| Ser19-Pro20s | 2.5 | 32.4±0.7 | 3.9 | 82 | DSC | 99M11 |
| | 3.5 | 53.0±0.4 | 3.3 | 160 | DSC | 99M11 |
| | 7.0 | 56.0±0.3 | 3.2 | 174 | DSC | 99M11 |
| | | 25.0 | | 59.4 | DSC (2) | 99M11 |
| Ser19-Pro20s1G | 2.5 | 31.1±0.7 | 3.9 | 85 | DSC | 99M11 |
| | 3.5 | 55.7±0.3 | 3.3 | 176 | DSC | 99M11 |
| | 7.0 | 61.1±0.3 | 3.1 | 192 | DSC | 99M11 |
| Ser19-Pro20s3G | 3.5 | 52.6±0.4 | 3.4 | 164 | DSC | 99M11 |
| | 7.0 | 57.6±0.3 | 3.2 | 187 | DSC | 99M11 |
| Ser19-Pro20s5G | 3.5 | 52.1±0.4 | 3.4 | 161 | DSC | 99M11 |
| | 7.0 | 56.5±0.3 | 3.2 | 173 | DSC | 99M11 |
| Asn47-Asp48s | 2.0 | 30.0±0.7 | 4.4 | 99 | DSC | 99M11 |
| | 2.5 | 42.0±0.5 | 3.8 | 142 | DSC | 99M11 |
| | 3.5 | 57.1±0.3 | 3.1 | 203 | DSC | 99M11 |
| | 7.0 | 52.8±0.4 | 3.3 | 195 | DSC | 99M11 |
| | | 25.0 | | 76.3 | DSC (2) | 99M11 |
| Asn47-Asp48s1G | 2.5 | 35.8±0.5 | 4.2 | 127 | DSC | 99M11 |
| | 3.5 | 56.6±0.3 | 3.2 | 195 | DSC | 99M11 |
| | 7.0 | 56.5±0.4 | 3.2 | 201 | DSC | 99M11 |
| Asn47-Asp48s3G | 3.5 | 55.1±0.3 | 3.2 | 182 | DSC | 99M11 |
| | 7.0 | 53.9±0.3 | 3.3 | 191 | DSC | 99M11 |
| Asn47-Asp48s5G | 3.5 | 54.3±0.3 | 3.3 | 180 | DSC | 99M11 |
| | 7.0 | 52.4±0.4 | 3.4 | 185 | DSC | 99M11 |

Explanations:

| | |
|---|---|
| wild type | for the sequence see Ref. 99M11 |
| pseudo-w.t. | the second and third residues were substituted by a Gly |
| Ser19-Pro20s | circular permutant, the pseudo-w.t. sequence was interrupted between Ser19 and Pro20, the previous N- and C-termini have been joined by the linker Lys60-Leu61-Asp62-Ser2-Gly3-Thr4-Gly5-Lys6, "s" refers to a Ser in the linker which was not present in shorter versions (see Ref. 95V5) |
| Asn47-Asp48s | circular permutant analogous to Ser19-Pro20s, the pseudo-w.t. sequence was interrupted between Asn47 and Asp48 |
| 1G, 3G, 5G | the connecting sequence was elongated with one, three, or five extra-Gly residues |

Remarks:

(1) data taken from Ref. 94V2

(2) data extrapolated to 25°C by a polynomial expression (see Ref. 99M11) which implies temperature dependence of ΔCp as follows,

pseudo-w.t.: $\Delta Cp = d\Delta H/dT = 27.49 - 0.1 \times T + (8.15 \times 10^{-5}) \times T^2$,

Ser19-Pro20s: $\Delta Cp = d\Delta H/dT = -5.95 + 0.087 \times T - (1.795 \times 10^{-4}) \times T^2$,

Asn47-Asp48s: $\Delta Cp = d\Delta H/dT = 27.17 - 0.1 \times T + (8.15 \times 10^{-5}) \times T^2$

*b) src SH3 domain*

Chicken src SH3 domain in $D_2O$

| | pD | $T_{ref}$ | $\Delta Cp$ | $\Delta H$ | Approach/Remarks | Ref |
|---|---|---|---|---|---|---|
| | 6 | 22 | 2.9±0.1 | 28.9±0.4 | heat/GuHCl (1,2) | 97G11 |

Remarks:
(1) heat denaturation, global fit of the temperature-GuHCl denaturation surface
(2) measured in 50 mM phosphate, pD 6, data at reference temperature of 22°C

*c) SH3 domains of Btk, Itk, and Tec*

SH3 Domains of the nonreceptor protein-tyrosin kinases Btk, Itk, and Tec

| Protein | pH | $T_{trs}$ | $\Delta Cp$ | $\Delta H$ | Approach/Remarks | | Ref |
|---|---|---|---|---|---|---|---|
| Btk-SH3 | 5.0 | 77.4 | 3.68 | 205.5 | DSC | | 98K10 |
| Btk-SH3 | 6.0 | 77.7 | 1.69 | 207.2 | DSC | | 98K10 |
| Btk-SH3 | 7.0 | 74.0 | 3.26 | 200.9 | DSC | | 98K10 |
| Btk-SH3 | 5.1–6.2 | 80 | 3.1±0.22 | 196 | DSC | (1,2) | 98K10 |
| Itk-SH3 | 4.1–6.0 | 69 | 3.4±0.25 | 178 | DSC | (1,2) | 98K10 |
| Tec-SH3 | 5.9–7.5 | 71 | 2.9±0.19 | 169 | DSC | (1,2) | 98K10 |
| Sso7d | 7.0 | 99 | 2.7±0.15 | 274 | DSC, heat (3) | | 96K5 |

Remarks:
(1) pH range where the protein achieves maximal thermostability, data taken from Fig. 5 in Ref. 98K10
(2) $\Delta Cp$ from $\Delta H$ versus $T_{trs}$
(3) Sso7d from the hyperthermophile *Sulfolobus solfataricus* is topologically identical to the eukaryotic SH3 domains, data from Ref. 96K5

**SH3 Domains,** see also Tyrosine Kinase

*d) SH3 domains of α-spectrin, Fyn, and Abl*

SH3 domains from α-spectrin (Spc-SH3), Fyn (Fyn-SH3), and Abl protein (Abl-SH3)

| Protein | pH | $T_{trs}$ | $\Delta Cp$ | $\Delta H$ | Appr./Rem. | | Ref |
|---|---|---|---|---|---|---|---|
| Spc-SH3 | 7.0 | 65.8±0.2 | 3.4±0.1 | 197±10 | DSC | (1) | 99F6 |
| Spc-SH3 | 3.5 | 62.8±0.2 | 3.4±0.1 | 188±10 | DSC | (1) | 99F6 |
| Abl-SH3 | 7.0 | 68.3±0.2 | 3.3±0.4 | 194±10 | DSC | (1) | 99F6 |
| Fyn-SH3 | 7.0 | 70.4±0.2 | 3.3±0.4 | 233±15 | DSC | (1) | 99F6 |

Remark:
(1) measured in 50 mM sodium phosphate

SH3 domain from the Fyn tyrosine kinase, mutants that affect a buried hydrogen bond between the side chains of Glu24 and Ser41

| Mutant | pH | $T_{trs}$ | $\Delta H$ | Approach/Remarks | | Ref |
|---|---|---|---|---|---|---|
| wild type | 8.0 | 80.1±0.7 | 215.1±4.2 | heat | (1–3) | 98M9 |
| Glu24→Ala | 8.0 | 72.5 | 174.5 | heat | (1,2) | 98M9 |
| Glu24→Arg | 8.0 | 45.6 | 95.4 | heat | (1,2) | 98M9 |
| Glu24→Asp | 8.0 | 82.2 | 222.6 | heat | (1,2) | 98M9 |
| Glu24→Gln | 8.0 | 74.1 | 190.8 | heat | (1,2) | 98M9 |
| Glu24→His | 8.0 | 73.6 | 187.9 | heat | (1,2) | 98M9 |
| Glu24→Ile | 8.0 | 55.2 | 120.5 | heat | (1,2) | 98M9 |

SH3 domain from the Fyn tyrosine kinase, mutants that affect a buried hydrogen bond between the side chains of Glu24 and Ser41 (continued)

| Mutant | pH | $T_{trs}$ | $\Delta H$ | Approach/Remarks | | Ref |
|---|---|---|---|---|---|---|
| Glu24→Leu | 8.0 | 53.0 | 110.9 | heat | (1,2) | 98M9 |
| Glu24→Lys | 8.0 | 73.5 | 182.8 | heat | (1,2) | 98M9 |
| Glu24→Phe | 8.0 | 66.6 | 153.1 | heat | (1,2) | 98M9 |
| Glu24→Pro | 8.0 | 64.2 | 153.1 | heat | (1,2) | 98M9 |
| Glu24→Ser | 8.0 | 67.8 | 167.4 | heat | (1,2) | 98M9 |
| Glu24→Thr | 8.0 | 74.4 | 192.5 | heat | (1,2) | 98M9 |
| Glu24→Tyr | 8.0 | 61.2 | 124.7 | heat | (1,2) | 98M9 |
| Glu24→Val | 8.0 | 65.8 | 150.6 | heat | (1,2) | 98M9 |
| Ser41→Ala | 8.0 | 73.9 | 176.1 | heat | (1,2) | 98M9 |
| Ser41→Arg | 8.0 | 82.6 | 212.1 | heat | (1,2) | 98M9 |
| Ser41→Asn | 8.0 | 77.5 | 173.2 | heat | (1,2) | 98M9 |
| Ser41→Asp | 8.0 | 59.4 | 128.9 | heat | (1,2) | 98M9 |
| Ser41→Gly | 8.0 | 58.6 | 114.6 | heat | (1,2) | 98M9 |
| Ser41→His | 8.0 | 74.8 | 182.4 | heat | (1,2) | 98M9 |
| Ser41→Ile | 8.0 | 73.3 | 175.7 | heat | (1,2) | 98M9 |
| Ser41→Leu | 8.0 | 75.6 | 187.4 | heat | (1,2) | 98M9 |
| Ser41→Lys | 8.0 | 85.5 | 223.0 | heat | (1,2) | 98M9 |
| Ser41→Phe | 8.0 | 61.4 | 120.9 | heat | (1,2) | 98M9 |
| Ser41→Thr | 8.0 | 65.6 | 141.8 | heat | (1,2) | 98M9 |
| Ser41→Tyr | 8.0 | 61.4 | 128.0 | heat | (1,2) | 98M9 |
| Ser41→Val | 8.0 | 70.3 | 163.2 | heat | (1,2) | 98M9 |
| double mutants: | | | | | | |
| (Glu24→Ala and Ser41→Ala) | | | | | | |
| | 8.0 | 68.2 | 166.1 | heat | (1,2) | 98M9 |
| (Glu24→Ala and Ser41→Leu) | | | | | | |
| | 8.0 | 77.1 | 188.7 | heat | (1,2) | 98M9 |
| (Glu24→Ala and Ser41→Lys) | | | | | | |
| | 8.0 | 71.4 | 163.6 | heat | (1,2) | 98M9 |
| (Glu24→Arg and Ser41→Asp) | | | | | | |
| | 8.0 | 70.8 | 168.6 | heat | (1,2) | 98M9 |
| (Glu24→Leu and Ser41→Ala) | | | | | | |
| | 8.0 | 51.4 | 101.7 | heat | (1,2) | 98M9 |

Remarks:
(1) transition monitored by CD at 220 nm
(2) buffer: 10 mM Tris-HCl, pH 8.0, 0.2 mM EDTA, 250 mM KCl
(3) averaged from eight measurements performed at protein concentrations from 10 to 100 µM

SH3 domain of human Fyn tyrosine kinase

| Protein | pH | $T_{trs}$ | $\Delta Cp$ | $\Delta H$ | Approach/Remarks | Ref |
|---|---|---|---|---|---|---|
| Fyn-SH3 | | | 3.6 | | (1,2) | 95A10 |

Remarks:
(1) Fyn-SH3 consists of 67 amino acid residues without disulfide bonds and prosthetic groups
(2) the $\Delta Cp$ value is cited in Ref. 98P11

## Sox-5 HMG Box

HMG Box from mouse Sox-5, and its interaction with DNA

| Protein/Complex | pH | $T_{trs}$ | $\Delta Cp$ | $\Delta H$ | Approach/Remarks | | Ref |
|---|---|---|---|---|---|---|---|
| Sox-5 without DNA | | | | | | | |
| trans. (1) | 6.0 | 34.4±0.5 | 1.7±0.4 | 88±6 | DSC | (1–5) | 98C11 |
| trans. (2) | 6.0 | 45.7±0.3 | 2.4±0.4 | 144±6 | DSC | (1–5) | 98C11 |
| complex Sox-5 + 10 bp DNA | | | | | | | |
| 58 μM | 6.0 | 61.5 | | 654 | DSC | (3,6,7) | 99P11 |
| 129 μM | 6.0 | 62.1 | | 611 | DSC | (3,6,7) | 99P11 |
| 135 μM | 6.0 | 62.4 | | 630 | DSC | (3,6,7) | 99P11 |
| complex Sox-5 + 12 bp DNA | | | | | | | |
| 59 μM | 6.0 | 66.6 | | 735 | DSC | (3,6,7) | 99P11 |
| 105 μM | 6.0 | 66.9 | | 760 | DSC | (3,6,7) | 99P11 |
| 187 μM | 6.0 | 67.0 | | 806 | DSC | (3,6,7) | 99P11 |
| complex Sox-5 + 16 bp DNA | | | | | | | |
| 101 μM | 6.0 | 57.9 | | 784 | DSC | (3,6,7) | 99P11 |
| 127 μM | 6.0 | 57.7 | | 767 | DSC | (3,6,7) | 99P11 |
| 152 μM | 6.0 | 57.3 | | 838 | DSC | (3,6,7) | 99P11 |
| 166 μM | 6.0 | 57.5 | | 860 | DSC | (3,6,7) | 99P11 |

Remarks:

(1) averaged values from DSC heating experiments at varying protein conc.

(2) results of deconvolution

(3) measured in 10 mM potassium phosphate, 100 mM KCl

(4) temperature-induced changes monitored by tryptophan fluorescence at 390 nm yield $T_{trs}$ = 43°C and $\Delta H$ = 150 kJ/mol

(5) temperature-induced changes monitored by tyrosine fluorescence at 300 nm yield $T_{trs}$ = 38°C and $\Delta H$ = 105 kJ/mol

(6) $\Delta H$ is the peak heat absorption defined by lines extrapolated from the pre- and posttransition zones into the transition region; for more detailed data, see Ref. 99P11

(7) given is the complex conc. (μM) and the base pair DNA duplex (xx bp)

## α-Spectrin

Human erythrocyte spectrin and recombinant spectrin peptides (Spα)

| Peptide | Buffer | $T_{trs}$ | $\Delta S$ | Approach/Remarks | Ref |
|---|---|---|---|---|---|
| Spα52–156 | 5P7.4 | 48±4 | 358±100 | heat, FL (1,2) | 98L7 |
| | PBS7.4 | 51±2 | 424±83 | heat, FL (1,2) | 98L7 |
| | 5P7.4 | 45±3 | 507±91 | heat, CD (1,3) | 98L7 |
| | PBS7.4 | 44 | 524 | heat, CD (1,3) | 98L7 |
| Spα157–262 | 5P7.4 | 58±0 | 890±25 | heat, FL (1,2) | 98L7 |
| | PBS7.4 | 54±2 | 806±33 | heat, FL (1,2) | 98L7 |
| | 5P7.4 | 55±2 | 981±17 | heat, CD (1,3) | 98L7 |
| | PBS7.4 | 53±0 | 798±83 | heat, CD (1,3) | 98L7 |
| Spα262–368 | 5P7.4 | 59±1 | 1264±158 | heat, FL (1,2) | 98L7 |
| | PBS7.4 | 63±1 | 1214±33 | heat, FL (1,2) | 98L7 |
| | 5P7.4 | 52±2 | 657±141 | heat, CD (1,3) | 98L7 |
| | PBS7.4 | 57±2 | 607±75 | heat, CD (1,3) | 98L7 |
| Spα52–262 | 5P7.4 | 51±2 | 615±33 | heat, FL (1,2) | 98L7 |
| | PBS7.4 | 52±1 | 765±42 | heat, FL (1,2) | 98L7 |
| | 5P7.4 | 50±1 | 615±25 | heat, CD (1,3) | 98L7 |
| | PBS7.4 | 50±1 | 840±33 | heat, CD (1,3) | 98L7 |
| Spα157–368 | 5P7.4 | 61±2 | 1064±25 | heat, FL (1,2) | 98L7 |
| | PBS7.4 | 59±2 | 1097±67 | heat, FL (1,2) | 98L7 |
| | 5P7.4 | 59±1 | 1039±67 | heat, CD (1,3) | 98L7 |

Human erythrocyte spectrin and recombinant spectrin peptides (Spα) (continued)

| Peptide | Buffer | $T_{trs}$ | $\Delta S$ | Approach/Remarks | Ref |
|---|---|---|---|---|---|
| | PBS7.4 | 57±1 | 1048±67 | heat, CD (1,3) | 98L7 |
| Spα52–368 | 5P7.4 | 52±3 | 757±100 | heat, FL (1,2) | 98L7 |
| | PBS7.4 | 53±2 | 1081±75 | heat, FL (1,2) | 98L7 |
| | 5P7.4 | 54±1 | 765±42 | heat, CD (1,3) | 98L7 |
| | PBS7.4 | 54±0 | 981±67 | heat, CD (1,3) | 98L7 |
| spectrin dimer | | | | | |
| | 5P7.4 | 46±0 | 1430±125 | heat, FL (1,2) | 98L7 |
| | PBS7.4 | 47±0 | 3658±524 | heat, FL (1,2) | 98L7 |
| | 5P7.4 | 48±0 | 624±17 | heat, CD (1,3) | 98L7 |
| | PBS7.4 | 47 | 948 | heat, CD (1,3) | 98L7 |
| Spα1 | PBS7.4 | 53.4 | | DSC        (4) | 97D3 |

Explanations:

     5P7.4    : 5 mM phosphate buffer

     PBS7.4  : 5 mM phosphate buffer, 150 mM NaCl

Remarks:

(1) $\Delta S$ in J/mol/K

(2) transition monitored by fluorescence spectroscopy

(3) transition monitored by CD at 222 nm

(4) Spα1 (residues 50–158) yields $\Delta H$ = 272 kJ/mol at 53.4°C

**α-Spectrin**, see also SH3 Domain

## Spherulin

Spherulin 3a from slime mold *Physarum polycephalum* in the presence and absence of calcium

| GuHCl conc. | pH | $T_{trs}$ | $\Delta Cp$ | $\Delta H$ | Appr./Rem. | Ref |
|---|---|---|---|---|---|---|
| spherulin 3a without Ca$^{2+}$: | | | | | | |
| 0.47 M | 7.0 | 57.9 | | 682 | DSC  (1,2) | 99K16 |
| 0.53 M | 7.0 | 56.3 | | 673 | DSC  (1,2) | 99K16 |
| 0.60 M | 7.0 | 55.8 | | 636 | DSC  (1,2) | 99K16 |
| 0.77 M | 7.0 | 51.8 | | 585 | DSC  (1,2) | 99K16 |
| 0.85 M | 7.0 | 51.3 | | 523 | DSC  (1,2) | 99K16 |
| 0.87 M | 7.0 | 49.7 | | 553 | DSC  (1,2) | 99K16 |
| 1.02 M | 7.0 | 49.0 | | 461 | DSC  (1,2) | 99K16 |
| 0.47–1.02 | | 49–58 | 18±2 | | DSC  (1–3) | 99K16 |
| 0.0 | 7.0 | 75 | | 800 | DSC  (1,2,5) | 99K16 |
| spherulin 3a with 3 mM Ca$^{2+}$: | | | | | | |
| 0.84 M | 7.0 | 68.7 | | 1020 | DSC  (2,4) | 99K16 |
| 0.98 M | 7.0 | 67.5 | | 954 | DSC  (2,4) | 99K16 |
| 0.98 M | 7.0 | 66.2 | | 937 | DSC  (2,4) | 99K16 |
| 1.32 M | 7.0 | 65.8 | | 947 | DSC  (2,4) | 99K16 |
| 1.40 M | 7.0 | 60.0 | | 820 | DSC  (2,4) | 99K16 |
| 1.79 M | 7.0 | 50.8 | | 675 | DSC  (2,4) | 99K16 |
| 0.84–1.79 | 7.0 | 51–69 | 18±2 | | DSC  (2–4) | 99K16 |
| 0.0 | 7.0 | 92 | | 1280 | DSC  (2,3,5) | 99K16 |

Remarks:

(1) buffer: 25 mM sodium cacodylate, 1 mM EDTA, pH 7.0

(2) data per mole of dimer, maximal error in $T_{trs}$ ±2°C and in $\Delta H^{cal}$ ±62 kJ/mol

(3) $\Delta Cp$ from $\Delta H^{cal}$ versus $T_{trs}$

(4) buffer: 25 mM sodium cacodylate, 3 mM CaCl$_2$, pH 7.0

(5) extrapolated value from $T_{trs}$ and $\Delta H$ versus GuHCl conc.

**Sso7d Protein**, see DNA-Binding Protein

## Sterol Carrier Protein

Recombinant human sterol carrier protein 2, wild type and mutant

| Protein | pH | $T_{trs}$ | $\Delta H$ | Approach/Remarks | | Ref |
|---|---|---|---|---|---|---|
| wild type | 6.8 | 70.5±0.5 | 183±10 | heat | (1,2) | 99J8 |
| Cys71→Ser | 6.8 | 59.9±0.9 | 164±22 | heat | (1,2) | 99J8 |

Remarks:
(1) transition monitored by far-UV CD
(2) buffer: 15 mM $K_2HPO_4$, 15 mM $KH_2PO_4$, 1 mM EDTA, pH 6.8

## Streptokinase

Streptokinase (SK) from various strains

| Transition | pH | $T_{trs}$ | $\Delta Cp$ | $\Delta H$ | Appr./Rem. | | Ref |
|---|---|---|---|---|---|---|---|
| SK from *Streptococcus equisimilis* strain H46A: | | | | | | | |
| trans. (1) | 7.5 | 46.0 | | 423 | DSC | (1) | 97W3 |
| trans. (2) | 7.5 | 60.9 | | 310 | DSC | (1) | 97W3 |
| SK from *Streptococcus pyogenes* strain A374: | | | | | | | |
| trans. (1) | 7.5 | 50.0 | | 452 | DSC | (1) | 97W3 |
| trans. (2) | 7.5 | 58.8 | | 297 | DSC | (1) | 97W3 |
| trans. (3) | 7.5 | 68.8 | | 272 | DSC | (1) | 97W3 |
| SK from *Streptococcus pyogenes strain* AT27: | | | | | | | |
| trans. (1) | 7.5 | 47.6 | | 393 | DSC | (1) | 97W3 |
| trans. (2) | 7.5 | 63.3 | | 259 | DSC | (1) | 97W3 |
| trans. (3) | 7.5 | 65.1 | | 201 | DSC | (1) | 97W3 |

Remark:
(1) the relative error of $\Delta H$ amounts to ±5%, the relative error of $T_{trs}$ amounts to ±0.5°C

Streptokinase from *Streptococcus equisimilis*, intact recombinant domain A (rSK-A, residues 1–146)

| Protein | pH | $T_{trs}$ | $\Delta Cp$ | $\Delta H$ | Appr./Rem. | | Ref |
|---|---|---|---|---|---|---|---|
| RSK-A | 7.0 | 52.4±0.5 | 7.9±0.5 | 278±20 | DSC | (1) | 99A5 |

Remark:
(1) measured in 20 mM phosphate buffer, pH 7.0

## Subtilisin BPN'

Subtilisin BPN' pro-domain, stabilization obtained by phage display selection

| Protein | pH | $T_{trs}$ | $\Delta Cp$ | $\Delta H$ | Approach/Remarks | | Ref |
|---|---|---|---|---|---|---|---|
| pro-R1 | 5.0 | 66 | 3.47 | 204 | DSC | (1–4) | 98R13 |

Remarks:
(1) the pro-domain consists of 77 amino acid residues compared with 275 of the mature protein
(2) the isolated pro-domain is to about 97% unfolded at 25°C, see also Ref. 95B12
(3) pro-R1 is the multiple mutant (Ala23→Cys, Lys27→Glu, Val37→Leu, and Gln40→Cys) obtained by phage display selection
(4) measured in 100 mM sodium acetate buffer, pH 5.0

## Subtilisin Inhibitor

Subtilisin inhibitor from *Streptomyces* (SSI) and mutants

| Mutant | pH | T | Δ(ΔH) | Approach/Remarks | | Ref |
|---|---|---|---|---|---|---|
| wild type | | 25 | 0.0 | DSC | (1,3,4,7) | 98T2 |
| Val13→Ala | | 25 | 37.6 | DSC | (2–4) | 98T2 |
| Val13→Gly | | 25 | −98.6 | DSC | (2–4) | 98T2 |
| Val13→Phe | | 25 | −73.6 | DSC | (2–4) | 98T2 |
| Val13→Ile | | 25 | −14.2 | DSC | (2–4) | 98T2 |
| Val13→Leu | | 25 | −25.1 | DSC | (2–4) | 98T2 |
| Val13→Met | | 25 | −20.1 | DSC | (2–4) | 98T2 |
| Met73→Lys | | 25 | −4.2 | DSC | (2–4) | 98T2 |
| Met73→Asp | | 25 | −27.6 | DSC | (2–4) | 98T2 |
| Met73→Glu | | 25 | −0.4 | DSC | (2–4) | 98T2 |
| Met73→Gly | | 25 | 1.7 | DSC | (2–4) | 98T2 |
| Met73→Ala | | 25 | −16.3 | DSC | (2–4) | 98T2 |
| Met73→Val | | 25 | 11.3 | DSC | (2–4) | 98T2 |
| Met73→Leu | | 25 | 3.8 | DSC | (2–4) | 98T2 |
| Met73→Ile | | 25 | 18.0 | DSC | (2–4) | 98T2 |
| Met103→Gly | | 25 | 22.6 | DSC | (2–4) | 98T2 |
| Met103→Ala | | 25 | 30.1 | DSC | (2–4) | 98T2 |
| Met103→Val | | 25 | 14.2 | DSC | (2–4) | 98T2 |
| Met103→Ile | | 25 | −13.0 | DSC | (2–4) | 98T2 |
| Met103→Leu | | 25 | 5.4 | DSC | (2–4) | 98T2 |
| wild type | 2.51 | 39.9 | | DSC | (1,5) | 98T2 |
| | 2.51 | 8.0 | | DSC | (1,6) | 98T2 |
| Val13→Leu | 2.51 | 9.2 | | DSC | (2,6) | 98T2 |
| | 2.99 | 44.1 | | DSC | (2,5) | 98T2 |
| Met73→Asp | 2.51 | 44.8 | | DSC | (2,5) | 98T2 |
| | 2.51 | 7.5 | | DSC | (2,6) | 98T2 |
| Met103→Leu | 2.51 | 43.5 | | DSC | (2,5) | 98T2 |
| | 2.51 | 7.6 | | DSC | (2,6) | 98T2 |
| | 2.99 | 46.7 | | DSC | (2,5) | 98T2 |

Remarks:
(1) wild-type SSI was obtained by cultivating *Streptomyces albogriseolus* S-3253
(2) SSI mutant proteins were harvested from *Streptomyces lividans* 66
(3) the Δ(ΔH) values are based on data from Refs. 94T3, 95T2, and 95T3
(4) Ref. 98T2 contains additional Δ(ΔH) values calculated at 100°C
(5) peak temperature of heat denaturation in 25 mM glycine buffer
(6) peak temperature of cold denaturation in 25 mM glycine buffer
(7) reference values for the wild-type protein at 25°C are ΔH = 98.2 kJ/mol, and ΔG = 89.9 kJ/mol, see Refs. 98T2 and 94T3

## Sunflower Albumin

Sunflower albumin 8 (SFA-8)

| Protein | pH | $T_{trs}$ | ΔCp | ΔH | Approach/Remarks | Ref |
|---|---|---|---|---|---|---|
| SFA-8 | 7.2 | 10 | 4.0±0.2 | −66.1±3.3 | heat/GuHSCN (1–3) | 99P3 |
| | 7.2 | 25 | 4.0±0.2 | −6.3±1.3 | heat/GuHSCN (1–3) | 99P3 |

Remarks:
(1) data from denaturant-induced unfolding carried out at different temperatures (10–55°C)
(2) buffer: 50 mM Tris-HCl
(3) Ref. 99P3 contains additional kinetic data on a compact misfolded state

## Superoxide Dismutase

Cu,Zn superoxide dismutase (Cu,Zn-SOD) from various species

| Protein | pH | $T_{trs}$ | $\Delta H$ | Approach/Remarks | Ref |
|---|---|---|---|---|---|
| ox | 7.8 | 88.0 | | (1,2) | 98B5 |
| sheep | 7.8 | 87.1 | | (1,2) | 98B5 |
| human | 7.8 | 83.6 | | (1,2) | 98B5 |
| shark | 7.8 | 84.1 | | (1,2) | 98B5 |
| yeast | 7.8 | 73.1 | | (1,2) | 98B5 |
| *X. laevis A* | 7.8 | 71.1 | | (1,2) | 98B5 |
| *X. laevis B* | 7.8 | 76.8 | | (1,2) | 98B5 |
| *P. leignathi* | 7.8 | 71.0 | | (1,2) | 98B5 |
| *E. coli* | 7.8 | 65.9 | | (1,2) | 98B5 |

Remarks:

(1) all data refer to Cu,Zn-SODs scanned under identical experimental conditions (100 mM phosphate buffer, pH 7.8, scan rate 60 K/h)

(2) the data were taken from literature, see Ref. 98B5

Cu,Zn superoxide dismutase (Cu,Zn-SOD) from various species

| Protein/Transition | pH | $T_{trs}$ | $\Delta H$ | Appr./Rem. | | Ref |
|---|---|---|---|---|---|---|
| bovine Cu,Zn-SOD, holoprotein | | | | | | |
| trans. (1) | 5.5 | 89 | | DSC | (1,2) | 88R4 |
| trans. (2) | 5.5 | 96 | | DSC | (1,2) | 88R4 |
| bovine Cu,Zn-SOD, apoprotein | | | | | | |
| | 5.5 | 57 | | DSC | (1,4) | 88R4 |
| bovine, Cu,Zn-SOD, $Zn^{2+}$ stabilized form | | | | | | |
| | 5.5 | 79 | 1172±63 | DSC | (1,4) | 88R4 |
| yeast Cu,Zn-SOD, holoprotein | | | | | | |
| trans. (1) | 5.5 | 77 | | DSC | (1) | 88R4 |
| trans. (2) | 5.5 | 82 | | DSC | (1) | 88R4 |
| equine Cu,Zn-SOD, holoprotein | | | | | | |
| | 5.5 | 93 | | DSC | (1,3) | 88R4 |

Remarks:

(1) measured in 0.1 M sodium acetate buffer, pH 5.5

(2) $T_{trs}$ and $\Delta H$ vary with scan rate

(3) melting and precipitation in a single peak

(4) single transition

Cu,Zn superoxide dismutase from *E. coli*

| Protein/Transition | | pH | $T_{trs}$ | $\Delta Cp$ | $\Delta H$ | Appr./Rem. | | Ref |
|---|---|---|---|---|---|---|---|---|
| holo | trans. (1) | 7.8 | 52.6 | 4.2 | 235 | DSC | (1,2) | 98B5 |
| | trans. (2) | 7.8 | 65.9 | −2.1 | 391 | DSC | (1,2) | 98B5 |
| holo | trans. (1) | 7.4 | 53.1 | 4.6 | 215 | DSC | (1,2) | 98B5 |
| | trans. (2) | 7.4 | 70.2 | −12.1 | 384 | DSC | (1,2) | 98B5 |
| holo | trans. (1) | 7.0 | 54.4 | 6.3 | 210 | DSC | (1,2) | 98B5 |
| | trans. (2) | 7.0 | 72.6 | −4.2 | 382 | DSC | (1,2) | 98B5 |
| holo | trans. (1) | 6.5 | 58.5 | 1.3 | 225 | DSC | (1,2) | 98B5 |
| | trans. (2) | 6.5 | 76.0 | −3.3 | 374 | DSC | (1,2) | 98B5 |
| holo | trans. (1) | 6.0 | 58.8 | 4.2 | 190 | DSC | (1,2) | 98B5 |
| | trans. (2) | 6.0 | 79.3 | −11.3 | 358 | DSC | (1,2) | 98B5 |
| apo | trans. (1) | 7.8 | 53.6 | 4.2 | 255 | DSC | (1,2) | 98B5 |
| apo | trans. (1) | 6.5 | 58.1 | 2.9 | 276 | DSC | (1,2) | 98B5 |

Remarks:

(1) results of deconvolution according to a two-state model ($\Delta H^{cal} = \Delta H^{v.H.}$)

(2) measured in 0.1 M potassium phosphate

Cu,Zn superoxide dismutase from *E. coli*, zinc-reconstituted protein

| Zinc/Protein | Transition | pH | $T_{trs}$ | $\Delta H$ | Appr./Rem. | | Ref |
|---|---|---|---|---|---|---|---|
| 0.45 | trans. (1) | 6.5 | 61.2 | 103 | DSC | (1,2) | 98B5 |
| | trans. (2) | 6.5 | 69.7 | 203 | DSC | (1,2) | 98B5 |
| 0.90 | trans. (1) | 6.5 | 62.9 | 128 | DSC | (1,2) | 98B5 |
| | trans. (2) | 6.5 | 69.1 | 169 | DSC | (1,2) | 98B5 |
| 1.35 | trans. (1) | 6.5 | 43.7 | 64 | DSC | (1,2) | 98B5 |
| | trans. (2) | 6.5 | 70.7 | 337 | DSC | (1,2) | 98B5 |

Remark:

(1) results of deconvolution according to a non-two-state model ($\Delta H^{cal} \neq \Delta H^{v.H.}$)

(2) measured in 0.1 M potassium phosphate

Bovine Cu,Zn superoxide dismutase (Cu,Zn-SOD), wild type and mutant

| Protein | Transition | pH | $T_{trs}$ | $\Delta H$ | Appr./Rem. | | Ref |
|---|---|---|---|---|---|---|---|
| wild type | trans. (1) | 7.8 | 82.8 | 473 | DSC | (1,2) | 90M6 |
| | trans. (2) | 7.8 | 89.5 | 674 | DSC | (1,2) | 90M6 |
| Cys6→Ala | trans. (1) | 7.8 | 80.7 | 515 | DSC | (1,2) | 90M6 |
| | trans. (2) | 7.8 | 85.8 | 519 | DSC | (1,2) | 90M6 |

Remarks:

(1) measured in 100 mM potassium phosphate buffer, pH 7.8

(2) results of deconvolution

Human Cu,Zn superoxide dismutase (SOD), wild type and circularly permuted variants

| Protein | Component | pH | $T_{trs}$ | $\Delta H$ | Appr./Rem. | Ref |
|---|---|---|---|---|---|---|
| native | minor | 7.8 | 81.2±0.3 | 494±38 | DSC (1,2) | 97B10 |
| | major | 7.8 | 84.3±0.3 | 795±84 | DSC (1,2) | 97B10 |
| cp(−1+1) | minor | 7.8 | 83.5±0.5 | 439±33 | DSC (1,2) | 97B10 |
| | major | 7.8 | 86.2±0.1 | 669±42 | DSC (1,2) | 97B10 |
| cp(−2+2) | minor | 7.8 | 80.9±1.0 | 385±84 | DSC (1,2) | 97B10 |
| | major | 7.8 | 83.5±0.8 | 665±17 | DSC (1,2) | 97B10 |
| cp(−3+3) | minor | 7.8 | 77.5±0.3 | 385±125 | DSC (1,2) | 97B10 |
| | major | 7.8 | 82.7±0.5 | 544±125 | DSC (1,2) | 97B10 |

Remarks:
(1) measured in 100 mM potassium phosphate, pH 7.8
(2) components resolved by deconvolution

Human manganese superoxide dismutase, mutants at Tyr34 and Glu143

| Protein | Component | pH | $T_{trs}$ | $\Delta H$ | Appr./Rem. | Ref |
|---|---|---|---|---|---|---|
| native | B | 7.8 | 70 | 594 | DSC (1–3) | 98G19 |
| | C | 7.8 | 88.9 | 379 | DSC (1–4) | 98G19 |
| Tyr34→Phe | B | 7.8 | 85.9 | 1469 | DSC (1–5) | 98G19 |
| Tyr34→Phe | C | 7.8 | 95.6 | 824 | DSC (1–5) | 98G19 |
| Gln143→Asn | A | 7.8 | 57.4 | 418 | DSC (1–4) | 98H10 |
| Gln143→Asn | B | 7.8 | 85.3 | 328 | DSC (1–4) | 98H10 |
| Gln143→Asn | C | 7.8 | 90.7 | 657 | DSC (1–4) | 98H10 |

Remarks:
(1) $\Delta H$ is given per mole of tetramer
(2) measured in 2 mM potassium phosphate buffer, pH 7.8
(3) for labeling of the transitions see also Ref. 96B9
(4) component C is the main transition
(5) five peaks were resolvable for mutant Tyr34→Phe

**Telluromethionine Containing Proteins**, see Annexin V

**Tenascin**

Fibronectin type III domain of human tenascin

| pH | $T_{trs}$ | $\Delta Cp$ | $\Delta H$ | Appr./Rem. | Ref |
|---|---|---|---|---|---|
| 5.0 | 56.7 | 6.11 | 279 | DSC (1) | 97C9 |
| | | 4.98 | | DSC (2) | 97C9 |
| 5.0 | 55±1 | | | heat (3) | 97C9 |
| 7.0 | 44±1 | | | heat (3) | 97C9 |

Remarks:
(1) $\Delta Cp$ from $\Delta H^{cal}$ versus $T_{trs}$
(2) $\Delta Cp$ from $\Delta H^{v.H}$ versus $T_{trs}$
(3) from both far-UV CD at 230 nm and near-UV CD at 284 nm

## Thermolysin

Mutants of thermolysin-like proteases (TLP) derived from *Bacillus stearothermophilus*

| Protein | pH | $T_{trs}$ | $\Delta T$ | Approach/Remarks | | Ref |
|---|---|---|---|---|---|---|
| TLP-tur1 | 5.3 | 71.1 | 0.0 | heat | (1–3) | 98V7 |
| Ala315→Val and insert 248–257 | | | | | | |
| | 5.3 | 73.4 | 2.3 | heat | (1,2) | 98V7 |
| double mutant (Thr63→Phe and Ala69→Pro) | | | | | | |
| | 5.3 | 75.2 | 4.1 | heat | (1,2) | 98V7 |
| multiple mutant (Thr63→Phe, Ala69→Pro, and Ala315→Val) and insert 248–257 | | | | | | |
| | 5.3 | 85.7 | 14.6 | heat | (1,2,4) | 98V7 |

Remarks:

(1) measured by activity after incubating in 20 mM sodium acetate, pH 5.3, 5 mM $CaCl_2$, 0.5% (v/v) isopropanol, 62.5 mM NaCl

(2) Ref. 98V7 contains additional data from a quasi-thermodynamic fit of the rest activity after standardized autolysis

(3) TLP-tur1 is TLP with two unfolding regions

(4) overadditivity of stabilizing effects

## Thioredoxin

Thioredoxin from *Bacillus acidocaldarius* (BacTrx), wild type and mutants

| Protein | pH | $T_{trs}$ | $\Delta H$ | Approach/Remarks | Ref |
|---|---|---|---|---|---|
| BacTrx | 7.0 | 102.6 | 288.6 | heat, v.H. (1) | 99P5 |
| Lys18→Gly | 7.0 | 90.6 | 276.0 | heat, v.H. (1) | 99P5 |
| Arg82→Glu | 7.0 | 90.6 | 301.6 | heat, v.H. (1) | 99P5 |
| double mutant (Lys18→Gly and Arg82→Glu) | | | | | |
| | 7.0 | 88.7 | 196.2 | heat, v.H. (1) | 99P5 |
| Asp102→X | 7.0 | 86.1 | 339.2 | heat, v.H. (1,2) | 99P5 |

Remarks:

(1) transition monitored by CD

(2) mutant Asp102→X, the last four amino acids were deleted

Thioredoxin from *E. coli* (TRX) studied by H/D exchange and electrospray ionization mass spectrometry (ESI-MS)

| Protein | pH | $T_{trs}$ | $\Delta H$ | Approach/Remarks | | Ref |
|---|---|---|---|---|---|---|
| TRX | 2.8–3.1 | 60 | 326 | heat | (1,2) | 99M1 |
| | 2.8–3.1 | 64 | 310 | heat | (1,3) | 99M1 |
| | 2.8–3.1 | 65 | 226 | heat | (1,4) | 99M1 |

Remarks:

(1) measured in 2% acetic acid-$d_1$, pH 2.8–3.1

(2) transition monitored by near-UV CD at 280 nm

(3) transition monitored by ESI-MS, ion peak intensity ratio

(4) transition monitored by ESI-MS, average charge state

Thioredoxin from *E. coli*, mutants

| Protein | Conc. | pH | $T_{trs}$ | $\Delta Cp$ | $\Delta H$ | Appr./Rem. | | Ref |
|---|---|---|---|---|---|---|---|---|
| wild type* | 0.090 mM | 7.0 | 84.1±0.7 | 10.3±5.4 | 371±11 | DSC | (1) | 99C4 |
| | 0.206 mM | 7.0 | 83.6±0.2 | 12.0±2.5 | 423±4 | DSC | (1) | 99C4 |
| | 0.360 mM | 7.0 | 83.7±0.2 | 8.7±2.1 | 439±3 | DSC | (1) | 99C4 |
| | | 7.0 | 25.0 | 5.9±1.0 | 84±18 | GuHCl | (2) | 99C4 |
| | | 7.0 | 25.0 | 7.5±2.5 | | GuHCl | (3) | 99C4 |
| double m. | 0.080 mM | 7.0 | 83.7±0.5 | 10.3±4.2 | 403±9 | DSC | (1) | 99C4 |
| | 0.112 mM | 7.0 | 83.6±0.7 | 9.3±8.8 | 456±13 | DSC | (1) | 99C4 |
| | 0.183 mM | 7.0 | 82.4±0.5 | 10.0±5.0 | 431±9 | DSC | (1) | 99C4 |
| | 0.238 mM | 7.0 | 81.9±0.5 | 16.9±7.1 | 456±16 | DSC | (1) | 99C4 |
| | | 7.0 | 25.0 | 6.7±1.0 | 16±2 | GuHCl | (2) | 99C4 |
| | | 7.0 | 25.0 | 8.8±1.3 | | GuHCl | (3) | 99C4 |

Explanations:

wild type* – pseudo-wild type, mutant Met37→Leu
double m. – double mutant Met37→Leu and Pro40→Ser

Remarks:
(1) buffer: 50 mM phosphate, pH 7.0
(2) data from GuHCl denaturation, $\Delta G$ versus T (see also Table 1)
(3) data from GuHCl denaturation (conc. dependence, see also Ref. 99C4)

Thioredoxin from *E. coli*, nonconservative amino acid substitutions

| Mutant/Protein conc. | $T_{trs}$ | $\Delta Cp$ | $\Delta H$ | Approach/Remarks | | Ref |
|---|---|---|---|---|---|---|
| Leu78→Cys: | | | | | | |
| 330 µM | 65.24 | | 262 | DSC | (1) | 97O1 |
| 162 µM | 65.20 | | 354 | DSC | (1) | 97O1 |
| 120 µM | 66.91 | | 363 | DSC | (1) | 97O1 |
| 120 µM | 66.30 | | 310 | DSC | (1) | 97O1 |
| Leu78→Cys, methyl derivative: | | | | | | |
| 346 µM | 69.33 | 7.8 | 420 | DSC | (1,2) | 97O1 |
| 310 µM | 69.36 | 7.8 | 447 | DSC | (1,2) | 97O1 |
| 198 µM | 69.74 | 4.0 | 395 | DSC | (1,2) | 97O1 |
| Leu78→Cys, ethyl derivative: | | | | | | |
| 467 µM | 69.34 | 6.1 | 393 | DSC | (1,2) | 97O1 |
| 333 µM | 69.65 | 5.2 | 408 | DSC | (1,2) | 97O1 |
| 258 µM | 69.69 | 6.2 | 406 | DSC | (1,2) | 97O1 |
| 134 µM | 69.80 | 6.5 | 405 | DSC | (1,2) | 97O1 |
| Leu78→Cys, propyl derivative: | | | | | | |
| 226 µM | 67.30 | 12.8 | 280 | DSC | (1,2) | 97O1 |
| 83 µM | 69.25 | 7.8 | 320 | DSC | (1,2) | 97O1 |
| 37 µM | 70.34 | 1.7 | 427 | DSC | (1,2) | 97O1 |
| Leu78→Cys, butyl derivative: | | | | | | |
| 327 µM | 63.07 | 7.8 | 224 | DSC | (1,2) | 97O1 |
| 232 µM | 63.72 | 9.2 | 221 | DSC | (1,2) | 97O1 |
| 110 µM | 65.49 | 9.1 | 236 | DSC | (1,2) | 97O1 |
| Leu78→Cys, n-pentyl derivative: | | | | | | |
| 97 µM | 62.63 | 11.7 | 300 | DSC | (1,2) | 97O1 |
| 73 µM | 62.43 | 10.3 | 213 | DSC | (1,2) | 97O1 |
| Leu78→Cys, cyclopentyl derivative: | | | | | | |
| 73 µM | 69.59 | 3.6 | 354 | DSC | (1,2) | 97O1 |
| 31 µM | 69.56 | 6.4 | 339 | DSC | (1,2) | 97O1 |
| average value | | 7.99±3.35 | | DSC | | 97O1 |

Remarks:
(1) buffer: 50 mM sodium phosphate, pH 7.0
(2) introduction of thiolsulfonated alkyl groups into the hydrophobic core, the modification was made at residue 78

Oxidized thioredoxin from *E. coli*, (Trx), cleaved and uncleaved Trx

| Protein/Conc. | | pH | $T_{trs}$ | $\Delta H$ | Approach/Remarks | | Ref |
|---|---|---|---|---|---|---|---|
| Trx | 15 µM | 5.7 | 88.1±0.1 | 507±20 | heat | (1,2) | 99G5 |
| | 40 µM | 5.7 | 88.7±0.5 | 488±18 | heat | (1,2) | 99G5 |
| | 80 µM | 5.7 | 88.3±0.5 | 437±13 | heat | (1,2) | 99G5 |
| | 80 µM | 5.7 | 87.9±0.2 | 527±18 | heat | (2,3) | 99G5 |
| | 160 µM | 5.7 | 87.0±0.4 | 461±13 | heat | (1,2) | 99G5 |
| | 15–160 | 5.7 | 88.0±0.6 | 474±26 | heat | (4) | 99G5 |
| NC | 12.5 µM | 5.7 | 62.1±0.2 | 347±12 | heat | (1,2,5) | 99G5 |
| | 20 µM | 5.7 | 63.5±0.2 | 344±6 | heat | (1,2,5) | 99G5 |
| | 40 µM | 5.7 | 65.2±0.1 | 368±2 | heat | (1,2,5) | 99G5 |
| | 80 µM | 5.7 | 66.4±0.1 | 328±4 | heat | (1,2,5) | 99G5 |
| | 80 µM | 5.7 | 66.5±0.1 | 363±5 | heat | (1,2,5) | 99G5 |
| | 160 µM | 5.7 | 67.3±0.2 | 352±10 | heat | (1,2,5) | 99G5 |
| | 450 µM | 5.7 | 69.2±0.5 | 349±26 | heat | (1,2,5) | 99G5 |

Remarks:

(1) transition monitored by far-UV CD

(2) measured in 10 mM potassium phosphate, pH 5.7

(3) transition monitored by near-UV CD

(4) average value

(5) NC = cleaved Trx (1–73, 74–108)

## Tobacco Mosaic Virus

Tobacco mosaic virus (strain U1) and its coat protein Ts mutant [remark (1)]

| Protein | pH | $T_{trs}$ | $\Delta H$ | Approach/Remarks | | Ref |
|---|---|---|---|---|---|---|
| intact U1 | 5.6 | 87.6 | 613 | DSC | | 98O6 |
| | 7.0 | 80.5 | 537 | DSC | | 98O6 |
| | 8.0 | 71.8 | 309 | DSC | | 98O6 |
| intact Ts21–66 | 5.6 | 84.6 | 614 | DSC | (2) | 98O6 |
| | 7.0 | 80.8 | 498 | DSC | (2) | 98O6 |
| | 8.0 | 71.9 | 404 | DSC | (2) | 98O6 |
| U1 RP | 5.6 | 73.2 | 415 | DSC | (3) | 98O6 |
| Ts21–66 RP | 5.6 | 70.9 | 397 | DSC | (2,3) | 98O6 |
| U1 A-protein | 8.0 | 40.9 | 251 | DSC | (4) | 98O6 |
| Ts21–66 A-protein | 8.0 | 37.8 | 269 | DSC | (2,4) | 98O6 |

Remarks:

(1) for tobacco mosaic virus see also Ref. 92M8

(2) Ts21–66 = tobacco mosaic virus carrying two amino acid substitutions (Ile21→Thr and Asp66→Gly) in the coat protein

(3) RP = repolymerized protein

(4) A-protein = small (3 or 4S) protein aggregates

## Toxins

Equinatoxin II (EqTxII) from the sea anemone *Actinia equina* L.

| pH | $T_{trs}$ | $\Delta H^{v.H.}$ | $\Delta H^{cal}$ | Approach | Remarks | Ref |
|---|---|---|---|---|---|---|
| transition monitored by UV absorption: | | | | | | |
| 2.0 | 39.9 | 293 | | heat, UV 232 nm | | 97P13 |
| 3.0 | 52.3 | 363 | | heat, UV 232 nm | | 97P13 |
| 3.5 | 59.0 | 402 | | heat, UV 232 nm | | 97P13 |
| 5.5–6.0 | 65.4 | 441 | | heat, UV 232 nm | | 97P13 |

Equinatoxin II (EqTxII) from the sea anemone *Actinia equina* L. (continued)

| pH | $T_{trs}$ | $\Delta H^{v.H.}$ | $\Delta H^{cal}$ | Approach | Remarks | Ref |
|---|---|---|---|---|---|---|
| transition monitored by DSC: | | | | | | |
| 2.0 | 44.4 | 240 | 117 | DSC | (1) | 97P13 |
| 3.0 | 54.9 | 380 | 310 | DSC | (1) | 97P13 |
| 3.5 | 61.9 | 403 | 371 | DSC | (1) | 97P13 |
| 5.5–6.0 | 65.8 | 430 | 441 | DSC | (1) | 97P13 |
| transition monitored by CD: | | | | | | |
| 2.0 | 38.3 | 264 | | heat, CD | (2) | 97P13 |
| 3.0 | 51.4 | 370 | | heat, CD | (2) | 97P13 |
| 3.5 | 59.3 | 350 | | heat, CD | (2) | 97P13 |
| 5.5–6.0 | 66.0 | 433 | | heat, CD | (2) | 97P13 |

Remarks:
(1) $\Delta Cp$ from $\Delta H^{v.H}$ versus $T_{trs}$ obtained by UV melting curves amounts to 5.8 kJ/mol/K; $\Delta Cp$ from DSC single calorimetric recordings amounts to 5.7 kJ/mol/K
(2) transition monitored by near-UV spectra (250–300 nm)

Cry3A δ-endotoxin from *Bacillus thuringiensis* var. *tenebrionis* and ist 55 kDa fragment

| Protein | pH | $T_{trs}$ | $\Delta Cp$ | $\Delta H$ | Approach/Remarks | | Ref |
|---|---|---|---|---|---|---|---|
| δ-endotoxin | 3.5 | 70.4 | | 1330 | DSC | (1,2b) | 99P10 |
| | 3.0 | 67.2 | | 1300 | DSC | (1,2a) | 99P10 |
| | 3.0 | 65.4 | | 1200 | DSC | (1,2b) | 99P10 |
| | 3.0 | 63.8 | | 1210 | DSC | (1,2c) | 99P10 |
| | 3.0 | 60.3 | | 945 | DSC | (1,2d) | 99P10 |
| | 2.8 | 59.9 | | 849 | DSC | (1,2b) | 99P10 |
| | 2.8 | 58.5 | | 840 | DSC | (1,2d) | 99P10 |
| | 2.5 | 59.3 | | 924 | DSC | (1,2b) | 99P10 |
| | 2.0 | 49.5 | | 644 | DSC | (1,2b) | 99P10 |
| | 2–3.5 | 49–70 | 33.0 | | DSC | (1,3) | 99P10 |
| 55 kDa fragment | 3.0 | 70.1 | | 832 | DSC | (1,2b) | 99P10 |
| | 2.75 | 61.1 | | 815 | DSC | (1,2b) | 99P10 |
| | 2.5 | 57.5 | | 727 | DSC | (1,2b) | 99P10 |
| | 2.0 | 49.0 | | 652 | DSC | (1,2b) | 99P10 |
| | 2–3 | 49–70 | 12.8 | | DSC | (1,3) | 99P10 |

Remarks:
(1) measured in 25 mM glycine-HCl
(2) heating rate: (2a) 2 K/min, (2b) 1 K/min, (2c) 0.5 K/min, (2d) 0.125 K/min
(3) $\Delta Cp$ from $\Delta H^{cal}$ versus $T_{trs}$

Heat-labile enterotoxin from *E.coli*, LT-I, subunit A with an engineered disulfide bond

| Mutant | pH | $T_{trs}$ | $\Delta H$ | Approach/Remarks | | Ref |
|---|---|---|---|---|---|---|
| wild-type LT-I | 7.5 | 54 | | DSC | (1) | 97A3 |
| double mutant (Asn40→Cys and Gly166→Cys) | | | | | | |
| | 7.5 | 60 | | DSC | (2,3) | 97A3 |

Remarks:
(1) the protein is similar to cholera toxin secreted by *Vibrio cholerae*
(2) mutant with disulfide bridge between residues 40 and 166
(3) $T_{trs}$ of the B pentamers of wild-type LT-I and double mutant amounts to 86°C

## Transglutaminase

Human erythrocyte transglutaminase

| pH | $T_{trs1}$ | $\Delta H_{trs1}$ | $T_{trs2}$ | $\Delta H_{trs2}$ | Approach/Remarks | | Ref |
|---|---|---|---|---|---|---|---|
| 6.1 | 51 | 850 | 52 | 290 | DSC | (1) | 99B8 |
| 7.5 | 51 | 880 | 60 | 150 | DSC | (1) | 99B8 |
| 9.5 | 41 | 280 | ~57 | 140 | DSC | (1) | 99B8 |

Remark:
(1) results of deconvolution

## tRNA Synthetase

Glutamyl-prolyl-tRNA synthetase, EPRS repeat (EPSR-R)

| Protein | pH | $T_{trs}$ | $\Delta H$ | Approach/Remarks | Ref |
|---|---|---|---|---|---|
| EPSR-R | 5.0 | 59 | 213 | DSC | 98R5 |

## Trypsin Inhibitor (BPTI)

Bovine pancreatic trypsin inhibitor (BPTI), 30–51 cysteine substitutions

| Mutant | pH | $T_{trs}$ | $\Delta Cp$ | $\Delta H$ | Approach/Remarks | | Ref |
|---|---|---|---|---|---|---|---|
| double mutant (Cys30→Ala and Cys51→Ala) | | | | | | | |
| | 7.0 | 66±0.5 | 3.5 | 257±13 | heat | (1–3,5) | 97L11 |
| | 7.0 | 71±0.5 | 2.7±0.8 | 248±17 | DSC | (2,7,9) | 97L11 |
| double mutant (Cys30→Val and Cys51→Ala) | | | | | | | |
| | 7.0 | 67±0.5 | 2.1 | 239±10 | heat | (1,2,4,5) | 97L11 |
| | 7.0 | 72±0.5 | 2.5±0.8 | 249±17 | DSC | (2,7,9) | 97L11 |
| double mutant (Cys30→Gly and Cys51→Ala) | | | | | | | |
| | 7.0 | 53±0.5 | 1.0 | 203±10 | heat | (1–3,5) | 97L11 |
| | 7.0 | 57±0.5 | 1.1±0.8 | 217±17 | DSC | (2,7,9) | 97L11 |
| double mutant (Cys30→Thr and Cys51→Ala) | | | | | | | |
| | 7.0 | 58±0.5 | 1.46 | 211±13 | heat | (1,2,4,5) | 97L11 |
| | 7.0 | 62±0.5 | 1.4±0.8 | 219±17 | DSC | (2,8,9) | 97L11 |
| double mutant (Cys30→Ser and Cys51→Ala) | | | | | | | |
| | 7.0 | 57±0.5 | 2.1 | 210±13 | heat | (1–3,5) | 97L11 |
| | 7.0 | 61±0.5 | 2.5±0.8 | 233±15 | DSC | (2,7,9) | 97L11 |
| double mutant (Cys30→Ala and Cys51→Ser) | | | | | | | |
| | 7.0 | 52±0.5 | 2.4 | 185±13 | heat | (1–3,5) | 97L11 |
| | 7.0 | 56±0.5 | 1.9±0.8 | 211±17 | DSC | (2,7,9) | 97L11 |
| double mutant (Cys30→Ser and Cys51→Ser) | | | | | | | |
| | 7.0 | 46±0.5 | 2.1 | 186±13 | heat | (1,2,4,6) | 97L11 |
| double mutant (Cys30→Gly and Cys51→Met) | | | | | | | |
| | 7.0 | 40±0.5 | 2.1 | 162±13 | heat | (1,2,4,6) | 97L11 |

Remarks:
(1) buffer: 10 mM potassium phosphate, 0.1 M NaCl, 0.2 mM EDTA, pH 7.0
(2) data from thermally induced unfolding monitored by CD at 222 and 205 nm
(3) $T_{trs}$ and van't Hoff heat at $T_{trs}$ from multiple measurements
(4) $T_{trs}$ and van't Hoff heat at $T_{trs}$ from single experiments
(5) $\Delta Cp$ from $\Delta H^{v.H}$ versus $T_{trs}$ obtained between pH 2 and 7
(6) $\Delta Cp$ estimated value
(7) $T_{trs}$, $\Delta H^{cal}$ and $\Delta Cp$ from multiple measurements
(8) $T_{trs}$, $\Delta H^{cal}$ and $\Delta Cp$ from single measurements
(9) reference data for the wild-type protein, see also Refs. 87S1, 93M4, and 97L11

Partially folded bovine pancreatic trypsin inhibitor (BPTI), chemically synthesized variant with the 14–38 disulfide bond intact [14–38]$_{Abu}$ and cysteines 5, 30, 51, and 55 replaced by α-amino-n-butyric acid (Abu)

| Residue | pH | T$_{trs}$ | ΔH | Approach/Remarks | Ref |
|---|---|---|---|---|---|
| Phe4 | 5.0 | 16.9(16.4–17.4) | 163(142–180) | heat/NMR (1–4) | 97B4 |
| Leu6 | 5.0 | 16.2(15.6–17.0) | 251(201–305) | heat/NMR (1–4) | 97B4 |
| Phe22 | 5.0 | 15.0(14.0–16.0) | 218(172–264) | heat/NMR (1–4) | 97B4 |
| Ala25 | 5.0 | 15.4(14.0–16.6) | 201(142–259) | heat/NMR (1–4) | 97B4 |
| Ala27 | 5.0 | 15.1(14.0–16.0) | 180(151–205) | heat/NMR (1–4) | 97B4 |
| Gly28 | 5.0 | 14.8(14.2–15.3) | 435(322–636) | heat/NMR (1–4) | 97B4 |
| Phe33 | 5.0 | 14.8(13.0–16.6) | 176(117–238) | heat/NMR (1–4) | 97B4 |
| Phe45 | 5.0 | 14.3(14.0–14.6) | 247(222–272) | heat/NMR (1–4) | 97B4 |
| Ala48 | 5.0 | 15.0(13.0–17.0) | 163(105–222) | heat/NMR (1–4) | 97B4 |
| all folded | 5.0 | 15.1(14.6–15.7) | 197(167–226) | heat/NMR (1–5) | 97B4 |
| all data sets (folded and unfolded) | | | | | |
| | 5.0 | 14.6(13.3–15.8) | 151(105–184) | heat/NMR (1–4,6) | 97B4 |
| [14–38]$_{Abu}$ | 6.0 | 19 | | heat/CD (7,8) | 95F6 |
| [14–38]$_{Abu}$ | 5.0 | 18.7 | | heat/CD (7,9) | 97B4 |
| [14–38]$_{Abu}$ | 6.4 | 19.2 | | heat/CD (7,10) | 97B4 |

Remarks:

(1)  $^{15}$N-bound $^1$H reporters at nine positions distributed along the backbone

(2)  measured was the disappearance and appearance with increasing temperature of signals of the folded and unfolded protein [14–38]$_{Abu}$

(3)  measured in 50 mM deuterated sodium acetate buffer

(4)  numbers in parentheses are the 67% confidence intervals from the fits

(5)  simultaneous fit based on all signals representing the folded form

(6)  simultaneous fit based on all signals representing the folded and unfolded form

(7)  data from equilibrium unfolding monitored by CD at 220 and 227 nm

(8)  thermal unfolding measured in 10 mM sodium phosphate buffer

(9)  thermal unfolding measured in 50 mM sodium acetate buffer

(10) thermal unfolding measured at pH 6.4 in water


**Trypsin Inhibitor**, Hen Egg White

Trypsin inhibitor from hen egg white, in aqueous polyol solution

| Cosolvent | Conc. | pH | T$_{trs}$ | ΔH | Appr./Rem. | Ref |
|---|---|---|---|---|---|---|
| control | | 7.0 | 59.0 | 236.4 | heat (1,2) | 98K3 |
| mannitol | 1.00 M | 7.0 | 64.8 | 298.3 | heat (1,2) | 98K3 |
| inositol | 0.75 M | 7.0 | 64.5 | 297.9 | heat (1,2) | 98K3 |
| xylitol | 2.00 M | 7.0 | 67.3 | 297.9 | heat (1,2) | 98K3 |
| adonitol | 2.00 M | 7.0 | 67.3 | 300.4 | heat (1,2) | 98K3 |
| sorbitol | 2.00 M | 7.0 | 71.8 | 305.4 | heat (1,2) | 98K3 |

Remarks:

(1) transition monitored by optical absorption at 287 nm

(2) buffer: 20 mM phosphate or MOPS, pH 7.0, 1.5 M GuHCl

## Tryptophan Represor

Tryptophan repressor from *E. coli*, analysis of melting curves obtaines at varying protein concentration, the presence of a dimeric intermediate

| Protein/Transition | pH | $T_{ref}$ | $\Delta C_p$ | $\Delta H_{ref}$ | Appr./Rem. | Ref |
|---|---|---|---|---|---|---|
| wild type | | | | | | |
| $N_2 \rightarrow I_2$ | 7.6 | 79.9±0.5 | 0.4±0.4 | 293±17 | heat  (1–4) | 97G7 |
| $I_2 \rightarrow 2U$ | 7.6 | 120±6 | 7.1±2.9 | 523±105 | heat  (1–4) | 97G7 |
| [2–66]$_2$ dimeric Trp repressor fragment (5) | | | | | | |
| $N_2 \rightarrow I_2$ | 7.6 | 69±2 | 0.8±0.8 | 146±13 | heat  (1–4) | 97G7 |
| $I_2 \rightarrow 2U$ | 7.6 | 105±5 | 5.4±1.7 | 368±75 | heat  (1–4) | 97G7 |

Remarks:
(1) buffer: 10 mM potassium phosphate, 0.1 mM EDTA, pH 7.6
(2) parameters were obtained from global fits of scans monitored by CD at 222 nm and optical absorption at 292 nm
(3) $T_{ref}$ is the reference temperature at which $\Delta G° = 0$ and is independent of monomer concentration
(4) $\Delta H_{ref}$ is $\Delta H^{v.H.}$ at $T_{ref}$
(5) [2–66]$_2$ dimeric Trp repressor fragment containing residues 2–66 of wild type

**Tryptophan repressor**, see also Repressor Proteins

## Tubulin

Bovine tubulin, isotypes, influence of decay at 0°C on $T_{trs}$ and $\Delta H$

| Protein | Time (1) | pH | $T_{trs}$ | $\Delta H$ | Approach/Remarks | | Ref |
|---|---|---|---|---|---|---|---|
| αβII | 0 h | 7.0 | 60.3±0.1 | 753±38 | DSC | (1,2) | 98S7 |
| αβII | 10 h | 7.0 | 60.1±0.1 | 536±25 | DSC | (1,2) | 98S7 |
| αβII | 20 h | 7.0 | 60.5±0.1 | 360±17 | DSC | (1,2) | 98S7 |
| αβIII | 0 h | 7.0 | 61.4±0.1 | 703±33 | DSC | (1,2) | 98S7 |
| αβIII | 10 h | 7.0 | 61.2±0.1 | 632±33 | DSC | (1,2) | 98S7 |
| αβIII | 20 h | 7.0 | 61.2±0.1 | 523±25 | DSC | (1,2) | 98S7 |

Remarks:
(1) time at 0°C in hours
(2) buffer: 0.05 M PIPES, pH 7.0, 1 mM EGTA, 0.5 mM MgCl$_2$, 1 mM GTP at 0°C

## Tumour Suppressor

Tumour suppressor protein p53, peptide (Ser303-Asp393), unphosphorylated (SS) and phosphorylated (SP) tetramers

| Form | pH | $T_{trs}$ | $\Delta C_p$ | $\Delta H$ | Approach/Remarks | | Ref |
|---|---|---|---|---|---|---|---|
| SS | 7.5 | 79.0 | 6.7 | 261 | DSC | (1) | 97S1 |
| SS | 7.5 | | 6.7±0.8 | | ITC | (1,2) | 97S1 |
| SP | 7.5 | 80.8 | 9.2 | 338 | DSC | (1) | 97S1 |
| SP | 7.5 | | 8.8±0.8 | | ITC | (1,2) | 97S1 |

Remarks:
(1) buffer: 50 mM sodium phosphate, 0.1 M NaCl, pH 7.5
(2) from $\Delta H$ of dissociation versus T

**Tyrosine Kinase**, see SH3 Domain

# U1A

N-terminal RNA binding domain of human U1A protein (RBD1), variant containing amino acids 1–95, RBD1(95A)

| Protein | pH | $T_{trs}$ | $\Delta Cp$ | $\Delta H$ | Approach/Remarks | | Ref |
|---------|------|-----------|-------------|------------|------------------|------|-------|
| RBD1(95A) | 2.00 | 55.3 | 4.87 | 242 | DSC | (1–3) | 97L13 |
| | 2.30 | 57.2 | 4.88 | 244 | DSC | (1–3) | 97L13 |
| | 2.50 | 58.9 | 5.33 | 256 | DSC | (1–3) | 97L13 |
| | 2.70 | 60.9 | 4.82 | 262 | DSC | (1–3) | 97L13 |
| | 2.85 | 62.8 | 3.56 | 275 | DSC | (1–3) | 97L13 |
| | 3.00 | 64.8 | 2.58 | 276 | DSC | (1–3) | 97L13 |
| | 3.20 | 69.0 | 0.51 | 290 | DSC | (1–3) | 97L13 |
| | 3.40 | 70.5 | 1.08 | 292 | DSC | (1–3) | 97L13 |
| | 2.0–2.85 | 55–63 | 5.02±0.42 | | DSC | (4,5) | 97L13 |

Remarks:
(1) buffer: 40 mM glycine HCl
(2) $\Delta H(T_{trs})$ and $T_{trs}$ from the fit of the DSC data to a two-state model
(3) errors in $\Delta H$ estimated at 5%, and at 15–20% for $\Delta Cp$, based on multiple experiments
(4) $\Delta Cp$ from $\Delta H$ versus $T_{trs}$
(5) the value is regarded as the most reliable one and used as a reference value for comparison with other RBD variants

N-terminal RNA binding domain of human U1A protein (RBD1), variants

| Protein | pH | $T_{trs}$ | $\Delta Cp$ | $\Delta H$ | Approach/Remarks | | Ref |
|---------|-----|-----------|-------------|------------|------------------|-------|-------|
| RBD1(102A) | 2.3 | 56.6 | 4.11 | 243 | DSC | (1) | 97L13 |
| RBD1(95A) | 2.3 | 57.0 | 4.88 | 245 | DSC | (2) | 97L13 |
| RBD1(8–99) | 2.3 | 47.0 | 3.98 | 169 | DSC | (3,4) | 97L13 |
| RBD1(Δloop3) | 2.3 | 65.4 | 3.39 | 240 | DSC | (5,6) | 97L13 |

Remarks:
(1) RBD1(102A), the construct containing 1–102 amino acids of the human U1A protein is referred to as wild type
(2) RBD1(95A), the construct containing residues 1–95
(3) RBD1(8–99), the construct containing residues 8–99 with Arg7→Met and Asn8→Ala
(4) $\Delta Cp$ at $T_{trs}$, $\Delta Cp$ is assumed to be temperature dependent, reference value $\Delta Cp(56°C) = 3.43$ kJ/mol/K
(5) RBD1(Δloop3) is RBD1(102A) with the six amino acids of loop3 (Ser46-Arg47-Ser48-Leu49-Lys50-Met51) replaced with four residues (Val-Pro-Gly-Arg)
(4) $\Delta Cp$ at $T_{trs}$, $\Delta Cp$ is assumed to be temperature dependent, reference value $\Delta Cp(56°C) = 3.93$ kJ/mol/K

# Ubiquitin

Yeast and bovine ubiquitin, results of thermal versus GuHCl-induced unfolding

| Protein | pH | $T_{trs}$ | $\Delta H$ | Approach/Remarks | | Ref |
|---------|-----|-----------|------------|------------------|-------|-------|
| bovine ubiquitin | 2.0 | 57 | 200 | DSC | (1–5) | 99I1 |
| | 3.0 | 75 | 255 | DSC | (1–5) | 99I1 |
| | 4.0 | 91 | 295 | DSC | (1–5) | 99I1 |
| yeast ubiquitin | 2.0 | 45 | 130 | DSC | (2–5) | 99I1 |
| | 2.4 | 50 | 145 | DSC | (2–5) | 99I1 |

Yeast and bovine ubiquitin, results of thermal versus GuHCl-induced unfolding (continued)

| Protein | pH | $T_{trs}$ | $\Delta H$ | Approach/Remarks | | Ref |
|---|---|---|---|---|---|---|
| | 2.9 | 63 | 200 | DSC | (2–5) | 99I1 |
| | 3.5 | 70 | 220 | DSC | (2–5) | 99I1 |
| | 4.0 | 80 | 240 | DSC | (2–5) | 99I1 |

Remarks:

(1) in agreement with data from Ref. 94W5

(2) buffer: 10 mM sodium acetate from pH 3.5 to 4.5, and 10 mM glycine from pH 1.5 to 3.5

(3) selected values taken from Fig. 6 from Ref. 99I1

(4) the pH dependencies of $\Delta G$ from DSC and GuHCl-induced unfolding do not coincide (see Table 1); the data are analyzed in terms of charge-charge interactions in Ref. 99I1

(5) $\Delta Cp$ for DSC in the presence of GuHCl was found to be 3.76±0.80 kJ/mol/K

Human ubiquitin in the presence of salts

| Salt | Conc. | pH | $T_{trs}$ | $\Delta H$ | Approach/Remarks | | Ref |
|---|---|---|---|---|---|---|---|
| NaCl | 0 mM | 2.0 | 57.1 | 201 | DSC | (1–3) | 98M2 |
| NaCl | 0 mM | 2.0 | 55.6 | 200 | DSC | (1–3) | 98M2 |
| NaCl | 0 mM | 2.0 | 56.0 | 198 | DSC | (1–3) | 98M2 |
| NaCl | 50 mM | 2.0 | 63.4 | 236 | DSC | (1–3) | 98M2 |
| NaCl | 100 mM | 2.0 | 66.3 | 252 | DSC | (1–3) | 98M2 |
| NaCl | 150 mM | 2.0 | 68.7 | 260 | DSC | (1–3) | 98M2 |
| NaCl | 200 mM | 2.0 | 71.1 | 274 | DSC | (1–3) | 98M2 |
| NaCl | 300 mM | 2.0 | 74.4 | 277 | DSC | (1–3) | 98M2 |
| NaCl | 500 mM | 2.0 | 79.6 | 282 | DSC | (1–3) | 98M2 |
| $CaCl_2$ | 25 mM | 2.0 | 61.7 | 225 | DSC | (1–3) | 98M2 |
| $CaCl_2$ | 50 mM | 2.0 | 65.8 | 247 | DSC | (1–3) | 98M2 |
| $CaCl_2$ | 75 mM | 2.0 | 68.5 | 259 | DSC | (1–3) | 98M2 |
| $CaCl_2$ | 100 mM | 2.0 | 70.5 | 272 | DSC | (1–3) | 98M2 |
| $MgCl_2$ | 25 mM | 2.0 | 62.5 | 240 | DSC | (1–3) | 98M2 |
| $MgCl_2$ | 50 mM | 2.0 | 66.3 | 252 | DSC | (1–3) | 98M2 |
| $MgCl_2$ | 100 mM | 2.0 | 71.2 | 279 | DSC | (1–3) | 98M2 |
| $MgCl_2$ | 150 mM | 2.0 | 73.1 | 281 | DSC | (1–3) | 98M2 |
| $MgCl_2$ | 200 mM | 2.0 | 76.2 | 294 | DSC | (1–3) | 98M2 |
| $Na_2SO_4$ | 50 mM | 2.0 | 74.4 | 288 | DSC | (1–3) | 98M2 |
| $NaClO_4$ | 100 mM | 2.0 | 69.9 | 280 | DSC | (1–3) | 98M2 |
| NaBr | 50 mM | 2.0 | 63.9 | 253 | DSC | (1–3) | 98M2 |
| $NaH_2PO_4$ | 0 mM | 2.0 | 53.8 | 177 | DSC | (2–4) | 98M2 |
| $NaH_2PO_4$ | 10 mM | 2.0 | 59.4 | 212 | DSC | (2–4) | 98M2 |
| $NaH_2PO_4$ | 50 mM | 2.0 | 66.4 | 243 | DSC | (2–4) | 98M2 |
| $NaH_2PO_4$ | 100 mM | 2.0 | 70.4 | 259 | DSC | (2–4) | 98M2 |
| GuHCl | 25 mM | 2.0 | 59.5 | 214 | DSC | (1–3) | 98M2 |
| GuHCl | 50 mM | 2.0 | 62.8 | 236 | DSC | (1–3) | 98M2 |
| GuHCl | 100 mM | 2.0 | 65.3 | 249 | DSC | (1–3) | 98M2 |
| GuHCl | 150 mM | 2.0 | 67.2 | 254 | DSC | (1–3) | 98M2 |
| GuHCl | 250 mM | 2.0 | 68.7 | 257 | DSC | (1–3) | 98M2 |
| GuHCl | 500 mM | 2.0 | 70.8 | 256 | DSC | (1–3) | 98M2 |
| GuHCl | 1000 mM | 2.0 | 71.2 | 248 | DSC | (1–3) | 98M2 |
| GuHCl | 2000 mM | 2.0 | 63.7 | 218 | DSC | (1–3) | 98M2 |

Remarks:

(1) buffer: 10 mM glycine if not otherwise noted

(2) the calorimetric enthalpy change $\Delta H$ agrees with $\Delta H^{fit}$, the enthalpy of the fit of a calorimetric profile to a two-state model

(3) estimated error in $T_{trs}$ is ±0.2°C, estimated error in $\Delta H$ is 5–7%

(4) pH adjusted by phosphoric acid

## Uridine Phosphorylase

Uridine phosphorylase from *E. coli* K-12

| pH | $T_1^*$ | $\Delta H_1$ | $T_2^*$ | $\Delta H_2$ | Appr./Rem. | Ref |
|---|---|---|---|---|---|---|
| 7.55 | 62.7±0.07 | 1098±464 | 61.7±0.04 | 2391±22 | DSC  (1–5) | 98L9 |

Remarks:

(1) scan rate-dependent irreversible denaturation analyzed in terms of various models deduced from the Lumry-Eyring theory

(2) the model which involves to consecutive irreversible steps was found to be the most satisfactory one, see also Ref. 98L10

(3) measured in 50 mM HEPES-NaOH, pH 7.55

(4) $T_1^*$ and $T_2^*$ are the temperatures at which the rate constants $k_1$ and $k_2$ are equal to 1 min$^{-1}$

(5) the data refer to a scan rate of 1 K/min

## Xylanase

Xylanase from *Bacillus circulans*, wild type and disulfide-bridge containing mutant (Ser100→Cys and Asn148→Cys)

| Protein | pH | $T_{trs}$ | $\Delta Cp$ | $\Delta H$ | Approach/Remarks | | Ref |
|---|---|---|---|---|---|---|---|
| wild type | 6.0 | 51.2 | 10.5 | 133±9 | DSC | (1–4) | 98D9 |
| disulfide m. | 6.0 | 58.6 | 10.5 | 138±11 | DSC | (1–4) | 98D9 |

Remarks:

(1) measured in the presence of 2.5 M urea at a scan rate of 58.0 K/h

(2) in the presence of low (nondenaturing) concentrations of urea, reversible calorimetric transitions (reheating) were observed whereas the transitions are irreversible in the absence of urea

(3) the data show scan rate dependence of the transition temperature in the case of reversible denaturation (see also J.R. Lepock et al., Biochemistry 31, 1992, 12706–12712)

(4) estimated value of $\Delta Cp = 10.5$ kJ/mol/K

Xylanase, β-1,4-glycanase (Cex) of *Cellulomonas fimi* (xylanase/exoglucanase)

| | pH | $T_{trs}$ | $\Delta Cp$ | $\Delta H$ | Appr./Rem. | Ref |
|---|---|---|---|---|---|---|
| | 7.0 | 64.0±0.3 | 24.6±2.8 | 1380±50 | DSC | 97N8 |

Xylanase from *Streptomyces halstedii* JM8 and variants, stability of domains

| Variant/Transition | pH | $T_{trs}$ | $\Delta Cp$ | $\Delta H$ | Appr./Rem. | | Ref |
|---|---|---|---|---|---|---|---|
| xylanase type Xys1S (1): | | | | | | | |
| trans. (1) | 7.5 | 60.6±0.2 | 9.2±0.8 | 423±29 | DSC | (1,4) | 98R14 |
| trans. (2) | 7.5 | 63.9±0.2 | 28.0±2.1 | 778±33 | DSC | (1,4) | 98R14 |
| xylanase variant Xys1Δ (2): | | | | | | | |
| trans. (1) | 7.5 | 49.1±0.2 | 7.9±1.3 | 431±17 | DSC | (2,4) | 98R14 |
| trans. (2) | 7.5 | 53.1±0.2 | 10.0±3.8 | 628±17 | DSC | (2,4) | 98R14 |
| xylanase variant Xys1VW (3): | | | | | | | |
| trans. (1) | 7.5 | 55.6±0.1 | 7.1±0.8 | 481±25 | DSC | (3,4) | 98R14 |
| trans. (2) | 7.5 | 58.4±0.1 | 12.0±4.2 | 774±25 | DSC | (3,4) | 98R14 |
| xylanase type Xys1S (1): | | | | | | | |
| trans. (1) | 4.1 | 43.1±0.3 | 9.2±0.8 | 238±13 | DSC | (1,4) | 98R14 |
| trans. (2) | 4.1 | 50.2±0.2 | 28.0±2.1 | 418±17 | DSC | (1,4) | 98R14 |

Xylanase from *Streptomyces halstedii* JM8 and variants, stability of domains (continued)

| Variant/Transition | pH | T$_{trs}$ | ΔCp | ΔH | Appr./Rem. | | Ref |
|---|---|---|---|---|---|---|---|
| xylanase variant Xys1Δ (2): | | | | | | | |
| trans. (1) | 4.1 | 42.4±0.2 | 7.9±1.3 | 414±17 | DSC | (2,4) | 98R14 |
| trans. (2) | 4.1 | 45.2±0.2 | 10.0±3.8 | 481±17 | DSC | (2,4) | 98R14 |
| xylanase variant Xys1VW (3): | | | | | | | |
| trans. (1) | 4.1 | 41.9±1.0 | 7.1±0.8 | 360±13 | DSC | (3,4) | 98R14 |
| trans. (2) | 4.1 | 47.9±1.0 | 12.0±4.2 | 594±17 | DSC | (3,4) | 98R14 |
| xylanase type Xys1S (1): | | | | | | | |
| trans. (1) | 8.9 | 46.5±0.2 | 9.2±0.8 | 301± 8 | DSC | (1,4) | 98R14 |
| trans. (2) | 8.9 | 53.7±0.1 | 28.0±2.1 | 569±33 | DSC | (1,4) | 98R14 |
| xylanase variant Xys1Δ (2): | | | | | | | |
| trans. (1) | 8.9 | 41.7±0.2 | 7.9±1.3 | 356± 8 | DSC | (2,4) | 98R14 |
| trans. (2) | 8.9 | 46.4±0.1 | 10.0±3.8 | 515± 8 | DSC | (2,4) | 98R14 |
| xylanase variant Xys1VW (3): | | | | | | | |
| trans. (1) | 8.9 | 50.0±0.2 | 7.1±0.8 | 435±29 | DSC | (3,4) | 98R14 |
| trans. (2) | 8.9 | 53.4±0.1 | 12.0±4.2 | 665±17 | DSC | (3,4) | 98R14 |

Remarks:

(1) *Streptomyces halstedii* JM8 secrets two xylanase forms, Xys1L and Xys1S, see also Ref. 94R5

(2) xylanase variant Xys1Δ was obtained by deletion of a Gly-rich linker region at the C-terminus

(3) xylanase Xys1VW was obtained by Val399→Trp point mutation

(4) ΔCp from ΔH versus T$_{trs}$ (at twelve different pH values)


# Zn-α$_2$-Glycoprotein

Zn-α$_2$-glycoprotein (Zn-α$_2$-gp) from the Cohn's fraction VI of human blood serum

| Protein | pH | T$_{trs}$ | ΔH | Approach/Remarks | Ref |
|---|---|---|---|---|---|
| Zn-α$_2$-gp | 7.4 | 66 | 113 | heat, v.H. (1,2) | 97K2 |

Remarks:

(1) measured in 50 mM Tris buffer

(2) transition monitored by CD at 218 nm

**Table 3.**
**Enthalpy and Heat Capacity Changes –**
**Specific** Values

## Adenylate Kinase

Adenylate kinase (AK) from the archaeon *Sulfolobus acidocaldarius*

| Protein | pH | $T_{trs}$ | $\Delta c_p$ | $\Delta h$ | Appr./Rem. | Ref |
|---------|-----|-----------|--------------|------------|------------|------|
| AK | 7.0 | 89 | 0.19 | 12.7 | GuHCl (1-3) | 98B1 |

Remarks:
(1) data from GuHCl-induced unfolding of AK in the temperature range from 5 to 70°C, see also Table 1
(2) data treatment by linear extrapolation, LEM-SB
(3) data per trimer

## Adrenodoxin

Recombinant bovine adrenodoxin (Adx) and preadrenodoxin (Padx)

| Protein | NaCl | pH | $T_{trs}$ | $\Delta c_p$ | Approach/Remarks | | Ref |
|---------|--------|-----|-------------|--------------|------------------|-----|------|
| Adx | 0 mM | 8.5 | 51.4±0.1 | 25.3±0.9 | DSC | (1,2) | 98G9 |
| | 50 mM | 8.5 | 55.6±0.3 | 22.8±1.6 | DSC | (1) | 98G9 |
| Padx | 0 mM | 8.5 | 47.8±0.3 | 14.9±0.8 | DSC | (1) | 98G9 |
| | 50 mM | 8.5 | 48.4±0.4 | 14.2±0.6 | DSC | (1) | 98G9 |
| | 100 mM | 8.5 | 49.5±0.2 | 16.8±0.8 | DSC | (1) | 98G9 |
| | 400 mM | 8.5 | 55.4 | 16.2 | DSC | (1) | 98G9 |
| | 0 mM | 8.5 | 47.8±0.2 | 15.9±0.8 | DSC | (3) | 98G9 |

Remarks:
(1) buffer: 40 mM glycine, pH 8.5, 20 mM sodium sulfide, 1 mM ascorbate, 10 mM 2-mercaptoethanol
(2) data from Ref. 95B9
(3) the above buffer with variation of sulfide concentration from 0 to 20 mM

## Alcohol Dehydrogenase

Alcohol dehydrogenase from yeast in the presence and absence of sucrose

| Sucrose conc. | pH | $T_{trs}$ | $\Delta h$ | Approach/Remarks | Ref |
|---------------|-----|-----------|------------|------------------|------|
| 0.0 % (w/w) | 7.5 | 63 | 4.2 | DSC | 98N1 |
| 44.4 % (w/w) | 7.5 | 70 | 6.9 | DSC | 98N1 |

## Annexin

Annexin I, recombinant porcine protein, and annexin V, recombinant human protein

| Protein | pH | $T_{trs}$ | $\Delta c_p$ | $\Delta h$ | Approach/Remarks | | Ref |
|---------|---------|-----------|--------------|------------|------------------|-----|--------|
| annexin I | 7.2 | 52 | | 18.8 | DSC | (1) | 98R18 |
| | 3.2-8.0 | 41-63 | 0.49 | | DSC | (2) | 98R18 |
| | 8.0 | 58.5 | | 21.3 | DSC | (2) | 98R18 |
| | 6.0 | 61.8 | | 21.3 | DSC | (2) | 98R18 |
| annexin V | 6.9 | 58 | | 16.1 | DSC | (1) | 98R18 |
| | 8.0 | 54.4 | | 19.0 | DSC | (2) | 98R18 |
| | 6.0 | 50.5 | | 17.3 | DSC | (2) | 98R18 |
| | | | 0.29 | | DSC | | 98R18 |

Remarks:
(1) measured in 50 mM Tris-HCl
(2) from a series of measurements performed in 50 mM sodium phosphate at a heating rate of 2 K/min, see also Table 2

## Apolipoprotein

Apolipoprotein B100, low-density lipoprotein (LDL) subspecies during copper-mediated oxidation

| Process | ox. stage | pH | $T_{trs}$ | $\Delta h$ | Approach/Remarks | | Ref |
|---|---|---|---|---|---|---|---|
| core melting | 1 | 7.4 | 28.6 | 3.3±0.8 | DSC | (1,2) | 98P13 |
| | 2 | 7.4 | 29.2 | 2.9±0.3 | DSC | (1,2) | 98P13 |
| | 3 | 7.4 | 29.2 | 3.0±0.3 | DSC | (1,2) | 98P13 |
| | 4 | 7.4 | 29.1 | 3.2±0.4 | DSC | (1,2) | 98P13 |
| | 5 | 7.4 | 29.5 | 3.4±0.5 | DSC | (1,2) | 98P13 |
| protein denat. | 1 | 7.4 | 79.3±0.8 | 2.4±0.2 | DSC | (1,2) | 98P13 |
| | 2 | 7.4 | 78.1±1.2 | 1.4±0.5 | DSC | (1,2) | 98P13 |
| | 3 | 7.4 | 77.8±1.3 | 1.0±0.1 | DSC | (1,2) | 98P13 |

Remarks:
(1) average values represent the means for analyses of LDL from six different donors
(2) for details of the oxidation see Ref. 98P13
(3) the half-width of the transition increases with inreasing oxidation
(4) the cooperative ratio CR = $\Delta H^{cal}/\Delta H^{v.H.}$ increases from about 1.0±0.1 at oxidation stage 1 to 2.0±0.3 at oxidation stage 3

## Aspartate Transcarbamoylase

Aspartate transcarbamoylase (ATCase), catalytic (C) and regulatory (R) subunits in the presence of ligands

| Sample | pH | $T_{trs}$ | $\Delta h$ | Approach/Remarks | | Ref |
|---|---|---|---|---|---|---|
| intact ATCase | | | | | | |
| ATCase | 7.0 | 82 | 21.13±0.25 | DSC | (1) | 78V1 |
| ATCase + PALA | 7.0 | 89.5 | 22.76±0.42 | DSC | (1,2) | 78V1 |
| ATCase + CTP | 7.0 | 79.5 | 20.21 | DSC | (1) | 78V1 |
| ATCase + ATP | 7.0 | 77.5 | 20.00±0.92 | DSC | (1) | 78V1 |
| catalytic subunit: | | | | | | |
| C, isolated | 7.0 | 80 | 16.48±0.42 | DSC | (1) | 78V1 |
| C, isolated, + PALA | 7.0 | 85.5 | 20.71±0.42 | DSC | (1,2) | 78V1 |
| C in ATCase | 7.0 | 82 | 22.72±0.75 | DSC | (1,3) | 78V1 |
| C in ATCase + PALA | 7.0 | 89.5 | 27.49±0.59 | DSC | (1-3) | 78V1 |
| regulatory subunit: | | | | | | |
| R, isolated | 7.0 | 55 | 8.24±0.17 | DSC | (1) | 78V1 |
| R, isolated, + CTP | 7.0 | 60 | 12.97±0.21 | DSC | (1) | 78V1 |
| R, isolated, + ATP | 7.0 | 59 | 12.89±0.50 | DSC | (1) | 78V1 |
| R in ATCase | 7.0 | 72.5 | 17.99±1.13 | DSC | (1,3) | 78V1 |
| R in ATCase + PALA | 7.0 | 74.5 | 13.22±0.38 | DSC | (1-3) | 78V1 |

Remarks:
(1) measured in 40 mM potassium phosphate, 2 mM 2-mercaptoethanol, 0.2 mM EDTA
(2) PALA = $N$-(phosphonacetyl)-L-aspartate
(3) data from curve analysis

## Chymotrypsinogen

α-chymotrypsinogen A, heat-induced denatured states (see also Table 2)

| pH | $T_{trs}$ | $\Delta c_p$ | $\Delta h$ | Approach/Remarks | | Ref |
|---|---|---|---|---|---|---|
| 1.3-3.6 | 35-57 | 0.38±0.08 | | DSC | (1) | 97C6 |

Remark:
(1) $\Delta c_p$ from $\Delta h$ versus $T_{trs}$ using $\Delta H^{cal}$ from DSC

# Collagen

Type I collagen fibrils from white rat skin, separation of two transitions

| pH | $T_{trs1}$ | $\Delta h_{trs1}$ | $T_{trs2}$ | $\Delta h_{trs2}$ | $\Delta h_{total}$ | $\Delta c_p$ | Appr./Rem. | | Ref |
|---|---|---|---|---|---|---|---|---|---|
| 3.0 | 40.7 | 72.3 | | | 72.3 | | DSC | | 98T6 |
| 6.0 | 40.7 | 71.5 | | | 71.5 | | DSC | | 98T6 |
| 6.3 | 40.9 | 72.3 | | | 72.3 | | DSC | | 98T6 |
| 6.5 | 40.8 | 73.6 | | | 73.6 | | DSC | | 98T6 |
| 6.75 | 40.7 | 63.5 | 48.5 | 10.9 | 74.4 | 0.38 | DSC | | 98T6 |
| 7.0 | 40.7 | 58.9 | 49.3 | 18.0 | 76.9 | 0.63 | DSC | | 98T6 |
| 7.15 | 39.9 | 53.1 | 49.6 | 28.8 | 81.9 | 1.05 | DSC | | 98T6 |
| 7.3 | 40.0 | 48.5 | 50.0 | 37.2 | 85.7 | 1.25 | DSC | | 98T6 |
| 7.55 | 40.5 | 30.5 | 50.7 | 60.6 | 96.1 | 1.46 | DSC | | 98T6 |
| 7.44 | | | 53.0 | 97.4 | 97.4 | 1.60 | DSC | (1) | 98T6 |

Remark:
(1) the collagen solution was kept for 50 min at 26°C at pH 7.44

Collagen from white rat skin and rat tendon, dependence of thermodynamic data from scan rate

| Scan rate (K/min) | pH | $T_{trs}$ | $\Delta c_p$ | $\Delta h$ | Approach/Remarks | | Ref |
|---|---|---|---|---|---|---|---|
| 0.0 | 3.7 | 36.8±0.5 | | 60±5 | DSC | (1,2) | 97B13 |
| 0.0 | 3.7 | | 0.12±0.05 | | DSC | (1,2) | 97M11 |
| 0.125 | 3.7 | | 0.17 | | DSC | (2) | 97M11 |
| 0.25 | 3.7 | | 0.22 | | DSC | (2) | 97M11 |
| 0.5 | 3.7 | | 0.30 | | DSC | (2) | 97M11 |
| 1.0 | 3.7 | | 0.50 | | DSC | (2) | 97M11 |
| 2.0 | 3.7 | | 0.72 | | DSC | (2) | 97M11 |
| 3.0 | | | 0.42±0.04 | | DSC | (3) | 97T5 |
| 3.0 | | | 0.54±0.12 | | DSC | (4) | 97T5 |

Remarks:
(1) 'equilibrium value' from scan rate dependent measurements, extrapolated to zero scan rate
(2) measured on collagen from rat skin in diluted protein solution (2 to 2.5 mg/ml) in 0.1 M citrate buffer, pH 3.7
(3) collagen from rat tendon at high protein concentration ranging from 10 to 50%, denaturation transition of the native state
(4) collagen from rat tendon at high protein concentration, transition from the glass state into the high viscoelastic state

Collagen from rat tail after UV irradiation

| Dose of radiation | $T_{trs}$ | $\Delta c_p$ | $\Delta h$ | Appr./Rem. | | Ref |
|---|---|---|---|---|---|---|
| 0 J/cm² | 112.6 | | 324.7 | DSC | (1) | 99S11 |
| 32 J/cm² | 108.3 | | 322.5 | DSC | (1) | 99S11 |
| 128 J/cm² | 106.6/92.2 | | 290.4 | DSC | (1) | 99S11 |

Remark:
(1) measured on film of collagen

Triple-helical domains of type IX collagen

| Form | pH | $T_{trs}$ | $\Delta h_{res}$ | Approach/Remarks | | Ref |
|---|---|---|---|---|---|---|
| COL1 | | 41.3 | 6.22 | DSC | (1,2) | 98M18 |
| COL2 | | 35.9,39.9 | 7.12 | DSC | (1,2) | 98M18 |
| COL3 | | 48.0 | 6.93 | DSC | (1,2) | 98M18 |

Remarks:
(1) measured in 0.05 M acetic acid
(2) $\Delta h_{res}$ in kJ/mol res.

## Concanavalin

Concanavalin A in the presence of $Mn^{2+}$ and $Ca^{2+}$

| $Mn^{2+}$ | $Ca^{2+}$ | pH | $T_{trs}$ | $\Delta h$ | Appr./Rem. | Ref |
|---|---|---|---|---|---|---|
| 0.18 | 0.55 | 5 | 101.0 | 31.0 | DSC (1,2) | 81Z1 |
| 0.06 | 0.23-2.0 | 5 | 73.5-74.5 | 13.4±0.8 | DSC (1,2) | 81Z1 |
| 10.0 | 12.4 | 5 | 103.0 | 28.5 | DSC (1,2) | 81Z1 |
| 0.01 | 14.0 | 5 | 93.0 | 21.3 | DSC (1-3) | 81Z1 |
| 13.4 | 0.2-0.3 | 5 | 94.5 | 22.2 | DSC (1,2) | 81Z1 |
| 0.56 | 0.23 | 5 | 75.5,93.5 | 12.1 | DSC (1,2) | 81Z1 |
| 0.56 | 0.50 | 5 | 76.0,95.0 | 13.8 | DSC (1,2) | 81Z1 |
| 0.56 | 1.00 | 5 | 76.0,96.0 | 15.1 | DSC (1,2) | 81Z1 |
| 0.56 | 1.82 | 5 | 77.0,97.0 | | DSC (1,2) | 81Z1 |
| 1.07 | 0.23 | 5 | 77.5,85.0,94.5 | 12.1 | DSC (1,2) | 81Z1 |
| 1.07 | 0.50 | 5 | 77.0,83.5,97.5 | 14.6 | DSC (1,2) | 81Z1 |
| 1.07 | 1.00 | 5 | 97.5 | 22.2 | DSC (1,2) | 81Z1 |
| 2.14 | 0.23 | 5 | 88.0,96.0 | 12.1 | DSC (1,2) | 81Z1 |

Remarks:

(1) $Mn^{2+}$ and $Ca^{2+}$ are given in mol/mol concanavalin-A monomer

(2) measured at a heating rate of 10 K/min

(3) excess $Ca^{2+}$ concentration

## Ferritin

Horse spleen apoferritin and human recombinant apoferritin

| Protein | pH | $T_{trs}$ | $\Delta c_p$ | $\Delta h$ | Approach/Remarks | Ref |
|---|---|---|---|---|---|---|
| horse spleen | 3.5 | 42.1 | 0.8 | 9.2 | DSC (1-3) | 96S17 |
| human recombinant | 3.5 | 50.0 | 0.4 | 8.8 | DSC (1-3) | 96S17 |

Remarks:

(1) dimeric subunit obtained by treatment at pH 1.8 followed by dialysis versus 40 mM glycine-HCl, pH 3.5

(2) data calculated for molar mass of 2×19.5 and 2×21.0 kDa for horse spleen and human recombinant protein, respectively

(3) for molar values see Table 2

Recombinant L subunit of human ferritin

| | pH | $T_{trs}$ | $\Delta c_p$ | $\Delta h$ | Approach/Remarks | Ref |
|---|---|---|---|---|---|---|
| | 2.0-2.8 | 37.2-79.8 | 0.23 | | DSC | 98M8 |

## Fetuin

Bovine serum fetuin, glycosylated (native) and deglycosylated

| | pH | $T_{trs}$ | $\Delta c_p$ | $\Delta h$ | Approach/Remarks | Ref |
|---|---|---|---|---|---|---|
| | 5.5-8.8 | 45-60 | 0.17 | | DSC (1) | 96W3 |

Remark:

(1) $\Delta c_p$ from $\Delta h$ versus $T_{trs}$ of both glycoslyated and deglycosylated protein

## Glucoamylase

Glucoamylase from *Aspergillus niger*, glycosylated (native) and deglycosylated

| pH | $T_{trs}$ | $\Delta c_p$ | $\Delta h$ | Approach/Remarks | | Ref |
|---|---|---|---|---|---|---|
| 4.5-8.0 | 55-70 | 0.42 | | DSC | (1) | 96W3 |

Remark:
(1) $\Delta c_p$ from $\Delta h$ versus $T_{trs}$ of both glycoslyated and deglycosylated protein

## Glutamine Synthetase

Glutamine synthetase (GS)

| Transition | pH | $T_{trs}$ | $\Delta h$ | Approach/Remarks | | Ref |
|---|---|---|---|---|---|---|
| Mn-GS: | | | | | | |
| trans. (1) | 7 | 42 | 0.9 | DSC | (1) | 97N12 |
| trans. (2) | 7 | 81 | 23.4 | DSC | (1) | 97N12 |
| Mn-GS in the presence of 100 mM KCl: | | | | | | |
| trans. (1) | 7 | 51.6 | 1.42 | DSC | (1,2) | 91G2 |
| Mn-GS in the presence of 3 M urea: | | | | | | |
| | 7 | 65 | 16.8±2.1 | DSC | (3) | 97N12 |

Remarks:
(1) dodecameric GS measured in 20 mM HEPES and 0.1 mM $MnCl_2$
(2) the second transition is not observable due to aggregation at about 67°C, see also molar values in Table 2, Refs. 91G2, 97N12
(3) monomeric form in 3 M urea, $\Delta h$ corrected for urea binding (see Ref. 95Z3) amounts to 58±4 J/g

## Glutathione Reductase

Glutathione reductase from cyanobacterium *Spirulina maxima*

| pH | $T_{trs}$ | $\Delta h$ | Approach/Remarks | | Ref |
|---|---|---|---|---|---|
| 7.0 | 76 | 18.8 | DSC | (1-4) | 97R7 |

Remarks:
(1) scan rate dependent $T_{trs}$, $T_{trs}$ = 76°C refers to a scan rate of 1.5 K/min
(2) $T_{trs}$ was taken from Fig. 1 in Ref. 97R7
(3) $\Delta h$ is an extrapolated value for infinite scan rate
(4) buffer: 0.1 M potassium phosphate, 1 mM EDTA

## Immunoglobulin

Mouse monoclonal immunolobulin G (IgG, isotype 1)

| Protein | pH | $T_{trs}$ | $\Delta h$ | Approach/Remarks | | Ref |
|---|---|---|---|---|---|---|
| IgG | 6.0 | 74 | 12.7 | DSC | (1) | 98V4 |

Remark:
(1) measured in 5 mM phosphate buffer, pH 6.0

Monoclonal immunoglobulin M (IgM) compared with rheumatoid immunoglobulin M (IgM-RF) and fragments

| Protein | pH | $T_{trs}$ | $\Delta h$ | Approach/Remarks | | Ref |
|---|---|---|---|---|---|---|
| IgM | 7.0 | 67.1 | 19.5 | DSC | (1) | 97P16 |
| IgM-RF | 7.0 | 71.1 | 19.4 | DSC | (1) | 97P16 |
| Fab(IgM) | 7.0 | 73.7 | 26.0 | DSC | (1) | 97P16 |
| Fab(IgM-RF) | 7.0 | 73.3 | 27.7 | DSC | (1) | 97P16 |
| Fc(IgM) | 7.0 | 71.2 | 17.3 | DSC | (1) | 97P16 |
| FC(IgM-RF) | 7.0 | 70.7 | 17.0 | DSC | (1) | 97P16 |

Remark:
(1) measured in 0.01 M potassium phosphate buffer containing 0.15 M NaCl

## Invertase

Yeast external invertase, glycosylated (native) and deglycosylated

| pH | $T_{trs}$ | $\Delta c_p$ | $\Delta h$ | Approach/Remarks | | Ref |
|---|---|---|---|---|---|---|
| 5.5-8.8 | 44-65 | 0.88 | | DSC | (1) | 96W3 |

Remark:
(1) $\Delta c_p$ from $\Delta h$ versus $T_{trs}$ of both glycoslyated and deglycosylated protein

## α-Lactalbumin

Bovine recombinant α-lactalbumin, effects of amino acid substitutions in the N-terminus

| Protein | pH | $T_{trs}$ | $\Delta h$ | Approach/Remarks | | Ref |
|---|---|---|---|---|---|---|
| apo-state: | | | | | | |
| des-Met α-LA | 8.06 | 33.0 | 5.0 | DSC | (1,2) | 99V2 |
| native α-LA | 8.06 | 25.9 | 12.6 | DSC | (1,3) | 99V2 |
| ΔGlu1 α-LA | 8.06 | 37.4 | 15.6 | DSC | (1,4) | 99V2 |
| $Ca^{2+}$-loaded state: | | | | | | |
| recomb. wt. α-LA | 8.06 | 60.1 | 17.5 | DSC | (1) | 99V2 |
| des-Met α-LA | 8.06 | 67.0 | 18.0 | DSC | (1,2) | 99V2 |
| native α-LA | 8.06 | 65.5 | 24.9 | DSC | (1,3) | 99V2 |
| ΔGlu1 α-LA | 8.06 | 69.8 | 27.0 | DSC | (1,4) | 99V2 |

Remarks:
(1) measured in 50 mM borate buffer, pH 8.06, in the presence of 2 mM EDTA (apo-state) or 1 mM $CaCl_2$ ($Ca^{2+}$-loaded state)
(2) recombinant protein, N-terminal Met enzymatically removed
(3) native protein from bovine milk
(4) recombinant protein, the N-terminal Met is placed in the position of Glu1 in the native protein

## Lipase

Lipase from *Pseudomonas cepacia* in the absence and presence of alcohols

| Alcohol conc. | pH | $T_{trs}$ | $\Delta h$ | $\Delta cp$ | Appr./Rem. | | Ref |
|---|---|---|---|---|---|---|---|
| without alcohol | 7.0 | 75.8-80.9 | 25.6±0.8 | 0.055±0.075 | DSC | (1-3) | 98T4 |
| MeOH 0.24-1.12 M | 7.0 | 76.9-74.6 | 25.3±0.5 | | DSC | (3-5) | 98T4 |
| EtOH 0.17-0.86 M | 7.0 | 76.7-73.9 | 25.3±0.9 | | DSC | (3-5) | 98T4 |
| PrOH 0.13-0.67 M | 7.0 | 76.9-70.8 | 23.7±1.8 | | DSC | (3-5) | 98T4 |
| BuOH 0.11,0.22 M | 7.0 | 75.6-74.0 | 25.0,22.5 | | DSC | (3-5) | 98T4 |

Remarks:

(1) measured at protein concentration from 29.3 to 236 $\mu$M in the absence of alcohol, $T_{trs}$ varies from 75.8 to 80.9°C without significant change in $\Delta h$

(2) analysis of the DSC tracings is consistent with the mechanism $N.Ca^{2+} \to D + Ca^{2+}$, see also Table 2

(3) abbreviations: MeOH, methanol; EtOH, ethanol; PrOH, propanol; BuOH, butanol

(4) buffer: 20 mM phosphate, pH 7.0

(5) measured at protein conc. of 98 $\mu$M

## Luciferase

Firefly luciferase, thermal unfolding in the presence of ethanol

| Ethanol conc. | pH | $T_{trs}$ | $\Delta h$ | Approach/Remarks | | Ref |
|---|---|---|---|---|---|---|
| 0.00 M | 7.8 | 41.7 | 6.7 | DSC | (1) | 94C13 |
| 0.15 M | 7.8 | 40.8 | | DSC | (1) | 94C13 |
| 0.30 M | 7.8 | 39.6 | | DSC | (1) | 94C13 |
| 0.60 M | 7.8 | 38.2 | | DSC | (1) | 94C13 |
| 1.20 M | 7.8 | 35.2 | | DSC | (1) | 94C13 |
| 2.30 M | 7.8 | 30.6 | | DSC | (1) | 94C13 |

Remark:

(1) measured in 100 mM glycylglycine buffer, pH 7.

## Maltose-Binding Protein

Maltose-binding protein from *E. coli*, in the presence and absence of maltose

| Maltose conc. | pH | $T_{trs}$ | $\Delta c_p$ | $\Delta h$ | Approach/Remarks | | Ref |
|---|---|---|---|---|---|---|---|
| 0-50 mM | 2.5-8.3 | 40-72 | 0.67±0.08 | | DSC | (1) | 97N13 |
| 0 mM | 2.5-3.4 | 70 | 0.63 | | DSC | (2) | 97N13 |

Remarks:

(1) $\Delta c_p$ from $\Delta h$ versus $T_{trs}$

(2) $\Delta c_p$ from extrapolated heat capacity between unfolded and folded states

## Plant Seed Proteins

11 S globulins from three seeds

| Species | pH | $T_{trs}$ | $\Delta c_p$ | $\Delta h$ | Appr./Rem. | Ref |
|---|---|---|---|---|---|---|
| soya bean | 8.0 | 92±0.3 | 0.2±0.02 | 25±1 | DSC (1,2) | 89G6 |
| faba bean | 8.0 | 94±0.3 | 0.3±0.03 | 23±1 | DSC (1,3) | 89G6 |
| sunflower | 8.0 | 95±1.4 | 0.4±0.1 | 32±1 | DSC (1,4) | 89G6 |
| mean | 8.0 | 94±1 | 0.3±0.07 | 27±3 | DSC | 89G6 |

Remarks:

(1) measured in at pH 8.0 in the presence of 0.3 M NaCl

(2) soya bean = *Glycine max* L

(3) faba bean = *Vicia faba* L

(4) sunflower = *Helianthus annuus* L

## Plasminogen Activator

Recombinant tissue plasminogen activator (rt-PA) and mutants, expressed in various systems

| Protein | pH | $T_{trs}$ | $\Delta h$ | Approach/Remarks | | Ref |
|---|---|---|---|---|---|---|
| rt-PA (CHO) | 4.0 | 59.0±0.8 | 10.0±0.5 | DSC | (1,2) | 94V5 |
| BM 06.022 | 4.0 | 58.4±0.2 | 11.3±0.4 | DSC | (1,3) | 94V5 |
| rt-PA (CHO) | 7.4 | 71.9±0.2 | 16.8±0.3 | DSC | (2,4) | 94V5 |
| BM 06.022 | 7.4 | 74.7±0.2 | 18.6±0.3 | DSC | (3,4) | 94V5 |
| rt-PA [remark (5)] | 7.4 | 72.8±0.8 | 17.6±1.8 | DSC | (5) | 94V5 |
| rt-PA [remark (6)] | 7.4 | 71.8±1.0 | 17.0±1.5 | DSC | (6) | 94V5 |
| rt-PA [remark (7)] | 7.4 | 73.1±1.0 | 16.5±1.8 | DSC | (7) | 94V5 |
| rt-PA [remark (8)] | 7.4 | 72.8±1.2 | 14.2±1.2 | DSC | (8) | 94V5 |
| r-EAEAYV[K2$_{t-PA}$]SR(H)$_6$ | | 71.5 | | DSC | (9) | 97N9 |

Remarks:

(1) buffer: 50 mM sodium citrate/250 mM L-arginine/$H_3PO_4$, pH 4.0

(2) native recombinant t-PA expressed in CHO cells

(3) Boehringer product BM 06.022, a domain deletion mutant of t-PA which contains only the kringle-2 and proteinase module

(4) buffer: 20 mM HEPES/NaOH + 250 mM L-arginine/HCl, pH 7.4

(5) rt-PA, Genentech Inc., CHO-expressed, data from Ref. 88R1

(6) rt-PA, codon: human-melanoma-cell-expressed, data from Ref. 88R1

(7) rt-PA, Monsanto Co., mouse-C127-mammary-cell-expresed, data from Ref. 88R1

(8) rt-PA, Wyeth-Ayerst, mouse-C127-mammary-cell-expresed, low concentration measurement, data from Ref. 88R1

(9) kringle-2 (K2$_{t-PA}$) domain of t-PA expressed in *Pichia pastoris* cell lines GS115 and KM71, sequence Glu-Ala-Glu-Ala-Tyr-Val-[K2$_{t-PA}$]-Ser-Arg-(His)$_6$

**Table 4.**
**Protein Denaturation by Trifluoroethanol (TFE) and Other Alcohol-Based Cosolvents**

The data area arranged as follows:

1) $\Delta G$, $c_{1/2}$, and m from protein denaturation by trifluoroethanol (TFE) and other alcohol-based cosolvents, compared with other denaturants
2) $\Delta G$, $c_{1/2}$, and m determined by urea and GuHCl-induced unfolding in the presence of TFE
3) Enthalpy and heat capacity changes obtained in the presence of TFE and other alcohols
4) Further data, for the most part referred to in Table 2:

1) $\Delta G$, $c_{1/2}$, and m from protein denaturation by trifluoroethanol (TFE) and other alcohol-based cosolvents, compared with other denaturants:

## Barnase

Barnase fragment 1–36 (B(1–36)), TFE-induced unfolding

| Protein | pH | T | $\Delta G$ | $c_{1/2}$ | m | Approach/Remark | | Ref |
|---|---|---|---|---|---|---|---|---|
| B(1–36) | 7 | 5 | 4.6 | | | TFE | (1–3) | 92S15 |

Remarks:
(1) linear extrapolation, see also Ref. 94J5
(2) transition monitored by far-UV CD at 222 nm
(3) measured in 2 mM sodium phosphate, pH 7

Barnase peptide 1–22, wild-type and variants, TFE-induced unfolding

| Protein | pH | T | $\Delta G$ | $c_{1/2}$ | m | Approach/Remark | | Ref |
|---|---|---|---|---|---|---|---|---|
| wild-type | 5.3 | 25 | 8.20±0.50 | | 159±8 | TFE | (1–3) | 94K13 |
| | 5.8 | 25 | 8.41±0.38 | | 167±17 | TFE | (1–3) | 94K13 |
| | 6.3 | 25 | 8.95±0.59 | | 151±17 | TFE | (1–3) | 94K13 |
| | 6.7 | 25 | 9.75±0.54 | | 167±13 | TFE | (1–3) | 94K13 |
| | 7.0 | 25 | 10.50±0.59 | | 167±17 | TFE | (1–3) | 94K13 |
| | 7.3 | 25 | 11.17±0.59 | | 151±13 | TFE | (1–3) | 94K13 |
| | 7.8 | 25 | 11.92±0.46 | | 146±13 | TFE | (1–3) | 94K13 |
| Thr16→Ser | 5.3 | 25 | 8.74±0.29 | | 121±17 | TFE | (1–3) | 94K13 |
| | 5.8 | 25 | 8.95±0.42 | | 130±17 | TFE | (1–3) | 94K13 |
| | 6.3 | 25 | 9.29±0.29 | | 138±13 | TFE | (1–3) | 94K13 |
| | 7.0 | 25 | 10.63±0.21 | | 138±13 | TFE | (1–3) | 94K13 |
| | 7.8 | 25 | 11.30±0.42 | | 138±17 | TFE | (1–3) | 94K13 |
| Tyr17→Ala | 5.3 | 25 | 9.37±0.46 | | 121±13 | TFE | (1–3) | 94K13 |
| | 5.8 | 25 | 9.50±0.42 | | 130±13 | TFE | (1–3) | 94K13 |
| | 6.3 | 25 | 9.83±0.50 | | 113±13 | TFE | (1–3) | 94K13 |
| | 6.7 | 25 | 10.29±0.63 | | 134±21 | TFE | (1–3) | 94K13 |
| | 7.0 | 25 | 10.63±0.63 | | 138±17 | TFE | (1–3) | 94K13 |
| | 7.8 | 25 | 11.13±0.59 | | 117±17 | TFE | (1–3) | 94K13 |
| Tyr13→Ala | 5.3 | 25 | 9.50±0.38 | | 134±8 | TFE | (1–3) | 94K13 |
| | 5.8 | 25 | 9.67±0.33 | | 113±8 | TFE | (1–3) | 94K13 |
| | 6.3 | 25 | 9.92±0.59 | | 138±13 | TFE | (1–3) | 94K13 |
| | 7.0 | 25 | 10.63±0.63 | | 126±17 | TFE | (1–3) | 94K13 |
| | 7.8 | 25 | 11.05±0.59 | | 121±17 | TFE | (1–3) | 94K13 |
| double mutant (Tyr13→Ala and Tyr13→Ala) | | | | | | | | |
| | 5.3 | 25 | 10.04±0.38 | | 109±8 | TFE | (1–3) | 94K13 |
| | 5.8 | 25 | 10.13±0.29 | | 92±8 | TFE | (1–3) | 94K13 |
| | 6.3 | 25 | 10.38±0.29 | | 109±8 | TFE | (1–3) | 94K13 |

Barnase peptide 1–22, wild-type and variants, TFE-induced unfolding(continued)

| Protein | pH | T | $\Delta G$ | $c_{1/2}$ | m | Approach/Remark | | Ref |
|---|---|---|---|---|---|---|---|---|
| | 7.0 | 25 | 10.67±0.46 | | 109±13 | TFE | (1–3) | 94K13 |
| | 7.8 | 25 | 11.00±0.29 | | 92±8 | TFE | (1–3) | 94K13 |
| His18→Gly | 5.3 | 25 | 11.42±0.38 | | 155±13 | TFE | (1–3) | 94K13 |
| | 6.3 | 25 | 11.59±0.50 | | 146±17 | TFE | (1–3) | 94K13 |
| | 7.8 | 25 | 11.63±0.33 | | 142±8 | TFE | (1–3) | 94K13 |

Remarks:

(1) linear extrapolation, see also Ref. 94J5

(2) transition monitored by far-UV CD at 222 nm

(3) measured in 5 mM MES buffer (pH 5.3-7) and 5 mM MOPS (pH 7.3, 7.8)


## Barstar

Barstar peptides, TFE-induced unfolding

| Peptide | pH | T | $\Delta G$ | m | Approach/Remarks | | Ref |
|---|---|---|---|---|---|---|---|
| LHL-wt | 6.3 | 5 | 9.5±2.3 | 39.3±8.5 | TFE | (1–3) | 97S13 |
| H2-wt | 6.3 | 5 | 16.4±1.6 | 61.2±6.0 | TFE | (1–3) | 97S13 |
| LH2-wt | 6.3 | 5 | 11.0±0.8 | 51.3±3.8 | TFE | (1–3) | 97S13 |
| H1H2-wt | 6.3 | 5 | 7.1±0.5 | 48.0±2.7 | TFE | (1–3) | 97S13 |
| LHL-3A | 6.3 | 5 | 5.3±0.7 | 13.8±1.5 | TFE | (1–3) | 97S13 |
| H2-Ma | 6.3 | 5 | 1.9±0.7 | 9.8±1.0 | TFE | (1–3) | 97S13 |
| H2-ER | 6.3 | 5 | 5.3±1.7 | 16.9±3.4 | TFE | (1–3) | 97S13 |
| LH2-CP | 6.3 | 5 | 6.6±2.8 | 22.0±7.3 | TFE | (1–3) | 97S13 |

Explanations:

L = loop, H1 = helix 1, H2 = helix 2, for details of the sequences see Ref. 97S13

Remarks:

(1) linear extrapolation, see also Ref. 94J5

(2) transition monitored by far-UV CD

(3) measured in 5 mM MES, pH 6.3


## Glycodelin

Human glycodelin A (GdA), 2-propanol-induced transition

| Protein | pH | T | $\Delta G$ | $c_{1/2}$ | m | Appr./Rem. | Ref |
|---|---|---|---|---|---|---|---|
| GdA | | 25 | 22.6±1.3 | 3.5±0.1 | 6.3±0.4 | PrOH (1,2) | 99G4 |

Remarks:

(1) 2-propanol-induced transition monitored by CD at 208 nm

(2) linear extrapolation

### α-Lactalbumin

Bovine holo α-lactalbumin, ethanol-induced transitions

| | pH | T | ΔG | Approach/Remarks | | Ref |
|---|---|---|---|---|---|---|
| | 8.0 | 20 | 19.25±0.25 | DSC | (1) | 98G17 |

Remark:
(1) from DSC measurements conducted in the presence of ethanol, ΔG is an extrapolated value for α-lactalbumin unfolding in water

### β-Lactoglobulin

Bovine β-lactoglobulin A (β-LG), unfolding by TFE and HFIP compared with various alcohols

| Protein | pH | T | ΔG | $c_{1/2}$ | m | Approach/Remarks | | Ref |
|---|---|---|---|---|---|---|---|---|
| β-LG | 2.0 | 20 | 23.0 | 11.5 | 2.0 | MetOH | (1,2) | 97H6 |
| | 2.0 | 20 | 23.2 | 5.4 | 4.3 | EtOH | (1,2) | 97H6 |
| | 2.0 | 20 | 23.1 | 3.5 | 6.6 | 2-PrOH | (1,2) | 97H6 |
| | 2.0 | 20 | 23.0 | 2.3 | 10.0 | TFE | (1,2) | 97H6 |
| | 2.0 | 20 | 23.8 | 0.66 | 35.0 | HFIP | (1,2) | 97H6 |

Remarks:
(1) linear extrapolation
(2) transition monitored by ellipticity at 222 nm

β-lactoglobulin A (β-LG), 2-propanol-induced transition

| Protein | pH | T | ΔG | $c_{1/2}$ | m | Appr./Rem. | | Ref |
|---|---|---|---|---|---|---|---|---|
| β-LG | | 25 | 15.9±2.1 | 4.0±0.2 | 3.8±0.8 | PrOH | (1,2) | 99G4 |

Remarks:
(1) 2-propanol-induced transition monitored by CD at 208 nm
(2) linear extrapolation

### Lysozyme HEW

Hen egg white lysozyme, TFE-induced unfolding compared with GuHCl-induced unfolding

| Protein | pH | T | ΔG | $c_{1/2}$ | m | Approach/Remark | | Ref |
|---|---|---|---|---|---|---|---|---|
| lysozyme | 5.2 | 25 | 41.8 | 5.0 | 8.4 | TFE | (1,2) | 96L6 |
| | 5.2 | 25 | 41.8 | 3.6 | 11.7 | GuHCl | (1,2) | 96L6 |

Remarks:
(1) linear extrapolation
(3) the data were taken from Ref. 98B10

## Mellitin

Mellitin, unfolding by TFE and HFIP compared with various alcohols

| Protein | pH | T | $\Delta G$ | $c_{1/2}$ | m | Approach/Remarks | | Ref |
|---|---|---|---|---|---|---|---|---|
| melittin | 2.0 | 20 | 9.5 | 10.0 | 0.95 | MetOH | (1,2) | 97H6 |
| | 2.0 | 20 | 9.7 | 5.1 | 1.9 | EtOH | (1,2) | 97H6 |
| | 2.0 | 20 | 8.6 | 2.4 | 3.6 | 2-PrOH | (1,2) | 97H6 |
| | 2.0 | 20 | 9.6 | 1.6 | 6.0 | TFE | (1,2) | 97H6 |
| | 2.0 | 20 | 9.4 | 0.41 | 23.0 | HFIP | (1,2) | 97H6 |

Remarks:

(1) linear extrapolation

(2) transition monitored by ellipticity at 222 nm

## Tendamistat

Tendamistat, wild-type and disulfide mutants, TFE-induced unfolding compared with GuHCl-induced unfolding

| Protein | pH | T | $\Delta G$ | $c_{1/2}$ | m | Approach/Remarks | | Ref |
|---|---|---|---|---|---|---|---|---|
| wild-type | 2 | 25 | 27.1 | 5.7 | 6.9 | TFE | (1,2) | 96S16 |
| | 2 | 25 | 28.4 | | 7.0 | GuHCl | (1,2) | 96S16 |
| double mutant (Cys11→Ala and Cys27→Thr) | | | | | | | | |
| | 2 | 25 | | 2.6 | | TFE | (1,2) | 96S16 |
| double mutant (Cys45→Ala and Cys73→Ala) | | | | | | | | |
| | 3 | 25 | | 3.6 | | TFE | (1,2) | 96S16 |

Remarks:

(1) linear extrapolation, LEM-SB

(2) transition monitored by ellipticity at 222 nm

*2) $\Delta G$, $c_{1/2}$ and m determined by urea and GuHCl-induced unfolding in the presence of TFE:*

## Acylphosphatase

Recombinant human muscle acylphosphatase, urea denaturation in the presence of TFE

| Solvent | pH | T | $\Delta G$ | $c_{1/2}$ | m | Appr./Rem. | Ref |
|---|---|---|---|---|---|---|---|
| no additives | 5.5 | 28 | 18.8 | 3.97 | 4.56 | urea (1–3,6) | 98C7 |
| no additives | 5.5 | 28 | 19.0 | 3.75 | 5.07 | urea (3,4,6) | 98C7 |
| 3.4% (v/v) TFE | 5.5 | 28 | 19.1 | 3.88 | 4.92 | urea (3–6) | 98C7 |

Remarks:

(1) data from equilibrium unfolding, transition monitored by fluorescence at 335 nm

(2) linear extrapolation, LEM-SB

(3) buffer: 50 mM acetate buffer, pH 5.5

(4) data from kinetics of unfolding, transition monitored by fluorescence at 335 nm

(5) TFE = 2,2,2-trifluoroethanol

(6) for further data obtained in the presence of glucose and phosphate see Table 1

## FKBP12

FKBP12, FK506-binding protein (immunophilin), unfolding of wild-type FKBP12 in the presence and absence of TFE by urea and GuHCl

| Protein | pH | T | $\Delta G$ | $c_{1/2}$ | m | Appr./Rem. | | Ref |
|---|---|---|---|---|---|---|---|---|
| data from equilibrium unfolding: | | | | | | | | |
| FKBP12 | 7.5 | 25 | 23.14±0.50 | 3.87±0.01 | 5.98±0.13 | urea | (1–3) | 99M2 |
| + 3.6% TFE | 7.5 | 25 | 24.69±1.30 | 4.40±0.03 | 5.61±0.29 | urea | (1–3) | 99M2 |
| + 9.6% TFE | 7.5 | 25 | 32.59±1.76 | 4.58±0.03 | 7.11±0.38 | urea | (1–3) | 99M2 |
| + 17 % TFE | 7.5 | 25 | 25.65±0.63 | 3.65±0.01 | 7.03±0.17 | urea | (1–3) | 99M2 |
| FKBP12 | 7.5 | 25 | 21.46±0.88 | 0.78±0.01 | 21.61±1.13 | GuHCl (1–3) | | 99M2 |
| FKBP12 | 7.5 | 25 | 34.10±3.39 | 26.80±0.20 | 1.26±0.13 | TFE | (1–4) | 99M2 |
| data from kinetics of unfolding and refolding: | | | | | | | | |
| FKBP12 | 7.5 | 25 | 24.3±1.3 | | 7.1±2.1 | urea | (3,5) | 99M2 |
| + 0.0% TFE | 7.5 | 25 | 24.7±0.8 | | 7.1±2.5 | urea | (3,5) | 99M2 |
| + 3.6% TFE | 7.5 | 25 | 28.9±0.8 | | 6.3±2.5 | urea | (3,5) | 99M2 |
| + 9.6% TFE | 7.5 | 25 | 32.2±0.8 | | 6.3±2.9 | urea | (3,5) | 99M2 |
| + 17 % TFE | 7.5 | 25 | 26.8±1.3 | | 6.7±2.9 | urea | (3,5) | 99M2 |
| FKBP12 | 7.5 | 25 | 20.9±0.4 | | 37.2±0.8 | GuHCl (3,5) | | 99M2 |

Remarks:

(1) nonlinear least-squares fit of the transition curve assuming a linear dependence of $\Delta G$ on the denaturant conc.

(2) transition monitored by fluorescence spectroscopy

(3) buffer: 50 mM Tris-HCl, 1 mM DTT, pH 7.5

(4) TFE denaturation, units for m are kJ/mol/%

(5) from denaturant dependence of kinetics of unfolding and refolding, for further data at various temperatures and in the presence of rapamycin see Table 1

FKBP12, FK506-binding protein (immunophilin), urea-induced unfolding in the presence of TFE

| Mutant | pH | T | $\Delta G$ | $c_{1/2}$ | m | Appr./Rem. | Ref |
|---|---|---|---|---|---|---|---|
| data at 0% TFE: | | | | | | | |
| wild-type | 7.5 | 25 | 25.56±0.33 | 3.87±0.02 | 5.98±0.21 | urea (1–4) | 99M3 |
| Arg57→Ala | 7.5 | 25 | 22.22±0.29 | 3.36±0.02 | 6.53±0.29 | urea (1–4) | 99M3 |
| Arg57→Gly | 7.5 | 25 | 16.07±0.25 | 2.43±0.02 | 6.95±0.33 | urea (1–4) | 99M3 |
| Glu60→Ala | 7.5 | 25 | 16.74±0.29 | 2.53±0.03 | 7.03±0.42 | urea (1–4) | 99M3 |
| Glu60→Gly | 7.5 | 25 | 13.77±0.38 | 2.08±0.05 | 6.28±0.50 | urea (1–4) | 99M3 |
| Glu61→Ala | 7.5 | 25 | 22.09±0.29 | 3.34±0.02 | 6.15±0.21 | urea (1–4) | 99M3 |
| Glu61→Gly | 7.5 | 25 | 15.19±0.25 | 2.30±0.02 | 6.57±0.25 | urea (1–4) | 99M3 |
| Val63→Ala | 7.5 | 25 | 13.22±0.25 | 2.00±0.03 | 7.07±0.42 | urea (1–4) | 99M3 |
| Ile7→Val | 7.5 | 25 | 21.76±0.29 | 3.29±0.02 | 5.90±0.29 | urea (1–4) | 99M3 |
| Ile76→Val | 7.5 | 25 | 22.43±0.33 | 3.39±0.03 | 6.95±0.42 | urea (1–4) | 99M3 |
| Leu97→Ala | 7.5 | 25 | 10.79±0.29 | 1.63±0.04 | 7.91±0.54 | urea (1–4) | 99M3 |
| Val98→Ala | 7.5 | 25 | 16.61±0.46 | 2.51±0.06 | 6.19±0.71 | urea (1–4) | 99M3 |
| Val101→Ala | 7.5 | 25 | 14.14±0.21 | 2.14±0.02 | 6.82±0.17 | urea (1–4) | 99M3 |
| data at 9.6% TFE: | | | | | | | |
| wild-type | 7.5 | 25 | 31.80±0.21 | 4.58±0.03 | 7.11±0.38 | urea (1–4) | 99M3 |
| Arg57→Ala | 7.5 | 25 | 29.71±0.17 | 4.28±0.02 | 6.57±0.25 | urea (1–4) | 99M3 |
| Arg57→Gly | 7.5 | 25 | 23.81±0.21 | 3.43±0.03 | 6.49±0.42 | urea (1–4) | 99M3 |
| Glu60→Ala | 7.5 | 25 | 25.15±0.21 | 3.62±0.03 | 6.07±0.38 | urea (1–4) | 99M3 |
| Glu60→Gly | 7.5 | 25 | 21.25±0.29 | 3.06±0.04 | 6.90±0.54 | urea (1–4) | 99M3 |
| Glu61→Ala | 7.5 | 25 | 28.95±0.21 | 4.17±0.03 | 6.44±0.42 | urea (1–4) | 99M3 |
| Glu61→Gly | 7.5 | 25 | 21.59±0.29 | 3.11±0.04 | 6.69±0.54 | urea (1–4) | 99M3 |
| Val63→Ala | 7.5 | 25 | 20.96±0.17 | 3.02±0.02 | 7.61±0.38 | urea (1–4) | 99M3 |

FKBP12, FK506-binding protein (immunophilin), urea-induced unfolding in the presence of TFE (continued)

| Mutant | pH | T | $\Delta G$ | $c_{1/2}$ | m | Appr./Rem. | Ref |
|--------|-----|-----|-------------|------------|------------|-------------|-------|
| Ile7→Val | 7.5 | 25 | 27.03±0.21 | 3.89±0.03 | 7.53±0.54 | urea (1–4) | 99M3 |
| Ile76→Val | 7.5 | 25 | 29.37±0.17 | 4.23±0.02 | 7.07±0.29 | urea (1–4) | 99M3 |
| Leu97→Ala | 7.5 | 25 | 16.19±0.59 | 2.33±0.08 | 7.03±0.88 | urea (1–4) | 99M3 |
| Val98→Ala | 7.5 | 25 | 21.97±0.08 | 3.16±0.01 | 7.15±0.25 | urea (1–4) | 99M3 |
| Val101→Ala | 7.5 | 25 | 20.50±0.17 | 2.95±0.02 | 8.08±0.42 | urea (1–4) | 99M3 |
| data at 3.6% TFE: | | | | | | | |
| wild-type | 7.5 | 25 | 31.55±0.04 | 4.55±0.02 | 7.11±0.38 | urea (1–4) | 99M3 |
| data at 17% TFE: | | | | | | | |
| wild-type | 7.5 | 25 | 25.36±0.04 | 3.65±0.02 | 7.03±0.30 | urea (1–4) | 99M3 |

Remarks:

(1) linear extrapolation, for further details, see Ref. 98M1

(2) transition monitored by fluorescence

(3) buffer: 50 mM Tris-HCl, pH 7.5, 1 mM DTT

(4) $\Delta G$ was calculated in 0% TFE using an average m value of 6.65±0.08 and in 9.6% TFE 6.82±0.17 kJ/mol/M

## β-Lactoglobulin

β-lactoglobulin, GuHCl-induced transitions in the presence of TFE that lead to an intermediate with nonnative α-helical structure

| TFE conc. | Transition | pH | T | $\Delta G$ | m | Appr./Rem. | Ref |
|-----------|------------|-----|-----|------------|------------|-------------|-------|
| transition monitored by far-UV CD: | | | | | | | |
| 0% | N → U | 2 | 4 | 41.8±1.2 | 17.1±0.5 | GuHCl (1,2) | 97H1 |
| 0% | I → U | 2 | 4 | 8.0±0.7 | 3.4±0.1 | GuHCl (1,2) | 97H1 |
| 0% | N → I | 2 | 4 | 33.8±1.8 | 13.7±0.6 | GuHCl (1,2) | 97H1 |
| 9.8% | N → U | 2 | 4 | 30.6±0.4 | 12.4±0.5 | GuHCl (1–3) | 97H1 |
| 9.8% | I → U | 2 | 4 | 18.3±0.4 | 6.7±0.1 | GuHCl (1–3) | 97H1 |
| 9.8% | N → I | 2 | 4 | 12.3±0.8 | 5.7±0.3 | GuHCl (1–3) | 97H1 |
| transition monitored by near-UV CD: | | | | | | | |
| 0% | N → I | 2 | 4 | 34.9±2.2 | 14.5±0.9 | GuHCl (1,2) | 97H1 |
| 9.8% | N → I | 2 | 4 | 10.7±1.6 | 5.8±0.8 | GuHCl (1–3) | 97H1 |

Remark:

(1) the approach assumes a linear combination of the contributions of the three states to the observed ellipticity

(2) the approach is based on a linear dependence of $\Delta G_i$ on denaturant concentration

(3) measured in the presence of 2,2,2-trifluoro-ethanol (TFE)

*3) Enthalpy and heat capacity changes obtained in the presence of TFE and other alcohols:*

## α–Lactalbumin

Bovine holo α-lactalbumin, DSC in the presence of ethanol

| EtOH (%) | pH | $T_{trs}$ | $\Delta Cp$ | $\Delta H$ | Appr./Rem. | Ref |
|---|---|---|---|---|---|---|
| 0.0 | 8.0 | 65.4±0.10 | 5.3±0.1 | 310.5±0.5 | DSC (1,2) | 98G17 |
| 5.0 | 8.0 | 61.4±0.20 | 5.3±0.3 | 312.5±1.0 | DSC (1,2) | 98G17 |
| 10.0 | 8.0 | 57.2±0.10 | 5.6±0.1 | 288.3±0.4 | DSC (1,2) | 98G17 |
| 20.0 | 8.0 | 48.3±0.10 | 6.8±0.1 | 269.7±1.3 | DSC (1,2) | 98G17 |
| 30.0 | 8.0 | 36.6±0.02 | 8.3±0.1 | 237.4±0.4 | DSC (1,2) | 98G17 |

Remarks:

(1) ethanol conc. in % (v/v)

(2) the heat capacity function of holo α-lactalbumin in the presence of ethanol follows a linear dependence according to

$\Delta Cp(T) = \Delta Cp(T_{trs}) + \Delta Cp'(T- T_{trs})$ with

| EtOH (v/v) | $T_{trs}$ | $\Delta Cp$ | $\Delta Cp'$ (kJ/mol/K²) |
|---|---|---|---|
| 0.0 | 65.4 | 5.3±0.1 | −0.050±0.004 |
| 5.0 | 61.4 | 5.3±0.3 | −0.036±0.009 |
| 10.0 | 57.2 | 5.6±0.1 | −0.085±0.002 |
| 20.0 | 48.3 | 6.8±0.1 | −0.020±0.010 |
| 30.0 | 36.6 | 8.3±0.1 | −0.022±0.003 |

## Lipase

Lipase from *Pseudomonas cepacia* in the absence and presence of alcohols

| Alcohol conc. | pH | $T_{trs}$ | $\Delta H$ | Approach/Remarks | | Ref |
|---|---|---|---|---|---|---|
| without alcohol | 7.0 | 75.8–80.9 | 849±25 | DSC | (1,2) | 98T4 |
| MeOH 0.24–1.12 M | 7.0 | 76.9–74.6 | 836±15 | DSC | (3,4) | 98T4 |
| EtOH 0.17–0.86 M | 7.0 | 76.7–73.9 | 837±30 | DSC | (3,4) | 98T4 |
| PrOH 0.13–0.67 M | 7.0 | 76.9–70.8 | 784±60 | DSC | (3,4) | 98T4 |
| BuOH 0.11,0.22 M | 7.0 | 75.6–74.0 | 828,745 | DSC | (3,4) | 98T4 |

Remarks:

(1) measured at protein concentrations from 29.3 to 236 μM in the absence of alcohol, $T_{trs}$ varies from 75.8 to 80.9°C without significant change in $\Delta H$, see also specific heat capacity change $\Delta Cp = 0.055±0.075$ J/g/K in Table 3

(2) analysis of the DSC tracings is consistent with the mechanism $N.Ca^{2+} \leftrightarrow D + Ca^{2+}$

(3) buffer: 20 mM phosphate, pH 7.0

(4) measured at protein conc. of 98 μM

## Luciferase

Firefly luciferase, thermal unfolding in the presence of ethanol

| Ethanol conc. | pH | $T_{trs}$ | $\Delta h$ | Approach/Remarks | | Ref |
|---|---|---|---|---|---|---|
| 0.00 M | 7.8 | 41.7 | 6.7 | DSC | (1) | 94C13 |
| 0.15 M | 7.8 | 40.8 | | DSC | (1) | 94C13 |
| 0.30 M | 7.8 | 39.6 | | DSC | (1) | 94C13 |
| 0.60 M | 7.8 | 38.2 | | DSC | (1) | 94C13 |
| 1.20 M | 7.8 | 35.2 | | DSC | (1) | 94C13 |
| 2.30 M | 7.8 | 30.6 | | DSC | (1) | 94C13 |

Remark:

(1) measured in 100 mM glycylglycine buffer, pH 7.8

**Lysozyme HEW**

Hen egg white lysozyme, $\Delta H$ obtained by ITC and DSC

| Protein | Alcohol | pH | $T_{trs}$ | $x_A^d$ | $\Delta H$ | Appr./Rem. | Ref |
|---------|---------|-----|-----------|---------|------------|------------|------|
| lysozyme | none | 2.0 | 40 | | 270 | DSC | 97W4 |
| | TFE | 2.0 | 30 | 0.050 | 320 | ITC | 97W4 |
| | TFE | 2.0 | 30 | 0.046 | 350 | DSC | 97W4 |
| | TFE | 6.0 | 20 | 0.125 | 180 | ITC | 97W4 |
| | MetOH | 2.0 | 40 | 0.170 | 400 | ITC | 97W4 |
| | EtOH | 2.0 | 40 | 0.103 | 430 | ITC | 97W4 |
| | 1-PrOH | 2.0 | 40 | 0.033 | 375 | ITC | 97W4 |

Explanation:

$x_A^d$ is the mole fraction of alcohol at which half conversion is observed

Remarks:

(1) measured by DSC

(2) measured by ITC

Hen egg white lysozyme in the presence of 1-PrOH

| 1-PrOH (M) | pH | $T_{trs}$ | $\Delta H$ | $\Delta Cp$ | Appr./Rem. | | Ref |
|------------|-----|-----------|------------|-------------|------------|-----|------|
| 0.0 | 2.0 | 52.03±0.52 | 381.6±7.9 | 6.49±0.50 | DSC | (1,2) | 79V |
| 0.67 | 2.0 | 47.44±0.50 | 387.4±6.7 | 5.19±0.59 | DSC | (1,2) | 79V |
| 1.34 | 2.0 | 41.64±0.55 | 399.2±11.3 | 5.44±0.63 | DSC | (1,2) | 79V |
| 2.00 | 2.0 | 33.36±0.61 | 397.1±2.9 | 4.60±1.30 | DSC | (1,2) | 79V |
| 2.67 | 2.0 | 25.44±0.37 | 322.6±4.6 | 4.27±2.01 | DSC | (1,2) | 79V |
| 3.34 | 2.0 | 15.69±0.51 | 246.9±7.1 | 4.73±0.63 | DSC | (1,2) | 79V |

Remark:

(1) measured in 40 mM glycine buffer, pH 2.0

(2) Ref. 79V contains additional data for hen egg white lysozyme in aqueous mixtures of methanol and ethanol

Hen egg white lysozyme, DSC in the presence of 1-propanol

| PrOH (M) | pH | $T_{trs}$ | $\Delta Cp$ | $\Delta H$ | Appr./Rem. | | Ref |
|----------|-----|-----------|-------------|------------|------------|-----|------|
| 0.67 | 3.7 | 70.2 | | 541 | DSC | | 89S6 |
| 1.34 | 3.7 | 63.3 | | 506 | DSC | | 89S6 |
| 2.0 | 3.7 | 56.8 | | 473 | DSC | | 89S6 |
| 2.67 | 3.7 | 49.5 | | 435 | DSC | | 89S6 |
| 4.0 | 3.7 | 40 | | 393 | DSC | | 89S6 |
| 0.7–4.0 | 3.7 | 40–70 | 6.5 | 393–541 | DSC | (1) | 89S6 |

Remarks:

(1) $\Delta Cp$ from $\Delta H^{cal}$ versus $T_{trs}$ at varying PrOH conc.

Hen egg white lysozyme, crosslinked between Glu35 and Trp108, DSC in the presence of 1-propanol

| PrOH (M) | pH | $T_{trs}$ | $\Delta Cp$ | $\Delta H$ | Appr./Rem. | | Ref |
|----------|-----|-----------|-------------|------------|------------|-----|------|
| 1.34 | 3.7 | 80.5 | | 608 | DSC | | 89S6 |
| 2.0 | 3.7 | 73.9 | | 591 | DSC | | 89S6 |
| 2.67 | 3.7 | 68.2 | | 532 | DSC | | 89S6 |
| 4.0 | 3.7 | 59.7 | | 506 | DSC | | 89S6 |
| 1.3–4.0 | 3.7 | 60–80 | 6.5 | 506–608 | DSC | (1) | 89S6 |

Remark:

(1) $\Delta Cp$ from $\Delta H^{cal}$ versus $T_{trs}$ at varying PrOH conc.

**Nuclease** (Staphylococcal nuclease)

Staphylococcal nuclease, heat denaturation in the presence of methanol

| MetOH (%) | pH | $T_{trs}$ | $\Delta Cp$ | $\Delta H$ | Appr./Rem. | | Ref |
|---|---|---|---|---|---|---|---|
| 0 % (v/v) | | 53 | | 340 | heat | (1,2) | 90N1 |
| 15 % (v/v) | | 44 | | 360 | heat | (1,2) | 90N1 |
| 35 % (v/v) | | 32 | | 355 | heat | (1,2) | 90N1 |
| 50 % (v/v) | | 19 | | 305 | heat | (1,2) | 90N1 |
| 60 % (v/v) | | 4 | | 255 | heat | (1,2) | 90N1 |
| 70 % (v/v) | | –4 | | 175 | heat | (1,2) | 90N1 |

Remarks:
(1) $\Delta H$ from van't Hoff treatment
(2) the data were taken from Table 1 and Figs. 3–4 of Ref. 90N1

**Pepsinogen**

Pepsinogen in the presence and absence of ethanol

| EtOH (M) | pH | $T_{trs}$ | $\Delta Cp$ | $\Delta H$ | Appr./Rem. | | Ref |
|---|---|---|---|---|---|---|---|
| 0 % | 6.0 | 66.2 | 24.3±1.7 | 1063 | DSC | (1) | 95M1 |
| 0 % | 6.4 | 62.8 | 24.3±1.7 | 1038 | DSC | | 94M1 |
| 0 % | 6.4 | 62.8 | | 1038 | DSC | | 95M1 |
| 0 % | 7.2 | 56.1 | | 816 | DSC | | 95M1 |
| 0 % | 7.7 | 55.0 | | 724 | DSC | | 95M1 |
| 20 % | 5.9 | 55.8 | 17.6±1.7 | 1025 | DSC | (1) | 95M1 |
| 20 % | 6.4 | 52.1 | 17.6±1.7 | 1038 | DSC | | 94M1 |
| 20 % | 6.4 | 52.1 | | 1038 | DSC | | 95M1 |
| 20 % | 6.8 | 47.2 | | 925 | DSC | | 95M1 |
| 20 % | 7.3 | 39.8 | | 803 | DSC | | 95M1 |
| 20 % | 8.0 | 35.3 | | 707 | DSC | | 95M1 |
| 20 % | 8.2 | 33.8 | | 711 | DSC | | 95M1 |

Remark:
(1) $\Delta Cp$ values obtained from single calorimetric recordings coincide with $\Delta Cp$ from $\Delta H$ versus $T_{trs}$

*4) Further data, for the most part referred to in Tables 2 and 3:*

Methanol: for cytochrome $c$ in the presence of methanol see Ref. 92F4
For ribonuclease A in the presence of 50% methanol see Ref. 94R1
Ethanol: for bovine serum albumin in the presence of ethanol see Ref. 97Z3
Aromatic alcohols: for interleukin-1 receptor in the presence of benzyl alcohol, m-cresol, and phenol see Ref. 98R4

**References
and Index of Proteins**

**References** (Tables 1–4)

68P1    Pace, C.N., Tanford, C.: Biochemistry 7 (1968) 198–208.
71B1    Biltonen, R., Schwartz, A.T., Wadsö, I.: Biochemistry 10 (1971) 3417–3423.
71S1    Shiao, D.F., Lumry, R., Fahey, J.: J. Amer. Chem. Soc. 93 (1971) 2024–2035.
73D1    Donovan, J.W., Ross, K.D.: Biochemistry 12 (1973) 512–517.
76T2    Tsong, T.Y.: Biochemistry 15 (1976) 5467–5473.
78F2    Freire, E., Biltonen, R.L.: Biopolymers 17 (1978) 463–479.
78G1    Gudkov, A.T., Khechinashvili, N.N., Bushuev, V.N.: Eur. J. Biochem. 90 (1978) 313–318.
78V1    Vickers, L.P., Donovan, J.W., Schachman, H.K.: J. Biol. Chem. 253 (1978) 8493–8498.
79G2    Gumpen, S., Hegg, P.O., Martens, H.: Biochim. Biophys. Acta 574 (1979) 189–196.
79P7    Pabo, C.O., Sauer, R.T., Sturtevant, J.M., Ptashne, M.: Proc. Natl. Acad. Sci. USA 76 (1979) 1608–1612.
80L1    Luer, C.A., Wong, K.-P.: Biochemistry 19 (1980) 176–183.
81Z1    Zahnley, J.C.: J. Inorg. Biochem. 15 (1981) 67–78.
85A2    Adams, B., Burgess, R.J., Pain, R.H.: Eur. J. Biochem. 152 (1985) 715–720.
87M2    Monnot, M., Gilles, A.-M., Girons, I.S., Michelson, S., Bârzu, O., Fermandjian, S.: J. Biol. Chem. 262 (1987) 2502–2506.
88R4    Roe, J.A., Butler, A., Scholler, D.M., Valentine, J.S., Marky, L., Breslauer, K.J.: Biochemistry 27 (1988) 950–958.
89G6    Grinberg, Y.Ya., Danilenko, A.N., Burova, T.V., Tolstoguzov, V.B.: J. Sci. Food Agric. 49 (1989) 235–248.
90B11   Bedzyk, W.D., Weidner, K.M., Denzin, L.K., Johnson, L.S., Hardman, K.D., Pantoliano, M.W., Asel, E.D., Voss Jr, E.W.: J. Biol. Chem. 265 (1990) 18615–18620.
90B12   Bonaccorsi di Patti, M.C., Musci, G., Giartosio, A., D'Alessio, S., Calabrese, L.: J. Biol. Chem. 265 (1990) 21016–21022.
90M6    McRee, D.E., Redford, S.M., Getzoff, E.D., Lepock, J.R., Hallewell, R.A., Tainer, J.A.: J. Biol. Chem. 265 (1990) 14234–14241.
90P4    Pace, C.N.: Trends Biochem. Sci. 15 (1990) 14–17.
90P5    Pace, C.N.: Trends Biotechnol. 8 (1990) 93–98.
90R5    Reinstein, J., Vetter, I.R., Schlichting, I., Rösch, P., Wittinghofer, A., Goody, R.S.: Biochemistry 29 (1990) 7440–7450.
91C8    Copeland, R.A., Ji, H., Halfpenny, A.J., Williams, R.W., Thompson, K.C., Herber, W.K., Thomas, K.A., Bruner, M.W., Ryan, J.A., Marquis-Omer, D., Sanyal, G., Sitrin, R.D., Yamazaki, S., Middaugh, C.R.: Arch. Biochem. Biophys. 289 (1991) 53–61.
91F3    Fisher, M.T.: Biochemistry 30 (1991) 10012–10018.
91H5    Herning, T., Yutani, K., Taniyama, Y., Kikuchi, M.: Biochemistry 30 (1991) 9882–9891.
91J3    Johnson, C.M., Cooper, A., Brown, A.J.P.: Eur. J. Biochem. 202 (1991) 1157–1164.
91K7    Kanaya, S., Katsuda, C., Kimura, S., Nakai, T., Kitakuni, E., Nakamura, H., Katayanagi, K., Morikawa, K., Ikehara, M.: J. Biol. Chem. 266 (1991) 6038–6044.
91L10   LiCalsi, C., Crocenzi, T.S., Freire, E., Roseman, S.: J. Biol. Chem. 266 (1991) 19519–19527.
91Z3    Žerovnik, E., Lenarcic, B., Jerala, R., Turk, V.: Biochim. Biophys. Acta 1078 (1991) 313–320.
91Z4    Zhang, Z.-Y., Poorman, R.A., Maggiora, L.L., Heinrikson, R.L., Kézdy, F.J.: J. Biol. Chem. 266 (1991) 15591–15594.

92B10   Barone, G., Giancola, C., Verdoliva, A.: Thermochim. Acta 199 (1992) 197–205.

92B11   Barone, G., Del Vecchio, P., Fessas, D., Giancola, C., Graziano, G., Pucci, P., Riccio, A., Ruoppolo, M.: J. Thermal Anal. 38 (1992) 2791–2802.

92B12   Bonaccorsi di Patti, M.C., Galtieri, A., Giartosio, A., Musci, G., Calabrese, L.: Comp. Biochem. Physiol. [B] 103 (1992) 183–188.

92G7    Gooley, P.R., Caffrey, M.S., Cusanovich, M.A., MacKenzie, N.E.: Biochemistry 31 (1992) 443–450.

92K13   Kanaya, S., Itaya, M.: J. Biol. Chem. 267 (1992) 10184–10192.

92S15   Sancho, J., Neira, J.L., Fersht, A.R.: J. Mol. Biol. 224 (1992) 749–758.

92Z6    Žerovnik, E., Jerala, R., Kroon-Žitko, Pain, R.H., Turk, V.: J. Biol. Chem. 267 (1992) 9041–9046.

93C8    Cedergren, L., Andersson, R., Jansson, B., Uhlén, M., Nilsson, B.: Protein Engng. 6 (1993) 441–448.

93K12   Khorasanizadeh, S., Peters, I.D., Butt, T.R., Roder, H.: Biochemistry 32 (1993) 7054–7063.

93L10   Linse, S., Thulin, E., Sellers, P.: Protein Sci. 2 (1993) 985–1000.

93M19   Marmorino, J.L., Auld, D.S., Betz, S.F., Doyle, D.F., Young, G.B., Pielak, G.J.: Protein Sci. 2 (1993) 1966–1974.

93S13   Schwarz, F.P., Puri, K.D., Bhat, R.G., Surolia, A.: J. Biol. Chem. 268 (1993) 7668–7677.

93V4    Vergères, G., Chen, D.Y., Wu, F.F., Waskell, L.: Arch. Biochem. Biophys. 305 (1993) 231–241.

94C13   Chiou, J.-S., Ueda, I.: J. Pharmaceut. Biomed. Anal. 12 (1994) 969–975.

94F4    Feller, G., Payan, F., Theys, F., Qian, M., Haser, R., Gerday, C.: Eur. J. Biochem. 222 (1994) 441–447.

94G8    Gekko, K., Kunori, Y., Takeuchi, H., Ichihara, S., Kodama, M.: J. Biochem. 116 (1994) 34–41.

94H12   Hargrove, M.S., Krzywda, S., Wilkinson, A.J., Dou, Y., Ikeda-Saito, M., Olson, J.S.: Biochemistry 33 (1994) 11767–11775.

94J5    Jasanoff, A., Fersht, A.R.: Biochemistry 33 (1994) 2129–2135.

94K13   Kippen, A.D., Arcus, V.L., Fersht, A.R.: Biochemistry 33 (1994) 10013–10021.

94K14   Kravchuk, Z.I., Vlasov, A.P., Lyakhovich, G.V., Martsev, S.P.: Biokhimiya 59 (1994) 1458–1477.

94M24   Martsev, S.P., Kravchuk, Z.I., Vlasov, A.P.: Immunol. Lett. 43 (1994) 149–152.

94N4    Neet, K.E., Timm, D.E.: Protein Sci. 3 (1994) 2167–2174.

94V4    Vlassi, M., Steif, C., Weber, P., Tsernoglou, D., Wilson, K.S., Hinz, H.-J., Kokkinidis, M.: Struct. Biol. 1 (1994) 706–716.

94V5    Vogl, T., Hinz, H.-J.: Biotechnol. Appl. Biochem. 20 (1994) 1–22.

95A10   Azuaga, A.I.: Doctoral Thesis (1995), University of Granada, Spain, data cited in Ref. 98P11: Plaxco, K.W., Guijarro, J.I., Morton, C.J., Pitkeathly, M., Campbell, I.D., Dobson, C.M.: Biochemistry 37 (1998) 2529–2537.

95B12   Bryan, P., Wang, L., Hoskins, J., Ruvinov, S., Strausberg, S., Alexander, P., Almog, O., Gilliland, G., Gallagher, T.: Biochemistry 34 (1995) 10310–10318.

95C7    Cohen, D.S., Pielak, G.J.: J. Amer. Chem. Soc. 117 (1995) 1675–1677.

95D7    Dolla, A., Florens, L., Bruschi, M., Dudich, I.V., Makarov, A.A.: Biochem. Biophys. Res. Commun. 211 (1995) 742–747.

95F6    Ferrer, M., Barany, G., Woodward, C.: Struct. Biol. 2 (1995) 211–217.

95G14   Grauschopf, U., Winther, J.R., Korber, P., Zander, T., Dallinger, P., Bardwell, J.C.A.: Cell 83 (1995) 947–955.

95K15   Keck, J.L., Marqusee, S.: Proc. Natl. Acad. Sci. USA 92 (1995) 2740–2744.

95K16   Knappik, A., Plückthun, A.: Protein Engng. 8 (1995) 81–89.

95K17   Kragelund, B.B., Knudsen, J., Poulsen, F.M.: J. Mol. Biol. 250 (1995) 695–706.

95L14   La Rosa, C., Milardi, D., Grasso, D., Guzzi, R., Sportelli, L.: J. Phys. Chem. B 99 (1995) 14864–14870.
95P11   Parker, M.J., Spencer, J., Clarke, A.R.: J. Mol. Biol. 253 (1995) 771–786.
95V10   Vignais, M.-L., Corbier, C., Mulliert, G., Branlant, C., Branlant, G.: Protein Sci. 4 (1995) 994–1000.
96B12   Bam, N.B., Cleland, J.L., Randolph, T.W.: Biotechnol. Prog. 12 (1996) 801–909.
96C19   Chauvin, F., Fomenkov, A., Johnson, C.R., Roseman, S.: Proc. Natl. Acad. Sci. USA 93 (1996) 7028–7031.
96D3    Dodge, R.W., Scheraga, H.A.: Biochemistry 35 (1996) 1548–1559.
96D4    Dolgikh, D.A., Uversky, V.N., Gabrielian, A.E., Chemeris, V.V., Fedorov, A.N., Navolotskaya, E.V., Zav'yalov, V.P., Kirpichnikov, M.P.: Protein Engng. 9 (1996) 195–201.
96G11   Garda-Salas, A.L., Santamaria, R.I., Marcos, M.J., Zhadan, G.G., Villar, E., Shnyrov, V.L.: Biochem. Mol. Biol. Int. 38 (1996) 161–170.
96H8    Henzl, M.T., Hapak, R.C., Goodpasture, E.A.: Biochemistry 35 (1996) 5856–5869.
96K11   Kaiser, D.A., Pollard, T.D.: J. Mol. Biol. 256 (1996) 89–107.
96K12   Kawamura, S., Kakuta, Y., Tanaka, I., Hikichi, K., Kuhara, S., Yamasaki, N., Kimura, M.: Biochemistry 35 (1996) 1195–1200.
96K13   Kovrigin, E.L., Potekhin, S.A.: Biofizika 41 (1996) 1201–1206.
96L5    Liemann, S., Benz, J., Burger, A., Voges, D., Hofmann, A., Huber, R., Göttig, P.: J. Mol. Biol. 258 (1996) 555–561.
96L6    Lu, H.: Thesis, University of Oxford, 1996, cited in Ref. 98B10 Buck, M.: Quart. Rev. Biophys. 31 (1998) 297–355.
96M13   Meijberg, W., Schuurman-Wolters, G.K., Robillard, G.T.: Biochemistry 35 (1996) 2759–2766.
96M14   Milardi, D., Fasone, S., La Rosa, C., Grasso, D.: Il Nuovo Cimento 18 (1996) 1347–1353.
96M15   Mines, G.A., Pascher, T., Lee, S.C., Winkler, J.R., Gray, H.B.: Chemistry & Biology 3 (1996) 491–497.
96M16   Musci, G., Bonaccorsi di Patti, M.C., Petruzzelli, R., Giartosio, A., Calabrese, L.: Biometals 9 (1996) 66–72.
96N3    Nikolaeva, O.P., Orlov, V.N., Dedova, I.V., Drachev, V.A., Levitsky, D.I.: Biochem. Mol. Biol. Int. 40 (1996) 653–661.
96O5    Ohmae, E., Iriyama, K., Ichihara, S., Gekko, K.: J. Biochem. 119 (1996) 703–710.
96P4    Pascher, T., Chesick, J.P., Winkler, J.R., Gray, H.B.: Science 271 (1996) 1558–1560.
96P5    Poklar, N.: unpublished results cited in: Poklar, N., Vesnaver, G., Lapanje, S., Biophys. Chem. 57 (1996) 279–289.
96R4    Reinstädler, D., Fabian, H., Backmann, J., Naumann, D.: Biochemistry 35 (1996) 15822–15830.
96S15   Saburova, E.A., Khechinashvili, N.N., Elfimova, L.I.: Molecular Biology (Moscow) 30 (1996) 1219–1228.
96S16   Schönbrunner, N., Wey, J., Engels, J., Georg, H., Kiefhaber, T.: J. Mol. Biol. 260 (1996) 432–445.
96S17   Stefanini, S., Cavallo, S., Wang, C.-Q., Tataseo, P., Vecchini, P., Giartosio, A., Chiancone, E.: Arch. Biochem. Biophys. 325 (1996) 58–64.
96T5    Todorova, R.T., Rogov, V.V., Vasilenko, K.S., Permyakov, E.A.: Biophys. Chem. 62 (1996) 39–45.
96V6    Vlasov, A.P., Kravchuk, Z.I., Martsev, S.P.: Biokhimiya 61 (1996) 212–235.
96W3    Wang, C., Eufemi, M., Turano, C., Giartosio, A.: Biochemistry 35 (1996) 7299–7307.

488     References

96Z2    Zolkiewski, M., Redowicz, M.J., Korn, E.D., Ginsburg, A.: Biophys. Chem. 59
        (1996) 365–371.
97A1    Agashe, V., Schmid, F.X., Udgaonkar, J.B.: Biochemistry 36 (1997)
        12288–12295.
97A2    Akasako, A., Haruki, M., Oobatake, M., Kanaya, S.: J. Biol. Chem. 272 (1997)
        18686–18693.
97A3    Akker, F. van den, Feil, I.K., Roach, C., Platas, A.A., Merritt, E.A., Hol., W.G.J.:
        Protein Sci. 6 (1997) 2644–2649.
97A4    Andreotti, G., Cubellis, M.V., Di Palo, M., Fessas, D., Sannia, G., Marino, G.:
        Biochem. J. 323 (1997) 259–264.
97A5    Arnold, U., Ulbrich-Hofmann, R.: Biochemistry 36 (1997) 2166–2172.
97A6    Aronsson, G., Brorsson, A.-C., Sahlman, L., Johnsson, B.-H.: FEBS Lett. 411
        (1997) 359–364.
97A7    Arrington, C.B., Robertson, A.D.: Biochemistry 36 (1997) 8686–8691.
97B1    Bagel'ova, J., Antalik, M., Tomori, Z.: Biochem. Mol. Biol. Int. 43 (1997)
        891–899.
97B2    Bai, Y., Karimi, A., Dyson, H.J., Wright, P.E.: Protein Sci. 6 (1997) 1449–1457.
97B3    Banerjee, S., Shigematsu, N., Pannell, L.K., Ruvinov, S., Orban, J., Schwarz, F.,
        Herzberg, O.: Biochemistry 36 (1997) 10857–10866.
97B4    Barbar, E., LiCata, V.J., Barany, G., Woodward, C.: Biophys. Chem. 64 (1997)
        45–57.
97B5    Beaucamp, N., Hofmann, A., Kellerer, B., Jaenicke, R.: Protein Sci. 6 (1997)
        2159–2165.
97B6    Behravan, G., Lycksell, P.-O., Larsson, G.: Protein Engng. 10 (1997) 1327–1331.
97B7    Bertini, I., Cowan, J.A., Luchinat, C., Natarajan, K., Piccioli, M.: Biochemistry 36
        (1997) 9332–9339.
97B8    Betz, S.F., Liebman, P.A., DeGrado, W.F.: Biochemistry 36 (1997) 2450–2458.
97B9    Bhat, M.G., Ganley, L.M., Ledman, D.W., Goodman, M.A., Fox, R.O.:
        Biochemistry 36 (1997) 12167–12174.
97B10   Boissinot, M., Karnas, S., Lepock, J.R., Cabelli, D.E., Tainer, J.A., Getzoff, E.D.,
        Hallewell, R.A.: EMBO J. 16 (1997) 2171–2178.
97B11   Boudker, O., Todd, M.J., Freire, E.: J. Mol. Biol. 272 (1997) 770–779.
97B12   Browne, J.P., Strom, M., Martin, S.R., Bayley, P.M.: Biochemistry 36 (1997)
        9550–9561.
97B13   Burdzhanadze, T.V., Metreveli, N.O., Mdzinarashvili, T.D., Mrevlishvili, G.M.:
        Biofizika 42 (1997) 75–77.
97C1    Calabrese, L., Musci, G.: in: Multi-Copper Oxidases, ed. by A. Messerschmidt,
        World Scientific Publ. Co., Singapore, New Jersey, London, Hong Kong, 1997,
        pp. 307–354.
97C2    Carra, J.H., Privalov, P.L.: Biochemistry 36 (1997) 526–535.
97C3    Catanzano, F., Gambuti, A., Graziano, G., Barone, G.: J. Biochem. 121 (1997)
        568–577.
97C4    Catanzano, F., Graziano, G., Cafaro, V., D'Alessio, G., Di Donato, A., Barone, G.:
        Biochemistry 36 (1997) 14403–14408.
97C5    Catanzano, F., Graziano, G., Capasso, S., Barone, G.: Protein Sci. 6 (1997)
        1682–1693.
97C6    Chalikian, T.V., Völker, J., Anafi, D., Breslauer, K.J.: J. Mol. Biol. 274 (1997)
        237–252.
97C7    Chan, C.-K., Hu, Y., Takahashi, S., Rousseau, D.L., Eaton, W.A., Hofrichter, J.:
        Proc. Natl. Acad. Sci. USA 94 (1997) 1779–1784.
97C8    Chang, Y., Zajicek, J., Castellino, F.J.: Biochemistry 36 (1997) 7652–7663.
97C9    Clarke, J., Hamill, S.J., Johnson, C.M.: J. Mol. Biol. 270 (1997) 771–778.

97C10  Colón, W., Wakem, L.P., Sherman, F., Roder, H.: Biochemistry 36 (1997) 12535–12541.

97D1   DeKoster, G.T., Robertson, A.D.: Biochemistry 36 (1997) 2323–2331.

97D2   DeKoster, G.T., Robertson, A.D.: Biophys. Chem. 64 (1997) 59–68.

97D3   DeSilva, T.M., Harper, S.L., Kotula, L., Hensley, P., Curtis, P.J., Otvos Jr., L., Speicher, D.W.: Biochemistry 36 (1997) 3991–3997.

97E1   Eberstadt, M., Huang, B., Olejniczak, E.T., Fesik, S.W.: Nature Struct. Biol. 4 (1997) 983–985.

97E2   Eftink, M.R., Ramsay, G.D., Beavers, A.: Proteins: Structure, Function, and Genetics 28 (1997) 227–240.

97F1   Farruggia, B., Garcia, G., D'Angelo, C., Picó, G.: Int. J. Biol. Macromol. 20 (1997) 43–51.

97F2   Fetrow, J.S., Horner, S.R., Oehrl, W., Schaak, D.L., Boose, T.L., Burton, R.E.: Protein Sci. 6 (1997) 197–210.

97F3   Fink, A.L., Oberg, K.A., Seshadri, S.: Folding & Design 3 (1997) 19–25.

97F4   Forsyth, Robertson, A.D., personal communication, cited in: Arrington, C.B., Robertson, A.D.: Biochemistry 36 (1997) 8686–8691.

97F5   Frye, K.J., Royer, C.A.: Protein Sci. 6 (1997) 789–793.

97G1   Ganesh, C., Shah, A.N., Swaminathan, C.P., Surolia, A., Varadarajan, R.: Biochemistry 36 (1997) 5020–5028.

97G2   Gasset, M., Saiz, J.L., Laynez, J., Sanz, L., Gentzel, M., Töpfer-Petersen, E., Calvete, J.J.: Eur. J. Biochem. 250 (1997) 735–744.

97G3   Gast, K., Zirwer, D., Damaschun, H., Hahn, U., Müller-Frohne, M., Wirth, M., Damaschun, G.: FEBS Lett. 403 (1997) 245–248.

97G4   Gegg, C.V., Bowers, K.E., Matthews, C.R.: Protein Sci. 6 (1997) 1885–1892.

97G5   Gerday, C., Aittaleb, M., Arpigny, J.L., Baise, E., Chessa, J.-P.,Garsoux, G., Petrescu, I., Feller, G.: Biochim. Biophys. Acta 1342 (1997) 119–131.

97G6   Giancola, C., De Sena, C., Fessas, D., Graziano, G., Barone, G.: Int. J. Biol. Macromol. 20 (1997) 193–204.

97G7   Gloss, L.M., Matthews, C.R.: Biochemistry 36 (1997) 5612–5623.

97G8   Godbole, S., Bowler, B.E.: J. Mol. Biol. 268 (1997) 816–821.

97G9   Godbole, S., Dong, A., Garbin, K., Bowler, B.E.: Biochemistry 36 (1997) 119–126.

97G10  Goedken, E.R., Raschke, T.M., Marqusee, S.: Biochemistry 36 (1997) 7256–7263.

97G11  Grantcharova, V.P., Baker, D.: Biochemistry 36 (1997) 15685–15692.

97G12  Grimsley, J.K., Scholtz, J.M., Pace, C.N., Wild, J.R.: Biochemistry 36 (1997) 14366–14374.

97G13  Gu, H., Kim, D., Baker, D.: J. Mol. Biol. 274 (1997) 588–596.

97G14  Guddat, L.W., Bardwell, J.C.A., Glockshuber, R., Huber-Wunderlich, M., Zander, T., Martin, J.L.: Protein Sci. 6 (1997) 1893–1900.

97H1   Hamada, D., Goto, Y.: J. Mol. Biol. 269 (1997) 479–487.

97H2   Hennecke, J., Spleiss, C., Glockshuber, R.: J. Biol. Chem. 272 (1997) 189–195.

97H3   Herbst, R., Schäfer, U., Seckler, R.: J. Biol. Chem. 272 (1997) 7099–7105.

97H4   Herrmann, L.M., Bowler, B.: Protein Sci. 6 (1997) 657–665.

97H5   Hiller, R., Zhou, Z.H., Adams, M.W.W., Englander, S.W.: Proc. Natl. Acad. Sci. USA 94 (1997) 11329–11332.

97H6   Hirota, N., Mizuno, K., Goto, Y.: Protein Sci. 6 (1997) 416–421.

97H7   Hoshino, M., Yumoto, N., Yoshikawa, S., Goto, Y.: Protein Sci. 6 (1997) 1396–1404.

97H8   Huang, H., Yuan, C.-S., Borchardt, R.T.: Protein Sci. 6 (1997) 1482–1490.

97I1   Ibarra-Molero, B., Sanchez-Ruiz, J.M.: Biochemistry 36 (1997) 9616–9624.

97I2   Ikura, T., Tsurupa, G.P., Kuwajima, K.: Biochemistry 36 (1997) 6529–6538; and corrections: Biochemistry 36 (1997) 11050.

490    References

97I3    Ionescu, R.M., Eftink, M.R.: Biochemistry 36 (1997) 1129–1140.
97I4    Itzhaki, L.S., Neira, J.L., Fersht, A.R.: J. Mol. Biol. 270 (1997) 89–98.
97J1    Jacobi, A., Huber-Wunderlich, M., Hennecke, J., Glockshuber, R.: J. Biol. Chem.
        272 (1997) 21692–21699.
97J2    Jain, S., Ahluwalia, J.C.: Thermochim. Acta 302 (1997) 17–24.
97J3    Jana, R., Hazbun, T.R., Mollah, A.K.M.M., Mossing, M.C.: J. Mol. Biol. 273
        (1997) 402–416.
97J4    Johnson, C.M., Oliveberg, M., Clarke, J., Fersht, A.R.: J. Mol. Biol. 268 (1997)
        198–208.
97J5    Juminaga, D., Wedemeyer, W.J., Garduño-Júarez, R., McDonald, M.A., Scheraga,
        H.A.: Biochemistry 36 (1997) 10131–10145.
97K1    Kaplan, W., Hüsler, P., Klump, H., Erhardt, J., Sluis-Cremer, N., Dirr, H.: Protein
        Sci. 6 (1997) 399–406.
97K2    Karpenko, V., Kaupová, M., Kodícek, M.: Biophys. Chem. 69 (1997) 209–217.
97K3    Kawamura, S., Tanaka, I., Yamasaki, N., Kimura, M.: J. Biochem. 121 (1997)
        448–455.
97K4    Kiefhaber, T., Bachmann, A., Wildegger, G., Wagner, C.: Biochemistry 36 (1997)
        5108–5112.
97K5    Kim, K., Ramanathan, R., Frieden, C.: Protein Sci. 6 (1997) 364–372.
97K6    Kobayashi, C., Suga, Y., Yamamoto, K., Yoma, T., Ogasahara, K., Yutani, K.,
        Urabe, I.: J. Biol. Chem. 272 (1997) 23011–23016.
97K7    Koh, J.T., Cornish, V.W., Schultz, P.G.: Biochemistry 36 (1997) 11314–11322.
97K8    Kohn, W.D., Kay, C.M., Hodges, R.S.: J. Mol. Biol. 267 (1997) 1039–1052.
97K9    Konno, T., Kamatari, Y.O., Kataoka, M., Akasaka, K.: Protein Sci. 6 (1997)
        2242–2249.
97K10   Kornilaev, B.A., Kurganov, B.I., Eronina, T.B., Chebotareva, N.A., Livanova,
        N.B., Orlov, V.N., Chernyak, V.Ya.: Molekularnaya Biol. 31 (1997) 98–107.
97K11   Kornilaev, B.A., Kurganov, B.I., Livanova, N.B., Eronina, T.B., Orlov, V.N.,
        Chernyak, V.Ya., Poglazov, B.F.: Doklady Akad. Nauk 352 (1997) 256–258.
97K12   Kurganov, B.I., Lyubarev, A.E., Sanchez-Ruiz, J.M., Shnyrov, V.L.: Biophys.
        Chem. 69 (1997) 125–135.
97L1    Laity, J.H., Lester, C.C., Shimotakahara, S., Zimmermann, D.E., Montelione, G.,
        Scheraga, H.A.: Biochemistry 36 (1997) 12683–12699.
97L2    Lamotte-Guéry, F. de, Pruvost, C., Minard, P., Delsuc, M.-A., Miginiac-Maslow,
        M., Schmitter, J.-M., Stein, M., Decottignies, P.: Protein Engng. 10 (1997)
        1425–1432.
97L3    Laurents, D.V., Baldwin, R.L.: Biochemistry 36 (1997) 1496–1504.
97L4    Lazar, G.A., Desjarlais, J.R., Handel, T.M.: Protein Sci. 6 (1997) 1167–1178.
97L5    Leckner, J., Bonander, N., Wittung-Stafshede, P., Malmström, B.G., Karlsson,
        B.G.: Biochim. Biophys. Acta 1342 (1997) 19–27.
97L6    Leckner, J., Wittung, P., Bonander, N., Karlsson, B.G., Malmström, B.G.: J. Biol.
        Inorg. Chem. 2 (1997) 368–371.
97L7    Ledent, P., Duez, C., Vanhove, M., Lejeune, A., Fonzé, E., Charlier, P., Rhazi-
        Filali, F., Thamm, I., Guillaume, G., Samyn, B., Devreese, B., Van Beeunem, J.,
        Lamotte-Brasseur, J., Frère, J.-M.: FEBS Lett. 413 (1997) 194–196.
97L8    Li, X., Lopez-Guisa, J.M., Ninan, N., Weiner, E.J., Rauscher III, F.J.,
        Marmorstein, R.: J. Biol. Chem. 272 (1997) 27324–27329.
97L9    Li, Y., Reilly, P.J., Ford, C.: Protein Engng. 10 (1997) 1199–1204.
97L10   Lima, L.M.T.R., de Prat-Gay, G.: J. Biol. Chem. 272 (1997) 19295–19303.
97L11   Liu, Y., Breslauer, K., Anderson, S.: Biochemistry 36 (1997) 5323–5335.
97L12   López-Hernández, E., Cronet, P., Serrano, L., Muñoz, V.: J. Mol. Biol. 266 (1997)
        610–620.
97L13   Lu, J., Hall, K.B.: Biophys. Chem. 64 (1997) 111–119.

97L14    Luo, Y., Kay, M.S., Baldwin, R.L.: Nature Struct. Biol. 4 (1997) 925–930.

97M1    Magdaleno, L., Gasset, M., Varea, J., Schambony, A.M., Urbanke, C., Raida, M., Töpfer-Petersen, E., Calvete, J.J.: FEBS Lett. 420 (1997) 179–185.

97M2    Malecki, J., Wasylewski, Z.: Eur. J. Biochem. 243 (1997) 660–669.

97M3    Matsubara, K., Ando, Y., Irie, T., Uekama, K.: Pharmaceutical Res. 14 (1997) 1401–1405.

97M4    Maurus, R., Overall, C.M., Bogumil, R., Luo, Y., Mauk, A.G., Smith, M., Brayer, G.D.: Biochim. Biophys. Acta 1341 (1997) 1–13.

97M5    Mayr, E.-M., Jaenicke, R., Glockshuber, R.: J. Mol. Biol. 269 (1997) 260–269.

97M6    Mei, G., Di Venere, A., Buganza, M., Vecchini, P., Rosato, N., Finazzi-Agro, A.: Biochemistry 36 (1997) 10917–10922.

97M7    Mollah, A.K.M.M., Mossing, M.C.: unpublished results, cited in Ref. 97J3: Jana, R., Hazbun, T.R., Mollah, A.K.M.M., Mossing, M.C.: J. Mol. Biol. 273 (1997) 402–416.

97M8    Mombelli, E., Afshar, M., Fusi, P., Mariani, M., Tortora, P., Connelly, J.P., Lange, R.: Biochemistry 36 (1997) 8733–8742.

97M9    Motoshima, H., Mine, S., Masumoto, K., Abe, Y., Iwashita, H., Hashimoto, Y., Chijiiwa, Y., Ueda, T., Imoto, T.: J. Biochem. 121 (1997) 1076–1081.

97M10    Motoshima, H., Ueda, T., Masumoto, K., Hashimoto, Y., Chijiiwa, Y., Imoto, T.: J. Biochem. 122 (1997) 25–31.

97M11    Mrevlishvili, G.M., Metreveli, N.O., Mdzinarashvili, T.D.: Biofizika 42 (1997) 78–81.

97M12    Mullins, L.S., Pace, C.N., Raushel, F.M.: Protein Sci. 6 (1997) 1387–1395.

97M13    Myers, J.K., Pace, C.N., Scholtz, J.M.: Biochemistry 36 (1997) 10923–10929.

97N1    Nagi, A.D., Regan, L.: Folding & Design 2 (1997) 67–75.

97N2    Nakaya, M., Kakinuma, M., Watabe, S.: Biochemistry 36 (1997) 9179–9184.

97N3    Narhi, L.O., Aoki, K.H., Philo, J.S., Arakawa, T.: J. Protein Chem. 16 (1997) 213–225.

97N4    Nath, U., Udgaonkar, J.B.: Biochemistry 36 (1997) 8602–8610.

97N5    Neira, J.L., Itzhaki, L.S., Otzen, D.E., Davis, B., Fersht, A.R.: J. Mol. Biol. 270 (1997) 99–110.

97N6    Nichols, J.C., Matthews, K.S.: J. Biol. Chem. 272 (1997) 18550–18557.

97N7    Nieba, L., Honegger, A., Krebber, C., Plückthun, A.: Protein Engng. 10 (1997) 435–444.

97N8    Nikolova, P.V., Creagh, A.L., Duff, S.J.B., Haynes, C.A.: Biochemistry 36 (1997) 1381–1388.

97N9    Nilsen, S.L., DeFord, M.E., Prorok, M., Chibber, B.A.K., Bretthauer, R.K., Castellino, F.J.: Biotechnol. Appl. Biochem. 25 (1997) 63–74.

97N10    Nölting, B., Golbik, R., Neira, J.L., Soler-Gonzalez, A.S., Schreiber, G., Fersht, A.R.: Proc. Natl. Acad. Sci. USA 94 (1997) 826–830.

97N11    Nölting, B., Golbik, R., Soler-González, A.S., Fersht, A.R.: Biochemistry 36 (1997) 9899–9905.

97N12    Nosworthy, N.J., Ginsburg, A.: Protein Sci. 6 (1997) 2617–2623.

97N13    Novokhatny, V., Ingham, K.: Protein Sci. 6 (1997) 141–146.

97O1    O'Brien, R., Wynn, R., Driscoll, P.C., Davis, B., Plaxco, K.W., Sturtevant, J.M., Ladbury, J.E.: Protein Sci. 6 (1997) 1325–1332.

97O2    Ogasahara, K., Yutani, K.: Biochemistry 36 (1997) 932–940.

97O3    Ohage, E.C., Graml, W., Walter, M.M., Steinbacher, S., Steipe, B.: Protein Sci. 6 (1997) 233–241.

97O4    Ohmura, T., Ueda, T., Motoshima, H., Tamura, T., Imoto, T.: J. Biochem. 122 (1997) 512–517.

97P1    Pace, C.N., personal communication, cited in: L.S. Mullins et al., Protein Sci. 6 (1997) 1387–1395.

97P2    Pace, C.N., Scholtz, J.M.: in: Protein Structure: a practical approach, ed. by T.E. Creighton, 2nd. ed., Oxford University Press, New York, 1997, pp. 299–321.

97P3    Padmanabhan, S., Jiménez, M.A., González, C., Sanz, J.M., Gimémez- Gallego, G., Rico, M.: Biochemistry 36 (1997) 6424–6434.

97P4    Park, S.-H., O'Neil, K.T., Roder, H.: Biochemistry 36 (1997) 14277–14283.

97P5    Parker, M.J., Clarke, A.R.: Biochemistry 36 (1997) 5786–5794.

97P6    Parker, M.J., Dempsey, C.E., Lorch, M., Clarke, A.R.: Biochemistry 36 (1997) 13396–13405.

97P7    Perrett, S., Zahn, R., Stenberg, G., Fersht, A.R.: J. Mol. Biol. 269 (1997) 892–901.

97P8    Peters, K., Hinz, H.-J., Cesareni, G.: Biol. Chem. 378 (1997) 1141–1152.

97P9    Pfeil, W., Gesierich, U., Kleemann, G.R., Sterner, R.: J. Mol. Biol. 272 (1997) 591–596.

97P10   Picó, G.: Int. J. Biol. Macromol. 20 (1997) 63–73.

97P11   Pierce, M.M., Nall, B.T.: Protein Sci. 6 (1997) 618–627.

97P12   Plaxco, K.W., Spitzfaden, C., Campbell, I.D., Dobson, C.M.: J. Mol. Biol. 270 (1997) 763–770.

97P13   Poklar, N., Lah, J., Salobir, M., Macek, P., Vesnaver, G.: Biochemistry 36 (1997) 14345–14352.

97P14   Prieto, J., Wilmans, M., Jiménez, M.A., Rico, M., Serrano, L.: J. Mol. Biol. 268 (1997) 760–778.

97P15   Protasevich, I., Ranjbar, B., Lobachov, V., Makarov, A., Gilli, R., Briand, C., Lafitte, D., Haiech, J.: Biochemistry 36 (1997) 2017–2024.

97P16   Protasevich, I.I., Ranjbar, B., Varlamova, E.Yu., Cherkasov, I.A., Lapuk, V.A.: Biochemistry (Moscow) 62 (1997) 914–918.

97Q1    Qu, K., Vaughn, J.L., Sienkiewicz, A., Scholes, C.P., Fetrow, J.S.: Biochemistry 36 (1997) 2884–2897.

97R1    Randzhbar, B., Protasevich, I.I., Shul'ga, A.A., Kurbanov, F.T., Lobachev, V.M., Kirpichnikov, M.P., Makarov, A.A.: Molekularnaya Biol. 31 (1997) 492–499.

97R2    Raschke, T.M., Marqusee, S.: Nature Struct. Biol. 4 (1997) 298–304.

97R3    Riddle, D.S., Santiago, J.V., Bray–Hall, S.T., Doshi, N., Grantcharova, V.P., Yi, Q., Baker, D.: Nature Struct. Biol. 4 (1997) 805–809.

97R4    Robinson, C.R., Liu, Y., Thomson, J.A., Sturtevant, J.M., Sligar, S.G.: Biochemistry 36 (1997) 16141–16146.

97R5    Robinson, C.R., Rentzeperis, D., Silva, J.L., Sauer, R.T.: J. Mol. Biol. 273 (1997) 692–700.

97R6    Rogers, D.P., Brouilette, C.G., Engler, J.A., Tendian, S.W., Roberts, L., Mishra, V.K., Anantharamaiah, G.M., Lund-Katz, S., Phillips, M.C., Ray, M.J.: Biochemistry 36 (1997) 288–300.

97R7    Rojo-Dominguez, A., Hernández-Arana, A., Mendoza-Hernández, G., Rendón, J.L.: Biochem. Mol. Biol. Int. 42 (1997) 631–639.

97R8    Ruvinov, S., Wang, L., Ruan, B., Almog, O., Gilliland, G.L., Eisenstein, E., Bryan, P.N.: Biochemistry 36 (1997)10414–10421.

97S1    Sakaguchi, K., Sakamoto, H., Lewis, M.S., Anderson, C.W., Erickson, J.W., Appella, E., Xie, D.: Biochemistry 36 (1997) 10117–10124.

97S2    Scalley, M.L., Baker, D.: Proc. Natl. Acad. Sci. USA 94 (1997) 10636–10640.

97S3    Scalley, M.L., Yi, Q., Gu, H., McCormack, A., Yates III, J.R., Baker, D.: Biochemistry 36 (1997) 3373–3382.

97S4    Schafmeister, C.E., LaPorte, S.L., Miercke, L.J.W., Stroud, R.M.: Nature Struct. Biol. 4 (1997) 1039–1046.

97S5    Schönbrunner, N., Koller, K.-P., Kiefhaber, T.: J. Mol. Biol. 268 (1997) 526–538.

97S6    Schöppe, A., Hinz, H.-J., Agashe, V.R., Ramachandran, S., Udgaonkar, J.B.: Protein Sci. 6 (1997) 2196–2202.

97S7    Schreiber, G., Frisch, C., Fersht, A.R.: J. Mol. Biol. 270 (1997) 111–122.

97S8    Shao, X., Hensley, P., Matthews, C.R.: Biochemistry 36 (1997) 9941–9949.

97S9    Sherman, M.A., Chen, Y., Mas, M.T.: Protein Sci. 6 (1997) 882–891.

97S10   Shimotakahara, S., Rios, C.B., Laity, J.H., Zimmerman, D.E., Scheraga, H.A., Montelione, G.T.: Biochemistry 36 (1997) 6915–6929.

97S11   Shnyrov, V.L., Zhadan, G.G., Cobaleda, C., Sagrera, A., Muñoz–Barroso, I., Villar, E.: Arch. Biochem. Biophys. 341 (1997) 89–97.

97S12   Silow, M., Oliveberg, M.: Proc. Natl. Acad. Sci. USA 94 (1997) 6084–6086.

97S13   Soler-González, A.S., Fersht, A.R.: Eur. J. Biochem. 249 (1997) 724–732.

97S14   Spitzfaden, C., Grant, R.P., Mardon, H.J., Campbell, I.D.: J. Mol. Biol. 265 (1997) 565–579.

97S15   Surin, A.K., Kotova, N.V., Kashparov, I.A., Marchenkov, V.V., Marchenkova, S.Yu., Semisotnov, G.V.: FEBS Lett. 405 (1997) 260–262

97S16   Swietnicki, W., Petersen, R., Gambetti, P., Surewicz, W.K.: J. Biol. Chem. 272 (1997) 27517–27520.

97S17   Szilák, L., Moitra, J., Vinson, C.: Protein Sci. 6 (1997) 1273–1283.

97T1    Takano, K., Funahashi, J., Yamagata, Y., Fujii, S., Yutani, K.: J. Mol. Biol. 274 (1997) 132–142.

97T2    Takano, K., Yamagata, Y., Fujii, S., Yutani, K.: Biochemistry 36 (1997) 688–698.

97T3    Tan, Y.-J., Oliveberg, M., Otzen, D.E., Fersht, A.R.: J. Mol. Biol. 269 (1997) 611–622.

97T4    Todorova, R.: Int. J. Biochem. Cell Biol. 29 (1997) 841–848.

97T5    Tsereteli, G.I., Belopol'skaya, T.V., Mel'nik, T.N.: Biofizika 42 (1997) 68–74.

97U1    Uhlmann, H., Iametti, S., Vecchio, G., Bonomi, F., Bernhardt, R.: Eur. J. Biochem. 248 (1997) 897–902.

97V1    Valladares, M.H., Felici, A., Weber, G., Adolph, H.W., Zeppezauer, M., Rossolini, G.M., Amicosante, G., Frère, J.-M., Galleni, M.: Biochemistry 36 (1997) 11534–11541.

97V2    Vanhove, M., Guillaume, G., Ledent, P., Richards, J.H., Pain, R.H., Frère, J.-M.: Biochem. J. 321 (1997) 413–417.

97V3    Viguera, A.R., Villegas, V., Avilés, F.X., Serrano, L.: Folding & Design 2 (1996/97) 23–33.

97V4    Villaverde, J., Cladera, J., Padrós, E., Rigaud, J.-L., Duñach, M.: Eur. J. Biochem. 24 (1997) 441–448.

97V5    Vogl, T., Jatzke, C., Hinz, H.-J., Benz, J., Huber, R.: Biochemistry 36 (1997) 1657–1668.

97W1    Walkenhorst, W.F., Green, S.M., Roder, H.: Biochemistry 36 (1997) 5795–5805.

97W2    Wang, A., Bolen, D.W.: Biochemistry 36 (1997) 9101–9108.

97W3    Welfle, K., Misselwitz, R., Schaup, A., Gerlach, D., Welfle, H.: Proteins: Structure, Function, and Genetics 27 (1997) 26–35.

97W4    Westh, P., Koga, Y.: J. Phys. Chem. B 101 (1997) 5755–5758.

97W5    Wildegger, G., Kiefhaber, T.: J. Mol. Biol. 270 (1997) 294–304.

97W6    Wittung-Stafshede, P., Gray, H.B., Winkler, J.R.: J. Amer. Chem. Soc. 119 (1997) 9562–9563.

97W7    Wong, C.-Y., Eftink, M.R.: Protein Sci. 6 (1997) 689–697.

97X1    Xie, G., Timasheff, S.N.: Biophys. Chem. 64 (1997) 25–43.

97X2    Xie, G., Timasheff, S.N.: Protein Sci. 6 (1997) 211–221.

97X3    Xie, G., Timasheff, S.N.: Protein Sci. 6 (1997) 222–232.

97Y1    Yang, J., Spek, E.J., Gong, Y., Zhou, H., Kallenbach, N.R.: Protein Sci. 6 (1997) 1264–1272.

97Y2    Yao, P., Xie, Y., Wang, Y.-H., Sun, Y.-L., Huang, Z.-X., Xiao, G.T., Wang, S.-D.: Protein Engng. 10 (1997) 575–581.

97Y3    Yi, Q., Scalley, M.L., Simons, K.T., Gladwin, S.T., Baker, D.: Folding & Design 2 (1997) 271–280.

97Z1    Zaidi, F.N., Nath, U., Udgaonkar, J.B.: Nature Struct. Biol. 4 (1997) 1016–1024.

97Z2    Zarnt, T., Tradler, T., Stoller, G., Scholz, C., Schmid, F.X., Fischer, G.: J. Mol. Biol. 271 (1997) 827–837.

97Z3    Zavodnik, I.B., Lapshina, E.A.: Biofizika 42 (1997) 1035–1039.

97Z4    Zhang, H., Stöckel, J., Melhorn, I., Groth, D., Baldwin, M.A., Prusiner, S.B., James, T.L., Cohen, F.E.: Biochemistry 36 (1997) 3543–3543.

97Z5    Zhang, J.-G., Matthews, J.M., Ward, L.D., Simpson, R.J.: Biochemistry 36 (1997) 2380–2389.

97Z6    Zhang, J.-X., Goldenberg, D.P.: Protein Sci. 6 (1997) 1563–1576.

97Z7    Zolkiewski, M., Redowicz, M.J., Korn, E.D., Hammer, III, J.A., Ginsburg, A.: Biochemistry 36 (1997) 7876–7883.

98A1    Adamek, D., Popovic, A., Blaber, S., Blaber, M.: in: Biocalorimetry: Applications of Calorimetry in the Biological Sciences, ed. by J.E. Ladbury and B.Z. Chowdhry, John Wiley & Sons, Chichester, New York, Weinheim, Brisbane, Singapore, Toronto, 1998, pp. 235– 241.

98A2    Ahmad, N., Srinivas, V.R., Reddy, G.B., Surolia, A.: Biochemistry 37 (1998) 16765–16772.

98A3    Allen, D.L., Pielak, G.J.: Protein Sci. 7 (1998) 1262–1263.

98A4    Apenten, R.K.O.: Int. J. Biol. Macromol. 23 (1998) 19–25.

98A5    Aphasizheva, I.Yu., Dolgikh, D.A., Abdullaev, Z.K., Uversky, V.N., Kirpichnikov, M.P., Ptitsyn, O.B.: FEBS Lett. 425 (1998) 101–104.

98A6    Arai, M., Ikura, T., Semisotnov, G.V., Kihara, H., Amemiya, Y., Kuwajima, K.: J. Mol. Biol. 275 (1998) 149–162.

98A7    Arnesano, F., Banci, L., Bertini, I., Koulougliotis, D.: Biochemistry 37 (1998) 17082–17092.

98B1    Backmann, J., Schäfer, G., Wyns, L., Bönisch, H.: J. Mol. Biol. 284 (1998) 817–833.

98B2    Bam, N., Cleland, J.L., Yang, J., Manning, M.C., Carpenter, J.F., Kelley, R.F., Randolph, T.W.: J. Pharm. Sci. 87 (1998) 1554–1559.

98B3    Baskakov, I., Bolen, D.W.: J. Biol. Chem. 273 (1998) 4831–4834.

98B4    Baskakov, I.V., Bolen, D.W.: Biochemistry 37 (1998) 18010–18017.

98B5    Battistoni, A., Folcarelli, S., Cervoni, L., Polizio, F., Desideri, A., Giartosio, A., Rotilio, G.: J. Biol. Chem. 273 (1998) 5655–5661.

98B6    Bhuyan, A.K., Udgaonkar, J.B.: Biochemistry 37 (1998) 9147–9155.

98B7    Bhuyan, A.K., Udgaonkar, J.B.: Proteins: Structure, Function, and Genetics 32 (1998) 241–247.

98B8    Bigotti, M.G., Allocatelli, C.T., Staniforth, R.A., Arese, M., Cutruzzolà, F., Brunori, M.: FEBS Lett. 425 (1998) 385–390.

98B9    Bryson, J.W., Desjarlais, J.R., Handel, T.M., DeGrado, W.F.: Protein Sci. 7 (1998) 1404–1414.

98B10   Buck, M.: Quart. Rev. Biophys. 31 (1998) 297–355.

98B11   Budisa, N., Huber, R., Golbik, R., Minks, C., Weyher, E., Moroder, L.: Eur. J. Biochem. 253 (1998) 1–9.

98B12   Burlacu-Miron, S., Perrier, V., Gilles, A.-M., Pistotnik, E., Craescu, C.T.: J. Biol. Chem. 273 (1998) 19102–19107.

98B13   Burns, L.L., Dalessio, P.M., Ropson, I.J.: Proteins: Structure, Function, and Genetics 33 (1998) 107–118.

98B14   Burova, T.V., Choiset, Y., Tran, V., Haertlé, T.: Protein Engng. 11 (1998) 1065-1073.

98C1    Catanzano, F., Graziano, G., Cafaro, V., D'Alessio, G., Di Donato, A., Barone, G.: Int. J. Biol. Macromol. 23 (1998) 277–285.

98C2    Catanzano, F., Graziano, G., Fusi, P., Tortora, P., Barone, G.: Biochemistry 37 (1998) 10493–10498.

98C3 Catanzano, F., Graziano, G., De Paola, B., Barone, G., D'Auria, S., Rossi, M., Nucci, R.: Biochemistry 37 (1998) 14484–14490.

98C4 Chamberlain, A.K., Marqusee, S.: Biochemistry 37 (1998) 1736–1742.

98C5 Chiti, F., Van Nuland, N.A.J., Taddei, N., Magherini, F., Stefani, M., Ramponi, G., Dobson, C.M.: Biochemistry 37 (1998) 1447–1455.

98C6 Chiti, F., Magherini, F., Taddei, N., Ilardi, C., Stefani, M., Bucciantini, M., Dobson, C.M., Ramponi, G.: Protein Engng. 11 (1998) 557–561.

98C7 Chiti, F., Taddei, N., van Nuland, N.A.J., Magherini, F., Stefani, M., Ramponi, G., Dobson, C.M.: J. Mol. Biol. 283 (1998) 893–903.

98C8 Choe, S.E., Matsudaira, P.T., Osterhout, J., Wagner, G., Shakhnovich, E.I.: Biochemistry 37 (1998) 14508–14518.

98C9 Chumanevich, A.A., Kravchuk, Z.I., Vlasov, A.P., Zhorov, O.V., Martsev, S.P.: Biokhimiya 63 (1998) 563–572.

98C10 Constans, A.J., Mayer, M.R., Sukits, S.F., Lecomte, J.T.J.: Protein Sci. 7 (1998) 1983–1993.

98C11 Crane–Robinson, C., Read, C.M., Cary, P.D., Driscoll, P.C., Dragan, A.I., Privalov, P.L.: J. Mol. Biol. 281 (1998) 705–717.

98C12 Creagh, A.L., Koska, J., Johnson, P.E., Tomme, P., Joshi, M.D., McIntosh, L.P., Kilburn, D.G., Haynes, C.A.: Biochemistry 37 (1998) 3529–3537.

98D1 Dalby, P.A., Clarke, J., Johnson, C.M., Fersht, A.R.: J. Mol. Biol. 276 (1998) 647–656.

98D2 Dalby, P.A., Oliveberg, M., Fersht, A.R.: J. Mol. Biol. 276 (1998) 625–646.

98D3 Dalessio, P.M., Ropson, I.J.: Arch. Biochem. Biophys. 359 (1998) 199–208.

98D4 Dall'Acqua, W., Simon, A.L., Mulkerrin, M.G., Carter, P.: Biochemistry 37 (1998) 9266–9273.

98D5 Das, B.K., Liang, J.J.–N.: Int. J. Biol. Macromol. 23 (1998) 191–197.

98D6 Das, T.K., Mazumdar, S., Mitra, S.: Eur. J. Biochem. 254 (1998) 662–670.

98D7 D'Auria, S., Rossi, M., Barone, G., Catanzano, F., Del Vecchio, P., Graziano, G., Nucci, R.: J. Biochem. 120 (1996) 292–300.

98D8 Davis-Searles, P.R., Morar, A.S., Saunders, A.J., Erie, D.A., Pielak, G.J.: Biochemistry 37 (1998) 17048–17053.

98D9 Davoodi, J., Wakarchuk, W.W., Surewicz, W., Carey, P.R.: Protein Sci. 7 (1998) 1538–1544.

98D10 De Filippis, V., De Antoni, F., Frigo, M., de Laureto, P.P., Fontana, A.: Biochemistry 37 (1998) 1686–1696.

98D11 Den Blaauwen, T., Driessen, A.J.M.: in: Biocalorimetry: Applications of Calorimetry in the Biological Sciences, ed. by J.E. Ladbury and B.Z. Chowdhry, John Wiley & Sons, Chichester, New York, Weinheim, Brisbane, Singapore, Toronto, 1998, pp. 253–265.

98D12 De Vos, S., Doumen, J., Langhorst, U., Steyaert, J.: J. Mol. Biol. 275 (1998) 651–661.

98E1 Eads, J.C., Mahoney, N.M., Vorobiev, S., Bresnick, A.R., Wen, K.-K., Rubenstein, P.A., Haarer, B.K., Almo, S.C.: Biochemistry 37 (1998) 11171–11181.

98F1 Fan, Y.-X., Zhou, J.-M., Kihara, H., Tsou, C.-L.: Protein Sci. 7 (1998) 2631–2641.

98F2 Foord, R.L., Leatherbarrow, R.J.: Biochemistry 37 (1998) 2969–2978.

98F3 Freskgård, P.-O., Petersen, L.C., Gabriel, D.A., Li, X., Persson, E.: Biochemistry 37 (1998) 7203–7212.

98F4 From, N.B., Bowler, B.E.: Biochemistry 37 (1998) 1623–1631.

98F5 Frye, K.J., Royer, C.A.: Protein Sci. 7 (1998) 2217–2222.

98F6 Fuentes, E.J., Wand, A.J.: Biochemistry 37 (1998) 3687–3698.

98F7 Fuentes, E.J., Wand, A.J.: Biochemistry 37 (1998) 9877–9883.

98G1    Gajiwala, K.: unpublished results, cited in Ref. 98H7: Hebert, E.J., Giletto, A.,
        Sevcik, J., Urbanikova, L., Wilson, K.S., Dauter, Z., Pace, C.N.: Biochemistry 37
        (1998) 16192–16200.

98G2    Gao, J., Yin, D.H., Yao, Y., Sun, H., Qin, Z., Schöneich, C., Williams, T.D.,
        Squier, T.C.: Biophys. J. 74 (1998) 1115–1134.

98G3    Garnier, C., Protasevich, I., Gilli, R., Tsvetkov, P., Lobachov, V., Peyrot, V.,
        Briand, C., Makarov, A.: Biochem. Biophys. Res. Commun. 249 (1998) 197–201.

98G4    Gast, K., Zirwer, D., Müller-Frohne, M., Damaschun, G.: Protein Sci. 7 (1998)
        2004–2011.

98G5    Georgescu, R.E., Li, J.-H., Goldberg, M.E., Tasayco, M.L., Chaffotte, A.F.:
        Biochemistry 37 (1998) 10286–10297.

98G6    Giver, L., Gershenson, A., Freskgard, P.-O., Arnold, F.H.: Proc. Natl. Acad. Sci.
        USA 95 (1998) 12809–12813.

98G7    Gloss, L.M., Matthews, C.R.: Biochemistry 37 (1998) 15990–15999.

98G8    Gloss, L.M., Matthews, C.R.: Biochemistry 37 (1998) 16000–16010.

98G9    Goder, V., Beckert, V., Pfeil, W., Bernhardt, R.: Arch. Biochem. Biophys. 359
        (1998) 31–41.

98G10   Goedken, E.R., Marqusee, S.: Proteins: Structure, Function, and Genetics 33
        (1998) 135–143.

98G11   Golbik, R., Zahn, R., Harding, S.E., Fersht, A.R.: J. Mol. Biol. 276 (1998)
        505–515.

98G12   Gomez-Orellana, I., Variano, B., Miura-Fraboni, J., Milstein, S., Paton, D.R.:
        Protein Sci. 7 (1998) 1352–1358.

98G13   Grättinger, M., Dankesreiter, A., Schurig, H., Jaenicke, R.: J. Mol. Biol. 280
        (1998) 525–533.

98G14   Grantcharova, V.P., Riddle, D.S., Santiago, J.V., Baker, D.: Nature Struct. Biol. 5
        (1998) 714–720.

98G15   Grinberg, A., Bernhardt, R.: Biochem. Biophys. Res. Commun. 249 (1998)
        933–937.

98G16   Grinberg, A.V., Bernhardt, R.: Protein Engng. 11 (1998) 1057–1064.

98G17   Grinberg, V.Ya., Grinberg, N.V., Burova, T.V., Dalgalarrondo, M., Haertlé, T.:
        Biopolymers 46 (1998) 253–265.

98G18   Gu, H., Kim, D., Baker, D.: J. Mol. Biol. 274 (1998) 588–596.

98G19   Guan, Y., Hickey, M.J., Borgstahl, G.E.O., Hallewell, R.A., Lepock, J.R.,
        O'Connor, D., Hsieh, Y., Nick, H.S., Silverman, D.N., Tainer, J.A.: Biochemistry
        37 (1998) 4722–4730.

98G20   Guijarro, J.I., Morton, C.J., Plaxco, K.W., Campbell, I.D., Dobson, C.M.: J. Mol.
        Biol. 276 (1998) 657–667.

98G21   Gursky, O., Atkinson, D.: Biochemistry 37 (1998) 1283–1291.

98H1    Haezebrouck, P., Noyelle, K., Van Dael, H.: Biochemistry 37 (1998) 6772–6780.

98H2    Hamburger, J.B., Chen, E., Narhi, L.O., Wu, G.-M., Brems, D.N.: Proteins:
        Structure, Function, and Genetics 32 (1998) 495–503.

98H3    Hamill, S.J., Meekhof, A.E., Clarke, J.: Biochemistry 37 (1998) 8071–8079.

98H4    Hammack, B., Attfield, K., Clayton, D., Dec, E., Dong, A., Sarisky, C., Bowler,
        B.E.: Protein Sci. 7 (1998) 1789–1795.

98H5    Hansen, P.E., Zhang, W., Lauritzen, C., Bjørn, S., Petersen, L.C., Norris, K.,
        Olsen, O.H., Betzel, C.: Biochemistry 37 (1998) 3645–3653.

98H6    Hawrot, E., Xiao, Y., Shi, Q., Norman, D., Kirkitadze, M., Barlow, P.N.: FEBS
        Lett. 432 (1998) 103–108.

98H7    Hebert, E.J., Giletto, A., Sevcik, J., Urbanikova, L., Wilson, K.S., Dauter, Z.,
        Pace, C.N.: Biochemistry 37 (1998) 16192–16200.

98H8    Hillier, B.J., Rodriguez, H.M., Gregoret, L.M.: Folding & Design 3 (1998) 87–93.

98H9    Hirano, N., Haruki, M., Morikawa, M., Kanaya, S.: Biochemistry 37 (1998) 12640–12648.

98H10   Hsieh, Y., Guan, Y., Chingkuang, T., Bratt, P.J., Angerhofer, A., Lepock, J.R., Hickey, M.J., Tainer, J.A., Nick, H.S., Silverman, D.N.: Biochemistry 37 (1998) 4731–4739.

98H11   Huang, S.-M., Chou, W.-Y., Lin, S.-I., Chang, G.-G.: Proteins: Structure, Function, and Genetics 31 (1998) 61–73.

98H12   Hung, H.-C., Chang, G.-G.: Proteins: Structure, Function, and Genetics 33 (1998) 49–61.

98I1    Iametti, S., Uhlmann, H., Ragg, E., Sala, N., Grinberg, A., Beckert, V., Bernhardt, R., Bonomi, F.: Eur. J. Biochem. 251 (1998) 673–681.

98I2    Ikeguchi, M., Fujino, M., Kato, M., Kuwajima, K., Sugai, S.: Protein Sci. 7 (1998) 1564–1574.

98I3    Ishikawa, N., Chiba, T., Chen, L.T., Shimizu, A., Ikeguchi, M., Sugai, S.: Protein Engng. 11 (1998) 333–335.

98I4    Iwakura, M., Nakamura, T.: Protein Engng. 11 (1998) 707–713.

98I5    Iwakura, M., Takenawa, T., Nakamura, T.: J. Biochem. 124 (1998) 769–777.

98J1    Johansson, J.S., Gibney, B.R., Rabanal, F., Reddy, K.S., Dutton, P.L.: Biochemistry 37 (1998) 1421–1429.

98J2    Johansson, J.S., Gibney, B.R., Skalicky, J.J., Wand, A.J., Dutton, P.L.: J. Amer. Chem. Soc. 120 (1998) 3881–3886.

98J3    Julenius, K., Thulin, E., Linse, S., Finn, B.E.: Biochemistry 37 (1998) 8915–8925.

98J4    Juminaga, D., Wedemeyer, W.J., Scheraga, H.A.: Biochemistry 37 (1998) 11614–11620.

98K1    Kakinuma, M., Nakaya, M., Hatanaka, A., Hirayama, Y., Watabe, S., Maeda, K., Ooi, T., Suzuki, S.: Biochemistry 37 (1998) 6606–6613.

98K2    Kasimova, M.R., Milstein, S.J., Freire, E.: J. Mol. Biol. 277 (1998) 409–418.

98K3    Kaushik, J.K., Bhat, R.: J. Phys. Chem. B 102 (1998) 7058–7066.

98K4    Kawamura, S., Abe, Y., Ueda, T., Masumoto, K., Imoto, T., Yamasaki, N., Kimura, M.: J. Biol. Chem. 273 (1998) 19982–19987.

98K5    Kay, M.S., Baldwin, R.L.: Biochemistry 37 (1998) 7859–7868.

98K6    Kern, G., Handel, T., Marqusee, S.: Protein Sci. 7 (1998) 2164–2174.

98K7    Kikuchi, M., Kawano, K., Nitta, K.: Protein Sci. 7 (1998) 2150–2155.

98K8    Kim, K., Frieden, C.: Protein Sci. 7 (1998) 1821–1828.

98K9    Kim, D.E., Yi, Q., Gladwin, S.T., Goldberg, J.M., Baker, D.: J. Mol. Biol. 284 (1998) 807–815.

98K10   Knapp, S., Mattson, P.T., Christova, P., Berndt, K.D., Karshikoff, A., Vihinen, M., Smith, C.I.E., Ladenstein, R.: Proteins: Structure, Function, and Genetics 31 (1998) 309–319.

98K11   Kohn, W.D., Kay, C.M., Hodges, R.S.: J. Mol. Biol. 283 (1998) 993–1012.

98K12   Kortemme, T., Ramirez-Alvarado, M., Serrano, L.: Science 281 (1998) 253–256.

98K13   Koshiba, T., Tsumoto, K., Masaki, K., Kawano, K., Nitta,. K., Kumagai, I.: Protein Engng. 11 (1998) 683–690.

98K14   Kragelund, B.B., Heinemann, B., Knudsen, J., Poulsen, F.M.: Protein Sci. 7 (1998) 2237–2248.

98K15   Kravchuk, Z.I., Chumanevich, A.A., Vlasow, A.P., Martsev, S.P.: J. Immunological Methods 217 (1998) 131–141.

98K16   Kuhlman, B., Boice, J.A., Fairman, R., Raleigh, D.P.: Biochemistry 37 (1998) 1025–1032.

98K17   Kuhlman, B., Luisi, D.L., Evans, P.A., Raleigh, D.P.: J. Mol. Biol. 284 (1998) 1661–1670.

98K18   Kuhlman, B., Raleigh, D.P.: Protein Sci. 7 (1998) 2405–2412.

98K19   Kuroki, R., Yutani, K.: J. Biol. Chem. 273 (1998) 34310–34315.

98K20   Kwok, S.C., Tripet, B., Man, J.H., Chana, M.S., Lavigne, P., Mant, C.T., Hodges, R.S.: Biopolymers 47 (1998) 101–123.
98L1    Lassalle, M.W., Hinz, H.-J.: Biochemistry 37 (1998) 8465–8472.
98L2    Lassalle, M.W., Hinz, H.-J., Wenzel, H., Vlassi, M., Kokkinidis, M., Cesareni, G.: J. Mol. Biol. 279 (1998) 987–1000.
98L3    Levitsky, D.I., Ponomarev, M.A., Geeves, M.A., Shnyrov, V.L., Manstein, D.J.: Eur. J. Biochem. 251 (1998) 275–280.
98L4    Li, W., Grayling, R.A., Sandman, K., Edmondson, S., Shriver, J.W., Reeve, J.N.: Biochemistry 37 (1998) 10563–10572.
98L5    Lipscomb, L.A., Gassner, N.C., Snow, S.D., Eldridge, A.M., Baase, W.A., Drew, D.L., Matthews, B.W.: Protein Sci. 7 (1998) 765–773.
98L6    Llinás, M., Marqusee, S.: Protein Sci. 7 (1998) 96–104.
98L7    Lusitani, D., Menhart, N., Keiderling, T.A., Fung, L.W.-M.: Biochemistry 37 (1998) 16546–16554.
98L8    Lyubarev, A.E., Kurganov, B.I.: Biokhimiya 63 (1998) 516–523.
98L9    Lyubarev, A.E., Kurganov, B.I., Burlakova, A.A., Orlov, V.N.: Biophys. Chem. 70 (1998) 247–257.
98L10   Lyubarev, A.E., Kurganov, B.I., Burlakova, A.A., Orlov, V.N., Poglazov, B.F.: Doklady Akad. Nauk 358 (1998) 830–832.
98M1    Main, E.R.G., Fulton, K.F., Jackson, S.E.: Biochemistry 37 (1998) 6145–6153.
98M2    Makhatadze, G.I., Lopez, M.M., Richardson III, J.M., Thomas, S.T.: Protein Sci. 7 (1998) 689–697.
98M3    Malakauskas, S.M., Mayo, S.L.: Nature Struct. Biol. 5 (1998) 470–475.
98M4    Maldonado, S., Jiménez, M.Á., Langdon, G.M., Sancho, J.: Biochemistry 37 (1998) 10589–10596.
98M5    Mandelman, D., Schwarz, F.P., Li, H., Poulos, T.L.: Protein Sci. 7 (1998) 2089–2098.
98M6    Martinez, J.C., Pisabarro, M.T., Serrano, L.: Nature Struct. Biol. 5 (1998) 721–729.
98M7    Martsev, S.P., Kravchuk, Z.I., Chumanevich, A.A., Vlasov, A.P., Dubnovitsky, A.P., Bespalov, I.A., Arosio, P., Deyev, S.M.: FEBS Lett. 441 (1998) 458–462.
98M8    Martsev, S.P., Vlasov, A.P., Arosio, P.: Protein Engng. 11 (1998) 377–381.
98M9    Maxwell, K.L., Davidson, A.R.: Biochemistry 37 (1998) 16172–16182.
98M10   McCrary, B.S., Bedell, J., Edmondson, P., Shriver, J.W.: J. Mol. Biol. 276 (1998) 203–224.
98M11   McGee, W.A., Nall, B.T.: Protein Sci. 7 (1998) 1071–1082.
98M12   McLean, M.A., Maves, S.A., Weiss, K.E., Krepich, S., Sligar, S.G.: Biochem. Biophys. Res. Commun. 252 (1998) 166–172.
98M13   Meeker, A.K., Sack Jr., G.H.: Proteins: Structure, Function, and Genetics 31 (1998) 381–387.
98M14   Meijberg, W., Schuurman–Wolters, G.K., Boer, H., Scheek, R.M., Robillard, G.T.: J. Biol. Chem. 273 (1998) 20785–20794.
98M15   Merabet, E.K., Burz, D.S., Ackers, G.K.: Methods Enzymol. 295 (1998) 450–467.
98M16   Merkel, J.S., Regan, L.: Folding & Design 3 (1998) 449–455.
98M17   Milardi, D., La Rosa, C., Grasso, D., Guzzi, R., Sportelli, L., Fini, C.: Eur. Biophys. J. 27 (1998) 273–282.
98M18   Miles, C.A., Knott, L., Sumner, I.G., Bailey, A.J.: J. Mol. Biol. 277 (1998) 135–144.
98M19   Miller, S., Schuler, B., Seckler, R.: Biochemistry 37 (1998) 9160–9168.
98M20   Milne, J.S., Mayne, L., Roder, H., Wand, A.J., Englander, S.W.: Protein Sci. 7 (1998) 739–745.
98M21   Mizuguchi, M., Arai, M., Ke, Y., Nitta, K., Kuwajima, K.: J. Mol. Biol. 283 (1998) 265–277.

98M22 Mössner, E., Huber-Wunderlich, M., Glockshuber, R.: Protein Sci. 7 (1998) 1233–1244.

98M23 Marmorino, J.L., Lethi, M., Pielak, G.J.: J. Mol. Biol. 275 (1998) 379–388.

98N1 Nath, S., Satpathy, G.R., Mantri, R., Deep, S., Ahluwalia, J.C.: Thermochim. Acta 309 (1998) 193–196.

98N2 Nguyen, D.M., Schleif, R.F.: J. Mol. Biol. 282 (1998) 751–759.

98N3 Nosworthy, N.J., Peterkofsky, A., König, S., Seok, Y.-J., Szcepanowski, R.H., Ginsburg, A.: Biochemistry 37 (1998) 6718–6726.

98O1 Oakley, M.G., Kim, P.S.: Biochemistry 37 (1998) 12603–12610.

98O2 Ogasahara, K., Lapshina, E.A., Sakai, M., Izu, Y., Tsunasawa, S., Kato, I., Yutani, K.: Biochemistry 37 (1998) 5939–5946.

98O3 Ogasahara, K., Nakamura, M., Nakura, S., Tsunasawa, S., Kato, I., Yoshimoto, T., Yutani, K.: Biochemistry 37 (1998) 17537–17544.

98O4 Ohmae, E., Iriyama, K., Ichihara, S., Gekko, K.: J. Biochem. 123 (1998) 33–41.

98O5 Oliveberg, M., Tan, Y.-J., Silow, M., Fersht, A.R.: J. Mol. Biol. 277 (1998) 933–943.

98O6 Orlov, V.N., Kust, S.V., Kalmykov, P.V., Krivosheev, V.P., Dobrov, E.N., Drachev, V.A.: FEBS Lett. 433 (1998) 307–311.

98O7 Otzen, D.E., Fersht, A.R.: Biochemistry 37 (1998) 8139–8146.

98P1 Pace, C.N., Hebert, E.J., Shaw, K.L., Schell, D., Both, V., Krajcikova, D., Sevcik, J., Wilson, K.S., Dauter, Z., Hartley, R.W., Grimsley, G.R.: J. Mol. Biol. 279 (1998) 271–286.

98P2 Pace, C.N., Scholtz, J.M.: Biophys. J. 75 (1998) 422–427.

98P3 Panse, V.G., Udgoankar, J.B., Varadarajan, R.: Biochemistry 37 (1998) 14477–14483.

98P4 Parker, M.H., Hefford, A.A.: Biotechnol. Appl. Biochem. 28 (1998) 69–76.

98P5 Parker, M.J., Lorch, M., Sessions, R.B., Clarke, A.R.: Biochemistry 37 (1998) 2538–2545.

98P6 Pavlov, D.A., Sobieszek, A., Levitsky, D.I.: Biokhimiya 63 (1998) 1116–1128.

98P7 Perl, D., Welker, C., Schindler, T., Schröder, K., Marahiel, M.A., Jaenicke, R., Schmid, F.X.: Nature Struct. Biol. 5 (1998) 229–235.

98P8 Perrier, V., Burlacu-Miron, S., Bourgeois, S., Surewicz, W.K., Gilles, A.-M.: J. Biol. Chem. 273 (1998) 19097–19101.

98P9 Pfeil, W.: Proteins: Structure, Function, and Genetics 30 (1998) 43–48.

98P10 Plaxco, K.W., Baker, D.: Proc. Natl. Acad. Sci. USA 95 (1998) 13591–13596.

98P11 Plaxco, K.W., Guijarro, J.I., Morton, C.J., Pitkeathly, M., Campbell, I.D., Dobson, C.M.: Biochemistry 37 (1998) 2529–2537.

98P12 Popovic, A., Adamek, D.H., Blaber, S.I., Blaber, M.: in: Biocalorimetry: Applications of Calorimetry in the Biological Sciences, ed. by J.E. Ladbury and B.Z. Chowdhry, John Wiley & Sons, Chichester, New York, Weinheim, Brisbane, Singapore, Toronto, 1998, pp. 277–282.

98P13 Prassl, R., Schuster, B., Laggner, P., Flamant, C., Nigon, F., Chapman, M.J.: Biochemistry 37 (1998) 938–944.

98P14 Prehoda, K.E., Mooberry, E.S., Markley, J.L.: Biochemistry 37 (1998) 5785–5790.

98Q1 Quian, W., Sun, Y.-L., Wang, Y.-H., Zhuang, J.-H., Xie, Y., Huang, Z.-X.: Biochemistry 37 (1998) 14137–14150.

98Q2 Quirk, D.J., Park, C., Thompson, J.E., Raines, R.T.: Biochemistry 37 (1998) 17958–17964.

98R1 Raffy, S., Sassoon, N., Hofnung, M., Betton, J.-M.: Protein Sci. 7 (1998) 2136–2142.

98R2 Reid, K.L., Rodriguez, H.M., Hillier, B.J., Gregoret, L.M.: Protein Sci. 7 (1998) 470–479.

98R3    Reiersen, H., Clarke, A.R., Rees, A.R.: J. Mol. Biol. 283 (1998) 255–264.

98R4    Remmele Jr., R.L., Nightlinger, N.S., Srinivasan, S., Gombotz, W.R.: Pharmaceut. Res. 15 (1998) 200–208.

98R5    Rho, S.B., Lee, J.S., Jeong, E.-J., Kim, K.-S., Kim, Y.G., Kim, S.: J. Biol. Chem. 273 (1998) 11267–11273.

98R6    Rietveld, A.W.M., Ferreira, S.T.: Biochemistry 37 (1998) 933–937.

98R7    Rishi, V., Anjum, F., Ahmad, F., Pfeil, W.: Biochem. J. 329 (1998) 137–143.

98R8    Robinson, C.R., Liu, Y., O'Brien, R., Sligar, S.G., Sturtevant, J.M.: Protein Sci. 7 (1998) 961–965.

98R9    Rogers, D.P., Roberts, L.M., Lebowitz, J., Engler, J.A., Brouilette, C.G.: Biochemistry 37 (1998) 945–955.

98R10   Rösgen, J., Hallerbach, B., Hinz, H.-J.: Biophys. Chem. 74 (1998) 153–161.

98R11   Rosso, S.B., Gonzalez, M., Bagatolli, L.A., Duffard, R.O., Fidelio, G.D.: Life Sci. 63 (1998) 2343–2351.

98R12   Rousseau, F., Schymkowitz, J.W.H., Sánchez del Pino, M., Itzhaki, L.S.: J. Mol. Biol. 284 (1998) 503–519.

98R13   Ruan, B., Hoskins, J., Wang, L., Bryan, P.N.: Protein Sci. 7 (1998) 2345–2553.

98R14   Ruiz-Arribas, A., Zhadan, G.G., Kutyshenko, V.P., Santamaria, R.I., Cortijo, M., Villar, E., Fernandez-Abalos, J.M., Calvete, J.J., Shnyrov, V.L.: Eur. J. Biochem. 253 (1998) 462–468.

98S1    Sauder, J.M., Roder, H.: Folding & Design 3 (1998) 293–301.

98S2    Schindler, T., Perl, D., Graumann, P., Sieber, V., Marahiel, M.A., Schmid, F.X.: Proteins: Structure, Function, and Genetics 31 (1998) 401–406.

98S3    Schuler, B., Seckler, R.: J. Mol. Biol. 281 (1998) 227–234.

98S4    Schulga, A., Kurbanov, F., Kirpichnikov, M., Protasevich, I., Lobachov, V., Ranjbar, B., Chekhov, V., Polyakov, K., Engelborghs, Y., Makarov, A.: Protein Engng. 11 (1998) 775–782.

98S5    Schultz, L.W., Hargraves, S.R., Klink, T.A., Raines, R.T.: Protein Sci. 7 (1998) 1620–1625.

98S6    Schwarz, F.P., Ahmed, H., Bianchet, M.A., Amzel, L.M., Vasta, G.R.: Biochemistry 37 (1998) 5867–5877.

98S7    Schwarz, P.M., Liggins, J.R., Ludueña, R.F.: Biochemistry 37 (1998) 4687–4692.

98S8    Schwehm, J.M., Kristyanne, E.S., Biggers, C.C., Stites, W.E.: Biochemistry 37 (1998) 6939–6948.

98S9    Sedlák, E., Antalik, M.: Biopolymers 46 (1998) 145–154.

98S10   Segel, D.J., Fink, A.L., Hodgson, K.O., Doniach, S.: Biochemistry 37 (1998) 12443–12451.

98S11   Shao, X., Matthews, C.R.: Biochemistry 37 (1998) 7850–7858.

98S12   Shifman, J.M., Moser, C.C., Kalsbeck, W.A., Bocian, D.F., Dutton, P.L.: Biochemistry 37 (1998) 16815–16827.

98S13   Sieber, V., Plückthun, A., Schmid, F.X.: Nature Biotechnol. 16 (1998) 955–960.

98S14   Sohl, J.L., Jaswal, S.S., Agard, D.A.: Nature 395 (1998) 817–819.

98S15   Soumillion, P., Fastrez, J.: Protein Engng. 11 (1998) 213–217.

98S16   Spector, S., Kuhlman, B., Fairman, R., Wong, E., Boice, J.A., Raleigh, D.P.: J. Mol. Biol. 276 (1998) 479–489.

98S17   Spek, E.J., Bui, A.H., Lu, M., Kallenbach, N.R.: Protein Sci. 7 (1998) 2431–2437.

98S18   Srinivas, V.R., Singha, N.C., Schwarz, F.P., Surolia, A.: FEBS Lett. 425 (1998) 57–60.

98S19   Staniforth, R.A., Bigotti, M.G., Cutruzzolà, F., Allocatelli, C.T., Brunori, M.: J. Mol. Biol. 275 (1998) 133–148.

98S20   Steensma, E., van Mierlo, C.P.M.: J. Mol. Biol. 282 (1998) 653–666.

98S21   Stevens, J.M., Hornby, J.A.T., Armstrong, R.N., Dirr, H.W.: Biochemistry 37 (1998) 15534–15541.

98S22   Swietnicki, W., Petersen, R.B., Gambetti, P., Surewicz, W.K.: J. Biol. Chem. 273 (1998) 31048–31052.

98S23   Syto, R., Murgolo, N.J., Braswell, E.H., Mui, P., Huang, E., Windsor, W.T.: Biochemistry 37 (1998) 16943–16951.

98T1    Takano, K., Yamagata, Y., Yutani, K.: J. Mol. Biol. 280 (1998) 749–761.

98T2    Tamura, A.: Thermochim. Acta 308 (1998) 35–40.

98T3    Tan, P.H., Sandmaier, B.M., Stayton, P.S.: Biophys. J. 75 (1998) 1473–1482.

98T4    Tanaka, A.: J. Biochem. 123 (1998) 289–293.

98T5    Telford, J.R., Wittung-Stafshede, P., Gray, H.B., Winkler, J.R.: Acc. Chem. Res. 31 (1998) 755–763.

98T6    Tiktopulo, E.I., Kajava, A.V.: Biochemistry 37 (1998) 8147–8152.

98T7    Tischenko, V.M., Abramov, V.M., Zav'yalov, V.P.: Biochemistry 37 (1998) 5576–5581.

98T8    Tishchenko, V.M., Lund, J., Goodall, M., Jefferis, R.: in: Biocalorimetry: Applications of Calorimetry in the Biological Sciences, ed. by J.E. Ladbury and B.Z. Chowdhry, John Wiley & Sons, Chichester, New York, Weinheim, Brisbane, Singapore, Toronto, 1998, pp. 267–275.

98T9    Todd, M.J., Semo, N., Freire, E.: J. Mol. Biol. 283 (1998) 475–488.

98T10   Trevino, R.J., Tsalkova, T., Kramer, G., Hardesty, B., Chirgwin, J.M., Horowitz, P.M.: J. Biol. Chem. 273 (1998) 27841–27847.

98T11   Tsaprailis, G., Chan, D.W.S., English, A.M.: Biochemistry 37 (1998) 2004–2016.

98T12   Tzoneva, R.D., Mishonova-Alexova, E.I.: Biochim. Biophys. Acta 1364 (1998) 420–424.

98U1    Ueda, I., Suzuki, A.: Biochim. Biophys. Acta 1380 (1998) 313–319.

98U2    Uversky, V.N., Karnoup, A.S., Segel, D.J., Seshadri, S., Doniach, S., Fink, A.L.: J. Mol. Biol. 278 (1998) 879–894.

98V1    Van Mierlo, C.P.M., Van Dongen, W.M.A.M., Vergeldt, F., Van Berkel, W.J.H., Steensma, E.: Protein Sci. 7 (1998) 2331–2344.

98V2    Van Nuland, N.A.J., Chiti, F., Taddei, N., Raugei, G., Ramponi, G., Dobson, C.M.: J. Mol. Biol. 283 (1988) 883–891.

98V3    Van Nuland, N.A.J., Meijberg, W., Warner, J., Forge, V., Scheek, R.M., Robillard, G.T., Dobson, C.M.: Biochemistry 37 (1998) 622–637.

98V4    Vermeer, A.W.P., Bremer, M.G.E.G., Norde, W.: Biochim. Biophys. Acta 1425 (1998) 1–12.

98V5    Villaverde, J., Cladera, J., Hartog, A., Berden, J., Padrós, Duñach, M.: Biophys. J. 75 (1998) 1980–1988.

98V6    Villegas, V., Martinez, J.C., Avilés, F.X., Serrano, L.: J. Mol. Biol. 283 (1998) 1027–1036.

98V7    Vriend, G., Berendsen, H.J.C., van den Burg, B., Venema, G., Eijsink, V.G.H.: J. Biol. Chem. 273 (1998) 35074–35077.

98W1    Waldner, J.C., Lahr, S.J., Edgell, M.H., Pielak, G.J.: Analyt. Biochem. 263 (1998) 116–118.

98W2    Wallace, L.A., Blatch, G.L., Dirr, H.W.: Biochem. J. 336 (1998) 413–418.

98W3    Wallace, L.A., Sluis-Cremer, N., Dirr, H.W.: Biochemistry 37 (1998) 5320–5328.

98W4    Wang, C., Lascu, I., Giartosio, A.: Biochemistry 37 (1998) 8457–8464.

98W5    Wang, L., Kallenbach, N.R.: Protein Sci. 7 (1998) 2460–2464.

98W6    Wendt, H., Thomas, R.M., Ellenberger, T.: J. Biol. Chem. 273 (1998) 5735–5743.

98W7    Wenk, M., Jaenicke, R., Mayr, E.-M.: FEBS Lett. 438 (1998) 127–130.

98W8    Wheeler, K.A., Hawkins, A.R., Pain, R., Virden, R.: Proteins: Structure, Function, and Genetics 33 (1998) 550–557.

98W9    Wisz, M.S., Garrett, C.Z., Hellinga, H.W.: Biochemistry 37 (1998) 8269–8277.

98W10   Wittung-Stafshede, P.: Biochim. Biophys. Acta 1382 (1998) 324–332.

98W11  Wittung-Stafshede, P., Gomez, E., Öhman, A., Aasa, R., Villahermosa, R.M., Leckner, J., Karlsson, B.G., Sanders, D., Fee, J.A., Winkler, J.R., Malmström, B.G., Gray, H.B., Hil, M.G.: Biochim. Biophys. Acta 1388 (1998) 437–443.

98W12  Wittung-Stafshede, P., Malmström, B.G., Sanders, D., Fee, J.A., Winkler, J.R., Gray, H.B.: Biochemistry 37 (1998) 3172–3177.

98W13  Wong, C.-Y., Eftink, M.R.: Biochemistry 37 (1998) 8947–8953.

98X1   Xu, J., Baase, W.A., Baldwin, E., Matthews, B.W.: Protein Sci. 7 (1998) 158–177.

98X2   Xu, Y., Mayne, L., Englander, S.W.: Nature Struct. Biol. 5 (1998) 774–778.

98Y1   Yamagata, Y., Kubota, M., Sumikawa, Y., Funahashi, J., Takano, K., Fujii, S., Yutani, K.: Biochemistry 37 (1998) 9355–9362.

98Y2   Yokota, A., Takenaka, H., Oh, T., Noda, Y., Segawa, S.-I.: Protein Sci. 7 (1998) 1717–1727.

98Z1   Zaiss, K., Schurig, H., Jaenicke, R.: in: Biocalorimetry: Applications of Calorimetry in the Biological Sciences, ed. by J.E. Ladbury and B.Z. Chowdhry, John Wiley & Sons, Chichester, New York, Weinheim, Brisbane, Singapore, Toronto, 1998, pp. 283– 293.

98Z2   Zakharov, S.D., Lindeberg, M., Griko, Y., Salamon, Z., Tollin, G., Prendergast, F.G., Cramer, W.A.: Proc. Natl. Acad. Sci. 95 (1998) 4282–4287.

98Z3   Závodszky, P., Kardos, J., Svingor, Á., Petsko, G.A.: Proc. Natl. Acad. Sci. USA 95 (1998) 7406–7411.

98Z4   Zhang, J., Matthews, C.R.: Biochemistry 37 (1998) 14881–14890.

98Z5   Zhang, J., Matthews, C.R.: Biochemistry 37 (1998) 14891–14899.

98Z6   Zheng, L., Hogue, C.W.V., Brennan, J.D.: Biophys. Chem. 71 (1998) 157–172.

99A1   Alexandrescu, A.T., Jaravine, V.A., Dames, S.A., Lamour, F.P.: J. Mol. Biol. 289 (1999) 1041–1054.

99A2   Anderson, D.E., Peters, R.J., Wilk, B., Agard, D.A.: Biochemistry 38 (1999) 4728–4735.

99A3   Arrington, C.B., Teesch, L.M., Robertson, A.D.: J. Mol. Biol. 285 (1999) 1265–1275.

99A4   Axe, D.D., Foster, N.W., Fersht, A.R.: J. Mol. Biol. 286 (1999) 1471–1485.

99A5   Azuaga, A.I., Woodruff, N.D., Conejero-Lara, F., Cox, V.F., Smith, R.A.G., Dobson, C.M.: Protein Sci. 8 (1999) 443–446.

99B1   Bai, Y.: Proc. Natl. Acad. Sci. USA 96 (1999) 477–480.

99B2   Barry, J.K., Matthews, K.S.: Biochemistry 38 (1999) 6520–6528.

99B3   Baskakov, I.V., Bolen, D.W.: Protein Sci. 8 (1999) 1314–1319.

99B4   Baxter, S.M., Fetrow, J.S.: Biochemistry 38 (1999) 4493–4503.

99B5   Beadle, B.M., Baase, W.A., Wilson, D.B., Gilkes, N.R., Shoichet, B.K.: Biochemistry 38 (1999) 2570–2576.

99B6   Beadle, B.M., McGovern, S.L., Patera, A., Shoichet, B.K.: Protein Sci. 8 (1999) 1816–1824.

99B7   Beldarrain, A., López-Lacomba, J.L., Furrazola, G., Barberia, D., Cortijo, M.: Biochemistry 38 (1999) 7865–7873.

99B8   Bergamini, C.M., Dean, M., Matteucci, G., Hanau, S., Tanfani, F., Ferrari, C., Boggian, M., Scatturin, A.: Eur. J. Biochem. 266 (1999) 575–582.

99B9   Bhattacharya, S., Falzone, C.J., Lecomte, J.T.J.: Biochemistry 38 (1999) 2577–2589.

99B10  Bhattacharyya, R.P., Sosnick, T.R.: Biochemistry 38 (1999) 2601–2609.

99B11  Bhuyan, A.K., Udgaonkar, J.B.: Biochemistry 38 (1999) 9158–9168.

99B12  Bieri, O., Wildegger, G., Bachmann, A., Wagner, C., Kiefhaber, T.: Biochemistry 38 (1999) 12460–12470.

99B13  Bilsel, O., Zitzewitz, J.A., Bowers, K.E., Matthews, C.R.: Biochemistry 38 (1999) 1018–1029.

99B14  Bishop, S.M., Ross, J.B.A., Kohanski, R.A.: Biochemistry 38 (1999) 3079–3089.

99B15   Blaber, S.I., Culajay, J.F., Khurana, A., Blaber, M.: Biophys. J. 77 (1999)
470–477.

99B16   Borén, K., Andersson, P., Larsson, M., Carlsson, U.: Biochim. Biophys. Acta 1430
(1999) 111–118.

99B17   Bortoleto, R.K., Ward, R.J.: FEBS Lett. 459 (1999) 438–442.

99B18   Branchu, S., Forbes, R.T., York, P., Petrén, S., Nyquist, H., Camber, O.: J. Pharm.
Sci. 88 (1999) 905–911.

99B19   Brown, B.M., Sauer, R.T.: Proc. Natl. Acad. Sci. USA 96 (1999) 1983–1988.

99B20   Burova, T.V., Choiset, Y., Jankowski, C.K., Haertlé, T.: Biochemistry 38 (1999)
15043–15051.

99B21   Bushueva, T.L., Teplova, M.V., Bushuev, V.N., Kudriashov, D.S., Vorotnikov,
A.V., Shirinskii, V.P.: Mol. Biol. (Moscow) 33 (1999) 227–236.

99C1    Cadieux, E., Powlowski, J.: Biochemistry 38 (1999) 10714–10722.

99C2    Capaldi, A.P., Ferguson, S.J., Radford, S.E.: J. Mol. Biol. 286 (1999) 1621–1632.

99C3    Cavagnero, S., Dyson, H.J., Wright, P.E.: J. Mol. Biol. 285 (1999) 269–282.

99C4    Chakrabarti, A., Srivastava, S., Swaminathan, C.P., Surolia, A., Varadarajan, R.:
Protein Sci. 8 (1999) 2455–2459.

99C5    Chamberlain, A.K., Fischer, K.F., Reardon, D., Handel, T.M., Marqusee, S.:
Protein Sci. 8 (1999) 2251–2257.

99C6    Chang, Y., Nilsen, S.L., Castellino, F.J.: J. Peptide Res. 53 (1999) 656–664.

99C7    Chaudhuri, T.K., Horii, K., Yoda, T., Arai, M., Nagata, S., Terada, T.P.,
Uchiyama, H., Ikura, T., Tsumoto, K., Kataoka, H., Matsushima, M., Kuwajima,
K., Kumagai, I.: J. Mol. Biol. 285 (1999) 1179–1194.

99C8    Chiti, F., Taddei, N., White, P.M., Bucciantini, M., Magherini, F., Stefani, M.,
Dobson, C.M.: Nature Struct. Biol. 6 (1999) 1005–1009.

99C9    Chivers, P.T., Sauer, R.T.: Protein Sci. 8 (1999) 2494–2500.

99C10   Christensen, T., Svensson, B., Sigurskjold, B.W.: Biochemistry 38 (1999)
6300–6310.

99C11   Clark, A.C., Frieden, C.: J. Mol. Biol. 285 (1999) 1765–1776.

99C12   Clark, A.C., Frieden, C.: J. Mol. Biol. 285 (1999) 1777–1788.

99C13   Coll, M.G., Protasevich, I.I., Torrent, J., Ribó, M., Lobachov, V.M., Makarov,
A.A., Vilanova, M.: Biochem. Biophys. Res. Commun. 265 (1999) 356–360.

99C14   Cregut, D., Civera, C., Macias, M.J., Wallon, G., Serrano, L.: J. Mol. Biol. 292
(1999) 389–401.

99D1    Dams, T., Jaenicke, R.: Biochemistry 38 (1999) 9169–9178.

99D2    Day, E.S., Wen, D., Garber, E.A., Hong, J., Avedissian, L.S., Rayhorn, P., Shen,
W., Zeng, C., Bailey, V.R., Reilly, J.O., Roden, J.A., Moore, C.B., Williams, K.P.,
Galdes, A., Whitty, A., Baker, D.P.: Biochemistry 38 (1999) 14868–14880.

99D3    Demarest, S.J., Boice, J.A., Fairman, R., Raleigh, D.P.: J. Mol. Biol. 294 (1999)
213–221.

99D4    Desai, G., Panick, G., Zein, M., Winter, R., Royer, C.A.: J. Mol. Biol. 288 (1999)
461–475.

99D5    Dirr, H.W., Wallace, L.A. Biochemistry 38 (1999) 15631–15640.

99D6    Dolla, A., Arnoux, P., Protasevich, I., Lobachov, V., Brugna, M., Giudici-
Orticoni, M.T., Haser, R., Czjzek, M., Makarov, A., Bruschi, M.: Biochemistry 38
(1999) 33–41.

99D7    Dürr, E., Jelesarov, I., Bosshard, H.R.: Biochemistry 38 (1999) 870–880.

99E1    Epand, R.F., Epand, R.M., Jung, C.Y.: Biochemistry 38 (1999) 454–458.

99E2    Esser, D., Rudolph, R., Jaenicke, R., Böhm, G.: J. Mol. Biol. 291 (1999)
1135–1146.

99E3    Eubanks, S., Lu, M., Peyton, D., Breslow, E.: Biochemistry 38 (1999)
13530–13541.

99E4    Evrard, C., Fastrez, J., Soumillion, P.: FEBS Lett. 460 (1999) 442–446.

504     References

99F1    Farruggia, B., Nerli, B., Di Nuci, H., Rigatusso, Picó, G.: Int. J. Biol. Macromol. 26 (1999) 23–33.
99F2    Farruggia, B., Picó, G.A.: Int. J. Biol. Macromol. 26 (1999) 317–323.
99F3    Feller, G., d'Amico, D., Gerday, C.: Biochemistry 38 (1999) 4613–4619.
99F4    Feng, Y., Minnerly, J.C., Zurfluh, L.L., Joy, W.D., Hood, W.F., Abegg, A.L., Grabbe, E.S., Shieh, J.-J., Thurman, T.L., McKearn, J.P., McWherter, C.A.: Biochemistry 38 (1999) 4553–4563.
99F5    Ferguson, N., Capaldi, A.P., James, R., Kleanthous, C., Radford, S.E.: J. Mol. Biol. 286 (1999) 1597–1608.
99F6    Filimonov, V.V., Azuaga, A.I., Viguera, A.R., Serrano, L., Mateo, P.L.: Biophys. Chem. 77 (1999) 195–208.
99F7    Frankel, M., Bishop, S.M., Ablooglu, A.J., Han, Y.-P., Kohanski, R.A.: Protein Sci. 8 (1999) 2158–2165.
99F8    Frankenberg, N., Welker, C., Jaenicke, R.: FEBS Lett. 454 (1999) 299–302.
99F9    Frayser, M., Sato, A.K., Xu, L., Stern, L.J.: Protein Expr. Purif. 15 (1999) 105–114.
99F10   Fuertes, M.A., Berberich, C., Lozano, R.M., Gimenez-Gallego, G., Alonso, C.: Eur. J. Biochem. 260 (1999) 559–567.
99F11   Fujiwara, K., Arai, M., Shimizu, A., Ikeguchi, M., Kuwajima, K., Sugai, S.: Biochemistry 38 (1999) 4455–4463.
99F12   Fulton, K.F., Main, E.R.G., Daggett, V., Jackson, S.E.: J. Mol. Biol. 291 (1999) 445–461.
99F13   Funahashi, J., Takano, K., Yamagata, Y., Yutani, K.: Protein Engng. 12 (1999) 841–850.
99G1    Ganesh, C., Banerjee, A., Shah, A., Varadarajan, R.: FEBS Lett. 454 (1999) 307–311.
99G2    Gassner, N.C., Baase, W.A., Hausrath, A.C., Matthews, B.W.: J. Mol. Biol. 294 (1999) 17–20.
99G3    Gassner, N.C., Baase, W.A., Lindstrom, J.D., Lu, J., Dahlquist, F.W., Matthews, B.W.: Biochemistry 38 (1999) 14451–14460.
99G4    Gaudiano, M.C., Pala, A., Barteri, M.: Biochim. Biophys. Acta 1431 (1999) 451–461.
99G5    Georgescu, R.E., Braswell, E.H., Zhu, D., Tasayco, M.L.: Biochemistry 38 (1999) 13355–13366.
99G6    Gibney, B.R., Dutton, P.L.: Protein Sci. 8 (1999) 1888–1898.
99G7    Giletto, A., Pace, C.N.: Biochemistry 38 (1999) 13379–13384.
99G8    Godbole, S., Bowler, B.E.: Biochemistry 38 (1999) 487–495.
99G9    Golbik, R., Fischer, G., Fersht, A.R.: Protein Sci. 8 (1999) 1505–1514.
99G10   Gorinstein, S., Zemser, M., Vargas-Albores, F., Ochoa, J.-L., Paredes-Lopez, O., Scheler, C., Aksu, S., Salnikov, J.: J. Protein Chem. 18 (1999) 239–247.
99G11   Greene, B., King, J.: J. Biol. Chem. 274 (1999) 16135–16140.
99G12   Greene, B., King, J.: J. Biol. Chem. 274 (1999) 16141–16146.
99G13   Greene, L.H., Grobler, J.A., Malinovskii, V.A., Tian, J., Acharya, K.R., Brew, K.: Protein Engng. 12 (1999) 581–587.
99G14   Griffiths-Jones, S.R., Maynard, A.J., Searle, M.S.: J. Mol. Biol. 292 (1999) 1051–1069.
99G15   Griko, Y.: J. Protein Chem. 18 (1999) 361–369.
99G16   Griko, Y., Remeta, D.P.: Protein Sci. 8 (1999) 554–561.
99G17   Grimsley, G.R., Shaw, K.L., Fee, L.R., Alston, R.W., Huyghues-Despointes, B.M.P., Thurlkill, R.L., Scholtz, J.M., Pace, C.N.: Protein Sci. 8 (1999) 1843–1849.
99G18   Gu, H., Doshi, N., Kim, D.E., Simons, K.T., Santiago, J.V., Nauli, S., Baker, D.: Protein Sci. 8 (1999) 2734–2741.

99G19 Gualfetti, P.J., Bilsel, O., Matthews, C.R.: Protein Sci. 8 (1999) 1623–1635.

99G20 Gualfetti, P.J., Iwakura, M., Lee, J.C., Kihara, H., Bilsel, O., Zitzewitz, J.A., Matthews, C.R.: Biochemistry 38 (1999) 13367–13378.

99G21 Gupta, R., Ahmad, F.: Biochemistry 38 (1999) 2471–2479.

99G22 Gursky, O.: Protein Sci. 8 (1999) 2055–2064.

99G23 Guzzi, R., Sportelli, L., La Rosa, C., Milardi, D., Grasso, D., Verbeet, M.Ph., Canters, G.W.: Biophys. J. 77 (1999) 1052–1063.

99G24 Ghirlando, R., Lund, J., Goodall, M., Jefferis, R.: Immunology Lett. 68 (1999) 47–52.

99H1 Haezebrouck, P., Noyelle, K., Joniau, M., Van Dael, H.: J. Mol. Biol. 293 (1999) 703–718.

99H2 Hansen, T., Urbanke, C., Leppänen, V.-M., Goldman, A., Brandenburg, K., Schäfer, G.: Arch. Biochem. Biophys. 363 (1999) 135–147.

99H3 Hasegawa, J., Shimahara, H., Mizutani, M., Uchiyama, S., Arai, H., Ishii, M., Kobayashi, Y., Ferguson, S.J., Sambongi, Y., Igarashi, Y.: J. Biol. Chem. 274 (1999) 37533–37537.

99H4 Hayes, M.V., Sessions, R.B., Brady, R.L., Clarke, A.R.: J. Mol. Biol. 285 (1999) 1857–1867.

99H5 Heitz, A., Le-Nguyen, D., Chiche, L.: Biochemistry 38 (1999) 10615–10625.

99H6 Henkel, W., Vogl, T., Echner, H., Voelter, W., Urbanke, C., Schleuder, D., Rauterberg, J.: Biochemistry 38 (1999) 13610–13622.

99H7 Hennecke, J., Sebbel, P., Glockshuber, R.: J. Mol. Biol. 286 (1999) 1197–1215.

99H8 Henzl, M.T., Graham, J.S.: FEBS Lett. 442 (1999) 241–245.

99H9 Higurashi, T., Nosaka, K., Mizobata, T., Nagai, J., Kawata, Y.: J. Mol. Biol. 291 (1999) 703–713.

99H10 Hollien, J., Marqusee, S.: Biochemistry 38 (1999) 3831–3836.

99H11 Honda, Y., Fukamizo, T., Okajima, T., Goto, S., Boucher, I., Brzezinski, R.: Biochim. Biophys. Acta 1429 (1999) 365–376.

99H12 Honda, S., Kobayashi, N., Munekata, E., Uedaira, H.: Biochemistry 38 (1999) 1203–1213.

99H13 Honda, S., Uedaira, H., Vonderviszt, F., Kidokoro, S. Namba, K.: J. Mol. Biol. 293 (1999) 719–732.

99H14 Hostetter, D.R., Weatherly, G.T., Beasley, J.R., Bortone, K., Cohen, D.S., Finger, S.A., Hardwidge, P., Kakouras, D.S., Saunders, A.J., Trojak, S.K., Waldner, J.C., Pielak, G.J.: J. Mol. Biol. 289 (1999) 639–644.

99H15 Huang, S., Ratliff, K.S., Schwartz, M.P., Spenner, J.M., Matouschek, A: Nature Struct. Biol. 6 (1999) 1132–1138.

99H16 Huyghues-Despointes, B.M.P., Langhorst, U., Steyaert, J., Pace, C.N., Scholtz, J.M.: Biochemistry 38 (1999) 16481–16490.

99I1 Ibarra-Molero, B., Loladze, V.V., Makhatadze, G.I., Sanchez-Ruiz, J.M.: Biochemistry 38 (1999) 8138–8149.

99I2 Ibarra-Molero, B., Makhatadze, G.I., Sanchez-Ruiz, J.M.: Biochim. Biophys. Acta 1429 (1999) 384–390.

99I3 Isaacson, R.L., Weeds, A.G., Fersht, A.R.: Proc. Natl. Acad. Sci. USA 96 (1999) 11247–11252.

99I4 Isogai, Y., Ota, M., Fujisawa, T., Izuno, H., Mukai, M., Nakamura, H., Iizuka, T., Nishikawa, K.: Biochemistry 38 (1999) 7431–7443.

99I5 Iwaoka, M., Wedemeyer, W.J., Scheraga, H.A.: Biochemistry 38 (1999) 2805–2815.

99J1 Jackson, G.S., Hill, A.F., Joseph, C., Hosszu, L., Power, A., Waltho, J.P., Clarke, A.R., Collinge, J.: Biochim. Biophys. Acta 1431 (1999) 1–13.

99J2 Jacob, M., Geeves, M., Holtermann, G., Schmid, F.X.: Nature Struct. Biol. 6 (1999) 923–926.

99J3    Jacob, M., Holtermann, G., Perl, D., Reinstein, J., Schindler, T., Geeves, M.A.,
        Schmid, F.X.: Biochemistry 38 (1999) 2882–2891.
99J4    Jäger, M., Plückthun, A.: J. Mol. Biol. 285 (1999) 2005–2019.
99J5    Jäger, M., Plückthun, A.: FEBS Lett. 462 (1999) 307–312.
99J6    James, E.L., Whisstock, J.C., Gore, M.G., Bottomley, S.P.: J. Biol. Chem. 274
        (1999) 9482–9488.
99J7    Jamin, M., Yeh, S.-R., Rousseau, D.L., Baldwin, R.L.: J. Mol. Biol. 292 (1999)
        731–740.
99J8    Jatzke, C., Hinz, H.-J., Seedorf, U., Assmann, G.: Biochim. Biophys. Acta 1432
        (1999) 265–274.
99J9    Jerala, R., Žerovnik, E.: J. Mol. Biol 291 (1999) 1079–1089.
99J10   Jin, L., Fukayama, J.W., Pelczer, I., Carey, J.: J. Mol. Biol. 285 (1999) 361–378.
99J11   Jung, S., Honegger, A., Plückthun, A.: J. Mol. Biol. 294 (1999) 163–180.
99K1    Kamiyama, T., Sadahide, Y., Nogusa, Y., Gekko, K.: Biochim. Biophys. Acta
        1434 (1999) 44–57.
99K2    Kardos, J., Bódi, A., Závodsky, P., Venekei, I., Gráf, L.: Biochemistry 38 (1999)
        12248–12257.
99K3    Kaushik, J.K., Bhat. R.: Protein Sci. 8 (1999) 222–233.
99K4    Kay, M.S., Ramos, C.H.I., Baldwin, R.L.: Proc. Natl. Acad. Sci. USA 96 (1999)
        2007–2012.
99K5    Kirkitadze, M.D., Dryden, D.T.F., Kelly, S.M., Price, N.C., Wang, X., Krych, M.,
        Atkinson, J.P., Barlow, P.N.: FEBS Lett. 459 (1999) 133–138.
99K6    Kirkitadze, M.D., Krych, M., Uhrin, D., Dryden, D.T.F., Smith, B.O., Cooper, A.,
        Wang, X., Hauhart, R., Atkinson, J.P., Barlow, P.N.: Biochemistry 38 (1999)
        7019–7031.
99K7    Kleanthous, C., Kühlmann, U.C., Pommer, A.J., Ferguson, N., Radford, S.E.,
        Moore, G.R., James, R., Hemmings, A.M.: Nature Struct. Biol. 6 (1999) 243–252.
99K8    Kleppe, R., Uhlemann, K., Knappskog, P.M., Haavik, J.: J. Biol. Chem. 274
        (1999) 33251–33258.
99K9    Knapp, S., Ladenstein, R., Galinski, E.A.: Extremophiles 3 (1999) 191–198.
99K10   Kobashigawa, Y., Sakurai, M., Nitta, K.: Protein Sci. 8 (1999) 2765–2772.
99K11   Kobayashi, N., Honda, S., Munekata, E.: Biochemistry 38 (1999) 3228–3234.
99K12   Koepf, E.K., Petrassi, H.M., Sudol, M., Kelly, J.W.: Protein Sci. 8 (1999)
        841–853.
99K13   Koshiba, T., Hayashi, T., Miwako, I., Kumagai, I., Ikura, T., Kawano, K., Nitta,
        K., Kuwajima, K.: Protein Engng. 12 (1999) 429–435.
99K14   Kragelund, B.B., Osmark, P., Neergaard, T.B., Schiødt, J., Kristiansen, K.,
        Knudsen, J., Poulsen, F.M.: Nature Struct. Biol. 6 (1999) 594–600.
99K15   Kragelund, B.B., Poulsen, K., Andersen, K.V., Baldursson, T., Krøll, J.B.,
        Neergård, T.B., Jepsen, J., Roepstorff, P., Kristiansen, K., Poulsen, F.M.,
        Knudsen, J.: Biochemistry 38 (1999) 2386–2394.
99K16   Kretschmar, M., Jaenicke, R.: J. Mol. Biol. 291 (1999) 1147–1153.
99K17   Kretschmar, M., Mayr, E.-M., Jaenicke, R.: J. Mol. Biol. 289 (1999) 701–705.
99K18   Krupakar, J., Swaminathan, C.P., Das, P.K., Surolia, A., Podder, S.K.: Biochem. J.
        338 (1999) 273–279.
99K19   Kuhlman, B., Luisi, D.L., Young, P., Raleigh, D.P.: Biochemistry 38 (1999)
        4896–4903.
99K20   Kuwajima, K., Arai, M., Mizuguchi, M., Koshiba, T., Nitta, K.: in: Old and New
        Views of Protein Folding, ed. by K.Kuwajima and M. Arai, Elsevier Science B.V.,
        Amsterdam, Lausanne, New York, Oxford, Shannon, Singapore, Tokyo, 1999,
        pp.135–144.
99L1    Laity, J.H., Montelione, G.T., Scheraga, H.A.: Biochemistry 38 (1999)
        16432–16442.

99L2    Lesieur, C., Frutiger, S., Hughes, G., Kellner, R., Pattus, F., van der Goot, F.G.: J. Biol. Chem. 274 (1999) 36722–36728.

99L3    Lett, C.M., Rosu-Myles, M.D., Frey, H.E., Guillemette, J.G.: Biochim. Biophys. Acta 1432 (1999) 40–48.

99L4    Levashov, P., Orlov, V., Boschi–Muller, S., Talfournier, F., Asryants, R., Bulatnikov, I., Muronetz, V., Branlant, G., Nagradova, N.: Biochim. Biophys. Acta 1433 (1999) 294–306.

99L5    Liemann, S., Glockshuber, R.: Biochemistry 38 (1999) 3258–3267.

99L6    Liggins, J.R., Lo, T.P., Brayer, G.D., Nall, B.T.: Protein Sci. 8 (1999) 2645–2654.

99L7    Llinás, M., Gillespie, B., Dahlquist, F.W., Marqusee, S.: Nature Struct. Biol. 6 (1999) 1072–1078.

99L8    Loladze, V.V., Ibarra-Molero, B., Sanchez-Ruiz, J.M., Makhatadze, G.I.: Biochemistry 38 (1999) 16419–16423.

99L9    Lorch, M., Mason, J.M., Clarke, A.R., Parker, M.J.: Biochemistry 38 (1999) 1377–1385.

99L10   Luisi, D.L., Kuhlman, B., Sideras, K., Evans, P.A., Raleigh, D.P.: J. Mol. Biol. 289 (1999) 167–174.

99L11   Luo, Y., Baldwin, R.L.: Proc. Natl. Acad. Sci. USA 96 (1999) 11283–11287.

99L12   Lyubarev, A.E., Kurganov, B.I.: Biokhimiya (Moscow) 64 (1999) 990–997.

99L13   Lyubarev, A.E., Kurganov, B.I., Orlov, V.N., Zhou, H.-M.: Biophys. Chem. 79 (1999) 199–204.

99M1    Maier, C.S., Schimerlik, M.I., Deinzer, M.L.: Biochemistry 38 (1999) 1136–1143.

99M2    Main, E.R.G., Fulton, K.F., Jackson, S.E.: J. Mol. Biol. 291 (1999) 429–444.

99M3    Main, E.R.G., Jackson, S.E.: Nature Struct. Biol. 6 (1999) 831–835.

99M4    Maki, K., Ikura, T., Hayano, T., Takahashi, N., Kuwajima, K.: Biochemistry 38 (1999) 2213–2223.

99M5    Malvezzi-Campeggi, F., Stroppolo, M.E., Mei, G., Rosato, N., Desideri, A.: Arch. Biochem. Biophys. 370 (1999) 201–207.

99M6    Manyusa, S., Mortuza, G., Whitford, D.: Biochemistry 38 (1999) 14352–14362.

99M7    Manyusa, S., Whitford, D.: Biochemistry 38 (1999) 9533–9540.

99M8    Marcos, M.J., Chehin, R., Arrondo, J.L., Zhadan, G.G., Villar, E., Shnyrov, V.L.: FEBS Lett. 443 (1999) 192–196.

99M9    Markovic-Housley, Z., Stolz, B., Lanz, R., Erni, B.: Protein Sci. 8 (1999) 1530–1535.

99M10   Martinez, J.C., Serrano, L.: Nature Struct. Biol. 6 (1999) 1010–1016.

99M11   Martinez, J.C., Viguera, A.R., Berisio, R., Wilmanns, M., Mateo, P.L., Filimonov, V.V., Serrano, L.: Biochemistry 38 (1999) 549–559.

99M12   Masson, P., Fortier, P.-L., Albaret, C., Cléry, C., Guerra, P., Lockridge, O.: Chem.-Biol. Interactions 119–120 (1999) 17–27.

99M13   Medved, L.V., Migliorini, M., Mikhailenko, I., Barrientos, L.G., Llinás, M., Strickland, D.K.: J. Biol. Chem. 274 (1999) 717–727.

99M14   Mei, G., Di Venere, A., Campeggi, F.M., Gilardi, G., Rosato, N., De Matteis, F., Finazzi-Agrò, A.: Eur. J. Biochem. 265 (1999) 619–626.

99M15   Merkel, J.S., Sturtevant, J.M., Regan, L.: Structure 7 (1999) 1333–1343.

99M16   Milne, J.S., Xu, Y., Mayne, L.C., Englander, S.W.: J. Mol. Biol. 290 (1999) 811–822.

99M17   Minks, C., Huber, R., Moroder, L., Budisa, N.: Biochemistry 38 (1999) 10649–10659.

99M18   Mizuguchi, M., Masaki, K., Nitta, K.: J. Mol. Biol. 292 (1999) 1137–1148.

99M19   Mohana-Borges, R., Silva, J.L., de Prat-Gay, G.: J. Biol. Chem. 274 (1999) 7732–7740.

99M20   Morii, H., Uedaira, H., Ogata, K., Ishii, S., Sarai, A.: J. Mol. Biol. 292 (1999) 909–920.

99M21   Motono, C., Yamagishi, A., Oshima, T.: Biochemistry 38 (1999) 1332–1337.

99M22   Moutiez, M., Burova, T.V., Haertlé, T., Quéméneur, E.: Protein Sci. 8 (1999) 106–112.

99M23   Myers, J.K., Oas, T.G.: Biochemistry 38 (1999) 6761–6768.

99N1    Neira, J.L., Fersht, A.R.: J. Mol. Biol. 285 (1999) 1309–1333.

99N2    Neira, J.L., Fersht, A.R.: J. Mol. Biol. 287 (1999) 421–432.

99N3    Neira, J.L., Sevilla, P., Menéndez, M., Bruix, M., Rico, M.: J. Mol. Biol. 285 (1999) 627–643.

99N4    Noland, B.W., Dangott, L.J., Baldwin, T.O.: Biochemistry 38 (1999) 16136–16145.

99O1    Odaert, B., Jean, F., Boutillon, C., Buisine, E., Melnyk, O., Tartar, A., Lippens, G.: Protein Sci. 8 (1999) 2773–2783.

99O2    Ohage, E., Steipe, B.: J. Mol. Biol. 291 (1999) 1119–1128.

99O3    Ohage, E., Wirtz, P., Barnikow, J., Steipe, B.: J. Mol. Biol. 291 (1999) 1129–1134.

99O4    Otzen, D.E., Christiansen, L., Schülein, M.: Protein Sci. 8 (1999) 1878–1887.

99O5    Otzen, D.E., Kristensen, O., Proctor, M., Oliveberg, M.: Biochemistry 38 (1999) 6499–6511.

99P1    Pace, C.N., Grimsley, G.R., Thomas, S.T., Makhatadze, G.I.: Protein Sci. 8 (1999) 1500–1504.

99P2    Padmanabhan, S., Laurents, D.V., Fernández, A.M., Elias-Arnanz, M., Ruiz-Sanz, J., Mateo, P.L., Rico, M., Filimonov, V.V.: Biochemistry 38 (1999) 15536–15547.

99P3    Pandya, M.J., Williams, P.B., Dempsey, C.E., Shewry, P.R., Clarke, A.R.: J. Biol. Chem. 274 (1999) 26828–26837.

99P4    Panick, G., Vidugiris, G.J.A., Malessa, R., Rapp, G., Winter, R., Royer, C.A.: Biochemistry 38 (1999) 4157–4164.

99P5    Pedone, E., Cannio, R., Saviano, M., Rossi, M., Bartolucci, S.: Biochem. J. 339 (1999) 309–317.

99P6    Perrett, S., Freeman, S.J., Butler, P.J.G., Fersht, A.R.: J. Mol. Biol. 290 (1999) 331–345.

99P7    Peterson, R.W., Nicholson, E.M., Thapar, R., Klevit, R.E., Scholtz, J.M.: J. Mol. Biol. 286 (1999) 1609–1619.

99P8    Plaxco, K.W., Millett, I.S., Segel, D.J., Doniach, S., Baker, D.: Nature Struct. Biol. 6 (1999) 554–556.

99P9    Poklar, N., Petrovcic, N., Oblak, M., Vesnaver, G.: Protein Sci. 8 (1999) 832–840.

99P10   Potekhin, S.A., Loseva, O.I., Tiktopulo, E.I., Dobritsa, A.P.: Biochemistry 38 (1999) 4121–4127.

99P11   Privalov, P.L., Jelesarov, I., Read, C.M., Dragan, A.I., Crane-Robinson, C.: J. Mol. Biol. 294 (1999) 997–1013.

99P12   Protasevich, I.I., Schulga, A.A., Vasilieva, L.I., Polyakov, K.M., Lobachov, V.M., Hartley, R.W., Kirpichnikov, M.P., Makarov, A.A.: FEBS Lett. 445 (1999) 384–388.

99R1    Ragona, L., Confalonieri, L., Zetta, L., DeKruif, K.G., Mammi, S., Peggion, E., Longhi, R., Molinari, H.: Biopolymers 49 (1999) 441–450.

99R2    Ragona, L., Fogolari, F., Romagnoli, S., Zetta, L., Maubois, J.L., Molinari, H.: J. Mol. Biol. 293 (1999) 953–969.

99R3    Ramos, C.H.I., Kay, M.S., Baldwin, R.L.: Biochemistry 38 (1999) 9783–9790.

99R4    Ramstein, J., Locker, D., Bianchi, M.E., Leng, M.: Eur. J. Biochem. 260 (1999) 692–700.

99R5    Raschke, T.M., Kho, J., Marqusee, S.: Nature Struct. Biol. 6 (1999) 825–831.

99R6    Reddy, G.B., Bharadwaj, S., Surolia, A.: Biochemistry 38 (1999) 4464–4470.

99R7    Redfield, C., Schulman, B.A., Milhollen, M.A., Kim, P.S., Dobson, C.M.: Nature Struct. Biol. 6 (1999) 948–952.

99R8    Reinstädler, D., Fabian, H., Naumann, D.: Proteins: Structure, Function, and Genetics 34 (1999) 303–316.

99R9    Remmele Jr., R.L., Bhat, S.D., Phan, D.H., Gombotz, W.R.: Biochemistry 38 (1999) 5241–5247.

99R10   Rentzeperis, D., Jonsson, T., Sauer, R.T.: Nature Struct. Biol. 6 (1999) 569–573.

99R11   Richardson, J.M., McMahon, K.W., MacDonald, C.C., Makhatadze, G.I.: Biochemistry 38 (1999) 12869–12875.

99R12   Riddle, D.S., Grantcharova, V.P., Santiago, J.V., Alm, E., Ruczinski, I., Baker, D.: Nature Struct. Biol. 6 (1999) 1016–1024.

99R13   Roberge, M., Shareck, F., Morosoli, R., Kluepfel, D., Dupont, C.: Protein Engng. 12 (1999) 251–257.

99R14   Rochdi, A., Foucat, L., Renou, J.P.: Biopolymers 50 (1999) 690–696.

99R15   Rochu, D., Ducret, G., Masson, P.: J. Chromatography 838 (1999) 157–165.

99R16   Roesler, K.R., Rao, A.G.: Protein Engng. 12 (1999) 967–973.

99R17   Rosengarth, A., Rösgen, J., Hinz., H.-J.: Eur. J. Biochem. 264 (1999) 989–995.

99R18   Rosengarth, A., Rösgen, J., Hinz, H.-J., Gerke, V.: J. Mol. Biol. 288 (1999) 1013–1025.

99R19   Royer, C.A., Hinck, A.P., Loh, S.N., Prehoda, K.E., Peng, X., Jonas, J., Markley, J.L.: Biochemistry 32 (1993) 5222–5232.

99R20   Ruan, B., Hoskins, J., Bryan, P.N.: Biochemistry 38 (1999) 8562–8571.

99R21   Ruiz-Sanz, J., Simoncsits, A., Törö, I., Pongor, S., Mateo, P.L., Filimonov, V.V.: Eur. J. Biochem. 263 (1999) 246–253.

99S1    Sadqi, M., Casares, S., Abril, M.A., López-Mayorga, O., Conejero- Lara, F.: Biochemistry 38 (1999) 8899–8906.

99S2    Sasahara, K., Nitta, K.: Protein Sci. 8 (1999) 1469–1474.

99S3    Sato, S., Kuhlman, B., Wu, W.-J., Raleigh, D.P.: Biochemistry 38 (1999) 5643–5650.

99S4    Schnappinger, D., Schubert, P., Berens, C., Pfleiderer, K., Hillen, W.: J. Biol. Chem. 274 (1999) 6405–6410.

99S5    Schwerdtfeger, R.M., Chiaraluce, R., Consalvi, V., Scandurra, R., Antranikian, G.: Eur. J. Biochem. 264 (1999) 479–487.

99S6    Sedlák, E., Antalik, M.: Biochim. Biophys. Acta 1434 (1999) 347–355.

99S7    Segel, D.J., Bachmann, A., Hofrichter, J., Hodgson, K.O., Doniach, S., Kiefhaber, T.: J. Mol. Biol. 288 (1999) 489–499.

99S8    Segel, D.J., Eliezer, D., Uversky, V., Fink, A.L., Hodgson, K.O., Doniach, S.: Biochemistry 38 (1999) 15352–15359.

99S9    Sheshadri, S., Lingaraju, G.M., Varadarajan, R.: Protein Sci. 8 (1999) 1689–1695.

99S10   Silow, M., Tan, Y.-J., Fersht, A.R., Oliveberg, M.: Biochemistry 38 (1999) 13006–13012.

99S11   Sionkowska, A., Kaminska, A.: Int. J. Biol. Macromol. 24 (1999) 337–340.

99S12   Sivaraman, T., Kumar, T.K.S., Tu, Y.T., Peng, H.J., Yu, C.: Arch. Biochem. Biophys. 363 (1999) 107–115.

99S13   Spector, S., Young, P., Raleigh, D.P.: Biochemistry 38 (1999) 4128–4136.

99S14   Spector, S., Raleigh, D.P.: J. Mol. Biol. 293 (1999) 763–768.

99S15   Starovasnik, M.A., O'Connell, M.P., Fairbrother, W.J., Kelley, R.F.: Protein Sci. 8 (1999) 1423–1431.

99S16   Steer, B.A., Dinardo, A.A., Merrill, A.R.: Biochem. J. 340 (1999) 631–638.

99S17   Storch, E., Daggett, V., Atkins, W.M.: Biochemistry 38 (1999) 5054–5064.

99S18   Sudharshan, E., Appu Rao, A.G.: J. Biol. Chem. 274 (1999) 35351–35358.

99S19   Sun, T.-X., Akhtar, N.J., Liang, J.J.-N.: J. Biol. Chem. 274 (1999) 34067–34071.

99T1    Taddei, N., Chiti, F., Paoli, P., Fiaschi, T., Bucciantini, M., Stefani, M., Dobson, C.M., Ramponi, G.: Biochemistry 38 (1999) 2135–2142.

510    References

99T2    Takano, K., Ota, M., Ogasahara, K., Yamagata, Y., Nishikawa, K., Yutani, K.:
        Protein Engng. 12 (1999) 663–672.
99T3    Takano, K., Tsuchimori, K., Yamagata, Y., Yutani, K.: Eur. J. Biochem. 266
        (1999) 675–682.
99T4    Takano, K., Yamagata, Y., Funahashi, J., Hioki, Y., Kuramitsu, S., Yutani, K.:
        Biochemistry 38 (1999) 12698–12708.
99T5    Takano, K., Yamagata, Y., Kubota, M., Funahashi, J., Fujii, S., Yutani, K.:
        Biochemistry 38 (1999) 6623–6629.
99T6    Tang, K.S., Guralnick, B.J., Wang, W.K., Fersht, A.R., Itzhaki, L.S.: J. Mol. Biol.
        285 (1999) 1869–1886.
99T7    Tanner, J.W., Eckenhoff, R.G., Liebman, P.A.: Biochim. Biophys. Acta 1430
        (1999) 46–56.
99T8    Taylor, J.W., Greenfield, N.J., Wu, B., Privalov, P.L.: J. Mol. Biol. 291 (1999)
        965–976.
99T9    Ternström, T., Mayor, U., Akke, M., Oliveberg, M.: Proc. Natl. Acad. Sci. USA
        96 (1999) 14854–14859.
99T10   Thies, M.J.W., Mayer, J., Augustine, J.G., Frederick, C.A., Lilie, H., Buchner, J.:
        J. Mol. Biol. 293 (1999) 67–79.
99T11   Tommos, C., Skalicky, J.J., Pilloud, D.L., Wand, A.J., Dutton, P.L.: Biochemistry
        38 (1999) 9495–9507.
99T12   Torrent, J., Connelly, J.P., Coll, M.G., Ribó, M., Lange, R., Vilanova, M.:
        Biochemistry 38 (1999) 15952–15961.
99T13   Travaglini-Allocatelli, C., Cutruzzolà, F., Bigotti, M.G., Staniforth, R.A., Brunori,
        M.: J. Mol. Biol. 289 (1999) 1459–1467.
99T14   Trevino, R.J., Gliubich, F., Berni, R., Cianci, M., Chirgwin, J.M., Zanotti, G.,
        Horowitz, P.M.: J. Biol. Chem. 274 (1999) 13938–13947.
99T15   Tsuji, T., Yoshida, K., Satoh, A., Kohno, T., Kobayashi, K., Yanagawa, H.: J.
        Mol. Biol. 268 (1999) 1581–1596.
99U1    Uedaira, H., Morii, H., Ishimura, M., Taniguchi, H., Namba, K., Vonderviszt, F.:
        FEBS Lett. 445 (1999) 126–130.
99U2    Uversky, V.N., Abdullaev, Z.Kh., Arseniev, A.S., Bocharov, E.V., Dolgikh, D.A.,
        Latypov, R.F., Melnik, T.N., Vassilenko, K.S., Kirpichnikov, M.P.: Biochim.
        Biophys. Acta 1432 (1999) 324–332.
99V1    Vallée, B., Teyssier, C., Maget–Dana, R., Ramstein, J., Bureaud, N., Schoentgen,
        F.: Eur. J. Biochem. 266 (1999) 40–52.
99V2    Veprintsev, D.B., Narayan, M., Permyakov, S.E., Uversky, V.N., Brooks, C.L.,
        Cherskaya, A.M., Permyakov, E.A., Berliner, L.J.: Proteins: Structure, Function,
        and Genetics 37 (1999) 65–72.
99W1    Wagschal, K., Tripet, B., Hodges, R.S.: J. Mol. Biol. 285 (1999) 785–803.
99W2    Wagschal, K., Tripet, B., Lavigne, P., Mant, C., Hodges, R.S.: Protein Sci. 8
        (1999) 2312–2329.
99W3    Waldner, J.C., Lahr, S.J., Edgell, M.H., Pielak, G.J.: Biopolymers 49 (1999)
        471–479.
99W4    Wall, J., Schell, M., Murphy, C., Hrncic, R., Stevens, F.J., Solomon, A.:
        Biochemistry 38 (1999) 14101–14108.
99W5    Wang, Q., Buckle, A.M., Foster, N.W., Johnson, C.M., Fersht, A.R.: Protein Sci. 8
        (1999) 2186–2193.
99W6    Wassenberg, D., Welker, C., Jaenicke, R.: J. Mol. Biol. 289 (1999) 187–193.
99W7    Welfle, K., Misselwitz, R., Hausdorf, G., Höhne, W., Welfle, H.: Biochim.
        Biophys. Acta 1431 (1999) 120–131.
99W8    Welker, C., Böhm, G., Schurig, H., Jaenicke, R.: Protein Sci. 8 (1999) 394–403.
99W9    Wenk, M., Baumgartner, R., Holak, T.A., Huber, R., Jaenicke, R., Mayr, E.-M.: J.
        Mol. Biol. 286 (1999) 1533–1545.

99W10 Wenk, M., Jaenicke, R.: J. Mol. Biol. 293 (1999) 117–124.

99W11 Wildegger, G., Liemann, S., Glockshuber, R.: Nature Struct. Biol. 6 (1999) 550–553.

99W12 Wirtz, P., Steipe, B.: Protein Sci. 8 (1999) 2245–2250.

99W13 Wittung-Stafshede, P.: Biochim. Biophys. Acta 1432 (1999) 401–405.

99W14 Wittung-Stafshede, P.: Protein Sci. 8 (1999) 1523–1529.

99W15 Wittung-Stafshede, P., Lee, J.C., Winkler, J.R., Gray, H.B.: Proc. Natl. Acad. Sci. USA 96 (1999) 6587–6590.

99W16 Wrabl, J., Shortle, D.: Nature Struct. Biol. 6 (1999) 876–883.

99W17 Wu, J.-W., Wang, Z.-X.: Protein Sci. 8 (1999) 2090–2097.

99X1 Xue, L.-L., Wang, Y.-H., Xie, Y., Yao, P., Wang, W.-H., Qian, W., Huang, Z.-X.: Biochemistry 38 (1999) 11961–11972.

99Y1 Yang, M., Liu, D., Bolen, D.W.: Biochemistry 38 (1999) 11216–11222.

99Y2 Yokota, A., Noda, Y., Tachibana, H., Segawa, S.: in: Old and New Views of Protein Folding, ed. by K.Kuwajima and M. Arai, Elsevier Science B.V., Amsterdam, Lausanne, New York, Oxford, Shannon, Singapore, Tokyo, 1999, pp.163–172.

99Y3 Yu, A., Ballard, L., Smillie, L., Pearlstone, J., Foguel, D., Silva, J., Jonas, A., Jonas, J.: Biochim. Biophys. Acta 1431 (1999) 53–63.

99Y4 Yuan, C., Li, J., Selby, T.L., Byeon, I.-J.L., Tsai, M.-D.: J. Mol. Biol. 294 (1999) 201–211.

99Z1 Zaiss, K., Jaenicke, R.: Biochemistry 38 (1999) 4633–4639.

99Z2 Zarutskie, J.A., Sato, A.K., Rushe, M.M., Chan, I.C., Lomakin, A., Benedek, G.B., Stern, L.J.: Biochemistry 38 (1999) 5878–5887.

99Z3 Žerovnik, E., Janjic, V., Francky, A., Mozetic-Francky, B.: Eur. J. Biochem. 260 (1999) 609–618.

99Z4 Zhong, H., Gilmanshin, R., Callender, R.: J. Phys. Chem. B 103 (1999) 3947–3953.

99Z5 Zimmerman, A.W., Rademacher, M., Rüterjans, H., Lücke, C., Veerkamp, J.H.: Biochem. J. 344 (1999) 495–501.

99Z6 Zitzewitz, J.A., Gualfetti, P.J., Perkons, I.A., Wasta, S.A., Matthews, C.R.: Protein Sci. 8 (1999) 1200–1209.

99Z7 Zitzewitz, J.A., Matthews, C.R.: Biochemistry 38 (1999) 10205–10214.

## Index of Proteins

The manufacturer's authorised representative in the EU is Springer
Nature Customer Service Centre GmbH, Europaplatz 3, 69115 Heidelberg,
Germany. If you have any concerns regarding our products, please
contact ProductSafety@springernature.com

Printed and bound by CPI Group (UK) Ltd, Croydon, CR0 4YY

28/04/2026

02098548-0003